Non-Linear Optical Properties of Matter

From Molecules to Condensed Phases

Edited by

Manthos G. Papadopoulos
National Hellenic Research Foundation, Greece

Andrzej J. Sadlej
Nicolaus Copernicus University, Poland

and

Jerzy Leszczynski
Jackson State University, USA

A C.I.P. Catalogue record for this book is available from the Library of Congress.

ISBN-10 1-4020-4849-1 (HB)
ISBN-13 978-1-4020-4849-4 (HB)
ISBN-10 1-4020-4850-5 (e-book)
ISBN-13 978-1-4020-4851-3 (e-book)

Published by Springer,
P.O. Box 17, 3300 AA Dordrecht, The Netherlands.

www.springer.com

Printed on acid-free paper

All Rights Reserved
© 2006 Springer
No part of this work may be reproduced, stored in a retrieval system, or transmitted
in any form or by any means, electronic, mechanical, photocopying, microfilming, recording
or otherwise, without written permission from the Publisher, with the exception
of any material supplied specifically for the purpose of being entered
and executed on a computer system, for exclusive use by the purchaser of the work.

CONTENTS

Preface ix
Manthos G. Papadopoulos, Jerzy Leszczynski and Andrzej J. Sadlej

Introduction xv
Mark A. Ratner

1. Microscopic Theory of Nonlinear Optics 1
   Patrick Norman and Kenneth Ruud

2. Accurate Nonlinear Optical Properties for Small Molecules 51
   Ove Christiansen, Sonia Coriani, Jürgen Gauss, Christof Hättig, Poul Jørgensen, Filip Pawłowski, and Antonio Rizzo

3. Determination of Vibrational Contributions to Linear and Nonlinear Optical Properties 101
   B. Kirtman and J.M. Luis

4. SOS Methods in Calculations of Electronic NLO Properties 129
   Wojciech Bartkowiak and Robert Zaleśny

5. Kohn–Sham Time-Dependent Density Functional Theory with Applications to Linear and Nonlinear Properties 151
   Dan Jonsson, Olav Vahtras, Branislav Jansik, Zilvinas Rinkevicius, Paweł Sałek, and Hans Ågren

6. Non-Linear Pulse Propagation in Many-Photon Active Isotropic Media 211
   A. Baev, S. Polyutov, I. Minkov, F. Gel'mukhanov and H. Ågren

7. Collective and Cooperative Phenomena in Molecular Functional Materials 251
   Anna Painelli and Francesca Terenziani

8. Multiconfigurational Self-Consistent Field-Molecular Mechanics Response Methods 283
   Kurt V. Mikkelsen

9. Solvatochromism and Nonlinear Optical Properties of
Donor-Acceptor π-Conjugated Molecules 299
Wojciech Bartkowiak

10. Symmetry Based Approach to the Evaluation of Second Order
NLO Properties of Carbon Nanotubes 319
L. De Dominicis and R. Fantoni

11. Atomistic Molecular Modeling of Electric Field Poling
of Nonlinear Optical Polymers 337
Megan R. Leahy-Hoppa, Joseph A. French, Paul D. Cunningham, and L. Michael Hayden

12. Nonlinear Optical Properties of Chiral Liquids 359
Peer Fischer and Benoît Champagne

13. Recent Progress in Molecular Design of Ionic Second-order
Nonlinear Optical Materials 383
Paresh Chandra Ray

14. Characterization Techniques of Nonlinear Optical Materials 419
Inge Asselberghs, Javier Pérez-Moreno and Koen Clays

15. Third-Order Nonlinear Optical Response of Metal Nanoparticles 461
Bruno Palpant

16. From Dipolar to Octupolar Phthalocyanine Derivatives:
The Example of Subphthalocyanines 509
Christian G. Claessens, Gema de la Torre and Tomás Torres

17. NLO Properties of Metal Alkynyl and Related Complexes 537
Joseph P.L. Morrall, Mark G. Humphrey, Gulliver T. Dalton, Marie P. Cifuentes and Marek Samoc

18. Ruthenium Complexes as Versatile Chromophores with Large,
Switchable Hyperpolarizabilities 571
Benjamin J. Coe

19. Linear and Nonlinear Optical Properties of Selected
Rotaxanes and Catenanes 609
Jacek Niziol, Kamila Nowicka and Francois Kajzar

20. Second Harmonic Generation from Gold and Silver Nanoparticles in Liquid Suspensions 645
Jérôme Nappa, Guillaume Revillod, Gaelle Martin, Isabelle Russier-Antoine, Emmanuel Benichou, Christian Jonin and Pierre-François Brevet

Index 671

PREFACE

Many areas of science and technology have benefited from the tremendous advance in computational techniques over the last two decades. In a number of cases theoretical predictions have directly impacted the design of new and improved materials with unique characteristics. An understanding of the details of molecular structures and their relationship with the desired properties of such materials has always been a key factor for successful interplay between theory and experiment. Among such important characteristics are the non-linear optical (NLO) properties of matter.

There is currently an immense interest in NLO properties of materials, since they provide valuable information about important aspects of matter. For example, they are related to electronic and vibrational structures as well as to intra- and intermolecular interactions. On the other hand, NLO properties are of great practical importance for the design of materials and devices, which have numerous and important applications (e.g. optical devices for the transfer and storage of data). Taking into account the widespread interest in the NLO properties/materials and the rapid progress of both the scientific and technological aspects of this field, we believe that it is timely and useful to select some topics of particular interest (this is subjective, of course) and invite leading experts to review current progress. This idea underlies the composition of the present volume. Obviously, the selection of topics was strongly influenced by our research interests. They may not cover all important areas of chemistry and physics of the NLO systems. However, in spite of the possible deficiencies, we believe that the reviews presented here cover a relatively wide class of problems and will be of interest for researchers and students in this area. Particular care was taken to impartially address both theoreticians and experimentalists to increase the extent of the mutual interaction between them.

This book starts with an introduction written by M. A. Ratner and then there is an opening chapter that introduces readers to the theory of linear and NLO properties. Norman and Ruud review the basic physical processes and their connection with the formulas derived from time-dependent perturbation theory. They briefly discuss the vibrational properties as well as the connection between the microscopic and macroscopic NLO properties.

Christiansen et al. review the coupled cluster (CC) response methods for the calculation of the electronic contributions to first and second hyperpolarizabilities and some magneto-optic NLO properties. The latter include the magneto-optical activity (Faraday effect, magneto-circular dichroism), Buckingham and Cotton–Mouton effects as well as Jones and magneto-electric birefringence. They discussed the basis set convergence for the properties of interest as well as the convergence of the properties with the wavefunction model. Christiansen et al. compare theoretical with experimental results and show that the CC approaches, connected with the hierarchy of correlation-consistent basis sets, can lead to very accurate results for small molecules. They also point out some of the present challenges for the accurate calculation of the properties of interest (e.g. computation of NLO properties of open shell molecules).

Kirtman and Luis review some of the theoretical/computational methods which have been proposed over the past fifteen years for the calculation of vibrational contributions to the linear and NLO properties. They discuss: (i) the time-dependent sum-over-states perturbation theory and the alternative nuclear relaxation/curvature approach, (ii) the static field-induced vibrational coordinates which reduce the number of nth-order derivatives to be evaluated, (iii) the convergence behavior of the perturbation series, (iv) an approach to treat large amplitude (low frequency) vibrations, (v) the effect of the basis set and electron correlation on the vibrational properties, and (vi) techniques to compute the linear and NLO properties of infinite polymers.

Bartkowiak and Zaleśny discuss the sum-over-states (SOS) method which is used for the calculation of NLO properties (electronic contribution) and multi-photon absorption. They comment on the various approximations, including the widely used few-states models, and the exact sum-over-states formulas. They show that one of the main advantages of the many variants of this approach is the interpretation of the NLO properties in terms of contributions from excited states. They comment on the limited utility of the SOS technique for small molecules, aggregates and clusters, but they point out, that it is still a very attractive tool for large molecules.

Jonsson et al. review the Kohn-Sham density functional theory (DFT) for time-dependent (TD) response functions. They describe the derivation of the working expressions. They also review recent progress in the application of TD-DFT to open shell systems. They reported results on several properties: (i) hyperpolarizabilities (e.g. para-nitroaniline, benzene, C_{60} fullerene), (ii) excited state polarizabilities (e.g. pyrimidine), (iii) three-photon absorption and (iv) EPR spin Hamiltonian parameters.

Baev et al. review a theoretical framework which can be useful for simulations, design and characterization of multi-photon absorption-based materials which are useful for optical applications. This methodology involves quantum chemistry techniques, for the computation of electronic properties and cross-sections, as well as classical Maxwell's theory in order to study the interaction of electromagnetic fields with matter and the related properties. The authors note that their dynamical method, which is based on the density matrix formalism, can be useful for both fundamental and applied problems of non-linear optics (e.g. self-focusing, white light generation etc).

Painelli and Terenziani discuss the cooperative and collective behavior resulting from classical electrostatic intermolecular interactions in molecular materials with negligible intermolecular overlap. The simple model they employ for clusters of push-pull chromophores neglects intermolecular overlap and describes them using a two-state model. They comment on the excitonic approximation which is expected

Preface

to work well for clusters of molecules with low polarizability. These authors reviewed the mean-field approximation, which has been introduced for the calculation of the polarizability of molecular crystals and films. Painelli and Terenziani describe: (i) the optical susceptibilities of some representative clusters and (ii) the excited states in a cluster of polar and polarizable molecules.

Mikkelsen reviews recent advances of the MCSCF/MM method. This approach has been developed in order to obtain frequency-dependent molecular properties for a solute perturbed by solvent interactions. He defines the Hamiltonian for the total system. It involves three components: the first describes the quantum mechanical (QM) system, the second, the classical (MM) system and the third their interaction. He describes the energy functional, the MCSCF wave function and the linear and quadratic response functions.

Bartkowiak reviewed the connection between the NLO response and solvatochromic behavior of donor-acceptor π-conjugated molecules. The marked NLO properties of these molecules are associated with an intra-molecular charge-transfer excited state. This author points out that the environmental interactions may have a very significant effect on the hyperpolarizabilities (they may even lead to a change of sign). Bartkowiak shows that a simple two-state model combined with the solvatochromic methods may allow the prediction of changes of in molecular hyperpolarizabilities as a function of the solvent polarity.

Dominicis and Fantoni present a method for the computation of the electronic first hyperpolarizability of chiral carbon nano-tubes (CNTs). The CNT eigenstates are computed by an algorithm reported by Damnjanovic et al. They discuss the symmetry properties of CNTs and selection rules for electronic transitions and demonstrate that the use of symmetry reveals the state-to-state transitions, which contribute to the first hyperpolarizability of CNTs. The latter is related to particular state-to-state transitions. The principles for predicting the magnitude of the first hyperpolarizability and its relation to the topology of CNTs are also discussed.

Leahy-Hoppa et al. review the results for an electric field-poled guest-host NLO polymer system, computed by using Monte Carlo and atomistic molecular dynamics simulations. They discuss the work of the groups investigating this topic. The Monte Carlo studies provide results in good agreement with experimental data and valuable information concerning the optimal loading concentration for a chromophore of a given shape and dipole moment. The findings of the teams, which have used molecular dynamics are also described. The alignment of dopants is studied by employing $<\cos\theta>$ and $<\cos^3\theta>$. Leahy-Hoppa et al. noted that the orientational order in the system is strongly dependent on the strength of the poling field.

Fischer and Champagne present an overview of linear and nonlinear optical properties of chiral molecules in isotropic media. The authors state the general symmetry requirements of chiroptical processes, and show that nonlinear chiral spectroscopies can arise within the electric dipole approximation. The authors describe sum-frequency–generation experiments at second order and demonstrate how nonlinear optics can be used to determine the absolute conformation of a chiral molecule in solution. This is discussed with recourse to electric-field induced sum-frequency generation, a third-order phenomenon. Aspects of BioCARS at fourth order are also discussed. The chapter includes a survey of computations that address the nonlinear optical properties of chiral molecules. Studies using semi-empirical as well as *ab initio* methods, such as time-dependent Hartree Fock theory and sum-over-states schemes, are shown to be helpful in linking chiral molecular structure to the extraordinary optical properties of chiral molecules. Examples include computations of the first hyperpolarizability for some helical molecules, such as helicenes and heliphenes.

Ray reviews some recent developments concerning the design of novel materials with large NLO effects. He considers a series of organic salts and various organometallic derivatives where it was found that metal-to-ligand charge-transfer has a dominant contribution to the second-order NLO response. He also discusses the first hyperpolarizability of several retinal derivatives, ionic octupolar molecules and zwitterionic derivativatives. Solvent effects on NLO properties are also reviewed.

Asselberghs et al. review the experimental techniques which are used to characterize the NLO response. These involve for the second order properties: second-harmonic generation (SHG), electric field-induced second harmonic generation (EFISH) and hyper-Rayleigh scattering (HRS). Experimental methods which are also used for the third-order nonlinear responses: the Maker fringe technique and the wedge-shaped technique. They review techniques which are frequently used to characterize third-order response: third harmonic generation (THG), degenerate four-wave mixing (DFWM), optical phase conjugation, optical Kerr-gate, self-focusing methods, nonlinear Fabry-Perot methods etc. The authors discuss the four most frequently used conventions in defining the non-linear polarization and the interacting fields in the frequency domain.

Palpant reviews the literature concerning the third-order NLO response of nanocomposite media consisting of noble metal (Cu, Ag, Au) nanoparticles surrounded by a dielectric host. He first considers the theoretical background used to describe the linear and NLO properties of noble metal nanoparticles and nanocomposite media. In these sections the influence of the surface plasmon resonance in nanoparticles as well as the effects of the interactions between neighboring particles is especially investigated. Subsequently, the main experimental results regarding the optical Kerr effect in nanocomposite media are presented. Palpant reviews the different NLO phenomena observed in such materials, as well as the intrinsic third-order non-linear susceptibility of noble metal nanoparticles in different host media. He also considers the dependence of the nonlinear properties on morphological parameters (particle size and shape, matrix kind, metal concentration) and laser excitation characteristics (intensity, wavelength, pulsewidth). Finally, the role of thermal effects in the NLO response of such materials is discussed.

Claessens et al. reviewed the recent progress in studies of NLO properties of boron-subphthalocyanines (SubPcs). These phthalocyanine derivatives consist of three isoindole units N-fused around a central boron atom, which bear an axial ligand. These authors noted that the optical response of these nearly octupolar derivatives is associated to the charge transfer inside the macrocycle π surface. They

Preface

considered NLO properties of phthalocyanines and related macrocycles and noted the role of dimensionality on these properties. They reviewed the NLO properties of subphthalocyanines: (i) in solution; they discussed the effect of various substituents on the properties, which have been measured by employing HRS, EFISH and THG, and (ii) in condensed phases.

Morrall et al. review NLO properties of iron, ruthenium, osmium, nickel, and gold alkynyl complexes, which the authors have prepared. HRS (at 1.064 μm) and Z-scan (at 0.8 μm) measurements have been employed. Structure-property relations have been established. Static first hyperpolarizability values have also been computed employing the two-state model. They relate the NLO coefficients to several factors and properties (e.g. ease of oxidation, π-system length, "dimensionality").

Coe describes a large number of ruthenium complexes commenting on their quadratic and cubic optical nonlinearities. He notes that the mechanisms which lead to large NLO effects are generally similar to those observed in metal-free organic molecules. However, some unusual effects have also been found, for example the decreasing first hyperpolarizability value with π-conjugation extension in pyridyl polyene complexes. Also of note, it has been found that the metal-to-ligand charge-transfer absorption and first hyperpolarizability response of the pentaammine complexes can be decreased by $Ru^{II} \to Ru^{III}$ oxidation. This facile redox-induced switching of the NLO response is fully reversible. Various experimental methods (e.g. HRS, Z-scan, DFWM) and computational techniques (e.g. TD-DFT, ZINDO) have been used to determine the NLO properties of the reviewed complexes. Several of these metallochromophores are found to have very large NLO responses.

Niziol et al. review the linear and NLO properties of some catenanes and rotaxanes studied in solutions or thin films. Techniques like UV-Vis spectrometry, second and third harmonic generation in thin films and electro-optic Kerr effects in solution have been employed. They review the synthesis and material processing of these derivatives. Niziol et al. describe how the rotation rate of the macrocycle in catenane solutions is more than an order of magnitude larger than in rotaxanes. They comment on the factors on which the rate of rotation depends. This new class of molecules, with mobile subparts, is very likely to have useful applications including the construction of synthetic molecular machines and all-optical switching elements.

Nappa et al present a review of studies on the second harmonic (SH) light scattered from aqueous suspensions of small gold and silver particles. Initially this work concerns the SH response from arbitrary particles, with minimum restrictions on their size or shape. They show that, in the case of metallic particles, the excitation fields should be considered as superpositions of the incident and polarization fields, because of the large (hyper)polarizabilities of the metallic particles and the possibility of resonance enhancements through surface plasmon excitations. Nappa et al. employing the hyper-Rayleigh intensities from small metallic particle suspensions, demonstrate that the NLO response originates from the breaking of centrosymmetry at the surface of the particles.

We would like to thank the authors for their excellent contributions and fruitful collaboration on this book. The very efficient assistance of all the reviewers and their critical and constructive comments as well as the help of Heribert Reis and Aggelos Avramopoulos in preparing the index of this book are greatly appreciated. We would value feedback from all readers of this book. Your comments are very important to us, so please feel free to e-mail your suggestions to: mpapad@eie.gr, teoajs@chem.uni.torun.pl, jerzy@ccmsi.us.

MGP   AJS   JL
May 2006

INTRODUCTION
Molecular Nonlinear Optics in 2005

MARK A. RATNER

Department of Chemistry, Northwestern University, Evanston, IL

Because science is by its nature an experimental discipline, only experimental data can make conceptual notions real. Simple ideas of nonlinear responses go back to the ancients, in their observations of magnetism. Systemic study of nonlinear properties really began in the nineteenth century, with the investigations of workers like Faraday and Seebeck on responses of materials to several applied fields. Maxwell's development of the theory of the electromagnetic field permits the correct mathematical description of optical and electromagnetic response, and therefore underlies formal approaches to responses both linear and nonlinear. The expansion of the molecular polarization in a Taylor series in the applied electromagnetic field is an obvious step to take. Considerations of the fine structure constant and the requirements for sum rules suggest that higher-order responses for strictly-limited magnitudes. The actual detailed investigation of materials' nonlinearities, in particular $\chi^{(n)}$, only began when the intense field permitted by use of laser light made it possible to observe such responses systematically. The first nonlinear process to be advantaged by laser measurements was Raman spectroscopy. In the last four decades, major progress has been made on all nonlinear optical response properties.

Interpretation of nonlinear molecular measurements on molecules, and indeed our intuitive understandings of any polarization, is almost always based on a state model of the molecule: the applied fields mix the levels of the molecular Hamiltonian so that spectral analysis (in the sense of sums over states, or SoS) becomes a very useful description. While more recent and more sophisticated electronic structure calculations have important direct-response methods, the SoS techniques, like the very simple two-level formula of Oudar and Chemla, have tremendous advantages in terms of generality and understanding.

Experimentally, there have been substantial advances in measurement, going beyond the earlier electric-field induced second harmonic generation (EFISH) techniques to include hyper-Rayleigh scattering, z-scan and other schemes. Modeling methods have become more nuances. Synthetic study of molecular and nonlinear optics has gone beyond early work on donor/acceptor π-systems. Applications and technological advances are appearing, and molecules are being used as modulators as well as materials for light manipulation.

Nevertheless, some goals remain unattained. Molecular materials were not yet routinely used for nonlinear technologies, essentially because of inadequate materials properties. The field still needs more systematic measurement and modeling approaches. Of equal importance, the extensions both of the knowledge base of experiments and of concepts for guiding our intuition about nonlinear response need to be strengthened. Clearly, the goal must be to utilize the great capabilities of synthetic chemistry to prepare molecular nonlinear materials by design, just as we can now prepare dyes, pharmaceuticals and polymers (although to different degrees of exactness and simplicity!) to have particular designed properties.

Some examples of intuition-based or modeling-based advances in nonlinear properties, specifically the first hyperpolarizability response β, have been developed in the past three decades. These include the bond-length alternation motif, the idea of octopolar molecular structures, the stronger responses of excited states and the use of purposely twisted π-electron molecules to modify the admixture of quinoid and aromatic structures.

Despite extensive research, major problems still remain unsolved in both of the crucial areas required for employing effectively nonlinear materials based on molecules: the molecular hyperpolarizability properties (β, γ and to a very limited extent higher-order responses) are still not either predictable or preparable using any meaningful structure/function understandings. Moreover, utilizing molecules to prepare actual materials with designed nonlinearities $\chi^{(2)}, \chi^{(3)}$ ... remains a very difficult problem.

In my view, this book contains the most in-depth and broad-based discussion of molecular nonlinear materials yet available. While it does not in itself discuss all the issues (there is relatively little on molecular crystals or on local-field effects), the combination of theoretical and experimental presentations makes the book of unique value to any investigators in the general of molecular nonlinear optics.

The sketch below indicates some of the major themes in nonlinear response that are addressed in this book. In the modeling area, many of the major themes are addressed, including fundamentals, environmental effects, processing and particular special topics. There is a good deal of attention to electronic-structure methods, including such issues as scaling properties, collective excitations, resonant excitations and vibrational nonlinear effects. One major underlying theme here is the development of electronic-structure themes for calculating molecular (as opposed to materials) response. Methods include density-functional theory approaches, time-dependant analyses, *ab initio* based (systematically improvable) approaches, and semi-empirical analyses. Each of these has both strengths and weaknesses, and those are brought out clearly.

There is also analysis of environmental effects from solvents and of solvatechromic phenomena, and discussion of poling as a processing technique for nonlinear structures.

The second half of the book concerns measurements and particular molecular structures. After a general overview, this section concentrates on specifics, ranging from complex molecules like catanenes and rotaxanes through ruthenium complexes and organometallics. More materials-oriented contributions on carbon nanotubes,

Introduction xvii

metal-dot conjugates, and ionic materials are also given. One chapter is devoted to the $\chi^{(3)}$ response, which remains much less investigated for molecular materials.

The book in general presents a balanced and informative description of progress in molecular non-linear optics in 2005. While it is clear that the field has progressed substantially, and that there is deep understanding in some promising areas for experimental and technological application, it is still true that some major themes remain challenging. These include the preparation of stable molecular systems exhibiting very large (but more particularly, predictable) β and γ properties, building of a knowledge base and an intuitive understanding of structure/function relationships in simple molecular entitites. Perhaps even more challenging (and less extensively addressed) is the design of actual nonlinear materials. Here it is necessary to go beyond design at the molecular level, to deal with an actual materials system. This problem is both less-well addressed in this book, and less-well addressed by the community. Poling methods are discussed here (and are indeed one of the standard ways in which molecular nonlinear materials are prepared and measured). Poled materials have substantial inherent disadvantages, arising both from fundamental statistical mechanics (the Boltzmann penalty is required to pole the material) and from kinetic long-term instabilities. Other approaches, particularly self-assembly and covalent cross-linking structures, will certainly play a major role in this area.

The editors are to be congratulated on the remarkable quality and completeness of this book, that shows both the advances and some of the remaining challenges in the general area of nonlinear materials based on molecular response.

CHAPTER 1

MICROSCOPIC THEORY OF NONLINEAR OPTICS

PATRICK NORMAN[1] AND KENNETH RUUD[2]

[1] *Department of Physics, Chemistry and Biology, Linköping University, SE-581 83 Linköping, Sweden*
[2] *Department of Chemistry, University of Tromsø, N-9037 Tromsø, Norway*

Abstract: In this chapter we give an introduction to the theory of linear and nonlinear optics. We show how the response of a molecule to an external oscillating electric field can be described in terms of intrinsic properties of the molecules, namely the (hyper)polarizabilities. We outline how these properties are described in the case of exact states by considering the time-development of the exact state in the presence of a time-dependent electric field. Approximations introduced in theoretical studies of nonlinear optical properties are introduced, in particular the separation of electronic and nuclear degrees of freedom which gives rise to the partitioning of the (hyper)polarizabilities into electronic and vibrational contributions. Different approaches for calculating (hyper)polarizabilities are discussed, with a special focus on the electronic contributions in most cases. We end with a brief discussion of the connection between the microscopic responses of an individual molecule to the experimentally observed responses from a molecular ensemble

Keywords: two-photon absorption; three-photon absorption; multi-photon absorption; nonlinear optics; polarizability; hyperpolarizability

1. INTRODUCTION

The field of molecular nonlinear optics has been growing since the first prediction of the nonlinear optical process, and a strong boost was given to the field with the experimental observation of nonlinear optical effects made by Franken et al. in 1961 [29]. The development of the modern laser had provided scientists with a source of the high-intensity fields needed for nonlinear optical processes to become effective. However, one could not observe a synchronous development in the quantum mechanical modeling of these processes, and there are several reasons for this delay. The molecules of experimental interest are in most cases large, and, in general, it is not until recently that standard first principle quantum chemical methods have been able to address the compounds of interest. Furthermore, the calculation of the higher-order molecular properties that determine the nonlinear optical responses require a development of highly sophisticated theoretical models, and it is a challenge to design efficient computational schemes for these models.

The majority of the early nonlinear optical materials were based on inorganic crystals. More recently, however, focus has shifted toward organic molecules due to the much greater design flexibility in molecular compounds, which allows for a fine tuning of the microscopic properties and thus the linear and nonlinear optical behavior of the materials. Molecular compounds can also have a narrow band absorption, one can re-orient their optical axis, they can display bistable electronic states of different spin symmetries, etc. This diversity makes molecular compounds suitable for specific target areas in light-control applications. In addition, organic molecules have favorable mechanical and thermal properties, which allow them to be used for a wider range of applications.

During the last 10–20 years, a large number of efficient theoretical methods for the calculation of linear and nonlinear optical properties have been developed—this development includes semi-empirical, highly correlated *ab initio*, and density functional theory methods. Many of these approaches will be reviewed in later chapters of this book, and applications will be given that illustrate the merits and limitations of theoretical studies of linear and nonlinear optical processes. It will become clear that theoretical studies today can provide valuable information in the search for materials with specific nonlinear optical properties. First, there is the possibility to screen classes of materials based on cost and time effective calculations rather then labor intensive synthesis and characterization work. Second, there is the possibility to obtain a microscopic understanding for the performance of the material—one can investigate the role of individual transition channels, dipole moments, etc., and perform systematic model improvements by inclusion of the environment, relativistic effects, etc.

The purpose of this chapter is to introduce the fundamentals of the theory of linear and nonlinear optical processes, and our focus will be on the general features of the theory. We will primarily restrict our discussion to the framework of exact-state theories, and focus on the occurrence of linear and nonlinear optical processes from a physical point of view. However, in order to set a frame of reference we provide a brief outline of the most common classes of methods in approximate-state theories. We will discuss the partitioning of molecular properties into electronic and vibrational contributions, and close the chapter with a brief discussion of the comparison of the microscopic properties with those of the bulk. We wish to stress that other chapters of this book will cover these latter aspects in greater detail.

The reader is also advised to consult previous reviews and books. Previous reviews in this field have been written by Ward [58], Buckingham [19], Buckingham and Orr [20], Bogaard and Orr [14], Dykstra et al. [26], Bishop [3, 4, 6], Hasanein [30], Shelton and Rice [55], Brédas et al. [18], Luo et al. [39], Bartlett and Sekino [2], Kirtman and Champagne [34], Nakano and Yamaguchi [43], Wolff and Wortmann [59], Bishop and Norman [11], and Champagne and Kirtman [22]. A selection of books concerning nonlinear optics are: *The elements of nonlinear optics* by Butcher and Cotter [21], *Introduction to nonlinear optical effects in molecules and polymers* by Prasad and Williams [49], *Nonlinear optics* by Boyd [17], and *Linear and nonlinear optical properties of molecules* by Wagniére [57].

2. MOLECULAR NONLINEAR OPTICAL PROPERTIES

Molecular nonlinear optics is the description of the change of the molecular optical properties by the presence of an intense light field. Since light either can be considered a classical electromagnetic wave or as a stream of photons, we may describe the interaction between light and matter in two apparently different ways, and we will start by considering how linear and nonlinear optical phenomena can be described in these two frameworks.

The discussion in this chapter, as well as in the rest of this book, will be concerned with electromagnetic radiation in the visible and infrared spectral regions. In the energy scale of molecules, this frequency range corresponds to the energies required for vibrational and electronic excitations. For these energies, the light-matter interaction can be regarded as scattering of photons by the electrons of the molecule. Each photon carries a linear momentum $p = E/c$ which is partially transferred to the molecule since the electron remains bound, but the great mass of the molecule effectively prohibits energy transfer so that the scattered photons will have, for all practical purposes, identical frequency as the incoming ones, a process known as elastic scattering. At the instant of interaction, the photon can be regarded as absorbed and the molecule as being in a virtual excited state, intermediate in energy to the stationary states of the system. However, the time-scale τ for this interaction, or the lifetime of the virtual state, is short enough not to violate the time-energy uncertainty relation

$$\text{(1)} \quad \tau \Delta E \sim \frac{\hbar}{2}$$

where ΔE denotes the energy difference between the nearest electronically excited state $|1\rangle$ and the virtual state. The de-excitation of the system from the virtual state back to the ground state $|0\rangle$ is associated with the emission of the scattered photon, a process referred to as linear optics, see Fig. 1.

A high intensity of the incident radiation enhances the probability for simultaneous multi-photon interactions with a single molecule, i.e. two or more photons are annihilated and absorbed by the molecule in a single quantum mechanical process. The frequency of the scattered photon does in such cases not have to be equal that of the absorbed photons, *e.g.* two quanta with frequency ω may be annihilated, creating a third photon with frequency 2ω. As indicated in Fig. 2, the system returns to its ground state $|0\rangle$ after the interaction has taken place, and the intermediate virtual state is separated from the first excited state by an energy ΔE. This is an example of a nonlinear optical process known as second-harmonic generation, which can

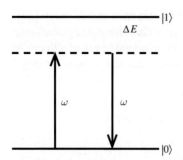

Figure 1. Elastic scattering of incident photons of frequency ω

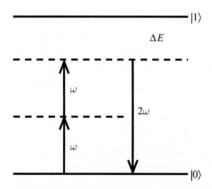

Figure 2. Second-harmonic generation involving two incident photons of frequency ω and a sum-frequency generated photon of frequency 2ω

be used to accomplish frequency conversion of light. From these basic principles, a large number of multi-photon interactions can be envisaged when considering incident photons of different frequencies. We also note that when the laser detuning decreases, i.e. when ΔE becomes smaller, the process can no longer be considered as an instantaneous scattering since the molecular state $|1\rangle$ will become absorbing.

In the complementary view of light–matter interaction in optics, the laser field is described as a electro-magnetic plane wave in which the molecular system resides, and the coupling between the two is, to a first approximation, the classical electric-dipole coupling. In the presence of the external electric field all charged particles in the molecule, electrons as well as atomic nuclei, will experience a force that perturbs their motions. In a classical sense, one would expect the charges to follow the time-oscillations of the electric field and thereby act as small antennas from which electro-magnetic radiation would be transmitted; the frequency of the transmitted wave would of course be the same as that of the external field. We can compare this classical picture to the elastic scattering process described in Fig. 1: The oscillating charges in the molecule give rise to an induced dipole moment, and the degree to

Microscopic Theory of Nonlinear Optics

which the external field $E(t)$ manages to set the charges in motion is, to first order, expressed in terms of the linear electric polarizability α. In general, the response of the charges depends on the frequency of the electric field, and the polarizability is therefore frequency dependent as we shall see later. The time-dependent polarization becomes

$$\mu(t) = \mu^0 + \alpha E(t) \tag{2}$$

where μ^0 is the permanent electric dipole moment of the molecule. We are concerned with optical fields, so the wavelength of $E(t)$ is at least in the order of a few hundred nanometers whereas the size of the molecule is no greater than a few nanometers. In most cases, the electric field is therefore taken to be uniform over the molecule.

Now there seems to be one immediate and important question to be addressed: if the particle and wave pictures of light are two versions of the same thing, how can we understand nonlinear light–matter interactions when the electric field is represented by a plane wave? The answer lies in a generalization of Eq. (2) which lets the polarization be expressed as a Taylor series in the electric field strength

$$\mu(t) = \mu^0 + \alpha E(t) + \frac{1}{2}\beta E^2(t) + \frac{1}{6}\gamma E^3(t) + \cdots \tag{3}$$

This equation introduces the first-order (nonlinear) hyperpolarizability β, the second-order hyperpolarizability γ, and so forth. Just as the linear polarizability, the nonlinear coupling constants depend on the frequency of the applied field. It is clear that the time-dependent polarization can have frequency components separate from those of the external field due to its power dependence on the electric field strength, and the molecule can thus emit sum-frequency-generated radiation in correspondence with for example Fig. 2. In optics, Eq. (3) provides the fundamental origin of nonlinearities, and, at the microscopic level, it is the expansion coefficients in this equation, or the hyperpolarizabilities, that govern the nonlinear optical performance of the material. Hence, theoretical modeling of nonlinear optical properties is concerned with the determination of these quantities given the structure of the system.

Let us now examine Eq. (3) in some detail. The electric field is vectorial and generally considered to be composed of a static component and one or more time-oscillating components according to

$$E_\alpha(t) = \sum_\omega E_\alpha^\omega e^{-i\omega t} \tag{4}$$

where E_α^ω are the Fourier amplitudes of the electric field along the molecular axis α. We note that the use of Greek subscripts for tensors in nonlinear optics follows the original notation of Buckingham [19]. The summation includes both positive and negative frequencies, and, since the external field is real, we have $E^\omega = [E^{-\omega}]^*$. Furthermore, a vectorial electric field implies that the linear polarizability is a second-rank tensor, the first-order hyperpolarizability is a third-rank tensor, and so forth. We insert the expression for the time-dependent electric field in Eq. (4) into Eq. (3) to obtain

$$\mu_\alpha(t) = \mu_\alpha^0 + \sum_\omega \alpha_{\alpha\beta}(-\omega;\omega)E_\beta^\omega e^{-i\omega t} \tag{5}$$

$$+ \frac{1}{2}\sum_{\omega_1,\omega_2} \beta_{\alpha\beta\gamma}(-\omega_\sigma;\omega_1,\omega_2)E_\beta^{\omega_1}E_\gamma^{\omega_2}e^{-i\omega_\sigma t}$$

$$+ \frac{1}{6}\sum_{\omega_1,\omega_2,\omega_3} \gamma_{\alpha\beta\gamma\delta}(-\omega_\sigma;\omega_1,\omega_2,\omega_3)E_\beta^{\omega_1}E_\gamma^{\omega_2}E_\delta^{\omega_3}e^{-i\omega_\sigma t} + \cdots$$

where ω_σ denotes the sum of optical frequencies: for terms involving β then $\omega_\sigma = (\omega_1 + \omega_2)$ and for terms involving γ then $\omega_\sigma = (\omega_1 + \omega_2 + \omega_3)$. The Einstein summation convention for repeated subscripts is assumed here and elsewhere. We note that any pairwise interchange of the indices and frequencies $\{\beta,\omega_1\}, \{\gamma,\omega_2\}$, and $\{\delta,\omega_3\}$ can be made without altering the physically observable polarization $\mu(t)$. It is therefore customary, but not necessary, to demand that the individual tensor elements are intrinsically symmetric

$$\beta_{\alpha\beta\gamma}(-\omega_\sigma;\omega_1,\omega_2) = \beta_{\alpha\gamma\beta}(-\omega_\sigma;\omega_2,\omega_1) \tag{6}$$

$$\gamma_{\alpha\beta\gamma\delta}(-\omega_\sigma;\omega_1,\omega_2,\omega_3) = \gamma_{\alpha\beta\delta\gamma}(-\omega_\sigma;\omega_1,\omega_3,\omega_2) \tag{7}$$

$$= \gamma_{\alpha\gamma\beta\delta}(-\omega_\sigma;\omega_2,\omega_1,\omega_3) = \gamma_{\alpha\gamma\delta\beta}(-\omega_\sigma;\omega_2,\omega_3,\omega_1)$$

$$= \gamma_{\alpha\delta\gamma\beta}(-\omega_\sigma;\omega_3,\omega_2,\omega_1) = \gamma_{\alpha\delta\beta\gamma}(-\omega_\sigma;\omega_3,\omega_1,\omega_2)$$

Furthermore, since the molecular polarization $\mu(t)$ as well as the electric field $E(t)$ are real, we have

$$\alpha(\omega;-\omega) = [\alpha(-\omega;\omega)]^* \tag{8}$$

$$\beta(\omega_\sigma;-\omega_1,-\omega_2) = [\beta(-\omega_\sigma;\omega_1,\omega_2)]^* \tag{9}$$

$$\gamma(\omega_\sigma;-\omega_1,-\omega_2,-\omega_3) = [\gamma(-\omega_\sigma;\omega_1,\omega_2,\omega_3)]^* \tag{10}$$

The frequency ω_σ is that of the generated molecular polarization, and, since the summations in Eq. (5) run over both positive and negative frequency components, the nonlinear hyperpolarizabilities will create both sum-frequency as well as difference-frequency generated polarization.

It is illustrative to consider a few specific examples. For instance, let two lasers A and B, which operate at frequencies ω_A and ω_B, respectively, interact. The external electric field experienced by the molecular system will in this case become

$$E_\alpha(t) = E_\alpha^{\omega_A}\cos(\omega_A t) + E_\alpha^{\omega_B}\cos(\omega_B t) \tag{11}$$

Figure 3. Frequency decomposition of two interfering laser fields

with a frequency decomposition that appears as in Fig. 3. According to Eq. (5) and including terms up to second-order in the field, the time-dependent polarization will be

$$\mu_\alpha(t) = \mu_\alpha^0 + \alpha_{\alpha\beta}[E_\beta^{\omega_A}\cos(\omega_A t) + E_\beta^{\omega_B}\cos(\omega_B t)] \tag{12}$$

$$+ \frac{1}{2}\beta_{\alpha\beta\gamma}[E_\beta^{\omega_A}\cos(\omega_A t) + E_\beta^{\omega_B}\cos(\omega_B t)][E_\gamma^{\omega_A}\cos(\omega_A t) + E_\gamma^{\omega_B}\cos(\omega_B t)]$$

With the use of the trigonometric identity

$$\cos u \cos v = \frac{1}{2}[\cos(u+v) + \cos(u-v)] \tag{13}$$

we can rewrite the polarization as

$$\mu_\alpha(t) = \mu_\alpha^0 + \sum_{\omega=\{\omega_A,\omega_B\}} \left[\alpha_{\alpha\beta}(-\omega;\omega)E_\beta^\omega \cos(\omega t) \right. \tag{14}$$

$$+ \frac{1}{2}\beta_{\alpha\beta\gamma}(0;\omega,-\omega)E_\beta^\omega E_\gamma^\omega + \frac{1}{2}\beta_{\alpha\beta\gamma}(-2\omega;\omega,\omega)E_\beta^\omega E_\gamma^\omega \cos(2\omega t) \bigg]$$

$$+ \beta_{\alpha\beta\gamma}(-(\omega_A+\omega_B);\omega_A,\omega_B)E_\beta^{\omega_A}E_\gamma^{\omega_B}\cos([\omega_A+\omega_B]t)$$

$$+ \beta_{\alpha\beta\gamma}(-(\omega_A-\omega_B);\omega_A,-\omega_B)E_\beta^{\omega_A}E_\gamma^{\omega_B}\cos([\omega_A-\omega_B]t)$$

The frequency decomposition of the polarization is illustrated in Fig. 4, and, in this figure, we recognize the linear polarization at the frequencies of the external field and the second-harmonic generation frequencies as a result from two-photon absorption (see also Fig. 2). In addition to these frequencies, we see that the induced molecular polarization will contain the frequencies $(\omega_A + \omega_B)$ (sum-frequency generation) and $(\omega_A - \omega_B)$ (difference-frequency generation) as well as a static component $\omega = 0$. Sum-frequency generation is similar to second-harmonic generation in that

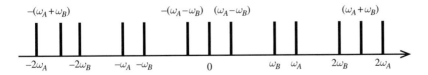

Figure 4. Frequency decomposition of the molecular polarization

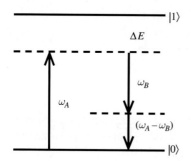

Figure 5. Difference-frequency generation involving photons of frequencies ω_A and ω_B

it involves the simultaneous absorption of two photons, although in this case one photon from each light source. Difference-frequency generation, on the other hand, has a fundamentally different microscopic origin. The corpuscular light–matter interaction in this nonlinear optical process is described in Fig. 5. Since the energy is conserved in this process as well as in all others, the light quanta of frequency ω_A is annihilated and those of frequencies ω_B and $(\omega_A - \omega_B)$ are both created at this instant (two-photon emission). The creation of the photon with frequency ω_B is a result of stimulated emission induced by laser B, and the electric field amplitude E^{ω_B} is thereby enhanced by this process. It is thus in principle possible to amplify a weak light signal B with a pump laser A.

Let us analyze another example, namely when the external electric field is composed of a single laser field with frequency ω and amplitude E^{ω} in addition to a static field with amplitude E^0:

$$(15) \quad E_\alpha(t) = E_\alpha^0 + E_\alpha^\omega \cos(\omega t)$$

Including terms up to third order in the electric field, the polarization in this case become

$$(16) \quad \mu_\alpha(t) = \mu_\alpha^0 + \alpha_{\alpha\beta}[E_\beta^0 + E_\beta^\omega \cos(\omega t)]$$
$$+ \frac{1}{2}\beta_{\alpha\beta\gamma}[E_\beta^0 + E_\beta^\omega \cos(\omega t)][E_\gamma^0 + E_\gamma^\omega \cos(\omega t)]$$
$$+ \frac{1}{6}\gamma_{\alpha\beta\gamma\delta}[E_\beta^0 + E_\beta^\omega \cos(\omega t)][E_\gamma^0 + E_\gamma^\omega \cos(\omega t)][E_\delta^0 + E_\delta^\omega \cos(\omega t)]$$

We make repeated use of the trigonometric identity in Eq. (13) and rewrite the polarization as

$$(17) \quad \mu_\alpha(t) = \tilde{\mu}_\alpha^0 + \tilde{\mu}_\alpha^\omega \cos(\omega t) + \tilde{\mu}_\alpha^{2\omega} \cos(2\omega t) + \tilde{\mu}_\alpha^{3\omega} \cos(3\omega t)$$

where the Fourier amplitudes of the polarization become

$$\tilde{\mu}_\alpha^0 = \mu_\alpha^0 + \alpha_{\alpha\beta}(0;0)E_\beta^0 + \frac{1}{2}\beta_{\alpha\beta\gamma}(0;0,0)E_\beta^0 E_\gamma^0 \tag{18}$$

$$+ \frac{1}{6}\gamma_{\alpha\beta\gamma\delta}(0;0,0,0)E_\beta^0 E_\gamma^0 E_\delta^0$$

$$+ \frac{1}{4}\beta_{\alpha\beta\gamma}(0;-\omega,\omega)E_\beta^\omega E_\gamma^\omega + \frac{1}{4}\gamma_{\alpha\beta\gamma\delta}(0;-\omega,\omega,0)E_\beta^0 E_\gamma^\omega E_\delta^\omega$$

$$\tilde{\mu}_\alpha^\omega = \alpha_{\alpha\beta}(-\omega;\omega)E_\beta^\omega + \beta_{\alpha\beta\gamma}(-\omega;\omega,0)E_\beta^\omega E_\gamma^0 \tag{19}$$

$$+ \frac{1}{2}\gamma_{\alpha\beta\gamma\delta}(-\omega;\omega,0,0)E_\beta^\omega E_\gamma^0 E_\delta^0$$

$$+ \frac{1}{8}\gamma_{\alpha\beta\gamma\delta}(-\omega;\omega,-\omega,\omega,)E_\beta^\omega E_\gamma^\omega E_\delta^\omega$$

$$\tilde{\mu}_\alpha^{2\omega} = \frac{1}{4}\beta_{\alpha\beta\gamma}(-2\omega;\omega,\omega)E_\beta^\omega E_\gamma^\omega \tag{20}$$

$$+ \frac{1}{4}\gamma_{\alpha\beta\gamma\delta}(-2\omega;\omega,\omega,0)E_\beta^\omega E_\gamma^\omega E_\delta^0$$

$$\tilde{\mu}_\alpha^{3\omega} = \frac{1}{24}\gamma_{\alpha\beta\gamma\delta}(-3\omega;\omega,\omega,\omega)E_\beta^\omega E_\gamma^\omega E_\delta^\omega \tag{21}$$

An alternative summary of this result is sometimes expressed as an expansion of the polarization amplitudes in terms of the electric field amplitudes

$$\tilde{\mu}_\alpha^{\omega_\sigma} = \alpha(-\omega_\sigma;\omega_1)E^{\omega_1} + \frac{1}{2}K^{(2)}\beta(-\omega_\sigma;\omega_1,\omega_2)E^{\omega_1}E^{\omega_2} \tag{22}$$

$$+ \frac{1}{6}K^{(3)}\gamma(-\omega_\sigma;\omega_1,\omega_2,\omega_3)E^{\omega_1}E^{\omega_2}E^{\omega_3} + \cdots$$

where the factors $K^{(n)}$ are required for the polarization related to the molecular response of order n to have the same static limit. By a direct comparison with Eqs. (17)–(21), we are able to identify these factors for some common nonlinear optical processes, see Table 1.

We have seen how the molecular properties in nonlinear optics are defined by the expansion of the molecular polarization in orders of the external electric field, see Eq. (5); beyond the linear polarization this definition introduces the so-called nonlinear hyperpolarizabilities as coupling coefficients between the two quantities. The same equation also expresses an expansion in terms of the number of photons involved in simultaneous quantum-mechanical processes: α, β, γ, and so on involve emission or absorption of two, three, four, etc. photons. The cross section for multiphoton absorption or emission, which takes place in nonlinear optical processes, is in typical cases relatively small and a high density of photons is required for these to occur.

Table 1. Common nonlinear optical processes

Process	Frequencies	Factor
	Second-order processes	$K^{(2)}$
Static	0; 0, 0	1
EOPE[a]	$-\omega; \omega, 0$	2
SHG[b]	$-2\omega; \omega, \omega$	1/2
	Third-order processes	$K^{(3)}$
Static	0; 0, 0, 0	1
EOKE[c]	$-\omega; \omega, 0, 0$	3
IDRI[d]	$-\omega; \omega, -\omega, \omega$	3/4
ESHG[e]	$-2\omega; \omega, \omega, 0$	3/2
THG[f]	$-3\omega; \omega, \omega, \omega$	1/4

[a] Electro-optical Pockels effect.
[b] Second-harmonic generation.
[c] Electro-optical Kerr effect.
[d] Intensity-dependent refractive index.
[e] Electric field-induced second harmonic generation.
[f] Third-harmonic generation.

3. TIME-DEPENDENT PERTURBATION THEORY

In this section we shall derive explicit expressions for the response functions that describe the interaction of a quantum mechanical system and an external electric field. The first order response function was in the previous section referred to as the molecular polarizability and higher-order response functions as the molecular hyperpolarizabilities. We will use both these terminologies as if they were synonymous although the notion of response functions is more general and can be used in other applications than nonlinear optics. As tool we will use time-dependent perturbation theory and assume that the solutions to the eigenvalue problem of the unperturbed system are known. In practice we are of course not able to obtain exact solutions even to the unperturbed system, but our analysis will nevertheless highlight the dependence of the polarizabilities on other molecular parameters such as excitation energies and transition moments. A thorough introduction to various electronic structure methods for determining unperturbed reference states have been given in [31].

The time-evolution of the state vector is given by the time-dependent Schrödinger equation

$$(23) \quad i\hbar \frac{\partial}{\partial t} |\psi(t)\rangle = \hat{H} |\psi(t)\rangle$$

where $\hat{H}$ is the Hamiltonian which may be time-independent or time-dependent. In the former case, the problem of solving Eq. (23) reduces to solving the eigenvalue equation for the Hamiltonian, but in the latter case such simplifications are not possible. However, when the time-dependent part of $\hat{H}$ is a small perturbation, we can still express the solution in terms of the eigenstates of the unperturbed system. Let us separate the Hamiltonian according to

(24) $\quad \hat{H} = \hat{H}_0 + \hat{V}(t)$

where $\hat{H}_0$ is the molecular Hamiltonian of the unperturbed system and $\hat{V}(t)$ is a small perturbation. We assume that the solutions to the eigenvalue problem of $\hat{H}_0$ are known

(25) $\quad \hat{H}_0|n\rangle = E_n|n\rangle$

where $|n\rangle$ are the exact rovibronic eigenstates and E_n the respective energies. Before being exposed to the perturbation, we assume the molecule to be in a reference state $|0\rangle$—in most cases the molecular ground state—and we wish to determine the amplitudes for the molecule to be in another state at a later time. Since the set $\{|n\rangle\}$ is complete, we can at any point in time express the general molecular state $|\psi(t)\rangle$ as an expansion according to

(26) $\quad |\psi(t)\rangle = \sum_n c_n(t)|n\rangle$

where $c_n(-\infty) = \delta_{0n}$ (Kronecker delta function) as a result of the system being initially in the ground state, and our task is to find the time-dependent expansion coefficients $c_n(t)$. We note that if the perturbation had not been present then $c_n(t) = c_n(0)\exp(-iE_n t/\hbar)$, and it is therefore appropriate to rewrite Eq. (26) as

(27) $\quad |\psi(t)\rangle = \sum_n d_n(t) e^{-iE_n t/\hbar}|n\rangle$

because changes of d_n in time are due only to the small perturbation. The coefficients d_n can be written in a power series of $\hat{V}$:

(28) $\quad d_n(t) = d_n^{(0)} + d_n^{(1)}(t) + d_n^{(2)}(t) + \cdots$

Operating on both sides of Eq. (27) with $i\hbar\partial/\partial t - \hat{H}$ from the left and multiplying from the left with the bra vector $\langle m|\exp(iE_m t/\hbar)$ gives us an equation from which we can determine $d_m(t)$

(29) $\quad i\hbar\frac{\partial}{\partial t}d_m^{(N)} = \sum_n \langle m|\hat{V}(t)|n\rangle e^{i\omega_{mn}t} d_n^{(N-1)}$

where ω_{mn} is the transition frequency $(E_m - E_n)/\hbar$, and $d_n^{(0)}$ is δ_{0n}. We see that Eq. (29) is recursive in that by feeding it with $d_n^{(0)}$ we can determine the next order response, $d_n^{(1)}$, etc. by straightforward time-integration. In accordance with the expansion of the coefficients d_n in orders of the perturbation $\hat{V}$, it is customary to write the wave function as

(30) $\quad |\psi(t)\rangle = |\psi^{(0)}(t)\rangle + |\psi^{(1)}(t)\rangle + |\psi^{(2)}(t)\rangle + \cdots$

where

$$|\psi^{(N)}(t)\rangle = \sum_n d_n^{(N)} e^{-iE_n t/\hbar} |n\rangle \qquad (31)$$

In general, expectation values will be time-dependent due to the time-dependence of the Hamiltonian. However, the expectation value of the molecular electric dipole operator $\hat{\mu}$ is of special interest to us, since it corresponds to the molecular polarization and therefore also to the polarizabilities and hyperpolarizabilities. With help of Eq. (30), the expectation value of $\hat{\mu}$ becomes

$$\langle \psi(t)|\hat{\mu}|\psi(t)\rangle = \langle\hat{\mu}\rangle^{(0)} + \langle\hat{\mu}\rangle^{(1)} + \langle\hat{\mu}\rangle^{(2)} + \langle\hat{\mu}\rangle^{(3)} + \cdots \qquad (32)$$

where the various orders of the time-dependent polarizations are

$$\langle\hat{\mu}\rangle^{(0)} = \langle 0|\hat{\mu}|0\rangle \qquad (33)$$

$$\langle\hat{\mu}\rangle^{(1)} = \langle\psi^{(0)}|\hat{\mu}|\psi^{(1)}\rangle + \langle\psi^{(1)}|\hat{\mu}|\psi^{(0)}\rangle \qquad (34)$$

$$\langle\hat{\mu}\rangle^{(2)} = \langle\psi^{(0)}|\hat{\mu}|\psi^{(2)}\rangle + \langle\psi^{(1)}|\hat{\mu}|\psi^{(1)}\rangle + \langle\psi^{(2)}|\hat{\mu}|\psi^{(0)}\rangle \qquad (35)$$

$$\langle\hat{\mu}\rangle^{(3)} = \langle\psi^{(0)}|\hat{\mu}|\psi^{(3)}\rangle + \langle\psi^{(1)}|\hat{\mu}|\psi^{(2)}\rangle + \langle\psi^{(2)}|\hat{\mu}|\psi^{(1)}\rangle + \langle\psi^{(3)}|\hat{\mu}|\psi^{(0)}\rangle \qquad (36)$$

We recognize the first term as the permanent electric dipole moment of the molecule, and from the latter terms we will shortly be able to identify expressions for the polarizabilities and hyperpolarizabilities.

In the electric dipole approximation the interaction between the molecule and the electric field is described by the operator

$$\hat{V}(t) = -\hat{\mu}_\alpha E_\alpha(t) e^{\epsilon t} \qquad (37)$$

where $\hat{\mu}_\alpha$ is the electric dipole operator along the molecular axis α, and $E(t)$ is the classical electric field which can be decomposed in the frequency domain according to Eq. (4). The promotion of the dipole moment to become an operator while leaving the electric field as an amplitude corresponds to the quantum mechanical treatment of the molecule and the classical treatment of the external field. The exponential factor with the small positive infinitesimal ϵ ensures that the perturbation is switched on adiabatically and that it vanishes for $t = -\infty$. Inserting $\hat{V}(t)$ and $d_n^{(0)}$ into Eq. (29) and performing the time integration, the first order response becomes

$$d_m^{(1)}(t) = -\frac{1}{i\hbar} \int_{-\infty}^{t} \sum_n \langle m|\hat{\mu}_\alpha \sum_{\omega_1} E_\alpha^{\omega_1} e^{-i\omega_1 t'} e^{\epsilon t'} |n\rangle e^{i\omega_{mn} t'} \delta_{0n} dt' \qquad (38)$$

$$= \frac{1}{\hbar} \sum_{\omega_1} \frac{\langle m|\hat{\mu}_\alpha|0\rangle E_\alpha^{\omega_1}}{\omega_{m0} - \omega_1 - i\epsilon} e^{i(\omega_{m0}-\omega_1)t} e^{\epsilon t}$$

where it is recognized that the contribution from the lower integration limit vanishes due to ϵ. Since the coefficients $d_m^{(1)}(-\infty)$ are zero due to the positive infinitesimal, the result is consistent with the molecule being in its ground state at $t = -\infty$.

Microscopic Theory of Nonlinear Optics

The second order response is obtained from Eq. (29) by insertion of the result for the first order response:

$$(39) \quad d_m^{(2)}(t) = -\frac{1}{i\hbar} \int_{-\infty}^{t} \sum_n \langle m|\hat{\mu}_\alpha \sum_{\omega_1} E_\alpha^{\omega_1} e^{-i\omega_1 t'} e^{\epsilon t'} |n\rangle e^{i\omega_{mn} t'}$$

$$\times \frac{1}{\hbar} \sum_{\omega_2} \frac{\langle n|\hat{\mu}_\beta|0\rangle E_\beta^{\omega_2}}{\omega_{n0} - \omega_2 - i\epsilon} e^{i(\omega_{n0}-\omega_2)t'} e^{\epsilon t'} \, dt'$$

$$= \frac{1}{\hbar^2} \sum_{\omega_1 \omega_2} \sum_n \frac{\langle m|\hat{\mu}_\alpha|n\rangle\langle n|\hat{\mu}_\beta|0\rangle E_\alpha^{\omega_1} E_\beta^{\omega_2}}{(\omega_{m0} - \omega_1 - \omega_2 - i2\epsilon)(\omega_{n0} - \omega_2 - i\epsilon)} e^{i(\omega_{m0}-\omega_1-\omega_2)t} e^{2\epsilon t}$$

Repeating the procedure once more, we obtain the third order response as

$$(40) \quad d_m^{(3)}(t) = -\frac{1}{i\hbar} \int_{-\infty}^{t} \sum_n \langle m|\hat{\mu}_\alpha \sum_{\omega_1} E_\alpha^{\omega_1} e^{-i\omega_1 t'} e^{\epsilon t'} |n\rangle e^{i\omega_{mn} t'}$$

$$\times \frac{1}{\hbar^2} \sum_{\omega_2 \omega_3} \sum_p \frac{\langle n|\hat{\mu}_\beta|p\rangle\langle p|\hat{\mu}_\gamma|0\rangle E_\beta^{\omega_2} E_\gamma^{\omega_3}}{(\omega_{n0} - \omega_2 - \omega_3 - i2\epsilon)(\omega_{p0} - \omega_3 - i\epsilon)} e^{i(\omega_{n0}-\omega_2-\omega_3)t'} e^{2\epsilon t'} \, dt'$$

$$= \frac{1}{\hbar^3} \sum_{\omega_1 \omega_2 \omega_3} \sum_{np} \frac{\langle m|\hat{\mu}_\alpha|n\rangle\langle n|\hat{\mu}_\beta|p\rangle\langle p|\hat{\mu}_\gamma|0\rangle E_\alpha^{\omega_1} E_\beta^{\omega_2} E_\gamma^{\omega_3}}{(\omega_{m0} - \omega_1 - \omega_2 - \omega_3 - i3\epsilon)(\omega_{n0} - \omega_2 - \omega_3 - i2\epsilon)(\omega_{p0} - \omega_3 - i\epsilon)}$$

$$\times e^{i(\omega_{m0}-\omega_1-\omega_2-\omega_3)t} e^{3\epsilon t}$$

We are now in a position to determine the molecular polarization including terms of at the most third order in the perturbation through Eqs. (33)–(36).

3.1 Linear Polarizability

The first-order polarization is obtained from Eq. (34) by insertion of the zeroth-order as well as the first-order correction to the wave function, where $|\psi^{(0)}\rangle = \exp(-iE_0 t/\hbar)|0\rangle$ and $|\psi^{(1)}\rangle$ is given by a combination of Eqs. (38) and (31):

$$(41) \quad \langle \hat{\mu}_\alpha \rangle^{(1)} = \langle \psi^{(0)}|\hat{\mu}_\alpha|\psi^{(1)}\rangle + \langle \psi^{(1)}|\hat{\mu}_\alpha|\psi^{(0)}\rangle$$

$$= \langle 0|e^{iE_0 t/\hbar} \hat{\mu}_\alpha \sum_n \frac{1}{\hbar} \sum_{\omega_1} \frac{\langle n|\hat{\mu}_\beta|0\rangle E_\beta^{\omega_1}}{\omega_{n0} - \omega_1 - i\epsilon} e^{i(\omega_{n0}-\omega_1)t} e^{\epsilon t} e^{-iE_n t/\hbar}|n\rangle$$

$$+ \sum_n \frac{1}{\hbar} \sum_{\omega_1} \frac{\langle 0|\hat{\mu}_\beta|n\rangle [E_\beta^{\omega_1}]^*}{\omega_{n0} - \omega_1 + i\epsilon} e^{-i(\omega_{n0}-\omega_1)t} e^{\epsilon t} e^{iE_n t/\hbar} \langle n|\hat{\mu}_\alpha e^{-iE_0 t/\hbar}|0\rangle$$

$$= \sum_{\omega_1} \frac{1}{\hbar} \sum_n \frac{\langle 0|\hat{\mu}_\alpha|n\rangle\langle n|\hat{\mu}_\beta|0\rangle}{\omega_{n0} - \omega_1 - i\epsilon} E_\beta^{\omega_1} e^{-i\omega_1 t} e^{\epsilon t}$$

$$+ \sum_{\omega_1} \frac{1}{\hbar} \sum_n \frac{\langle 0|\hat{\mu}_\beta|n\rangle\langle n|\hat{\mu}_\alpha|0\rangle}{\omega_{n0} - \omega_1 + i\epsilon} [E_\beta^{\omega_1}]^* e^{i\omega_1 t} e^{\epsilon t}$$

$$= \sum_{\omega_1} \frac{1}{\hbar} \sum_n \left[\frac{\langle 0|\hat{\mu}_\alpha|n\rangle\langle n|\hat{\mu}_\beta|0\rangle}{\omega_{n0} - \omega_1 - i\epsilon} + \frac{\langle 0|\hat{\mu}_\beta|n\rangle\langle n|\hat{\mu}_\alpha|0\rangle}{\omega_{n0} + \omega_1 + i\epsilon} \right] E_\beta^{\omega_1} e^{-i\omega_1 t} e^{\epsilon t}$$

In the last step we have used that $[E^{\omega_1}]^* = E^{-\omega_1}$ and the fact that ω_1 runs over both positive and negative frequencies. By a direct comparison to Eq. (5), we are able to identify the resulting quantum mechanical formula for the linear polarizability as

$$(42) \qquad \alpha_{\alpha\beta}(-\omega;\omega) = \frac{1}{\hbar}\sum_n\left[\frac{\langle 0|\hat{\mu}_\alpha|n\rangle\langle n|\hat{\mu}_\beta|0\rangle}{\omega_{n0}-\omega-i\epsilon} + \frac{\langle 0|\hat{\mu}_\beta|n\rangle\langle n|\hat{\mu}_\alpha|0\rangle}{\omega_{n0}+\omega+i\epsilon}\right]$$

It is clear that this formula can be used directly as it stands for practical computations of the optical polarization once the excitation energies and transition moments of the system are known. More importantly, however, it elucidates the dependence of the linear polarizability on the quantum mechanical properties of the molecule.

For the two terms in the sum-over-states expression in Eq. (42) that involve the ground state $n = 0$, the transition frequency ω_{n0} is zero. The two terms are of opposite sign and will therefore cancel, and it is common practice to exclude the ground state from the summation and to use a primed summation symbol $\sum'_n$ for the sum over excited states.

We have made a point of carrying along the positive infinitesimal ϵ in the perturbation not only to avoid singularities or divergencies in the time-integration step of the expansion coefficients d_n above but also to avoid divergences in the linear response function itself. With a reasonable laser detuning, however, the imaginary term in the denominator of Eq. (42) can safely be neglected in the calculation of $\alpha(-\omega;\omega)$, and, since the set of eigenstates $\{|n\rangle\}$ can be chosen as real, without loss of generality, the linear response function is real in the nonresonant region. The working formula will in this case take the form

$$(43) \qquad \alpha_{\alpha\beta}(-\omega;\omega) = \frac{1}{\hbar}\sum_n{}'\left[\frac{\langle 0|\hat{\mu}_\alpha|n\rangle\langle n|\hat{\mu}_\beta|0\rangle}{\omega_{n0}-\omega} + \frac{\langle 0|\hat{\mu}_\beta|n\rangle\langle n|\hat{\mu}_\alpha|0\rangle}{\omega_{n0}+\omega}\right]$$

We note that after having excluded the ground state from the summation, the polarizability is convergent in the limit of nonoscillating fields also without the imaginary term in the denominator. The singularities that occur for static fields with $\epsilon = 0$ and inclusion of the ground state in the summation are called secular divergences, since they can be removed by mathematical manipulations of the sum-over-states expression. We will later see that secular divergences appear for the hyperpolarizabilities as well. However, with $\epsilon = 0$, the linear polarizability, or the linear polarization propagator as it is sometimes denoted, will also be divergent for perturbation frequencies equal to the transition frequencies of the system. These singularities are true divergences and are known as resonances of the system. This fact is frequently utilized in approximative calculations in order to find estimates of the molecular excitation energies as poles of the polarization propagator. The residues of the propagator can be used to identify the absolute value of the transition moments between the ground and excited states and thereby describe the intensities in the linear absorption spectrum. The residue of the linear polarizability is seen from Eq. (43) to equal

$$(44) \quad \lim_{\omega \to \omega_{f0}} (\omega_{f0} - \omega) \, \alpha_{\alpha\beta}(-\omega; \omega) = \langle 0|\hat{\mu}_\alpha|f\rangle\langle f|\hat{\mu}_\beta|0\rangle$$

where $|f\rangle$ denotes the final state in the one-photon absorption process.

We have treated the perturbing electric field as classical, which means that the lifetimes of the eigenstates $|n\rangle$ of the unperturbed Hamiltonian are infinite. The linear absorption from the ground state to the excited states is in this case described by Dirac delta functions peaked at the excitation energies, and monochromatic light sources need to be exactly tuned for transitions to occur. In reality, and when treating also the electric field quantum mechanically, there is no such thing as an exact zero external field, as well as there are no infinite lifetimes of excited states even of perfectly isolated systems. A phenomenological way to incorporate the situation of finite excited state lifetimes into our description is to introduce population decay rates Γ_{n0} that equal the inverse lifetime, or the lifetime broadening, of the excited state $|n\rangle$. The expression for the linear polarizability is then written as

$$(45) \quad \alpha_{\alpha\beta}(-\omega; \omega) = \frac{1}{\hbar}\sum_n{}' \left[\frac{\langle 0|\mu_\alpha|n\rangle\langle n|\mu_\beta|0\rangle}{\omega_{n0} - \omega - i\Gamma_{n0}/2} + \frac{\langle 0|\mu_\beta|n\rangle\langle n|\mu_\alpha|0\rangle}{\omega_{n0} + \omega + i\Gamma_{n0}/2} \right]$$

In general, and for the nonlinear hyperpolarizabilities to be derived below, one introduces Γ_{mn} for the transition between states $|m\rangle$ and $|n\rangle$. In effect the imaginary term $i\Gamma_{n0}/2$ takes the place of $i\epsilon$ in Eq. (42). The linear absorption spectrum, which corresponds to the imaginary part of Eq. (45), will be built from "smeared out" Dirac delta functions of Lorentzian shape, i.e. the frequency-integrated absorption will remain constant regardless of the value of the lifetime broadening. The real part of the polarizability is related to the refractive index n of the sample

$$(46) \quad n_2 = \frac{1 + 2\alpha N_A/3\varepsilon_0}{1 - \alpha N_A/3\varepsilon_0} \approx 1 + \frac{N_A}{2\varepsilon_0}\alpha$$

where ε_0 is the permittivity in vacuum, and N_A is Avogadro's number.

3.2 First-order Hyperpolarizability

Just as an explicit formula for the linear polarizability is identified from the linear polarization, we are able to retrieve the corresponding formula for the first-order hyperpolarizability from the second-order polarization. We obtain the second-order polarization from Eq. (35) by insertion of the first-order and second-order corrections to the wave function, where $\psi^{(1)}$ and $\psi^{(2)}$ are given by Eqs. (38) and (39), respectively, in combination with Eq. (31). Let us first consider the three terms in Eq. (35) one at a time:

(47) $\langle\psi^{(0)}|\hat{\mu}_\alpha|\psi^{(2)}\rangle = \langle 0|e^{iE_0 t/\hbar}\hat{\mu}_\alpha$

$$\times \sum_n \frac{1}{\hbar^2} \sum_{\omega_1 \omega_2} \sum_p \frac{\langle n|\hat{\mu}_\beta|p\rangle\langle p|\hat{\mu}_\gamma|0\rangle E_\beta^{\omega_1} E_\gamma^{\omega_2}}{(\omega_{n0}-\omega_1-\omega_2-i2\epsilon)(\omega_{p0}-\omega_2-i\epsilon)}$$

$$\times e^{i(\omega_{n0}-\omega_1-\omega_2)t} e^{2\epsilon t} e^{-iE_n t/\hbar}|n\rangle$$

$$= \sum_{\omega_1 \omega_2} \frac{1}{\hbar^2} \sum_{np} \frac{\langle 0|\hat{\mu}_\alpha|n\rangle\langle n|\hat{\mu}_\beta|p\rangle\langle p|\hat{\mu}_\gamma|0\rangle E_\beta^{\omega_1} E_\gamma^{\omega_2}}{(\omega_{n0}-\omega_1-\omega_2-i2\epsilon)(\omega_{p0}-\omega_2-i\epsilon)} e^{-i(\omega_1+\omega_2)t} e^{2\epsilon t}$$

(48) $\langle\psi^{(1)}|\hat{\mu}_\alpha|\psi^{(1)}\rangle =$

$$\times \sum_n \frac{1}{\hbar} \sum_{\omega_1} \frac{\langle 0|\hat{\mu}_\beta|n\rangle [E_\beta^{\omega_1}]^*}{\omega_{n0}-\omega_1+i\epsilon} e^{-i(\omega_{n0}-\omega_1)t} e^{\epsilon t} e^{iE_n t/\hbar} \langle n|\hat{\mu}_\alpha$$

(49) $\times \sum_p \frac{1}{\hbar} \sum_{\omega_1} \frac{\langle p|\hat{\mu}_\gamma|0\rangle E_\gamma^{\omega_2}}{\omega_{p0}-\omega_2-i\epsilon} e^{i(\omega_{p0}-\omega_2)t} e^{\epsilon t} e^{-iE_p t/\hbar}|p\rangle$

$$= \sum_{\omega_1 \omega_2} \frac{1}{\hbar^2} \sum_{np} \frac{\langle 0|\hat{\mu}_\beta|n\rangle\langle n|\hat{\mu}_\alpha|p\rangle\langle p|\hat{\mu}_\gamma|0\rangle E_\beta^{\omega_1} E_\gamma^{\omega_2}}{(\omega_{n0}+\omega_1+i\epsilon)(\omega_{p0}-\omega_2-i\epsilon)} e^{-i(\omega_1+\omega_2)t} e^{2\epsilon t}$$

(50) $\langle\psi^{(2)}|\hat{\mu}_\alpha|\psi^{(0)}\rangle = \sum_n \frac{1}{\hbar^2} \sum_{\omega_1 \omega_2} \sum_p \frac{\langle p|\hat{\mu}_\beta|n\rangle\langle 0|\hat{\mu}_\gamma|p\rangle [E_\beta^{\omega_1}]^* [E_\gamma^{\omega_2}]^*}{(\omega_{n0}-\omega_1-\omega_2+i2\epsilon)(\omega_{p0}-\omega_2+i\epsilon)}$

$$\times e^{-i(\omega_{n0}-\omega_1-\omega_2)t} e^{2\epsilon t} e^{iE_n t/\hbar} \langle n|\hat{\mu}_\alpha e^{-iE_0 t/\hbar}|0\rangle$$

$$= \sum_{\omega_1 \omega_2} \frac{1}{\hbar^2} \sum_{np} \frac{\langle 0|\hat{\mu}_\gamma|p\rangle\langle p|\hat{\mu}_\beta|n\rangle\langle n|\hat{\mu}_\alpha|0\rangle E_\beta^{\omega_1} E_\gamma^{\omega_2}}{(\omega_{n0}+\omega_1+\omega_2+i2\epsilon)(\omega_{p0}+\omega_2+i\epsilon)} e^{-i(\omega_1+\omega_2)t} e^{2\epsilon t}$$

where we again have used that $[E^\omega]^* = E^{-\omega}$ and the fact that ω_1 and ω_2 are dummy summation indices that run over all positive and negative frequencies. We note that neither of the three equations above are symmetric in the indices β and γ, but we have already discussed in connection with Eq. (6) that we desire the hyperpolarizability tensors to be intrinsically symmetric. Remembering, however, that also β and γ are merely summation indices by the use of the Einstein summation convention, it is clear that we can force Eqs. (47)–(50) to be intrinsically symmetric without altering the physical polarization; we do so by operating with the operator $1/2\sum\mathcal{P}_{1,2}$ which performs the summation of terms obtained by permuting the pairs (β,ω_1) and (γ,ω_2). The factor of one half obviously causes the polarization to maintain its original value. The final expression for the second-order polarization thereby becomes

$$
(51) \quad \langle \hat{\mu} \rangle^{(2)} = \langle \psi^{(0)} | \hat{\mu} | \psi^{(2)} \rangle + \langle \psi^{(1)} | \hat{\mu} | \psi^{(1)} \rangle + \langle \psi^{(2)} | \hat{\mu} | \psi^{(0)} \rangle
$$

$$
= \frac{1}{2} \sum_{\omega_1 \omega_2} \frac{1}{\hbar^2} \sum \mathcal{P}_{1,2} \sum_{np} \left[\frac{\langle 0 | \hat{\mu}_\alpha | n \rangle \langle n | \hat{\mu}_\beta | p \rangle \langle p | \hat{\mu}_\gamma | 0 \rangle}{(\omega_{n0} - \omega_1 - \omega_2 - i2\epsilon)(\omega_{p0} - \omega_2 - i\epsilon)} \right.
$$

$$
+ \frac{\langle 0 | \hat{\mu}_\beta | n \rangle \langle n | \hat{\mu}_\alpha | p \rangle \langle p | \hat{\mu}_\gamma | 0 \rangle}{(\omega_{n0} + \omega_1 + i\epsilon)(\omega_{p0} - \omega_2 - i\epsilon)}
$$

$$
\left. + \frac{\langle 0 | \hat{\mu}_\gamma | p \rangle \langle p | \hat{\mu}_\beta | n \rangle \langle n | \hat{\mu}_\alpha | 0 \rangle}{(\omega_{n0} + \omega_1 + \omega_2 + i2\epsilon)(\omega_{p0} + \omega_2 + i\epsilon)} \right] E_\beta^{\omega_1} E_\gamma^{\omega_2} e^{-i(\omega_1 + \omega_2)t} e^{2\epsilon t}
$$

and, by a direct comparison to Eq. (5), we identify the expression for the first-order hyperpolarizability to be

$$
(52) \quad \beta_{\alpha\beta\gamma}(-\omega_\sigma; \omega_1, \omega_2) = \frac{1}{\hbar^2} \sum \mathcal{P}_{1,2} \sum_{np} \left[\frac{\langle 0 | \hat{\mu}_\alpha | n \rangle \langle n | \hat{\mu}_\beta | p \rangle \langle p | \hat{\mu}_\gamma | 0 \rangle}{(\omega_{n0} - \omega_\sigma - i2\epsilon)(\omega_{p0} - \omega_2 - i\epsilon)} \right.
$$

$$
+ \frac{\langle 0 | \hat{\mu}_\beta | n \rangle \langle n | \hat{\mu}_\alpha | p \rangle \langle p | \hat{\mu}_\gamma | 0 \rangle}{(\omega_{n0} + \omega_1 + i\epsilon)(\omega_{p0} - \omega_2 - i\epsilon)}
$$

$$
\left. + \frac{\langle 0 | \hat{\mu}_\gamma | p \rangle \langle p | \hat{\mu}_\beta | n \rangle \langle n | \hat{\mu}_\alpha | 0 \rangle}{(\omega_{n0} + \omega_\sigma + i2\epsilon)(\omega_{p0} + \omega_2 + i\epsilon)} \right]
$$

In connection with the linear response function, we discussed briefly the possibility of incorporating absorption of light through imaginary damping terms that parallels the positive infinitesimals in the expressions for polarizabilities and hyperpolarizabilities. With a reasonable detuning, we argued that the linear absorption is negligible and the linear response function is real. Analogously, if all one- and two-photon frequencies are nonresonant then the first-order hyperpolarizability will be real, which is equivalent to letting ϵ equal zero in Eq. (52). In the nonresonant region, therefore, the expression for the first-order hyperpolarizability can be written in a more compact form:

$$
(53) \quad \beta_{\alpha\beta\gamma}(-\omega_\sigma; \omega_1, \omega_2)
$$

$$
= \frac{1}{\hbar^2} \sum \mathcal{P}_{-\sigma,1,2} \sum_{np} \frac{\langle 0 | \hat{\mu}_\alpha | n \rangle \langle n | \hat{\mu}_\beta | p \rangle \langle p | \hat{\mu}_\gamma | 0 \rangle}{(\omega_{n0} - \omega_\sigma)(\omega_{p0} - \omega_2)}
$$

where $\sum \mathcal{P}_{-\sigma,1,2}$ denote the sum of the six terms one gets by permuting pairs of $(\alpha, -\omega_\sigma)$, (β, ω_1) and (γ, ω_2). The verification of Eq. (53) is done in a straightforward manner by a direct comparison of the six terms in Eq. (52) with $\epsilon = 0$. The first-order hyperpolarizability tensor is said to possess full permutation symmetry in the nonresonant region.

Before closing the derivation of the first-order hyperpolarizability, we wish to remove the apparent divergences of Eq. (53) in the limit of non-oscillating perturbing fields. Just as we did for $\alpha(-\omega; \omega)$, it is our intention to derive an equivalent expression for $\beta(-\omega_\sigma; \omega_1, \omega_2)$ that excludes terms involving the ground state in the summation. Let us begin by splitting Eq. (53) according to

$$(54) \quad \beta_{\alpha\beta\gamma}(-\omega_\sigma; \omega_1, \omega_2) = \frac{1}{\hbar^2} \sum \mathcal{P}_{-\sigma,1,2} \left[\sum_{np}{}' \frac{\langle 0|\hat{\mu}_\alpha|n\rangle\langle n|\hat{\mu}_\beta|p\rangle\langle p|\hat{\mu}_\gamma|0\rangle}{(\omega_{n0} - \omega_\sigma)(\omega_{p0} - \omega_2)} \right.$$

$$+ \sum_p{}' \frac{\langle 0|\hat{\mu}_\alpha|0\rangle\langle 0|\hat{\mu}_\beta|p\rangle\langle p|\hat{\mu}_\gamma|0\rangle}{-\omega_\sigma(\omega_{p0} - \omega_2)}$$

$$+ \sum_n{}' \frac{\langle 0|\hat{\mu}_\alpha|n\rangle\langle n|\hat{\mu}_\beta|0\rangle\langle 0|\hat{\mu}_\gamma|0\rangle}{-(\omega_{n0} - \omega_\sigma)\omega_2}$$

$$+ \left. \frac{\langle 0|\hat{\mu}_\alpha|0\rangle\langle 0|\hat{\mu}_\beta|0\rangle\langle 0|\hat{\mu}_\gamma|0\rangle}{\omega_\sigma\omega_2} \right]$$

Since

$$(55) \quad \frac{1}{\omega_\sigma\omega_2} = \frac{1}{\omega_1\omega_2} - \frac{1}{\omega_\sigma\omega_1}$$

the six permutations generated from the last term in Eq. (54) will cancel each other. Furthermore, with the full permutation operator on the outside, we are free to interchange any two pairs of indices in the respective terms in Eq. (54): in the second term we interchange $(\alpha, -\omega_\sigma)$ and (β, ω_1), and in the third term we interchange (γ, ω_2) and (β, ω_1). The expression for the first-order hyperpolarizability thereby appears as

$$(56) \quad \beta_{\alpha\beta\gamma}(-\omega_\sigma; \omega_1, \omega_2) = \frac{1}{\hbar^2} \sum \mathcal{P}_{-\sigma,1,2} \left[\sum_{np}{}' \frac{\langle 0|\hat{\mu}_\alpha|n\rangle\langle n|\hat{\mu}_\beta|p\rangle\langle p|\hat{\mu}_\gamma|0\rangle}{(\omega_{n0} - \omega_\sigma)(\omega_{p0} - \omega_2)} \right.$$

$$+ \sum_n{}' \langle 0|\hat{\mu}_\alpha|n\rangle\langle n|\hat{\mu}_\gamma|0\rangle\langle 0|\hat{\mu}_\beta|0\rangle$$

$$\left. \times \left(\frac{1}{\omega_1(\omega_{n0} - \omega_2)} - \frac{1}{(\omega_{n0} - \omega_\sigma)\omega_1} \right) \right]$$

$$= \frac{1}{\hbar^2} \sum \mathcal{P}_{-\sigma,1,2} \left[\sum_{np}{}' \frac{\langle 0|\hat{\mu}_\alpha|n\rangle\langle n|\hat{\mu}_\beta|p\rangle\langle p|\hat{\mu}_\gamma|0\rangle}{(\omega_{n0} - \omega_\sigma)(\omega_{p0} - \omega_2)} \right.$$

$$\left. - \sum_n{}' \frac{\langle 0|\hat{\mu}_\alpha|n\rangle\langle 0|\hat{\mu}_\beta|0\rangle\langle n|\hat{\mu}_\gamma|0\rangle}{(\omega_{n0} - \omega_\sigma)(\omega_{n0} - \omega_2)} \right]$$

$$= \frac{1}{\hbar^2} \sum \mathcal{P}_{-\sigma,1,2} \sum_{np}{}' \frac{\langle 0|\hat{\mu}_\alpha|n\rangle\langle n|\overline{\hat{\mu}}_\beta|p\rangle\langle p|\hat{\mu}_\gamma|0\rangle}{(\omega_{n0} - \omega_\sigma)(\omega_{p0} - \omega_2)}$$

where $\overline{\hat{\mu}}$ is the fluctuation dipole moment operator $\hat{\mu} - \langle 0|\hat{\mu}|0\rangle$. This equation represents an appropriate form of the quantum mechanical formula for the first-order

Microscopic Theory of Nonlinear Optics

hyperpolarizability in the nonresonant region and including non-oscillating external fields. The molecular parameters that enter this expression are the excitation energies, the ground to excited state transition moments, the excited to excited state transition moments, and the permanent dipole moment of the ground as well as the excited states. All this information is in principle contained in $\beta(-\omega_\sigma; \omega_1, \omega_2)$, and it can be extracted from the second-order response function by a residue analysis.

The first-order hyperpolarizability has both single and double residues. From Eq. (53), we see that one of the first-order residues becomes

$$(57) \quad \lim_{\omega_2 \to \omega_{f0}} (\omega_{f0} - \omega_2)\beta(-\omega_\sigma; \omega_1, \omega_2) =$$

$$= \frac{1}{\hbar^2} \sum_n \left[\frac{\langle 0|\hat{\mu}_\alpha|n\rangle\langle n|\hat{\mu}_\beta|f\rangle\langle f|\hat{\mu}_\gamma|0\rangle}{(\omega_{n0} - \omega_1 - \omega_{f0})} + \frac{\langle 0|\hat{\mu}_\beta|n\rangle\langle n|\hat{\mu}_\alpha|f\rangle\langle f|\hat{\mu}_\gamma|0\rangle}{(\omega_{n0} + \omega_1)} \right]$$

$$= \frac{1}{\hbar^2} \sum_n \left[\frac{\langle 0|\hat{\mu}_\alpha|n\rangle\langle n|\hat{\mu}_\beta|f\rangle}{(\omega_{nf} - \omega_1)} + \frac{\langle 0|\hat{\mu}_\beta|n\rangle\langle n|\hat{\mu}_\alpha|f\rangle}{(\omega_{n0} + \omega_1)} \right] \langle f|\hat{\mu}_\gamma|0\rangle$$

where it includes two terms due to the permutation of pairs of indices $(\alpha, -\omega_\sigma)$ and (β, ω_1). This residue, when evaluated for $\omega_1 = -\omega_{f0}/2$, is closely connected to the two-photon matrix element describing absorption of two monochromatic light quanta in the transition from the ground $|0\rangle$ to the excited state $|f\rangle$, in analogy with the correspondence between the one-photon matrix element and the residue of the linear polarizability. Turning to the double residues, we focus on the case when $\omega_1 = -\omega_{f0}$ and $\omega_2 = \omega_{g0}$. Since we are also interested in the situation when $|f\rangle = |g\rangle$, we choose the expression in Eq. (56) for the hyperpolarizability and where the secular divergences have been removed. This particular double residue will then become

$$(58) \quad \lim_{\omega_1 \to -\omega_{f0}} (\omega_{f0} + \omega_1) \left[\lim_{\omega_2 \to \omega_{g0}} (\omega_{g0} - \omega_2)\beta(-\omega_\sigma; \omega_1, \omega_2) \right] =$$

$$= \langle 0|\hat{\mu}_\beta|f\rangle\langle f|\overline{\hat{\mu}_\alpha}|g\rangle\langle g|\hat{\mu}_\gamma|0\rangle$$

When $|f\rangle \neq |g\rangle$, the matrix element in the middle will equal the one-photon transition moment between the two excited states $|f\rangle$ and $|g\rangle$. On the other hand, when $|f\rangle = |g\rangle$ the same matrix element will equal the difference between the dipole moment of the excited state $|f\rangle$ and the ground state $|0\rangle$. This provides a first example of the possibility of extracting lower-order properties of molecular excited states from higher-order ground state properties.

3.3 Second-order Hyperpolarizability

We must now once more return to the perturbation expansion of the molecular polarization and consider the third-order polarization in Eq. (36) from which we will identify a formula for the second-order hyperpolarizability in analogy with what we have achieved for the polarizability and the first-order hyperpolarizability. The first-, second-, and third-order corrections to the wave function are given by Eqs. (38), (39), and (40), respectively, in combination with Eq. (31). The four terms that contribute to the third-order polarization are:

$$(59) \quad \langle \psi^{(0)} | \hat{\mu}_\alpha | \psi^{(3)} \rangle = \langle 0 | e^{iE_0 t/\hbar} \hat{\mu}_\alpha \sum_n \frac{1}{\hbar^3} \sum_{\omega_1 \omega_2 \omega_3} \sum_{mp}$$

$$\times \frac{\langle n|\hat{\mu}_\beta|m\rangle \langle m|\hat{\mu}_\gamma|p\rangle \langle p|\hat{\mu}_\delta|0\rangle E_\beta^{\omega_1} E_\gamma^{\omega_2} E_\delta^{\omega_3}}{(\omega_{n0} - \omega_1 - \omega_2 - \omega_3 - i3\epsilon)(\omega_{m0} - \omega_2 - \omega_3 - i2\epsilon)(\omega_{p0} - \omega_3 - i\epsilon)}$$

$$\times e^{i(\omega_{n0} - \omega_1 - \omega_2 - \omega_3)t} e^{3\epsilon t} e^{-iE_n t/\hbar} | n \rangle$$

$$= \sum_{\omega_1 \omega_2 \omega_3} \frac{1}{\hbar^3} \sum_{nmp}$$

$$\times \frac{\langle 0|\hat{\mu}_\alpha|n\rangle \langle n|\hat{\mu}_\beta|m\rangle \langle m|\hat{\mu}_\gamma|p\rangle \langle p|\hat{\mu}_\delta|0\rangle E_\beta^{\omega_1} E_\gamma^{\omega_2} E_\delta^{\omega_3}}{(\omega_{n0} - \omega_1 - \omega_2 - \omega_3 - i3\epsilon)(\omega_{m0} - \omega_2 - \omega_3 - i2\epsilon)(\omega_{p0} - \omega_3 - i\epsilon)}$$

$$\times e^{-i(\omega_1 + \omega_2 + \omega_3)t} e^{3\epsilon t}$$

$$(60) \quad \langle \psi^{(1)} | \hat{\mu}_\alpha | \psi^{(2)} \rangle = \sum_n \frac{1}{\hbar} \sum_{\omega_1} \frac{\langle 0|\hat{\mu}_\beta|n\rangle [E_\beta^{\omega_1}]^*}{\omega_{n0} - \omega_1 + i\epsilon} e^{-i(\omega_{n0} - \omega_1)t} e^{\epsilon t} e^{iE_n t/\hbar} \langle n|\hat{\mu}_\alpha$$

$$\times \sum_m \frac{1}{\hbar^2} \sum_{\omega_2 \omega_3} \sum_p \frac{\langle m|\hat{\mu}_\gamma|p\rangle \langle p|\hat{\mu}_\delta|0\rangle E_\gamma^{\omega_2} E_\delta^{\omega_3}}{(\omega_{m0} - \omega_2 - \omega_3 - i2\epsilon)(\omega_{p0} - \omega_3 - i\epsilon)}$$

$$\times e^{i(\omega_{m0} - \omega_2 - \omega_3)t} e^{2\epsilon t} e^{-iE_m t/\hbar} | m \rangle$$

$$= \sum_{\omega_1 \omega_2 \omega_3} \frac{1}{\hbar^3} \sum_{nmp} \frac{\langle 0|\hat{\mu}_\beta|n\rangle \langle n|\hat{\mu}_\alpha|m\rangle \langle m|\hat{\mu}_\gamma|p\rangle \langle p|\hat{\mu}_\delta|0\rangle E_\beta^{\omega_1} E_\gamma^{\omega_2} E_\delta^{\omega_3}}{(\omega_{n0} + \omega_1 + i\epsilon)(\omega_{m0} - \omega_2 - \omega_3 - i2\epsilon)(\omega_{p0} - \omega_3 - i\epsilon)}$$

$$\times e^{-i(\omega_1 + \omega_2 + \omega_3)t} e^{3\epsilon t}$$

$$(61) \quad \langle \psi^{(2)} | \hat{\mu}_\alpha | \psi^{(1)} \rangle = \sum_n \frac{1}{\hbar^2} \sum_{\omega_1 \omega_2} \sum_m \frac{\langle m|\hat{\mu}_\beta|n\rangle \langle 0|\hat{\mu}_\gamma|m\rangle [E_\beta^{\omega_1}]^* [E_\gamma^{\omega_2}]^*}{(\omega_{n0} - \omega_1 - \omega_2 + i2\epsilon)(\omega_{m0} - \omega_2 + i\epsilon)}$$

$$\times e^{-i(\omega_{n0} - \omega_1 - \omega_2)t} e^{2\epsilon t} e^{iE_n t/\hbar} \langle n|\hat{\mu}_\alpha$$

$$\times \sum_p \frac{1}{\hbar} \sum_{\omega_3} \frac{\langle p|\hat{\mu}_\delta|0\rangle E_\delta^{\omega_3}}{\omega_{p0} - \omega_3 - i\epsilon} e^{i(\omega_{p0} - \omega_3)t} e^{\epsilon t} e^{-iE_p t/\hbar} | p \rangle$$

$$= \sum_{\omega_1 \omega_2 \omega_3} \frac{1}{\hbar^3} \sum_{nmp} \frac{\langle 0|\hat{\mu}_\gamma|m\rangle \langle m|\hat{\mu}_\beta|n\rangle \langle n|\hat{\mu}_\alpha|p\rangle \langle p|\hat{\mu}_\delta|0\rangle E_\beta^{\omega_1} E_\gamma^{\omega_2} E_\delta^{\omega_3}}{(\omega_{n0} + \omega_1 + \omega_2 + i2\epsilon)(\omega_{m0} + \omega_2 + i\epsilon)(\omega_{p0} - \omega_3 - i\epsilon)}$$

$$\times e^{-i(\omega_1 + \omega_2 + \omega_3)t} e^{3\epsilon t}$$

(62) $\langle\psi^{(3)}|\hat{\mu}_\alpha|\psi^{(0)}\rangle = \sum_n \frac{1}{\hbar^3} \sum_{\omega_1\omega_2\omega_3} \sum_{mp}$

$\times \frac{\langle m|\hat{\mu}_\beta|n\rangle\langle p|\hat{\mu}_\gamma|m\rangle\langle 0|\hat{\mu}_\delta|p\rangle [E_\beta^{\omega_1}]^* [E_\gamma^{\omega_2}]^* [E_\delta^{\omega_3}]^*}{(\omega_{n0} - \omega_1 - \omega_2 - \omega_3 + i3\epsilon)(\omega_{m0} - \omega_2 - \omega_3 + i2\epsilon)(\omega_{p0} - \omega_3 + i\epsilon)}$

$\times e^{-i(\omega_{n0} - \omega_1 - \omega_2 - \omega_3)t} e^{3\epsilon t} e^{iE_n t/\hbar} \langle n|\hat{\mu}_\alpha e^{-iE_0 t/\hbar}|0\rangle$

$= \sum_{\omega_1\omega_2\omega_3} \frac{1}{\hbar^3} \sum_{nmp}$

$\times \frac{\langle 0|\hat{\mu}_\delta|p\rangle\langle p|\hat{\mu}_\gamma|m\rangle\langle m|\hat{\mu}_\beta|n\rangle\langle n|\hat{\mu}_\alpha|0\rangle E_\beta^{\omega_1} E_\gamma^{\omega_2} E_\delta^{\omega_3}}{(\omega_{n0} + \omega_1 + \omega_2 + \omega_3 + i3\epsilon)(\omega_{m0} + \omega_2 + \omega_3 + i2\epsilon)(\omega_{p0} + \omega_3 + i\epsilon)}$

$\times e^{-i(\omega_1 + \omega_2 + \omega_3)t} e^{3\epsilon t}$

We have used that $[E^\omega]^* = E^{-\omega}$ and the fact that ω_1, ω_2, and ω_3 are dummy summation indices that run over both positive and negative frequencies. None of Eqs. (59)–(62) is symmetric in the tensor indices β, γ, and δ. As pointed out in connection with Eq. (7), we normally choose our hyperpolarizability tensors to possess intrinsic symmetry, and it is clear that we can accomplish this without changing the polarization of the molecule by taking the average of the six terms generated by permuting pairs of the dummy indices (β, ω_1), (γ, ω_2), and (δ, ω_3); we denote this operation with the symbol $1/6\sum\mathcal{P}_{1,2,3}$, where the factor of one sixth is required to maintain the same value of the polarization. The third-order polarization in Eq. (36) can then be written as

(63) $\langle\hat{\mu}\rangle^{(3)} = \langle\psi^{(0)}|\hat{\mu}|\psi^{(3)}\rangle + \langle\psi^{(1)}|\hat{\mu}|\psi^{(2)}\rangle + \langle\psi^{(2)}|\hat{\mu}|\psi^{(1)}\rangle + \langle\psi^{(3)}|\hat{\mu}|\psi^{(0)}\rangle$

$= \frac{1}{6} \sum_{\omega_1\omega_2\omega_3} \frac{1}{\hbar^3} \sum \mathcal{P}_{1,2,3} \sum_{nmp}$

$\times \Bigg[\frac{\langle 0|\hat{\mu}_\alpha|n\rangle\langle n|\hat{\mu}_\beta|m\rangle\langle m|\hat{\mu}_\gamma|p\rangle\langle p|\hat{\mu}_\delta|0\rangle}{(\omega_{n0} - \omega_1 - \omega_2 - \omega_3 - i3\epsilon)(\omega_{m0} - \omega_2 - \omega_3 - i2\epsilon)(\omega_{p0} - \omega_3 - i\epsilon)}$

$+ \frac{\langle 0|\hat{\mu}_\beta|n\rangle\langle n|\hat{\mu}_\alpha|m\rangle\langle m|\hat{\mu}_\gamma|p\rangle\langle p|\hat{\mu}_\delta|0\rangle}{(\omega_{n0} + \omega_1 + i\epsilon)(\omega_{m0} - \omega_2 - \omega_3 - i2\epsilon)(\omega_{p0} - \omega_3 - i\epsilon)}$

$+ \frac{\langle 0|\hat{\mu}_\gamma|m\rangle\langle m|\hat{\mu}_\beta|n\rangle\langle n|\hat{\mu}_\alpha|p\rangle\langle p|\hat{\mu}_\delta|0\rangle}{(\omega_{n0} + \omega_1 + \omega_2 + i2\epsilon)(\omega_{m0} + \omega_2 + i\epsilon)(\omega_{p0} - \omega_3 - i\epsilon)}$

$+ \frac{\langle 0|\hat{\mu}_\delta|p\rangle\langle p|\hat{\mu}_\gamma|m\rangle\langle m|\hat{\mu}_\beta|n\rangle\langle n|\hat{\mu}_\alpha|0\rangle}{(\omega_{n0} + \omega_1 + \omega_2 + \omega_3 + i3\epsilon)(\omega_{m0} + \omega_2 + \omega_3 + i2\epsilon)(\omega_{p0} + \omega_3 + i\epsilon)} \Bigg]$

$\times E_\beta^{\omega_1} E_\gamma^{\omega_2} E_\delta^{\omega_3} e^{-i(\omega_1 + \omega_2 + \omega_3)t} e^{3\epsilon t}$

and, by a direct comparison to Eq. (5), we identify the expression for the second-order hyperpolarizability as

$$\gamma_{\alpha\beta\gamma\delta}(-\omega_\sigma;\omega_1,\omega_2,\omega_3) = \frac{1}{\hbar^3}\sum \mathcal{P}_{1,2,3} \tag{64}$$

$$\times \sum_{nmp}\left[\frac{\langle 0|\hat{\mu}_\alpha|n\rangle\langle n|\hat{\mu}_\beta|m\rangle\langle m|\hat{\mu}_\gamma|p\rangle\langle p|\hat{\mu}_\delta|0\rangle}{(\omega_{n0}-\omega_\sigma-i3\epsilon)(\omega_{m0}-\omega_2-\omega_3-i2\epsilon)(\omega_{p0}-\omega_3-i\epsilon)}\right.$$

$$+\frac{\langle 0|\hat{\mu}_\beta|n\rangle\langle n|\hat{\mu}_\alpha|m\rangle\langle m|\hat{\mu}_\gamma|p\rangle\langle p|\hat{\mu}_\delta|0\rangle}{(\omega_{n0}+\omega_1+i\epsilon)(\omega_{m0}-\omega_2-\omega_3-i2\epsilon)(\omega_{p0}-\omega_3-i\epsilon)}$$

$$+\frac{\langle 0|\hat{\mu}_\gamma|m\rangle\langle m|\hat{\mu}_\beta|n\rangle\langle n|\hat{\mu}_\alpha|p\rangle\langle p|\hat{\mu}_\delta|0\rangle}{(\omega_{n0}+\omega_1+\omega_2+i2\epsilon)(\omega_{m0}+\omega_2+i\epsilon)(\omega_{p0}-\omega_3-i\epsilon)}$$

$$+\left.\frac{\langle 0|\hat{\mu}_\delta|p\rangle\langle p|\hat{\mu}_\gamma|m\rangle\langle m|\hat{\mu}_\beta|n\rangle\langle n|\hat{\mu}_\alpha|0\rangle}{(\omega_{n0}+\omega_\sigma+i3\epsilon)(\omega_{m0}+\omega_2+\omega_3+i2\epsilon)(\omega_{p0}+\omega_3+i\epsilon)}\right]$$

Just as for the lower-order responses, we have kept the imaginary terms in the denominators in order to maintain convergence of the second-order hyperpolarizability at all frequencies, as well as to indicate the possibility of including damping in the near-resonant and resonant regions of the frequency spectrum. In the nonresonant region, however, the imaginary terms may be left out, and the expression for $\gamma(-\omega_\sigma;\omega_1,\omega_2,\omega_3)$ can be written more compactly as

$$\gamma_{\alpha\beta\gamma\delta}(-\omega_\sigma;\omega_1,\omega_2,\omega_3) = \frac{1}{\hbar^3}\sum \mathcal{P}_{-\sigma,1,2,3} \tag{65}$$

$$\times \sum_{nmp}\frac{\langle 0|\hat{\mu}_\alpha|n\rangle\langle n|\hat{\mu}_\beta|m\rangle\langle m|\hat{\mu}_\gamma|p\rangle\langle p|\hat{\mu}_\delta|0\rangle}{(\omega_{n0}-\omega_\sigma)(\omega_{m0}-\omega_2-\omega_3)(\omega_{p0}-\omega_3)}$$

where $\sum \mathcal{P}_{-\sigma,1,2,3}$ denotes the sum of the 24 terms one gets by permuting pairs of $(\alpha,-\omega_\sigma)$, (β,ω_1), (γ,ω_2), and (δ,ω_3). A direct comparison of Eq. (64) with $\epsilon = 0$ and Eq. (65) verifies this result. The second-order hyperpolarizability tensor is said to possess full permutation symmetry in the nonresonant region.

The secular divergences in Eq. (65), i.e. the singularities due to terms in which one or more of the states equal $|0\rangle$, can be removed in a similar manner as we did for the lower-order properties. In fact, explicit nondivergent formulas have been derived for molecular hyperpolarizabilities up to fourth order [5], and for the second-order hyperpolarizability, the resulting formula can be written as

$$\gamma_{\alpha\beta\gamma\delta}(-\omega_\sigma;\omega_1,\omega_2,\omega_3) = \frac{1}{\hbar^3}\sum \mathcal{P}_{-\sigma,1,2,3} \tag{66}$$

$$\times\left[\sum_{nmp}{}'\frac{\langle 0|\hat{\mu}_\alpha|n\rangle\langle n|\overline{\hat{\mu}_\beta}|m\rangle\langle m|\overline{\hat{\mu}_\gamma}|p\rangle\langle p|\hat{\mu}_\delta|0\rangle}{(\omega_{n0}-\omega_\sigma)(\omega_{m0}-\omega_2-\omega_3)(\omega_{p0}-\omega_3)}\right.$$

$$\left.-\sum_{nm}{}'\frac{\langle 0|\hat{\mu}_\alpha|n\rangle\langle n|\hat{\mu}_\beta|0\rangle\langle 0|\hat{\mu}_\gamma|m\rangle\langle m|\hat{\mu}_\delta|0\rangle}{(\omega_{n0}-\omega_\sigma)(\omega_{m0}-\omega_3)(\omega_{m0}+\omega_2)}\right]$$

Microscopic Theory of Nonlinear Optics

This quantum mechanical expression is applicable when all one-, two-, and three-photon combinations of the fields are nonresonant, and it also applies to the case of static fields.

The second-order hyperpolarizability has single, double and triple residues. We will consider a few of these that correspond to certain ground- and excited-state properties. From Eq. (65), we see that one of the first-order residues becomes

(67) $\lim_{\omega_3 \to \omega_{f0}} (\omega_{f0} - \omega_3)\gamma(-\omega_\sigma; \omega_1, \omega_2, \omega_3)$

$$= \frac{1}{\hbar^3} \sum \mathcal{P}_{-\sigma,1,2} \sum_{nm} \frac{\langle 0|\hat{\mu}_\alpha|n\rangle\langle n|\hat{\mu}_\beta|m\rangle\langle m|\hat{\mu}_\gamma|f\rangle\langle f|\hat{\mu}_\delta|0\rangle}{(\omega_{n0} - \omega_\sigma)(\omega_{m0} - \omega_2 - \omega_{f0})}$$

$$= \frac{1}{\hbar^3} \sum \mathcal{P}_{-\sigma,1,2} \sum_{nm} \frac{\langle 0|\hat{\mu}_\alpha|n\rangle\langle n|\hat{\mu}_\beta|m\rangle\langle m|\hat{\mu}_\gamma|f\rangle}{(\omega_{nf} - \omega_1 - \omega_2)(\omega_{mf} - \omega_2)} \langle f|\hat{\mu}_\delta|0\rangle$$

This expression is closely related to the three-photon matrix element describing an excitation from the molecular ground state to the excited state $|f\rangle$ by simultaneous absorption of three photons. When the residue above is evaluated at $\omega_1 = \omega_2 = -\omega_{f0}/3$ it will provide the matrix element corresponding to absorption of three monochromatic photons. From the first residue of the first-order hyperpolarizability in Eq. (57) we could identify the matrix element describing two-photon absorption. If we now consider one higher order of response as well as one higher order residue, we can identify the expression for the matrix element describing two-photon absorption between two excited states, say $|g\rangle$ and $|f\rangle$. The following double residue of the second-order hyperpolarizability needs to be considered:

(68) $\lim_{\omega_2 \to -\omega_{g0}} (\omega_{g0} + \omega_2) \left[\lim_{\omega_3 \to \omega_{f0}} (\omega_{f0} - \omega_3)\gamma(-\omega_\sigma; \omega_1, \omega_2, \omega_3) \right]$

$$= \frac{1}{\hbar^3} \sum \mathcal{P}_{-\sigma,1} \sum_m \frac{\langle 0|\hat{\mu}_\gamma|g\rangle\langle g|\hat{\mu}_\beta|m\rangle\langle m|\hat{\mu}_\alpha|f\rangle\langle f|\hat{\mu}_\delta|0\rangle}{(\omega_{m0} + \omega_\sigma - \omega_{f0})}$$

$$= \frac{1}{\hbar^3} \sum \mathcal{P}_{-\sigma,1} \sum_m \frac{\langle g|\hat{\mu}_\beta|m\rangle\langle m|\hat{\mu}_\alpha|f\rangle}{(\omega_{mg} + \omega_1)} \langle 0|\hat{\mu}_\gamma|g\rangle\langle f|\hat{\mu}_\delta|0\rangle$$

$$= \frac{1}{\hbar^3} \sum_m \left[\frac{\langle g|\hat{\mu}_\alpha|m\rangle\langle m|\hat{\mu}_\beta|f\rangle}{(\omega_{mf} - \omega_1)} + \frac{\langle g|\hat{\mu}_\beta|m\rangle\langle m|\hat{\mu}_\alpha|f\rangle}{(\omega_{mg} + \omega_1)} \right] \langle 0|\hat{\mu}_\gamma|g\rangle\langle f|\hat{\mu}_\delta|0\rangle$$

Out of the 24 permutations in Eq. (65) for $\gamma(-\omega_\sigma; \omega_1, \omega_2, \omega_3)$ only those obtained by permuting the pairs $(\alpha, -\omega_\sigma)$ and (γ, ω_2) contribute to this double residue. The residue is to be evaluated at $\omega_1 = -\omega_{fg}/2$ in order to relate to monochromatic two-photon absorption in the excited state $|g\rangle$.

It is also possible to have $|g\rangle = |f\rangle$ in the double residue. However, this situation corresponds to a secular singularity in the second-order hyperpolarizability and it is therefore appropriate to use Eq. (66) as a starting point:

$$(69) \quad \lim_{\omega_2 \to -\omega_{f0}} (\omega_{f0} + \omega_2) \left[\lim_{\omega_3 \to \omega_{f0}} (\omega_{f0} - \omega_3) \gamma(-\omega_\sigma; \omega_1, \omega_2, \omega_3) \right]$$

$$= \frac{1}{\hbar^3} \sum \mathcal{P}_{-\sigma,1} \left[\sum_m{}' \frac{\langle 0|\hat{\mu}_\gamma|f\rangle \langle f|\overline{\hat{\mu}_\beta}|m\rangle \langle m|\overline{\hat{\mu}_\alpha}|f\rangle \langle f|\hat{\mu}_\delta|0\rangle}{(\omega_{m0} + \omega_\sigma - \omega_{f0})} \right.$$

$$- \sum_n{}' \frac{\langle 0|\hat{\mu}_\alpha|n\rangle \langle n|\overline{\hat{\mu}_\beta}|0\rangle \langle 0|\overline{\hat{\mu}_\gamma}|f\rangle \langle f|\hat{\mu}_\delta|0\rangle}{(\omega_{n0} - \omega_\sigma)}$$

$$\left. - \frac{\langle 0|\hat{\mu}_\gamma|f\rangle \langle f|\overline{\hat{\mu}_\beta}|0\rangle \langle 0|\overline{\hat{\mu}_\alpha}|f\rangle \langle f|\hat{\mu}_\delta|0\rangle}{(\omega_{f0} - \omega_\sigma)} \right]$$

The relevance of this double residue to molecular properties is not yet obvious, but let us nevertheless first mention how the surviving terms in the residue came about. By considering the permutation of the pairs $(\alpha, -\omega_\sigma)$ and (γ, ω_2) in the first summation in Eq. (66) we see that it gives rise to the first term of the double residue. The second and third terms in the residue both come from the second summation in Eq. (66). We recall that the primed summations exclude the ground state $|0\rangle$. However, the third term in the residue equals the contribution from the omitted ground state in the first summation, and we can therefore write the double residue as

$$(70) \quad \lim_{\omega_2 \to -\omega_{f0}} (\omega_{f0} + \omega_2) \left[\lim_{\omega_3 \to \omega_{f0}} (\omega_{f0} - \omega_3) \gamma(-\omega_\sigma; \omega_1, \omega_2, \omega_3) \right]$$

$$= \frac{1}{\hbar^3} \sum \mathcal{P}_{-\sigma,1} \left[\sum_m \frac{\langle 0|\hat{\mu}_\gamma|f\rangle \langle f|\overline{\hat{\mu}_\beta}|m\rangle \langle m|\overline{\hat{\mu}_\alpha}|f\rangle \langle f|\hat{\mu}_\delta|0\rangle}{(\omega_{mf} + \omega_1)} \right.$$

$$\left. - \sum_n{}' \frac{\langle 0|\hat{\mu}_\alpha|n\rangle \langle n|\hat{\mu}_\beta|0\rangle \langle 0|\hat{\mu}_\gamma|f\rangle \langle f|\hat{\mu}_\delta|0\rangle}{(\omega_{n0} - \omega_1)} \right]$$

where we have used that $\omega_1 = \omega_\sigma$ and $\langle 0|\overline{\hat{\mu}}|f\rangle = \langle 0|\hat{\mu}|f\rangle$. The first summation in Eq. (70) has a secular singularity at $\omega_1 = 0$ due to the term for which $|m\rangle = |f\rangle$. However, this term will cancel that generated by the permutation of the pairs $(\alpha, -\omega_\sigma)$ and (β, ω_1), and a nondivergent formula would therefore exclude the excited state $|f\rangle$ in the summation. We thus finally obtain

$$(71) \quad \lim_{\omega_2 \to -\omega_{f0}} (\omega_{f0} + \omega_2) \left[\lim_{\omega_3 \to \omega_{f0}} (\omega_{f0} - \omega_3) \gamma(-\omega_\sigma; \omega_1, \omega_2, \omega_3) \right]$$

$$= \frac{1}{\hbar^3} \sum \mathcal{P}_{-\sigma,1} \left[\sum_{m \neq f} \frac{\langle f|\hat{\mu}_\alpha|m\rangle \langle m|\hat{\mu}_\beta|f\rangle}{(\omega_{mf} - \omega_1)} - \sum_{n \neq 0} \frac{\langle 0|\hat{\mu}_\alpha|n\rangle \langle n|\hat{\mu}_\beta|0\rangle}{(\omega_{n0} - \omega_1)} \right]$$

$$\times \langle 0|\hat{\mu}_\gamma|f\rangle \langle f|\hat{\mu}_\delta|0\rangle$$

where we permuted the pairs $(\alpha, -\omega_\sigma)$ and (β, ω_1) in the first term. If we compare Eq. (71) with Eq. (43), we see that, apart from some factors, the double residue of the second-order hyperpolarizability equals the excited-to-ground state difference in the linear polarizability.

3.4 Higher-order Hyperpolarizabilities

It is clear that the route followed hitherto can be extended to an arbitrary order in the perturbation, and thus provide the description of general multi-photon interactions. However, since very high light intensities are needed in order to observe the nonlinear responses, the predominant interest in optics are focused at processes incorporated in the linear polarizability and the first- and second-order hyperpolarizability. The expression for the general-order nonresonant response $X^n_{\alpha\beta\ldots}(-\omega_\sigma; \omega_1, \omega_2, \ldots, \omega_n)$ can be written as

$$(72) \quad X^n_{\alpha\beta\ldots}(-\omega_\sigma; \omega_1, \omega_2, \ldots, \omega_n)$$
$$= \hbar^{-n} \sum \mathcal{P}_{\alpha,\beta,\ldots} \sum_{a_1} \sum_{a_2} \cdots \sum_{a_n} \langle 0|\hat{\mu}_\alpha|a_1\rangle\langle a_1|\hat{\mu}_\beta|a_2\rangle \cdots$$
$$\times \left[(\omega_{a_1} - \omega_\sigma)(\omega_{a_2} - \omega_\sigma + \omega_1)\cdots(\omega_{a_n} - \omega_n)\right]^{-1}$$

In this very general formula we are using the notation of Bishop [5], where conventionally $X^1_{\alpha\beta} = \alpha_{\alpha\beta}$ (linear polarizability), $X^2_{\alpha\beta\gamma} = \beta_{\alpha\beta\gamma}$ (first-order hyperpolarizability), and $X^3_{\alpha\beta\gamma\delta} = \gamma_{\alpha\beta\gamma\delta}$ (second-order hyperpolarizability).

3.5 Absorption

We have so far used time-dependent perturbation theory to compute the time-dependent corrections to the wave function of the unperturbed system $|0\rangle$ due to an external electric field, see Eq. (27). From the same equation, it is clear that the probability to find the system in state $|f\rangle$ at time t is

$$(73) \quad P_{0\to f} = |d_f(t)|^2$$

The transition from the ground $|0\rangle$ to the excited state $|f\rangle$ can occur by the absorption of either a single photon or many photons. One-photon absorption is governed by the first-order amplitudes $d^{(1)}$ in Eq. (38), two-photon absorption by the second-order amplitudes $d^{(2)}$ in Eq. (39), and so forth.

From Eq. (38) we see that the first-order amplitude can be written as

$$(74) \quad d_f^{(1)}(t) = \frac{1}{\hbar} \sum_{\omega_1} \langle f|\hat{\mu}_\beta|0\rangle E_\beta^{\omega_1} F(t, \omega_{f0} - \omega_1)$$

where $F(t, \omega_{f0} - \omega_1)$ is a dimensionless function that depends on the time t and the separation of the frequency of the perturbing field from the transition frequency of the system ω_{f0}. For t large compared to $(2\pi/\omega_{f0})$, the function F is sharply peaked at $\omega_{f0} = \omega_1$, which is the same as saying that one-photon absorption occurs only when the photon energy matches the excitation energy. The probability for one-photon transitions per unit time will be proportional to the square of the one-photon matrix element

$$M_{0f}^{(1)} = \langle 0|\hat{\mu}_\beta|f\rangle \tag{75}$$

We note that the probability for one-photon absorption, since it depends on the square of the amplitude and therefore also the square of the amplitude of the electric field, is linearly dependent on the intensity of the light. One-photon absorption is for that reason synonymous with linear absorption.

The second-order amplitude, taken from Eq. (39), is

$$d_f^{(2)}(t) = \frac{1}{\hbar^2} \sum_{\omega_1 \omega_2} \sum_n \frac{\langle f|\hat{\mu}_\beta|n\rangle \langle n|\hat{\mu}_\gamma|0\rangle}{\omega_{n0} - \omega_2} E_\beta^{\omega_1} E_\gamma^{\omega_2} F(t, \omega_{f0} - \omega_1 - \omega_2) \tag{76}$$

where the function $F(t, \omega_{f0} - \omega_1 - \omega_2)$ this time depends on the time t and the separation of the sum of the perturbing frequencies and the transition frequency of the system ω_{f0}. In analogy with the case of linear absorption: for t large compared to $(2\pi/\omega_{f0})$, the function F is sharply peaked at $\omega_{f0} = \omega_1 + \omega_2$, which means that two-photon absorption occurs only when the total energy of the two photons matches the excitation energy. We deduce that the two-photon transition matrix element is written as

$$M_{0f}^{(2)} = \frac{1}{\hbar} \sum_n \left[\frac{\langle 0|\hat{\mu}_\beta|n\rangle \langle n|\hat{\mu}_\gamma|f\rangle}{\omega_{n0} - \omega_1} + \frac{\langle 0|\hat{\mu}_\gamma|n\rangle \langle n|\hat{\mu}_\beta|f\rangle}{\omega_{n0} - \omega_2} \right] \tag{77}$$

where we have symmetrized the summations over dummy indices. This symmetrization obviously does not effect the value of the amplitude $d^{(2)}$ as long as we compensate with a factor of one-half in front. The complex conjugation associated with reversing the order of the states in the numerator is irrelevant since transition probabilities depend on the absolute square of the matrix elements. The reader should compare the expression for the two-photon matrix element to the first-order residue of the first-order hyperpolarizability in Eq. (57). One of the interesting characteristics of two-photon absorption is that since the square of the second-order amplitude, i.e. the probability of two-photon absorption, depends on the perturbing electric field to the fourth power, it will exhibit a square dependence on the light intensity. Two-photon absorption is therefore nonlinear to first order with respect to the intensity of the incident laser beam.

Carried out to yet a higher order in the perturbation, we will be able to obtain an expression for the three-photon absorption matrix element. The third-order amplitude, taken from Eq. (40), is

$$d_f^{(3)}(t) = \frac{1}{\hbar^3} \sum_{\omega_1 \omega_2 \omega_3} \sum_{mn} \frac{\langle f|\hat{\mu}_\beta|n\rangle \langle n|\hat{\mu}_\gamma|m\rangle \langle m|\hat{\mu}_\delta|0\rangle}{(\omega_{n0} - \omega_2 - \omega_3)(\omega_{m0} - \omega_3)} E_\beta^{\omega_1} E_\gamma^{\omega_2} E_\delta^{\omega_3}$$
$$\times F(t, \omega_{f0} - \omega_1 - \omega_2 - \omega_3) \tag{78}$$

Microscopic Theory of Nonlinear Optics

Table 2. Molecular properties described by the first-, second-, and third-order response functions

Response	Residue	Molecular property
$\alpha(-\omega;\omega)$	—	Linear electric dipole polarizability.
	$\omega = \omega_{f0}$	One-photon transition matrix elements between the ground state $|0\rangle$ and the excited state $|f\rangle$.
$\beta(-\omega_\sigma;\omega_1,\omega_2)$	—	First-order nonlinear electric dipole hyperpolarizability.
	$\omega_2 = \omega_{f0}$	Two-photon transition matrix element between the ground state $|0\rangle$ and the excited state $|f\rangle$.
	$\omega_1 = -\omega_{f0}$ $\omega_2 = \omega_{g0}$	One-photon transition matrix element between the excited states $|f\rangle$ and $|g\rangle$.
	$\omega_1 = -\omega_{f0}$ $\omega_2 = \omega_{f0}$	Permanent electric dipole moment of the excited state $|f\rangle$.
$\gamma(-\omega_\sigma;\omega_1,\omega_2,\omega_3)$	—	Second-order nonlinear electric dipole hyperpolarizability.
	$\omega_3 = \omega_{f0}$	Three-photon transition matrix element between the ground state $|0\rangle$ and the excited state $|f\rangle$.
	$\omega_2 = -\omega_{f0}$ $\omega_3 = \omega_{g0}$	Two-photon transition matrix element between the excited states $|f\rangle$ and $|g\rangle$.
	$\omega_2 = -\omega_{f0}$ $\omega_3 = \omega_{f0}$	Linear electric dipole polarizability of the excited state $|f\rangle$.

The three-photon absorption matrix element, symmetrized in the dummy indices, can thus be written as

$$(79) \quad M_{0f}^{(3)} = \frac{1}{\hbar^2} \sum \mathcal{P}_{1,2,3} \sum_{mn} \frac{\langle 0|\hat{\mu}_\beta|m\rangle\langle m|\hat{\mu}_\gamma|n\rangle\langle n|\hat{\mu}_\delta|f\rangle}{(\omega_{n0}-\omega_1-\omega_2)(\omega_{m0}-\omega_1)}$$

where the operator $\sum \mathcal{P}_{1,2,3}$ denote that the summation of terms obtained by permuting the pairs (β, ω_1), (γ, ω_2), and (δ, ω_3) is performed. The connection to the first-order residue of the second-order hyperpolarizability in Eq. (67) is verified by explicit comparison of the six terms in each case. We see that the probability of three-photon absorption depend on the light intensity to the third power. In Table 2 we have summarized some of the molecular properties that can be retrieved from the linear and nonlinear response functions.

4. THE SEPARATION OF ELECTRONIC AND NUCLEAR DEGREES OF FREEDOM

The expressions presented in the previous section apply to exact states. In practice, it is not possible to obtain the complete set of exact states and we therefore need ways to obtain approximate solutions. The most important step when developing strategies for evaluating linear and nonlinear optical properties is the introduction of the Born–Oppenheimer approximation [15, 16]. As we will briefly outline, and as described in much more detail in Chapter 3, the Born–Oppenheimer approximation leads to a partitioning of the (hyper)polarizabilities into three distinct contributions: One purely electronic contribution, one so-called pure vibrational contribution, and finally the zero-point vibrational contribution. There are also contributions to the (hyper)polarizabilities from the overall rotation of the molecule. However, for isotropic samples, the rotational contribution is to a large extent accounted for by doing a classical isotropic averaging [3]. The contributions that could be classified as pure rotational contributions to the (hyper)polarizabilities have been shown to be rather small [7].

In the next section we will discuss the general strategies for approximate calculations of nonlinear optical properties and describe in general terms in which way the different approaches can recover the different contributions to the (hyper)polarizabilities. The details of these different methods are, however, left for other chapters of this book.

The Born–Oppenheimer approximation states that the electrons are able to adjust themselves instantaneously to the motions of the nuclei. The motions of the nuclei are in this approximation therefore not able to induce electronic transitions, an assumption that is also known as the adiabatic approximation. The electrons thus create an effective electronic potential in which the nuclei move, and for a given electronic state the variation in the electronic energy with respect to the nuclear configuration defines a *potential energy surface* for the electronic state. The electronic Schrödinger equation can be written as

$$(80) \quad \hat{H}^{el}(\mathbf{R}, \mathbf{r}) \phi_K^{el}(\mathbf{R}, \mathbf{r}) = V_K^{el}(\mathbf{R}) \phi_K^{el}(\mathbf{R}, \mathbf{r})$$

where $V_K^{el}(\mathbf{R})$ is the electronic energy for electronic state K and $\phi_K^{el}(\mathbf{R}, \mathbf{r})$ the corresponding electronic wave function. We have explicitly indicated the dependence of these quantities on the nuclear configuration. The electronic Hamiltonian is given as

$$(81) \quad \hat{H}^{el}(\mathbf{R}, \mathbf{r}) = -\frac{\hbar^2}{2m_e} \sum_i \nabla_i^2 - \frac{e^2}{4\pi\epsilon_0} \sum_{i\mu} \frac{Z_\mu}{r_{i\mu}} + \frac{e^2}{4\pi\epsilon_0} \sum_{i>j} \frac{1}{r_{ij}}$$

We note that, in contrast to above, the nuclear repulsion is often included as an additional repulsive potential in the electronic Hamiltonian and thus in the electronic energy. From the solutions of the electronic Schrödinger equation, we obtain the potential that governs the nuclear motions, and the Scrödinger equation for the nuclei can then be solved for the potential provided by the electrons

$$(82) \quad \hat{H}_K^{nuc}(\mathbf{R}) \Psi_{K,k}^{nuc}(\mathbf{R}) = \mathcal{E}_{K,k} \Psi_{K,k}^{nuc}(\mathbf{R})$$

where $\mathcal{E}_{K,k}$ is the total vibronic energy and $\Psi_{K,k}^{nuc}(\mathbf{R})$ is the kth vibrational wave function for the Kth electronic state. In this equation, the nuclear Hamiltonian is given as

$$(83) \quad \hat{H}_K^{nuc}(\mathbf{R}) = -\sum_{\mu=1}^{N} \frac{\hbar^2}{2M_\mu} \nabla_\mu^2 + \frac{e^2}{4\pi\epsilon_0} \sum_{\mu>\nu} \frac{Z_\mu Z_\nu}{R_{\mu\nu}} + V_K^{el}(\mathbf{R})$$

Microscopic Theory of Nonlinear Optics 29

where the summations run over all nuclei in the molecule, and M_μ is the mass of nucleus μ. There are $3N$ degrees of freedom in this equation, of which three degrees of freedom correspond to the translation of the center of mass of the molecule and three (or two in the case of linear molecules) degrees of freedom correspond to the overall rotation of the molecule about the center of mass. The remaining $3N-6(5)$ coordinates describe the relative motions of the nuclei of the molecule, and constitute the vibrations of the molecule. The translation of the molecular center of mass is not quantized and therefore not of any interest to our discussion. To a good first approximation, the rotational motion of the molecule is decoupled from the vibrational motion, and we may thus treat this separately.

Considering the product form of the total molecular wave function, let us briefly return to the linear polarizability given in Eq. (43), and let us assume that the complete set of vibronic product states can be considered as the exact states of the molecular system. The linear polarizability can then be written as [9, 10, 28]

$$(84) \quad \alpha_{\alpha\beta}(-\omega;\omega) = \frac{1}{\hbar}\sum_{K,k}{}' \left[\frac{\langle 0,0|\hat{\mu}_\alpha|K,k\rangle\langle k,K|\hat{\mu}_\beta|0,0\rangle}{\omega_{K,k}-\omega} + \frac{\langle 0,0|\hat{\mu}_\beta|K,k\rangle\langle k,K|\hat{\mu}_\alpha|0,0\rangle}{\omega_{K,k}+\omega} \right]$$

where the prime indicates that the summation runs over all electronic K and vibrational states k apart from the vibronic ground state and $\hbar\omega_{K,k}$ denotes the energy difference between the intermediate and ground vibronic states.

To reduce the complexity of Eq. (84), we make the observation that energies involved in electronic transitions are in general much larger than the vibrational excitations that contribute significantly to the summation. We may therefore assume that $\omega_{K,k} \approx \omega_{K0}$, which allows us to partition the coupled summations in Eq. (84) into two contributions, one over electronic excited states involving only the ground vibrational state, and one over the vibrationally excited states involving only the electronic ground state

$$(85) \quad \alpha_{\alpha\beta}(-\omega;\omega) =$$

$$= \frac{1}{\hbar}\sum_{K\neq 0}\langle 0| \left[\frac{\langle 0|\hat{\mu}_\alpha|K\rangle\langle K|\hat{\mu}_\beta|0\rangle}{\omega_{K0}-\omega} + \frac{\langle 0|\hat{\mu}_\beta|K\rangle\langle K|\hat{\mu}_\alpha|0\rangle}{\omega_{K0}+\omega} \right] |0\rangle$$

$$+ \frac{1}{\hbar}\sum_{k\neq 0} \left[\frac{\langle 0|\mu_\alpha^{00}|k\rangle\langle k|\mu_\beta^{00}|0\rangle}{\omega_{k0}-\omega} + \frac{\langle 0|\mu_\beta^{00}|k\rangle\langle k|\mu_\alpha^{00}|0\rangle}{\omega_{k0}+\omega} \right]$$

where, in the former term, the closure over (k) has been carried out and, in the latter term, we have introduced the notation $\mu^{00} = \langle 0|\hat{\mu}|0\rangle$. The latter of the two terms only involves excitations within the vibrational manifold and is for this reason referred to as the *pure vibrational* contribution [3, 6]. In recent years, various efficient approaches have been developed for the calculation of pure vibrational contributions to linear and nonlinear polarizabilities—as discussed in Chapter 3—and it has been demonstrated that vibrational contributions to nonlinear optical properties can be large, and in many cases as large as the electronic contributions. The reason for their importance can to some extent be related to the fact that the excitations within the vibrational manifold requires smaller energies, leading an enhancement of the denominator of the second contribution in Eq. (85). However, since the commonly used laser wavelengths are much shorter than the wavelengths associated with the vibrational excitations, their importance is in general much reduced at optical frequencies since ω in this case will be much larger than ω_{k0}, thus leading to an increase in the energy denominators in Eq. (85).

The first term in Eq. (84) is the corresponding electronic contribution to the linear polarizability averaged over the ground vibrational state. It is customary to partition this contribution into two terms, one term corresponding to the purely electronic contribution as obtained at the equilibrium geometry, and one contribution which arises from the averaging of the polarizability over the vibrational ground state, a term referred to as the *zero-point vibrational averaging* correction to the polarizability.

To briefly summarize, it is customary, and also highly advantageous in order to develop efficient schemes for the calculation of the total molecular (hyper)polarizabilities, to partition the molecular property into three different contributions:

$$(86) \quad P = P^e + P^{\text{ZPVA}} + P^v$$

where P^e is the electronic contribution to the property calculated at the equilibrium geometry of the molecule, P^{ZPVA} the zero-point vibrational averaging contribution, and P^v the pure vibrational contribution.

The techniques and approximations involved in obtaining computationally tractable schemes for the calculation of the linear and nonlinear optical properties differ for the three contributions given in Eq. (86), and the different strategies will be presented and reviewed in the different chapters of this book. In the next section, we will briefly describe a few of the approximate methods used to calculate hyperpolarizabilities. Most of these methods will be directed toward the electronic contributions, but some of the approaches will also be able to extract information about the pure vibrational contributions.

5. COMPUTATIONAL STRATEGIES

We will focus attention on computational methods based on a quantum mechanical wave function treatment of the system, supplemented with a classical interaction with the external electric fields. The quantum mechanical system should be possible to describe with a time-independent Hamiltonian, whereas the external perturbation can be time-independent or time-dependent.

In cases when our system represents a subsystem embedded in a medium, we will not consider thermal interactions such as collisions or chemical interactions such as hydrogen bonding. However, the quantum mechanical approaches presented in this chapter can be used in conjunction with classical methods for a statistical sampling the phase space of a system that partially include the environment. There are also physical interactions between the quantum mechanical system and the environment that may be described by a dielectric continuum model and which can be included in a self-consistent field approach (see Chapter 4).

Let us return to the problem of solving the response of the quantum mechanical system to an external electric field. The zeroth-order wave function of the quantum mechanical system is obtained by use of any of the standard approximate methods in quantum chemistry and the coupling to the field is described by the electric dipole operator. There exist a number of ways to determine the response functions, some of which differ in formulation only, whereas others will be inherently different. We will give a short review of the characteristics of the most common formulations used for the calculation of molecular polarizabilities and hyperpolarizabilities. The survey begins with the assumption that the external perturbing fields are non-oscillatory, in which case we may determine molecular properties at zero frequencies, and then continues with the general situation of time-dependent fields and dynamic properties.

When a time-independent external perturbation is applied on a molecule the molecular charges re-orient; the nuclei relax into a new configuration and the electronic motion is altered, thereby representing the equilibrated ground state of the molecule in the presence of the field. The corresponding equilibrated wave function is stationary, as opposed to the case when a time-dependent perturbation is applied, and we can thus determine the molecular energy as the time-independent expectation value of the Hamiltonian

$$(87) \quad \hat{H} = \hat{H}_0 \; \hat{\mu}_\alpha E_\alpha$$

where H_0 denotes the unperturbed Hamiltonian and the perturbation is given by the coupling between the electric dipole moment operator $\hat{\mu}$ and the electric field strength E. It is clear that the equilibrated wave function $|\psi(E)\rangle$ as well as the molecular energy $\mathcal{E}(E)$ depend on the applied field, and that we, for a given field, have

$$(88) \quad \hat{H}|\psi(E)\rangle = \mathcal{E}(E)|\psi(E)\rangle$$

Let us next consider the derivative of the energy with respect to a component of the electric field

$$(89) \quad \frac{\partial \mathcal{E}(E)}{\partial E_\alpha} = \frac{\partial}{\partial E_\alpha} \langle \psi(E)|\hat{H}|\psi(E)\rangle$$

$$= \left\langle \frac{\partial \psi(E)}{\partial E_\alpha} \middle| \hat{H} \middle| \psi(E) \right\rangle + \left\langle \psi(E) \middle| \hat{H} \middle| \frac{\partial \psi(E)}{\partial E_\alpha} \right\rangle$$

$$+ \left\langle \psi(E) \middle| \frac{\partial \hat{H}}{\partial E_\alpha} \middle| \psi(E) \right\rangle$$

The first two terms correspond to the variation of the wave function with respect to the external field, but, since $\psi(E)$ is optimal for *any* variation of the wave function, we have

(90) $$\delta\langle\psi|\hat{H}|\psi\rangle = 0$$

where δ is used to symbolize general variations of the wave function (but not the Hamiltonian). Eq. (89) thus simplifies to

(91) $$\frac{\partial\mathcal{E}(E)}{\partial E_\alpha} = \left\langle\psi(E)\left|\frac{\partial\hat{H}}{\partial E_\alpha}\right|\psi(E)\right\rangle$$

which is a result known as the Hellmann–Feynman theorem [27, 32]. In our case one obtains for the first-order derivative of the energy with respect to the electric field

(92) $$\frac{\partial\mathcal{E}(E)}{\partial E_\alpha} = -\langle\psi(E)|\hat{\mu}_\alpha|\psi(E)\rangle = -\mu_\alpha(E)$$

i.e. minus the field-dependent molecular dipole moment. If this derivative is evaluated at zero field strength we obviously retrieve the permanent electric dipole moment of the molecule:

(93) $$\mu_\alpha^0 = -\left.\frac{\partial\mathcal{E}(E)}{\partial E_\alpha}\right|_{E=0}$$

Moreover, if we recall Eq. (3), which we used as a starting point for defining linear and nonlinear optical properties, then it is seen that for time-independent perturbations, we may equally well choose an expansion of the molecular energy for this purpose

(94) $$\mathcal{E}(E) = \mathcal{E}_0 - \mu_\alpha^0 E_\alpha - \frac{1}{2}\alpha_{\alpha\beta}E_\alpha E_\beta - \frac{1}{6}\beta_{\alpha\beta\gamma}E_\alpha E_\beta E_\gamma$$
$$- \frac{1}{24}\gamma_{\alpha\beta\gamma\delta}E_\alpha E_\beta E_\gamma E_\delta + \cdots$$

where $\mathcal{E}_0$ denotes the molecular energy of the unperturbed system. Another way to see this fact is to note that the expressions obtained by evaluation of the response functions at zero frequencies *equal* the energy corrections which are well known from time-independent perturbation theory. For the higher-order derivatives of the energy with respect to the perturbation we thus have

(95) $$\alpha_{\alpha\beta} = -\left.\frac{\partial^2\mathcal{E}(E)}{\partial E_\alpha \partial E_\beta}\right|_{E=0}$$

(96) $$\beta_{\alpha\beta\gamma} = -\left.\frac{\partial^3\mathcal{E}(E)}{\partial E_\alpha \partial E_\beta \partial E_\gamma}\right|_{E=0}$$

(97) $$\gamma_{\alpha\beta\gamma\delta} = -\left.\frac{\partial^4\mathcal{E}(E)}{\partial E_\alpha \partial E_\beta \partial E_\gamma \partial E_\delta}\right|_{E=0}$$

Microscopic Theory of Nonlinear Optics

When addressing properties in the static limit, we are thus left with two alternatives: (i) the polarization propagator approach or (ii) the energy-derivative approach. However, a word of caution is needed at this point regarding calculations involving approximate states due to the fact that Eq. (90), and thereby also the Hellmann–Feynman theorem, is only fulfilled for variationally optimized wave functions. In other cases, such as for instance computational methods based on Møller–Plesset perturbation theory (MP2, MP3, etc.) or truncated configuration interaction (CIS, CISD, etc.), the energy-derivative and polarization propagator techniques will not provide identical results even in the limit of static frequencies.

5.1 Finite-field Approaches

We have have now paved the way for computational strategies to take form. Immediately Eq. (94) invites to the basic and straightforward idea of numerical differentiation, which is known in the literature as the finite-field technique. For atomic calculations, this idea is easily implemented in an existing code for any electronic structure method, as it only involves the response of the electronic density to the external static field in accordance with Fig. 6. In the absence of the perturbation, the dipole moment is zero due to the spherical symmetry of the atom, but in the presence of the external field the Coulomb force acts on the electrons as well as the atomic nucleus thereby displacing them in opposite directions: the nucleus tends to move along with and the electrons opposed to the electric field E with an induced dipole moment μ as a result.

In the program it is only necessary to modify the matrix elements of the one-electron part of the Hamiltonian h_{ij} by adding the dipole moment integrals:

$$(98) \quad \langle \phi_i | \hat{h} | \phi_j \rangle \longrightarrow \langle \phi_i | \hat{h} | \phi_j \rangle + e \langle \phi_i | \hat{r}_\alpha | \phi_j \rangle E_\alpha$$

where $\hat{h}$ is the one-electron part of the Hamiltonian for the atomic system containing the kinetic energy of the electrons and the attractive nuclear potential.

This small modification makes the code able to determine the atomic energy in the presence of the electric field in accordance with Eq. (94), and simple polynomial fits allow for the determination of the energy expansion coefficients. In Fig. 7

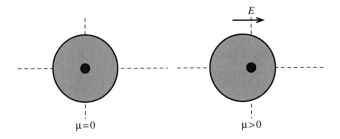

Figure 6. The induced dipole moment of an atom in an electric field

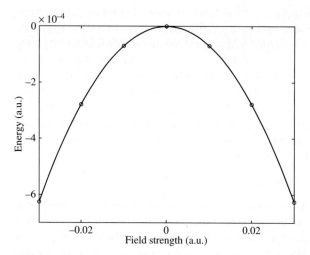

Figure 7. The total energy of helium in an external and uniform electric field is shown. The zero level of the energy is set to $\mathcal{E}_0$

we illustrate the dependence of the full configuration interaction (FCI) energy on an externally applied finite field with use of the triply augmented, correlation consistent, valence triple-ζ basis set of Dunning and co-workers [61].

In order to achieve numerical accuracy in the calculation of energy derivatives by means of finite field differentiation, the applied fields need to be small—we have chosen a grid size of 0.01 a.u. in this case—and, in principle, the smaller the grid size the better the accuracy. However, since the quartic power of the field multiplies γ, we need to converge the energy to at least 10^{-8} a.u. in order to compute this property. The first nonzero correction to the energy is of second order in the field and thereby in the order of 10^{-4} a.u., and it is noteworthy that the perturbation energy is about a hundred times smaller than the correlation energy (which, for $E = 0$ and in the present basis set, amounts to 3.9×10^{-2} a.u.). Despite this fact, the perturbation energy can be quite accurately determined at the Hartree–Fock level of theory.

The data points are fitted in a least-square sense to a fourth degree polynomial, and the properties thereby obtained are presented in Table 3. Since the atom possesses spherical symmetry there is only a single independent component of the α-tensor as well as the γ-tensor. The curvature of the energy, or the polarizability, at the SCF level differs by less than 5% compared to the FCI result, and the MP2 value captures slightly more than half of the correlation effect. Electron correlation plays a more important role in the determination of the fourth-order property γ. Again the MP2 method captures slightly more than half of the total contribution, which amounts to 21% at the FCI level of theory. The trends we have seen here in the example of the helium atom are more or less representative for closed-shell molecules in general.

Table 3. Electric properties of helium

	Finite field[a]			Analytical derivative	
	SCF	MP2	FCI	SCF	FCI
$\alpha_{zz}(0;0)$	1.3238	1.3643	1.3885	1.3238	1.3885
		3.1%	4.9%		4.9%
$\gamma_{zzzz}(0;0,0,0)$	35.181	39.967	42.563	34.692	41.925
		13.6%	21.0%		20.9%

[a] The finite field results are obtained by a polynomial fit to energy points at field strengths of 0.0, ±0.01, ±0.02, and ±0.03 a.u.

Before taking this example to its end, we point out some of the drawbacks with the finite-field method. In Table 3 we also present the analytical derivative results for α and γ, or, equivalently, the results obtained in the limit of a zero grid size. For the second-order derivative of the energy our numerical grid is able to provide us with five accurate digits, but for the fourth-order derivative, on the other hand, the accuracy is barely two digits. This may prove troublesome in the case of more extended molecules for which the energy is not always so easy to converge to a high degree of accuracy, especially with the demands for diffuse and polarizing basis sets that we have for these properties. In addition there is a practical concern in the calculation involved with the application of an external field. From Fig. 6 it is clear that one can no longer maintain the symmetry elements found in the plane orthogonal to the external field. Apart from producing a larger number of nonzero integrals in the calculation, this factor may also affect the convergence of the wave function negatively.

If we now turn to a situation where the system is comprised of a molecule rather than an atom, then the program modification needed is a bit more involved than what is shown in Eq. (98). The reason is of course that we must now not only consider the changes on the electronic motion but also the changes in the nuclear motion and equilibrium structure that are induced by the external electric field. In order to determine the field-dependent equilibrium structure Q_E, we need at least to be able to determine the molecular gradient, and care must also be taken in order not to let the molecule rotate and align its dipole moment with the external field [8]. It is advantageous to divide the field-dependent energy into three parts:

$$(99) \quad \mathcal{E}(Q_E, E) = \mathcal{E}^e(Q_0, E) + [\mathcal{E}^e(Q_E, E) - \mathcal{E}^e(Q_0, E)] + \mathcal{E}^{ZPV}(Q_E, E)$$

Here the first term corresponds to the electronic energy at the equilibrium geometry of the unperturbed molecule Q_0, the second contribution represents the difference in electronic energy between Q_E and Q_0, and, finally, the third term is the zero-point vibrational energy of the nuclei. From Eq. (99) we understand that the energy differentiation we perform to obtain the molecular properties will lead to three quite separate contributions. The first contribution represents the response of the electrons to the perturbation and thus corresponds to the electronic contribution to the properties. For the calculation of this electronic part of the property one only need to repeat what we have just done in the atomic case. The second and third terms in Eq. (99) will give rise to what is collectively known as the vibrational contributions to the properties (as already introduced in Section 4), or individually known as the nuclear relaxation and curvature contributions.

5.2 Analytic Derivative Approaches

The discussion of the pros and cons of the finite-field approach made it clear that we need an analytical formulation of the derivatives of the molecular energy with respect to the external fields in order to maintain computational efficiency and numerical stability. The phrase "analytical derivative approaches" is often used to denote methods where closed-form expressions have been derived for the part of the molecular properties that regards the motions of the electrons, i.e. the part related to the first term in Eq. (99).

In deriving such expressions we should recognize that, for approximate wave functions, the energy depends both explicitly and implicitly on the perturbing fields. We may formally write the energy as

$$(100) \quad \mathcal{E}(\lambda, E) = \langle \psi(\lambda) | \hat{H}_0 - \hat{\mu}_\alpha E_\alpha | \psi(\lambda) \rangle$$

where we identify the molecular Hamiltonian in the presence of the electric field (the explicit dependence) and λ collectively denote the wave function parameters for a given electronic structure method (e.g. the molecular orbital coefficients in the case of a Hartree–Fock calculation). For simplicity we have here assumed that all field dependence is contained in the Hamiltonian. In other applications, for instance those concerned with external magnetic field perturbations, it is advantageous to also have an explicit field dependence in the wave function parameters [38, 53, 60], but for electric properties the situation is, in almost all cases, characterized by Eq. (100). Attempts at including an explicit electric field dependence in the basis set have been taken by several authors [1, 24, 54], but the approach has not received much attention in studies of electric properties. We note, however, that if we are concerned with the analytical evaluation of the vibrational contributions to linear and nonlinear optical properties, an explicit dependence on the change in the nuclear positions exists in the basis set since the basis sets in general are attached to the nuclear centers [50, 51].

The implicit field dependence is yet not apparent, since the parameters λ in Eq. (100) are chosen freely. Let us therefore denote the set of parameters which are *optimized* in the presence of the field by λ_E. The implicit dependence then becomes obvious, as we will then have a corresponding equation to that of Eq. (88) for approximate wave functions. In the following derivations, we will use the notation

$$(101) \quad \left. \frac{\partial \mathcal{E}(\lambda, E)}{\partial \lambda} \right|_{\lambda = \lambda_E} = \frac{\partial \mathcal{E}(\lambda_E, E)}{\partial \lambda}$$

The first-order derivative of the energy is

$$\text{(102)} \quad \frac{d\mathcal{E}(\lambda_E, E)}{dE_\alpha} = \frac{\partial \mathcal{E}(\lambda_E, E)}{\partial \lambda} \frac{\partial \lambda_E}{\partial E_\alpha} + \frac{\partial \mathcal{E}(\lambda_E, E)}{\partial E_\alpha}$$

which is the analogue of Eq. (89) for approximate wave functions. We have already concluded that the first term vanishes for variational wave functions fulfilling the Hellmann–Feynman theorem [25, 44] and we will in the following only consider this case. The explicit formula for the dipole moment then becomes

$$\text{(103)} \quad \mu_\alpha = -\left.\frac{d\mathcal{E}(\lambda_E, E)}{dE_\alpha}\right|_{E=0} = \langle \psi(\lambda_0) | \hat{\mu}_\alpha | \psi(\lambda_0) \rangle$$

The second-order derivative of the energy is

$$\text{(104)} \quad \frac{d^2\mathcal{E}(\lambda_E, E)}{dE_\alpha dE_\beta} = \left[\frac{\partial}{\partial E_\alpha} + \frac{\partial \lambda_E}{\partial E_\alpha}\frac{\partial}{\partial \lambda}\right]\frac{\partial \mathcal{E}(\lambda_E, E)}{\partial E_\beta}$$

$$= \frac{\partial \lambda_E}{\partial E_\alpha}\frac{\partial^2 \mathcal{E}(\lambda_E, E)}{\partial \lambda \partial E_\beta}$$

where we have used the fact that the first term in the intermediate step vanishes since the Hamiltonian depends only linearly on the electric field. We see that in order to determine the second-order properties we need to determine the first-order response of the wave function parameters with respect to the perturbation. We can readily obtain a working formula for this response from the variational condition $\partial \mathcal{E}(\lambda_E, E)/\partial \lambda = 0$ [Eq. (90)] as

$$\text{(105)} \quad 0 = \frac{d}{dE_\alpha}\left[\frac{\partial \mathcal{E}(\lambda_E, E)}{\partial \lambda}\right] = \frac{\partial^2 \mathcal{E}(\lambda_E, E)}{\partial E_\alpha \partial \lambda} + \frac{\partial^2 \mathcal{E}(\lambda_E, E)}{\partial \lambda^2}\frac{\partial \lambda_E}{\partial E_\alpha}$$

It is convenient to introduce a short-hand notation for the various derivatives evaluated at zero field strength. We recognize the first-, second-, and higher-order derivatives of the energy with respect to λ as the electronic gradient, Hessian and so on, and we denote the n'th order derivative by $E^{[n]}$. In analogy, for minus the second derivative of the energy with respect to the field E and the parameters λ we speak about a property gradient. Minus the energy derivative once with respect to the field and n times with respect to the wave function parameters is denoted $A^{[n]}$, $B^{[n]}$, $C^{[n]}$, etc. for the various components of the dipole moment operator. The first-order response of the wave function parameters can therefore be written

$$\text{(106)} \quad \frac{\partial \lambda_0}{\partial E_\alpha} = -\left(E^{[2]}\right)^{-1} A^{[1]}$$

and the explicit formula for the linear polarizability is

$$(107) \quad \alpha_{\alpha\beta} = -\frac{d^2\mathcal{E}(\lambda_E, E)}{dE_\alpha dE_\beta}\bigg|_{E=0} = -A^{[1]}\left(E^{[2]}\right)^{-1} B^{[1]} = -A^{[1]} N^B$$

where the so-called response vector N^B has been introduced to further shorten the notation. Computationally it is the inverse of the Hessian which is the expensive part in this expression due to the possibly large number of wave function parameters.

The third-order derivative of the energy is

$$(108) \quad \frac{d^3\mathcal{E}(\lambda_E, E)}{dE_\alpha dE_\beta dE_\gamma} = \left[\frac{\partial}{\partial E_\alpha} + \frac{\partial \lambda_E}{\partial E_\alpha}\frac{\partial}{\partial \lambda}\right]\left[\frac{\partial}{\partial E_\beta} + \frac{\partial \lambda_E}{\partial E_\beta}\frac{\partial}{\partial \lambda}\right]\frac{\partial \mathcal{E}(\lambda_E, E)}{\partial E_\gamma}$$

$$= \left[\frac{\partial^2 \lambda_E}{\partial E_\alpha \partial E_\beta}\frac{\partial}{\partial \lambda} + \frac{\partial \lambda_E}{\partial E_\alpha}\frac{\partial \lambda_E}{\partial E_\beta}\frac{\partial^2}{\partial \lambda^2}\right]\frac{\partial \mathcal{E}(\lambda_E, E)}{\partial E_\gamma}$$

$$= \frac{\partial^2 \lambda_E}{\partial E_\alpha \partial E_\beta}\frac{\partial^2 \mathcal{E}(\lambda_E, E)}{\partial \lambda \partial E_\gamma} + \frac{\partial \lambda_E}{\partial E_\alpha}\frac{\partial \lambda_E}{\partial E_\beta}\frac{\partial^3 \mathcal{E}(\lambda_E, E)}{\partial \lambda^2 \partial E_\gamma}$$

and it is seen that the third-order property depends on the first- and second-order response of the wave function parameters with respect to the perturbation. A working formula for the second-order response of the wave function is obtained, like the first-order response, from the variational condition:

$$(109) \quad 0 = \frac{d^2}{dE_\alpha dE_\beta}\left[\frac{\partial \mathcal{E}(\lambda_E, E)}{\partial \lambda}\right]$$

$$= \left[\frac{\partial}{\partial E_\alpha} + \frac{\partial \lambda_E}{\partial E_\alpha}\frac{\partial}{\partial \lambda}\right]\left[\frac{\partial}{\partial E_\beta} + \frac{\partial \lambda_E}{\partial E_\beta}\frac{\partial}{\partial \lambda}\right]\frac{\partial \mathcal{E}(\lambda_E, E)}{\partial \lambda}$$

$$= \frac{\partial^2 \lambda_E}{\partial E_\alpha \partial E_\beta}\frac{\partial^2 \mathcal{E}(\lambda_E, E)}{\partial \lambda^2} + \frac{\partial \lambda_E}{\partial E_\beta}\frac{\partial^3 \mathcal{E}(\lambda_E, E)}{\partial \lambda^2 \partial E_\alpha}$$

$$+ \frac{\partial \lambda_E}{\partial E_\alpha}\frac{\partial^3 \mathcal{E}(\lambda_E, E)}{\partial \lambda^2 \partial E_\beta} + \frac{\partial \lambda_E}{\partial E_\alpha}\frac{\partial \lambda_E}{\partial E_\beta}\frac{\partial^3 \mathcal{E}(\lambda_E, E)}{\partial \lambda^3}$$

and the second-order response of the wave function thus becomes

$$(110) \quad \frac{\partial^2 \lambda_0}{\partial E_\alpha \partial E_\beta} = -\left(E^{[2]}\right)^{-1}\left[N^B A^{[2]} + N^A B^{[2]} + N^A N^B E^{[3]}\right]$$

The explicit formula for the first-order hyperpolarizability thus becomes

$$(111) \quad \beta_{\alpha\beta\gamma} = -\frac{d^3\mathcal{E}(\lambda_E, E)}{dE_\alpha dE_\beta dE_\gamma}\bigg|_{E=0}$$

$$= -\left[N^B A^{[2]} N^C + N^A B^{[2]} N^C + N^A N^B E^{[3]} N^C + N^A N^B C^{[2]}\right]$$

and we see that it is necessary to solve three linear response equations in order to obtain the three response vectors corresponding to the respective components of the hyperpolarizability tensor. In addition the calculation involves the contraction of the third-rank generalized Hessian matrix $E^{[3]}$. We note that we only need the first-order response of the wave function to determine the third derivative of the energy. This is an example of the $2n + 1$ rule of energy derivative theory and perturbation theory, which states that the perturbed energy of order $2n + 1$ can be determined from a knowledge of the perturbed wave function to order n.

The fourth derivative of the energy with respect to the electric fields is given by

$$(112) \quad \frac{d^4 \mathcal{E}(\lambda_E, E)}{dE_\alpha dE_\beta dE_\gamma dE_\delta} = \left[\frac{\partial}{\partial E_\alpha} + \frac{\partial \lambda_E}{\partial E_\alpha} \frac{\partial}{\partial \lambda} \right] \left[\frac{\partial}{\partial E_\beta} + \frac{\partial \lambda_E}{\partial E_\beta} \frac{\partial}{\partial \lambda} \right]$$
$$\times \left[\frac{\partial}{\partial E_\gamma} + \frac{\partial \lambda_E}{\partial E_\gamma} \frac{\partial}{\partial \lambda} \right] \frac{\partial \mathcal{E}(\lambda_E, E)}{\partial E_\delta}$$

which, when evaluated, will bottle down to a rather lengthy expression that involves first-, second-, and third-order responses of the wave function to the perturbation. Only the last one is unknown at this point and we can find it by returning to the variational condition:

$$(113) \quad 0 = \frac{d^3}{dE_\alpha dE_\beta dE_\gamma} \left[\frac{\partial \mathcal{E}(\lambda_E, E)}{\partial \lambda} \right]$$
$$= \left[\frac{\partial}{\partial E_\alpha} + \frac{\partial \lambda_E}{\partial E_\alpha} \frac{\partial}{\partial \lambda} \right] \left[\frac{\partial}{\partial E_\beta} + \frac{\partial \lambda_E}{\partial E_\beta} \frac{\partial}{\partial \lambda} \right] \left[\frac{\partial}{\partial E_\gamma} + \frac{\partial \lambda_E}{\partial E_\gamma} \frac{\partial}{\partial \lambda} \right] \frac{\partial \mathcal{E}(\lambda_E, E)}{\partial \lambda}$$

and we can thereafter evaluate the second-order hyperpolarizability as

$$(114) \quad \gamma_{\alpha\beta\gamma\delta} = - \left. \frac{d^4 \mathcal{E}(\lambda_E, E)}{dE_\alpha dE_\beta dE_\gamma dE_\delta} \right|_{E=0}$$

The explicit formula has been derived and implemented [45].

The main advantages of the energy derivative approach compared to the finite-field technique are numerical accuracy and computational speed. However, the approach is not readily extended to incorporate the contributions to the molecular properties from the motions of the nuclei. Furthermore, the analytical derivative approach is also restricted to time-independent perturbations, whereas the calculation of optical properties where significant dispersion is to be expected often is of greater interest.

5.3 Sum-over-states Approaches

An approach that *in principle* accounts for all contributions to linear and nonlinear optical properties is known as the sum-over-states (SOS) technique. Without having emphasized its use in practical calculations we have already met this formalism in the section on time-dependent perturbation theory. In the SOS method the working formulas for $\alpha(-\omega;\omega)$, $\beta(-\omega_\sigma;\omega_1,\omega_2)$, $\gamma(-\omega_\sigma;\omega_1,\omega_2,\omega_3)$ are given by Eqs. (43), (56), and (66), respectively. In the derivation it was assumed that the states in the summations were the true eigenstates of the molecular Hamiltonian $\hat{H}$. We understood $\hat{H}$ to be complete and including electronic as well as nuclear degrees of freedom, and in which case the states are the true nonadiabatic vibronic eigenstates of the system and hence the properties are the exact ones. Nothing prevents us, however, to introduce the adiabatic approximation and to assume the wave functions to be products of electronic and nuclear (vibrational) parts. In this case, the Born–Oppenheimer electronic plus vibrational properties will appear. We can even reduce the accuracy to the extent that we adopt the electronic Hamiltonian, work with the spectrum of electronic states, and thus extract the electronic part of the properties. In all these cases, the SOS property expressions remain unchanged.

The use of the SOS expressions in conjunction with approximate states (vibrational or electronic) provide us with methods to use in practical calculations. The electronic wave functions and transition moments can be determined with our standard electronic structure methods (SCF, CI, MCSCF, etc.) and the vibrational wave functions and transition moments (if considered) are determined using the potential energy surfaces. The SOS formulas may be formally separated into electronic and vibrational contributions to the properties (see Section 4), and this fact makes the SOS expressions pertinent in all calculations regardless of the choice of electronic structure method. Criticism of the SOS approach mainly concerns calculations of the electronic contributions to the properties as the SOS technique is often hampered by slow convergence with respect to the number of states that need to be included in the summations. It has been used with success only for very small systems, most notably by Bishop and co-workers in α- and γ-calculations of calibrational quality on helium [12] and H_2 [13].

Despite the discouraging performance of the SOS approach for the electronic contributions to properties of more extended systems, the SOS formulas are frequently used for reasons of interpretation. Strongly absorbing states in the linear absorption spectrum may be identified as main contributors to the linear and nonlinear response properties, and truncated SOS expressions can be formed with this consideration in mind. In certain cases, molecules may have a single strongly dominant one-photon transition, and so-called two-states-models (TSM) can then be applied. The two states in question are obviously the ground state $|0\rangle$ and the intense excited state $|f\rangle$.

A class of systems for which the TSM gives a reasonable description of the nonlinear responses is that of charge-transfer molecules, or push-pull systems, as demonstrated by Luo et al. [40] for *para*-nitroaniline (pNA). The reason for this success is the existence of an intense transition that involves the transfer of charge from the highest-occupied molecular orbital (HOMO), which is a molecular orbital localized on the donor group (the amino-group for pNA), to the lowest unoccupied molecular orbital (LUMO), which is a molecular orbital localized on the acceptor group (the nitro-group for pNA).

Applied to the diagonal components of the linear polarizability the two-states approximation of Eq. (43) is

(115) $\quad \alpha(-\omega; \omega) = \alpha(0; 0)D^{(1)}(\omega)$

(116) $\quad \alpha(0; 0) = \dfrac{2|\mu^{0f}|^2}{\hbar \omega_{f0}}$

(117) $\quad D^{(1)}(\omega) = \dfrac{1}{1 - \omega^2/\omega_{f0}^2}$

where $\mu^{0f} = \langle 0|\hat{\mu}|f\rangle$ denotes the transition dipole moment between the ground and the excited states. Furthermore, the dispersion $D^{(1)}(\omega)$ has been separated from the static value $\alpha(0; 0)$, and the TSM could therefore be applied to the dispersion part only and in conjunction with a more accurate static value. The corresponding set of formulas for the first-order hyperpolarizabilities were first presented by Oudar and Chemla [48] and applied to charge-transfer state contributions in nitroanilines. The two-states approximation of Eq. (56) for the diagonal components of $\beta(-\omega_\sigma; \omega_1, \omega_2)$ is

(118) $\quad \beta(-\omega_\sigma; \omega_1, \omega_2) = \beta(0; 0, 0)D^{(2)}(\omega_1, \omega_2)$

(119) $\quad \beta(0; 0, 0) = \dfrac{3|\mu^{0f}|^2(\mu^f - \mu^0)}{(\hbar \omega_{f0})^2}$

(120) $\quad D^{(2)}(\omega_1, \omega_2) = \dfrac{1 - \omega_1 \omega_2/\omega_{f0}^2}{(1 - \omega_1^2/\omega_{f0}^2)(1 - \omega_2^2/\omega_{f0}^2)(1 - (\omega_1 + \omega_2)^2/\omega_{f0}^2)}$

where the dispersion $D^{(2)}(\omega_1, \omega_2)$ again has been separated out from the static value $\beta(0; 0, 0)$, and μ^0 and μ^f denote the permanent dipole moment of the ground and the excited state, respectively. It is immediately seen from Eq. (119) that in the TSM the sign of β is dictated by the change in dipole moment between the ground and the excited states.

For the electro-optical Pockels effect (EOPE), the dispersion simplifies to

(121) $\quad D^{(2)}(\omega, 0) = \dfrac{1}{(1 - \omega^2/\omega_{f0}^2)^2}$

and for the second-harmonic generation process (SHG) it becomes

(122) $\quad D^{(2)}(\omega, \omega) = \dfrac{1}{(1 - \omega^2/\omega_{f0}^2)(1 - 4\omega^2/\omega_{f0}^2)}$

In Fig. 8 the EOPE and SHG dispersions are compared in the two-states model, and significant differences are observed for laser field frequencies exceeding 10% of the transition frequency ω_{f0}. The stronger dispersion for the SHG optical process is due to the two-photon resonance at $2\omega = \omega_{f0}$.

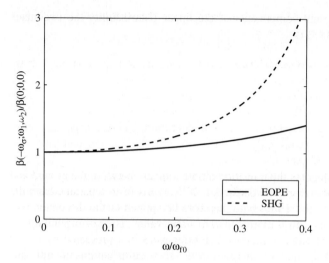

Figure 8. Dispersion of the first-order hyperpolarizability $\beta(-\omega_\sigma; \omega_1, \omega_2)$ in the two-states model

To date, the TSM is much less used for the second-order hyperpolarizability, but the corresponding TSM expressions for $\gamma(-\omega_\sigma; \omega_1, \omega_2, \omega_3)$ can easily be derived from Eq. (66).

5.4 Response Theory

When the external electric field is time-dependent, there is no well-defined energy of the molecular system in accordance with Eq. (100), and the wave function response can thus not be retrieved from a variational condition on the energy as in the analytic derivative approach described above. Instead the response parameters have to be determined from the time-dependent Schrödinger equation, a procedure which was illustrated in Section 3 for the exact state case. In approximate state theories, however, our wave function space only partially spans the N-electron Hilbert space, and the response functions that correspond to an approximate state wave function will clearly be separate from those of the exact state wave function. This fact is disregarded in the sum-over-states approach, and, apart from the computational aspect of slowly converging SOS expressions, it is of little concern when highly accurate wave function models are used. But for less flexible wave function models, the correct response functions should be used in the calculation of nonlinear optical properties.

The time-dependent Schrödinger equation may be disguised in different forms, and in exact state theory all formulations will result in identical results for the molecular properties. In the case of approximate state theories, this is no longer true, but one has to choose an appropriate formulation of wave mechanics in order to get results consistent with the theory. Over the years, *response theory* has come to denote formulations of time-dependent perturbation theory that are suitable for extension to approximate state theories.

The steps involved in response theory are:
1. Find a nonredundant parameterization of the wave function space at hand. It is a common procedure to use the exponential operator ansatz.
2. Choose an equation-of-motion, based on the time-dependent Schrödinger equation, from which the set of parameters can be determined by use of perturbation theory. Since the parameterizations are electronic structure dependent, the detailed working expressions that arise, and thus also the implementations, will have to differ with different wave function methods.
3. Owing to the fact that the state vectors are not eigenstates of the unperturbed molecular Hamiltonian, the linear differential equations for the parameters will be coupled. A Fourier transformation leads to matrix equations in the frequency domain to be solved for the Fourier amplitudes. These matrix equations are often solved with iterative techniques due to their large sizes.

In this chapter, we will not be concerned with the detailed expressions of the response functions that we find for the standard electronic structure methods in theoretical chemistry. However, we will briefly outline the basic elements in two alternative formulations of response theory, namely the *polarization propagator* and the *quasi-energy derivative* approaches.

5.4.1 Polarization propagator approaches

An introduction to the polarization propagator is made through Eq. (5). As the name suggests, the polarization propagator describes the propagation in time of the molecular polarization, and the framework is general and can in principle be applied to any kind of perturbations to a reference state. A pioneering work that has been the foundation for the development of propagator methods in quantum chemistry is the book by Linderberg and Öhrn [36] (the book has recently been re-written [37]).

From the point of view of a computational chemist, one of the most appreciated strengths of the polarization propagator approach is that, although being generally applicable to many fields in physics, it also delivers efficient, computationally tractable formulas for specific applications. Today we see implementations of the theory for virtually all standard electronic structure methods in quantum chemistry, and the implementations include both linear and nonlinear response functions. The double-bracket notation is the most commonly used one in the literature, and, in analogy with Eq. (5), the response functions are defined by the expansion

$$(123) \quad \langle \psi(t)|\hat{\mu}_\alpha|\psi(t)\rangle = \langle 0|\hat{\mu}_\alpha|0\rangle - \sum_\omega \langle\langle \hat{\mu}_\alpha; \hat{\mu}_\beta \rangle\rangle_\omega E_\beta^\omega e^{-i\omega t}$$

$$+ \frac{1}{2} \sum_{\omega_1,\omega_2} \langle\langle \hat{\mu}_\alpha; \hat{\mu}_\beta, \hat{\mu}_\gamma \rangle\rangle_{\omega_1,\omega_2} E_\beta^{\omega_1} E_\gamma^{\omega_2} e^{-i\omega_\sigma t}$$

$$- \frac{1}{6} \sum_{\omega_1,\omega_2,\omega_3} \langle\langle \hat{\mu}_\alpha; \hat{\mu}_\beta, \hat{\mu}_\gamma, \hat{\mu}_\delta \rangle\rangle_{\omega_1,\omega_2,\omega_3} E_\beta^{\omega_1} E_\gamma^{\omega_2} E_\delta^{\omega_3} e^{-i\omega_\sigma t} + \cdots$$

A straight-forward comparison between Eqs. (123) and (5) makes it clear that the connecting formulas between the molecular properties of interest in this work and the response functions are

(124) $$\alpha_{\alpha\beta}(-\omega;\omega) = -\langle\langle\hat{\mu}_\alpha; \hat{\mu}_\beta\rangle\rangle_\omega$$

(125) $$\beta_{\alpha\beta\gamma}(-\omega_\sigma;\omega_1,\omega_2) = \langle\langle\hat{\mu}_\alpha; \hat{\mu}_\beta, \hat{\mu}_\gamma\rangle\rangle_{\omega_1,\omega_2}$$

(126) $$\gamma_{\alpha\beta\gamma\delta}(-\omega_\sigma;\omega_1,\omega_2,\omega_3) = -\langle\langle\hat{\mu}_\alpha; \hat{\mu}_\beta, \hat{\mu}_\gamma, \hat{\mu}_\delta\rangle\rangle_{\omega_1,\omega_2,\omega_3}$$

The alternation of the sign in the equations above are explained by the perturbation operator in the electric dipole approximation being equal to

(127) $$\hat{V}(t) = -\hat{\mu}_\alpha E_\alpha(t)$$

When atomic units are adopted, the electric dipole operators are to be replaced by the negative of the position operators, i.e., $\hat{\mu}_\alpha = -\hat{r}_\alpha$. So, if we refer to response functions with position operators, there is a sign difference compared to properties of *all* orders.

As mentioned above, the detailed computationally tractable formulas for the response functions for any given electronic structure methods are quite involved. If $|0\rangle$ denotes a wave function in the absence of the external field, the time-dependent reference state is expressed using an exponential operator

(128) $$|\psi(t)\rangle = \exp\left[\sum_n \lambda_n(t)\hat{\Lambda}_n\right]|0\rangle$$

For variational methods, such as Hartree–Fock (HF), multi-configurational self-consistent field (MCSCF), and Kohn–Sham density functional theory (KS-DFT), the initial values of the parameters are equal to zero and $|0\rangle$ thus corresponds to the reference state in the absence of the perturbation. The $\hat{\Lambda}$ operators are the non-redundant state-transfer or orbital-transfer operators, and carries no time-dependence (the sole time-dependence lies in the complex λ parameters). Furthermore, the operator $\sum_n \lambda_n(t)\hat{\Lambda}_n$ is anti-Hermitian, and the exponential operator is thus explicitly unitary so that the norm of the reference state is preserved. Perturbation theory is invoked in order to solve for the time-dependence of the parameters, and we expand the parameters in orders of the perturbation

(129) $$\lambda(t) = \lambda^{(1)} + \lambda^{(2)} + \cdots$$

We have used the fact that $\lambda^{(0)} = 0$, in accordance with the zeroth-order wave function being equal to $|0\rangle$. It was demonstrated by Olsen and Jørgensen that the time-dependence of the λ parameters could be determined by applying the Ehrenfest theorem to the operators $\hat{\Lambda}_n$, and require the equation to hold for each order of the perturbation and for each Fourier frequency [46].

Microscopic Theory of Nonlinear Optics

For the coupled cluster methods, which are non-variational, the initial values of the λ's are nonzero, and $|0\rangle$ does not correspond to the unperturbed reference state but, in most applications, to the Hartree–Fock state. The initial values of the parameters are found in an iterative optimization of the coupled cluster state, and the time-dependent values of the parameters were determined from the coupled-cluster time-dependent Schrödinger equation by Koch and Jørgensen [35]. The coupled cluster state is not norm conserving, but the inner product of the coupled cluster state vector $|CC(t)\rangle$ and a constructed dual vector $\langle\overline{CC}(t)|$ remains a constant of time

$$(130) \quad \langle\overline{CC}(t)|CC(t)\rangle = 1$$

In a truncated coupled cluster approach, the two vectors are not connected by the adjoint operation; but without truncations a representation of the exact state situation is retrieved and one state is the adjoint of the other. The generalized Hellmann–Feynman theorem is proven to hold

$$(131) \quad \frac{\partial}{\partial E_\alpha}\langle\overline{CC}(t)|\hat{H}_0 - \hat{\mu}_\alpha E_\alpha|CC(t)\rangle = \langle\overline{CC}(t)|\hat{\mu}_\alpha|CC(t)\rangle$$

and the coupled cluster response functions can be identified from the time-development of the transition expectation value $\langle\overline{CC}(t)|\hat{\mu}_\alpha|CC(t)\rangle$ rather than from the time-development of the true expectation value in Eq. (123).

5.4.2 Quasi-energy derivative approaches

One somewhat displeasing detail in the approximate polarization propagator methods discussed in the previous section is the fact that concern needs to be made as to which formulation of wave mechanics that is used. This point has been elegantly resolved by Christiansen et al. in their quasi-energy formulation of response theory [23], in which a general and unified theory is presented for the evaluation of response functions for variational as well as nonvariational electronic structure methods.

We have seen that the analytic energy derivative technique provides an alternative for determining molecular properties in the static limit. When the system is subjected to a time-dependent perturbation, on the other hand, there is no well-defined molecular energy. However, for time-periodic perturbations, one may introduce the quasi-energy $Q(t)$ as the time-averaged expectation value of the operator $\hat{H}_0 - i\hbar\partial/\partial t$. The quasi-energy plays the same role as the molecular energy in the time-independent situation, and response functions can be obtained with the derivative technique discussed earlier in the time-independent case with Eq. (100) as starting point.

The time evolution of the wave function parameters $\lambda(t)$ is determined from the time-averaged variational condition

$$(132) \quad \delta Q(t) = 0$$

and the response functions are identified from the time-averaged Hellmann–Feynman theorem. To the extent possible, the final expressions for the approximate state response functions maintain the symmetry relations valid in the exact theory. For variational state theory, the computational results are identical to those obtained from the propagator formalism, but for nonvariational state theories this is not necessarily true. If different in performance, the authors argue that the Fourier component variational perturbation theory is the best alternative [23].

6. BULK PROPERTIES

Our discussion has so far been concerned with the microscopic response of a molecule to an external electric field, and thus with an expansion of the molecular energy in orders of the response with respect to the external field, giving rise to the molecular (hyper)polarizabilities. Although experimental data for nonlinear optical properties of molecules in the gas phase do exist [55], the majority of experimental measurements are done in the liquid or solid states, as these states also are the ones that are of greatest interest with respect to developing materials with specifically tailored (non)linear optical properties.

In the macroscopic case, we will instead consider the polarization $P(t)$ of the medium by a time-dependent electric field, and we may expand the polarization in orders of the applied external field as

$$(133) \quad P_\alpha(t) = P_\alpha^{(0)}(t) + P^{(1)}(t) + P^{(2)}(t) + \cdots$$

where $P_\alpha^{(1)}(t)$ is linear in the electric field, $P_\alpha^{(2)}(t)$ is quadratic in the electric field and so on. We note the very close correspondence to the expansion of the molecular dipole moment in Eq. (3). Using now the Fourier transformation of the time-dependent electric field to express the field in terms of a finite number of frequency components Eq. (4) and using $\chi^{(n)}$ to denote a macroscopic susceptibility tensor of rank $n+1$, we can rewrite Eq. (133) as

$$(134) \quad P(t) = P^{(0)}(t) + \varepsilon_0 \sum_\omega \chi_{\alpha\beta}^{(1)}(-\omega;\omega) E_\beta^\omega$$

$$+ \varepsilon_0 \sum_{\omega_1,\omega_2} \chi_{\alpha\beta\gamma}^{(2)}(-\omega_\sigma;\omega_1,\omega_2) E_\beta^{\omega_1} E_\gamma^{\omega_2} + \cdots$$

A comparison of Eq. (5) and Eq. (134) shows that the macroscopic susceptibility is, assuming the electric fields to be the same, directly proportional to the microscopic (hyper)polarizabilities, and we could thus define from the experimental data a microscopic hyperpolarizability by

$$(135) \quad \chi_{\text{mic}}^{(1)} = \alpha/\varepsilon_0$$

$$(136) \quad \chi_{\text{mic}}^{(2)} = \frac{1}{2}\beta/\varepsilon_0$$

$$(137) \quad \chi_{\text{mic}}^{(3)} = \frac{1}{6}\gamma/\varepsilon_0$$

and so on for higher-order responses, assuming that we have calculated the microscopic susceptibility from the observed data by dividing by the number of molecules in the sample.

The microscopic susceptibility will however still not be directly related to the (hyper)polarizabilities as given by Eqs. (135)–(137) due to the fact that the electric field experienced by the individual molecule in a macroscopic sample is not the same as the applied external electric field (which would be experienced by an isolated molecule in the gas phase). Instead, the molecule in a macroscopic sample experiences a *local field* E_{loc}^{ω}. This field is different from the macroscopically applied field because of the polarization of the surrounding molecules.

A direct comparison between experimentally observed susceptibilities and calculated microscopic (hyper)polarizabilities is thus hampered by the fact that whereas the macroscopic electric field used in experiment is well known, the local field experienced by the molecule is in most cases unknown, and only when corrections for the local field effects can be made will a direct comparison between theory and experiment be made.

It is also important to realize that the nonlinear optical properties of a molecule in solution or in the solid state will differ from that of the isolated molecule due to polarization effects caused by the surrounding molecules. In theoretical calculations of molecules in the liquid phase, these effects may be modeled using for instance dielectric continuum models [33, 41, 42, 52, 56]. The use of such schemes for estimating the polarization of the solute by the solvent does not resolve the issue of local field factors.

The determination of local field factors is a difficult topic, even for such simplified models as the dielectric continuum models, and for this reason we will not go into further detail about the determination of local field factors in this chapter, referring instead to Chapter 4. We note, however, that for the case of a spherical cavity, Onsager demonstrated that the local field is related to the macroscopic field through [47].

$$(138) \quad E_{\text{loc}}^{\omega} = \frac{(\varepsilon_{\infty}+2)\varepsilon_{\omega}}{\varepsilon_{\infty}+2\varepsilon_{\omega}} E^{\omega}$$

where ε_{∞} and ε_{ω} is the dielectric constant of the medium for a static and an optical field, respectively.

7. CONCLUDING REMARKS

In this chapter we have introduced the basic elements of the theory of nonlinear optical properties. Emphasis has been laid on the basic physical processes involved and how these processes are reflected in the basic formulas derived from time-dependent perturbation theory. We have also briefly outlined the strategies for developing efficient computational methods for the calculation of linear and nonlinear optical properties. A brief discussion of the contributions to the nonlinear optical properties arising from the nuclear motions, as well as the connection between the molecular microscopic properties and the bulk properties most often observed in experiment. As such, this chapter should provide the reader with the background needed to understand in more detail these various aspects of the theoretical modeling of nonlinear optical properties as will be discussed in the remainder of this book.

ACKNOWLEDGEMENTS

This work has received support from the Nordic Research Academy through the project QMMM (Grant No 030262). KR has been supported by the Norwegian Research Council through a Strategic University Program in theoretical chemistry (Grant No 154011/420), an YFF grant (Grant No 162746/V00) and a NANOMAT grant (Grant No 158538/431).

REFERENCES

[1] Baranowska, A., Sadlej, A.J.: Chem. Phys. Lett., **398**, 270 (2004)
[2] Bartlett, R.J., Sekino, H.: In Karna, S.P., Yeates, A.T., (eds) Nonlinear Optical Materials, ACS Symposium Ser. **628**, p. 23. American Chemical Society, Washington (1996)
[3] Bishop, D.M.: Rev. Mod. Phys., **62**, 343 (1990)
[4] Bishop, D.M.: Adv. Quantum Chem., **25**, 1 (1994)
[5] Bishop, D.M.: J. Chem. Phys., **100**, 6535 (1994)
[6] Bishop, D.M.: Adv. Chem. Phys., **104**, 1 (1998)
[7] Bishop, D.M., Cheung, L.M., Buckingham, A.D.: Mol. Phys., **41**, 1225 (1980)
[8] Bishop, D.M., Hasan, M., Kirtman, B.: J. Chem. Phys., **103**, 4157 (1995)
[9] Bishop, D.M., Kirtman, B.: J. Chem. Phys., **95**, 2646 (1991)
[10] Bishop, D.M., Kirtman, B.: J. Chem. Phys., **97**, 5255 (1992)
[11] Bishop, D.M., Norman, P.: In Nalwa, H.S. (ed.) Handbook of Advanced Electronic and Photonic Materials, Academic, San Diego (2000)
[12] Bishop, D.M., Pipin, J.: J. Chem. Phys., **91**, 3549 (1989)
[13] Bishop, D.M., Pipin, J., Cybulski, S.M.: Phys. Rev. A, **43**, 4845 (1991)
[14] Bogaard, M.P., Orr, B.J.: In Buckingham, A.D. (ed.) MTP International Review of Science, Physical Chemistry, vol. 2 of 2, Chap 5, p. 149. Butterworth, London (1975)
[15] Born, M., Huang, K.: Clarendon Press, Oxford. Appendix 8 (1954)
[16] Born, M., Oppenheimer, R.: Ann. d. Phys., **84**, 457 (1927)
[17] Boyd, R.W.: Nonlinear Optics. 2nd edition Academic, Inc., San Diego (2003)
[18] Brédas, J.L., Adant, C., Tackx, P., Persoons, A.: Chem. Rev., **94**, 243 (1994)
[19] Buckingham, A.D.: Adv. Chem. Phys., **12**, 107 (1967)
[20] Buckingham, A.D., Orr, B.J.: Quart. Rev., **21**, 195 (1967)
[21] Butcher, P.N., Cotter, D.: The elements of nonlinear optics. Cambridge University Press, Cambridge (1990)
[22] Champagne, B., Kirtman, B.: In: Nalwa, H.S., (ed.) Handbook of Advanced Electronic and Photonic Materials. Academic, San Diego (2000)
[23] Christiansen, O., Jørgensen, P., Hättig, C.: Int. J. Quantum Chem., **68**, 1 (1998)
[24] Darling, C.L., Schlegel, H.B.: J. Phys. Chem., **98**, 5855 (1994)
[25] Diercksen, G.H.F., Roos, B.O., Sadlej, A.J.: Chem. Phys., **59**, 29 (1981)
[26] Dykstra, C.E., Liu, S.-Y., Malik, D.J.: Adv. Chem. Phys., **75**, 37 (1989)
[27] Feynman, R.P.: Phys. Rev., **56**, 340 (1939)
[28] Flytzanis, C.: In Rabin, H., Tang, C.L., (eds) Quantum Electronics, vol. IA, p. 9. Academic, New York (1975)
[29] Franken, P.A., Hill, A.E., Peters, C.W., Weinreich, G.: Phys. Rev. Lett., **7**, 118 (1961)

[30] Hasanein, A.A.: Adv. Chem. Phys., **85**, 415 (1993)
[31] Helgaker, T., Jørgensen, P., Olsen, J.: Molecular Electronic-Structure Theory, Wiley (2000)
[32] Hellmann, H.: Einführung in die Quantenchemie. Deuticke, Leipzig (1937)
[33] Kirkwood, J.G.: J. Chem. Phys., **2**, 351 (1934)
[34] Kirtman, B., Champagne, B.: Int. Rev. Phys. Chem., **16**, 389 (1997)
[35] Koch, H., Jørgensen, P.: J. Chem. Phys., **93**, 3333 (1990)
[36] Linderberg, J., Öhrn, Y.: Propagators in Quantum Chemistry. Academic, London (1973)
[37] Linderberg, J., Öhrn, Y.: Propagators in Quantum Chemistry. Wiley, Weinheim (2004)
[38] London, F.: J. Phys. Radium, **8**, 397 (1937)
[39] Luo, Y., Ågren, H., Jørgensen, P., Mikkelsen, K.V.: Adv. Quantum Chem., **26**, 165 (1995)
[40] Luo, Y., Ågren, H., Vahtras, O., Jørgensen, P.: Chem. Phys. Lett., **207**, 190 (1993)
[41] Miertuš, S., Scrocco, E., Tomasi, J.: Chem. Phys., **55**, 117 (1981)
[42] Mikkelsen, K.V., Dalgaard, E., Swanstrøm, P.: J. Phys. Chem., **91**, 3081 (1987)
[43] Nakano, M., Yamaguchi, K.: Trends in Chemical Physics, **5**, 87 (1997)
[44] Nerbrant, P.O., Roos, B.O., Sadlej, A.J.: Int. J. Quantum Chem., **15**, 135 (1979)
[45] Norman, P., Jonsson, D., Vahtras, O., Ågren, H.: Chem. Phys., **203**, 23 (1996)
[46] Olsen, J., Jørgensen, P.: J. Chem. Phys., **82**, 3235 (1985)
[47] Onsager, L.: J. Am. Chem. Soc., **58**, 1456 (1936)
[48] Oudar, J.L., Chemla, D.S.: J. Chem. Phys., **66**, 2664 (1977)
[49] Prasad, P.N., Williams, D.J.: Introduction to nonlinear optical effects in molecules and polymers. Wiley, Inc., New York (1991)
[50] Pulay, P.: Mol. Phys., **18**, 473 (1970)
[51] Pulay, P.: In Yarkony, D.R., (ed.) Modern Electronic Structure Theory, p. 1191. World Scientific (1995)
[52] Rinaldi, D., Rivail, J.-L.: Theor. Chim. Acta, **32**, 57 (1973)
[53] Ruud, K., Helgaker, T., Bak, K.L., Jørgensen, P., and Jensen, H.J. Aa. J. Chem. Phys., **99**, 3847 (1993)
[54] Sadlej, A.J.: Chem. Phys. Lett., **47**, 50 (1977)
[55] Shelton, D.P., Rice, J.E.: Chem. Rev., **94**, 3 (1994)
[56] Tomasi, J., Persico, M.: Chem. Rev., **94**, 2027 (1994)
[57] Wagniére, G.H.: Linear and nonlinear optical properties of molecules. Verlag Helvetica Chimica Acta, Basel (1993)
[58] Ward, J.F.: Rev. Mod. Phys., **37**, 1 (1965)
[59] Wolff, J.J., Wortmann, R.: Adv. Phys. Org. Chem., **32**, 121 (1999)
[60] Wolinski, K., Hinton, J.F., Pulay, P.: J. Am. Chem. Soc., **112**, 8251 (1990)
[61] Woon, D.E., Dunning Jr., T.H.: J. Chem. Phys., **100**, 2975 (1994)

CHAPTER 2

ACCURATE NONLINEAR OPTICAL PROPERTIES FOR SMALL MOLECULES
Methods and results

OVE CHRISTIANSEN[1], SONIA CORIANI[2], JÜRGEN GAUSS[3], CHRISTOF HÄTTIG[4], POUL JØRGENSEN[1], FILIP PAWŁOWSKI[5], AND ANTONIO RIZZO[6]

[1] *Department of Chemistry, University of Århus, Langelandsgade 140, DK-8000 Århus C, Denmark*
[2] *Dipartimento di Scienze Chimiche, Università degli Studi di Trieste, Via L. Giorgieri 1, I-34127 Trieste, Italy*
[3] *Institut für Physikalische Chemie, Universität Mainz, D-55099 Mainz, Germany*
[4] *Forschungszentrum Karlsruhe, Institute of Nanotechnology, P.O. Box 3640, D-76021 Karlsruhe, Germany*
[5] *Department of Chemistry, University of Oslo, P.O. Box 1033 Blindern, N-0315 Oslo, Norway*
[6] *Istituto per i Processi Chimico-Fisici del Consiglio Nazionale delle Ricerche, Area della Ricerca, via G. Moruzzi 1, loc. S. Cataldo, I-56124 Pisa, Italy*

Abstract: During the last decade it became possible to calculate by quantum chemical *ab initio* methods not only static but also frequency-dependent properties with high accuracy. Today, the most important tools for such calculations are coupled cluster response methods in combination with systematic hierarchies of correlation consistent basis sets. Coupled cluster response methods combine a computationally efficient treatment of electron correlation with a qualitatively correct pole structure and frequency dispersion of the response functions. Both are improved systematically within a hierarchy of coupled cluster models.

The present contribution reviews recent advances in the highly accurate calculation of frequency-dependent properties of atoms and small molecules, electronic structure methods, basis set convergence and extrapolation techniques. Reported applications include first and second hyperpolarizabilities, Faraday, Buckingham and Cotton–Mouton effects as well as Jones and magneto-electric birefringence

Keywords: coupled cluster, CCSD, CC3, response theory, quasi-energy Lagrangian, time-dependent perturbation theory, frequency-dependent properties, hyperpolarizabilities, basis set convergence, magnetic optical rotation, magnetic circular dichroism, Verdet constant, Faraday effect, Buckingham effect, Cotton–Mouton effect, Jones birefringence, magneto-electric birefringence

1. INTRODUCTION

The developments of laser technology have in the last decades made the field of nonlinear optics (NLO) increasingly important. Today lasers that produce light with well-defined properties and high intensities are in widespread use. In recent years it has become possible to generate in addition very strong magnetic fields. Thus a wealth of NLO processes that arise in the presence of strong static and dynamic electric and magnetic fields became accessible. Accordingly many modern experiments are related in one or the other way to NLO processes. Another motivation for studying NLO properties is the quest for new materials with specific NLO properties for creating new generations of optical devices. This quest has initiated a significant activity in the development of theoretical methods for describing NLO processes.

The theoretical prediction of NLO properties using high accuracy *ab initio* methods has been primarily focused on small molecules, where there has been a significant interplay between theory and experiment [1–39]. Among NLO properties, the interest in birefringences and their absorptive counterparts (dichroisms) has increased during the last decade with the progress in optics and detection techniques on the experimental side, and with the fast advance of methods and computational power from the theoretical side [40]. The interplay between theory and experiment has also been essential for illuminating the role of various physical effects in relation to NLO processes and for the design of molecules, chromophores, functional groups, etc., with specific NLO properties.

The calculation of NLO properties with high accuracy is challenging and requires consideration of many different issues. While electron correlation is known to be important for accurate predictions of essentially all molecular properties, it is often of extreme importance for NLO properties. Rather extensive one-electron basis sets must be used to obtain reasonably well-converged results. NLO properties are related to optical experiments with external frequencies different from zero. The calculation of frequency-dependent properties can nowadays be achieved in theoretical methods for example with the response theoretical approach [41–44] discussed in this review. However, it cannot easily be obtained via the finite field techniques that are often used in the calculation of static properties. Molecular vibrations have also a significant effect on NLO properties. They cause an averaging of the electronic contribution over the vibrational motion leading to a so-called zero-point vibrational (ZPV) correction to the electronic contribution calculated at the equilibrium geometry. For many NLO properties there are also significant pure vibrational contributions. We refer to later chapters of this book for a detailed discussion of these contributions.

Relativistic effects have also been considered. Though they may be important for heavier elements, we shall not discuss this issue in detail here, since most of the applications presented are for molecules containing only light atoms. Our review will focus on the prediction of electronic contributions to NLO properties of molecules in the gas phase discussing the above issues in relation to theory and to actual calculations on small molecules. It will be shown that with benchmark calculations it is nowadays possible to predict NLO effects with unprecedented accuracy in a variety of systems and in this way to assess the quality of experiments.

The *ab initio* calculation of NLO properties has been a topic of research for about three decades. In particular, response theory has been used in combination with a number of electronic structure methods to derive so-called response functions [41–48]. The latter describe the response of a molecular system for the specific perturbation operators and associated frequencies that characterize a particular experiment. For example, molecular hyperpolarizabilities can be calculated from the quadratic and cubic response functions using electric dipole operators. From the frequency-dependent response functions one can also determine expressions for various transition properties (e.g. for multi-photon absorption processes) and properties of excited states [42].

One of the advantages of response theory is that it can be applied for most quantum chemical methods, including Hartree-Fock self-consistent-field (HF-SCF) [42, 45], multi-configurational SCF (MCSCF) [42, 49, 50], and coupled cluster (CC) [44, 46, 51]. In recent years density functional theory (DFT) based response theory [52, 53] has also received considerable interest. Response theory based upon HF-SCF and DFT has the advantage of being applicable to rather large systems, but on the other hand is not always appropriate for detailed and quantitatively accurate comparisons with experiment. Since dynamic electron correlation is typically very important for the predictions of NLO properties, efficient and accurate treatments of dynamic correlation are required for obtaining quantitative results. Due to the exponential parameterization and the accompanying correct scaling with the size of the system, CC methods have the possibility to give accurate results for systems with both few and many electrons as long as the calculations are feasible.

The high accuracy of CC methods is well established for a number of molecular properties including structures, vibrations, NMR shielding etc. In particular CC models with an approximate treatment of triple excitations as e.g. the CCSD(T) method [54] provide a high accuracy. Despite all its qualities there is an inherent problem in the use of CCSD(T) in relation to NLO properties: while static properties can be obtained with high accuracy using CCSD(T) the description of the frequency dependence is dubious within this approach. A similar problem holds for second-order Møller-Plesset Perturbation theory (MP2). These methods have a frequency dependence that is corrupted by the presence of poles due to the HF-SCF reference state. However, a hierarchy of CC methods has been developed [55–71] for the calculation of response properties including CC2 [57], coupled cluster singles and doubles (CCSD) [72], and CC3 [56, 58], etc. The advantage of CC2 and CC3, as compared to MP2 and CCSD(T), is that they have a frequency dependence that matches the quality of the static properties. Thus in the CC2, CCSD and CC3 sequence both static and frequency-dependent molecular properties as well as electronic excitation energies are obtained with increasing accuracy at the cost of increasing computational effort. For these methods linear [60–64], quadratic [65–67] and cubic response functions [67, 68] have been implemented and allow today accurate and systematic calculations of NLO properties including hyperpolarizabilities

[30, 32, 36, 37, 73], magneto-optical effects [38, 39, 70, 71, 74], two-photon absorption [75, 76], and so on. In principle, the hierarchy of CC methods also extends further to the full coupled cluster singles, doubles and triples (CCSDT) method and beyond, but at this level the calculation begins to be too demanding using present day standards for computer equipment and such methods have been little used for the calculation of NLO properties besides benchmark studies [34, 39, 64, 77, 78]. On the other hand CC2 and in some cases even CCSD are not accurate enough for detailed studies of NLO properties of small molecules. The experience with CC3 has been that this is a very accurate method also for NLO properties. Indeed, CC3 is probably the most accurate methodology for frequency-dependent properties applicable nowadays to systems with more than a few electrons. In this paper the focus is on high accuracy NLO calculations and the role of CC3 is in this context extremely important since it often gives the final increase in accuracy that allows us to support, interpret, predict, or reject experimental results with confidence.

In the next section we summarize the theoretical background for coupled cluster response theory and discuss certain issues related to their actual implementation. In Sections 3 and 4 we describe the application of quadratic and cubic response in calculations of first and second hyperpolarizabilities. The use of response theory to calculate magneto-optical properties as e.g. the Faraday effect, magnetic circular dichroism, Buckingham effect, Cotton–Mouton effect or Jones birefringence is discussed in Section 5. Finally we give some conclusions and an outlook in Section 6.

2. COUPLED CLUSTER RESPONSE THEORY

In standard time-independent coupled cluster theory the wavefunction is parameterized as

$$\text{(1)} \qquad |\text{CC}\rangle = \exp(\hat{T})|\text{Ref}\rangle$$

where $\hat{T} = \sum_\mu t_\mu \hat{\tau}_\mu$ is the cluster operator, $|\text{Ref}\rangle$ a reference state – usually taken as the Hartree-Fock wavefunction $|\text{HF}\rangle$ – and $\hat{\tau}_\mu$ excitation operators [79] which applied to the reference state generate excited determinants or configurations $|\Phi_\mu\rangle = \hat{\tau}_\mu|\text{Ref}\rangle$. The amplitudes t_μ of the cluster operator are determined by projecting the Schrödinger equation onto the bra states $\langle \text{Ref}|\hat{\tau}_\mu^\dagger \exp(-\hat{T})$:

$$\text{(2)} \qquad 0 = \left\langle \text{Ref} \left| \hat{\tau}_\mu^\dagger \exp(-\hat{T}) \hat{H} \exp(\hat{T}) \right| \text{Ref} \right\rangle$$

and the energy is obtained as

$$\text{(3)} \qquad E_{\text{CC}} = \langle \text{Ref}|\hat{H}|\text{CC}\rangle.$$

By truncating the cluster operator $\hat{T}$ after single, double, triple, ... replacements the hierarchy of coupled cluster models CCS, CCSD, CCSDT, etc. is obtained. In this series the accuracy of the results – but also the computational costs – converge rapidly to those of Full Configuration Interaction (FCI). This motivated the development of several intermediate methods which are derived from the above standard models by either

- augmenting the CC energy of Eq. (3) with perturbative corrections (e.g. CCSD[T] [80, 81], CCSD(T) [54], CCSDT(Q_f) [82], ...)
- or introducing approximations into the cluster equations (2) – (e.g. CC2 [57], CC3 [56], CCSDT-1a [83, 84], ...)

For a detailed review of coupled cluster methods for ground state energies and static properties the reader is referred to [79, 85–87].

2.1 Including a Time-Dependent Perturbation

In response theory one considers a quantum mechanical system described by the time-independent Hamiltonian $\hat{H}^{(0)}$ which is perturbed by a time-dependent perturbation $\hat{V}(t, \epsilon)$

$$(4) \quad \hat{H}(t, \epsilon) = \hat{H}^{(0)} + \hat{V}(t, \epsilon).$$

We assume that the perturbation $\hat{V}$ can be expanded in a sum over monochromatic Fourier components as

$$(5) \quad \hat{V}(t, \epsilon) = \sum_j \epsilon_j(\omega_j) \exp(-i\omega_j t) \hat{X}_j,$$

where $\hat{X}_j$ are hermitian time-independent one-electron operators and ϵ_j are the real amplitudes of the associated field-strengths. (The letter ϵ without index is above and in the following used as abbreviation for all field strengths $\{\epsilon_j\}$ included in $\hat{V}$.) We shall furthermore assume for simplicity that the perturbations in Eq. (5) have a common time period and that $\hat{V}$ is hermitian, which implies that for each frequency ω_j also its negative is included in the summation.

The time-dependent ground-state coupled cluster wavefunction for such a system is conveniently parameterized in a form, where the oscillating phase factor caused by the so-called level-shift [43–45, 88, 89] or time-dependent quasi-energy $\mathcal{W}(t, \epsilon)$ (vide infra) is explicitly isolated [42–46, 90, 91]:

$$(6) \quad \left|\overline{CC}(t, \epsilon)\right\rangle = \exp\left(-i \int_{t_0}^{t} \mathcal{W}(t', \epsilon) dt'\right) \left|CC(t, \epsilon)\right\rangle,$$

with

$$(7) \quad |CC(t, \epsilon)\rangle = \exp(\hat{T}(t, \epsilon)) |\text{Ref}\rangle.$$

Note that in the time-independent limit $|CC(t, \epsilon)\rangle$ becomes the usual time-independent CC wavefunction, Eq. (1), while the full time-dependent wavefunction $|\overline{CC}(t, \epsilon)\rangle$ goes to $\exp(iE_{CC}(t_0 - t))|CC\rangle$. The reference state should be chosen such that it is time-independent – for the following it will be assumed that $|\text{Ref}\rangle$ is the Hartree-Fock wavefunction of the unperturbed molecule $|\text{HF}_0\rangle$. By keeping the uncorrelated reference state fixed in the presence of the perturbation, a two–step approach, which would introduce artificial poles into the correlated CC wavefunction due to the response of the orbitals, is avoided. Inserting the above ansatz into the time-dependent Schrödinger equation and projecting on the bra states $\langle \text{HF}_0 | \hat{\tau}_\mu^\dagger \exp(-\hat{T})$, the time-dependent cluster equations become

$$(8) \quad 0 = \left\langle \text{HF}_0 \left| \hat{\tau}_\mu^\dagger \exp(-\hat{T}) \hat{H}(t, \epsilon) \exp(\hat{T}) \right| \text{HF}_0 \right\rangle - i\frac{d}{dt} t_\mu = e_\mu(t, \epsilon) - i\frac{d}{dt} t_\mu.$$

The energy, which in the time-dependent case is no longer a constant of motion, has to be generalized to the time-dependent quasi-energy:

$$(9) \quad \mathcal{W}(t, \epsilon) = \left\langle \text{HF}_0 \left| \exp(-\hat{T}) \hat{H}(t, \epsilon) \exp(\hat{T}) \right| \text{HF}_0 \right\rangle.$$

Frequency-dependent higher-order properties can now be obtained as derivatives of the real part[1] of the time-average of the quasi-energy $\{\mathcal{W}\}_T$ with respect to the field strengths of the external perturbations. To derive computational efficient expressions for the derivatives of the coupled cluster quasi-energy, which obey the $2n+1$- and $2n+2$-rules of variational perturbation theory [44, 45, 93], the (quasi-)energy is combined with the cluster equations to a Lagrangian:

$$(10) \quad L(t, \epsilon) = \mathcal{W}(t, \epsilon) + \sum_\mu \bar{t}_\mu \left(e_\mu(t, \epsilon) - i\frac{d}{dt} t_\mu \right).$$

The time-average of the quasi-energy Lagrangian

$$(11) \quad \{L(t, \epsilon)\}_T = \lim_{t_0 \to \infty} \frac{1}{2t_0} \int_{-t_0}^{+t_0} L(t, \epsilon)\, dt$$

is required to be stationary with respect to the cluster amplitudes t_μ and the Lagrange multipliers $\bar{t}_\mu$ [43–45, 48, 88, 94]:

$$(12) \quad \delta \{L(t, \epsilon)\}_T = 0.$$

---

[1] The imaginary part of $\mathcal{W}(t, \epsilon)$ is connected to the time derivative of the wavefunction norm and does not contain information about physical observables. For normalized wavefunctions and bra and ket state conjugated to each other the quasi-energy is real [44, 45]. However, the CC energy is calculated from a projection expression and in is general not real. Thus with CC wavefunctions it is important to consider only the real part of the quasi-energy [44, 65, 92].

Accurate NLO Properties for Small Molecules

The last equation is a variation principle for the coupled cluster quasi-energy and wavefunction within oscillating harmonic external fields. If one inserts a perturbation and Fourier expansion as ansatz for the cluster amplitudes

$$(13) \quad t_\mu(t, \epsilon) = t_\mu^{(0)} + \sum_{n=1}^{\infty} \frac{1}{n!} \sum_{j_1} \cdots \sum_{j_n} t_\mu^{X_{j_1}\ldots X_{j_n}}(\omega_{j_1}, \ldots, \omega_{j_n})$$

$$\times \prod_{m=1}^{n} \epsilon_{j_m}(\omega_{j_m}) \exp(-i\omega_{j_m} t)$$

and an analogous ansatz for the Lagrange multipliers Eq. (12) becomes a set of linear equations for the individual expansion coefficients of t_μ and $\bar{t}_\mu$, apart from the unperturbed cluster amplitudes, $t_\mu^{(0)}$, for which the nonlinear cluster equations, Eq. (2), are recovered. The response functions, i.e. the frequency-dependent higher-order properties, are obtained as derivatives of the real part of the time-dependent quasi-energy Lagrangian [44, 65, 92, 95] with respect to the field strengths:

$$(14) \quad \langle\langle X_1; X_2, \ldots, X_n \rangle\rangle_{\omega_2,\ldots,\omega_n} = \left(\frac{d^n \{\frac{1}{2}L(t, \epsilon) + \frac{1}{2}L(t, \epsilon)^*\}_T}{d\epsilon_1(\omega_1) \ldots d\epsilon_n(\omega_n)} \right)_0$$

$$(15) \quad = \frac{1}{2} \hat{C}^{\pm} \left(\frac{d^n \{L(t, \epsilon)\}_T}{d\epsilon_1(\omega_1) \ldots d\epsilon_n(\omega_n)} \right)_0$$

where $\hat{C}^{\pm}$ symmetrizes with respect to an inversion of the signs of all frequencies and simultaneous complex conjugation.[2] Note that as a consequence of the time-averaging of the quasi-energy Lagrangian, the derivative in the last equation gives only a non-vanishing result if the frequencies of the external fields fulfill the matching condition $\sum_i \omega_i = 0$.

2.2 The Issue of Orbital Relaxation

As pointed out above, in the time-dependent case the correlation treatment cannot be based on *time-dependent* Hartree-Fock orbitals – at least not on the real frequency axis in the vicinity of poles of the response functions. Thus, the polarization of the wavefunction must be described through the variables of the correlation method, i.e. for the CC approach by means of the cluster amplitudes. This has important implications on the choice or suitability of correlation methods. As it is apparent from the sum-over-states expression for the n-th response function [96]

$$(16) \quad \langle\langle X_{n+1}; X_n, \ldots, X_1 \rangle\rangle_{\omega_n,\ldots,\omega_1} = \hat{P}^{X_{n+1}X_n\ldots X_1} \sum_{l_1\ldots l_n} \langle \Psi_0 | \hat{X}_{n+1} | \Psi_{l_n} \rangle$$

$$\times \prod_{m=1}^{n} \frac{\langle \Psi_{l_m} | \hat{X}_m | \Psi_{l_{m-1}} \rangle}{E_0 - E_{l_m} + \sum_{i=1}^{m} \omega_i}$$

---

[2] The operator $\hat{C}^{\pm}$ is defined by $\hat{C}^{\pm} f(\omega_1, \omega_2, \ldots) = f(\omega_1, \omega_2, \ldots) + f(-\omega_1, -\omega_2, \ldots)^*$.

(where $\hat{P}^{X_{n+1}X_n\cdots X_1}$ is a symmetrization operator generating all possible permutations of the operators together with the accompanied frequencies and the summations are over all states Ψ_{l_m}) a qualitatively correct description requires up to $(n+1)/2$-tuple excitations, i.e. single excitations for linear and quadratic response functions, double excitations for cubic and quartic response functions and so on. To account simultaneously for dynamic correlation also at least the next two higher excitation levels need to be included, i.e. at least triple excitations for the linear and quadruple excitations for the cubic response function. These excitations are not necessarily connected and are therefore most efficiently parameterized via the exponential coupled cluster ansatz, Eq. (1). In a configuration interaction (CI) picture, however, these high excitations need to be explicitly included. For this reason, truncated CI methods have never been applied very successfully for the calculation of frequency-dependent properties of many-electron systems, even though implementations have been reported in the literature [97–100].

Also for coupled cluster methods the use of unrelaxed orbitals has some implications. A perturbation theoretical analysis based on a Møller-Plesset like partitioning of the Hamiltonian as $\hat{H}(t, \epsilon) = \hat{F} + \hat{U} + \hat{V}(t, \epsilon)$, where $\hat{U}$ is the electron fluctuation potential and $\hat{F} + \hat{V}(t, \epsilon)$ is used as zeroth-order Hamiltonian, leads to the conclusion that in the presence of "unrelaxed" fields the results of coupled cluster methods are in general not correct through the same order in the fluctuation potential as they are in the absence of such fields [59]. Thus, CCSD, for example, gives response functions which are only correct through second order in $\hat{U}$, while CCSD ground state energies are correct through third order (compare Table 1). However, it is unclear

Table 1. Hierarchy of coupled cluster methods for response calculations. The table summarizes to which order in the electron fluctuation potential ground state and single excitation energies and response functions are obtained correctly at a given level of the correlation treatment. The analysis is based on a Møller-Plesset like partitioning of the Hamiltonian as $\hat{H}(t, \epsilon) = \hat{F} + \hat{U} + \hat{V}(t, \epsilon)$, where $\hat{U}$ is the electron fluctuation potential [58, 59]

Method	Scaling of comp. effort[a]	Ground state energy[b]	Single excitation poles	Response functions
HF-SCF	$< N^4$ (iter.)	1	1	1
CCS	$< N^4$ (iter.)	1	1	0
MP2	nN^4	2	—	—
CC2	nN^4 (iter.)	2	2	1
CCSD	n^2N^4 (iter.)	3	2	2
CCSD(T)	n^3N^4	4	—	—
CC3	n^3N^4 (iter.)	4	3	3
CCSDT-1a/b	n^3N^4 (iter.)	4	2	2
CCSDT	n^4N^4 (iter.)	4	4	4

[a] n and N are, respectively, the number of electrons and basis functions, an addition (iter.) indicates that the highest cost appear in iterative steps. Given are the formal scaling without exploiting screening; the exact costs depend heavily on the implementation.
[b] for the unperturbed system, i.e. no fields added after the HF calculation.

how the loss of one order in perturbation theory affects the accuracy of the results. It is well established for static properties like dipole moments and polarizabilities that the orbital–relaxed and orbital–unrelaxed results differ substantially at the CCSD level. But the results are sometimes on different sides of the FCI values, with deviations of similar magnitudes depending on the molecule and the CC model [34, 39, 77, 101–103]. We shall return to this issue in the numerical studies in Sections 4.2 and 5.3.

2.3 Implementation in Quantum Chemistry Programs

The approach outlined above combines the calculation of response functions (i.e. of frequency-dependent properties) with the theory of analytic derivatives developed for static higher-order properties. In the limit of a static perturbation all equations above reduce to the usual equations for (unrelaxed) coupled cluster energy derivatives. This is an invaluable advantage for the implementation of frequency-dependent properties in quantum chemistry programs.

The CC response functions are conveniently expressed in terms of partial derivatives of $L(t, \epsilon)$ with respect to cluster amplitudes, Lagrange multipliers and field strengths, which are given for the lowest orders in Table 2. Assuming that all perturbations included in $\hat{V}(t, \epsilon)$ are described by one-electron operators and that the basis set does not depend on the external fields (as it would e.g. with so-called *gauge including atomic orbitals* (GIAOs) in the case of magnetic perturbations [71, 104–108]) the quadratic response function for example is obtained as:

$$(17) \quad \langle\langle A; B, C \rangle\rangle_{\omega_B,\omega_C} = \frac{1}{2}\hat{C}^{\pm\omega}\hat{P}^{ABC}\left\{\frac{1}{6}Gt^A t^B t^C + \frac{1}{2}F^A t^B t^C + \frac{1}{2}\bar{t}^A \mathbf{B} t^B t^C + \bar{t}^A \mathbf{A}^B t^C \right\}.$$

In the above equation the operator $\hat{P}^{ABC}$ symmetrizes with respect to permutations of the perturbations A, B, and C together with the accompanied frequencies and

Table 2. Partial derivatives of the Lagrange functional for the coupled cluster quasi-energy

$$\eta_\mu^{(0)} = \left(\frac{\partial W}{\partial t_\mu}\right)_{\epsilon=0} \qquad A_{\mu\nu} = \left(\frac{\partial^2 L}{\partial \bar{t}_\mu \partial t_\nu}\right)_{\epsilon=0}$$

$$F_{\mu\nu} = \left(\frac{\partial^2 L}{\partial t_\mu \partial t_\nu}\right)_{\epsilon=0} \qquad B_{\mu\nu\gamma} = \left(\frac{\partial^3 L}{\partial \bar{t}_\mu \partial t_\nu \partial t_\gamma}\right)_{\epsilon=0}$$

$$G_{\mu\nu\gamma} = \left(\frac{\partial^3 L}{\partial t_\mu \partial t_\nu \partial t_\gamma}\right)_{\epsilon=0} \qquad \xi_\mu^X = \left(\frac{\partial^2 L}{\partial \epsilon_X \partial \bar{t}_\mu}\right)_{\epsilon=0}$$

$$\eta_\mu^X = \left(\frac{\partial^2 L}{\partial \epsilon_X \partial t_\mu}\right)_{\epsilon=0} \qquad A_{\mu\nu}^X = \left(\frac{\partial^3 L}{\partial \epsilon_X \partial \bar{t}_\mu \partial t_\nu}\right)_{\epsilon=0}$$

$\hat{C}^{\pm\omega}$ is the symmetrization defined in Section 2.1. The first-order responses of the cluster amplitudes and Lagrange multipliers are denoted as, respectively, t^X and $\bar{t}^X$. In addition to these and the unperturbed cluster amplitudes $t^{(0)}$ also the zeroth-order Lagrange multipliers $\bar{t}^{(0)}$ are needed. All these parameters are obtained from linear so-called response equations in which the Jacobian $\mathbf{A}$ as stability matrix enters as a central object:

(18) $$\bar{t}^{(0)}\mathbf{A} = -\eta^{(0)}$$

(19) $$[\mathbf{A} - \omega_X \mathbf{1}] t^X(\omega_X) = -\xi^X$$

(20) $$\bar{t}^X(\omega_X)[\mathbf{A} + \omega_X \mathbf{1}] = -\left(\eta^X + \mathbf{F}t^X(\omega_X)\right).$$

The above equations are relatively straightforward to implement in a "direct" way without explicitly setting up and storing the matrices (and higher-order tensors) $\mathbf{A}$, $\mathbf{F}$, $\mathbf{G}$, $\mathbf{A}^X$, etc. Only intermediates of at most the size of the cluster amplitudes – i.e. $(nN)^m$, where n is the number of electrons, N the number of basis functions and m the highest excitation level – need to be stored in memory or on disk. Actually, for some approximate CC models one can reduce this further, as we shall see in the following for the CC3 model.

2.4 The Approximate Triples Model CC3

For nonlinear (magneto-) optical properties, calculations of an accuracy close to that of modern gas phase experiments require – similar to what has also been found for other properties like structures [79, 109], reaction enthalpies [79, 110, 111], vibrational frequencies [112, 113], NMR chemical shifts [114], etc. – at least an approximate inclusion of connected triple excitations in the wavefunction. This has been known for years now from calculations of static hyperpolarizabilities with the CCSD(T) approximation [9–13]. CCSD(T) accounts rather efficiently for connected triples through a perturbative correction on top of CCSD. For the reasons pointed out in Section 2.1 CCSD(T) is, as a two–step approach, not suitable for the calculation of frequency-dependent properties. Therefore, the CC3 model has been proposed [56, 58] as an alternative to CCSD(T) especially designed for use in connection with response theory. CC3 is an approximation to CCSDT – alike CCSDT-1a and related methods – where the triples equations are truncated such that the scaling of the computational efforts with system size is reduced to $\mathcal{O}(n^3 N^4)$ as for CCSD(T), but the iterative character of the method is kept. The CC3 response functions are correct through third order in the fluctuation potential.

An important aspect of CC3 and other approximate triples methods is that due to the approximations made in the triples equations, it is possible to avoid the storage of the triples parts of the cluster amplitudes and Lagrange multipliers. This is essential since otherwise the storage and not the CPU requirements would limit the applicability of these methods.

To reduce the storage requirements, however, a more complicated partitioned formulation must be used which exploits the particular simple structure of the triples equations and the diagonality of the triples–triples block of the Jacobian. It allows to express the triples parts of the cluster amplitudes and Lagrange multipliers in terms of the singles and doubles parts and to reformulate the ground state cluster equations and the response equations that have to be solved to calculate properties, into effective equations in the space spanned by just the single and double excitation manifolds. Thereby the storage of full sets of triples amplitudes on disk or in core memory can be avoided. Instead the triples parts are calculated whenever needed "on-the-fly" from the singles and doubles parts and some integral intermediates. However, in the orbital-unrelaxed response approach needed for frequency-dependent properties some partial derivatives of $L(t,\epsilon)$ that appear in the right-hand-side vectors (e.g. ξ^X, η^X, and $\mathbf{A}^X$) induce an off-diagonal coupling between the triples amplitudes or multipliers of different order in response. This leads in higher order to quite involved expressions and algorithms, in particular if the optimal formal scaling of the computational costs with $\mathcal{O}(n^4 N^3)$ should be conserved, and eventually this coupling will limit the applicability of a fully partitioned formulation.

What gives CC3 special prominence in response calculations is that this method is the simplest CC wavefunction model that yields response functions correct through third order in the fluctuation potential [56, 58]. The first implementations of the linear and quadratic response functions at the CC3 level that exploit the partitioning to eliminate the storage requirements for triples amplitudes were reported by Christiansen, Gauss and Stanton [62, 66] based on the triples code of the ACES II program package [115]. While the iterative solution of the response equations with given right hand sides proceeded with $\mathcal{O}(\mathcal{N}^7)$ scaling, these authors allowed in some non-iterative terms for higher-order scalings of the computational costs to obtain a close to optimal performance for small molecules. Hald et al. [63, 69] and Pawłowski et al. [67, 73] reported an alternative implementation of the linear and quadratic response functions of CC3 and a first implementation of the cubic response function which conserves strictly the $\mathcal{O}(\mathcal{N}^7)$ scaling of the CC3 model at the expense of somewhat larger costs for small molecules. The latter implementation is based on the integral-direct coupled cluster code [116, 117] included in the Dalton program [118].

3. FREQUENCY-DEPENDENT FIRST HYPERPOLARIZABILITIES

The conceptually simplest NLO property is the electric first dipole hyperpolarizability β. Nevertheless, it is a challenging property from both the theoretical and experimental side, which is related to the fact that, as third-rank tensor, it is a purely anisotropic property. Experimentally this means that β in isotropic media (gas or liquid phase) cannot be measured directly as such, but only extracted from the temperature dependence of the third-order susceptibilities $\chi^{(3)}$. In calculations anisotropic properties are often subject to subtle cancellations between different contributions and accurate final results are only obtained with a carefully balanced treatment of all important contributions.

Table 3. Basis set convergence of first hyperpolarizabilities $\beta_\parallel$ (in a.u.) at the (unrelaxed) CCSD level. For FH, CO, and H$_2$O results are given for the static limit $\beta_\parallel(0)$, while for HCl the results are for $\beta_\parallel(-\omega,\omega,0)$ with $\omega = 0.072003$ a.u. (632.8 nm). In all calculations a frozen core approximation with the electrons in the 1s shells of non-hydrogen atoms kept inactive in the correlation and response calculations is employed

Basis	FH[a]	HCl[b]	CO[c]	H$_2$O[d]
aug-cc-pVDZ	−9.32	19.53	27.95	−16.27
aug-cc-pVTZ	−9.80	12.42	26.34	−18.40
aug-cc-pVQZ	−8.63	8.77	25.37	−18.02
aug-cc-pV5Z	−8.29	9.49	25.12	−18.24
d-aug-cc-pVDZ	−6.01	10.96	26.78	−14.09
d-aug-cc-pVTZ	−7.68	9.71	25.43	−17.73
d-aug-cc-pVQZ	−7.73	8.95	25.06	−18.07
d-aug-cc-pV5Z	−7.77	8.94	24.99	−18.11
t-aug-cc-pVDZ	−6.29		26.77	−16.02
t-aug-cc-pVTZ	−7.93	9.48	25.26	−18.48
t-aug-cc-pVQZ	−7.87		25.04	−18.28
q-aug-cc-pVDZ	−6.40		27.00	−15.94
q-aug-cc-pVTZ	−7.93	9.34	25.34	−18.50
q-aug-cc-pVQZ	−7.84			

[a] $R_{FH} = 1.7328$ bohr; Z axis points from F to H; [65, 119].
[b] $R_{HCl} = 1.27455$ Å; Z axis points is from H to Cl; [37] (including some previously unpublished numbers).
[c] $R_{CO} = 2.132$ bohr; Z axis points from O to C; present work and [36].
[d] $R_{OH} = 0.957$ Å, $\angle(HOH) = 104.5°$; Z axis points from O to the center of mass; present work and [36].

3.1 Basis Set Convergence

According to the $2n + 1$-rule first hyperpolarizabilities require similar to the linear polarizabilities only the calculation of the first-order response of the wavefunction. Thus, the minimal basis set requirements for a qualitative correct description of β_{ijk} are similar to those for the components of the linear polarizability α_{ij}, i.e. for calculations at a correlated level the basis set should be at least of triple-ζ quality augmented with diffuse functions for a better description of the long range part of the electron density and its response. However, for the first hyperpolarizability the experimentally observed properties (in most cases $\beta_\parallel$ or $\beta_\perp$) are anisotropic tensors and therefore usually more sensitive to the quality of the wavefunction than the isotropic polarizability $\bar{\alpha}$. Within the hierarchy of correlation-consistent basis sets [120–123] augmentation with a single set of diffuse functions (aug-cc-pVXZ) is often sufficient for the calculation of polarizabilities of polyatomic molecules [101, 123]. For hyperpolarizabilities, in particular for small molecules, the doubly-augmented basis sets (d-aug-cc-pVXZ) give much better convergence with the cardinal number X, while the addition of even more diffuse functions as in the t-aug-cc-pVXZ sets does not give any further systematic improvements (see Table 3).

3.2 Convergence with Correlation Treatment

The situation is somewhat different for the convergence with the wavefunction model, i.e. the treatment of electron correlation. As an anisotropic and nonlinear property the first dipole hyperpolarizability is considerably more sensitive to the correlation treatment than linear dipole polarizabilities. Uncorrelated methods like HF-SCF or CCS yield for $\beta_\parallel$ results which are for small molecules at most qualitatively correct. Also CC2 is for higher-order properties not accurate enough to allow for detailed quantitative studies. Thus the CCSD model is the lowest level which provides a consistent and accurate treatment of dynamic electron correlation effects for frequency-dependent properties. With the CC3 model which also includes the effects of connected triples the electronic structure problem for $\beta_\parallel$ seems to be solved with an accuracy that surpasses that of the latest experiments (vide infra).

3.3 Comparison with Experiment: CO, H_2O, FH, and HCl

3.3.1 Carbon monoxide

Table 4 shows results from [36] for the static and the second harmonic generation hyperpolarizabilities $\beta_\parallel^{SHG}$ of CO at 694.3 nm. The electronic contributions were obtained from CC3/d-aug-cc-pVTZ calculations carried out at $R_{CO} = 2.132$ bohr. These were approximately corrected for remaining basis set errors by adding the difference between CCSD/d-aug-cc-pVQZ and CCSD/d-aug-cc-pVTZ results for the same frequency and internuclear distance. For CO the triples correction for $\beta_\parallel(0)$ is 1.72 a.u. or $\approx 6\%$. At a wavelength of 694.3 nm the triples correction is already 2.35 a.u. or $\approx 7\%$. Thus, there is in this case a notable triples effect on the frequency dispersion. Since there is no information available about correlation contributions beyond CC3, it is difficult to assess the accuracy of these results.

The experimental result for $\beta_\parallel^{SHG}$ at 694.3 nm is 30.2 ± 3.2 a.u. [1, 16]. This is in good agreement with the best theoretical estimate for the electronic contribution

Table 4. The first hyperpolarizability of CO (from [36], results in a.u.)

Method	$\beta_\parallel^{static}$	$\beta_\parallel^{SHG}$ (694.3 nm)
CCSD/d-aug-cc-pVQZ[a]	25.06	29.09
CC3/d-aug-cc-pVTZ[a]	27.15	31.91
CC3/d-aug-cc-pVQZ[a,b]	26.78	31.44
CCSD(T)[c]	23.5	27.00
experimental value[d]		30.2 ± 3.2

[a] [36]; $R_{CO} = 2.132$ bohr.
[b] estimated from CC3/d-aug-cc-pVTZ and CCSD results.
[c] [8] static CCSD(T) result with Sadlej POL basis; for $\beta_\parallel^{SHG}$ scaled with HF-SCF dispersion.
[d] [1] rescaled in [16].

(31.44 a.u.). However, a direct comparison of the two values is hampered by the fact that presently no results are available for the zero-point vibrational correction as well as for the pure vibrational contribution.

3.3.2 Water

The same approach leads for water, H_2O, at the equilibrium geometry of $R_{OH} = 0.957$ Å and $\angle(HOH) = 104.5°$ to a best estimate of -21.28 a.u. for the electronic second harmonic generation hyperpolarizability $\beta_\parallel^{SHG}$ at 694.3 nm [36]. Also in this case connected triples are essential to obtain quantitative accuracy (they amount to ≈4%), but they have only a minor effect on the dispersion (0.01 a.u. at 694.3 nm). For both the zero-point vibrational (ZPV) correction and the pure vibrational contribution, there are results available from previous theoretical investigations at the HF-SCF [15], MCSCF [14] and MP2 [15] levels. It is difficult to judge which of these results, which differ considerably, are most accurate. Probably the MP2 results of [15] are most consistent with the coupled cluster calculations. At this level the ZPV correction has been estimated to be of the order of -0.95 a.u. The pure vibrational contribution has been calculated to about -0.21 a.u. at a frequency of 0.07 a.u. This frequency corresponds roughly to the experimental wavelength of 694.3 nm. Adding the MP2 vibrational corrections to the coupled cluster result for the electronic hyperpolarizability one obtains as a best estimate for the final result at 694.3 nm a value of -22.4 a.u. as summarized in Table 5. This is in excellent agreement with the experimental result for $\beta_\parallel^{SHG}$ of -22.0 ± 0.9 a.u. reported in [1, 16] for this wavelength.

A comparison with an experimental result for a wavelength of 1064 nm of -19.2 ± 0.9 a.u. [17] is limited by missing data for the pure vibrational contribution at this wavelength. However, the best estimate for the electronic contribution including the aforementioned MP2 value for the ZPV correction obtained in [36] is with -19.81 a.u. in agreement with the experimental value.

Table 5. The first hyperpolarizability of H_2O (from [36], results in a.u.)

Method	$\beta_\parallel^{static}$	$\beta_\parallel^{SHG}$ (694.3 nm)
CCSD/d-aug-cc-pVQZ[a]	-18.07	-21.98
CC3/d-aug-cc-pVTZ[a]	-17.04	-21.02
CC3/d-aug-cc-pVQZ[a,b]	-17.38	-21.28
zero point vibr. correct.[c]		-0.95
pure vibr. contrib.[c]		-0.21
best theoretical estimate		-22.4
experimental value[d]		-22.0 ± 0.9

[a] [36]; $R_{OH} = 0.957$ Å, $\angle(HOH) = 104.5°$.
[b] Estimated from CC3/d-aug-cc-pVTZ and CCSD results.
[c] [15] MP2 results. ZPV correction was calculated for static limit.
[d] [1] rescaled in [16].

3.3.3 Hydrogen fluoride

A somewhat special case is $\beta_\parallel^{SHG}$ of hydrogen fluoride (FH). Since its experimental determination by Dudley and Ward [7] about twenty years ago and its first *ab initio* investigation by Sekino and Bartlett [13], there has been a discussion in the literature whether the experimental result is in agreement with theory or not [8, 13, 16, 18, 26–28, 31, 33, 65, 66, 124]. However, both the experimental and the calculated results have relatively large uncertainties of 5–10%. For the theoretical value the main source of uncertainty are the vibrational contributions. The electronic contribution is today known with good accuracy from CC3 calculations [66] in a t-aug-cc-pVTZ basis and estimates for the remaining basis set error from CCSD calculations in large basis sets [65, 119] and for higher-order correlation effects from FCI calculations [34] in the aug-cc-pVDZ basis. The zero-point vibrational correction and the pure vibrational contributions have been calculated from (unrelaxed) CCSD/t-aug-cc-pVTZ results for the property curves and a potential energy curve evaluated at the CCSD(T)/t-aug-cc-pVTZ level [119]. Combining these results, one arrives for $\beta_\parallel^{SHG}$(694.3 nm) at a best estimate of -9.45 ± 0.5 a.u., as shown in Table 6. This estimate includes for the remaining correlation and basis set errors an uncertainty of 0.2 a.u. and for the ZPV correction and the pure vibrational contributions uncertainties of, respectively, 0.1 and 0.2 a.u.

The experimental value is with -10.9 ± 1.0 a.u. somewhat larger, but has also a relatively large uncertainty of $\approx 10\%$. It is thus difficult to conclude from these numbers that there exists a significant disagreement between the experimental and theoretical results. However, as suggested previously, an experimental

Table 6. Second harmonic generation hyperpolarizability $\beta_\parallel^{SHG}$ of FH at 694.3 nm (in a.u.)

	$\beta_\parallel^{SHG}$ (694.3 nm)
CC3/t-aug-cc-pVTZ[a] at R_e	−8.32
est. remaining basis effects[b]	+0.16 ± 0.1
est. remaining correl. effects[c]	+0.07 ± 0.1
zero point vibr. correct.[d,e]	−0.72 ± 0.1
pure vibr. contrib.[e]	−0.64 ± 0.2
best theoretical estimate	−9.45 ± 0.5
experimental value[f]	−10.9 ± 1.0

[a] [66], $R_e = 1.7328$ bohr.
[b] diff. between CCSD/d-aug-cc-pV5Z and CCSD/d-aug-cc-pVTZ, [65, 119].
[c] diff. between CC3/aug-cc-pVDZ and FCI/aug-cc-pVDZ, [34].
[d] diff. between vibrational averaged and single point (at $R_{FH} = 1.7328$ bohr) results, [119].
[e] calculated with CCSD/t-aug-cc-pVTZ frequency-dependent properties and a CCSD(T)/t-aug-cc-pVTZ potential, [119].
[f] [7] based on CF_4 as reference.

Table 7. Second harmonic generation hyperpolarizability $\beta_\parallel^{SHG}$ of HCl at 694.3 nm (in a.u.)

	$\beta_\parallel^{SHG}$ (694.3 nm)
CC3/d-aug-cc-pVTZ[a] at R_e	9.84
est. remaining basis effects[b]	−0.94
zero point vibr. correct.[c]	−0.16
pure vibr. contrib.[d]	+0.4
best theoretical estimate	9.14
experimental value[e]	9.76 ± 0.96

[a] [37], $R_{HCl} = 1.27455$ Å.
[b] diff. between CCSD/d-aug-cc-pV5Z and CCSD/d-aug-cc-pVTZ, [37].
[c] static result from CC3/d-aug-cc-pVTZ property and CCSD(T)/aug-cc-pVQZ potential, [37].
[d] from CC3/d-aug-cc-pVTZ frequency-dependent properties and CCSD(T)/aug-cc-pVQZ potential, [37].
[e] [7] based on CF_4 as reference gas.

reinvestigation of $\beta_\parallel^{SHG}$ for FH, would be needed to resolve the remaining discrepancy, in particular since the measurement of [7] was carried out with CF_4 as reference gas, for which the second hyperpolarizability is not known very accurately.

3.3.4 Hydrogen chloride

For hydrogen chloride (see Table 7.) a very similar approach results in considerably better agreement between the experimental and the *ab initio* calculated result. The electronic contribution has been obtained from calculations at the CC3/d-aug-cc-pVTZ level supplemented by CCSD/d-aug-cc-pV5Z for remaining basis set effects. The ZPV correction and the pure vibrational contributions were calculated from CC3/d-aug-cc-pVTZ results for the properties and a CCSD(T)/aug-cc-pVQZ potential energy curve. The final best estimate obtained in [37] for $\beta_\parallel^{SHG}$(694.3 nm) by combining these results is with 9.14 a.u. in very good agreement with the experimental value of 9.76 ± 0.96 a.u. from [7].

4. FREQUENCY-DEPENDENT SECOND HYPERPOLARIZABILITIES

Naively, one would expect that second hyperpolarizabilities γ are theoretically and experimentally more difficult to obtain than first hyperpolarizabilities β. From a computational point of view the calculation of fourth-order properties requires, according to the $2n+1$-rule, second-order responses of the wavefunction and thus the solution of considerably more equations than needed for β (cf. Section 2.3). However, unlike β the second dipole hyperpolarizability γ has two isotropic tensor components ($\gamma_{\parallel}$ and $\gamma_{\perp}$) which are amenable to very accurate gas phase measurements [2–6, 16, 29] on atoms and small and highly symmetric molecules for which also very accurate calculations can be carried out.

4.1 Basis Set Convergence

Since second hyperpolarizabilities depend in addition to the first-order also on the second-order response of the wavefunction, the minimal requirements with respect to the choice of basis sets are for γ somewhat higher than for the linear polarizabilities α and the first hyperpolarizabilities β, in particular for atoms and small molecules. For the latter at least doubly-polarized basis sets augmented with a sufficient number of diffuse functions (e.g. d-aug-cc-pVTZ or t-aug-cc-pVTZ) are needed to obtain qualitatively correct results. Highly accurate results at a correlated level will in general only be obtained in quadruple-ζ or better basis sets.

Table 8 shows the results of basis set studies for the static isotropic second hyperpolarizabilities of Ne, N_2, and CH_4. These numbers illustrate the typical basis set

Table 8. Basis set convergence of $\gamma_{\parallel}(0)$ (in a.u.) in HF-SCF and (unrelaxed) CCSD

Basis set	HF-SCF			CCSD		
	Ne[a]	N_2[b]	CH_4[c]	Ne[a]	N_2[b]	CH_4[c]
aug-cc-pVDZ	20.3	456	1497	29.6	684	1931
aug-cc-pVTZ	28.3	560	1745	42.2	745	2199
aug-cc-pVQZ	38.6	654	1830	58.5	822	2285
aug-cc-pV5Z	50.1	685		76.8	844	
d-aug-cc-pVDZ	49.3	672	1774	83.9	938	2328
d-aug-cc-pVTZ	57.3	697	1868	94.2	895	2364
d-aug-cc-pVQZ	62.0	710	1864	100.2	881	2330
d-aug-cc-pV5Z	67.5	714		108.0	873	
d-aug-cc-pV6Z	68.6	714			869	
t-aug-cc-pVDZ	53.2	669	1815	94.4	935	2378
t-aug-cc-pVTZ	66.5	710	1869	110.3	903	2365
t-aug-cc-pVQZ	67.8	713	1868	110.0	883	
t-aug-cc-pV5Z	68.6	714		110.2	873	
t-aug-cc-pV6Z	68.9	715		110.0		
q-aug-cc-pVDZ	53.3	680	1816	94.5	943	2380
q-aug-cc-pVTZ	68.9	717	1871	114.5	909	
q-aug-cc-pVQZ	68.5	715		111.2	884	
q-aug-cc-pV5Z	68.7			110.2		
q-aug-cc-pV6Z	68.9			110.0		
limit	68.8	715				

[a] [32, 119], CCSD calculations with frozen core approximation for 1s shell.
[b] $R_{NN} = 2.074$ bohr; [32, 35].
[c] $R_{CH} = 2.052$ bohr; [32], CCSD calculations with frozen core approximation for 1s shell at C.

convergence of this property, which apart from the higher minimal needs, resembles that of the isotropic polarizability $\bar{\alpha}$. As observed before by other authors [123], the results converge slowly with the cardinal number (or ζ-level) if the augmentation level is low. This reflects the important role of diffuse functions for these properties. For atoms (as Ne) one obtains both at the HF-SCF and at the correlated level an acceptable convergence first in the triply-augmented series t-aug-cc-pVXZ. The fourth set of diffuse functions improves slightly the convergence, but for correlated calculations an increase in the cardinal number X is more important. For molecules the basis set convergence is somewhat faster, in particular with respect to diffuse functions. For N_2 and CH_4 already in the doubly-augmented series an increase of the cardinal number X is of the same importance as triple augmentation.

At the Hartree-Fock level the hyperpolarizabilities usually increase if the augmentation level and also if the cardinal number X are increased. For the correlated contribution to $\gamma_{||}(0)$ the convergence pattern is dominated by different effects: At the CCSD level an increase of $\gamma_{||}(0)$ with the cardinal number beyond T is only found for the lower augmentation levels. In particular for molecules we observe, as illustrated in Table 8 for N_2 and CH_4, a monotonic decrease of the second hyperpolarizability when the correlation treatment is improved in the series X = T, Q, 5, etc. The results for X = T typically overestimate the correlation contribution to $\gamma_{||}(0)$ by a few percent. Many correlated hyperpolarizability calculations in the literature were performed with basis sets of triple-ζ or similar quality and basis set convergence was often only explored with respect to augmentation with diffuse functions. From the above observations one may conclude that many of these studies obtained too large results for $\gamma_{||}(0)$.

4.2 Convergence with Correlation Treatment

Coupled cluster response calculations are usually based on the HF-SCF wavefunction of the unperturbed system as reference state, i.e. they correspond to so-called orbital-unrelaxed derivatives. In the static limit this becomes equivalent to finite field calculations where the perturbation is added to the Hamiltonian after the HF-SCF step, while in the orbital-relaxed approach the perturbation is included already in the HF-SCF calculation. For frequency-dependent properties the orbital-relaxed approach leads to artificial poles in the correlated results whenever one of the involved frequencies becomes equal to an HF-SCF excitation energy. However, in the static limit both unrelaxed and relaxed coupled cluster calculations can be used and for both approaches the hierarchy CCS (HF-SCF), CC2, CCSD, CC3, ... converges in the limit of a complete cluster expansion to the Full CI result.[3] Thus, the question arises, whether for second hyperpolarizabilities one

---

[3] Note that if a frozen core approximation is used a small difference will persist in the Full CI limit since the unrelaxed series converges to a limit with the core orbitals constrained to be those of the unperturbed system whereas in the relaxed series the relaxation contribution of the core orbitals is included. For electric (hyper)polarizabilities this effect is usually negligible.

Table 9. Convergence of $\gamma_{\parallel}(0)$ with the wavefunction model. The basis sets are for Ne and Ar t-aug-cc-pV5Z and for N_2 and CH_4 t-aug-cc-pVTZ for HF-SCF, CCS and CC2 and d-aug-cc-pVTZ for CCSD and CCSD(T). Coupled cluster results with frozen–core approximation[a] with the exception of Ne and Ar, where all–electron CCSD(T) and CC3 results are given. Where not stated otherwise, the results are taken from [32]

	Orbital–relaxed			Orbital–unrelaxed			
	HF-SCF	CCSD	CCSD(T)	CCS	CC2	CCSD	CC3
Ne	68.6	99.1	107.1	78.2	135.8	110.2	108.0[b]
Ar	961	1106	1159	1154	1247	1178	1172[c]
N_2	710	854	911	825	992	895	
CH_4	1869	2176	2294	2228	2557	2364	

[a] frozen core corresponding to 1s on C, N, Ne and $1s2s2p_1$ on Ar.
[b] [73].
[c] [125].

of the two approaches is superior, i.e., converges faster to the FCI limit. A theoretical analysis of the response functions [59] based on a Møller-Plesset like partitioning of the Hamiltonian indicates that the unrelaxed hierarchy converges at least initially slower. It is therefore sometimes assumed in the literature that the orbital-relaxed hierarchy is more reliable [31, 126, 127], although numerical investigations [34, 77, 102, 103] on small molecules do not support this supposition.

In order to compare the performance of the relaxed and unrelaxed hierarchies for second hyperpolarizabilities, the static limit $\gamma_{\parallel}(0)$ has been calculated for Ne, Ar, N_2, and CH_4 using relaxed HF-SCF, CCSD and CCSD(T) and unrelaxed CCS, CC2, CCSD and CC3 (see Table 9). The lowest levels, HF-SCF and CCS, give results of similar accuracy, although the CCS values are somewhat higher and for all these cases between the HF-SCF and the correlated results. This indicates that the singles excitation manifold in the cluster expansion accounts implicitly for most of the orbital relaxation effects. CC2 overestimates the correlation contributions to response properties and does not improve systematically upon HF-SCF and CCS – most likely a consequence of the fact that CC2 response functions are, as the HF-SCF ones, only correct through first order in the electron-electron interaction. This effect is less severe for (orbital-relaxed) MP2, which usually gives quite good results for second hyperpolarizabilities [8, 11, 16, 123], but this method cannot be used in the unrelaxed response approach needed for frequency-dependent properties. At the CCSD level the unrelaxed results have errors which are by a factor 2 to 4 smaller than those obtained with orbital-relaxed CCSD, which, similar as HF-SCF, systematically underestimates isotropic hyperpolarizabilities. This has also been observed for other molecules and (isotropic) properties [34, 55, 59, 128–130] and it seems to be a common trend. After including the effects of connected triple excitations at the CCSD(T) or CC3 level, the differences between the relaxed and unrelaxed approaches are small and, in the basis set limit, both provide results close to experimental accuracies. Thus, there is no indication that the implicit treatment of orbital relaxation via the singles and higher cluster excitations impairs the accuracy of coupled cluster response methods compared to the calculation of orbital-relaxed derivatives.

4.3 Comparison with Experiment: Ne and N_2

4.3.1 Neon

The ESHG hyperpolarizability of Ne has been the subject of several experimental investigations and numerous *ab initio* calculations. The 'ups' and 'downs' of the theoretical and experimental estimates for γ_0^{Ne} have been described in detail in reviews by Bishop [24], Shelton and Rice [16] as well as in recent work by Hättig and Jørgensen [32] and Pawłowski, Jørgensen and Hättig [73]. After the latest ESHG experiments by Shelton and Donley [23], measurements at four frequencies were presumed to be accurate within $\pm 0.4 - 0.8$ a.u. A fit of the expression $\gamma_\parallel^{ESHG}(\omega) = \gamma_0(1 + 6A\omega^2 + 36B\omega^4)$ to these four points with B constrained to 5.50 a.u. (a value arbitrarily chosen between the results of a MP2 [12, 25] and a CASSCF [19] calculation) led for γ_0^{Ne} to an estimate of 108 ± 2.2 a.u. However, the dispersion curve obtained by this fit does not overlap with the error bars of the two ESHG measurements at 1319 and 1064 nm (see Fig. 1). As already pointed out by Shelton and Donley [23], this indicated that one of these two points has a larger error than conveyed by its statistical uncertainty. From the available data it was not possible to determine which of the two points is inaccurate.

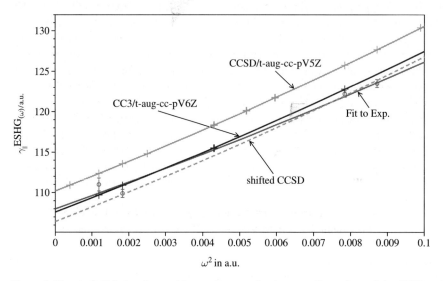

Figure 1. Electric field induced second harmonic generation in neon. Comparison of the CCSD and CC3 results from [73] with the experimental data from [6, 23]. For "shifted CCSD" the dispersion curve obtained with CCSD has been shifted to match best the experimental points for the two highest frequencies

A CCSD response calculation in a large basis set for the ESHG dispersion curve [30] showed that the inaccurate point is the one at 1319 nm. A revised extrapolation based on the CCSD dispersion curve led to a new static limit of $\gamma_0^{Ne} = 106.4 \pm 2.2$ a.u.

In [73] the non-relativistic CC3 basis set limit of γ_0^{Ne} was determined to 107.0 ± 0.4 a.u. The corresponding CCSD(T) value is about 0.8 a.u. smaller. For the first-order relativistic correction to γ_0^{Ne} Klopper et al. [131] obtained at the CCSD(T)/q-aug-cc-pCV6Z level a value of 0.59 a.u. The largest uncertainty in the calculations is presently due to correlation effects beyond CC3, since FCI calculations are only possible in small basis sets. The best available FCI result [125] for γ_0^{Ne}, calculated in a d-aug-cc-pVDZ basis, is -0.17 a.u. below the corresponding CC3 value. It is not known how well this result reproduces the basis set limit of the higher-order correlation contributions and we include therefore in our best estimate an uncertainty of 0.6 a.u. Adding the relativistic and the FCI–CC3 correction to the CC3 results leads to a best estimate for γ_0^{Ne} of 107.4 ± 1.0 a.u. in good agreement with the value extrapolated from the experimental results (106.4 ± 2.2 a.u.). For a comparison with the frequency-dependent experimental results see Fig. 1.

4.3.2 Nitrogen

For this molecule, ESHG measurements are available for a much larger number of frequencies. Thus, there is little doubt about the accuracy of the experimental values and a dispersion curve could be fitted to the measured data (corrected for rovibrational contributions) without referring to calculated dispersion coefficients [2, 3, 6].

For N_2, there is a sizeable triples contribution to the lowest dipole allowed excitation energy of about 0.07 eV or 0.7%. As a consequence of this (unrelaxed) CCSD underestimates the absolute value of the isotropic hyperpolarizability and its dispersion. The electronic contribution to the static limit $\gamma_0^{N_2}$ has been calculated at the CCSD/t-aug-cc-pVTZ level [32] to be 903.0 a.u. However, as indicated above, the triple-ζ level is often too low for the calculation of second hyperpolarizabilities and the $\gamma_0^{N_2}$ obtained at the CCSD/t-aug-cc-pVTZ level turned out to be about 40 a.u. above the CCSD basis-set limit result. The latter has been calculated to be 863.3 ± 3.3 a.u. [35]. Before comparing this result to the value extrapolated from experimental results, it has to be corrected for the ZPV contribution, which has been obtained in [32] to 12.0 a.u., thereby yielding 875.3 ± 3.3 a.u. as the best estimate for the electronic contribution to $\gamma_0^{N_2}$ at the CCSD level. Shelton [6] obtained an experimental value of $\gamma_0^{N_2}$, 917 ± 9 a.u., from the extrapolation of the results in [2] corrected for the pure rotational and vibrational contributions. The discrepancy between this experimental value and the CCSD best estimate is as large as 42 a.u. and makes very clear the importance of the triples contribution.

After inclusion of connected triples at the CC3 level and accounting for the zero-point vibrational contribution at this level and for remaining basis set effects, a theoretical value of 912.9 ± 4.3 a.u. is obtained as best estimate for the electronic

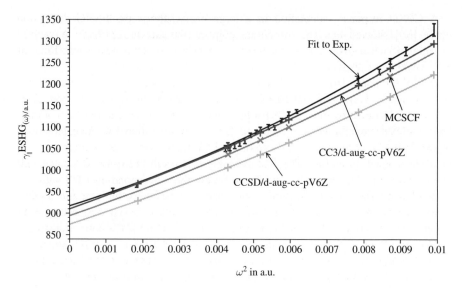

Figure 2. Electric field induced second harmonic generation in nitrogen N_2 (in a.u.). Comparison of the CCSD [32] and CC3 [35] results with a previous MCSCF calculation [20] and the experimental data from [2, 3, 6]

$\gamma_0^{N_2}$ [35]. The latter value is in excellent agreement with the value of 917 ± 9 a.u. extrapolated from the ESHG experiments [6].

A comparison of the CCSD and CC3 results obtained in the d-aug-cc-pV6Z basis with the experimental results is shown in Fig. 2. While the CCSD results for $\gamma_\parallel^{ESHG}(\omega)$ differ from the experimental results typically by 6–8%, the CC3 results are generally in perfect agreement with experiment. Only for the highest frequencies the deviations become larger than the error bars of the measurements. These deviations are attributed mainly to remaining correlation effects, which are enhanced for the highest frequencies. The relativistic contribution to $\gamma_\parallel^{ESHG}(\omega)$ of N_2 is small (about 1.0 a.u.) [35].

5. MIXED ELECTRIC AND MAGNETIC PROPERTIES: BIREFRINGENCES AND DICHROISMS

In the previous section we discussed pure electric–dipole hyperpolarizabilities, in particular second harmonic generation. Another important class of NLO processes includes birefringences and dichroisms which can be rationalized (at least to lowest orders in perturbation theory) in terms of response functions involving, besides the electric–dipole, also magnetic–dipole and electric–quadrupole operators. Prominent examples related to quadratic response functions are:
- magneto-optical activity (MOA). It comprises two effects, both originating from a differential interaction of the right and left circularly polarized components of linearly polarized light with matter due to the presence of a magnetic field

parallel to the direction of propagation. The first is the magnetic optical rotation (MOR, also known as Faraday effect) [21], i.e. an induced rotation of the plane of polarization (or circular birefringence). The other is an induced circular dichroism referred to as magnetic-circular dichroism (MCD) [21, 22, 132, 133].
- electric-field gradient (EFG) induced birefringence, also known as Buckingham effect (BE). This birefringence is observed when linearly polarized light interacts with matter in the presence of an electric-field gradient [134, 135]. The relationship with MOA and in particular MOR will be discussed later.

Examples of NLO processes that involve also cubic response functions are:
- Cotton-Mouton effect (CME) [136, 137], the linear birefringence which arises in a sample with a magnetic induction field perpendicular to the linearly polarized light beam. The CME is the magnetic analogue of the Kerr effect (KE) [138, 139] which describes the birefringence induced by an electric field. The optical axes in CME are the same as in the magneto-electric birefringence (MEB), the latter being observed in the presence of both an electric and a magnetic field perpendicular to each other and to the direction of propagation. In experimental investigations, the three birefringences usually superimpose, but their effects can be separated via their different dependences on the field strengths, i.e., KE $\propto E^2$, CME $\propto B^2$, and MEB $\propto EB$.
- Jones birefringence (JB) [140–143]: this birefringence is observed when linearly polarized light traverses a sample in the presence of both an electric and a magnetic field perpendicular to the direction of propagation but (unlike in the case of MEB) parallel to each other. The theoretical expression for JB is identical to that for MEB, but it should be noted that the two effects correspond to different experiments with the optical axes in the JB directed at $\pm 45°$ with respect to those of the other three birefringences (KE, CME and MEB).

In the following, we will discuss how these effects can be investigated using quantum chemical methods and present results from high-level *ab initio* calculations that have been reported in the last few years.

An important issue for optical processes that involve a magnetic field as perturbation is the possible gauge-dependence of the results when obtained by means of approximate quantum chemical calculations. The dependence on the chosen origin for the vector potential (usually termed in the quantum chemical community the gauge-origin dependence) is unphysical and entirely due to the approximations imposed on the calculations (in particular, but not only, the use of finite one-electron basis sets). In many cases, the gauge-origin dependence can successfully be dealt with by using explicitly magnetic-field dependent basis functions such as the gauge-including atomic orbitals (GIAOs, also known as London atomic orbitals (LAOs)) [104, 105, 144]. However, some complications arise in the coupled-cluster treatment for frequency-dependent properties. Straightforward solutions to the gauge-origin problem are here only possible if the corresponding response functions can be written as derivatives of lower-order response functions with respect to the applied magnetic field components (for a detailed discussion, see [71]).

5.1 Magneto-optical Activity

The key quantity for the Faraday effect [21, 145, 146] is the Verdet constant [147] $V(\omega)$ which for molecules in the gas phase is given by

$$(21) \quad V(\omega) \propto i\varepsilon_{\alpha\beta\gamma} \langle\langle \hat{\mu}_\alpha; \hat{m}_\beta, \hat{\mu}_\gamma \rangle\rangle_{0,\omega}$$

i.e., as the isotropic average over mixed electric- and magnetic-dipole quadratic response function components. In Eq. (21), $\varepsilon_{\alpha\beta\gamma}$ denotes the Levi-Civita tensor and $\hat{\mu}$ and $\hat{m}$ indicate the electric dipole and magnetic dipole operators, respectively.

MCD is quantitatively analyzed in terms of three magnetic rotatory strengths, the so-called Faraday A, B, and C terms [21]. The B terms are particularly important for detecting hidden transitions, even though they are in general small and difficult to measure [148]. Therefore, theoretical determinations of MCD may play an important role for a correct interpretation of experimental data. For a transition between the states a and j, the B term is given as the residue of the quadratic response function entering $V(\omega)$ in Eq. (21), i.e.

$$(22) \quad B(a \to j) \propto i\varepsilon_{\alpha\beta\gamma} \left[\lim_{\omega \to \omega_{aj}} (\omega - \omega_{aj}) \langle\langle \hat{\mu}_\gamma; \hat{m}_\beta, \hat{\mu}_\alpha \rangle\rangle_{0,\omega} \right]$$

with $\hbar\omega_{aj}$ as the corresponding transition energy.

5.1.1 Basis set convergence: MOA

Verdet constants have been investigated in [74] and [38] for a few centro-symmetric molecules using coupled-cluster response-theory techniques. For atoms, corresponding results can be found in [149].

The basis set convergence in the calculations of $V(\omega)$ is illustrated by discussing H_2 for which the CCSD calculations yield FCI results [74]. As seen from Table 10 augmentation with diffuse functions is essential to obtain reliable results. With the first augmentation level (aug-cc-pVXZ), convergence within 3 decimals is obtained at the quadruple- and quintuple-ζ level. Double and triple augmentation on the other hand barely affect the results with remaining changes in the order of less than 0.1%. Comparison with the explicitly-correlated results from Bishop and Cybulski [150] proves the 'benchmark' quality of the corresponding CCSD results. It is also worth noting that the basis set convergence is more or less independent of the frequency.

For other molecules (nitrogen (N_2), acetylene (C_2H_2), and methane (CH_4)), the basis set studies from [74] indicate again the importance of diffuse functions. However, compared to H_2, it is found that the second augmentation is slightly more important, while for a sufficiently large valence basis set the effects due to triple augmentation are almost negligible.

5.1.2 Convergence with electron correlation treatment: MOA

The convergence with respect to the electron correlation treatment has been analyzed in [74] for N_2, C_2H_2, and CH_4 using the CCS, CC2, CCSD hierarchy. Additional

CC3 results have been reported in [38]; those calculations were motivated by the rather large discrepancies to experiment seen at the CCSD level.

The most significant observation with respect to electron correlation is the surprisingly large triples contribution for N_2 and C_2H_2. While the CCS, CC2, and CCSD calculations indicate smooth convergence with the usual reduction in the correlation contributions at each consecutive level and, thus, predict that triple excitation contributions should be rather small, the opposite is seen in the actual calculations [38]. The effect due to triple excitations as obtained at the CC3 level was found to increase the CCSD results by 4% and 16%. Apparently there seems to be no guarantee that triple excitation effects are small when the CC2 model is a good approximation to CCSD. On the other hand, triple excitation effects are found to be rather small for CH_4 (only 1.5% of the CCSD results) [38] consistent with the prediction based on the corresponding CCSD-CC2 difference [74].

5.1.3 Comparison with experiment: MOA

Verdet constants for H_2, N_2, C_2H_2, and CH_4 have been measured at various frequencies (at 1 atm and 298 K) by Ingersoll and Liebenberg [151] in 1956 and are considered to be accurate within 1%. The claimed accuracy has been confirmed in the case of H_2 – by the explicitly correlated results of Bishop and Cybulski [150] and again by the FCI/d-aug-cc-pV5Z results of [74] (see Table 10).

For N_2, C_2H_2, and CH_4, Fig. 3 shows the deviation (in percent) from experiment for the calculated Verdet constants. As already discussed, for both N_2 and C_2H_2 triples corrections are substantial, as their inclusion brings the computational results within 1% of the experimental values. However, in the case of N_2 consideration of vibrational effects (about 1.3%, obtained at the CCSD/t-aug-cc-pVTZ level using

Table 10. FCI Verdet constants (a.u. $\times 10^7$) for H_2 at four different frequencies[a]

Basis Set	ω (au)			
	0.11391	0.08284	0.06509	0.05360
cc-pVQZ	0.22640	0.11607	0.07070	0.04761
aug-cc-pVQZ	0.45643	0.22998	0.13912	0.09335
d-aug-cc-pVQZ	0.45679	0.23002	0.13911	0.09334
t-aug-cc-pVQZ	0.45665	0.22995	0.13907	0.09331
cc-pV5Z	0.28236	0.14421	0.08773	0.05903
aug-cc-pV5Z	0.45597	0.22965	0.13887	0.09319
d-aug-cc-pV5Z	0.45562	0.22944	0.13876	0.09310
[150][b]	0.45556	0.22946	0.13877	0.09312
[150][c]	0.50527	0.25361	0.15316	0.10270
Exp. [151]	0.501	0.251	0.150	0.103

[a] From [74].
[b] Explicitly correlated results at equilibrium geometry.
[c] Rovibrationally and thermally averaged values.

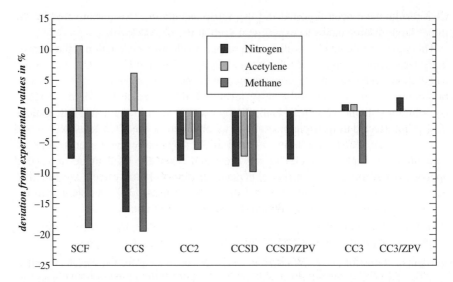

Figure 3. Deviation (in percent) from experiment [151] for the calculated Verdet constant $V(\omega)$ of N_2, C_2H_2, and CH_4. The calculations have been carried out in the aug-cc-pVQZ (N_2) and d-aug-cc-pVTZ (C_2H_2 and CH_4) basis sets and at frequency $\omega = 0.11391$ a.u. ZPV corrections for N_2 amount to about 1.3% and have been determined at the CCSD level (for details see text and [38])

a CCSD(T)/aug-cc-pVQZ potential curve [38]) slightly worsens the agreement with experiment, although the result remains within three times the experimental uncertainty (i.e., the 99.9% confidence interval of the experiment). Whereas the agreement between experimental and calculated Verdet constants can be considered satisfactorily for N_2 and C_2H_2, the situation is quite different for CH_4. The CC3 results for $V(\omega)$ are still about 8–9% lower than the corresponding experimental values (see Fig. 3) and, thus, are clearly outside three times the experimental uncertainty. A possible reason for this discrepancy might be the large magnitude of the ZPV effects, whose consideration has been proven important earlier for other properties of methane [114, 152–154]. ZPV correction have been estimated for CH_4 by performing calculations (at the HF-SCF level) for both the equilibrium bond distance ($R_e = 1.0858$ Å) [155] and the vibrationally averaged bond distance $R_0 = 1.09397$ Å [156]. The values obtained for the latter geometry are, for all considered frequencies, by about 2.8% larger than the corresponding equilibrium results, thus indicating the importance of ZPV averaging in the case of CH_4.

5.1.4 Gauge issues: MOA and MCD

Both the Verdet constant $V(\omega)$ of MOR and the B term of MCD suffer from gauge-origin dependence when computed for non-centrosymmetric molecules using approximate quantum chemical models. The gauge-origin dependence can be dealt with for these properties by using GIAOs/LAOs and recasting both quantities in the form of derivatives of a property with respect to an external magnetic field [70, 71].

Accurate NLO Properties for Small Molecules

The Verdet constant can be written as a total derivative of the frequency-dependent dipole polarizability with respect to the strength of an external magnetic field:

$$(23) \quad V(\omega) \propto \omega \varepsilon_{\alpha\beta\gamma} \Im \left(\frac{d\alpha_{\alpha\beta}(-\omega; \omega)}{dB_\gamma} \right)_0$$

and the B term can be reformulated as the corresponding derivative of the one-photon dipole transition strength $S_{\alpha\beta}^{aj}$:

$$(24) \quad B(a \to j) \propto \varepsilon_{\alpha\beta\gamma} \left(\frac{d\Im(\langle a|\mu_\alpha|j\rangle\langle j|\mu_\beta|a\rangle)}{dB_\gamma} \right)_0 = \varepsilon_{\alpha\beta\gamma} \Im \left(\frac{dS_{\alpha\beta}^{aj}}{dB_\gamma} \right)_0$$

However, the equivalence of the response functions to the property derivatives is in approximate methods not always strict, as, for example, CC response functions as defined in Section 2 do not involve contributions due to orbital relaxation while property derivatives usually do. The incorporation of orbital relaxation effects in the property derivatives is mandatory when perturbation-dependent basis functions such as GIAOs/LAOs are used. Applying the above reformulation to the expressions for $\alpha(-\omega, \omega)$ and S^{aj} obtained from the CC response functions takes only relaxation with respect to the (static) external magnetic field into account [70, 71]. The frequency-dependent electric fields are treated in an unrelaxed manner, which avoids spurious poles due to orbital relaxation (see Section 2.2).

Table 11. CCSD results for the total Verdet constant at $\omega = 0.11391$ a.u. in the case of hydrogen fluoride. Results labeled as "Unrelaxed" refer to the use of the unrelaxed (one-electron) magnetic dipole moment operator together with the usual magnetic-field independent basis sets. Results labeled "Relaxed" include additional contributions due to orbital relaxation in the presence of the magnetic field. Results labeled "LAO" are those obtained when using GIAOs/LAOs[a]

Basis	Gauge[b] origin	$V(\omega) \times 10^8$ a.u.		
		Unrelaxed	Relaxed	LAO
cc-pVDZ	CM	1.3365	1.3313	1.2259
	H	0.8713	0.8822	1.2259
aug-cc-pVDZ	CM	2.6116	2.6070	3.0381
	H	2.4365	2.4561	3.0381
aug-cc-pVTZ	CM	2.7843	2.7805	3.0883
	H	2.6473	2.6699	3.0883
d-aug-cc-pVDZ	CM	3.2737	3.2697	3.3473
	H	3.3233	3.2504	3.3473
d-aug-cc-pVTZ	CM	3.2190	3.2153	3.2126
	H	3.1936	3.2186	3.2126

[a] All results taken from [71].
[b] CM indicates the center of mass.

Table 12. Effect of a gauge-origin shift on the B term for the five lowest dipole-allowed (singlet) transitions from the ground state XA_1 of formaldehyde (CH_2O) obtained at the CCSD/aug-cc-pVTZ level. The molecule is placed on the xz-plane with the C_2 axis along z and the center of mass (CM) as origin of the coordinate system. Direction of transition in parentheses and excitation energies in atomic units (from [70])

Sym.	exc. energy	trans. strength	Gauge origin at CM		Gauge origin at $z = 25$ bohr	
			Unrelaxed	LAO	Unrelaxed	LAO
B_1	0.2659	0.0189 (x)	−1.6844	−1.6819	−1.6325	−1.6819
B_1	0.2987	0.0394 (x)	−31.796	−34.190	−29.899	−34.190
A_1	0.3019	0.0543 (z)	30.776	33.939	32.865	33.939
B_2	0.3433	0.0006 (y)	−0.4109	−0.3640	−0.6172	−0.3640
A_1	0.3578	0.1360 (z)	−10.730	−12.323	−3.3815	−12.323

The advantage of using GIAOs/LAOs is nicely demonstrated in the case of hydrogen fluoride and formaldehyde (CH_2O). In Table 11, results for the Verdet constants $V(\omega)$ of HF are summarized. While those obtained without using GIAOs/LAOs exhibit a distinct gauge-origin dependence (which however becomes smaller for the larger basis sets) gauge-origin independence is preserved in the GIAO/LAO calculations. In addition, it would appear that the use of GIAOs/LAOs leads to a faster basis set convergence as well-known for the calculation of other magnetic properties [157].

Table 12 collects first CCSD results for the various contributions to the B term due to the five lowest dipole-allowed electronic transitions in formaldehyde. The lacking gauge-origin independence is here illustrated via calculations for two different gauge origins. Again, the use of GIAOs/LAOs resolves this problem and is thus generally recommended.

5.2 Buckingham Effect

For closed-shell systems, the temperature dependence of the BE is given by [40, 134, 135, 158, 159]

$$(25) \quad \Delta n = A_0 + \frac{A_1}{T}$$

with $\Delta n = n_\parallel - n_\perp$ as the corresponding anisotropy of the refractive index,

$$(26) \quad A_0 \propto b(\omega) = \frac{1}{15}\{B_{\alpha\beta,\alpha\beta} - \mathcal{B}_{\alpha,\alpha\beta,\beta}\} - \frac{5}{\omega}\varepsilon_{\alpha\beta\gamma}J'_{\alpha,\beta,\gamma}$$

$$(27) \quad A_1 \propto -\Theta_{\alpha\beta}\alpha_{\alpha\beta} - \mu_\alpha\left(A_{\beta,\alpha\beta} + \frac{5}{\omega}\varepsilon_{\alpha\beta\gamma}G'_{\beta,\gamma}\right)$$

and

(28) $B_{\alpha\beta,\gamma\delta}(-\omega;\omega,0) \propto \Re\langle\langle\hat{\mu}_\alpha;\hat{\mu}_\beta,\hat{\Theta}_{\gamma\delta}\rangle\rangle_{\omega,0}$

(29) $\mathcal{B}_{\alpha,\beta\gamma,\delta}(-\omega;\omega,0) \propto \Re\langle\langle\hat{\mu}_\alpha;\hat{\mu}_\delta,\hat{\Theta}_{\beta\gamma}\rangle\rangle_{0,\omega}$

(30) $J'_{\alpha,\beta,\gamma}(-\omega;\omega,0) \propto \Im\langle\langle\hat{\mu}_\alpha;\hat{m}_\beta,\hat{\mu}_\gamma\rangle\rangle_{\omega,0}$

(31) $\alpha_{\alpha\beta}(-\omega;\omega) \propto \Re\langle\langle\hat{\mu}_\alpha;\hat{\mu}_\beta\rangle\rangle_\omega$

(32) $A_{\alpha,\beta\gamma}(-\omega;\omega) \propto \Re\langle\langle\hat{\mu}_\alpha;\hat{\Theta}_{\beta\gamma}\rangle\rangle_\omega$

(33) $G'_{\alpha,\beta}(-\omega;\omega) \propto \Im\langle\langle\hat{\mu}_\alpha;\hat{m}_\beta\rangle\rangle_\omega.$

In the equations above, μ and Θ indicate the permanent dipole and (traceless) quadrupole moments, respectively. In the case of non-dipolar molecules the birefringence becomes

(34) $\Delta n \propto b(\omega) + \dfrac{2\Theta_{\alpha\beta}\alpha_{\alpha\beta}}{15kT} = b(\omega) + \dfrac{F}{T}$

and for spherical systems it reduces to [159, 160]

(35) $\Delta n \propto b(\omega).$

The connection between BE and MOA are evident, as the response function in Eq. (30) also appears in the corresponding expression for the Verdet constant $V(\omega)$, Eq. (21), though with a different frequency argument. The response function in Eq. (30) plays also a central role in the BE, as for low frequencies it constitutes the most important contribution to the temperature-independent term $b(\omega)$ and as it is solely responsible for the BE in the case of spherical systems. A more detailed discussion concerning the relationship between $V(\omega)$ and $b(\omega)$ can be found in [160] and [149].

5.2.1 Basis set convergence

The basis set convergence in the quantum chemical calculation of the temperature-independent contribution to the BE has been investigated for several different small molecules in [149, 161–164]. Due to the different operators involved in the response functions required for the calculation of $b(\omega)$, it is difficult to put forward theoretical arguments or simple rules concerning the basis set convergence. Nevertheless, smooth convergence with extension of the basis set is observed in all cases investigated so far. To some extent this is probably due to some error cancellation as the involved quadrupole hyperpolarizabilities are usually known to be quite demanding with respect to the size of basis sets.

Overall, it seems that – similar to the calculation of hyperpolarizabilities – (doubly) augmented basis sets of at least triple-ζ or better quadruple-ζ quality are required to obtain reasonable results, as the use of non-augmented basis sets might even lead to qualitatively wrong results, i.e., a wrong sign of $b(\omega)$. For N_2

monotonic convergence is observed for both the singly augmented and the doubly augmented series and the two series approach the basis set limit from different sides [161]. The good convergence achieved in such calculations is, for example, seen by the fact that for X = Q the remaining differences are with 0.2% rather small. Similar trends are also seen for the other molecules [161, 162], although the convergence pattern is for Cl_2 somewhat less regular [163].

5.2.2 Convergence with correlation treatment

The temperature-independent contribution $b(\omega)$ as well as the other BE related terms have so far mostly been computed either at HF-SCF and/or CCSD levels. For Ne and Ar CCS and CC2 results have also been reported [149]. CO is the only case for which CC3 results are available [165]. Overall, it is seen that correlation effects are not negligible and in some cases even larger than the remaining basis set effects. For example, they amount (based on CCSD results) to about 0.5% for N_2, 15% for acetylene, and 11% for methane [161]. In the case of CO inclusion of triple excitation effects (at the CC3 level) lead to a substantial decrease (about 8 to 10%) in $b(\omega)$ in comparison to the CCSD results. Their inclusion thus seems to be essential when aiming at a rigorous comparison with experiment.

5.2.3 Comparison with experiment: Quadrupolar molecules

The experimental importance of the BE is due to the fact that it allows the experimental determination of molecular electric quadrupole moments Θ. The procedure is based on Eq. (34) and was first proposed by Buckingham [134] in the late fifties. With the assumption that the molecular polarizability is known, Θ can be extracted from measurements of Δn. In order to separate the A_0 from the A_1 contribution, it is necessary to perform BE measurements over an appropriately large temperature range.

Nevertheless, in several cases measurements were carried out only at a single temperature and the contribution due to A_0 was assumed to be much smaller than F/T and consequently neglected. Quadrupole moments obtained in this way are referred to as "apparent" quadrupole moments and are not necessarily identical to the actual quadrupole moment of the molecule.

In other cases [166, 167] literature values for (supposedly) $b(\omega)$ taken from quantum chemical calculations were used to separate the A_0 and A_1 contributions. While this is in principle preferred over the first approach, the use of computational data leads to questions concerning their reliability, as low-level data might easily deteriorate the accuracy of the quadrupole moments obtained in this way. In addition, there has been some confusion in the earlier literature concerning the hyperpolarizability correction term $b(\omega)$. In the original papers by Buckingham and coworkers [134, 168] this term was called B and was referred to as a quadrupole hyperpolarizability. This was apparently misunderstood and led to the incorrect use of the averaged static dipole-dipole-quadrupole hyperpolarizability $\frac{2}{15}B_{\alpha\beta\alpha\beta}$ as correction term instead of $b(\omega)$ in the determination of the quadrupole moment [166, 167].

The most striking example of the consequences of these different assumptions and choices is the N_2 case. In 1998 Graham and co-workers [167] published an experimental result for the quadrupole moment of N_2 based on a single temperature measurement of the birefringence. The measured apparent quadrupole moment (obtained by neglecting the A_0 term) was found to be $(-5.25 \pm 0.08) \times 10^{-40}\,\text{Cm}^2$. After revision with an incorrect A_0 value taken from the literature, the quadrupole moment of N_2 was determined to $(-4.65 \pm 0.08) \times 10^{-40}\,\text{Cm}^2$, i.e., noticeably lower than previous experimental and theoretical estimates.

Using our CCSD results [161] for $b(\omega)$, we redetermined the experimental moment from the birefringence measurement of Graham and coworkers to $(-5.01 \pm 0.08) \times 10^{-40}\,\text{Cm}^2$ – now in excellent agreement with the best currently available theoretical results $((-4.93 \pm 0.03) \times 10^{-40}\,\text{Cm}^2$ [169], see Table 13).

Table 13. Theoretical estimates and experimental values for the quadrupole moment Θ, the orientational term F/T, and temperature independent correction $b(\omega)$ (all quantities in atomic units and for $T = 273.15\,\text{K}$) for N_2, C_2H_2, Cl_2, CO_2, and CS_2

	Θ		F/T	$b(\omega)$	
	Theory	Exp.	Theory	Theory	Exp.
N_2	-1.098 ± 0.007[a]	-1.11 ± 0.04[b] -1.12 ± 0.02[d]	-785.1[c]	-36.13[c]	-76.62 ± 23.57[b]
C_2H_2	4.79 ± 0.03[e]	4.66 ± 0.19[f]	9257[c]	-193.3[c]	
Cl_2	2.327 ± 0.010[g]	$2.31/2.36 \pm 0.04$[h]	-6479.9[i]	-213.6[i]	
CO_2	-3.185 ± 0.020[j]	-3.187 ± 0.13[k] -3.18 ± 0.14[m]	-7536.7[l]	-54.46[l]	-159 ± 77[k] -118 ± 178[n]
CS_2	2.338[o]	2.56 ± 0.11[k]	25500[p]	-410[p]	-1179 ± 766[k]

[a] [169]. CCSD(T) basis set limit including ZPV corrections, estimates of correlation beyond CCSD(T), and relativistic effects.
[b] [170].
[c] [161]. CCSD/d-aug-cc-pVQZ/frozen-core equilibrium results.
[d] the apparent moment $\Theta^{eff} = -1.170$ in [167] after revision with b from [161].
[e] [171]. ZPV corrected CCSD(T) basis set limit including estimates for correlation beyond CCSD(T) and relativistic effects.
[f] the apparent moment $\Theta^{eff} = -4.55 \pm 0.22$ in [173] after revision with b from [161].
[g] [163]. ZPV corrected CCSD(T) basis set limit including estimates for correlation beyond CCSD(T) and relativistic contributions.
[h] [163]. Revision of the apparent quadrupole moment (2.23 ± 0.04) in [167] with the calculated $b(\omega)$.
[i] [163]. ZPV corrected CCSD/d-aug-cc-pVQZ result including relativistic and pure vibrational corrections.
[j] [162]. ZPV corrected CCSD(T) basis set limit including estimates for correlation beyond CCSD(T) and relativistic effects.
[k] [172].
[l] [162]. CCSD/d-aug-cc-pVQZ/frozen-core equilibrium results.
[m] [167].
[n] [174].
[o] [162]. CCSD(T)/d-aug-cc-pVQZ/frozen-core equilibrium value.
[p] [162]. CCSD/aug-cc-pVQZ/frozen-core equilibrium value.

More recently, Ritchie, Watson and Keir [170] carried out new experiments on the BE of N_2 for a variety of different temperatures, thus enabling a proper separation of the temperature-independent contribution from the orientational term. These authors also performed new measurements for the polarizability anisotropy. Combining all their experimental data, they were able to extract a new experimental value of $(-4.97 \pm 0.16) \times 10^{-40}\,\text{Cm}^2$ for the quadrupole moment of N_2. The good agreement with the previous (with the CCSD result) revised experimental value as well as the corresponding theoretical results (see Table 13) clearly eliminates any remaining doubts concerning the experimental quadrupole moment of N_2.

Revisions of the apparent quadrupole moments of quadrupolar systems based on *ab initio* results for $b(\omega)$ were undertaken in the last few years for Cl_2 [163] and C_2H_2 [161, 171] (see Table 13). Direct comparison of the hyperpolarizability terms $b(\omega)$ calculated *ab initio* and derived from measurements at various temperatures, such as in the case of CO_2, CS_2 [172] and the most recent results for N_2 [170] turns out to be difficult in all cases due to the very large error bars associated with the experimentally derived results (see Table 13).

5.2.4 Comparison with experiment: Atoms and spherical systems

From a theoretical point of view, rare gas atoms are ideal test systems. This is also the case for the BE, as for these systems the BE is entirely due to the hyperpolarizability term $b(\omega)$. Table 14 collects our "best theoretical results" for the hyperpolarizability term $b(\omega)$ of all atomic and spherical systems investigated so far together with the available experimental data in the literature.

From the data in Table 14, it is possible to derive estimates for the anisotropy Δn under some hypothetical standard experimental conditions for pressure, temperature, path length and electric field gradient, see [40, 70, 159]. For both helium and neon, it appears that the BE is too small to be detected with current experimental devices (for a discussion, see [175]). For krypton, CH_4, and SF_6 theory predicts birefringences of approximatively the same size and about twice as large as for argon. Experimentally, it has so far not been possible to detect the BE for argon and SF_6 [176]. For CH_4 theory and experiment agree on the sign of the effect, but the theoretical value is much smaller, although still within three standard deviations of the experiment. Clearly the computational results need to be improved and (as already discussed for the Verdet constant of methane) ZPV effects need to be considered before a final judgment of the experimental value seems appropriate. To our knowledge, no experimental data is available for krypton.

5.2.5 Comparison with experiment: Dipolar species

Experimental measurements of the BE can also be used to determine quadrupole moments for dipolar systems. The values obtained in this way refer to an origin which is denoted as the effective quadrupole center (EQC) [135]. The EQC is defined as the point at which the combination of dipole-quadrupole and dipole-magnetic dipole polarizabilities multiplied with the dipole moment given in Eq. (27)

Table 14. Best theoretical estimates and experimental values of the hyperpolarizability contribution $b(\omega)$ at 632.8 nm (in atomic units)

System	Theory	Experiment
Helium	−3.15[a]	
Neon	−5.89[b]	
Argon	−58.0[c]	0 ± 109[d]
Krypton	−117.90[e]	
CH_4	−97.3[f]	−262 ± 66[d]
SF_6	−95[g]	0 ± 109[d]

[a] [149]. FCI/d-aug-cc-pV6Z results.
[b] [149].
[c] [149]. CCSD/d-aug-cc-pV5Z results.
[d] [176].
[e] [40]. CCSD/d-aug-cc-pV5Z results.
[f] [161]. CCSD/d-aug-cc-pVQZ results.
[g] [149]. CCSD/d-aug-cc-pVDZ results.

vanishes so that an expression for the BE similar to the one for quadrupolar molecules, Eq. (34), is recovered.

The position of the effective quadrupole center, R_z, cannot be determined from the BE experiment and has to be identified by means of calculations. This is usually done by determining the origin shift which is necessary to make the dipolar contribution in the A_1 term to zero.

Two semiclassical theories existed in the literature for the BE of dipolar species, one due to Buckingham and coworkers (BLH) [134, 135, 160], and one due to Imrie and Raab (IR) [158]. In [159, 162–165], it was shown by means of quantum chemical calculations that these two theories led to different numerical results for the induced anisotropy in a variety of cases. The differences could not be ascribed to deficiencies in the computational approach, as they persisted even for two-electron systems (such as helium and H_2 [159]), for which practically "exact" results are available.

The most striking disagreement was observed for the EQC and the resulting quadrupole moment of CO [164, 165]. In our first study [164] CCSD(T) calculations for the quadrupole moment at the center of mass and CCSD calculations for the EQC were used to determine the quadrupole moment with respect to the EQC within both theories. Our results slightly favored the BLH theory, but the computations were not of sufficient accuracy to decide unambiguously which of the two theories could be correct.

In [165], a second analysis was carried out focusing on the remaining deficiencies in the used computational approach. The main aspect was that the EQCs were now determined using the CC3 approach with an inclusion of triple excitations. The obtained results for the EQC are summarized in Table 15.

The CC3 data clearly favored the BLH theory, but did not supply any explanation for the discrepancy between the two theories. Triggered by these results, a revision

Table 15. Computed and experimentally derived effective quadrupole centers (EQC, R_z, in bohr) with respect to the center of mass for CO at 632.8 nm. The molecule is oriented along the z axis with positive direction pointing from C to O, so that the dipole moment is positive. CCSD/d-aug-cc-pVQZ results are from [164]. CC3/d-aug-cc-pVQZ results are from [165]. The computed shifts are ro-vibrationally corrected

Property	BLH	IR
R_z^{CCSD}	4.52±0.5	2.50±0.5
R_z^{CC3}	5.96±0.5	3.06±0.5
$R_z^{exp,app}$ [a]	7.7±0.5	
$R_z^{exp,CCSD}$ [b]	6.25±0.51	5.78±0.51
$R_z^{exp,CC3}$ [c]	6.16±0.51	5.69±0.51

[a] "Apparent" experimentally derived shift obtained from $\Theta^{(R_z)} = \Theta^{(CM)} - 2R_z\mu$ combining the experimental values for the dipole moment μ, the apparent quadrupole moment Θ^{eff} from Buckingham birefringence and the quadrupole moment at the center of mass Θ^{CM} from molecular beam electric resonance.
[b] The experimentally derived value where Θ^{eff} from Buckingham birefringence has been corrected for $b(\omega)$ computed at the CCSD level. See [164, 165] for details.
[c] The experimentally derived value where Θ^{eff} from Buckingham birefringence has been corrected for $b(\omega)$ computed at the CC3 level. See [164, 165] for details.

of both theories has been recently undertaken and an error in the IR approach was identified and corrected (for details, see [177, 178]).

5.3 Cotton–Mouton Effect

The general expression of CME [136, 137] has for diamagnetic fluids the form of Eq. (25), where now A_0 and A_1 involve the following properties [179, 180]

$$(36) \quad A_0 \propto \Delta\eta(\omega) = \frac{1}{5}\left(\eta_{\alpha\beta,\alpha\beta} - \frac{1}{3}\eta_{\alpha\alpha,\beta\beta}\right)$$

$$(37) \quad A_1 \propto \left(\alpha_{\alpha\beta}\xi_{\alpha\beta} - \frac{1}{3}\alpha_{\alpha\alpha}\xi_{\beta\beta}\right)$$

$\xi_{\alpha\beta}$ is the magnetizability and $\eta_{\alpha\beta,\gamma\delta}$ the hypermagnetizability tensor

$$(38) \quad \eta_{\alpha\beta,\gamma\delta} = \eta^{dia}_{\alpha\beta,\gamma\delta}(-\omega;\omega,0) + \eta^{para}_{\alpha\beta,\gamma\delta}(-\omega;\omega,0,0)$$

$$(39) \quad \eta^{para}_{\alpha\beta,\gamma\delta}(-\omega;\omega,0,0) \propto \Re\langle\langle\hat{\mu}_\alpha;\hat{\mu}_\beta,\hat{m}_\gamma,\hat{m}_\delta\rangle\rangle_{\omega,0,0}$$

$$(40) \quad \eta^{dia}_{\alpha\beta,\gamma\delta}(-\omega;\omega,0) \propto \Re\langle\langle\hat{\mu}_\alpha;\hat{\mu}_\beta,\hat{\xi}^{dia}_{\gamma\delta}\rangle\rangle_{\omega,0}$$

where $\hat{\xi}^{dia}$ is the diamagnetic susceptibility operator, and the other quantities were already defined above. As for BE above, the inverse-temperature contribution dies out for spherical systems, where thus the anisotropy is entirely given by the higher order contribution A_0

(41) $\qquad \Delta n \propto \Delta \eta(\omega)$

5.3.1 Convergence with the basis set

Table 16 shows the results of a basis set study from [39] for the anisotropy of the hypermagnetizability $\Delta \eta$ and the all–parallel component $\eta_{zz,zz}$ at the CC3 level for Ne and Ar. As it can be inferred from the table, the convergence of η with the increase in the cardinal number X and the augmentation level n of the n-aug-cc-pVXZ series is mostly monotonic and smooth, especially for neon. Similar to the purely electric second hyperpolarizability $\gamma_\|$, the results converge very slowly with the singly-augmented basis sets and triply augmentation is needed for atoms to obtain good convergence with the cardinal number. As for $\gamma_\|$, at least triple-ζ basis sets are required to obtain a qualitatively correct description. With the t-aug-cc-pVQZ and t-aug-cc-pV5Z basis sets the results are expected to be converged is within a few percent, both for the individual components and for the anisotropy. When core–valence basis sets are employed in all–electron calculations the differences are for Ne in the order of 0.03 a.u. or $\approx 1\%$ for X = T, dropping to ≈ 0.01 a.u. or $\approx 0.3\%$ for X = Q. For argon the effect of the core functions is somewhat larger, in particular for the anisotropy $\Delta \eta$.

Table 16. Cotton–Mouton effect in neon and argon. CC3 response results for the static limit (from [39], all–electron calculations, results in a.u.)

Basis Set	Neon		Argon	
	$\eta_{zz,zz}$	$\Delta \eta$	$\eta_{zz,zz}$	$\Delta \eta$
aug-cc-pVDZ	−1.61	0.50	−16.8	5.2
aug-cc-pVTZ	−2.13	0.96	−20.8	11.7
aug-cc-pVQZ	−2.60	1.47	−24.0	17.3
aug-cc-pV5Z	−2.87	2.04	−21.7	19.9
d-aug-cc-pVDZ	−3.67	1.60	−32.3	13.6
d-aug-cc-pVTZ	−3.38	2.21	−24.1	22.6
d-aug-cc-pVQZ	−3.15	2.54	−22.7	23.5
d-aug-cc-pV5Z	−2.96	2.86	−21.7	24.6
t-aug-cc-pVDZ	−3.92	1.72	−32.1	13.5
t-aug-cc-pVTZ	−3.07	2.83	−22.3	25.5
t-aug-cc-pVQZ	−2.94	2.87	−21.6	25.1
t-aug-cc-pV5Z	−2.93	2.91		
t-aug-cc-pCVTZ	−3.05	2.82	−22.1	25.1
t-aug-cc-pCVQZ	−2.94	2.87	−21.5	24.8

5.3.2 Convergence with electron correlation treatment

The need for an accurate account of electron correlation for the high-order properties involved in CME, JB and MEB is particularly evident in systems of spherical symmetry, where these properties provide the whole contribution to the observable.

Table 17 summarizes the results of a benchmark study for the convergence of $\Delta\eta$ for neon with the correlation treatment in a series of CC methods from [39]. Whereas CCSD yields results still ca. 3% too low (orbital-relaxed) or 1.4% too high (orbital-unrelaxed), the CCSDT results differs from the FCI limit by only 0.002 a.u. or ≈0.1%. Thus CCSDT provides for the Cotton–Mouton effect of Ne results which are converged within approximately 0.1%.

Figure 4 shows the convergence behavior of the (static) hypermagnetizability anisotropy $\Delta\eta(0)$ of neon and argon up to CCSDT in the fairly large basis sets d-aug-cc-pV5Z (Ne) and d-aug-cc-pVQZ (Ar). The orbital-relaxed results were obtained for HF-SCF, MP2, CCSD, CCSD(T), CC3 and CCSDT by numerical differentiation

Table 17. Convergence of the anisotropy of the static hypermagnetizability $\Delta\eta(0)$ of neon with the correlation treatment (from [39]; finite field orbital–relaxed results (unless specified) for d-aug-cc-pVDZ, frozen core approximation for $1s$ shell, numbers in atomic units.)

CCSD	CCSD (unrel.)	CCSDT	CCSDTQ	CCSDTQP	FCI
1.546	1.614	1.594	1.593	1.593	1.592

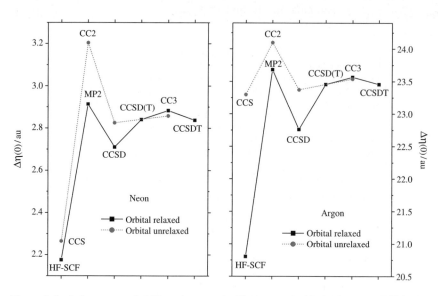

Figure 4. Static hypermagnetizability anisotropy, $\Delta\eta(0)$, computed with the d-aug-cc-pV5Z basis set (Neon) and d-aug-cc-pVQZ basis set (Argon). Orbital-relaxed results obtained with a finite field approach from analytically evaluated magnetizabilities are compared to those obtained from orbital-unrelaxed quadratic and cubic response functions

of analytically computed magnetizabilities. A second series of orbital-unrelaxed results was obtained via analytic quadratic and cubic response functions for the CCS, CC2, CCSD and CC3 wavefunction models. Figure 4 illustrates how the two series converge to a common limit as the level of electron correlation is increased. For instance for neon, the CCSD results obtained in the two approaches differ by almost 10%, while at the CC3 level the difference is five times smaller. These percentages are sensibly smaller for argon. It is expected that at the CCSDT level the orbital-relaxed and unrelaxed approaches give essentially the same results. In the orbital-relaxed hierarchy the contribution of connected triples (CCSDT vs. CCSD) increases the anisotropy by ≈3–5%. In the unrelaxed hierarchy the triples effect seems to be somewhat smaller. The latter hierarchy was used to calculate the CME at the experimental wavelength.

5.3.3 Comparison with experiment

Table 18, taken from [39] (see also [40]), shows a comparison between theory and experiment for the CME of neon and argon gas. For the best non-relativistic theoretical results the absolute uncertainties can be placed around 0.1 a.u. for Ne and (probably) around 1 a.u. for Ar, i.e. in the order of 2–3%. They substantiate the claim that, at least in this case, the computational results should be considered the reference nowadays. For instance the hypermagnetizability of argon is of a magnitude that allows a relatively easy measurement and can be computed with an accuracy more than one order of magnitude better than presently afforded by experiment. Thus, currently it appears advantageous to use the computed value for argon as a standard for calibration of the optical apparata employed for CME measurements.

Table 18. Comparison of experimental and computational results for the hypermagnetizability anisotropy $\Delta\eta(\omega)$ of neon and argon. Results in a.u.

	λ	Theory	Experiment
Neon	∞	2.89(10)[a,b]	
	790 nm		3.2 ± 1.2[c]
	514.5 nm		1.4 ± 0.1[d]
Argon	∞	24.7(10)[a,b]	
	790 nm		27 ± 5[c]
	514.5 nm		30.0 ± 4.5[e]

[a] The number in parenthesis indicates the estimated absolute error on the last digit.
[b] [39]. Static value estimated at CCSDT level; frequency dependence estimated using CC3 analytic response (and found to be negligible).
[c] [181], pressure of 1.52×10^5 Pa, temperatures in the range 285–293 K. The value is estimated for an average of 289 K.
[d] [182], temperature of 298.15 K and extrapolated at a pressure of 1 atm.
[e] [183].

5.4 Jones and Magnetoelectric Birefringence

As for BE and CME, also for JB and MEB the birefringence can be expressed with a general relationship such as Eq. (25). In this case it is [141, 143, 184]:

$$A_0 \propto 3G_{\alpha,\beta\alpha\beta} + 3G_{\alpha,\beta\beta,\alpha} - 2G_{\alpha,\alpha\beta,\beta} \tag{42}$$
$$- \frac{\omega}{2}\varepsilon_{\alpha\beta\gamma}\left(a'_{\alpha,\beta\delta,\delta,\gamma} + a'_{\alpha,\beta\delta,\gamma,\delta}\right)$$

$$A_1 \propto \mu_\alpha \left(3G_{\alpha,\beta\beta} + 3G_{\beta,\alpha\beta} - 2G_{\beta,\beta\alpha}\right) \tag{43}$$
$$- \frac{\omega}{2}\varepsilon_{\alpha\beta\gamma}\left(\mu_\gamma a'_{\alpha,\beta\delta,\delta} + \mu_\delta a'_{\alpha,\beta\delta,\gamma}\right)$$

and now

$$a'_{\alpha,\beta\gamma,\delta,\epsilon}(-\omega;\omega,0,0) \propto \Im\langle\langle \hat{\mu}_\alpha; \hat{q}_{\beta\gamma}, \hat{m}_\delta, \hat{\mu}_\epsilon \rangle\rangle_{\omega,0,0} \tag{44}$$

$$G_{\alpha,\beta\gamma,\delta}(\omega) = G^{para}_{\alpha,\beta\gamma,\delta}(\omega) + G^{dia}_{\alpha,\beta\gamma,\delta}(\omega) \tag{45}$$

$$G^{para}_{\alpha,\beta\gamma,\delta}(-\omega;\omega,0,0) \propto \Re\langle\langle \hat{\mu}_\alpha; \hat{m}_\beta, \hat{m}_\gamma, \hat{\mu}_\delta \rangle\rangle_{\omega,0,0} \tag{46}$$

$$G^{dia}_{\alpha,\beta\gamma,\delta}(-\omega;\omega,0) \propto \Re\langle\langle \hat{\mu}_\alpha; \hat{\xi}^{dia}_{\beta\gamma}, \hat{\mu}_\delta \rangle\rangle_{\omega,0} \tag{47}$$

$$a'_{\alpha,\beta\gamma,\delta}(-\omega;\omega,0) \propto \Im\langle\langle \hat{\mu}_\alpha; \hat{q}_{\beta\gamma}, \hat{m}_\delta \rangle\rangle_{\omega,0} \tag{48}$$

$$G_{\alpha,\beta\gamma}(\omega) = G^{para}_{\alpha,\beta\gamma}(\omega) + G^{dia}_{\alpha,\beta\gamma}(\omega) \tag{49}$$

$$G^{para}_{\alpha,\beta\gamma}(-\omega;\omega,0) \propto \Re\langle\langle \hat{\mu}_\alpha; \hat{m}_\beta, \hat{m}_\gamma \rangle\rangle_{\omega,0} \tag{50}$$

$$G^{dia}_{\alpha,\beta\gamma}(-\omega;\omega) \propto \Re\langle\langle \hat{\mu}_\alpha; \hat{\xi}^{dia}_{\beta\gamma} \rangle\rangle_\omega \tag{51}$$

where $\hat{q}$ is the traced quadrupole moment operator.

5.4.1 Convergence with the basis set

The basis set convergence of the response functions that enter the expressions for JB and MEB are illustrated in Figs. 5 and 6 with CC3 results for Ne and Ar from [185]. As demonstrated before in [184], the non- and singly-augmented basis sets (cc-pVXZ and aug-cc-pVXZ) are inadequate for mixed electric and magnetic field birefringences. An acceptable rate of convergence requires at least doubly-augmented basis sets, but for the atoms Ne and Ar the third set of diffuse functions in the t-aug-cc-pVXZ series still improves the convergence. As for other third- and fourth-order magneto-electric properties double-ζ basis sets give rise to rather large errors, while beyond the DZ level the changes with the cardinal number are smooth. With respect to convergence with the one-electron basis set the most demanding contributions to $\Lambda_\omega = G_{jjii} + \frac{\omega}{2}a'_{jkiii}$ – and thus to the observable Δn – are G^{dia} and a', i.e. the quadratic response functions which involve both the electric dipole and quadrupole operators. Similar as for $\gamma_\parallel$ and η, at least t-aug-cc-pVTZ basis sets appear to be needed for atoms to guarantee a good level of confidence that basis

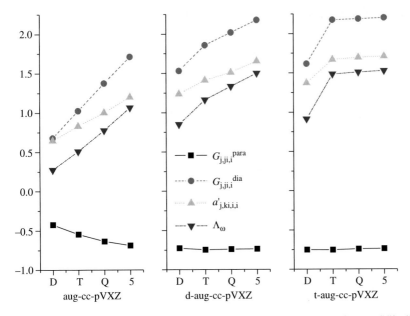

Figure 5. Contributions to the Jones and magneto-electric birefringences of neon. CC3 all–electron response results in atomic units

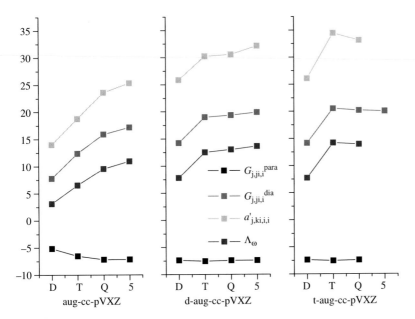

Figure 6. Contributions to the Jones and magneto-electric birefringences of argon. CC3 all–electron response results in atomic units

set convergence is approached for these contributions. With these basis sets the results seem to be converged within a few percent. (For further information on the basis set convergence for JB and MEB for other atoms and molecules the reader is refereed to [184].)

5.4.2 Convergence with electron correlation treatment

Figure 7 illustrates the convergence of the response properties involved in JB and MEB with the improvement of the electron correlation treatment in the hierarchy HF-SCF, CCS, CC2, CCSD, CC3. The results are those obtained for neon and argon, using the d-aug-cc-pV5Z and d-aug-cc-pVQZ basis sets, respectively, in all electron calculations within an orbital-unrelaxed approach. Again, the convergence is somewhat slower for neon than for argon, with HF-SCF and CCS on one side and CC2 on the other side of CCSD and CC3. After some initial oscillations, all properties seem to be quite well converged at the CCSD level, with CC3 yielding only small corrections. The triples effect on the observable, measured by comparing CCSD and CC3 results, are for Ne and Ar in the order of 1%. In [185] it is shown that these percentages increase to ≈2–3% for nitrogen and even higher (ca. 6%) for CO and can in some cases be notably higher for the individual response properties than for the total observable.

5.4.3 Comparison with experiment

Measurements of JB and MEB are a very recent occurrence, and they have been limited so far to systems in the condensed phase [186–188], where measurable birefringences can be observed in optical paths of the order of centimeters.

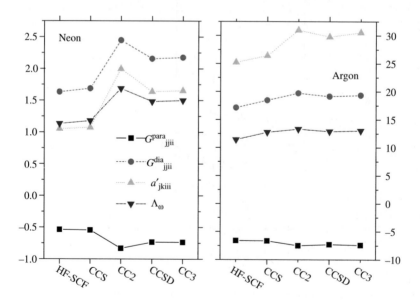

Figure 7. JB and MEB of neon and argon. Analytical response result, all electron, unrelaxed orbitals

Table 19. Comparison of estimated JB and MEB with the values computed for CME. ϕ is the retardance ($\phi = 2\pi l \Delta n/\lambda$), i.e. the property which is actually measured. $\lambda = 632.8$ nm, path length $l = 1$ m, pressure $P = 1$ bar, temperature $T = 273.15$ K, electric field strength $E = 2.6 \times 10^6$ V m^{-1}, magnetic induction field $B = 3$ T. The acronym nXZ is a shorthand notation for the correlation consistent n-aug-cc-pVXZ basis set. "Wf" is the wave function model employed. From [40]

	Jones			CME		
	$\Delta n/10^{-15}$	Basis/Wf	ϕ/nrad	$\Delta n/10^{-12}$	Basis/Wf	ϕ/nrad
Ne	−0.035	t5Z/CC3	−0.35	0.00581	t5Z/CCSDT	5.8×10^1
Ar	−0.30	t5Z/CC3	−3.0	0.0497	d5Z/CC3	4.9×10^2
N$_2$	−0.75	dQZ/CCSD	−7.4	−2.84	tTZ/B3LYP	-2.8×10^4
C$_2$H$_2$	−3.0	aQZ/CCSD	−30	−1.50	aTZ/MCSCF	-1.5×10^4
CO	−1.2	dQZ/CCSD	−12	−1.80	tTZ/B3LYP	-1.8×10^4

Computational analysis has been limited, on the other hand to the gas phase [40, 184, 185, 189, 190]. However, optical apparata designed for measurements in the gas phase of other birefringences, as the Kerr and Cotton-Mouton effects, may in principle be adapted to measure JB and MEB. It is thus probably useful to provide reliable predictions of the magnitude of the effect. Table 19 shows a comparison from [40] of predicted birefringences and corresponding retardances for the two birefringences discussed in this section, i.e. CME and JB (or MEB).

For these systems Jones and magneto-electric birefringences yield retardances which are under the usual experimental conditions about two orders of magnitude weaker than those observed for CME. For the rare gases the predicted retardances for the JB and MEB are comparable to those of BE. The current detection limit for retardances of ≈ 2 nrad [175] makes their observation in these systems rather challenging, but they should be measurable for N$_2$, CO and even better for C$_2$H$_2$.

6. CONCLUSIONS AND OUTLOOK

The calculation of NLO properties with high accuracy is a very challenging task requiring consideration of many different issues, including the electronic, vibrational, and condensed phase contributions. In this review we have described the application of CC response methods for the calculation of the electronic contributions to first and second hyperpolarizabilities as well as some magneto-optical NLO properties. As a result of many years work on implementing CC response theory [46, 47, 58–63, 65–71, 191–203] we can nowadays employ CC methods for detailed and accurate studies of many different molecular properties. A cornerstone has been the CC3 wavefunction model. The development of efficient implementations for these methods has required a significant effort. Yet, actual applications are today still limited to small systems. The pay-back is, however, that high accuracy can indeed be obtained with CC3. Since electron correlation is typically of extreme importance for NLO properties the CC3 model stands central for such calculations. With the accuracy obtained at this level the interpretation of experiments can be discussed with confidence.

Looking into the future for the calculation of NLO properties for small molecules some challenges remain. A basic problem is that the standard CC wavefunction models including CCSD and CC3 are single-reference methods. While for multi-reference cases MCSCF response theory can be applied, for more than a few electrons it will still be very difficult to achieve an accuracy comparable to that obtained in CC calculations on closed–shell single-reference molecules. A more practical restriction in the present implementations of CC3 response theory is that it does not include open–shell molecules even when they can be described by single reference theory. However, very recently an open–shell CC3 implementation (using one additional approximation) was presented for response calculations of excitation energy [204].

The example of neon, where relativistic contributions account for as much as $\approx 0.5\%$ of $\gamma_\parallel$, shows that relativistic effects can turn out to be larger for high-order NLO properties and need to be included if aiming at high accuracy. Some efforts to implement linear and nonlinear response functions for two- and four-component methods and to account for relativity in response calculations using relativistic direct perturbation theory or the Douglas-Kroll-Hess Hamiltonian have started recently [131, 205, 206]. But presently, only few numerical investigations are available and it is unclear when it will become important to include relativistic effects for the frequency dispersion.

An important issue for magneto-optical properties is gauge origin independence. GIAOs/LAOs are a widely applied approach to ensure gauge origin independence in *ab initio* calculation of properties that involve static magnetic fields. For frequency-dependent magnetic fields it is a still unsolved problem how gauge origin independence can be ensured in approximate calculations. GIAOs include the magnetic field already at the level of basis functions and thus necessarily in the uncorrelated reference function. This is incompatible with the requirement that the frequency-dependence of the wavefunction should solely be determined through a single–step method to ensure a correct pole structure.

Another important question in electronic structure calculations for NLO properties is what method to use for larger molecules. While DFT can be used for rather large systems it fails to have the accuracy and systematic behavior of CC theory. In particular problems have been reported for calculations of NLO properties [52, 53, 207, 208]. On the other hand it is unlikely that accurate CC methods including triple excitations like CC3 will become applicable to large systems within the next few years. In recent years the so-called local coupled cluster approach extended successfully the applicability of CC methods for ground state energies to larger molecules. However, the local coupled cluster ansatz faces in its present formulation severe problems when applied in the framework of response theory to excitation energies and to frequency-dependent properties if orbital relaxation has to be treated implicitly through the cluster operator [209, 210]. A complementary approach to reduce the computational costs involves inner projection methods for electron repulsion integrals, as Cholesky decomposition and the resolution-of-the-identity approximation. These have recently been successfully applied for excitation energies and linear response properties at the CC2 level [202, 211]. The operation count for some computational steps can be reduced with inner projection methods by orders of magnitude, but only in special cases the scaling of the costs with system size $\mathcal{N}$ is reduced.

Going beyond the narrow framework of electronic structure theory other major challenges become apparent. The rovibrational contributions to NLO properties are known to be significant. This includes both the vibrational averaging as well as pure (ro)vibrational contributions. The development of general, accurate and efficient methods for the calculation of vibrational contributions is an important area for future research. It requires input from electronic structure calculations and the theoretical interface between electronic and vibrational structure theory is an important issue in this research. For a more detailed discussion we refer to a later Chapter of this book.

Proceeding from the gas to the condensed phase many new issues appear. For NLO properties several additional complications arise when an environment interacts with the system under investigation and external fields need to be considered. Already for the gas phase the proper definition of local field factors and the pressure dependence of (magneto-) optical properties is a difficult issue. In condensed phase, an important question is the proper definition of solute properties and the solvent effect for the electronic property itself. We refer the reader to a comprehensive discussion of solvent effects on NLO properties in a later Chapter. Some of the schemes for modeling solvent effects have been employed in connection with calculation of electronic NLO properties, also recently at the CC level [212, 213]. This is still an area where much progress is expected in the coming years.

ACKNOWLEDGEMENTS

The authors acknowledge support from the EU training and research network NANOQUANT ("Understanding Nanomaterials from a Quantum Perspective", contract No. MRTN-CT-2003-506842), the Deutsche Forschungsgemeinschaft, the Fonds der Chemischen Industrie, the Danish Research Agency, and the Danish Center for Scientific Computing. Many thanks are also due to the long list of collaborators who contributed over many years to the work reviewed here.

REFERENCES

[1] Ward, J.F., Miller, C.K.: Phys. Rev. A. **19**, 826 (1979)
[2] Mizrahi, V., Shelton, D.P.: Phys. Rev. Letters. **55**, 696 (1985)
[3] Mizrahi, V., Shelton, D.P.: Phys. Rev. A. **31**, 3145 (1985)

[4] Shelton, D.P., Mizrahi, V.: Phys. Rev. A. **33**, 72 (1986)
[5] Mizrahi, V., Shelton, D.P.: Phys. Rev. A. **33**, 1396 (1986)
[6] Shelton, D.P.: Phys. Rev. A. **42**, 2578 (1990)
[7] Dudley, J., Ward, J.: J. Chem. Phys. **82**, 4673 (1985)
[8] Sekino, H., Bartlett, R.J.: J. Chem. Phys. **98**, 3022 (1993)
[9] Taylor, P.R., Lee, T.J., Rice, J.E., Almlöf, J.: Chem. Phys. Lett. **163**, 359 (1989)
[10] Maroulis, G., Thakkar, A.J.: Chem. Phys. Lett. **156**, 87 (1989)
[11] Rice, J.E., Taylor, P.R., Lee, T.J., Almlöf, J.: J. Chem. Phys. **94**, 4972 (1991)
[12] Rice, J.E., Scuseria, G.E., Lee, T., Taylor, P.R., Almlöf, J.: Chem. Phys. Lett. **191**, 23 (1992)
[13] Sekino, H., Bartlett, R.J.: J. Chem. Phys. **84**, 2726 (1986)
[14] Luo, Y., Ågren, H., Vahtras, O., Jørgensen, P., Spirko, V., Hettema, H.: J. Chem. Phys. **98**, 7159 (1993)
[15] Bishop, D.M., Kirtmann, B., Kurtz, H.A., Rice, J.E.: J. Chem. Phys. **98**, 8024 (1993)
[16] Shelton, D.P., Rice, J.E.: Chem. Rev. **94**, 3 (1994)
[17] Kaatz, P., Donley, E.A., Shelton, D.P.: J. Chem. Phys. **108**, 849 (1998)
[18] Chong, D.P., Langhoff, S.R.: J. Chem. Phys. **93**, 570 (1990)
[19] Jensen, H.J.Å., Jørgensen, P., Hettema, H., Olsen, J.: Chem. Phys. Lett. **187**, 387 (1991)
[20] Luo, Y., Vahtras, O., Ågren, H., Jørgensen, P.: Chem. Phys. Lett. **205**, 555 (1993)
[21] Buckingham, A.D., Stephens, P.J.: Ann. Rev. Phys. Chem. **17**, 399 (1966)
[22] Schatz, P.N., McCaffery, A.J.: Q. Rev. **23**, 552 (1969)
[23] Shelton, D.P., Donley, E.A.: Chem. Phys. Lett. **195**, 591 (1992)
[24] Bishop, D.M.: Adv. Quant. Chem. **25**, 1 (1994)
[25] Rice, J.E.: J. Chem. Phys. **96**, 7580 (1992)
[26] Jaszuński, M., Jørgensen, P., Rizzo, A.: Theor. Chim. Acta. **90**, 291 (1995)
[27] Bartlett, R.J., Sekino, H.: In: Karna, S.P., Yeates, A.T. (eds.) Non Linear Optical Materials: Theory and Modeling, American Chemical Society Symposium Series. Vol. 628 American Chemical Society, Washington (1996)
[28] Papadopoulos, M.G., Waite, J., Buckingham, A.D.: J. Chem. Phys. **102**, 371 (1995)
[29] Shelton, D.P., Palubinskas, J.J.: J. Chem. Phys. **104**, 2482 (1996)
[30] Hättig, C., Jørgensen, P.: Chem. Phys. Lett. **283**, 109 (1998)
[31] Rozyczko, P., Bartlett, R.J.: J. Chem. Phys. **107**, 10823 (1997)
[32] Hättig, C., Jørgensen, P.: J. Chem. Phys. **109**, 2762 (1998)
[33] Franke, R., Müller, H., Noga, J.: J. Chem. Phys. **114**, 7746 (2001)
[34] Larsen, H., Olsen, J., Hättig, C., Jørgensen, P., Christiansen, O., Gauss, J.: J. Chem. Phys. **111**, 1917 (1999)
[35] Pawłowski, F., Jørgensen, P., Hättig, C.: Chem. Phys. Lett. **413**, 272 (2005)
[36] Christiansen, O., Gauss, J., Stanton, J.F.: Chem. Phys. Lett. **305**, 147 (1999)
[37] Rizzo, A., Coriani, S., Fernández, B., Christiansen, O.: Phys. Chem. Chem. Phys. **4**, 2884 (2002)
[38] Coriani, S., Jørgensen, P., Christiansen, O., Gauss, J.: Chem. Phys. Lett. **330**, 463 (2000)
[39] Rizzo, A., Kállay, M., Gauss, J., Pawłowski, F., Jørgensen, P., Hättig, C.: J. Chem. Phys. **121**, 9461 (2004)
[40] Rizzo, A., Coriani, S.: Birefringences: a challenge for both theory and experiment. Adv. Quantum Chem **50**, 143 (2005)
[41] Oddershede, J.: Propagators Methods, in Ab initio Methods in Quantum Chemistry, Part II, K.P. Lawley (ed.), Adv. Chem. Phys., vol. 69 pp. 201–239. Wiley, New York (1987)
[42] Olsen, J., Jørgensen, P.: Time-Dependent Response Theory with Applications to Self-Consistent Field and Multiconfigurational Self-Consistent Field Wave Functions, in Modern Electronic Structure Theory, Yarkony, D.R., (ed.) vol. 2, chapter 13, pp. 857–990. World Scientific, Singapore (1995)
[43] Sasagane, K., Aiga, F., Itoh, R.: J. Chem. Phys. **99**, 3738 (1993)
[44] Christiansen, O., Jørgensen, P., Hättig, C.: Int. J. Quantum Chem. **68**, 1 (1998)
[45] Langhoff, P.W., Epstein, S.T., Karplus, M.: Rev. Mod. Phys. **44**, 602 (1972)
[46] Koch, H., Jørgensen, P.: J. Chem. Phys. **93**, 3333 (1990)

[47] Stanton, J.F., Bartlett, R.J.: J. Chem. Phys. **98**, 7029 (1993)
[48] Hättig, C., Heß, B.A.: Chem. Phys. Lett. **233**, 359 (1995)
[49] Yeager, D.L., Jørgensen, P.: Chem. Phys. Lett. **65**, 77 (1979)
[50] Jørgensen, P., Jensen, H.J.A., Olsen, J.: J. Chem. Phys. **89**, 3654 (1988)
[51] Dalgaard, E., Monkhorst, H.J.: Phys. Rev. A. **28**, 1217 (1983)
[52] van Gisbergen, S.J.A., Snijders, J.G., Baerends, E.J.: J. Chem. Phys. **109**, 10644 (1998)
[53] Sałek, P., Vahtras, O., Helgaker, T., Ågren, H.: J. Chem. Phys. **117**, 9630 (2002)
[54] Raghavachari, K., Trucks, G.W., Pople, J.A., Head-Gordon, M.: Chem. Phys. Lett. **157**, 479 (1989)
[55] Koch, H., Christiansen, O., Jørgensen, P., Olsen, J.: Chem. Phys. Lett. **244**, 75 (1995)
[56] Koch, H., Christiansen, O., Jørgensen, P., Sánchez de Merás, A., Helgaker, T.: J. Chem. Phys. **106**, 1808 (1997)
[57] Christiansen, O., Koch, H., Jørgensen, P.: Chem. Phys. Lett. **243**, 409 (1995)
[58] Christiansen, O., Koch, H., Jørgensen, P.: J. Chem. Phys. **103**, 7429 (1995)
[59] Christiansen, O.: A hierarchy of coupled cluster models for accurate calculations of molecular properties. Ph.D. Thesis, University of Aarhus (1997)
[60] Kobayashi, R., Koch, H., Jørgensen, P.: Chem. Phys. Lett. **219**, 30 (1994)
[61] Christiansen, O., Halkier, A., Koch, H., Jørgensen, P., Helgaker, T.: J. Chem. Phys. **108**, 2801 (1998)
[62] Christiansen, O., Gauss, J., Stanton, J.F.: Chem. Phys. Lett. **292**, 437 (1998)
[63] Hald, K., Pawłowski, F., Jørgensen, P., Hättig, C.: J. Chem. Phys. **118**, 1292 (2003)
[64] Kállay, M., Gauss, J. Mol Struct. (THEOCH EM), in press (2006)
[65] Hättig, C., Christiansen, O., Koch, H., Jørgensen, P.: Chem. Phys. Lett. **269**, 428 (1997)
[66] Gauss, J., Christiansen, O., Stanton, J.F.: Chem. Phys. Lett. **296**, 117 (1998)
[67] Pawłowski, F.: Development and implementation of CC3 response theory for calculation of frequency-dependent molecular properties. Benchmarking of static molecular properties. Ph.D. Thesis, University of Aarhus (2004)
[68] Hättig, C., Christiansen, O., Jørgensen, P.: Chem. Phys. Lett. **282**, 139 (1998)
[69] Hald, K.: Molecular properties in Coupled-Cluster theory. Ph.D. Thesis, University of Aarhus (2002)
[70] Coriani, S.: Ab initio determination of molecular properties. Ph.D. Thesis, University of Aarhus (2000)
[71] Coriani, S., Hättig, C., Jørgensen, P., Helgaker, T.: J. Chem. Phys. **113**, 3561 (2000)
[72] Purvis III, G.D., Bartlett, R.J.: J. Chem. Phys. **76**, 1910 (1982)
[73] Pawłowski, F., Jørgensen, P., Hättig, C.: Chem. Phys. Lett. **391**, 27 (2004)
[74] Coriani, S., Hättig, C., Jørgensen, P., Halkier, A., Rizzo, A.: Chem. Phys. Lett. **281**, 445 (1997). Erratum, ibid. **293** 324 (1998)
[75] Hättig, C., Christiansen, O., Jørgensen, P.: J. Chem. Phys. **108**, 8355 (1998)
[76] Thomsen, C.L., Madsen, D., Keiding, S.R., Thogersen, J., Christiansen, O.: J. Chem. Phys. **110**, 3453 (1999)
[77] Koch, H., Harrison, R.J.: J. Chem. Phys. **95**, 7479 (1991)
[78] Larsen, H., Hättig, C., Olsen, J., Jørgensen, P.: Chem. Phys. Lett. **291**, 536 (1998)
[79] Helgaker, T., Jørgensen, P., Olsen, J.: Molecular Electronic-Structure Theory. Wiley, New York (2000)
[80] Urban, M., Noga, J., Cole, S.J., Bartlett, R.J.: J. Chem. Phys. **83**, 4041 (1985)
[81] Noga, J., Bartlett, R.J., Urban, M.: Chem. Phys. Lett. **134**, 126 (1987)
[82] Kucharski, S.A., Bartlett, R.J.: J. Chem. Phys. **108**, 9221 (1998)
[83] Lee, Y.S., Bartlett, R.J.: J. Chem. Phys. **80**, 4371 (1984)
[84] Lee, Y.S., Kucharski, S.A., Bartlett, R.J.: J. Chem. Phys. **81**, 5906 (1984)
[85] Gauss, J.: In von Ragué Schleyer, P., The Encyclopedia of Computational Chemistry. Wiley, New York (1998)
[86] Bartlett, R.J. (ed.): Recent Advances in Coupled-Cluster Methods. World Scientific, Singapore (1997)

[87] Bartlett, R.J.: Coupled-Cluster Theory: An Overview of Recent Developments. In: Yarkony, D.R., (ed.) Modern Electronic Structure Theory, vol. 2, chapter 16, pp. 1047–1131. World Scientific, Singapore (1995)
[88] Kutzelnigg, W.: Theor. Chim. Acta. **83**, 263 (1992)
[89] Rice, J.E., Handy, N.C.: J. Chem. Phys. **94**, 4959 (1991)
[90] Sambe, H.: Phys. Rev. A. **7**, 2203 (1973)
[91] Olsen, J., Jørgensen, P.: J. Chem. Phys. **82**, 3235 (1985)
[92] Pedersen, T.B., Koch, H.: J. Chem. Phys. **106**, 8059 (1997)
[93] Helgaker, T., Jørgensen, P.: Calculation of geometrical derivatives in molecular electronic structure theory. In: Wilson, S., Diercksen, G.H.F., (eds) Methods in Computational Physics, pp. 513–421. Plenum Press, New York (1992)
[94] Kramer, P., Saraceno, M.: Geometry of the Time-Dependent Variatonal Principle in Quantum Mechanics. Number 140 in Lecture Notes in Physics. Springer-Verlag, Berlin (1981)
[95] Hättig, C., Christiansen, O., Jørgensen, P.: J. Chem. Phys. **108**, 8331 (1998)
[96] Butcher, P.N., Cotter, D.: The Elements of Nonlinear Optics. Cambridge University, Cambridge (1990)
[97] Nesbet, R.K.: Phys. Rev. A. **16**, 1 (1977)
[98] Visser, F., Wormer, P.E.S., Jacobs, W.P.J.H.: J. Chem. Phys. **82**, 3753 (1985)
[99] Wormer, P.E.S., Rijks, W.: Phys. Rev. A. **33**, 2928 (1986)
[100] Spelsberg, D., Lorenz, T., Meyer, W.: J. Chem. Phys. **99**, 7845 (1993)
[101] Christiansen, O., Hättig, C., Gauss, J.: J. Chem. Phys. **109**, 4745 (1998)
[102] Salter, E.A., Sekino, H., Bartlett, R.J.: J. Chem. Phys. **87**, 502 (1987)
[103] Kobayashi, R., Koch, H., Jørgensen, P., Lee, T.J.: Chem. Phys. Lett. **211**, 94 (1993)
[104] London, F.: J. Phys. Radium. **8**, 397 (1937)
[105] Hameka, H.F.: Mol. Phys. **1**, 203 (1958)
[106] Ditchfield, R.: J. Chem. Phys. **56**, 5688 (1972)
[107] Ditchfield, R.: Mol. Phys. **27**, 789 (1974)
[108] Helgaker, T., Jørgensen, P.: J. Chem. Phys. **95**, 2595 (1991)
[109] Bak, K.L., Gauss, J., Jørgensen, P., Olsen, J., Helgaker, T., Stanton, J.F.: J. Chem. Phys. **114**, 6548 (2001)
[110] Bak, K.L., Jørgensen, P., Olsen, J., Helgaker, T., Klopper, W.: J. Chem. Phys. **112**, 9229 (2000)
[111] Bak, K.L., Halkier, A., Jørgensen, P., Olsen, J., Helgaker, T., Klopper, W.: J. Mol. Struct. **567**, 375 (2001)
[112] Martin, J.M.L.: J. Chem. Phys. **100**, 8186 (1994)
[113] Ruden, T.A., Helgaker, T., Jørgensen, P., Olsen, J.: J. Chem. Phys. **121**, 5874 (2004)
[114] Auer, A.A., Gauss, J., Stanton, J.F.: J. Chem. Phys. **118**, 10407 (2003)
[115] ACES2 (Mainz-Austin-Budapest version), a quantum- chemical program package for high-level calculations of energies and properties by J.F.Stanton et al., see http://www.aces2.de.
[116] Koch, H., Sánchez de Merás, A., Helgaker, T., Christiansen, O.: J. Chem. Phys. **104**, 4157 (1996)
[117] Koch, H., Christiansen, O., Kobayashi, R., Jørgensen, P., Helgaker, T.: Chem. Phys. Lett. **228**, 233 (1994)
[118] Helgaker, T., Jensen, H.J. Å., Jørgensen, P., Olsen, J., Ruud, K., Ågren, H., Auer, A.A., Bak, K.L., Bakken, V., Christiansen, O., Coriani, S., Dahle, P., Dalskov, E.K., Enevoldsen, T., Fernandez, B., Hättig, C., Hald, K., Halkier, A., Heiberg, H., Hettema, H., Jonsson, D., Kirpekar, S., Kobayashi, R., Koch, H., Mikkelsen, K.V., Norman, P., Packer, M.J., Pedersen, T.B., Ruden, T.A., Sanchez, A., Saue, T., Sauer, S.P.A., Schimmelpfennig, B., Sylvester-Hvid, K.O., Taylor, P.R., Vahtras, O.: DALTON – an electronic structure program, release 1.2 (2001)
[119] Hättig, C.: Coupled-Cluster-Methoden zur Berechnung nichtlinearer optischer Eigenschaften und angeregter Zustände von Molekülen. Habil. Thesis, University of Karlsruhe (2003)
[120] Dunning, T.H.: J. Chem. Phys. **90**, 1007 (1989)
[121] Kendall, R.A., Dunning, T.H., Harrison, R.J.: J. Chem. Phys. **96**, 6796 (1992)
[122] Woon, D.E., Dunning, T.H.: J. Chem. Phys. **98**, 1358 (1993)

[123] Woon, D.E., Dunning, T.H.: J. Chem. Phys. **100**, 2975 (1994)
[124] Hättig, C., Koch, H., Jørgensen, P.: J. Chem. Phys. **109**, 3293 (1998)
[125] Høst, S., Jørgensen, P., Köhn, A., Pawłowski, F., Klopper, W., Hättig, C.: J. Chem. Phys. **123**, 094303 (2005)
[126] Koch, H., Kobayashi, R., Jørgensen, P.: Int. J. Quantum Chem. **49**, 835 (1994)
[127] Aiga, F., Sasagane, K., Itoh, R.: Int. J. Quantum Chem. **51**, 87 (1994)
[128] Christiansen, O., Koch, H., Jørgensen, P., Olsen, J.: Chem. Phys. Lett. **256**, 185 (1996)
[129] Larsen, H., Hald, K., Olsen, J., Jørgensen, P.: J. Chem. Phys. **115**, 3015 (2001)
[130] Larsen, H., Olsen, J., Jørgensen, P., Christiansen, O.: J. Chem. Phys. **113**, 6677 (2000). Erratum ibid. **114** (2001) 10985
[131] Klopper, W., Coriani, S., Helgaker, T., Jørgensen, P.: J. Phys. B. **37**, 3753 (2004)
[132] Stephens, P.J.: Ann. Rev. Phys. Chem. **25**, 201 (1974)
[133] Stephens, P.J.: Adv. Chem. Phys. **35**, 197 (1976)
[134] Buckingham, A.D.: J. Chem. Phys. **30**, 1580 (1959)
[135] Buckingham, A.D., Longuet-Higgins, H.C.: Mol. Phys. **14**, 63 (1968)
[136] Cotton, A., Mouton, M.: Compt. Rend. **141**, 317 (1905)
[137] Buckingham, A.D., Pople, J.A.: Proc. Phys. Soc. B. **69**, 1133 (1956)
[138] Kerr, J.: Phil. Mag. **4 (50)**, 337 (1875)
[139] Kerr, J.: Phil. Mag. **4 (50)**, 416 (1875)
[140] Jones, R.C.: J. Opt. Soc. Am. **38**, 671 (1948)
[141] Graham, E.B., Raab, R.E.: Proc. R. Soc. Lond. A. **390**, 73 (1983)
[142] Pockels, F.: Radium. **10**, 152 (1913)
[143] Graham, E.B., Raab, R.E.: Mol. Phys. **52**, 1241 (1984)
[144] Wolinski, K., Hinton, J., Pulay, P.: J. Am. Chem. Soc. **112**, 8251 (1990)
[145] Faraday, M.: Philos. Mag. **28**, 294 (1846)
[146] Faraday, M.: Philos. Trans. R. Soc. London. **136**, 1 (1846)
[147] Verdet, E.M.: Ann. Chimie (3rd Ser.). **41**, 370 (1854)
[148] Snyder, P.A., Hansen, R.W.C., Rowe, E.M.: J. Phys. Chem. **100**, 17756 (1996)
[149] Coriani, S., Hättig, C., Rizzo, A.: J. Chem. Phys. **111**, 7828 (1999)
[150] Bishop, D.M., Cybulski, S.M.: J. Chem. Phys. **93**, 590 (1990)
[151] Ingersoll, L.R., Liebenberg, D.H.: J. Opt. Soc. Am. **46**, 538 (1956)
[152] Bishop, D.M., Sauer, S.P.A.: J. Chem. Phys. **107**, 8502 (1997)
[153] Ruud, K., Åstrand, P.-O., Taylor, P.R.: J. Chem. Phys. **112**, 2668 (2000)
[154] Stanton, J.: Mol. Phys. **97**, 841 (1999)
[155] Gray, D.L., Robiette, A.G.: Mol. Phys. **37**, 1901 (1979)
[156] Herranz, J., Stoicheff, B.P.: J. Mol. Spectr. **10**, 448 (1963)
[157] Gauss, J., Stanton, J.: Adv. Chem. Phys. **123**, 355 (2003)
[158] Imrie, D.A., Raab, R.E.: Mol. Phys. **74**, 833 (1991)
[159] Coriani, S., Halkier, A., Rizzo, A.: The electric-field-gradient-induced birefringence and the determination of molecular quadrupole moments. In: Pandalai, G. (ed.) Recent Research Developments in Chemical Physics, Vol. 2, pg. 1, Transworld Scientific, Kerala, India (2001)
[160] Buckingham, A.D., Jamieson, M.J.: Mol. Phys. **22**, 117 (1971)
[161] Coriani, S., Hättig, C., Jørgensen, P., Rizzo, A., Ruud, K.: J. Chem. Phys. **109**, 7176 (1998)
[162] Coriani, S., Halkier, A., Rizzo, A., Ruud, K.: Chem. Phys. Lett. **326**, 269 (2000)
[163] Cappelli, C., Ekström, U., Rizzo, A., Coriani, S.: J. Comp. Meth. Sci. Eng. (JCMSE). **4**, 365 (2004)
[164] Rizzo, A., Coriani, S., Halkier, A., Hättig, C.: J. Chem. Phys. **113**, 3077 (2000)
[165] Coriani, S., Halkier, A., Jonsson, D., Gauss, J., Rizzo, A., Christiansen, O.: J. Chem. Phys. **118**, 7329 (2003)
[166] Russell, A.J., Spackman, M.A.: Mol. Phys. **88**, 1109 (1996)
[167] Graham, C., Imrie, D.A., Raab, R.E.: Mol. Phys. **93**, 49 (1998)
[168] Buckingham, A.D., Disch, R.L.: Proc. Roy. Soc. A. **273**, 275 (1963)

[169] Halkier, A., Coriani, S., Jørgensen, P.: Chem. Phys. Lett. **294**, 292 (1998)
[170] Ritchie, G.L.D., Watson, J.N., Keir, R.I.: Chem. Phys. Lett. **370**, 376 (2003)
[171] Halkier, A., Coriani, S.: Chem. Phys. Lett. **303**, 408 (1999)
[172] Watson, J.N., Craven, I.E., Ritchie, G.L.D.: Chem. Phys. Lett. **274**, 1 (1997)
[173] Keir, R.I., Lamb, D.W., Ritchie, G.L.D., Watson, J.N.: Chem. Phys. Lett. **279**, 22 (1997)
[174] Battaglia, M.R., Buckingham, A.D., Neumark, D., Pierens, R.K., Williams, J.H.: Mol. Phys. **43**, 1015 (1981)
[175] Ritchie, G.L.D.: Field-gradient induced birefringence: a direct route to molecular quadrupole moments. In Clary, D.C., Orr, B., (eds) Optical, Electric and Magnetic Properties of Molecules, pg. 67, Elsevier, Amsterdam, The Netherlands (1997)
[176] Buckingham, A.D., Disch, R.L., Dunmur, D.A.: J. Am. Chem. Soc. **90**, 3104 (1968)
[177] de Lange, O.L., Raab, R.E.: Mol. Phys. **102**, 125 (2004)
[178] Raab, R.E., de Lange, O.L.: Mol. Phys. **101**, 3467 (2003)
[179] Rizzo, C., Rizzo, A., Bishop, D.M.: Int. Rev. Phys. Chem. **16**, 81 (1997)
[180] Rizzo, A., Rizzo, C.: Mol. Phys. **96**, 973 (1999)
[181] Muroo, K., Ninomiya, N., Yoshino, M., Takubo, Y.: J. Opt. Soc. Am. B. **20**, 2249 (2003)
[182] Cameron, R., Cantatore, G., Melissinos, A.C., Rogers, J., Semertzidis, Y., Halama, H., Prodell, A., Nezrick, F.A., Rizzo, C., Zavattini, E.: J. Opt. Soc. Am. B. **8**, 520 (1991)
[183] Carusotto, S., Iacopini, E., Polacco, E., Scuri, F., Stefanini, G., Zavattini, E.: J. Opt. Soc. Am. B. **1**, 635 (1984)
[184] Rizzo, A., Coriani, S.: J. Chem. Phys. **119**, 11064 (2003)
[185] Pawłowski, F., Jørgensen, P., Rizzo, A., Hättig, C.: to be published
[186] Rikken, G.L.J.A., Raupach, E., Roth, T.: Physica B. **294–295**, 1 (2001)
[187] Roth, T., Rikken, G.L.J.A.: Phys. Rev. Lett. **85**, 4478 (2000)
[188] Roth, T.: Experimental verification of the Jones birefringence induced in liquids. Diploma thesis, Darmstadt University of Technology and Grenoble High Magnetic Field Laboratory (2000)
[189] Rizzo, A., Cappelli, C., Jansík, B., Jonsson, D., Sałek, P., Coriani, S., Ågren, H.: J. Chem. Phys. **121**, 8814 (2004)
[190] Rizzo, A., Cappelli, C., Jansík, B., Jonsson, D., Sałek, P., Coriani, S., Wilson, D.J.D., Helgaker, T., Ågren, H.: J. Chem. Phys. **122**, 234314 (2005)
[191] Koch, H., Jensen, H.J.A., Jørgensen, P., Helgaker, T.: J. Chem. Phys. **93**, 3345 (1990)
[192] Christiansen, O., Koch, H., Halkier, A., Jørgensen, P., Helgaker, T., Sánchez de Merás, A.M.: J. Chem. Phys. **105**, 6921 (1996)
[193] Christiansen, O., Koch, H., Jørgensen, P., Helgaker, T.: Chem. Phys. Lett. **263**, 530 (1996)
[194] Koch, H., Kobayashi, R., Sánchez de Merás, A., Jørgensen, P.: J. Chem. Phys. **100**, 4393 (1994)
[195] Datta, B., Sen, P., Mukherjee, D.: J. Phys. Chem. **99**, 6441 (1995)
[196] Stanton, J.F., Bartlett, R.J.: J. Chem. Phys. **99**, 5178 (1993)
[197] Hald, K., Jørgensen, P., Olsen, J., Jaszuński, M.: J. Chem. Phys. **115**, 671 (2001)
[198] Stanton, J.F., Gauss, J.: J. Chem. Phys. **100**, 4695 (1994)
[199] Gwaltney, S.R., Bartlett, R.J.: J. Chem. Phys. **110**, 62 (1999)
[200] Nooijen, M., Bartlett, R.J.: J. Chem. Phys. **102**, 3629 (1995)
[201] Watts, J.D., Bartlett, R.J.: Chem. Phys. Lett. **233**, 81 (1995)
[202] Hättig, C., Weigend, F.: J. Chem. Phys. **113**, 5154 (2000)
[203] Rozyczko, P., Perera, S.A., Nooijen, M., Bartlett, R.J.: J. Chem. Phys. **107**, 6736 (1997)
[204] Smith, C.E., King, R.A., Crawford, T.D.: J. Chem. Phys. **122**, 054110 (2004)
[205] Sałek, P., Helgaker, T., Saue, T.: Chem. Phys. **311**, 187 (2005)
[206] Norman, P., Schimmelpfennig, B., Ruud, K., Jensen, H.J.A., Ågren, H.: J. Chem. Phys. **116**, 6914 (2002)
[207] Cai, Z.L., Sendt, K., Reimers, J.R.: J. Chem. Phys. **117**, 5543 (2002)
[208] Sałek, P., Helgaker, T., Vahtras, O., Ågren, H., Jonsson, D., Gauss, J.: Mol. Phys. **103**, 439 (2005)

[209] Korona, T., Pflüger, K., Werner, H.-J.: Phys. Chem. Chem. Phys. **6**, 2059 (2004)
[210] Russ, N., Crawford, T.: Chem. Phys. Lett. **400**, 104 (2004)
[211] Pedersen, T.B., Sánchez de Merás, A.M., Koch, H.: J. Chem. Phys. **120**, 8887 (2004)
[212] Kongsted, J., Osted, A., Mikkelsen, K.V., Christiansen, O.: J. Chem. Phys. **119**, 10519 (2003)
[213] Kongsted, J., Osted, A., Mikkelsen, K.V., Christiansen, O.: J. Chem. Phys. **120**, 3787 (2004)

CHAPTER 3

DETERMINATION OF VIBRATIONAL CONTRIBUTIONS TO LINEAR AND NONLINEAR OPTICAL PROPERTIES

B. KIRTMAN AND J.M. LUIS

Department of Chemistry and Biochemistry, University of California, Santa Barbara, California 93106 and Institute of Computational Chemistry and Department of Chemistry, University of Girona, Campus de Montilivi, 17071 Girona, Catalonia, Spain

Abstract: A review of methods for calculating vibrational contributions to linear and nonlinear optical properties is presented. Our aim is to provide an overview of the various approaches that have been developed using illustrative equations supplemented with references to the detailed formulations. The treatment of electrical and mechanical anharmonicity is considered in some detail for resonant as well as non-resonant processes. Issues such as the choice of basis set, the treatment of electron correlation, and the convergence of perturbation expansions are examined. Although much of the presentation is general, there is a special emphasis on organic pi-conjugated systems

Keywords: nonlinear optical properties, vibrational hyperpolarizabilities, nuclear relaxation hyperpolarizabilities, curvature hyperpolarizabilities, field-induced coordinates, post-VSCF methods, mode-mode coupling, optimized effective potential, TPA, Franck-Condon factors, infinite periodic polymers

In this chapter we review some of the developments that have been made over the past fifteen years with regard to the calculation of vibrational contributions to linear and nonlinear (NLO) optical properties. Despite a number of advances it is important to recognize that more are needed since there is still no fully satisfactory general treatment for either resonant or non-resonant NLO processes in polyatomic molecules. Two major intertwining approaches to practical computations that include electrical and mechanical anharmonicity have emerged. The older approach is from the viewpoint of ordinary sum-over-states perturbation theory and it is presented in Section 1. The other approach, discussed in Section 2, is from what may be called the nuclear relaxation/curvature point of view. Even though there is an exact correspondence between the two, which has been exploited in developing the nuclear relaxation/curvature approach, the latter has spawned valuable new concepts and related computational procedures. One such offshoot, namely field-induced coordinates (FICs), is the subject of Section 3. The FICs have led to a substantial reduction in computational cost.

An important issue with regard to any perturbation treatment is the convergence behavior of the perturbation series. This is considered in Section 4 where problematic cases are identified. Then a potentially viable treatment of such cases, based on vibrational SCF and post-SCF procedures, is elaborated in Section 5. In Section 6 we turn to the practical issues of basis set requirements and treatment of electron correlation. Here the emphasis is on quasilinear pi-conjugated molecules and, for that case, we examine the difficulties encountered with the use of density functional theory.

In all of the above the focus is on non-resonant properties. We turn our attention in Section 7 to resonant phenomena, particularly one- and two-photon absorption (OPA/TPA). After outlining the perturbation theory approach to OPA/TPA, the problem of evaluating the key Franck-Condon (and related) integrals is examined. There is increasing activity in this area, but we limit ourselves to a new procedure that accounts for anharmonic mode-mode coupling along with Duschinsky rotation of the normal coordinates and the shift in equilibrium geometry. Then, future directions are indicated whereby calculations for resonant phenomena may catch up to those for non-resonant processes. Finally, in the last section of this chapter the specialized treatment required for infinite polymers, involving either extrapolation of finite oligomer results or application of periodic boundary conditions is considered.

1. PERTURBATION THEORY VIEWPOINT

We are interested here in the linear and nonlinear optical properties that determine the response of a chemical system to spatially uniform electric fields. The vibrational contribution to this response, which arises from vibronic coupling, can often be as important as the pure electronic contribution or even more important [1–14]. In addition, it is often inadequate in this context to treat the effect of vibrational motions at the harmonic level of approximation. The purpose of this review, then, is to show how the vibrational contribution to linear and nonlinear optical properties can be evaluated with both harmonic and anharmonic effects included.

1.1 Sum Over States Formulation

There are two major ways to view the vibrational contribution to molecular linear and nonlinear optical properties, i.e. to (hyper)polarizabilities. One of these is from the time-dependent sum-over-states (SOS) perturbation theory (PT) perspective. In the usual SOS-PT expressions [15], based on the adiabatic approximation, the intermediate vibronic states $|K, k\rangle$ are of two types. Either the electronic wavefunction

|K> refers to the ground state ($K = 0$) or it refers to an excited state. The electronic property is considered to arise entirely from $K \neq 0$ intermediate states. Hence, all terms containing one or more intermediate states with $K = 0$ are considered to be part of what has come to be known as the *pure vibrational* contribution.

1.1.1 Resonant vs. non-resonant

The energy denominators in SOS-PT contain one or more factors of the general form $\omega_{kK} \pm (i\Gamma_{kK}/2 + \omega)$ where ω_{kK} is the excitation frequency from the ground vibronic state to $|K, k>$, Γ_{kK} is the population decay rate for $|K, k>$, and ω is an optical frequency or a sum of the optical frequencies that characterize the particular process. Resonant, or near-resonant processes occur when $\omega_{kK} \pm \omega \approx 0$. Otherwise, the process is non-resonant and we neglect Γ_{kK}. The resonant case is discussed in Section 7, while the remainder of this review focuses on the non-resonant case.

1.2 Clamped Nucleus Approximation

The first step in simplifying the SOS-PT formulas is to apply the clamped nucleus approximation for the states $K \neq 0$ [16]. In this approximation the energy denominators $\hbar\omega_{kK}$ are replaced by the difference in electronic energies at a fixed nuclear configuration, i.e. by $E(K, \mathbf{R}) - E(0, \mathbf{R})$. The consequences of this approximation have been investigated and were found to be negligible [16].

1.3 Bishop and Kirtman Perturbation Treatment

In the Bishop and Kirtman (BK) perturbation treatment [17–19] two basic additional assumptions are made. First, when $|K>$ is an intermediate excited electronic state it is assumed that, under ordinary non-resonant conditions, one may ignore the optical frequency term $i\Gamma_{kK}/2 + \omega$ in the corresponding energy denominator as compared to the electronic excitation energy. Then, after summing over all intermediate states other than $K = 0$, one is left with the pure vibrational (hyper)polarizability, P^v. The latter may be expressed compactly in terms of so-called square bracket quantities. Thus,

$$(1) \quad \alpha^v(-\omega_\sigma; \omega_1) = [\mu^2]$$

$$(2) \quad \beta^v(-\omega_\sigma; \omega_1, \omega_2) = [\mu\alpha] + [\mu^3]$$

$$(3) \quad \gamma^v(-\omega_\sigma; \omega_1, \omega_2, \omega_3) = [\alpha^2] + [\mu\beta] + [\mu^2\alpha] + [\mu^4]$$

in which, for example,

$$(4) \quad [\mu\alpha] = \frac{1}{2\hbar} \sum P_{\alpha\beta\gamma} \sum_k{}' (\mu_\alpha)_{0k}(\alpha_{\beta\gamma})_{k0}[(\omega_k + \omega_\sigma)^{-1} + (\omega_k - \omega_\sigma)^{-1}]$$

where the prime on the summation indicates that $k = 0$ is excluded. A complete set of square bracket formulas is given in Table 1 of [19]. The total BK (hyper)polarizability is the sum of the pure clamped nucleus electronic term (evaluated at the electronic ground state equilibrium geometry) plus the pure vibrational contribution plus the zero-point vibrational averaging correction, P^{zpva}. In order to evaluate the quantities that depend upon nuclear motion BK assume that the instantaneous electrical properties (α, β, and the dipole moment μ), as well as the pure vibrational potential, may be expanded as a power series in the normal coordinate displacements about the equilibrium electronic ground state geometry.

1.3.1 Double harmonic model

The zeroth-order approximation in the BK perturbation treatment of pure vibrational NLO is the double harmonic model. As far as electrical properties are concerned this approximation includes just the terms in the instantaneous property expression that are linear in the normal coordinates (there is no vibrational contribution from the constant term). To these are added the quadratic terms in the pure vibrational (or mechanical) potential which constitute the usual harmonic approximation. Then, in zeroth-order roughly half of the square brackets vanish leaving:

$$(5) \quad \alpha^v(-\omega_\sigma;\omega_1) = [\mu^2]^0$$

$$(6) \quad \beta^v(-\omega_\sigma;\omega_1,\omega_2) = [\mu\alpha]^0$$

$$(7) \quad \gamma^v(-\omega_\sigma;\omega_1,\omega_2,\omega_3) = [\alpha^2]^0 + [\mu\beta]^0$$

The remaining anharmonic square bracket terms are obtained by BK using double perturbation theory with the definition of orders given below.

1.3.2 Anharmonicity

Quadratic terms in the property expansions are considered to be first-order in electrical anharmonicity, cubic terms are taken to be second-order, etc. Similarly, cubic terms in the vibrational potential are considered to be first-order in mechanical anharmonicity, quartic terms are second-order, and so forth. The notation (n, m) is used hereafter for the order of electrical (n) and mechanical (m) anharmonicity whereas the total order $(n+m)$ is denoted by I, II, ... Although our definition of orders is reasonable other choices are possible. Two key questions are: (1) How important are anharmonicity contributions to vibrational NLO properties and (2) What is the convergence behavior of the double perturbation series in electrical and mechanical anharmonicity? Both questions will be addressed later. Here we note that compact expressions, complete through order II in electrical plus mechanical anharmonicity, have been presented [19]. The formulas of order I contain either cubic force constants or second derivatives of the electrical properties with respect to the normal coordinates. Depending upon the level of calculation at least one order of numerical differentiation is ordinarily required to determine these anharmonicity parameters. For electrical properties, the additional normal coordinate derivative may be replaced by an electric field derivative using relations such as $\partial^2\mu/\partial Q_i\partial Q_j = -\partial^3 E/\partial F\partial Q_i\partial Q_j = -\partial k_{ij}/\partial F$ where F is the field and k_{ij} is a vibrational force constant. There are only two non-vanishing square brackets of order I: $[\mu^3]^I$ for β^v and $[\mu^2\alpha]^I$ for γ^v. Note that these are different from the square brackets that appear in zeroth-order. In fact, each square bracket contains only even-order or only odd-order contributions. For the formulas of order II a second numerical differentiation is necessary. In that case direct calculation of the required anharmonicity parameters can be quite tedious and subject to substantial round-off error. All zeroth-order square brackets (i.e. Eq. (5)–(7)) have corresponding second-order terms. In addition, γ^v contains a second-order $[\mu^4]^{II}$ square bracket which vanishes in zeroth-order.

1.4 Comparison of Vibrational and Electronic NLO

In assessing the importance of vibrational NLO properties it is necessary to distinguish between processes that involve at least one dc field and those that do not. The vibrational contribution is negligible for the latter category, which includes second and third harmonic generation. However, there are many other processes that do involve a dc-field. These include the dc-Pockels effect [dc-P; $\beta(-\omega; \omega, 0)$], the electro-optic Kerr effect [EOKE; $\gamma(-\omega; \omega, 0, 0)$], dc-second harmonic generation [dc-SHG; $\gamma(-2\omega; \omega, \omega, 0)$], and all static properties. The intensity-dependent refractive index [IDRI; $\gamma(-\omega; \omega, -\omega, \omega)$] may also be put in this category because the two optical frequencies of opposite phase in effect cancel one other. In fact, the vibrational contribution to IDRI is usually large in keeping with the general rule that this contribution becomes more important as the number of dc fields increases (See for instance [20]). An example is the π-conjugated donor-acceptor molecule $(NO_2)_2(CH=CH)_3(NH_2)_2$. Our calculated results [21] for the ratio P^v/P^e(static) (P^v is the vibrational property and P^e(static) the static electronic property) are: γ(static) = 87, β(static) = 6.4, γ(EOKE) = 5.8, and γ(IDRI) = 4.4. For all other processes smaller ratios were obtained. As a very rough rule of thumb, when the process involves one or more static fields the above ratio will be on the order of unity for the π-conjugated organic molecules that are of interest as NLO materials. It's clear, however, that much larger ratios can be obtained as well.

2. ALTERNATIVE NUCLEAR RELAXATION/CURVATURE VIEWPOINT

There exists an alternative to the BK perturbation approach for calculating pure vibrational contributions. It is based on determining the change in the electronic and zero-point vibrational energy (or, more generally, P^e and P^{zpva}) due to the distortion of the equilibrium geometry induced by a static external field [22–30]. From this viewpoint it is natural to divide the total *static* (hyper)polarizability into pure electronic (P^e), nuclear relaxation (P^{nr}) and curvature (P^c) contributions. On the one hand, P^e is due to the change in the electronic cloud caused by the electric field with the nuclei clamped at the field-free equilibrium geometry. On the other hand, P^{nr} arises from the change in the electronic energy caused by the field-induced relaxation of the equilibrium geometry. Finally, P^c is due to the change in zero-point vibrational energy caused by the change in the curvature, i.e., the shape, of the potential energy surface (PES) induced by the static external field. P^c can be divided into two terms: the zero-point vibrational averaging (*zpva*) contribution, P^{zpva}, and the remainder P^{c-zpva}.

2.1 Static Nuclear Relaxation and Curvature (hyper)Polarizabilities

Under the influence of a uniform static electric field, the electronic energy of a chemical system can be expressed as a double power series expansion in the normal coordinates and the electric field [24]:

$$(8) \quad V(\mathbf{Q},\mathbf{F}) = \sum_{n=0}^{3N-6} \sum_{i_1=1}^{3N-6} \cdots \sum_{i_n=1} \sum_{m=0} \sum_{\alpha_1=1}^{x,y,z} \cdots \sum_{\alpha_m=1}^{x,y,z} a_{nm}^{i_1\ldots i_n,\alpha_1\ldots\alpha_m} Q_{i_1}\ldots Q_{i_n} F_{\alpha_1}\ldots F_{\alpha_m}$$

where

$$(9) \quad a_{nm}^{i_1\ldots i_n,\alpha_1\ldots\alpha_m} = \frac{1}{n!m!} \left(\frac{\partial^{(n+m)} V(Q_1,\ldots,Q_{3N-6}, F_x, F_y, F_z)}{\partial Q_i \partial Q_j \ldots \partial F_{\alpha_1} \partial F_{\alpha_m} \ldots} \right)_{Q=0, F=0}$$

The P^{nr} and P^c contributions to the static property value may be obtained from this power series in the following manner. First, we impose the minimum condition (i.e. $\partial V(\mathbf{Q},\mathbf{F})/\partial Q_i = 0 \ \forall i$) on the potential of Eq. (8). This leads to analytical expressions for the field-dependent equilibrium geometry in terms of field-free normal coordinate displacements. Then, substitution of these displacements back into Eq. (8) yields a power series in the static electric field and that gives directly the nuclear relaxation contribution, P^{nr}, to the static (hyper)polarizability [24]. For instance, $\alpha_{\alpha\beta}^{nr}(0;0)$ and $\beta_{\alpha\beta\gamma}^{nr}(0;0,0)$ are obtained in this manner as:

$$(10) \quad \alpha_{\alpha\beta}^{nr}(0;0) = \frac{1}{2} \sum_{i=1}^{3N-6} P_{\alpha\beta}\, a_{11}^{i,\alpha} q_1^{i,\beta}$$

and

$$(11) \quad \beta_{\alpha\beta\gamma}^{nr}(0;0,0) = \sum_{i=1}^{3N-6} P_{\alpha\beta\gamma}\, a_{12}^{i,\alpha\beta} q_1^{i,\gamma} - \sum_{ij}^{3N-6} P_{\alpha\beta\gamma}\, a_{21}^{ij,\alpha} q_1^{i,\beta} q_1^{j,\gamma}$$

$$+ \sum_{ijk}^{3N-6} P_{\alpha\beta\gamma}\, a_{30}^{ijk} q_1^{i,\alpha} q_1^{j,\beta} q_1^{k,\gamma}$$

where the notation

$$(12) \quad q_1^{i,\alpha} = \frac{a_{11}^{i,\alpha}}{2\, a_{20}^{ii}}$$

has been introduced and $P_{\alpha\beta}$ indicates a sum over all permutations of the indices $\alpha\beta\ldots$. Identical expressions for P^{nr} may be obtained by substituting back into the μ^e expansion rather than the V expansion [23, 27].

The static P^c are derived in the same fashion as the static P^{nr} except that one substitutes back into the double power series expansion for zero-point energy instead of the electronic potential energy [22, 24]. In doing so it is important to take into account the fact that the field-dependent harmonic force constant matrix contains off-diagonal elements when expressed in terms of field-free normal coordinates. The order of anharmonicity included in P^c depends upon the order included in the zero-point energy, which is indicated by a roman superscript in parenthesis. $P^{c(I)}$ is derived from the first-order zero-point energy, $P^{c(III)}$ from the third-order zero-point energy, and so forth. The zero-point energy contains no even order corrections. As will be explained further in Section 2.3, apart from P^{zpva} the terms in $P^{c(I)}$ are all higher-order in anharmonicity than those in P^{nr}. This may be seen, for example, from the expression for $\alpha_{\alpha\beta\gamma}^{c(I)}(0;0)$:

$$(13) \quad \alpha_{\alpha\beta}^{c(I)}(0;0) = \frac{P_{\alpha\beta}}{4} \sum_{i=1}^{3N-6} \frac{1}{\sqrt{2a_{20}^{ii}}} \left[2a_{22}^{ii,\alpha\beta} - 6\sum_{j=1}^{3N-6} a_{30}^{iij} q_2^{j,\alpha\beta} - 6\sum_{j=1}^{3N-6} a_{31}^{iij,\alpha} q_1^{j,\beta} \right.$$

$$+ 12 \sum_{j,k=1}^{3N-6} a_{40}^{iijk} q_1^{j,\alpha} q_1^{k,\beta} + 6 \sum_{j,k=1}^{3N-6} \frac{a_{30}^{iij} a_{21}^{jk,\alpha}}{a_{20}^{j}} q_1^{k,\beta} - 9 \sum_{j,k,l=1}^{3N-6} \frac{a_{30}^{iij} a_{30}^{jkl}}{a_{20}^{j}} q_1^{k,\alpha} q_1^{l,\beta}$$

$$\left. - 4 \sum_{j=1}^{3N-6} \frac{a_{21}^{ij,\alpha} a_{21}^{ij,\beta}}{A_{20}^{ij}} + 24 \sum_{j,k=1}^{3N-6} \frac{a_{21}^{ij,\alpha} a_{30}^{ijk}}{A_{20}^{ij}} q_1^{k,\beta} - 36 \sum_{j,k,l=1}^{3N-6} \frac{a_{30}^{ijk} a_{30}^{ijl}}{A_{20}^{ij}} q_1^{k,\alpha} q_1^{l,\beta} \right]$$

where

$$(14) \quad q_2^{i,\alpha\beta} = \frac{a_{12}^{i,\alpha\beta}}{2a_{20}^{ii}}, \quad \text{and} \quad A_{20}^{ij} = 2\left(\sqrt{a_{20}^i} + \sqrt{a_{20}^j}\right)^2$$

Note that the first two terms on the *rhs* of Eq. (13), which are first-order, constitute the *zpva* contribution. All the remaining terms are second-order. The same $P^{c(I)}$ expressions may be obtained from the $[\mu^{zpva}]^I$ double power series expansion.

2.2 Connection with Perturbation Theory

There is a straightforward correspondence between the BK perturbation theory formulas and those obtained from the nuclear relaxation/curvature approach. P^{nr} contains the lowest-order BK term of each square bracket type in the expression for P^v, whereas P^c contains the remaining BK P^v terms [1, 23]. As indicated above P^c may be split into two components: P^{zpva} and $P^{c-zpva} = P^c - P^{zpva}$, from which one obtains $P^{c(I)} = [P^{zpva}]^I + P^{c-zpva(I)}$ and $P^{c(III)} = [P^{zpva}]^{III} + P^{c-zpva(III)}$, etc. $P^{c-zpva(I)}$ contains exactly the same type of square bracket terms as P^{nr} except that they are the next higher (non-vanishing) order. $P^{c-zpva(III)}$ contains the next higher-order terms beyond those in $P^{c-zpva(I)}$, etc. For instance, the static γ^{nr}, $\gamma^{c-zpva(I)}$ and $\gamma^{c-zpva(III)}$ are [22]:

$$\text{(15)} \quad \gamma^{nr}_{\alpha\beta\gamma\delta}(0;0,0,0) = [\alpha^2]^0 + [\mu\beta]^0 + [\mu^2\alpha]^I + [\mu^4]^{II}$$

$$\text{(16)} \quad \gamma^{c-zpva(I)}_{\alpha\beta\gamma\delta}(0;0,0,0) = [\alpha^2]^{II} + [\mu\beta]^{II} + [\mu^2\alpha]^{III} + [\mu^4]^{IV}$$

$$\text{(17)} \quad \gamma^{c-zpva(III)}_{\alpha\beta\gamma\delta}(0;0,0,0) = [\alpha^2]^{IV} + [\mu\beta]^{IV} + [\mu^2\alpha]^V + [\mu^4]^{VI}$$

2.3 Dynamic Infinite Optical Frequency Nuclear Relaxation and c-zpva (hyper)Polarizabilities

Under the infinite optical frequency approximation, which corresponds to the limit $\omega \to \infty$, the expression for the dynamic P^{nr} and P^{c-zpva} also can be obtained from the nuclear relaxation/curvature point of view. In general terms the criterion for validity of this approximation is that $(\omega_v/\omega)^2 \ll 1$ for all vibrational frequencies ω_v. For typical laser optical frequencies test calculations [31–33] confirm that the infinite optical frequency approximation is highly accurate, although this does not necessarily hold at lower frequencies [33].

The $\omega \to \infty$ limit of P^{nr} may be obtained by means of the same procedure as in Section 2.1, but expanding the static α^e and β^e as a double power series instead of the electronic energy or μ^e [23, 27]. This leads to expressions for $\beta^{nr}(-\omega;\omega,0)_{\omega\to\infty}$, $\gamma^{nr}(-\omega;\omega,0,0)_{\omega\to\infty}$ and $\gamma^{nr}(-2\omega;\omega,\omega,0)_{\omega\to\infty}$. In the $\omega \to \infty$ limit $\beta^{nr}(-2\omega;\omega,\omega)_{\omega\to\infty}$ and $\gamma^{nr}(-3\omega;\omega,\omega,\omega)_{\omega\to\infty}$ are zero. Properties such as $\beta^{nr}(0;\omega,-\omega)_{\omega\to\infty}$ and $\gamma^{nr}(-\omega;\omega,-\omega,\omega)_{\omega\to\infty}$ cannot be obtained by this procedure. Note that one or more of the higher-order square bracket terms that appear in the static P^{nr} expression will vanish in the infinite optical frequency limit. For instance, $\gamma^{nr}(-\omega;\omega,0,0)_{\omega\to\infty}$ is given by:

$$\text{(18)} \quad \gamma^{nr}_{\alpha\beta\gamma\delta}(-\omega;\omega,0,0)_{\omega\to\infty} = [\alpha^2]^0_{\omega\to\infty} + [\mu\beta]^0_{\omega\to\infty} + [\mu^2\alpha]^I_{\omega\to\infty}$$

Infinite optical frequency values for $P^{c-zpva(I)}$ can be obtained by expanding the static $[\alpha^{zpva}]^I$ and $[\beta^{zpva}]^I$ as a double power series [22]. The treatment is exactly analogous to that used to derive P^{nr} from the electronic α and β. Similarly, the next-highest order nonvanishing zpva corrections, $[\alpha^{zpva}]^{III}$ and $[\beta^{zpva}]^{III}$, yield the infinite optical frequency $P^{c-zpva(III)}$. However, if order III is required it is undoubtedly better to use other methods, to be described later, that give all orders simultaneously.

2.4 Finite Field Approach

The bottleneck in calculating P^{nr} and P^{c-zpva} from analytical expressions is due to the required evaluation of high-order derivatives with respect to the normal modes [34]. This problem can be circumvented by using Finite Field (FF) methods with the nuclear relaxation/curvature approach, which is the major advantage of the

Determination of Vibrational Contributions

latter. The FF calculations involve carrying out a geometry optimization for the molecule in the presence of a static electric field. Because of the shift in geometry the electronic and *zpva* properties will vary as a function of the field. Thus, we can define:

(19) $\quad \Delta P^e = P^e(\mathbf{R_F}, \mathbf{F}) - P^e(\mathbf{R_0}, 0)$

where $\mathbf{F}$ is the field and $\mathbf{R_F}$ is the field-dependent equilibrium geometry. An exactly analogous relation may be written for ΔP^{zpva}. Numerical differentiation of ΔE^e or $\Delta \mu^e$ (ΔE^{zpva} or $\Delta \mu^{zpva}$) with respect to the field yields the static P^{nr} (P^{c-zpva}), whereas numerical differentiation of $\Delta \alpha^e$ and $\Delta \beta^e$ ($\Delta \alpha^{zpva}$ and $\Delta \beta^{zpva}$) leads to the infinite optical frequency approximation for various P^{nr} (P^{c-zpva}) [22, 27]. For example, the first derivative of $\Delta \alpha^e$ with respect to the field gives $\beta^e_{\alpha\beta\gamma}(0; 0, 0) + \beta^{nr}_{\alpha\beta\gamma}(-\omega; \omega, 0)_{\omega \to \infty}$ (see [34] for other processes.) This method is the only feasible way to carry out ab initio calculations for large chemical systems, or for medium size molecules if accurate post-Hartree-Fock methods are applied.

2.4.1 Field-dependent geometry optimization and Eckart conditions

The key step in the FF determination of P^{nr} and P^{c-zpva} is the calculation of the field-dependent equilibrium geometry. In such calculations the field-free Eckart conditions must be enforced in order to prevent molecular reorientation during geometry optimization. Since the Eckart conditions are mass-dependent P^{nr} will exhibit an isotope effect. Although this feature is often not recognized, it is present in both FF and analytical calculations. We have found that the nuclear relaxation isotope effect is comparable to the zero-point vibrational averaging isotope effect, but with a different mass-dependence [35].

2.5 Evaluation of Anharmonicity Contributions not Included in Nuclear Relaxation

The evaluation of P^{c-zpva} using analytical expressions requires the calculation of high-order derivatives. For instance, the expression for the square bracket $[\mu^4]^{IV}$ contains sixth-order derivatives of the electronic energy with respect to the normal modes. That is why such calculations are computationally prohibitive for medium/large organic molecules and can only be done with the FF method. In fact, even the calculation of P^{zpva}, which must be done for $P = \alpha$ and β to obtain the dynamic P^{c-zpva}, can become quite expensive for medium-size or larger molecules. This cost can be dramatically reduced with the aid of field-induced coordinates as we will see in the next section. Finally, one might be tempted to assume that high-order anharmonic terms will make a negligible contribution to vibrational NLO. However, the results we have obtained for typical π-conjugated NLO molecules show that this may not be the case [34].

3. FIELD INDUCED COORDINATES

The main problem in the analytical evaluation of vibrational hyperpolarizabilities for medium size and larger molecules is the large number of nth-order derivatives with respect to normal modes that must be computed. This number is on the order of $(3N-6)^n$ with N being the number of normal modes. The static and infinite optical frequency P^{nr} can be computed using the FF procedure and the same is true of P^{c-zpva} assuming that P^{zpva} is available. However, in order to calculate P^{zpva}, or P^{nr} and P^{c-zpva} at arbitrary frequencies, the BK analytical expressions are currently the only available alternative. It turns out that one can circumvent this difficulty by introducing a set of static field-induced vibrational coordinates which radically reduce the number of nth-order derivatives to be evaluated [33, 35–39].

3.1 Definition and Determination of FICs

In the nuclear relaxation approach one determines the change in the equilibrium geometry induced by a static applied field. This change in geometry constitutes a displacement coordinate, which we call simply a field-induced coordinate (FIC) [35]. Obviously, such a coordinate can be expanded as a power series in the field giving rise to a first-order FIC, second-order FIC, etc. [36]. The first-order FIC contains only harmonic terms, but the higher-order FICs can be broken down further into harmonic and anharmonic components. As we will see below these FICs have remarkable properties. For example, the first-order FIC generated by a longitudinal field contains all the information necessary to compute the nuclear relaxation contribution to the static longitudinal β or the longitudinal dc-P effect. In other words, the perturbation theory expressions containing sums over 3N-6 normal coordinates can be reduced to formulas that involve only a single FIC. Thus, instead of having to determine on the order of $(3N-6)^3$ cubic force constants only a single cubic force constant need be obtained. As one might imagine this opens the possibility for major simplification of the calculations required to evaluate anharmonic contributions to static nuclear relaxation (hyper)polarizabilities. It turns out that the static FICs also yield relevant information to calculate the static P^{zpva} as well as P^{nr} and P^{c-zpva} in the infinite optical frequency approximation. Although the original definition of the FICs was based on static fields, the idea has been extended to the construction of frequency-dependent FICs so that one can account for the frequency dispersion of the vibrational NLO properties as well [33]. The FICs may be determined either analytically or by an FF method.

3.1.1 Finite field determination

In the finite field approach the FICs are generated simply by evaluating numerical derivatives of the change in equilibrium geometry induced by a finite field with respect to the magnitude of that field [35]. However, the FICs may also be determined analytically.

3.1.2 Analytical expressions

The value of the ith field-free normal coordinate displacement induced by a uniform static electric field, F, be written as a power series in the field:

$$(20) \quad Q_i^F(F_x, F_y, F_z) = \sum_\alpha^{x,y,z} \frac{\partial Q_i^F}{\partial F_\alpha} F_\alpha + \frac{1}{2} \sum_{\alpha,\beta}^{x,y,z} \frac{\partial^2 Q_i^F}{\partial F_\alpha \partial F_\beta} F_\alpha F_\beta + \cdots$$

Thus, the static first- and second-order FICs are defined by [36]:

$$(21) \quad \chi_1^\alpha = \sum_{i=1}^{3N-6} \frac{\partial Q_i^F}{\partial F_\alpha} Q_i = -\sum_{i=1}^{3N-6} q_1^{i,\alpha} Q_i \quad \text{and}$$

$$(22) \quad \chi_2^{\alpha\beta} = \frac{1}{2} \sum_{i=1}^{3N-6} \frac{\partial^2 Q_i^F}{\partial F_\alpha \partial F_\beta} Q_i$$

$$= \sum_{i=1}^{3N-6} \left[-q_2^{i,\alpha\beta} + \sum_{j=1}^{3N-6} \frac{a_{21}^{ij,\alpha}}{a_{20}^{ii}} q_1^{j,\beta} - \sum_{j,k=1}^{3N-6} \frac{3 a_{30}^{ijk}}{2 a_{20}^{ii}} q_1^{j,\alpha} q_1^{k,\beta} \right] Q_i$$

In a similar fashion, higher order FICs are defined from the higher order terms of Eq. (20).

As already mentioned the static FICs are useful for determining the static or infinite optical frequency P^{nr}, P^{zpva} and P^{c-zpva}. In order to perform an equivalent simplification for arbitrary frequencies we have also derived frequency-dependent FICs [33]. Their derivation is based on the relationship between the static and dynamic BK vibrational NLO expressions. For instance, the expression for the first-order frequency-dependent FICs is given by:

$$(23) \quad \chi_{1,|\omega|}^\alpha = -\sum_{i=1}^{3N-6} q_{1,|\omega|}^{i,\alpha} Q_i$$

where

$$(24) \quad q_{1,|\omega|}^{i,\alpha} = \frac{a_{11}^{i,\alpha}}{2 a_{20}^{ii} - \omega^2}$$

Notice that $\chi_{1,|\omega|}^\alpha$ correctly reduces to χ_1^α when $\omega = 0$.

3.1.3 Harmonic and anharmonic FICs

The first-order FICs (χ_1^α) depend only on the harmonic parameters a_{11} and a_{20}. On the contrary, the second-order FICs ($\chi_2^{\alpha\beta}$) depend also on the anharmonicity parameters a_{21} and a_{30}. Removing anharmonic terms from the $\chi_2^{\alpha\beta}$ expression one can define the harmonic second-order static FIC:

$$(25) \quad \chi_{2,har}^{\alpha\beta} = \frac{1}{2} \sum_{i=1}^{3N-6} \left(\frac{\partial^2 Q_i^F}{\partial F_\alpha \partial F_\beta} \right)_{har} Q_i = -\sum_{i=1}^{3N-6} q_2^{i,\alpha\beta} Q_i$$

An analogous harmonic expression can be defined for any n^{th}-order FIC.

3.2 Analytical Calculation of Vibrational NLO from FICs

The analytical expressions for P^{nr}, P^{zpva} and P^{c-zpva} can be written in terms of derivatives with respect to FICs rather than normal modes. Both sets of formulas are completely equivalent, although the number of derivatives that appear in the expression in terms of FICs is small and independent of molecular size. For instance, $\beta_{zzz}^{nr}(-\omega;\omega,0)_{\omega\to\infty}$ can be written as a function of a single first-order FIC. The first step in deriving this expression is to construct, in principle, a set of vibrational coordinates which are orthogonal linear combinations of the field-free normal coordinates. This set contains $\phi_1^\alpha = \chi_1^\alpha$ and its orthogonal complement $\{\phi_2, \phi_3, \ldots \phi_{3N-6}\}$. Then,

$$(26) \quad \phi_i = \sum_{j=1}^{3N-6} M_{ij} Q_j$$

where $\mathbf{M}$ is an orthogonal matrix and $M_{1i} = -q_1^{i,z} = \partial Q_i / \partial F_z$. Using the chain rule to express $\partial \alpha_{zz}/\partial Q_i$ in terms of $\partial \alpha_{zz}/\partial \phi_i$ we have:

$$(27) \quad \beta_{zzz}^{nr}(-\omega;\omega,0)_{\omega\to\infty} = \sum_{i=1}^{3N-6} \frac{\partial \alpha_{zz}}{\partial Q_i}\frac{\partial Q_i}{\partial F_z} = \sum_{i,j=1}^{3N-6} \frac{\partial \alpha_{zz}}{\partial \phi_j} M_{ji} M_{1i} = \frac{\partial \alpha_{zz}}{\partial \phi_1}\sum_{i=1}^{3N-6} M_{1i}^2$$

$$= \frac{\partial \alpha_{zz}}{\partial \phi_1}\sum_{i=1}^{3N-6} M_{1i}\frac{\partial Q_i}{\partial F_z} = \frac{\partial \alpha_L}{\partial \phi_1}\frac{\partial \phi_1}{\partial F_{zz}}$$

Whereas the normal mode expression for $\beta_{zzz}^{nr}(-\omega;\omega,0)_{\omega\to\infty}$ requires the evaluation of 3N-6 $\partial \alpha_{zz}/\partial Q_i$ derivatives, the FIC expression requires calculation of just the one property derivative $\partial \alpha_{zz}/\partial \phi_1$.

3.2.1 Nuclear relaxation contribution

For each diagonal component of the static or infinite optical frequency nuclear relaxation (hyper)polarizability tensor only one or two FICs are required. In the case of $\alpha_{\alpha\beta}^{nr}(0;0)$, $\beta_{\alpha\beta\gamma}^{nr}(0;0,0)$, $\beta_{\alpha\beta\gamma}^{nr}(-\omega;\omega,0)_{\omega\to\infty}$ and $\gamma_{\alpha\beta\gamma\delta}^{nr}(-2\omega;\omega,\omega,0)_{\omega\to\infty}$ one needs only χ_1^α; for $\gamma_{\alpha\beta\gamma\delta}^{nr}(-\omega;\omega,-\omega,\omega)_{\omega\to\infty}$ only $\chi_{2,har}^{\alpha\alpha}$; for $\gamma_{\alpha\beta\gamma\delta}^{nr}(-\omega;\omega,0,0)_{\omega\to\infty}$ only χ_1^α and either $\chi_{2,har}^{\alpha\alpha}$ or $\chi_2^{\alpha\alpha}$; and for $\gamma_{\alpha\beta\gamma\delta}^{nr}(0;0,0,0)$ only χ_1^α and $\chi_2^{\alpha\alpha}$ [36]. In order to obtain all components one needs all 3 χ_1^α, or/and all 6 $\chi_{2,har}^{\alpha\beta}$ (or all 6 $\chi_2^{\alpha\beta}$). Finally, for arbitrary frequencies the same type and number of FICs are required as for the corresponding static properties except that the static FICs are replaced by frequency-dependent FICs [33].

3.2.2 zpva contribution

In BK square bracket notation, the first-order zpva correction consists of two terms, i.e. $[P^{zpva}]^I = [P]^{0,1} + [P]^{1,0}$. The term $[P]^{1,0}$ contains the second derivatives $\partial^2 P^e / \partial Q_a^2$ which are evaluated by obtaining the appropriate electric field derivative of the diagonal element of the Hessian [37]. This is done without using FICs. On the other hand, $[P]^{0,1}$ contains first derivatives of the diagonal elements of the Hessian with respect to the set of normal modes and, in that case, FICs can be employed advantageously. Indeed, the expression for $[P]^{0,1}$ may be written as [37]:

$$(28) \quad [P]^{0,1} = -\frac{\hbar}{4} \left(\sum_{i}^{3N-6} \frac{1}{\omega_i} \frac{\partial \omega_i^2}{\partial \chi_P} \right) \frac{\partial P^e / \partial \chi_P}{\omega_{\chi_P}^2}$$

where ω_i and ω_{χ_P} are circular frequencies obtained from the diagonal element of the Hessian defined by Q_i and by the FIC χ_P. If P^e is the dipole moment, then $\chi_P = \chi_1^\alpha$. Similarly, for the polarizability $\chi_P = \chi_{2,har}^{\alpha\beta}$, whereas for the first and second hyperpolarizability $\chi_P = \chi_{3,har}^{\alpha\beta\gamma}$ and $\chi_P = \chi_{4,har}^{\alpha\beta\gamma\delta}$.

3.2.3 c-zpva contribution

The analytical formulas for P^{c-zpva} can also be simplified using FICs. However, the order of the derivatives involved is so high that their evaluation is feasible only for small molecules. For that reason these vibrational contributions are usually evaluated through the FF method of Kirtman, Luis and Bishop [22]. This method utilizes the analytical evaluation of P^{zpva} as described in the immediately preceding sub-section and, thus, the FICs have an important role in decreasing the cost of the calculations [34].

3.3 Reduction in Computational Cost of Vibrational NLO

The fact that the number of FICs needed to compute any vibrational hyperpolarizability does not depend upon the size of the molecule leads to important computational advantages. For instance, the calculation of the longitudinal component of the static γ^{nr} for 1,1-diamino-6,6-diphosphinohexa-1,3,5-triene requires quartic derivatives of the electronic energy with respect to vibrational displacements (i.e. quartic force constants) [34]. Such fourth derivatives may be computed by double numerical differentiation of the analytical Hessian matrix. With normal coordinates it is necessary to compute the Hessian matrix 3660 times, whereas using FICs only 6 Hessian calculations are required.

4. BEHAVIOR OF PERTURBATION SERIES

The convergence behavior of the BK double perturbation series has not been extensively studied. Recently, by using FICs in the FF procedures it has become possible to investigate the initial convergence of the perturbation series for some typical π-conjugated NLO molecules [21, 34]. One obvious way to monitor convergence is by looking at the square bracket terms of successively higher order. However, for quasilinear π-conjugated oligomers, the terms of order I and II are often larger than the zeroth-order double harmonic terms [36]. Further consideration shows that it is more appropriate to monitor the convergence by looking at two separate perturbation sequences.

4.1 Definition of Separate Electronic and zpva Perturbation Sequences

Based on the nuclear relaxation/curvature approach it is natural to divide the total hyperpolarizability into two different perturbation sequences [37]:

(A) $\quad P^e, [P^{zpva}]^{\text{I}}, [P^{zpva}]^{\text{III}}, \ldots$

(B) $\quad P^{nr}, P^{(c-zpva)(\text{I})}, P^{(c-zpva)(\text{III})}, \ldots$

All terms in each sequences are listed in increasing order of perturbation theory. As seen earlier, the successive terms in sequence (B) can be calculated from the effect of nuclear relaxation on the successive terms of sequence (A).

4.2 Initial Convergence of Perturbation Sequences

It is usually found, or assumed, that the perturbation series is, at least initially, convergent [1]. For typical π-conjugated NLO molecules, $[P^{zpva}]^{\text{I}}$ is small in comparison with P^e provided the computations are done at a sufficiently high level of theory (e.g. MP2/6-31+G(d)). However, the same is not always true for the first two terms of sequence (B) [34, 40]. $P^{(c-zpva)(\text{I})}/P^{nr}$ ratios as large as 0.68 have been calculated at the MP2/6-31G level for dynamic hyperpolarizabilities obtained in the infinite optical frequency approximation. For static hyperpolarizabilities larger ratios are found; they sometimes even exceed unity. This divergent behavior does not occur for dynamic properties because some of the large anharmonic perturbation terms contain the optical frequency in the denominator which makes the contribution approach zero as the frequency of the optical field increases.

4.2.1 Treatment of problematic cases

When the mechanical and/or electrical anharmonicity terms are large compared to the terms included in the double harmonic potential one can expect that the BK perturbation treatment will either diverge or converge slowly. For weakly bound systems, such as HF [41, 42] or H_2O [41] dimers, calculations reveal that the perturbation series diverges immediately. More generally, one can anticipate that difficulties will arise whenever there are large amplitude vibrations, such as a low frequency torsional mode. Even in less obvious cases, like some of the π-conjugated NLO molecules referred to above, problematic behavior can occur. In all these instances the approaches discussed thus far cannot be used in their current form. As an alternative, a new FF procedure for calculating vibrational (hyper)polarizabilities has been developed [43]. Instead of relying on the usual perturbation expansion, this procedure utilizes a self-consistent mean field solution of the vibrational Schrödinger equation. On top of that, the instantaneous mode-mode coupling may be taken into account by many-body perturbation theory, coupled cluster or configuration interaction techniques.

5. VARIATIONAL TREATMENT

As noted in Section 4 the BK perturbation series is not always well-behaved. Poor convergence behavior can, and does, occur when the anharmonicity is large. One must be particularly alert to this possibility in the case of van der Waals molecules or, in general, molecules that have large amplitude (low frequency) vibrations. In such circumstances, P^{nr} is not sufficient to accurately describe vibrational (hyper)polarizabilities and the addition of first-order *zpva* and *c-zpva* contributions is likely not to be sufficient either. Then, we need to focus on the entire curvature contribution which, in turn, is obtained from the entire zpva correction to the energy and electrical properties (see Section 3).

5.1 VSCF Method

In recent years there has been growing interest in extending the methodology of electronic structure calculations to vibrational problems. The starting point is a self-consistent-field (SCF) treatment of the coupling between normal modes leading to the vibrational SCF (VSCF) method [44–47]. Whereas electronic coupling is limited to pair interactions, there may be terms in the vibrational potential that couple all 3N-6 normal modes. However, one may expect the importance of the coupling to diminish rapidly as the number of coupled modes increases. Computer codes now exist for VSCF calculations that include coupling up to four modes at a time [48], which should almost always be adequate. By solving the VSCF equations numerically one accounts for intramode anharmonicity completely (see further below) and mode-mode anharmonic coupling within a mean field approximation. This approach has the advantage that it does not rely on an expansion in orders of perturbation theory. The treatment of mode mode coupling can subsequently be improved by applying various post-SCF methods just as in the case of the electronic structure problem.

As discussed in Section 2, the lowest-order effects due to anharmonicity are included as part of the nuclear relaxation contribution to vibrational NLO properties, whereas all other anharmonic effects are part of the curvature term. Assuming that nuclear relaxation can be evaluated as described above we focus here on curvature, which is determined by the zero-point vibrationally averaged properties, at least in the static and infinite optical frequency limits. Given the potential energy surface (PES), solution of the VSCF equations yield the mean field approximation for the zero-point vibrational energy. If the zero-point energy calculation is repeated in the presence of electric fields of different magnitude (but fixed direction), then numerical differentiation provides the *c-zpva* static electrical properties in the direction of the field. Note that, in principle, *all* anharmonic effects beyond lowest-order are thereby taken into account. A first set of calculations based on the VSCF approach has very recently been reported for a few small molecules [43]. In principle, a similar treatment may be carried out with the PES replaced by the dipole moment or (hyper)polarizability surface. For electrical property surfaces the zero-point average would be obtained by numerical integration using the VSCF wavefunction. The dipole moment yields the same information as the potential energy (with one less numerical differentiation required) whereas α and β give the infinite optical frequency vibrational hyperpolarizabilities.

In order to carry out a VSCF treatment a PES or corresponding property surface is required. For the VSCF and VMP2 methods an 'exact' (up to a given order in the number of coupled modes) numerical *ab initio* PES can be used. This is a very important advantage with respect to the BK formulation which relies on a power series expansion in the normal coordinates that is truncated at some relatively low order. Other post-SCF procedures discussed below currently use a normal coordinate expansion as well. This tremendously simplifies the calculations but may also introduce significant errors as found in trial VMP2 calculations [43]. In cases where there is a known large amplitude motion not along a normal coordinate, the expansion can be carried out at individual points along the 'reaction path' for that particular motion [49, 50].

5.2 Post-VSCF Methods

All of the methods available for improving the electronic SCF treatment are, in principle, available for improving vibrational wavefunctions and properties. The lowest level post-VSCF treatment is VMP2. A few VMP2 calculations have now been carried out using the 'exact' numerical PES with up to 3-mode coupling terms included [43]. In doing so it was found that the standard methodology employed in the GAMESS quantum chemistry program had to be modified (for both VMP2 and VSCF) in order to obtain satisfactory electric field derivatives. The VMP2 results turn out to be adequate as long as the anharmonic mode-mode coupling is not too large. No higher level treatments have been carried out for the 'exact' PES, although full vibrational CI (FVCI) calculations were done for a PES truncated at the quartic terms and, in some instances, further modified to ensure proper behavior of the potential. The key to extending the VCI [51–53] calculations so as to avoid truncation lies in finding a convenient way to truncate the number of states that are included without, of course, losing significant accuracy. Some progress along these lines has been recently been achieved in connection with the related problem of calculating Franck-Condon factors [54]. Other approaches that may eventually prove fruitful are coupled cluster methods (VCC) [55] and, especially for quasilinear molecules, the renormalization group method [56, 57].

6. BASIS SET REQUIREMENTS AND TREATMENT OF ELECTRON CORRELATION

6.1 Basis Set Requirements

The basis set that should be used for a given calculation depends, of course, on the desired accuracy. It also depends on the property, the nature of the system, and the level of treatment. For NLO applications π-conjugated organic molecules are of particular interest. These molecules are usually extended spatially in one dimension and the longitudinal component of the (hyper)polarizability tensor tends to be dominant. As a general rule, it has been found that the atomic basis necessary to achieve a given accuracy is smaller for the longitudinal electronic properties in such systems than it is for the perpendicular properties (or for the properties of more compact molecules) [58, 59]. The justifying argument is that basis functions on neighboring atoms compensate for deficiencies on any one atom. Other rough general rules for the electronic properties are that basis set requirements increase with the order of the property and the level of correlation treatment. For vibrational (hyper)polarizabilities only limited studies on medium-size organic molecules have been carried out [21], but the same generalizations appear to be valid. In addition, it seems that the basis set requirements increase with the order of anharmonicity. In the vibrational (hyper)polarizability studies it turns out that the 6-31G basis, often used in NLO calculations, does not always provide even qualitative accuracy. On the other hand, in almost all instances the 6-31+G(d) basis has given adequate results (10% accuracy) at the MP2 level for the total contribution and for each of the individual terms.

6.2   Treatment of Electron Correlation by *ab initio* Methods

The effect of electron correlation on electronic and vibrational (hyper)polarizabilities can be quite important. Typical increases or decreases in individual contributions to β and γ are about a factor of two, but range up to an order of magnitude [21, 34, 60, 61]. They originate indirectly from the change in geometry and directly from the change in the electronic charge distribution. In π-conjugated organic molecules a large geometry effect can occur because the properties are very sensitive to the bond length alternation. As compared to QCISD, the MP2 method adequately reproduces the effect of correlation on the relative magnitude of the various vibrational (and electronic) contributions to each property and, in addition, gives a reasonable prediction for the individual terms. However, higher level methods are necessary to obtain quantitative estimates for the latter. For medium-size molecules ($\sim$10 first row atoms) such calculations are feasible by taking advantage of field-induced coordinates [36, 37]. However, for the larger systems that are often of interest as NLO materials even an MP2 treatment may prove quite tedious. This suggests that it might be worthwhile to consider DFT as an alternative.

6.3   The DFT Alternative

It is now well-established [62–68] that there is a fundamental problem with DFT electrical property calculations done on extended chains using conventional functionals. Results obtained overshoot the correct values and the error grows dramatically, especially for nonlinear polarizabilities, as the chain length increases. The error has been traced to a poor treatment of electron exchange [63, 69, 70]; the DFT exchange potential should produce a counteracting field but conventional potentials do not. Although some progress has made at solving this problem by introducing the current density as an additional variable [66–68] the most promising approach at this time appears to be through the use of an optimized effective potential (OEP) [71, 72]. An efficient OEP procedure that has recently been developed [73, 74] has now been applied for exact exchange (OEP-EXX) [70, 75] and been shown to largely eliminate the dramatic overestimation problem. Of course, there remains a significant correlation effect which is not accounted for by this procedure. Unfortunately, conventional DFT correlation functionals are unsuitable for such purposes since their effect on calculated electrical properties is minimal. However, OEP-EXX does open the door for incorporating improved functionals in an efficient DFT procedure that will avoid much more costly *ab initio* treatments. Work is in progress towards that end.

7. RESONANT NLO PROPERTIES

In this section we consider the role of molecular vibrations in resonant L&NLO properties. These properties govern the intensity of light absorption (or emission) accompanying the transition between two vibronic energy levels. Of interest here is one- and two-photon absorption (OPA and TPA). Applications of TPA, in particular, include three-dimensional optical data storage and photodynamic therapy.

7.1 General Theory of OPA and TPA

We focus on absorption from the ground electronic state to some excited electronic state Λ. The OPA intensities are determined by the imaginary part of the linear polarizability whereas the TPA intensities are governed by the imaginary part of the second hyperpolarizability expression for the intensity-dependent refractive index (i.e. $\text{Im}\gamma_{\alpha\beta\gamma\delta}(-\omega;\omega,-\omega,\omega)$). In the former case the intensity will be significant only when the incident photon frequency (ω) coincides, or nearly coincides, with the energy difference between the ground state and the excited state Λ. For TPA the energy difference must nearly coincide with 2ω in order to have simultaneous absorption of two photons.

Expressions for vibrational OPA [76] and TPA [77–86] may be derived starting from the same SOS vibronic (hyper)polarizability formula used by BK for non-resonant processes except that now we are specifically interested in the imaginary component. Thus, using the linear polarizability as an example, we have:

$$(29) \quad \text{Im}\alpha_{\alpha\beta}(-\omega;\omega) = \hbar^{-1} {\sum_{K,k}}' \left(\frac{\langle 0,0|\hat{\mu}_\alpha|K,k\rangle\langle k,K|\hat{\mu}_\beta|0,0\rangle}{(\omega_{kK} - i\Gamma_{kK}/2 - \omega)} + \frac{\langle 0,0|\hat{\mu}_\alpha|K,k\rangle\langle k,K|\hat{\mu}_\beta|0,0\rangle}{(\omega_{kK} + i\Gamma_{kK}/2 + \omega)} \right)$$

Determination of Vibrational Contributions

where the primes indicate exclusion of the $|0, 0\rangle$ state in the sum over all vibronic states. The phenomenological damping terms $i\Gamma_{kK}/2$ that appear here allow us to deal with the resonant and near-resonant region, where singularities would otherwise occur. At or near a resonance there will be one term in the SOS for which $\omega_{kK} \approx \omega (K = \Lambda)$ while all other terms are considered to be negligible [77].

In the case of TPA we use the analogous expression for $\mathrm{Im}\gamma_{\alpha\beta\gamma\delta}(-\omega; \omega, -\omega, \omega)$ and the corresponding resonant condition is $\omega_{kK} \approx 2\omega (K = \Lambda)$. After eliminating the non-resonant states in the sum over K, we proceed as in the BK treatment of non-resonant NLO. Thus, for all remaining intermediate states, except those involving the electronic ground state, we use the clamped nucleus approximation for the energy denominators. This is valid for TPA as long as there are no electronic one-photon resonances or near-resonances at the optical frequency ω. If such resonances do occur, they require a treatment similar to ordinary OPA. That leaves a sum over vibrational states associated with the ground electronic state and with the two-photon resonant state. The imaginary part of this sum is the vibrational TPA. Using a curly bracket notation analogous to the square brackets of the BK treatment one obtains [77]:

$$(30) \quad \mathrm{Im}\gamma^v_{\alpha\beta\gamma\delta}(-\omega; \omega, -\omega, \omega) = \{\alpha^2\} + \{\mu^2\alpha\} + \{\mu^4\}$$

Note that, in contrast with the corresponding non-resonant process, there is no $\{\mu\beta\}$ term. Furthermore, in addition to ground state electrical properties the curly bracket expressions also contain transition dipole moments and polarizabilities. For example,

$$(31) \quad \{\mu^4\} = P_{\alpha\gamma}P_{\beta\delta}\frac{\hbar\Gamma}{2}\sum_{l,k,m}{}' \left\{ \begin{array}{c} \frac{\langle 0^0|\mu_\alpha^{00}(\mathbf{R})|k^0\rangle\langle k^0|\mu_\gamma^{0\Lambda}(\mathbf{R})|l^\Lambda\rangle}{\hbar^3\left((\omega_{l\Lambda}-2\omega)^2+\Gamma^2/4\right)(\omega_{k0}-\omega)} \\ \times \frac{\langle l^\Lambda|\mu_\beta^{\Lambda 0}(\mathbf{R})|m^0\rangle\langle m^0|\mu_\delta^{00}(\mathbf{R})|0^0\rangle}{\hbar(\omega_{m0}-\omega)} \end{array} \right\}$$

where the superscripts 0 or Λ have been used to indicate the electronic state.

7.2 Perturbation Treatment

As in the BK treatment of non-resonant properties double perturbation theory may be used to evaluate the curly bracket quantities. The curly brackets are more difficult to evaluate than square brackets due to the presence of transition, as well as ground state, electrical properties.

7.2.1 Overall procedure

As in the BK procedure, the electrical properties and the potential energy surface may be expanded as a Taylor series in the normal coordinates. Orders of perturbation theory are defined in the same way as for the non-resonant case. Electrical property terms that are quadratic, cubic, ... in the normal coordinates are taken to be first-order, second-order, ... ; terms in the potential energy function that are cubic, quartic, ... are defined to be first-order, second-order, ... For the electrical properties it is convenient to employ the electronic ground state normal coordinates in the expansion. Then the transition dipole moments and polarizabilities can be written as a sum of terms each of which involves an easily evaluated normal coordinate integral multiplied by a Franck-Condon (FC) integral. Hence, the calculation of FC integrals is the key step in the procedure.

7.2.2 Evaluation of FC and related integrals

In the literature one can find several methodologies to evaluate FC integrals at the harmonic level. Most of them [87–95] are based on the generating function approach of Sharp and Rosenstock [96], or the recursion relations of Doctorov, Malkin, and Man'ko [97]. However, including vibrational anharmonicity implies a large increase in computational cost and in the complexity of formulation. Current methods for including anharmonicity are either limited in practice to triatomic molecules [98–101] or assume separability of the normal modes, which means that only diagonal (in the normal modes) anharmonicity terms are taken into account [102, 103]. Recently, a simple new method has been developed for calculating accurate FC factors including non-diagonal (i.e. mode-mode) anharmonic coupling and Duschinsky rotations [104]. This method was used to successfully simulate the $C_2H_4^+ \tilde{X}^2 B_{3u} \leftarrow C_2H_4 \tilde{X}^1 A_g$ band in the photoelectron spectrum [54].

In formulating this new methodology one begins by taking the difference of the Schrödinger equations for nuclear motion in the ground and excited electronic states. Then, using the Hermitian property of the vibrational Hamiltonian, it is easy to show that [104]:

$$(32) \quad \left\langle \psi_{v_g}^g \middle| \hat{H}^g - \hat{H}^e \middle| \psi_{v_e}^e \right\rangle = \left(E_{v_g}^g - E_{v_e}^e \right) S_{v_g v_e}$$

In Eq. (32) $\hat{H}^g$, $\psi_{v_g}^g$ and $E_{v_g}^g$ are the vibrational Hamiltonian, wavefunction and energy of the ground electronic state; $\hat{H}^e$, $\psi_{v_e}^e$ and $E_{v_e}^e$ are their counterparts for the electronic excited state; and $S_{v_g v_e} = \left\langle \psi_{v_g}^g \middle| \psi_{v_e}^e \right\rangle$ is an FC overlap integral. Upon expanding $\psi_{v_g}^g$ in the complete set of vibrational eigenfunctions for the excited electronic state, and recognizing that the total nuclear kinetic energy operator is the same for $\hat{H}^g$ and $\hat{H}^e$, one obtains a set of homogenous linear simultaneous equations for a given v_g:

$$(33) \quad \sum_{\mu_e} S_{v_g \mu_e} \left[\left\langle \psi_{\mu_e}^e \middle| \hat{V}^g - \hat{V}^e \middle| \psi_{v_e}^e \right\rangle + \left(E_{v_e}^e - E_{v_g}^g \right) \delta_{\mu_e v_e} \right] = 0, \forall v_e$$

Eq. (33), together with the usual normalization condition, can readily be solved in principle for the FC S_{v_g} vector. In practice, however, this is an infinite set of equations that must be truncated (see further below) to obtain a solution.

In order to incorporate Duschinsky rotations and anharmonicity the potential energy difference $\hat{V}^g - \hat{V}^e$ is expanded as a power series in the excited electronic state normal coordinates, $\mathbf{Q}^e$ using the familiar relation $\mathbf{Q}^g = \mathbf{J}\mathbf{Q}^e + \mathbf{K}$ between the ground state normal coordinates, $\mathbf{Q}^g$, and $\mathbf{Q}^e$. Here $\mathbf{J}$ is the Duschinsky rotation matrix given by $\mathbf{J} = \mathbf{L}^{g\dagger}\mathbf{L}^e$ and $\mathbf{K} = \mathbf{L}^{g\dagger}\mathbf{R}$ arises because of the shift in equilibrium geometry. Thus $\mathbf{R}$ is the difference between the ground state equilibrium geometry (in mass-weighted Cartesian coordinates) and the excited state equilibrium geometry, while $\mathbf{L}^g$ and $\mathbf{L}^e$ are the unitary matrices that transform from mass-weighted Cartesian displacement coordinates to normal coordinates (i.e. $\mathbf{Q}^g = \mathbf{L}^{g\dagger}\mathbf{X}^g$ and $\mathbf{Q}^e = \mathbf{L}^{e\dagger}\mathbf{X}^e$). The effect of anharmonicity can be accounted for by perturbation theory. For instance, the first-order perturbation equation is given by:

$$(34) \quad \sum_{\mu_e} S^{(1)}_{v_g \mu_e} \left[\left\langle \psi^e_{\mu_e} | \hat{V}^g - \hat{V}^e | \psi^e_{v_e} \right\rangle + \left(E^e_{v_e} - E^g_{v_g} \right) \delta_{\mu_e v_e} \right]^{(0)}$$

$$+ \sum_{\mu_e} S^{(0)}_{v_g \mu_e} \left[\left\langle \psi^e_{\mu_e} | \hat{V}^g - \hat{V}^e | \psi^e_{v_e} \right\rangle + \left(E^e_{v_e} - E^g_{v_g} \right) \delta_{\mu_e v_e} \right]^{(1)} = 0$$

In Eq. (34) first-order corrections to the vibrational wavefunctions and energies are determined by the terms in $\hat{V}^g$ and $\hat{V}^e$ that are cubic in the normal coordinates. This equation is solved imposing the first-order normalization condition (i.e. $\mathbf{S}^{(1)\dagger}_{v_g} \mathbf{S}^{(0)}_{v_g} = 0$).

The most critical step in this new procedure is the truncation of the vibrational basis set for the excited electronic state. This basis set must contain all functions necessary to obtain accurate FC factors, but must also be small enough for the calculations to be efficient. The algorithm used involves an iterative build-up of the basis set by increasing the range of vibrational quantum numbers while, simultaneously, removing unnecessary functions [54].

7.3 Future Variational Treatment

Initial results obtained for TPA and for photoelectron spectra of small systems, show that anharmonicity must be included in the calculation of FC factors to reproduce experiment [54, 77, 104]. However, it is difficult to treat larger anharmonic systems by means of perturbation theory. Such systems can be handled by applying the variation/perturbation methods of electronic structure theory that have been, and continue to be, extended to the vibrational Schrödinger equation as discussed earlier. The FC integrals that appear in the equations for resonant (hyper)polarizabilities may be calculated employing approaches like VSCF, VMP2, VCI and VCC. That will allow us to include anharmonic contributions to all orders and thereby remove the intrinsic limitations of the perturbation expansion in terms of normal coordinates.

8. FROM MOLECULES TO INFINITE PERIODIC POLYMERS

8.1 Finite Oligomer Method

In the quest for systems having large NLO properties one is often interested in quasilinear stereoregular pi-conjugated polymers. Although such polymers in practice have a finite conjugation length, the infinite chain limit is a useful model.

Early calculations for infinite polymers were based on extrapolating the properties of finite oligomer chains. However, extrapolation can be difficult because hyperpolarizabilities converge slowly with chain length and the functional form for the dependence on the number of repeat units is unknown. In dealing with these problems perhaps the most important observation to make is that defining the property per repeat unit as $P(N) - P(N-1)$ leads to much more rapid convergence than the alternative, i.e. $P(N)/N$ [105, 106]. This is because chain end effects are largely cancelled in the former expression. In practice, however, it may not be possible in some cases to use that expression when numerical errors lead to erratic behavior as a function of N. Another useful 'trick' is illustrated by the following example. In calculations on linear polyene chains [107] it has been found that the longitudinal γ converges slowly with chain length at both the HF and MP2 levels. The ratio of the HF and MP2 values, on the other hand, converges a lot more rapidly. Thus, if HF calculations are feasible on chains of sufficient length to obtain a reliable extrapolation to the HF polymer result, then far shorter chains suffice to determine the corresponding MP2 value. This is fortunate considering the much longer computation times required for MP2, as opposed to HF, calculations. The ratio method just described is a general approach that has been successfully applied in a number of different circumstances.

As far as the extrapolations *per se* are concerned, a reasonably satisfactory 'stability' procedure has been developed [105, 108]. This procedure utilizes a variety of *ad hoc* fitting functions and data subsets to obtain a large number of extrapolated values. Then, those values that show large variations with respect to systematic extension of the fitting functions and/or data sets are eliminated leaving a reduced set. From the latter one can obtain both a mean value for the infinite chain property and an error estimate. There is also a mathematically more rigorous approach to extrapolation that is based on the use of sequence transformations to accelerate convergence [109]. Although sequence transformations are promising, further studies are necessary since they have been applied only to the field-free problem. Finally, we mention a novel Hartree-Fock 'elongation' algorithm [110–114] currently under development. This algorithm makes finite oligomer calculations feasible for longer chains than could otherwise be done by virtue of a linear scaling technique whereby monomers are successively added to a growing chain.

8.2 Periodic Boundary Conditions and Electric Field Polarization

For a long time the *finite oligomer* approach was the only method available for determining linear and nonlinear polarizabilities of infinite stereoregular polymers. Recently, however, the problem of carrying out electronic band structure (or crystal orbital) calculations in the presence of static or frequency-dependent electric fields has been solved [115, 116]. A related discretized Berry phase treatment of *static* electric field polarization has also been developed for 3D solid state systems

[117–120]. Difficulties arise in the band structure treatment for quasilinear periodic chains because the scalar dipole interaction potential is neither periodic nor bounded. These difficulties are overcome in the approach presented in [115] by using the time-dependent vector potential, A, instead of the scalar potential. In that formulation the momentum operator p is replaced by $\pi = p + (e/c)A$ while the corresponding quasi-momentum k becomes $\kappa = k + (e/c)A$. Then, a proper treatment of the time-dependence of κ, leads to the time-dependent self-consistent field Hartree-Fock (TDHF) equation [115]:

$$(35) \quad \left[\mathbf{F}(k) - i\mathbf{S}(k)\frac{\partial}{\partial t}\right]\mathbf{C}(k) + e\mathbf{E}(t)\left[\mathbf{M}(k) + i\mathbf{S}(k)\frac{\partial}{\partial k}\right]\mathbf{C}(k) = \mathbf{S}(k)\mathbf{C}(k)\varepsilon(k)$$

where $\mathbf{E}(t)$ is the time-dependent electric field. In Eq. (35) $\mathbf{F} = \mathbf{h} + \mathbf{D}(\mathbf{E})[2\mathbf{J} - \mathbf{K}]$, $\mathbf{S}$, $\mathbf{C}$ and $\boldsymbol{\varepsilon}$ are the usual k-dependent Fock, overlap, orbital coefficient and Lagrangian multiplier matrices. The term proportional to $\mathbf{E}$ on the *lhs* is the replacement for the usual scalar potential used in finite chain calculations. From the formulas given in [115] $\mathbf{M}(k)$ can be immediately recognized as a sawtooth dipole potential term, whereas the $\partial/\partial k$ term is associated with a flow of charge through the system. Taken together the two terms in square brackets represent the polarization of the infinite polymer induced by the field. In solving Eq. (35) by perturbation theory the non-canonical method (i.e. non-diagonal ε) due to Karna and Dupuis is utilized [123]. An exactly analogous treatment is applicable to Kohn-Sham density functional theory (KS-DFT) with the Fock matrix replaced by the corresponding KS matrix. However, as discussed above, KS-DFT cannot be recommended for this purpose until a satisfactory correlation potential is developed. Although it is clear that a time-dependent MP2 formulation can be developed along the same lines as above, the relevant equations have not been presented as yet.

8.3 Perturbation Treatment of Electric Fields

The quantities $\mathbf{F}(k)$, $\mathbf{C}(k)$ and $\boldsymbol{\varepsilon}(k)$ all depend upon the electric field(s). By expanding both sides of Eq. (35) as a power series in the field(s) and, then, equating terms of like power and frequency-dependence one obtains crystal orbital perturbation equations that are analogous to those of conventional (non-canonical) TDHF theory for finite molecules [123]. In analogy with that conventional treatment, the electrical properties can be obtained to all orders by evaluating the average value of the polarization term (i.e. the second term in square brackets on the *lhs* of Eq. (35)). Finally, by suitable algebraic manipulations one can derive a crystal orbital $2n+1$ rule that is analogous to the molecular $2n+1$ rule.

In zeroth-order ε is chosen to be diagonal and the result is the ordinary field-free Hartree-Fock crystal orbital equation. As usual $\mathbf{C}(k)$ is complex for arbitrary k. Apart from this aspect the most important difference between the crystal orbital and molecular TDHF perturbation equations is the presence of the $\partial \mathbf{C}/\partial k$ term in the former. Since $\partial \mathbf{C}/\partial \mathbf{k}$ is multiplied by $\mathbf{E}$, field-free derivatives of $\mathbf{C}(k)$ with respect to k appear for the first time in the first-order perturbation equations. These field-free derivatives can be obtained, for the most part, by analytically differentiating $S(k)$ and the zeroth-order $F(k)$ with respect to k. However, as shown in [116] $\partial C/\partial k$ also depends upon the choice made for the arbitrary phase factors that multiply the orbital eigenvectors. The corresponding phase angles must vary continuously with k in order to have well-defined derivatives. Although this requirement will not automatically be fulfilled in a crystal orbital calculation, a satisfactory (but not unique) procedure for imposing continuity has been provided in [116]. Even so, the zeroth-order phase angle is inherently determined only up to an arbitrary integer multiple of ka where a is the unit cell length. Beyond zeroth-order, perturbation corrections to $\partial C/\partial k$ can be handled in a normal manner without introducing any further ambiguity. The arbitrariness in the phase angle does not affect calculated (hyper)polarizabilities but the calculated dipole moment may differ from the true value by an integer multiple of a. Fortunately, the dipole moment is normally known well enough to eliminate the unintended contribution. This analysis accounts for the behavior of the calculated dipole moment in infinite polymers that had previously not been understood.

8.4 Finite Fields

In order to determine vibrational NLO properties efficiently it is necessary to carry out finite field geometry optimizations as we have seen. In principle, Eq. (35) can be used directly for this purpose. There are, however, practical considerations related to convergence of the self-consistent field (SCF) iterations. The most obvious iterative sequence is: (i) determine the zero-field solution; (ii) evaluate $\partial C/\partial k$; (iii) substitute $\partial C/\partial k$ from the previous step into the TDHF equation; (iv) solve for $C(k)$ and return to step (ii); etc. until convergence is achieved. In order to carry out step (iv) the normalization condition $C^\dagger SC = 1$ may be used to write $\partial C/\partial k = [(\partial C/\partial k)C^\dagger S]C$. Then the multiplicative form of the field-free equation is preserved and the polarization matrix will remain Hermitian for all iterations. Investigations are underway to test the convergence properties of the above iterative sequence and to determine how the convergence properties depend upon the magnitude of the field as well as the number of k-points that are sampled in the band structure treatment.

9. WHICH IS THE BEST METHOD FOR CALCULATING VIBRATIONAL CONTRIBUTIONS TO NONLINEAR OPTICAL PROPERTIES?

There are three main factors to consider in deciding on the best method to compute vibrational NLO properties: i) the frequency of the optical fields, ii) the size of the chemical system, and iii) the anharmonicity of the (static) field-dependent PES. When the frequencies of the optical fields are comparable to vibrational frequencies, the only option is the BK procedure presented in Section 1. Nevertheless, the computational cost of such calculations can be radically reduced by combining BK expressions with frequency-dependent field-induced coordinates (see Section 3).

In the static limit or at visible/near UV wavelengths, the nuclear relaxation/curvature procedure is suitable (see Section 2), but the BK method may still be the best choice if the molecule has less than a total of 20 atoms and the PES is nearly harmonic. For larger chemical systems the only feasible methods are the nuclear relaxation/curvature procedures based on the Finite Field technique and field-induced coordinates. These procedures are also preferred for small chemical systems when the anharmonicity is such that one has to go beyond the lowest-order terms. In general, this will occur whenever there are large amplitude vibrational motions and/or large electrical anharmonicities. Under such circumstances it may happen that the perturbation treatment is non-convergent (see Section 4), in which case the Finite Field technique must be combined with vibrational SCF and post-SCF methods to obtain the required zero-point vibrational average energies and electrical properties (see Section 5). Currently all approximate procedures for calculating the vibrational contribution to resonant NLO properties are based on perturbation theory (see Section 7). In this case, however, the PES of a second state is also involved. Hence, the likelihood of convergence problems is greater than for non-resonant processes. This is particularly so when evaluating Franck-Condon, and related, factors associated with highly excited vibrational states. Finally, for reasonable accuracy it is advisable to use MP2/6-31+G or a higher level treatment to evaluate the PES (see Section 6). Conventional DFT should not be employed for longitudinally extended systems pending the design of a satisfactory potential for electron correlation to accompany an optimized exchange potential.

Infinite stereoregular polymers require special consideration (see Section 8). Band structure techniques that parallel the methodologies formulated for ordinary molecules are evolving rapidly. However, the Finite Field procedures need further development and we recommend that, for the time being, extrapolation of finite oligomer calculations should be employed.

REFERENCES

[1] Bishop, D.M.: Rev. Mod. Phys. **62**, 343 (1990)
[2] Bishop, D.M.: Adv. Chem. Phys. **104**, 1 (1998)
[3] Kirtman, B., Champagne, B., Luis, J.M.: J. Comp. Chem. **21**, 1572 (2000)
[4] Champagne, B., Kirtman, B.: In: Nalwa, H.S. (ed.), Handbook of Advanced Electronic and Photonic Materials, Vol. 9, Chap. 2, p. 63. Academic, San Diego CA (2001)
[5] Bishop, D.M., Norman, P.: In: Nalwa, H.S. (ed.), Handbook of Advanced Electronic and Photonic Materials, vol. 9, Chap. 2, p. 1. Academic, San Diego CA (2001)
[6] Macak, P., Luo, Y., Norman, P., Ågren, H.: J. Chem. Phys. **113**, 7055 (2000)
[7] Jug, K., Chiodo, S., Calaminici, P., Avramopoulos, A., Papadopoulos, M.G.: J. Phys. Chem. **A107**, 4172 (2003)
[8] Jacquemin, D., André, J.M., Champagne, B.: J. Chem. Phys. **118**, 3956 (2003)
[9] Avramopoulos, A., Reis, H., Li, J., Papadopoulos, M.G.: J. Am. Chem. Soc. **126**, 6179 (2004)
[10] Ormendizo, P., Andrade, P., Aragão, A., Amaral, O.A.V., Fonseca, T.L., Castro, M.A.: Chem. Phys. Lett. **392**, 270 (2004)
[11] Acebal, P., Blaya, S., Carretero, L.: Chem. Phys. Lett. **382**, 489 (2003)
[12] Saal, A., Ouamerali, O.: Int. J. Quantum Chem. **96**, 333 (2004)
[13] Alparone, A., Millefiori, A., Millefiori, S.: J. Mol. Struct. (Theochem) **640**, 123 (2003)

[14] Squitieri, E.: J. Phys. Org. Chem. **17**, 131 (2004)
[15] Orr, B.J., Ward, J.F.: Mol. Phys. **20**, 513 (1971)
[16] Bishop, D.M., Kirtman, B., Champagne, B.: J. Chem. Phys. **107**, 5780 (1997)
[17] Bishop, D.M., Kirtman, B.: J. Chem. Phys. **95**, 2646 (1991)
[18] Bishop, D.M., Kirtman, B.: J. Chem. Phys. **97**, 5255 (1992)
[19] Bishop, D.M., Luis, J.M., Kirtman, B.: J. Chem. Phys. **108**, 10013 (1998)
[20] Perpète, E.A., André, J.-M., Champagne, B.: J. Chem. Phys. **109**, 4624 (1998)
[21] Torrent-Sucarrat, M., Solà, M., Duran, M., Luis, J.M., Kirtman, B.: J. Chem. Phys. **120**, 6346 (2004)
[22] Kirtman, B., Luis, J.M., Bishop, D.M.: J. Chem. Phys. **108**, 10008 (1998)
[23] Luis, J.M., Martí, J., Duran, M., Andrés, J.L., Kirtman, B.: J. Chem. Phys. **108**, 4123 (1998)
[24] Luis, J.M., Duran, M., Andrés, J.L.: J. Chem. Phys. **107**, 1501 (1997)
[25] Papadopoulos, G., Willets, A., Handy, N.C., Underhill, A.E.: Mol. Phys. **88**, 1063 (1996)
[26] Champagne, B., Vanderheoven, H., Perpète, E.A., André, J.M.: Chem. Phys. Lett. **248**, 301 (1996)
[27] Bishop, D.M., Hasan, M., Kirtman, B.: J. Chem. Phys. **103**, 4157 (1995)
[28] Cohen, M.J., Willets, A., Amos, R.D., Handy, N.C.: J. Chem. Phys. **100**, 4467 (1994)
[29] Andrés, J.L., Bertrán, J., Duran, M., Martí, J.: J. Phys. Chem. **98**, 2803 (1994)
[30] Adamowicz, L., Bartlett, R.J.: J. Chem. Phys. **84**, 4988 (1986)
[31] Bishop, D.M., Dalskov, E.K.: J. Chem. Phys. **104**, 1004 (1996)
[32] Quinet, O., Champagne, B.: J. Chem. Phys. **109**, 10594 (1998)
[33] Luis, J.M., Duran, M., Kirtman, B.: J. Chem. Phys. **115**, 4473 (2001)
[34] Torrent-Sucarrat, M., Solà, M., Duran, M., Luis, J.M., Kirtman, B.: J. Chem. Phys. **116**, 5363 (2002)
[35] Luis, J.M., Duran, M., Andrés, J.L., Champagne, B., Kirtman, B.: J. Chem. Phys. **111**, 875 (1999)
[36] Luis, J.M., Duran, M., Champagne, B., Kirtman, B.: J. Chem. Phys. **113**, 5203 (2000)
[37] Luis, J.M., Champagne, B., Kirtman, B.: Int. J. Quantum Chem. **80**, 471 (2000)
[38] Del Zoppo, M., Castiglioni, C., Veronelli, M., Zerbi, G.: Synth. Met. **57**, 3919 (1993)
[39] Castiglioni, C., Del Zoppo, M., Zerbi, G.: J. Raman Spectrosc. **24**, 485 (1993)
[40] Reis, H., Papadopoulos, M.G., Avramopoulos, A.: J. Phys. Chem. **A107**, 3907 (2003)
[41] Eckart, U., Sadlej, A.J.: Mol. Phys. **99**, 735 (2001)
[42] Bishop, D.M., Pipin, J., Kirtman, B.: J. Chem. Phys. **102**, 6778 (1995)
[43] Torrent-Sucarrat, M., Luis, J.M., Kirtman, B.: J. Chem. Phys. **122**, 204108 (2005).
[44] Bowman, J.M.: J. Chem. Phys. **68**, 608 (1978)
[45] Gerber, R.B., Ratner, M.A.: Chem. Phys. Lett. **68**, 195 (1979)
[46] Bowman, J.M.: Acc. Chem. Res. **19**, 202 (1986)
[47] Gerber, R.B., Ratner, M.A.: Adv. Chem. Phys. **70**, 97 (1988)
[48] Bowman, J.M., Carter, S., Huang, X.-C.: Int. Rev. Phys. Chem. **22**, 533 (2003)
[49] Miller, W.H., Handy, N.C., Adams, J.E.: J. Chem. Phys. **72**, 99 (1980)
[50] Carter, S., Handy, N.C.: J. Chem. Phys. **113**, 987 (2000)
[51] Bowman, J., Christoffel, K.M., Tobin, F.: J. Phys. Chem. **83**, 905 (1979)
[52] Christoffel, K.M., Bowman, J.M.: Chem. Phys. Lett. **85**, 220 (1982)
[53] Carter, S., Bowman, J.M., Handy, N.C.: Theor. Chim. Acta **100**, 191 (1998)
[54] Luis, J.M., Torrent-Sucarrat, M., Solà, M., Bishop, D.M., Kirtman, B.: J. Chem. Phys. **122**, 184104 (2005).
[55] Christiansen, O.: J. Chem. Phys. **120**, 2149 (2004)
[56] White, S.R.: Phys. Rev. Lett. **69**, 2863 (1992)
[57] White, S.R.: Phys. Rev. **B48**, 10345 (1993)
[58] Hasan, M., Kirtman, B.: J. Chem. Phys. **96**, 470 (1992)
[59] Jacquemin, D., Champagne, B., Hättig, C.: Chem. Phys. Lett. **319**, 327 (2000)
[60] Millefiori, S., Alparone, A.: Phys. Chem. Chem. Phys. **2**, 2495 (2000)
[61] Perpète, E.A., Champagne, B., André, J.M., Kirtman, B.: J. Mol. Struct. **425**, 115 (1998)

[62] Champagne, B., Perpète, E.A., van Gisbergen, S.J.A., Baerends, E.J., Snijders, J.G., Soubra-Ghaoui, C., Robins, K.A.: Kirtman, B.: J. Chem. Phys. **109**, 10489 (1998); erratum **110**, 11664 (1999)
[63] van Gisbergen, S.J.A., Schipper, P.R.T., Gritsenko, O.V., Baerends, E.J., Snijders, J.G., Champagne, B.: Kirtman, B.: Phys. Rev. Lett. **83**, 694 (1999)
[64] Champagne, B., Perpete, E.A.: Int. J. Quantum Chem. **75**, 441 (1999)
[65] Champagne, B., Perpète, E.A., Jacquemin, D., van Gisbergen, S.J.A., Baerends, E.J., Soubra-Ghaoui, C., Robins, K.A., Kirtman, B.: J. Phys. Chem. **A104**, 4755 (2000)
[66] Vignale, G., Kohn, W.: Phys. Rev. Lett. **77**, 2037 (1996)
[67] van Faassen, M., de Boeij, P.L., van Leeuwen, R., Berger, J.A., Snijders, J.G.: Phys. Rev. Lett. **88**, 186401 (2002)
[68] van Faassen, M., de Boeij, P.L., van Leeuwen, R., Berger, J.A., Snijders, J.G.: Phys. J. Chem. Phys. **118**, 1044 (2003)
[69] Kummel, S., Kronik, L., Perdew, J.P.: Phys. Rev. Lett. **93**, 213002 (2004)
[70] Bulat, F.A., Toro-Labbe, A., Champagne, B., Kirtman, B., Yang, W.: J. Chem. Phys. **123**, 014319 (2005)
[71] Krieger, J.B., Li, Y., Iafrate, G.J.: Phys. Rev. **A45**, 101 (1992)
[72] Gruning, M., Gritsenko, O., Baerends, E.J.: J. Chem. Phys. **116**, 6435 (2002)
[73] Yang, W., Wu, Q.: Phys. Rev. Lett. **89**, 143002 (2002)
[74] Wu, Q., Yang, W.: J. Theoret. Comp. Chem. **2**, 627 (2003)
[75] Mori-Sanchez, P., Wu, Q., Yang, W.: J. Chem. Phys. **119**, 11001 (2003)
[76] Norman, P., Bishop, D.M., Aa. Jensen, H.J., Oddershede, J.: J. Chem. Phys. **115**, 10323 (2001)
[77] Bishop, D.M., Luis, J.M., Kirtman, B.: J. Chem. Phys. **116**, 9729 (2002)
[78] Painelli, A., Del Freo, L., Terenziani, F.: Chem. Phys. Lett. **346**, 470 (2001)
[79] Macak, P., Luo, Y., Norman, P., Ågren, H.: J. Chem. Phys. **113**, 7055 (2000)
[80] Luo, Y., Norman, P., Macak, P., Ågren, H.: J. Phys. Chem. **A104**, 4718 (2000)
[81] Macak, P., Luo, Y., Norman, P., Ågren, H.: Chem. Phys. Lett. **330**, 447 (2000)
[82] Sundholm, D., Olsen, J., Jørgensen, P.: J. Chem. Phys. **102**, 4143 (1995)
[83] Luo, Y., Ågren, H., Knuts, S., Jørgensen, P.: Chem. Phys. Lett. **213**, 356 (1993)
[84] Luo, Y., Ågren, H., Knuts, S., Minaev, B.F., Jørgensen, P.: Chem. Phys. Lett. **209**, 513 (1993)
[85] Honig, B., Jortner, J.: J. Chem. Phys. **47**, 3698 (1967)
[86] Honig, B., Jortner, J., Szoke, A.: J. Chem. Phys. **46**, 2714 (1967)
[87] Kikuchi, H., Kubo, M., Watanabe, N., Suzuki, H.: J. Chem. Phys. **119**, 729 (2003)
[88] Ervin, K.M., Ramond, T.M., Davico, G.E., Schwartz, R.L., Casey, S.M., Lineberger, W.C.: J. Phys. Chem. **A105**, 10822 (2001)
[89] Islampour, R., Dehestani, M., Lin, S.H.: J. Mol. Spectrosc. **194**, 179 (1999)
[90] Ruhoff, P.T.: Chem. Phys. **186**, 355 (1994)
[91] Chen, P.: In: Ng, C.-Y., Baer, T., Powis, I. (ed.) Unimolecular and Bimolecular Reaction Dynamics, p. 371, Wiley, Chichester (1994)
[92] Schumm, S., Gerhards, M., Kleinermanns, K.: J. Phys. Chem. **A104**, 10648 (2000)
[93] Berger, R., Fischer, C., Klessinger, M.: J. Phys. Chem. **A102**, 7157 (1998)
[94] Callis, P.R., Vivian, J.T., Slater, L.S.: Chem. Phys. Lett. **244**, 53 (1995)
[95] Gruner, D., Nguyen, A., Brumer, P.: J. Chem. Phys. **101**, 10366 (1994)
[96] Sharp, T.E., Rosenstock, H.M.: J. Chem. Phys. **41**, 3453 (1964)
[97] Doctorov, E.V., Malkin, I.A., Man'ko, V.I.: J. Mol. Spectrosc. **64**, 302 (1977)
[98] Mok, D.K.W., Lee, E.P.F., Chau, F.-T., Wang, D., Dyke, J.M.: J. Chem. Phys. **113**, 5791 (2000)
[99] Barinova, G., Markovic, N., Nyman, G.: J. Chem. Phys. **111**, 6705 (1999)
[100] Serrano-Andrés, L., Forsberg, N., Malmqvist, P.-Å.: J. Chem. Phys. **108**, 7202 (1998)
[101] Takeshita, K., Shida, N.: Chem. Phys. **210**, 461 (1996)
[102] Hazra, A., Nooijen, M.: Phys. Chem. Chem. Phys. **7**, 1719 (2005)
[103] Hazra, A., Chang, H.H., Nooijen, M.: J. Chem. Phys. **121**, 2125 (2004)
[104] Luis, J.M., Bishop, D.M., Kirtman, B.: J. Chem. Phys. **120**, 813 (2004)
[105] Champagne, B., Jacquemin, D., André, J.-M., Kirtman, B.: J. Phys. Chem. **A101**, 3158 (1997)

[106] Kudin, K.N., Car, R., Resta, R.: J. Chem. Phys. **122**, 134907 (2005)
[107] Jacquemin, D., Champagne, B., Perpète, E.A., Luis, J.M., Kirtman, B.: J. Phys. Chem. **A105**, 9748 (2001)
[108] Kirtman, B., Toto, J.L., Robins, K.A., Hasan, M.: J. Chem. Phys. **102**, 5350 (1995)
[109] Weniger, E.J., Kirtman, B.: Computers & Math with Appl. **45**, 189 (2003)
[110] Korchowiec, J., Gu, F.L., Imamura, A., Kirtman, B., Aoki, Y.: Int. J. Quantum Chem. **102**, 785 (2005)
[111] Onishi, S., Gu, F.L., Naka, K., Imamura, A., Kirtman, B., Aoki, Y.: J. Phys. Chem. **A108**, 8478 (2004)
[112] Gu, F.L., Aoki, Y., Korchowiec, J., Imamura, A., Kirtman, B.: J. Chem. Phys. **121**, 10385 (2004)
[113] Imamura, A., Aoki, Y., Maekawa, K.: J. Chem. Phys. **95**, 5419 (1991)
[114] Aoki, Y., Imamura, A.: J. Chem. Phys. **97**, 8432 (1992)
[115] Kirtman, B., Gu, F.L., Bishop, D.M.: J. Chem. Phys. **113**, 1294 (2000)
[116] Bishop, D.M., Gu, F.L., Kirtman, B.: J. Chem. Phys. **114**, 7633 (2001)
[117] King-Smith, R.D., Vanderbilt, D.: Phys. Rev. **B47**, 1651 (1993)
[118] Vanderbilt, D., King-Smith, R.D.: Phys. Rev. **B48**, 4442 (1993)
[119] Resta, R.: Int. J. Quantum. Chem. **75**, 599 (1999)
[120] Gonze, X., Ghosez, Ph., Godby, R.W.: Phys. Rev. Lett. **74**, 4035 (1995)
[121] Souza, I., Iniguez, J., Vanderbilt, D.: Phys. Rev. Lett. **89**, 117602 (2002)
[122] Umari, P., Pasquarello, A.: Phys. Rev. Lett. **89**, 157602 (2002)
[123] Karna, S.P., Dupuis, M.: J. Comput. Chem. **12**, 487 (1991)

CHAPTER 4

SOS METHODS IN CALCULATIONS OF ELECTRONIC NLO PROPERTIES

WOJCIECH BARTKOWIAK[1] AND ROBERT ZALEŚNY[2]

[1] *Institute of Physical and Theoretical Chemistry, Wrocław University of Technology, Wyb. Wyspiańskiego 27, 50-370 Wrocław, Poland*
[2] *Department of Quantum Chemistry, Faculty of Chemistry, Nicolaus Copernicus University, Gagarina 7, 87-100 Toruń, Poland*

Abstract: The sum-over-states technique which is extensively used in calculations of nonlinear optical properties, is presented and discussed. We focus on the electronic contributions to first- and second-order hyperpolarizability. The SOS approach to the calculation of the multiphoton absorption is also discussed. The various approximations to exact sum-over-states formulae are presented. In particular, we describe the so-called few-levels models, which are widely used in qualitative analysis of nonlinear electrical properties

Keywords: First-order hyperpolarizability; second-order hyperpolarizability; sum-over states method; multiphoton absorption; nonlinear optics

1. INTRODUCTION

In the past two decades, a significant effort has been made towards development of reliable computational techniques for calculations of nonlinear optical properties of molecules. This has been reflected in many methods for calculations of first- (β) and second-order (γ) hyperpolarizabilities implemented in widely available quantum-chemical packages. However, the purely resonant properties, like two- and three-photon absorptivities have been coded in only a few of them. Also the inclusion of the influence of environment on NLO properties made a significant step forward in comparisons of theoretical and experimental data for large organic systems.

Among several different approaches to the problem of the evaluation of the nonlinear response of molecular systems, one can distinguish finite field (FF) approaches [28, 47], response theory [89, 90], and sum-over-states (SOS) methods [91, 113]. Those three approaches are available in many flavours. In this chapter we shall describe the last approach to the computation of resonant and non-resonant electronic nonlinear optical properties of molecules.

Obviously, there are numerous possible ways of classifying methods of computations of NLO properties. In the case of SOS techniques we classify the methods according to the quality of representation of electronic excited states.

One of the methods most frequently used in calculations of electronic excited states is the configuration interaction technique (CI). When combined with semiempirical Hamiltonians the CI method becomes an attractive method for investigations of electronic structure of large organic systems. Undoubtedly, it is the most popular method for calculations of electronic contributions to NLO properties based on the SOS formalism. The discussion of the CI/SOS techniques is presented in Section 4.

In early years of quantum chemistry, several theoretical papers were devoted to calculations of linear and nonlinear responses of molecules to the electric field perturbations using the Uncoupled Hartree-Fock (UCHF) method. In comparison with the CI ansatz, the UCHF is less accurate in the description of electronic structure of molecules. Since this method was of some interest in computations of NLO properties we present this method in Section 5.

Much of the interest in the field of molecular nonlinear optics has been focused on the so-called structure-property relationships. The most interesting systems are organic π-conjugated molecules. The experience gained in this area indicates that organic systems with low-lying energy states of charge-transfer character exhibit relatively large nonlinear optical response in comparison with similar compounds without such electronic states. The perturbation theory-based analysis gives a rational explanation to these observations. Since the SOS formulae express the nonlinear optical properties in terms of energies and dipole moments of excited states, transition moments between various excited states, it is logical and convenient to interpret the NLO properties in terms of these parameters. Such an approach underlies the so-called few-states models and is presented in Section 6.

The evaluation of the vibrational counterpart of nonlinear optical properties is not explicitly discussed in this review. This area is covered by appropriate references. Methods to compute the vibrational contributions to linear and nonlinear optical properties shall be presented by Kirtman and Luis in one of the chapters of the present book.

The aim of the present contribution is to review computational techniques based on the SOS formalism. We shall focus on electronic contributions to first- (β) and second-order (γ) hyperpolarizabilities. With the exception of the imaginary part of γ, the non-resonant properties shall be discussed. Some attention will be given to the so-called few-states models, as being very useful 'by-product' of the SOS formalism.

2. BASIC DEFINITIONS

In the presence of static uniform electric field, the total energy of molecular system can be expressed as a Taylor series:

$$E(F) = E(0) - \mu_\alpha F_\alpha - \frac{1}{2!}\alpha_{\alpha\beta}F_\alpha F_\beta - \frac{1}{3!}\beta_{\alpha\beta\gamma}F_\alpha F_\beta F_\gamma \tag{1}$$
$$- \frac{1}{4!}\gamma_{\alpha\beta\gamma\delta}F_\alpha F_\beta F_\gamma F_\delta - \ldots$$

where $E(0)$ denotes energy of a molecule without external perturbation. The Greek symbols $\alpha, \beta \ldots$, label tensor quantities. Single subscript denotes the first-rank tensor, double subscript stands for the second-rank tensor, etc. These subscripts are chosen to be Cartesian co-ordinates. In the whole chapter we adopted the Einstein summation convention. The above presented expansion is also known as the T-convention [115].

On the assumption that external electric field separately influences the electronic and nuclear motions, we may split the response of a molecule in two parts, namely electronic and vibrational:

$$\beta = \beta^e + \beta^{vib} \tag{2}$$
$$\gamma = \gamma^e + \gamma^{vib} \tag{3}$$

For the discussion of the vibrational counterpart of NLO properties we refer to original papers and reviews [5, 14, 16, 18, 19, 20, 26, 38, 60, 99].

The general SOS formula for the n-th order polarizability tensor component derived from time-dependent perturbation theory, $X^n_{\alpha\beta\ldots}(-\omega_\sigma; \omega_1, \omega_2, \ldots, \omega_n)$ with $\omega_\sigma = \sum_i \omega_i$ reads:

$$X^n_{\alpha\beta\ldots}(-\omega_\sigma; \omega_1, \omega_2, \ldots, \omega_n) \tag{4}$$
$$= \hbar^{-n} \sum \mathcal{P}_{-\sigma,1,2,\ldots,n} \sum_{a_1} \sum_{a_2} \ldots \sum_{a_n} \langle g|\hat{\mu}_\alpha|a_1\rangle \langle a_1|\hat{\mu}_\beta|a_2\rangle \ldots$$
$$\times [(\omega_{a_1} - \omega_\sigma)(\omega_{a_2} - \omega_\sigma + \omega_{a_1}) \ldots (\omega_{a_n} - \omega_n)]^{-1}$$

where $\mathcal{P}_{-\sigma,1,2,\ldots,n}$ denotes the operator permuting the pairs $(\alpha, -\omega_\sigma), (\beta, \omega_1) \ldots$. The subscripts $a_1, a_2, \ldots, a_n$ run over excited states of the system with energies $\hbar\omega_1, \hbar\omega_2, \ldots, \hbar\omega_n$, respectively. These states, depending on the level of aproximation, may be purely electronic, vibronic or rotational-vibrational-electronic. In the above formula we adopted the notation used by Bishop [15]. We find this notation compact and elegant. One immediately identifies $X^1_{\alpha\beta}, X^2_{\alpha\beta\gamma}$ and $X^3_{\alpha\beta\gamma\delta}$ as $\alpha_{\alpha\beta}, \beta_{\alpha\beta\gamma}$ and $\gamma_{\alpha\beta\gamma\delta}$, respectively.

3. ELECTRONIC NLO PROPERTIES

The sum-over-states expression for the electronic first-order hyperpolarizability (β) can be written as:

(5) $$\beta_{\alpha\beta\gamma}(-\omega_\sigma; \omega_1, \omega_2)$$

$$= \hbar^{-2} \sum \mathcal{P}_{-\sigma,1,2} {\sum_K}' {\sum_L}' \frac{\langle 0|\hat{\bar{\mu}}_\alpha|K\rangle\langle K|\hat{\bar{\mu}}_\beta|L\rangle\langle L|\hat{\bar{\mu}}_\gamma|0\rangle}{(\omega_K - \omega_\sigma)(\omega_L - \omega_2)}$$

where $\hat{\bar{\mu}}$ is the fluctuation dipole moment operator $\hat{\bar{\mu}} = \hat{\mu} - \langle 0|\hat{\mu}|0\rangle$ and $\omega_\sigma = \omega_1 + \omega_2$. The symbol $|K\rangle$ stands for the electronic state K. The primes appearing in the above equation indicate the exclusion of the ground electronic state from the summation. The energies in the denominators are defined with respect to the ground electronic state:

(6) $$\omega_K = \frac{E_K - E_0}{\hbar}$$

Formally, all electronic states should be included in Eq. (5). However, most of the quantum-chemical methods employ the LCAO-MO method where finite set of the so-called basis functions is used. Hence, only a finite set of electronic states can be used in the summation given by Eq. (5).

In order to compare theoretical data with the quantity experimentally available for polar molecules, the calculated components of the β_{ijk} tensor are to be transformed to the vector quantity defined as:

(7) $$\beta_\mu = \sum_{\eta=x,y,z} \frac{\mu_\eta \beta_\eta}{|\mu|}$$

where

(8) $$\beta_\eta = \frac{1}{5} \sum_{\xi=x,y,z} (\beta_{\eta\xi\xi} + \beta_{\xi\eta\xi} + \beta_{\xi\xi\eta})$$

and $|\mu|$ is the dipole moment.

It is important to stress the fact that most of the SOS calculations of first-order hyperpolarizbility for organic systems were performed using semiempirical methods combined with the various variants of the CI technique [1, 2, 6, 7, 11, 34, 41, 43, 44, 49, 52, 53, 57, 82, 83]. We will address this subject in more details in Section 4.

With all the symbols defined previously, the SOS equation defining the electronic second-order hyperpolarizability (γ) reads:

(9) $$\gamma_{\alpha\beta\gamma\delta}(-\omega_\sigma; \omega_1, \omega_2, \omega_3) = \gamma^{(+)}_{\alpha\beta\gamma\delta}(-\omega_\sigma; \omega_1, \omega_2, \omega_3) + \gamma^{(-)}_{\alpha\beta\gamma\delta}(-\omega_\sigma; \omega_1, \omega_2, \omega_3)$$

$$= \hbar^{-3} \sum \mathcal{P}_{-\sigma,1,2,3} \times \left\{ {\sum_K}' {\sum_L}' {\sum_M}' \frac{\langle 0|\hat{\bar{\mu}}_\alpha|K\rangle\langle K|\hat{\bar{\mu}}_\beta|L\rangle\langle L|\hat{\bar{\mu}}_\gamma|M\rangle\langle M|\hat{\mu}_\delta|0\rangle}{(\omega_K - \omega_\sigma)(\omega_L - \omega_2 - \omega_3)(\omega_M - \omega_3)} \right.$$

$$\left. - {\sum_K}' {\sum_L}' \frac{\langle 0|\hat{\bar{\mu}}_\alpha|K\rangle\langle K|\hat{\bar{\mu}}_\beta|0\rangle\langle 0|\hat{\bar{\mu}}_\gamma|L\rangle\langle L|\hat{\mu}_\delta|0\rangle}{(\omega_K - \omega_\sigma)(\omega_L - \omega_3)(\omega_L + \omega_2)} \right\}$$

The average value of second-order hyperpolarizability is defined as:

(10) $\langle \gamma \rangle = \frac{1}{15}(\gamma_{\xi\xi\eta\eta} + \gamma_{\xi\eta\xi\eta} + \gamma_{\xi\eta\eta\xi})$

In the present contribution the discussion of the NLO response is restricted to off-resonant case. The only exception is the purely resonant quantity, namely imaginary part of second-order hyperpolarizability in the resonant regime $(\text{Im}\gamma(-\omega; \omega, -\omega, \omega))$. This quantity describes the process of simultaneous absorption of two quanta. The two-photon absorption (TPA) process is much better understood than the three-photon absorption. The basic quantity associated with the two-photon absorption process is the two-photon absorption tensor (S^{0F}). In the most general case referring to two different photons (different polarizations $\vec{\zeta}_1 \neq \vec{\zeta}_2$ and different energies $\hbar\omega_1 \neq \hbar\omega_2$) $S^{0F}_{\alpha\beta}$ is given by [75, 81]:

(11) $S^{0F}_{\alpha\beta} = \hbar^{-1} \sum_K \left\{ \frac{\langle 0|\vec{\zeta}_1\hat{\mu}_\alpha|K\rangle\langle K|\vec{\zeta}_2\hat{\mu}_\beta|F\rangle}{\omega_K - \omega_1} + \frac{\langle 0|\vec{\zeta}_2\hat{\mu}_\alpha|K\rangle\langle K|\vec{\zeta}_1\hat{\mu}_\beta|F\rangle}{\omega_K - \omega_2} \right\}$

where $\hbar\omega_1 + \hbar\omega_2$ should satisfy the resonance condition and $\langle K|\vec{\zeta}_1\hat{\mu}|L\rangle$ is the transition moment between electronic states K and L, respectively.

Since in most experiments one source of photons is used, we can substitute the angular frequencies ω_1 and ω_2 for $0.5 \cdot \omega_F$.

The averaging procedure of the two-photon absorption tensor over all orientations of the absorbing molecule leads to [75, 81]:

(12) $\langle \delta^{0F} \rangle = \left\langle \left| S^{0F}_{\alpha\beta}(\vec{\zeta}_1, \vec{\zeta}_2) \right|^2 \right\rangle$

$= \frac{1}{30} \sum_{\alpha\beta} \left\{ S^{0F}_{\alpha\alpha} \left(S^{0F}_{\beta\beta} \right)^* F + S^{0F}_{\alpha\beta} \left(S^{0F}_{\alpha\beta} \right)^* G + S^{0F}_{\alpha\beta} \left(S^{0F}_{\beta\alpha} \right)^* H \right\}$

where $F = F(\vec{\zeta}_1, \vec{\zeta}_2), G = G(\vec{\zeta}_1, \vec{\zeta}_2), H = H(\vec{\zeta}_1, \vec{\zeta}_2)$ are the polarization variables. Since in the Eq. (11) we include both the ground $|0\rangle$ and the final $|F\rangle$ state and the tensor elements are real, the S^{0F} tensor is symmetric. In the case of two linearly polarized photons, all three polarization variables are equal to two and the two-photon absorptivity can be written as:

(13) $\langle \delta^{0F} \rangle = \frac{1}{15} \sum_{\alpha\beta} \left\{ S^{0F}_{\alpha\alpha} \left(S^{0F}_{\beta\beta} \right)^* + 2 S^{0F}_{\alpha\beta} \left(S^{0F}_{\alpha\beta} \right)^* \right\}$

The possibility of simultaneous absorption of two quanta was concluded on purely theoretical basis by Göppert-Mayer in 1931 [64]. First experimental observations of this process in organic systems were reported in early 1960's by Peticolas et al. [96, 97]. Basically, there are two methods of calculation of two-photon absorptivities. The first technique is based on response theory [90]. The two-photon absorption cross section can be determined by the single residue of the cubic response function. Alternatively, one can extract the two-photon transition moments from the single residue of the quadratic response function.

Most of the calculations of the two-photon absorptivities at the *ab initio* level were performed using the response theory [61, 63, 65, 66, 86, 87, 112]. Recently, Salek et al. [102] presented the implementation of the density-functional theory for the linear and the nonlinear response functions. In particular, in their most recent paper they reported on calculations of the two-photon absorption cross sections in terms of the single residue of the quadratic response function [101].

The second method commonly used for the evaluation of the two-photon absorptivities is the conventional sum-over-states technique. In general, the summation over excited states converges very slowly. However, the molecules with low lying excited state of charge-transfer character are frequently exception to this statement. Most of the calculations employing this method have been performed at the semiempirical level of theory [3, 10, 13, 35, 46, 67, 95, 127].

In their recent papers, Tretiak et al. proposed the technique for calculations of TPA properties which is to some extent the combination of the methods described above [45, 74, 108, 109]. The method proposed by Tretiak takes an advantage of the quantities that can be calculated within the linear response theory framework. The remaining quantities that appear in the expressions for the two-photon absorption cross section can be evaluated as the functional derivatives based on the time-dependent density functional (TDDFT) method. Although the response theory is involved in their evaluation, it is important to note that the TPA cross section is calculated via SOS formulae.

During the past decade, an increasing interest in multiphoton absorption of organic molecular systems has been observed. There were many attempts to establish the so-called structure-property relationships for organic systems using computational techniques [3, 10, 46, 51, 58, 66, 87, 95, 118, 119, 125, 126, 127].

4. CI METHOD AND SOS FORMALISM

As mentioned in section 1, the combination of the CI method and semiempirical Hamiltonians is an attractive method for calculations of excited states of large organic systems. However, some of the variants of the CI ansatz are not in practical use for large molecules even at the semiempirical level. In particular, this holds for full configuration interaction method (FCI). The truncated CI expansions suffer from several problems like the lack of size-consistency, and violation of Hellmann-Feynman theorem. Additionally, the calculations of NLO properties bring the problem of minimal level of excitation in CI expansion neccessary for the correct description of electrical response calculated within the SOS formalism.

In the case of calculations of electronic contributions to γ the problem of inclusion of double excitations is much more pronounced than in the case of computations of β. This can be illustrated as follows. Let the electronic states appearing in Eqs. (5) and (9) be classified as pure singly or doubly excited states. This is plausible under assumption that no mixing of singly and doubly excited configurations with the ground states occurs without the perturbation. According to the rules of evaluation

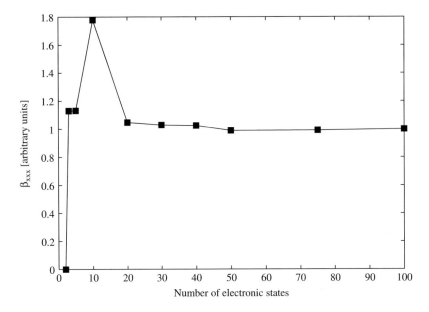

Figure 1. The dependence of the static first-order hyperpolarizability β_{zzz} on the number of electronic states included in Eq. (5)

of the matrix elements between determinants for one-electron operators (i.e., $\hat{\mu}$) no doubly excited configurations appear in Eq. (5) for β. However, in the case of Eq. (9), doubly excited configurations do make contribution to γ. Obviously, this picture is only an approximate one. Nevertheless, it illustrates the importance of inclusion of doubly excited configurations in the calculations of second-order hyperpolarizability using SOS approach. The problem of treating electron-electron interactions (EEI) via configuration interaction method (CI) and its influence on the γ values was addressed in details by Pierce [98] and Morley et al. [83].

The most frequent method of calculations of excited state wave functions appearing in Eq. (9) is the CI method. A given electronic state is obtained by the diagonalization of the CI matrix. Two variants of the CI method are most popular, namely configuration interaction with singles (CIS) and configuration interaction with singles and doubles (CISD). Mono-excited coniguration interaction method (MECI) or Tamm-Dancoff approximation are the synonyms of the former. As argued previously, in most cases the CIS method should be sufficient level of theory for description of first-order NLO response. The results of semiempirical calculations of β for organic systems confirm this supposition [8, 57]. Fig. 1 presents the results of calculations of β_{zzz} component for *p*-nitroaniline molecule (PNA) at the CIS level of theory using the GRINDOL Hamiltonian [55]. The values of the longitudinal component, β_{zzz}, are normalized with respect to the converged value, i.e. β calculated with inclusion of 100 electronic states. The electronic excited states were obtained by diagonalization of the CI matrix constructed from 600 singly excited configurations. As one can see, the convergence of β_{zzz} with respect to the number of electronic states is rather fast. 20 electronic excited states are sufficient to account for the most of the response of PNA through third order of external perturbation. The sign of β is consistent with the *ab initio* results [62, 104]. For most organic push-pull systems the SCI/SOS level of theory is satisfactory in reproducing the values of β.

The comparison of CIS/SOS and CISD/SOS results of calulations of γ leads to the opposite conclusions. The double and possibly higher order excitations are mandatory to obtain the second-order hyperpolarizability consistent with experimental data. The total SOS formula for γ can be divided into two parts, namely $\gamma^{(+)}$ and $\gamma^{(-)}$. The convergence of these terms with respect to the number of electronic states included in summation is presented in Figs. 2 and 3. The first figure show the longitudinal component (γ_{zzzz}), while the second figure present the average value $\langle\gamma\rangle$.

As previously, the GRINDOL/SCI/SOS level of theory was employed in order to illustrate the importance of double substitutions. In the first case, the γ_{zzzz} becomes negative for 300 excited electronic states. It is important to note that both $\gamma^{(+)}$ and $\gamma^{(-)}$ are of the same order of magniute, but with different signs. Hence, even

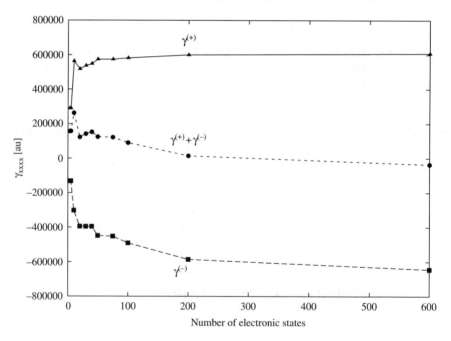

Figure 2. The plot of longitudinal component of the static second-order hyperpolarizability (γ_{zzzz}) as a function of number of electronic states included in summation. The triangles denote the first term in Eq. (9) and squares stand for the second term in Eq. (9). Full circles label the sum of the two contributions

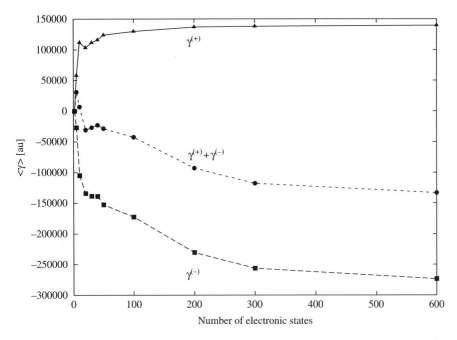

Figure 3. The plot of the static average second-order hyperpolarizability ($\langle\gamma\rangle$) as a function of number of electronic states included in summation. The triangles denote the first term in Eq. (9) and squares stand for the second term in Eq. (9). Full circles label the sum of the two contributions

small changes caused by the inclusion of high-lying electronic states may lead to wrong sign of γ_{zzzz}. In the second case presented in Fig. 3 the averaged second-order hyperpolarizability is negative in the whole range of excited states included in Eq. (9). Based on the data presented by Pierce [98] (Tables 1 and 3 in his paper) one may conclude what follows: The values of γ for most organic systems are positive. In order to reproduce the correct sign for γ one should include at least doubly substituted configurations in the CI expansion.

It has been observed that the semiempirical SOS results for β are much higher than the respective values obtained using other techniques (like FF approach) [57, 115]. One of the reasons of this discrepancy lies in the overestimated values of transition moments ($\langle K|\hat{\mu}|L\rangle$) [98, 115].

5. UNCOUPLED HARTREE-FOCK METHOD

The Uncoupled-Hartree-Fock method (UCHF) [31, 32, 50, 54, 85, 88, 100, 110] is also referred to as the sum-over-orbitals (SOO) method. In this technique, one takes the unperturbed Hamiltonian H^0 as a sum of one-particle Hamiltonians:

$$(14) \quad H^0 = \sum_i h^0(i)$$

with the perturbation being the sum of one-particle interactions:

(15) $\quad H' = \sum_i h'(i)$

The eigenfunctions of the unperturbed Hamiltonian H^0 are chosen to be the determinants of the form:

(16) $\quad |\Psi^0\rangle = |\chi_1^{(0)} \ldots \chi_a^{(0)} \ldots \chi_N^{(0)}\rangle$

where the subscript runs over the N spin orbitals. $a, b, c, \ldots$ denote the hole states, while $r, s, t, \ldots$ are used for the particle states. The superscripts label the spin orbitals being the solutions of the unperturbed eigenvalue problem:

(17) $\quad h^0 \chi_i^{(0)} = \epsilon_i^{(0)} \chi_i^{(0)}$

The ground state wave function $|\Psi^0\rangle$ is an eigenfunction of the Hamiltonian defined by Eq. (14):

(18) $\quad H^0|\Psi^0\rangle = \left(\sum_a \epsilon_a^{(0)}\right)|\Psi^0\rangle$

where the sum expands over all occupied orbitals a. Since the perturbation H' can be written as a sum of one-particle interactions, we may write analogously to Eq. (18):

(19) $\quad (H^0 + H')|\Phi^0\rangle = \left(\sum_a \epsilon_a\right)|\Phi^0\rangle$

where ϵ_a denote the energy of occupied orbital a in the presence of perturbation H'. The total energy of the perturbed system is:

(20) $\quad \mathcal{E} = \sum_a \epsilon_a$

We may formally write down the perturbation expansion for the energy $\mathcal{E}$ and introduce perturbation in explicit form, i.e. $H' = -\mathbf{F}\mu$. The expressions for first- (β) and second-order (γ) hyperpolarizabilities will be then given by:

(21) $\quad \beta_{\alpha\beta\gamma}(0) = \mathcal{S}(\alpha\beta\gamma) \left\{ \sum_{ars} \frac{\langle a|\mu_\alpha|r\rangle \langle r|\mu_\gamma|s\rangle \langle s|\mu_\beta|a\rangle}{(\epsilon_r - \epsilon_a)(\epsilon_s - \epsilon_a)} \right.$
$\left. - \sum_{abr} \frac{\langle a|\mu_\alpha|r\rangle \langle b|\mu_\gamma|a\rangle \langle r|\mu_\beta|b\rangle}{(\epsilon_r - \epsilon_a)(\epsilon_s - \epsilon_b)} \right\}$

(22) $\quad \gamma_{\alpha\beta\gamma\delta}(0) = \mathcal{S}(\alpha\beta\gamma\delta) \left\{ \sum_{arst} \frac{\langle a|\mu_\alpha|r\rangle \langle r|\mu_\delta|s\rangle \langle s|\mu_\gamma|t\rangle \langle t|\mu_\beta|a\rangle}{(\epsilon_r - \epsilon_a)(\epsilon_s - \epsilon_a)(\epsilon_t - \epsilon_a)} \right.$
$+ \sum_{abcr} \frac{\langle a|\mu_\alpha|r\rangle \langle b|\mu_\delta|a\rangle \langle c|\mu_\gamma|b\rangle \langle r|\mu_\beta|c\rangle}{(\epsilon_r - \epsilon_a)(\epsilon_r - \epsilon_b)(\epsilon_r - \epsilon_c)}$
$- \sum_{abrs} \frac{\langle a|\mu_\alpha|r\rangle \langle r|\mu_\delta|s\rangle \langle b|\mu_\gamma|a\rangle \langle s|\mu_\beta|b\rangle}{(\epsilon_r - \epsilon_a)(\epsilon_s - \epsilon_a)(\epsilon_s - \epsilon_b)}$
$- \sum_{abrs} \frac{\langle a|\mu_\alpha|r\rangle \langle b|\mu_\delta|a\rangle \langle r|\mu_\gamma|s\rangle \langle s|\mu_\beta|b\rangle}{(\epsilon_r - \epsilon_a)(\epsilon_r - \epsilon_b)(\epsilon_s - \epsilon_b)}$
$\left. - \sum_{abrs} \frac{\langle a|\mu_\alpha|r\rangle \langle b|\mu_\delta|s\rangle \langle r|\mu_\gamma|b\rangle \langle s|\mu_\beta|a\rangle}{(\epsilon_r - \epsilon_a)(\epsilon_r - \epsilon_b)(\epsilon_s - \epsilon_a)} \right\}$

In the above expressions S labels the appropriate symmetrizer. The way of deriving frequency-independent β and γ outlined above is also known as the orbital perturbation theory. Alternatively, one can start from general expressions for β and γ obtained within time-dependent perturbation theory. The sum over excited states follows then from the summation over excited determinants. The evaluation of matrix elements is done using the Slater rules. The neglect of the frequency-dependence in SOS expressions leads to Eqs. (21) and (22). The detailed description of this procedure can be found in the paper of Jacquemin et al. [42]. The resulting expressions for β and γ are only crude approximations to those arising from more advanced treatments. Instead of the energy differences between excited states of the system and the ground state energy, the orbital energy differences appear. For excited states with dominant singly substituted determinant $|\Psi_a^r\rangle$ the difference $(\epsilon_r - \epsilon_a)$ is usually a poor representation of the excitation energy. In Table 1 there are presented the results of calculations of NLO properties for the PNA molecule together with the values of energies of HOMO and LUMO. The excitation energy to the lowest-lying excited state of the charge-transfer character for PNA is about 4 eV. The CIS wave function for this state is dominated by configuration $|\Psi_{HOMO}^{LUMO}\rangle$. The excitation energy estimated as HOMO–LUMO difference gives the value about 10 eV. It can be a simple explanation of the fact, that UCHF hyperpolarizabilities are usually underestimated in comparison with more advanced treatments. Simple illustration of the importance of the energy differences appearing in denominators in Eqs. (21) and (22) is presented in Table 5. As it is seen, the values of β and γ at the DFT level, are overestimated over an order of magnitude in comparison with the HF results.

Hameka et al., has calculated polarizabilities up to third order for several organic systems [40, 76, 77, 78, 121, 122, 123, 124]. In his pioneering papers, the nonlinear polarizability was calculated by using sum over orbitals within the Hückel approximation. This approach was later improved by using extended Hückel method (EHM) [123, 124]. The quality of EHM third-order electric susceptibility[1] was tested by

Table 1. UCHF results of nonlinear optical properties of p-nitroaniline molecule lying in xy plane. All values are given in atomic units. The results are presented in the so-called T-convention

	HF/4-31G	B3LYP/4-31G
β_{zzz}	783	8870
γ_{zzzz}	48826	1288193
ϵ_{HOMO}	−0.3268	−0.2251
ϵ_{LUMO}	0.0554	−0.0697
$\Delta\epsilon$	0.3822	0.1554

---

[1] In the original papers by Hameka et al., the quantity γ is called susceptibility, but since they calculated pure molecular response it should be rather referred to as the third-order polarizability or the second-order hyperpolarizability.

comparison with more advanced treatment, namely in Pariser-Parr-Pople (PPP) calculations [121, 122, 123, 124]. The molecular excited states were calculated by using configuration interactions with singles method at the PPP level.

More recent studies include calculations of NLO properties of polyene series [4, 42] and various benzene derivatives [9, 24, 39, 105, 106, 107, 111]. Most recent calculations based on the Uncoupled Hartree-Fock scheme were devoted to large organic systems like fullerenes [103], nanotubes [116, 117], carbon cages [37], oligomers [25] and polymers [92].

6. FEW-STATES MODELS

The popularity of the SOS methods in calculations of non-linear optical properties of molecules is due to the so-called few-states approximations. The sum-over-states formalism defines the response of a system in terms of the spectroscopic parameters, like excitations energies and transition moments between various excited states. Depending on the level of approximation, those states may be electronic or vibronic or electronic-vibrational-rotational ones. Under the assumption that there are few states which contribute more than others, the summation over the whole spectrum of the Hamiltonian can be reduced to those states. In a very special case, one may include only one excited state which is assumed to dominate the molecular response through the given order in perturbation expansion. The first applications of two-level model to calculations of β date from late 1970s [93, 94]. The two-states model for first-order hyperpolarizability with only one excited state included can be written as:

$$(23) \quad \beta^e_{zzz}(0) = 6\hbar^{-2} \frac{(\langle K|\hat{\mu}_z|K\rangle - \langle 0|\hat{\mu}_z|0\rangle)\langle 0|\hat{\mu}_z|K\rangle^2}{\omega_K^2}$$

At first look, such an approximation seems to be very crude. However, there are molecular systems for which such an approximation works quite well. An example of such a molecule is p-nitroaniline. The results presented in Fig. 1 show that the inclusion of the third electronic singlet state (with excitation energy near 4 eV) gives the value of β close to the converged value for 100 electronic states in SOS method. This is partially due to fortuitous cancellation of contributions from higher-lying excited electronic states with excitation energies near 10 eV.

In a similar way it is possible to reduce the SOS expressions defining other nonlinear optical properties. The two-photon transition moment within two-level approximation reads:

$$(24) \quad S^{0K}_{zz} = 4 \frac{\langle 0|\hat{\mu}_z|K\rangle(\langle K|\hat{\mu}_z|K\rangle - \langle 0|\hat{\mu}_z|0\rangle)}{\omega_K}$$

where the summation over all intermediate electronic states is reduced to the final electronic state $|K\rangle$. The orientationally averaged two-photon absorptivity is then given by:

$$(25) \quad \langle \delta^{0K}\rangle = \frac{16}{5}\left(S^{0K}_{zz}\right)^2 = \frac{16}{5} \frac{\langle 0|\hat{\mu}_z|K\rangle(\langle K|\hat{\mu}_z|K\rangle - \langle 0|\hat{\mu}_z|0\rangle)}{\omega_K}$$

The formulae for generalized few-state models for two-photon absorption with numerical illustration at the *ab initio* level were given by Cronstrand et al. [29]. More recently, Cronstrand et al., have presented approximate expressions for three-photon absorption with application to PNA and LiH molecules [30].

There were also some attempts to derive the few-states models for second-order hyperpolarizability [23, 36, 48]. In the simplest case, i.e. within the two-level approximation, we may write:

$$(26) \quad \gamma_{zzzz}(0) = 24\hbar^{-3} \frac{\langle 0|\hat{\mu}_z|K\rangle^2(\langle K|\hat{\mu}_z|K\rangle - \langle 0|\hat{\mu}_z|0\rangle) - \langle 0|\hat{\mu}_z|K\rangle^4}{\omega_K^3}$$

Although this approximation seems to be very rough there is an indication that it works quite well for the betaine dye exhibiting strong negative solvatochromism [119].

6.1 Two-form Model

One of the most interesting applications of the few-state models in the interpretation of the NLO properties of molecules is the two-form model proposed by Barzoukas et al. [12, 21]. In the approach proposed by Barzoukas et al. [12, 21], the ground state $|0\rangle$ of the system under study is reperesented as a linear combination of two limiting resonant forms: neutral $|N\rangle$ and zwitterionic $|Z\rangle$ (see Fig. 4):

$$(27) \quad |0\rangle = \cos\frac{\theta}{2}|N\rangle + \sin\frac{\theta}{2}|Z\rangle$$

while the excited electronic state $|K\rangle$ is given by:

$$(28) \quad |K\rangle = -\sin\frac{\theta}{2}|N\rangle + \cos\frac{\theta}{2}|Z\rangle$$

Both $|0\rangle$ and $|K\rangle$ are assumed to be orthogonal. For small values of θ the ground state will be dominated by $|N\rangle$ and the excited state by $|Z\rangle$. The opposite holds for large values of θ ($\theta \sim \pi$). In the two-form model one introduces the coupling element t:

$$(29) \quad t = -\langle Z|H|N\rangle = \frac{1}{2}V\tan\theta$$

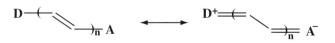

Figure 4. Schematic representation of two limiting resonant-forms, namely neutral and zwitterionic

where V is the energy gap between the two resonance forms is:

$$(30) \quad V = \langle Z|H|Z \rangle - \langle N|H|N \rangle$$

Eq. (29) defines parameter θ. Assuming transition dipole between two limiting forms to be: $\langle N|\mu|Z \rangle = 0$ one can express ground state dipole moment (μ_0), excited state dipole moment (μ_K), transition moment between states (μ_{0K}) and energy gap (E_{K0}) in terms of θ parameter:

$$(31) \quad \mu_0 = \langle 0|\mu|0 \rangle = \frac{\mu_N - \mu_Z - \Delta\mu \cos\theta}{2}$$

$$(32) \quad \mu_K = \langle K|\mu|K \rangle = \frac{\mu_N + \mu_Z + \Delta\mu \cos\theta}{2}$$

$$(33) \quad \mu_{0K} = \langle 0|\mu|K \rangle = \frac{\Delta\mu \sin\theta}{2}$$

$$(34) \quad E_{0K} = \frac{2t}{\sin\theta}$$

where $\Delta\mu$ stands for dipole moments difference between two resonant forms: $\mu_Z - \mu_N$. It is convenient to introduce one more parameter which describes the extent of mixing of limiting forms[2]:

$$(35) \quad \kappa = -\cos\theta = -\frac{V}{\sqrt{V^2 + 4t^2}}$$

Defined in such a way, the κ parameter is proportional to $\Delta\mu$ and hence it becomes a convenient quantity connecting molecular structure and amount of internal charge-transfer in push-pull systems. Once the expressions for μ_0, μ_K, μ_{0K} and E_{K0} are derived, one may relate the nonlinear optical properties (β, γ and δ) to κ parameter [10, 12, 21]:

$$(36) \quad \beta^e_{zzz}(0) = 6\hbar^{-2} \frac{(\langle K|\hat{\mu}_z|K \rangle - \langle 0|\hat{\mu}_z|0 \rangle)\langle 0|\hat{\mu}_z|K \rangle^2}{\omega_K}$$

$$= \frac{3\Delta\mu^3}{8t^2} \sin^4\theta \cos\theta = -\kappa(1-\kappa^2)^2 \frac{3\Delta\mu^3}{8t^2}$$

$$(37) \quad \gamma_{zzzz}(0) = 24 \frac{\langle 0|\hat{\mu}_z|K \rangle^2 (\langle K|\hat{\mu}_z|K \rangle \langle 0|\hat{\mu}_z|0 \rangle)^2 - \langle 0|\hat{\mu}_z|K \rangle^4}{(\omega_K)^3}$$

$$= \frac{3\Delta\mu^4}{16t^3} \sin^5\theta(5\cos^2\theta - 1) = (1-\kappa^2)^{\frac{5}{2}}(5\kappa^2 - 1)\frac{3\Delta\mu^4}{16t^3}$$

---

[2] We use the symbol κ here for the reason of compactness, although in the original papers the symbol *MIX* was used.

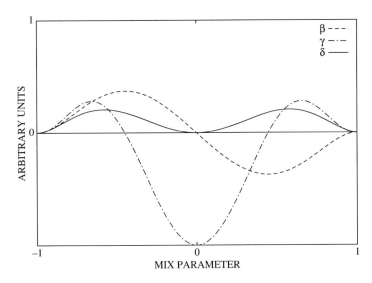

Figure 5. Evolution of non-linear optical properties (β, γ and δ) within two-state approximation as a function of κ parameter

In a similar way one may derive the expression for the two-photon absorptivity δ as a function of κ:

$$(38) \quad \langle \delta^{0F} \rangle = \frac{16}{5}(S_{zz}^{0F})^2 = \frac{16}{5} \frac{\langle 0|\hat{\mu}_z|F\rangle(\langle F|\hat{\mu}_z|F\rangle - \langle 0|\hat{\mu}_z|0\rangle)}{\omega_F} =$$

$$= \frac{\Delta\mu^4}{5t^2}\sin^4\theta\cos^2\theta = \kappa^2(1-\kappa^2)^2\frac{\Delta\mu^4}{5t^2}$$

The evolution of β, γ and δ as a function of κ parameter is presented in Fig. 5.

The two-form model has its roots in the valence-bond charge-transfer (VB-CT) model derived by Mulliken [84] and used with minor modifications by Warshel et al. for studying reactions in solutions [114]. Goddard et al. applied this VB-CT model to study the nonlinear optical properties of the charge-transfer systems. [27, 59]. The analysis of the relationship between electronic and vibrational components of the hyperpolarizabilities within the two-state valence-bond approach was presented by Bishop et al. [17]. Despite the limitations of the VB-CT model, it is very simple and gives some insight into mutual relationships between nonlinear optical responses through the various orders.

In his papers, Marder et al., have considered the relation between bond-legth alternation (BLA) and non-linear optical properties of π-conjugated systems [22, 33, 68, 69, 70, 71, 72, 73, 79, 80]. As it was shown, the NLO properties strongly depend on the BLA parameter. One can distinguish three forms of donor-π-acceptor system: neutral, polymethine-like and zwitterionic. The variations of β and γ for different forms can be substantial. It is well know, that solvent molecules generate electric field across the push-pull systems. Depending on the polarity of the solvent molecules, one may control the degree of internal charge-transfer in D-π-A system. Thus, in the case of charge-transfer species we may find the correspondence between the geometry (BLA) and internal charge-transfer which, due to the few-states model, is connected with NLO properties.

7. FINAL REMARKS

The SOS method presented in this chapter is still of practical use in calculations of large organic systems. In particular, many groups employ this technique for calculations of multiphoton absorption. Probably, one of the most advantageous features of this method is the possibility of analysis of the nonlinear optical properties of molecules in terms of contributions from electronic excited states. The most common implementation of SOS formulae is based on the configuration interaction expansion combined with semiempirical Hamiltonians. The fact, that any truncated CI method is not size-consistent makes the SOS method of small usefullness in calculations of NLO properties of aggregates and clusters. This is a serious limitation. Moreover, the same argument holds for the NLO properties in excited states even when calculated at the CIS level of theory [120]. One should also realize that the violation of the Hellmann-Feynman theorem (which is the case for any truncated CI expansions) may lead to the non-physical results of calculations of molecular electric properties [56]. After the words of scepticism, it should be pointed out that the SOS method gives a clear picture of the physics involved in the nonlinear optical response of molecular systems.

Although the SOS method, as a practical way of calculations, is considered nowadays to be obsolete for small molecules, for large systems it still seems to be sufficiently attractive to use at low cost of calculations.

REFERENCES

[1] Abe, J., Shirai, Y., Nemoto, N., Miyata, F., Nagase, Y.: Heterocyclic piridinium betaines, a new class of second-order nonlinear optical materials: Combined theoretical and experimental investigation of first-order hyperpolarizability through ab initio, INDO/S, and hyper-rayleigh scattering. J. Phys. Chem. B **101**, 576–582 (1997)

[2] Albert, I.D.L., Marks, T.J., Ratner, M.A.: Conformationally-induced geometric electron localization. Interrupted conjugation, very large hyperpolarizabilities, and sizable infrared absorption in simple twisted molecular chromophores. J. Am. Chem. Soc. **119**, 3155–3156 (1997)

[3] Albota, M., Beljonne, D., Brédas, J.-L., Ehrlich, J.E., Fu, J.-Y., Heikal, A.A., Hess, S.E., Kogej, T., Levin, M.D., Marder, S.R., McCord-Maughon, D., Perry, J.W., Röckel, H., Rumi, M.: Design of organic molecules with large two-photon absorption cross sections. Science **281**, 1653–1656 (1998)

[4] André, J.-M., Barbier, C., Bodart, V., Delhalle, J.: Trends in calculations of polarizabilities and hyperpolarizabilities of long molecules. In: Nonlinear Optical Properties of Organic Molecules and Crystals, vol. 2. Academc, New York (1987)

[5] Åstrand, P.-O., Ruud, K., Sundholm, D.: A modified variation-perturbation approach to zero-point vibrational motion. Theor. Chem. Acc. **103**, 365–373 (2000)

[6] Bartkowiak, W.: Theoretical study of hyperpolarizabilities of aminobenzodifuranone. Synth. Met. **109**, 109–111 (2000)

[7] Bartkowiak, W., Lipiński, J.: Conformation and solvent dependence of the first molecular hyperpolarizability of piridinium-N-phenolate betaine dyes. Quantum chemical calculations. J. Phys. Chem. A **102**, 5236–5240 (1998)

[8] Bartkowiak, W., Lipiński, J.: Solvent effect on the nonlinear optical properties of para-nitroaniline studied by Langevin dipoles-Monte Carlo (LD/MC) approach. Computers Chem. **22**, 31–37 (1998)

[9] Bartkowiak, W., Strasburger, K., Leszczynski, J.: Studies of molecular hyperpolarizabilities (β, γ) for 4-nitroaniline (PNA). The application of quantum mechanical/Langevin dipoles/Monte Carlo (QM/LD/MC) and sum-over-orbitals (SOO) methods. J. Mol. Struct. (THEOCHEM) **549**, 159–163 (2001)

[10] Bartkowiak, W., Zaleśny, R., Leszczynski, J.: Relation between bond-length alternation and two-photon absorption of push-pull conjugated molecules: A quantum-chemical study. Chem. Phys. **287**, 103–112 (2003)

[11] Bartkowiak, W., Zaleśny, R., Niewodniczański, W., Leszczynski, J.: Quantum chemical calculations of the first- and second-order hyperpolarizability of molecules in solutions. J. Phys. Chem. A **105**, 10702–10710 (2001)

[12] Barzoukas, M., Runser, C., Fort, A., Blanchard-Desce, M.: A two-state description of (hyper) polarizabilities of push-pull molecules based on a two-form model. Chem. Phys. Lett. **257**, 531–537 (1996)

[13] Birge, R.R., Pierce, B.M.: A theoretical analysis of the two-photon properties of linear polyenes and the visual chromophores. J. Chem. Phys. **70**, 165–178 (1979)

[14] Bishop, D.M.: Molecular vibrational and rotational motion in static and dynamic electric field. Rev. Mod. Phys. **62**, 343–374 (1990)

[15] Bishop, D.M.: Explicit nondivergent formulas for atomic and molecular dynamic hyperpolarizabilities. J. Chem. Phys. **100**, 6535–6542 (1994)

[16] Bishop, D.M.: Molecular vibration and nonlinear optics. Adv. Chem. Phys. **104**, 1–40 (1998)

[17] Bishop, D.M., Champagne, B., Kirtman, B.: Relationship between static vibrational and electronic hyperpolarizabilities of π-conjugated push-pull molecules within the two-state valence-bond charge-transfer model. J. Chem. Phys. **109**, 9987–9994 (1998)

[18] Bishop, D.M., Kirtman, B.: A peturbation method for calculating vibrational dynamic dipole polarizabilities and hyperpolarizabilities. J. Chem. Phys. **95**, 2646–2658 (1991)

[19] Bishop, D.M., Kirtman, B., Champagne, B.: Differences between the exact sum-over-states and the canonical approximation for the calculation of static and dynamic hyperpolarizabilities. J. Chem. Phys. **103**, 365–373 (1997)

[20] Bishop, D.M., Norman, P.: Calculations of dynamic hyperpolarizabilities for small and medium sized molecules. In: Handbook of Advanced Electronic and Photonic Materials and Devices, vol. 9. Academic, San Diego (2001)

[21] Blanchard-Desce, M., Barzoukas, M.: Two-form two-state analysis of polarizabilities of push-pull molecules. J. Opt. Soc. Am. B **15**, 302–307 (1998)

[22] Bourhill, G., Brédas, J.L., Cheng, L.-T., Marder, S.R., Meyers, F., Perry, J.W., Tiemann, B.G.: Experimental demonstration of the dependence of the first hyperpolarizability of donor-acceptor-substituted polyenes on the ground-state polarization and bond length alternation. J. Am. Chem. Soc. **116**, 2619–2620 (1994)

[23] Brédas, J.L., Adant, C., Tackx, P., Persoons, A.: Third-order nonlinear optical response in organic materials: Theoretical and experimental aspects. Chem. Rev. **94**, 243–278 (1994)

[24] Bursi, R., Lankhorst, M., Feil, D.: Uncoupled Hartree-Fock calculations of the polarizability and hyperpolarizabilities of nitrophenols. J. Comp. Chem. **16**, 545–562 (1995)

[25] Champagne, B., Jacquemin, D., André, J.-M., Kirtman, B.: Ab initio coupled Hartree-Fock investigation of the static first hyperpolarizability of model all-trans-polymethineimine oligomers of increasing size. J. Phys. Chem. A **101**, 3158–3165 (1997)

[26] Champagne, B., Kirtman, B.: Theoretical approach to the design of organic molecular and polymeric nonlinear optical materials. In: Handbook of Advanced Electronic and Photonic Materials and Devices, vol. 9. Academic, San Diego (2001)

[27] Chen, G., Lu, D., Goddard, III, W.A.: Valence-bond charge-transfer solvation model for nonlinear optical properties of organic molecules in polar solvents. J. Chem. Phys. **101**, 5860–5864 (1994)
[28] Cohen, H.D., Roothaan, C.C.J.: Electric dipole polarizability of atoms by the Hartree-Fock method. I. Theory for closed-shell systems. J. Chem. Phys. **43**, S34–S39 (1965)
[29] Cronstrand, P., Luo, Y., Ågren, H.: Generalized few-state models for two-photon absorption of conjugated molecules. Chem. Phys. Lett. **352**, 262–269 (2002)
[30] Cronstrand, P., Norman, P., Luo, Y., Ågren, H.: Few-states models for three-photon absorption. J. Chem. Phys. **121**, 2020–2029 (2004)
[31] Dalgarno, A.: Perturbation theory for atomic systems. Proc. Roy. Soc. (London) **A251**, 282–290 (1959)
[32] Dalgarno, A., McNamee, J.M.: Calculation of polarizabilities and shielding factors. J. Chem. Phys. **35**, 1517–1518 (1961)
[33] Dehu, C., Meyers, F., Hendrickx, E., Clays, K., Persoons, A., Marder, S.R., Brédas, J.L.: Solvent effects on the second-order nonlinear optical response of π-conjugated molecules: A combined evaluation through self-consistent reaction field calculations and hyper-Rayleigh scaterring measurements. J. Am. Chem. Soc. **117**, 10127–10128 (1995)
[34] Di Bella, S., Marks, T.J., Ratner, M.A.: Environmental efects on nonlinear optical chromophore performance. Calculation of molecular quadratic hyperpolarizabilities in solvating media. J. Am. Chem. Soc. **116**, 4440–4445 (1994)
[35] Dick, B., Hohlneicher, G.: Importance of initial and final states as intermediate states in two-photon spectroscopy of polar molecules. J. Chem. Phys. **76**, 5755–5760 (1982)
[36] Dirk, C.W., Cheng, L.-T., Kuzyk, M.G.: A simplified three-level model describing the molecular third-order nonlinear optical susceptibility. Int. J. Quant. Chem. **43**, 27–36 (1992)
[37] Fanti, M., Fowler, P.W., Orlandi, G., Zerbetto, F.: Ab initio scaling of the second hyperpolarizability of carbon cages. J. Chem. Phys. **107**, 5072–5075 (1997)
[38] Freo, L. Del, Terenziani, F., Painelli, A.: Static nonlinear optical susceptibilities: Testing approximation schemes against exact results. J. Chem. Phys. **116**, 755–761 (2002)
[39] Hamada, T.: Ab initio estimation of quadratic hyperpolarizabilities of organic molecules: Sum over states vs. coupled perturbed Hartree-Fock. Nonlinear Optics **16**, 279–289 (1996)
[40] Hameka, H.F.: Calculation of linear and nonlinear electric susceptibilities of conjugated hydrocarbon chains. J. Chem. Phys. **67**, 2935–2942 (1977)
[41] Hrobarik, P., Zahradnik, P., Fabian, W.M.F.: Computational design of benzothiazole-derived pushpull dyes with high molecular quadratic hyperpolarizabilities. Phys. Chem. Chem. Phys. **6**, 495–501 (2004)
[42] Jacquemin, D., Champagne, B., André, J.-M.: Molecular orbital expressions for approximate uncoupled Hartree-Fock second hyperpolarizabilities. A Pariser-Parr-Pople assessment for model polyacetylene chains. Chem. Phys. **197**, 107–127 (1995)
[43] Kanis, D.R., Marks, T.J., Ratner, M.A.: Calculation and electronic description of molecular quadratic hyperpolarizabilities employing the ZINDO-SOS quantum chemical formalism. Chromophore architecture and substituent effects. Nonlinear Optics **6**, 317–335 (1994)
[44] Keinan, S., Zojer, E., Brédas, J.-L., Ratner, M.A., Marks, T.J.: Twisted π-system electro-optic chromophores. A CIS vs. MRD-CI theoretical investigation. J. Mol. Struct. (Theochem) **633**, 227–235 (2003)
[45] Kobko, N., Masunov, A., Tretiak, S.: Calculations of the third-order nonlinear optical responses in push-pull chromophores with a time-dependent density functional theory. Chem. Phys. Lett. **392**, 444–451 (2004)
[46] Kogej, T., Beljonne, D., Meyers, F., Perry, J.W., Marder, S.R., Brédas, J.L.: Mechanism for enhancement of two-photon absorption in donor-acceptor conjugated chromophores. Chem. Phys. Lett. **298**, 1–6 (1998)
[47] Kurtz, H.J., Stewart, J.J.P., Dieter, K.M.: Calculation of the nonlinear optical properties of molecules. J. Comp. Chem. **11**, 82–87 (1990)
[48] Kuzyk, M.G., Dirk, C.W.: Effects of centrosymmetry on the nonresonant electronic third-order nonlinear optical susceptibility. Phys. Rev. A **41**, 5098–5109 (1990)

[49] Lalama, S.J., Garito, A.F.: Origin of the nonlinear second-order optical susceptibility of organic systems. Phys. Rev. A **20**, 1179–1194 (1979)
[50] Langhoff, P.W., Karplus, M., Hurst, R.P.: Approximations to Hartree-Fock perturbation theory. J. Chem. Phys. **44**, 505–514 (1966)
[51] Lee, W.-H., Lee, H., Kim, J.-A., Choi, J.-H., Cho, M., Jeon, S.-J., Cho, B.R.: Two-photon absorption and nonlinear optical properties of octupolar molecules. J. Am. Chem. Soc. **123**, 10658–10667 (2001)
[52] Li, D., Marks, T.J., Ratner, M.A.: π electron calculations for prediciting non-linear optical properties of molecules. Chem. Phys. Lett. **131**, 370–375 (1986)
[53] Li, D., Ratner, M., Marks, T.J.: Molecular and macromolecular nonlinear optical materials. Probing architecture/electronic structure/frequency doubling relationship via SCF-LCAO MECI π electron formalism. J. Am. Chem. Soc. **110**, 1707–1715 (1988)
[54] Liebmann, S.P., Moskowitz, J.W.: Polarizabilities and hyperpolarizabilities of small polyatomic molecules in the uncoupled Hartree-Fock approximation. J. Chem. Phys. **54**, 3622–3631 (1971)
[55] Lipiński, J.: Modified all-valence INDO/spd method for ground and excited state properties of isolated molecules and molecular complexes. Int. J. Quant. Chem. **34**, 423–434 (1988)
[56] Lipiński, J.: On the consequences of the violation of the Hellmann-Feynman theorem in calculations of electric properties of molecules. Chem. Phys. Lett. **363**, 313–318 (2002)
[57] Lipiński, J., Bartkowiak, W.: Conformation and solvent dependence of the first and second molecular hyperpolarizabilities of charge-transfer chromophores. Quantum-chemical calculations. Chem. Phys. **245**, 263–276 (1999)
[58] Liu, X.-J., Feng, J.-K., Ren, A.-M., Zhou, X.: Theoretical studies of the spectra and two-photon absorption cross section for porphyrin and carbaporphirin. Chem. Phys. Lett. **373**, 197–206 (2003)
[59] Lu, D., Chen, G., Perry, J.W., Goddard, III, W.A.: Valence-bond charge-transfer model for nonlinear optical properties of charge-transfer organic molecules. J. Am. Chem. Soc. **116**, 10679–10685 (1994)
[60] Luis, J.M., Martí, J., Duran, M., Andrés, J.L.: Nuclear relaxation and vibrational contributions to the static electrical properties of polyatomic molecules: Beyond the Hartree-Fock approximation. Chem. Phys. **217**, 29–42 (1997)
[61] Luo, Y., Ågren, H., Knuts, S., Minaev, B.F., Jørgensen, P.: Response theory calculations of the vibrationally induced $^1A_{1g} - {}^1B_{1u}$ two-photon spectrum of benzene. Chem. Phys. Lett. **209**, 513–518 (1993)
[62] Luo, Y., Ågren, H., Vahtras, O.: The hyperpolarizability dispersion of para-nitroaniline. Chem. Phys. Lett. **207**, 190–194 (1993)
[63] Luo, Y., Norman, P., Macak, P., Ågren, H.: Solvent-induced two-photon absorption of a push-pull molecule. J. Phys. Chem. A **104**, 4718–4722 (2000)
[64] Göppert-Mayer, M.: Über Elementarakte mit zwei Quantensprungen. Ann. Phys. **9**, 273–294 (1931)
[65] Macak, P., Luo, Y., Ågren, H.: Simulations of vibronic profiles in two-photon absorption. Chem. Phys. Lett. **330**, 447–456 (2000)
[66] Macak, P., Luo, Y., Norman, P., Ågren, H.: Electronic and vibronic contributions to two-photon absorption of molecules with multi-branched structures. J. Chem. Phys. **113**, 7055–7061 (2000)
[67] Marchese, F.T., Seliskar, C.J., Jaffé, H.H.: The use of CNDO in spectroscopy. XV. Two-photon absorption. J. Chem. Phys. **72**, 4149–4203 (1980)
[68] Marder, S.R., Beratan, D.N., Cheng, L.-T.: Approaches for optimizing the first electronic hyperpolarizability of conjugated organic molecules. Science **252**, 103–106 (1991)
[69] Marder, S.R., Cheng, L.-T., Tiemann, B.G., Friedli, A.C., Blanchard-Desce, M., Perry, J.W., Skindhøj, J.: Large first hyperpolarizability in push-pull polyenes by tuning of the bond length alternation and aromaticity. Science **263**, 511–514 (1994)
[70] Marder, S.R., Gorman, C.B., Meyers, F., Perry, J.W., Bourhill, G., Brédas, J.-L., Pierce, B.M.: A unified description of linear and nonlinear polarization in organic polymethine dyes. Science **265**, 632–635 (1994)
[71] Marder, S.R., Gorman, C.B., Tiemann, B.G., Cheng, L.-T.: Stronger acceptors can diminish nonlinear optical response in simple donor-acceptor polyenes. J. Am. Chem. Soc. **115**, 3006–3007 (1993)

[72] Marder, S.R., Perry, J.W., Bourhill, G., Gorman, C.B., Tiemann, B.G., Mansour, K.: Relation between bond-length alternation and second electronic hyperpolarizability of conjugated organic molecules. Science **261**, 186–189 (1993)
[73] Marder, S.R., Perry, J.W., Tiemann, B.G., Gorman, C.B., Gilmour, S., Biddle, S.L., Bourhill, G.: Direct observation of reduced bond length alternation in donor/acceptor polyenes. J. Am. Chem. Soc. **115**, 2524–2526 (1993)
[74] Masunov, A., Tretiak, S.: Prediction of two-photon absorption properties for organic chromophores using time-dependent density-functional theory. J. Phys. Chem. B **108**, 899–907 (2004)
[75] McClain, W.M., Harris, R.A.: Two-photon molecular spectroscopy in liquid and gases. In: Excited States, vol. 3. Academc, New York (1977)
[76] McIntyre, E.F., Hameka, H.F.: Calculation of nonlinear electric susceptibilities of aromatic hydrocarbon chains. J. Chem. Phys. **68**, 5534–5537 (1978)
[77] McIntyre, E.F., Hameka, H.F.: Extended basis set calculations of nonlinear susceptibilities of conjugated hydrocarbons. J. Chem. Phys. **69**, 4814–4820 (1978)
[78] McIntyre, E.F., Hameka, H.F.: Improved calculation of nonlinear electric susceptibilities of conjugated hydrocarbon chains. J. Chem. Phys. **68**, 3481–3484 (1978)
[79] Meyers, F., Brédas, J.L., Pierce, B.M., Marder, S.R.: Nonlinear optical properties of donor-acceptor polyenes: Frequency-dependent calculations of the relationship among molecular polarizabilities, α, β, and γ, and bond-length alternation. Nonlinear Optics **14**, 61–71 (1995)
[80] Meyers, F., Marder, S.R., Pierce, B.M., Brédas, J.L.: Electric field modulated nonlinear optical properties of donor-acceptor polyenes: Sum-over-states investigation of the relationship between molecular polarizabilities (α, β and γ) and bond length alternation. J. Am. Chem. Soc. **116**, 10703–10714 (1994)
[81] Monson, P.R., McClain, W.M.: Polarization dependence of the two-photon absorption of tumbling molecules with application to liquid 1-chloronaphtalene and benzene. J. Chem. Phys. **53**, 29–37 (1970)
[82] Morell, J.A., Albrecht, A.C.: Second-order hyperpolarizability of p-nitroaniline calculated from perturbation theory based expression using CNDO/S generated electronic states. Chem. Phys. Lett. **64**, 46–50 (1979)
[83] Morley, J.O., Pavlides, P., Pugh, D.: On the calculation of the hyperpolarizabilities of organic molecules by the sum over virtual excited states method. Int. J. Quant. Chem. **43**, 7–26 (1992)
[84] Mulliken, R.S.: Molecular compounds and their spectra. II. J. Am. Chem. Soc. **74**, 811–824 (1952)
[85] Musher, J.I.: Hartree-Fock theory of atomic properties. J. Chem. Phys. **46**, 369–372 (1967)
[86] Norman, P., Cronstrand, P., Ericsson, J.: Theoretical study of linear and nonlinear absorption in platinum-organic compounds. Chem. Phys. **285**, 207–220 (2002)
[87] Norman, P., Luo, Y., Ågren, H.: Large two-photon absorption cross sections in two-dimensional, charg-transfer, cumulene-containing aromatic molecules. J. Chem. Phys. **111**, 7758–7765 (1999)
[88] O'Hare, J.M., Hurst, R.P.: Hyperpolarizabilities of some polar diatomic molecules. J. Chem. Phys. **46**, 2356–2366 (1967)
[89] Olsen, J., Jørgensen, P.: Linear and nonlinear response functions for an exact state and for and MCSCF state. J. Chem. Phys. **82**, 3235–3264 (1985)
[90] Olsen, J., Jørgensen, P.: Time-dependent response theory with applications to self-consistent field and multiconfigurational self-consistent field wave functions. In: Modern Electronic Structure Theory, vol. 2. World Scientific, Singapore (1995)
[91] Orr, B.J., Ward, J.F.: Perturbation theory of the non-linear optical polarization of an isolated system. Mol. Phys. **20**, 513–526 (1971)
[92] Otto, P., Gu, F.L., Ladik, J.: Calculation of ab initio dynamic hyperpolarizabilities of polymers. J. Chem. Phys. **110**, 2717–2726 (1999)
[93] Oudar, J.L.: Optical nonlinearities of conjugated molecules. Stilbene derivatives and highly polar aromatic compounds. J. Chem. Phys. **67**, 446–457 (1977)
[94] Oudar, J.L., Chemla, D.S.: Hyperpolarizabilities of the nitroanilines and their relations to the excited state dipole moment. J. Chem. Phys. **66**, 2664–2668 (1977)

[95] Pati, S.K., Marks, T.J., Ratner, M.A.: Conformationally tuned large two-photon absorption cross sections in simple molecular chromophores. J. Am. Chem. Soc. **123**, 7287–7291 (2001)
[96] Peticolas, W.L., Goldsborough, J.P., Rieckhoff, K.E.: Double photon exctation in organic crystals. Phys. Rev. Lett. **10**, 43–45 (1963)
[97] Peticolas, W.L., Rieckhoff, K.E.: Double-photon excitation on organic molecules in dilute solution. J. Chem. Phys. **39**, 1347–1348 (1963)
[98] Pierce, B.M.: A theoretical analysis of third-order nonlinear optical properties of linear polyenes and benzene. J. Chem. Phys. **91**, 791–811 (1989)
[99] Quinet, O., Kirtman, B., Champagne, B.: Analytical time-dependent Hartree-Fock evaluation of the dynamic zero-point vibrationally averaged (ZPVA) first hyperpolarizability. J. Chem. Phys. **118**, 505–513 (2003)
[100] Sadlej, A.J.: Perturbation theory of the electron correlation effects for atomic and molecular properties. J. Chem. Phys. **75**, 320–331 (1981)
[101] Sałek, P., Vahtras, O., Guo, J., Luo, Y., Helgaker, T., Ågren, H.: Calculations of two-photon absorption cross sections by means of density-functional theory. Chem. Phys. Lett. **374**, 446–452 (2003)
[102] Sałek, P., Vahtras, O., Helgaker, T., Ågren, H.: Density-functional theory of linear and nonlinear time-dependent molecular properties. J. Chem. Phys. **117**, 9630–9645 (2002)
[103] Shuai, Z., Brédas, J.L.: Electronic structure and nonlinear optical properties of fullerenes C_{60} and C_{70}: A valence-effective-Hamiltonian study. Phys. Rev. B **46**, 16135–16141 (1992)
[104] Sim, F., Chin, S., Dupuis, M., Rice, J.E.: Electron correlation effects in hyperpolarizabilities of p-nitroaniline. J. Phys. Chem. **97**, 1158–1163 (1993)
[105] Tomonari, M., Ookubo, N., Takada, T.: Missing-orbital analysis of molecular hyperpolarizability β calculated by a simplified sum-over-states method. Enhancement of the off-diagonal component β_{zxx}. Chem. Phys. Lett. **215**, 45–52 (1993)
[106] Tomonari, M., Ookubo, N., Takada, T.: Missing-orbital analysis of molecular hyperpolarizability β calculated by a simplified sum-over-states method: "multi-state" model for β_{zxx}. J. Mol. Struct.: (Theochem) **311**, 221–232 (1994)
[107] Tomonari, M., Ookubo, N., Takada, T., Feyereisen, M.W., Almlöf, J.: Simplified sum-over-states calculations and missing-orbital analysis on hyperpolarizabilities of benzene derivatives. Chem. Phys. Lett. **203**, 603–610 (1993)
[108] Tretiak, S., Chernyak, V.: Resonant nonlinear polarizabilities in the time-dependent density functional theory. J. Chem. Phys. **119**, 8809–8823 (2003)
[109] Tretiak, S., Mukamel, S.: Density matrix analysis and simulation of electronic excitations in conjugated and aggregated molecules. Chem. Rev. **102**, 3171–3212 (2002)
[110] Tuan, D.F.-T., Epstein, S.T., Hirschfelder, J.O.: Improvements of uncoupled Hartree-Fock expectations values for physical properties. J. Chem. Phys. **44**, 431–433 (1966)
[111] Velders, G.J.M., Gillet, J.-M., Becker, P.J., Feil, D.: Electron density analysis of nonlinear optical materials. An ab initio study of different conformations of benzene derivatives. J. Phys. Chem. **95**, 8601–8606 (1991)
[112] Wang, C.-K., Macak, P., Luo, Y., Ågren, H.: Effects of π centers and symmetry on two-photon absorption cross sections of organic chromophores. J. Chem. Phys. **114**, 9813–9820 (2001)
[113] Ward, J.F.: Calculation of nonlinear optical susceptibilities using diagrammatic perturbation theory. Rev. Mod. Phys. **37**, 1–18 (1965)
[114] Warshel, A., Weiss, R.M.: An empirical valence bond approach for comparing reactions in solutions and in enzymes. J. Am. Chem. Soc. **102**, 6218–6226 (1980)
[115] Willets, A., Rice, J.E., Burland, D.M., Shelton, D.P.: Problems in the comparison of theoretical and experimental hyperpolarizabilities. J. Chem. Phys. **97**, 7590–7599 (1992)
[116] Xie, R.-H.: Empirical exponent law of the second-order hyperpolarizability in small armchair and zig-zag nanotubes. J. Chem. Phys. **108**, 3626–3629 (1998)
[117] Xie, R.-H., Jiang, J.: Large third-order optical nonlinearities of C_{60}–derived nanotubes in infrared. Chem. Phys. Lett. **280**, 66–72 (1997)

[118] Zaleśny, R., Bartkowiak, W., Leszczynski, J.: Theoretical study of the two-photon absorption in phtochromic fulgides. J. Luminescence **105**, 111–116 (2003)
[119] Zaleśny, R., Bartkowiak, W., Styrcz, S., Leszczynski, J.: Solvent effects on conformationally induced enhancement of the two-photon absorption cross section of a pyridinium-N-phenolate betaine dye. A quantum-chemical study. J. Phys. Chem. A **106**, 4032–4037 (2002)
[120] Zaleśny, R., Sadlej, A.J., Leszczynski, J.: Size-nonextensive contributions in singles-only CI. Struct. Chem. **15**, 379–384 (2004)
[121] Zamani-Khamiri, O., Hameka, H.F.: Polarizability calculations with the SCF method. I. Linear and dynamic polarizabilities of conjugated hydrocarbons and aromatics. J. Chem. Phys. **71**, 1607–1610 (1979)
[122] Zamani-Khamiri, O., Hameka, H.F.: Polarizability calculations with the SCF method. IV. Various conjugated hydrocarbons. J. Chem. Phys. **73**, 5693–5697 (1980)
[123] Zamani-Khamiri, O., McIntyre, E.F., Hameka, H.F.: Polarizability calculations with the SCF method. II. The benzene molecule. J. Chem. Phys. **72**, 1280–1284 (1980)
[124] Zamani-Khamiri, O., McIntyre, E.F., Hameka, H.F.: Polarizability calculations with the SCF method. III. Ethylene, butadiene, and hexatriene. J. Chem. Phys. **72**, 5906–5908 (1980)
[125] Zhou, X., Ren, A.-M., Feng, J.-K., Liu, X.-J.: A comparative study of the two-photon absorption properties of a new octupolar molecule – truxeone derivative and relative molecules. Chem. Phys. Lett. **373**, 167–175 (2003)
[126] Zhou, X., Ren, A.-M., Feng, J.-K., Liu, X.-J., Zhang, J., Liu, J.: One- and two-photon absorption properties of novel multi-branched molecules. Phys. Chem. Chem. Phys. **4**, 4346–4352 (2002)
[127] Zhou, Y.-F., Wang, X.-M., Zhao, X., Jiang, M.-H.: A quantum-chemical INDO/CI method for calculating two-photon cross sections. J. Phys. Chem. Sol. **62**, 1075–1079 (2001)

CHAPTER 5

KOHN–SHAM TIME-DEPENDENT DENSITY FUNCTIONAL THEORY WITH APPLICATIONS TO LINEAR AND NONLINEAR PROPERTIES

DAN JONSSON[1], OLAV VAHTRAS[2], BRANISLAV JANSIK[2], ZILVINAS RINKEVICIUS[2], PAWEŁ SAŁEK[2], AND HANS ÅGREN[2]

[1] *Department of Physics, Stockholm University, SE-10691 Stockholm, Sweden*
[2] *Laboratory of Theoretical Chemistry, The Royal Institute of Technology, SE-10044 Stockholm, Sweden*

Abstract: We review Kohn–Sham density-functional theory for time-dependent response functions up to and including cubic response. The working expressions are derived from an explicit exponential parametrization of the density operator and the Ehrenfest principle, alternatively the quasi-energy ansatz. While the theory retains the adiabatic approximation, implying that the time-dependency of the functional is obtained only implicitly—through the time-dependency of the density itself rather than through the form of the exchange-correlation functionals—our implementation generalizes previous time-dependent approaches in that arbitrary functionals can be chosen for the perturbed densities (energy derivatives or response functions). Thus, the response of the density can always be obtained using the stated density functional, or optionally different functionals can be applied for the unperturbed and perturbed densities, even different functionals for different response order. In particular, general density functionals beyond the local density approximation can be applied, such as hybrid functionals with exchange–correlation at the generalized gradient-approximation level and fractional exact Hartree–Fock exchange. We also review some recent progress in time-dependent density functional theory for open-shell systems, in particular spin-restricted and spin restricted-unrestricted formalisms for property calculations. We highlight a sample of applications of the theory

1. INTRODUCTION

The development of quantum modeling has by now reached a level where a wide range of microscopic phenomena taking place at different length and time scales can be studied. One important factor behind this development is the research in linear scaling technologies for the Coulomb interaction and in the algorithms for optimization of the density or the wave function. Another factor is the development of density functional theory, in particular, the improvement of exchange-correlation functionals that match asymptotic criteria of various kinds. Yet another factor refers to the general strive to augment electronic structure theory with concomitant theories or models for the calculations of properties that have taken place over the last two decades.

Analytic response theory, which represents a particular formulation of time-dependent perturbation theory, has constituted a core technology in much of the this development. Response functions provide a universal representation of the response of a system to perturbations, and are applicable to all computational models, density-functional as well as wave-function models, and to all kinds of perturbations, dynamic as well as static, internal as well as external perturbations. The analytical character of the theory with properties evaluated from analytically derived expressions at finite frequencies, makes it applicable for a large range of experimental conditions. The theory is also model transferable in that, once the computational model has been defined, all properties are obtained on an equal footing, without further approximations.

The vast success of density functional theory in chemistry on one hand and of (time-dependent) response theory on the other has made it desirable to merge the two into a modeling toolbox for general properties. A rigorous foundation of time-dependent DFT was given by Runge and Gross [1] already in 1984 and there has since been an ever increasing number of studies and applications concerning time-dependent response properties [2–40]. Most of the applications have been focused on linear properties, where in particular polarizabilities and excitation energies have been studied both for LDA and GGA type functionals. The generalization of linear time-dependent response to quadratic and to cubic order of response theory has recently been accomplished by the present authors [41, 42], motivated by the great number of properties that then can be addressed, for instance in the field of non-linear optics and magnetic resonance experiments. In parallel to generalizations to higher order we must also see generalizations to higher complexity in the operators; spin dependence in the operators which imply spin-dependent response in the wave function or density.

In order to accomplish such a goal we found it worthwhile to go back and edit the formulation of response theory based on the Ehrenfest theorem, of Olsen and Jorgensen in 1985, which has been very successful for ab initio property calculations, such as those based on Hartree–Fock, Multi-Configurational and Coupled Cluster wave functions, and generalize it to a density functional theory in which we make an exponential ansatz of the time-evolution of the density [41, 42]. The results of such an approach was that we obtained a direct analogue in density functional theory to what has been accomplished on the ab initio side, which also is of pedagogic and programming advantage. The other, perhaps more decisive advantage, is that we so obtained a time-dependent DFT methodology for linear and non-linear properties that transcends the ALDA—adiabatic local density—approximation in which the response of the density always is obtained from LDA functional derivatives even when the density itself is calculated using a finer approximation, such as the generalized-gradient approximation. By contrast, in our implementation, the response of the density is always obtained using the stated DFT functional. We still invoke the adiabatic approximation, however, meaning that the time-dependent functional is assumed to depend on the time-dependent density in the same manner that the time-independent functional depends on the time-independent density. This allows the application of basically all modern exchange-correlation functionals for the perturbed density, like hybrid functionals including exchange–correlation functionals at the general gradient-approximation level and fractional exact Hartree–Fock exchange. This means also that different functionals can be used for the unperturbed and perturbed densities, even different functionals for different orders of the perturbation (different orders of response functions or energy derivatives).

The purpose of this review is to highlight the recent theoretical development of density functional response theory, or more specifically Kohn–Sham time-dependent density functional theory, where we also address some recent progress in time-dependent DFT for open-shell systems.

2. THEORY

2.1 Response Functions

Response theory is perturbation theory with emphasis on properties rather than on states and energies, aiming to describe how the properties respond to internal or external perturbations. Consider the expectation value of a time-independent operator $\hat{A}$, which for the unperturbed system is given by $\langle 0|\hat{A}|0\rangle$. In the presence of a time-dependent perturbation $\hat{V}(t)$, the expectation value of $\hat{A}$ becomes time-dependent and may furthermore be expanded in powers of the perturbation:

(1) $$\langle t|\hat{A}|t\rangle = \langle t|\hat{A}|t\rangle^{(0)} + \langle t|\hat{A}|t\rangle^{(1)} + \langle t|\hat{A}|t\rangle^{(2)} + \langle t|\hat{A}|t\rangle^{(3)} + \cdots$$

Each term in this expansion has a Fourier representation of the form

(2) $$\langle t|\hat{A}|t\rangle^{(1)} = \int \langle\langle \hat{A}; \hat{V}^{\omega_1}\rangle\rangle_{\omega_1} \exp(-i\omega_1 t)\, d\omega_1$$

(3) $$\langle t|\hat{A}|t\rangle^{(2)} = \frac{1}{2}\iint \langle\langle \hat{A}; \hat{V}^{\omega_1}, \hat{V}^{\omega_2}\rangle\rangle_{\omega_1,\omega_2} \exp[-i(\omega_1+\omega_2)t]\, d\omega_1\, d\omega_2$$

(4) $$\langle t|\hat{A}|t\rangle^{(3)} = \frac{1}{6}\iiint \langle\langle \hat{A}; \hat{V}^{\omega_1}, \hat{V}^{\omega_2}, \hat{V}^{\omega_3}\rangle\rangle_{\omega_1,\omega_2,\omega_3}$$
$$\times \exp[-i(\omega_1+\omega_2+\omega_3)t]\, d\omega_1\, d\omega_2\, d\omega_3$$

and so on, where $\hat{V}^\omega$ is the Fourier transform of $\hat{V}(t)$

(5) $$\hat{V}(t) = \int_{-\infty}^{\infty} \hat{V}^\omega \exp(-i\omega t)\, d\omega$$

In practice, we work with monochromatic perturbations; the integrals in Eqs. (2)–(5) are then replaced by discrete sums.

In the following, we shall derive and implement expressions for the linear, quadratic, and cubic response functions $\langle\langle \hat{A}; \hat{V}^{\omega_1} \rangle\rangle_{\omega_1}$, $\langle\langle \hat{A}; \hat{V}^{\omega_1}, \hat{V}^{\omega_2} \rangle\rangle_{\omega_1,\omega_2}$, and $\langle\langle \hat{A}; \hat{V}^{\omega_1}, \hat{V}^{\omega_2}, \hat{V}^{\omega_3} \rangle\rangle_{\omega_1,\omega_2,\omega_3}$ within the framework of Kohn–Sham DFT. In this particular formulation of DFT, the electron density $\rho(\mathbf{r}, t)$ is parameterized in terms of a time-dependent reference determinant $|t\rangle$. There are thus no problems associated with the calculation of expectation values of one-electron operators (even differential ones), as assumed in Eq. (1).

2.2 The Spin-restricted Kohn–Sham Approach

2.2.1 Motivation

It is a fundamental fact of quantum mechanics, that a spin-independent Hamiltonian will have pure spin eigenstates. For approximate wave functions that do not fulfill this criterion, e.g. those obtained with various unrestricted methods, the expectation value of the square of the total spin angular momentum operator, $\langle S^2 \rangle$, has been used as a measure of the degree of spin contamination. S^2 is obviously a two-electron operator and the evaluation of its expectation value thus requires knowledge of the two-electron density matrix.

Density functional theory (DFT) is used to calculate the one-electron density of an interacting system by means of a fictitious non-interacting system with the same density. Within this framework it is thus not possible to evaluate two-electron properties without further approximations. Tempting as it may be to use the non-interacting two-electron density, because it is easily implemented in quantum chemistry programs, it is not a theoretically justified procedure. Wang et al. [43] expressed the two-electron density matrix in terms of one-electron density matrices, using Löwdin's relation [44], which is exact for a Slater determinant, but otherwise an approximation. The one-electron density matrix was further written in terms of the density and a correlation hole function, of which the latter was estimated within two simple approximations; a Gaussian function and a homogeneous electron gas approximation. It was shown that for the electron gas approximation the exact value $S(S+1)$ was a lower bound for the calculated value of $\langle S^2 \rangle$ and that spin contamination was absent whenever $\rho^\alpha(\mathbf{r}) > \rho^\beta(\mathbf{r})$ in all space. This obviously holds for spin-restricted density functional method where the non-interacting wave-function is a high-spin determinant.

One of our main motivations for pursuing the development of a density functional response theory for open-shell systems has been to calculate spin-Hamiltonian parameters which are fundamental to experimental magnetic resonance spectroscopy. It is only within the context of a state with well-defined spin we can speak of effective spin Hamiltonians. The relationship between microscopic and effective Hamiltonians rely on the Wigner-Eckart theorem for tensor operators of a specific rank and states which transform according to their irreducible representations [45].

2.2.2 Spin-restricted optimization

In Kohn–Sham theory [46] the ground state energy of an interacting system of N electrons in an external potential $v(\mathbf{r})$ is obtained from

$$(6) \quad E[\rho] = -\frac{1}{2}\int d\mathbf{r} \langle \Psi | \hat{\psi}(\mathbf{r})^\dagger \nabla^2 \hat{\psi}(\mathbf{r}) | \Psi \rangle + \frac{1}{2}\iint d\mathbf{r}_1 d\mathbf{r}_2 \frac{\rho(\mathbf{r}_1)\rho(\mathbf{r}_2)}{|\mathbf{r}_1 - \mathbf{r}_2|} + E_{xc}[\rho] + \int d\mathbf{r}\rho(\mathbf{r})v(\mathbf{r})$$

where the use of Kohn–Sham (KS) spin-orbitals $\{\phi_i\}$ is implied;

$$(7) \quad \rho(\mathbf{r}) = \sum_i^N \phi_i^* \phi_i \equiv \langle \Psi | \hat{\psi}(\mathbf{r})^\dagger \hat{\psi}(\mathbf{r}) \Psi \rangle \equiv \langle \Psi | \hat{\rho}(\mathbf{r}) \Psi \rangle$$

$$(8) \quad \hat{\psi}(\mathbf{r}) = \sum_i \phi_i(\mathbf{r}) a_i$$

which diagonalize the effective one-electron Hamiltonian

$$(9) \quad F(\mathbf{r}) = -\frac{1}{2}\nabla^2 + \int d\mathbf{r}' \frac{\rho(\mathbf{r}')}{|\mathbf{r} - \mathbf{r}'|} + \frac{\delta E_{xc}}{\delta \rho(\mathbf{r})} + v(\mathbf{r})$$

The expectation values in (6) and (7) are with respect to the KS determinant which represents the state of the non-interacting reference system

$$(10) \quad |\Psi\rangle = (\prod_{i=1}^N a_i^\dagger)|\text{vac}\rangle$$

The exact energy functional (and the exchange correlation functional) are indeed functionals of the total density, even for open-shell systems [47]. However, for the construction of approximate functionals of closed as well as open-shell systems, it has been advantageous to consider functionals with more flexibility, where the α- and β-densities can be varied separately, i.e. $E[\rho_\alpha, \rho_\beta]$. The variational search for a minimum of the $E[\rho_\alpha, \rho_\beta]$ functional can be carried out by unrestricted and spin-restricted approaches. The two methods differ only by the conditions of constraint imposed in minimization and lead to different sets of Kohn–Sham equations for the spin orbitals. The unrestricted Kohn–Sham approach is the one most commonly used and is implemented in various standard quantum chemistry software packages. However, this method has a major disadvantage, namely a spin contamination problem, and in recent years the alternative spin-restricted Kohn–Sham approach has become a popular contester [48–50].

Consider a trial Kohn–Sham (KS) determinant, either a closed shell determinant or an open-shell "high-spin" determinant where all singly occupied orbitals have α spin. It is parameterized by a real unitary exponential operator, and the purpose of the transformation is to transform the orbitals to a state of minimum energy

$$(11) \quad |\tilde{0}\rangle = e^{-\hat{\kappa}}|0\rangle$$

If the α and β orbitals are constructed from the same set and the transformation is of the form

$$(12) \quad \hat{\kappa} = \sum_{rs} \kappa_{rs}(a^{\dagger}_{r\alpha}a_{s\alpha} + a^{\dagger}_{r\beta}a_{s\beta})$$

we have a spin-restricted optimization. Thus, a variation of the energy then gives

$$(13) \quad \delta E[\rho_\alpha, \rho_\beta] = \sum_{\sigma=\alpha}^{\beta} \int d\mathbf{r} \frac{\delta E}{\delta \rho_\sigma(\mathbf{r})} \delta \rho_\sigma(\mathbf{r})$$

$$= \sum_{\sigma=\alpha}^{\beta} \int d\mathbf{r} \frac{\delta E}{\delta \rho_\sigma(\mathbf{r})} \langle [\delta\hat{\kappa}, \hat{\rho}_\sigma(\mathbf{r})] \rangle$$

$$= \langle [\delta\hat{\kappa}, \hat{F}] \rangle$$

where we have introduced the Fock operator

$$(14) \quad \hat{F} = \sum_{\sigma=\alpha}^{\beta} \int d\mathbf{r} \hat{\psi}^{\dagger}_\sigma(\mathbf{r}) \frac{\delta E}{\delta \rho_\sigma(\mathbf{r})} \hat{\psi}_\sigma(\mathbf{r}) = \sum_{\sigma=\alpha}^{\beta} \sum_{rs} F^{\sigma}_{rs} a^{\dagger}_{r\sigma} a_{s\sigma}$$

of which the the α- and β-parts have matrix representations corresponding to the KS Fock matrices. A restricted open-shell Roothan optimization which is based on the diagonalization of an effective Fock matrix can be designed in different ways—the main criterion is that the off-diagonal blocks (closed-open, open-virtual, closed-virtual) are zero for a converged state. If we translate $\langle [\hat{\kappa}, \hat{F}] \rangle$ to a matrix formulation in an atomic orbital basis and we use the overlap S, the closed- (c) and open-shell (o) Fock matrices

$$(15) \quad F^c = \frac{F^\alpha + F^\beta}{2}$$

$$(16) \quad F^o = F^\alpha$$

and density matrices

$$(17) \quad D^c = 2D^\beta$$

$$(18) \quad D^o = D^\alpha - D^\beta$$

we obtain a gradient of the form

$$(19) \quad S(D^c F^c + D^o F^o) - (F^c D^c + F^o D^o)S$$

Similarly, a specific choice of an adequate effective Fock matrix is the closed shell Fock matrix with a correction such that the off-diagonal blocks associated with the open-shell orbitals are adjusted to be proportional to the orbital gradient

$$(20) \quad F = F^c + F^v + (F^v)^T$$

$$F^v = SD^o(F^c - F^o)((D^c + D^o)S - 1)$$

The effective Fock matrix (20) is in our implementation [51] the quantity which is averaged in the optimization based on the direct inversion in the iterative subspace (DIIS) method [52].

2.2.3 Spin polarization in restricted theories

Hyperfine couplings, in particular the isotropic part which measures the spin density at the nuclei, puts special demands on spin-restricted wave-functions. For example, complete active space (CAS) approaches are designed for a correlated treatment of the valence orbitals, while the core orbitals are doubly occupied. This leaves little flexibility in the wave function for calculating properties of this kind that depend on the spin polarization near the nucleus. This is equally true for self-consistent field methods, like restricted open-shell Hartree–Fock (ROHF) or Kohn–Sham (ROKS) methods. On the other hand, unrestricted methods introduce spin contamination in the reference (ground) state resulting in overestimation of the spin-polarization.

The philosophy of the restricted-unrestricted (RU) approach is a physically motivated compromise between the restricted and unrestricted methods; to optimize the wave function with a spin-restricted approach and to account for perturbations with an unrestricted approach. That is, a ground state constructed from α and β spin-orbitals with common orbital parts is used with satisfies the variational condition

$$(21) \quad \frac{dE}{d\kappa} = 0$$

where E is the electronic energy and κ is a molecular (spatial) orbital parameter (like the parameters of Eq. (12) collected in a vector). Next, the properties are calculated with a set of parameters that allow the α- and β-orbitals to respond independently to the perturbation. This can be done by considering parameters for α and β-orbitals separately, or as in the work of Fernandez et al. [53] in terms of singlet and triplet rotations of the ground state orbitals $\kappa = (\kappa_s, \kappa_t)$.

Next, we outline the derivations of Fernandez et al. [53] and the generalization for DFT. The hyperfine coupling (HFC) Hamiltonian is scaled with a perturbation strength parameter x equation such that the total energy and the wave function parameters are functions of x

$$(22) \quad \hat{H}(x) = \hat{H}_0 + x\hat{H}_{\text{hfc}}$$

such that energy and the parameters are functions of x

$$(23) \quad E(x, \kappa) = \langle \kappa(x) | \hat{H}(x) | \kappa(x) \rangle$$

This gives a first-order change in the total electronic energy as

$$(24) \quad E_{\text{hfc}} = \left. \frac{dE}{dx} \right|_{x=0} = \frac{\partial E}{\partial x} + \sum_{\sigma=s,t} \left. \frac{\partial E}{\partial \kappa_\sigma} \frac{d\kappa_\sigma}{dx} \right|_{x=0}$$

$$= \langle \kappa(0) | \hat{H}_{\text{hfc}} | \kappa(0) \rangle + \left. \frac{\partial E}{\partial \kappa_t} \frac{d\kappa_t}{dx} \right|_{x=0}$$

where we used Eq. (21), the variational condition for the singlet parameters; if the wave function is assumed to be optimized for finite x, the first-order response of the singlet-parameters is by definition zero.

In order to determine the response of the triplet parameters Fernandez et al. [53] assume that the change in the non-zero energy gradient is negligible when the perturbation is turned on ($x \neq 0$); this gives an unbiased treatment of perturbed and un-perturbed systems and ensures a continuous energy. A second approximation involves neglecting the singlet contributions in the calculations of the polarization effects. This is the same as to neglect the singlet-triplet coupling in the second derivative of the energy. The problem is then reduced to solving conventional RPA-type equations and the energy shift becomes

$$(25) \quad E_{\text{hfc}} = \langle \kappa | \hat{H}_{\text{hfc}} | \kappa \rangle - \frac{\partial^2 E}{\partial x \partial \kappa_t} \left(\frac{\partial^2 E}{\partial \kappa_t^2} \right)^{-1} \frac{\partial E}{\partial \kappa_t} \bigg|_{x=0}$$

which using the notation for response functions is equivalent to

$$(26) \quad E_{\text{hfc}} = \langle \hat{H}_{\text{hfc}} \rangle + \langle\langle \hat{H}_0; \hat{H}_{\text{hfc}} \rangle\rangle_0$$

This somewhat odd linear response function containing the unperturbed Hamiltonian is thus not a physical property but a term which reflects the computational structure of this theory (it vanishes for exact and fully variational theories).

In DFT the energy is not the expectation value of a Hamiltonian, but rather a functional of the form

$$(27) \quad E[\rho_x, x] = E[\rho_x] + x \langle \hat{H}_{\text{hfc}} \rangle$$

The ground state density will depend on the field strength parameter x, which is denoted by a subscript. The functional has both an explicit and an implicit dependence on x and the first-order energy correction, the total derivative is formed by a partial derivative and a functional derivative

$$(28) \quad \frac{dE}{dx}\bigg|_{x=0} = \frac{\partial E}{\partial x} + \int d\mathbf{r} \, \frac{\delta E}{\delta \rho_x(\mathbf{r})} \frac{\partial \rho_x(\mathbf{r})}{\partial x}\bigg|_{x=0}$$

The first term is the expectation value and from the fact that the density change is the static linear response function

$$(29) \quad \frac{\partial \rho_x}{\partial x}\bigg|_{x=0} = \langle\langle \hat{\rho}(\mathbf{r}); \hat{H}_{\text{hfc}} \rangle\rangle$$

and identifying the Fock operator of Eq. (14) we have that the second part of (28) is the linear response function for the Fock operator and the perturbation giving

$$(30) \quad \frac{dE}{dx}\bigg|_{x=0} = \langle \hat{H}_{\text{hfc}} \rangle + \langle\langle \hat{F}; \hat{H}_{\text{hfc}} \rangle\rangle$$

For the special case that the exchange-correlation potential is the Hartree–Fock exchange, Eqs. (26) and (30) are identical.

2.3 Time-dependent Kohn–Sham Theory

In Kohn–Sham theory, we assume that the time-dependent density $\rho(\mathbf{r}, t)$ is represented in terms of a time-dependent reference Slater determinant $|t\rangle$. The Kohn–Sham energy is then written as a functional of this density in the following manner:

$$E[\rho, t] = T_s[\rho] + V_{\text{ext}}[\rho, t] + J[\rho] + E_{\text{xc}}[\rho] + V_{\text{NN}} \tag{31}$$

The first term is the kinetic energy evaluated as an expectation value

$$T_s[\rho] = -\frac{1}{2} \sum_i \langle t | \nabla_i^2 | t \rangle \tag{32}$$

whereas the second and third terms represent, respectively, the classical Coulomb interactions of the electron density with the external potential and with itself:

$$V_{\text{ext}}[\rho, t] = V_N[\rho] + V[\rho, t] = \int \rho(\mathbf{r}, t) (\sum_I \frac{Z_I}{|\mathbf{r} - \mathbf{R}_I|} + v(\mathbf{r}, t)) \, d\tau \tag{33}$$

$$J[\rho] = \frac{1}{2} \iint \frac{\rho(\mathbf{r}_1, t) \rho(\mathbf{r}_2, t)}{r_{12}} \, d\tau_1 \, d\tau_2$$

In Eq. (32), we have split the external potential into a static nuclear potential and an explicitly time-dependent perturbation. The exchange–correlation functional $E_{\text{xc}}[\rho]$ in Eq. (31) contains all two-electron interactions except the Hartree term $J[\rho]$—that is, it includes the effects of exchange and correlation. In addition, it corrects for the error made in the evaluation of the kinetic energy according to Eq. (32). The last term in Eq. (31) represents the classical nuclear–nuclear repulsion energy.

In the widely used adiabatic approximation, which we have adopted here, the time-dependence of the exchange–correlation energy is contained in the density—that is, the exchange–correlation functional is approximated using the same functional form in the time-dependent and time-independent cases. It is not obvious that this approximation holds for other than slowly varying external fields, but it has been verified that the adiabatic approximation is adequate for calculating excitation energies [10].

In Kohn–Sham theory, the time evolution of the spin orbitals is governed by the differential equation

$$[f(\mathbf{r}_1, t) + v(\mathbf{r}_1, t)] \phi_j(\mathbf{r}_1, t) = i \frac{d\phi_j(\mathbf{r}_1, t)}{dt} \tag{34}$$

where we have introduced the Kohn–Sham operator

$$f(\mathbf{r}_1, t) = h(\mathbf{r}_1) + j(\mathbf{r}_1, t) + v_{\text{xc}}(\mathbf{r}_1, t) \tag{35}$$

which we choose to define without the explicit perturbation term (it still depends implicitly on the perturbation through the density). The first term in Eq. (35) contains the one-electron combined kinetic and nuclear-attraction operators

$$(36) \quad h(\mathbf{r}_1) = -\frac{1}{2}\nabla_1^2 + \sum_I \frac{Z_I}{|\mathbf{r}_1 - \mathbf{R}_I|}$$

whereas the last two terms are the Coulomb and exchange–correlation potentials, respectively:

$$(37) \quad j(\mathbf{r}_1, t) = \int \frac{\rho(\mathbf{r}_2, t)}{|\mathbf{r}_1 - \mathbf{r}_2|} \, d\tau_2$$

$$(38) \quad v_{xc}(\mathbf{r}_1, t) = \left. \frac{\delta E_{xc}}{\delta \rho(\mathbf{r}_1)} \right|_{\rho(\mathbf{r}_1) = \rho(\mathbf{r}_1, t)}$$

Note that these terms are themselves functionals of the density.

2.4 Time Evolution of the Reference Determinant and the Density

The Kohn–Sham reference determinant satisfies the equation

$$(39) \quad (H + V)|t\rangle = i\frac{d}{dt}|t\rangle; \quad H = \sum_i f(\mathbf{r}_i, t)$$

where the total Kohn–Sham Hamiltonian has been partitioned into a term H implicitly dependent on the perturbation and the perturbation itself V (cf. Eq. (34)). In our second-quantization formulation of time-dependent Kohn–Sham theory, we adopt an exponential parameterization of the time-evolution operator, representing the time-dependent reference Kohn–Sham determinant in the following manner:

$$(40) \quad |t\rangle = \exp[-\hat{\kappa}(t)]|0\rangle, \quad |0\rangle = \prod_{k,\sigma} a_{k\sigma}^\dagger |\text{vac}\rangle$$

Here $|0\rangle$ is the unperturbed Kohn–Sham determinant, and $\hat{\kappa}(t)$ the anti-Hermitian operator

$$(41) \quad \hat{\kappa}(t) = \sum_{rs} \kappa_{rs}(t) E_{rs} = \sum_{rs} \kappa_{rs}(t) \sum_\sigma a_{r\sigma}^\dagger a_{s\sigma}; \quad \kappa_{rs}^*(t) = -\kappa_{sr}(t)$$

where $a_{r\sigma}^\dagger$ and $a_{s\sigma}$ are the creation and annihilation operators, respectively, of spin orbitals r and s of spin σ.

The fundamental variational parameters of our theory are the elements of the rotation matrix $\kappa_{rs}(t)$. As in Hartree–Fock theory, the non-redundant rotations are those between occupied and unoccupied orbitals. Equation (40) implies that the individual Kohn–Sham spin orbitals obey the transformation law

$$(42) \quad a_{k\sigma}^\dagger(t) = \exp[-\hat{\kappa}(t)] a_{k\sigma}^\dagger \exp[\hat{\kappa}(t)]$$

and that orthonormality is preserved.

To study the time-dependence of the electron density, we introduce the second-quantized density operator

$$\hat{\rho}(\mathbf{r}) = \sum_{pq} \phi_p^*(\mathbf{r})\phi_q(\mathbf{r}) E_{pq} \tag{43}$$

Recognizing that the density is an expectation value of this operator and given the parameterization in Eq. (40), we obtain the following expression for the time-dependent density:

$$\rho(\mathbf{r},t) = \langle t|\hat{\rho}(\mathbf{r})|t\rangle = \langle 0|\exp[\hat{\kappa}(t)]\hat{\rho}(\mathbf{r})\exp[-\hat{\kappa}(t)]|0\rangle \tag{44}$$

A Baker–Campbell–Hausdorff expansion of the exponential time-evolution operator gives for the density (and similarly for other operators)

$$\rho(\mathbf{r},t) = \rho(\mathbf{r},0) + \langle 0|[\hat{\kappa}(t), \hat{\rho}(\mathbf{r})]|0\rangle + \frac{1}{2}\langle 0|[\hat{\kappa}(t),[\hat{\kappa}(t),\hat{\rho}(\mathbf{r})]]|0\rangle \tag{45}$$

$$+ \frac{1}{6}\langle 0|[\hat{\kappa}(t),[\hat{\kappa}(t),[\hat{\kappa}(t),\hat{\rho}(\mathbf{r})]]]|0\rangle + \mathcal{O}(\kappa^4)$$

which will be used frequently in the following sections.

2.5 Perturbation Expansion of the Density and Kohn–Sham Operators

In the absence of the perturbation, the time-evolution operator produces only a dynamical phase factor, which cancels out for expectation values. In the presence of the perturbation, it is assumed that an expansion in powers of the perturbation exists such that

$$\hat{\kappa}(t) = \hat{\kappa}^{(1)}(t) + \hat{\kappa}^{(2)}(t) + \hat{\kappa}^{(3)}(t) + \cdots \tag{46}$$

with Fourier representations

$$\hat{\kappa}^{(1)}(t) = \int \hat{\kappa}^{\omega_1} \exp(-i\omega_1 t)\, d\omega_1 \tag{47}$$

$$\hat{\kappa}^{(2)}(t) = \frac{1}{2}\iint \hat{\kappa}^{\omega_1,\omega_2} \exp[-i(\omega_1+\omega_2)t]\, d\omega_1\, d\omega_2 \tag{48}$$

$$\hat{\kappa}^{(3)}(t) = \frac{1}{6}\iiint \hat{\kappa}^{\omega_1,\omega_2,\omega_3} \exp[-i(\omega_1+\omega_2+\omega_3)t]\, d\omega_1\, d\omega_2\, d\omega_3 \tag{49}$$

For monochromatic perturbations, the integrations are replaced by summations. Using the expansions Eqs. (45) and (46), we introduce the perturbed density matrices up to third order as

$$D_{pq}^{(0)} = \langle 0|E_{pq}|0\rangle \tag{50}$$

$$D_{pq}^{(1)} = \langle 0|[\hat{\kappa}^{(1)}, E_{pq}]|0\rangle \tag{51}$$

$$\text{(52)} \quad D^{(2)}_{pq} = \langle 0|[\hat{\kappa}^{(2)}, E_{pq}] + \frac{1}{2}[\hat{\kappa}^{(1)}, [\hat{\kappa}^{(1)}, E_{pq}]]|0\rangle$$

$$\text{(53)} \quad D^{(3)}_{pq} = \langle 0|[\hat{\kappa}^{(3)}, E_{pq}] + \frac{1}{2}[\hat{\kappa}^{(1)}, [\hat{\kappa}^{(2)}, E_{pq}]] + \frac{1}{2}[\hat{\kappa}^{(2)}, [\hat{\kappa}^{(1)}, E_{pq}]]$$
$$+ \frac{1}{6}[\hat{\kappa}^{(1)}, [\hat{\kappa}^{(1)}, [\hat{\kappa}^{(1)}, E_{pq}]]]|0\rangle$$

which allow us to write the n:th order correction to the density as

$$\text{(54)} \quad \rho^{(n)}(\mathbf{r}, t) = \sum_{pq} \phi_p^*(\mathbf{r})\phi_q(\mathbf{r}) D^{(n)}_{pq}(t)$$

We now expand the second-quantized Kohn–Sham Hamiltonian in orders of the perturbation

$$\text{(55)} \quad \hat{H} = \sum_n \hat{H}^{(n)}, \quad \hat{H}^{(n)} = \sum_{pq} f^{(n)}_{pq} E_{pq}$$

where

$$\text{(56)} \quad f^{(n)}_{pq} = \delta_{n0} h_{pq} + j^{(n)}_{pq} + v^{(n)}_{\text{xc},pq}$$

The first contribution to $f^{(0)}_{pq}$ is the one-electron integral over the kinetic-energy and nuclear-attraction operators of Eq. (36),

$$\text{(57)} \quad h_{pq} = \langle \phi_p| -\frac{1}{2}\nabla^2 + \sum_I \frac{Z_I}{|\mathbf{r} - \mathbf{R}_I|}|\phi_q\rangle$$

The electron-repulsion n:th order Coulomb interaction integrals

$$\text{(58)} \quad j^{(n)}_{pq} = \sum_{rs} g_{pqrs} D^{(n)}_{rs}$$

constitute the second contribution to the Kohn–Sham matrix element $f^{(n)}_{pq}$ and are obtained from the two-electron integrals

$$\text{(59)} \quad g_{pqrs} = \langle \phi_p(1)\phi_r(1)|\frac{1}{r_{12}}|\phi_q(2)\phi_s(2)\rangle$$

and from the time-dependent density-matrix elements in Eqs. (50)–(52). The last contribution to Eq. (56) is the integral over the exchange–correlation potential $v_{\text{xc}}[\rho]$,

$$\text{(60)} \quad v^{(n)}_{\text{xc},pq}(t) = \langle \phi_p|v^{(n)}_{\text{xc}}(\mathbf{r}, t)|\phi_q\rangle$$

Since the potential depends on the density ρ, this contribution to the Kohn–Sham matrix element depends on the perturbation,

$$v_{\text{xc}}^{(1)}(\mathbf{r}, t) = \int \frac{\delta v_{\text{xc}}(\mathbf{r})}{\delta \rho(\mathbf{r}')} \rho^{(1)}(\mathbf{r}', t) \, \mathrm{d}\tau' \tag{61}$$

$$v_{\text{xc}}^{(2)}(\mathbf{r}, t) = \int \frac{\delta v_{\text{xc}}(\mathbf{r})}{\delta \rho(\mathbf{r}')} \rho^{(2)}(\mathbf{r}', t) \, \mathrm{d}\tau' \tag{62}$$

$$+ \frac{1}{2} \iint \frac{\delta^2 v_{\text{xc}}(\mathbf{r})}{\delta \rho(\mathbf{r}') \delta \rho(\mathbf{r}'')} \rho^{(1)}(\mathbf{r}', t) \rho^{(1)}(\mathbf{r}'', t) \, \mathrm{d}\tau' \, \mathrm{d}\tau''$$

$$v_{\text{xc}}^{(3)}(\mathbf{r}, t) = \int \frac{\delta v_{\text{xc}}(\mathbf{r})}{\delta \rho(\mathbf{r}')} \rho^{(3)}(\mathbf{r}', t) \, \mathrm{d}\tau' \tag{63}$$

$$+ \iint \frac{\delta^2 v_{\text{xc}}(\mathbf{r})}{\delta \rho(\mathbf{r}') \delta \rho(\mathbf{r}'')} \rho^{(1)}(\mathbf{r}', t) \rho^{(2)}(\mathbf{r}'', t) \, \mathrm{d}\tau' \, \mathrm{d}\tau''$$

$$+ \frac{1}{6} \iiint \frac{\delta^3 v_{\text{xc}}(\mathbf{r})}{\delta \rho(\mathbf{r}') \delta \rho(\mathbf{r}'') \delta \rho(\mathbf{r}''')}$$

$$\times \rho^{(1)}(\mathbf{r}', t) \rho^{(1)}(\mathbf{r}'', t) \rho^{(1)}(\mathbf{r}''', t) \, \mathrm{d}\tau' \, \mathrm{d}\tau'' \, \mathrm{d}\tau'''$$

2.6 The Ehrenfest Method

Let us consider the time-development Eq. (41) of the expectation value of an operator $\hat{Q}$ parametrized by $\hat{\kappa}(t)$ [Eq. (41)]

$$\hat{Q}(t) = \exp[-\hat{\kappa}(t)] \hat{Q} \exp[\hat{\kappa}(t)] \tag{64}$$

Differentiating the expectation value $\langle t|\hat{Q}(t)|t\rangle = \langle 0|\hat{Q}|0\rangle$ [see Eq. (40)] with respect to time and invoking the Schrödinger equation (atomic units)

$$\left(\hat{H}(t) + \hat{V}(t) - \mathrm{i}\frac{\mathrm{d}}{\mathrm{d}t}\right)|t\rangle = 0 \tag{65}$$

we obtain the Ehrenfest theorem

$$\langle t|\mathrm{i}\dot{\hat{Q}}(t) + [\hat{Q}(t), \hat{H}(t) + \hat{V}(t)]|t\rangle = 0 \tag{66}$$

which forms the basis for the time-dependent variation principle [54]. Collecting the exponential operators in Eqs. (40) and (64) and using the identity

$$\exp[\hat{\kappa}(t)] \left(\frac{\mathrm{d}}{\mathrm{d}t} \exp[-\hat{\kappa}(t)] \hat{Q} \exp[\hat{\kappa}(t)]\right) \exp[-\hat{\kappa}(t)] \tag{67}$$

$$= [\exp[\hat{\kappa}(t)] \left(\frac{\mathrm{d}}{\mathrm{d}t} \exp[-\hat{\kappa}(t)]\right), \hat{Q}]$$

we find that the Ehrenfest theorem may be written in the more convenient form

$$(68) \quad \langle 0|[\hat{Q}, \exp[\hat{\kappa}(t)]\left(\hat{H}(t)+\hat{V}(t)-i\frac{d}{dt}\right)\exp[-\hat{\kappa}(t)]]|0\rangle = 0$$

This equation holds for any time-independent one-electron operator $\hat{Q}$. In particular, it holds for the spin-averaged excitation operators E_{pq} in the expansion of $\hat{\kappa}(t)$ in Eq. (41). Collecting these operators in the column vector $\hat{\mathbf{q}}$, we arrive at a set of nonlinear equations from which the time-dependence of $\hat{\kappa}(t)$ may be determined. In the following, we shall use these equations to determine the first- and second-order terms in Eq. (46) and thereby the linear, quadratic, and cubic response functions.

2.6.1 Linear response

The linear response equations are obtained by expanding Eq. (68) to first order, yielding a differential equation in time for $\hat{\kappa}^{(1)}$,

$$(69) \quad \langle 0|[\hat{\mathbf{q}}, [\hat{\kappa}^{(1)}, \hat{H}^{(0)}]+\hat{H}^{(1)}]|0\rangle + i\langle 0|[\hat{\mathbf{q}}, \dot{\hat{\kappa}}^{(1)}]|0\rangle = -\langle 0|[\hat{\mathbf{q}}, \hat{V}(t)]|0\rangle$$

In the frequency domain, this equation becomes an algebraic equation for $\hat{\kappa}^\omega$,

$$(70) \quad \langle 0|[\hat{\mathbf{q}}, [\hat{\kappa}^\omega, \hat{H}^{(0)}]+\hat{H}^\omega]|0\rangle + \omega\langle 0|[\hat{\mathbf{q}}, \hat{\kappa}^\omega]|0\rangle = -\langle 0|[\hat{\mathbf{q}}, \hat{V}^\omega]|0\rangle$$

We have here introduced the Fourier transform of the perturbed Kohn–Sham matrix

$$(71) \quad \hat{H}^\omega = \sum_{pq} f_{pq}^\omega E_{pq}$$

where

$$(72) \quad f_{pq}^\omega = j_{pq}^\omega + v_{\text{xc},pq}^\omega$$

$$= \sum_{rs} g_{pqrs} D_{rs}^\omega + \langle \phi_p| \int \frac{\delta v_{\text{xc}}}{\delta \rho(\mathbf{r}')}\rho^\omega(\mathbf{r}')\,d\tau'|\phi_q\rangle$$

$$(73) \quad D_{rs}^\omega = \langle 0|[\hat{\kappa}^\omega, E_{rs}]|0\rangle$$

$$(74) \quad \rho^\omega(\mathbf{r}) = \langle 0|[\hat{\kappa}^\omega, \hat{\rho}(\mathbf{r})]|0\rangle$$

To bring out the matrix structure of Eq. (70), we may express the first-order parameters in matrix form

$$(75) \quad \hat{\kappa}^\omega = (\cdots E_{pq} \cdots)\begin{pmatrix}\vdots\\ \kappa_{pq}^\omega \\ \vdots\end{pmatrix} \equiv \hat{\mathbf{q}}^\dagger \kappa^\omega$$

such that we formally solve the linear equations

$$(76) \quad (\mathbf{E}-\omega\mathbf{S})\kappa^\omega = \mathbf{V}^\omega$$

Kohn–Sham Time-Dependent Density Functional Theory

In Eq. (76), the matrix $\mathbf{E}$ has the structure of an electronic Hessian

(77) $\quad \mathbf{E} = \langle 0|[\hat{\mathbf{q}},[\hat{H}^{(0)},\hat{\mathbf{q}}^\dagger]]|0\rangle$

$\qquad + \sum_{pqrs} \langle 0|[\hat{\mathbf{q}}, E_{pq}]|0\rangle [\frac{1}{|\mathbf{r}-\mathbf{r}'|} + \frac{\delta v_{xc}(\mathbf{r})}{\delta \rho(\mathbf{r}')}]_{pqrs} \langle 0|[E_{rs}, \hat{\mathbf{q}}^\dagger]|0\rangle$

where, in the last term, we have introduced two-electron integrals defined by analogy with Eq. (59). In Eq. (76), we have introduced the generalized overlap matrix

(78) $\quad \mathbf{S} = \langle 0|[\hat{\mathbf{q}}, \hat{\mathbf{q}}^\dagger]|0\rangle$

and the right-hand side is the perturbation vector of Eq. (70)

(79) $\quad \mathbf{V}^\omega = \langle 0|[\hat{\mathbf{q}}, \hat{V}^\omega]|0\rangle$

From the solution $\hat{\kappa}^\omega$ to Eq. (76), the linear response function is easily obtained for any one-electron operator $\hat{A}$,

(80) $\quad \langle\langle \hat{A}; \hat{V}\rangle\rangle_\omega = \langle 0|[\hat{\kappa}^\omega, \hat{A}]|0\rangle = -\mathbf{A}^\dagger(\mathbf{E}-\omega\mathbf{S})^{-1}\mathbf{V}^\omega$

where

(81) $\quad \mathbf{A} = \langle 0|[\hat{\mathbf{q}}, \hat{A}]|0\rangle$

It should be noted the working equations of linear response theory resemble Eq. (70) more closely than Eq. (76) since $\mathbf{E}$ and $\mathbf{S}$ are never constructed explicitly and since, in the iterative algorithms we use, κ corresponds to trial vectors. Furthermore, the exchange–correlation potentials that we consider are local and never give rise to two-electron integrals—even the GGA functionals are local in this sense.

2.6.2 Quadratic response

The quadratic response to a perturbation requires the solution of $\hat{\kappa}^{(2)}$ from the second-order expansion of Eq. (68),

(82) $\quad \langle 0|[\hat{\mathbf{q}}, [\hat{\kappa}^{(2)}, \hat{H}^{(0)}] + \hat{H}^{(2)}]|0\rangle + i\langle 0|[\hat{\mathbf{q}}, \dot{\hat{\kappa}}^{(2)} + \frac{1}{2}[\hat{\kappa}^{(1)}, \dot{\hat{\kappa}}^{(1)}]]|0\rangle$

$\qquad = -\langle 0|\hat{\mathbf{q}}, [\hat{\kappa}^{(1)}, \hat{V}(t) + \hat{H}^{(1)}] + \frac{1}{2}[\hat{\kappa}^{(1)}, [\hat{\kappa}^{(1)}, \hat{H}^{(0)}]]]|0\rangle$

With the Fourier transform Eq. (48), this equation can be rearranged to give

(83) $\quad \langle 0|[\hat{\mathbf{q}}, [\hat{\kappa}^{\omega_1,\omega_2}, \hat{H}^{(0)}] + \hat{H}^{\omega_1,\omega_2}]|0\rangle + (\omega_1+\omega_2)\langle 0|[\hat{\mathbf{q}}, \hat{\kappa}^{\omega_1,\omega_2}]|0\rangle$

$\qquad = -\hat{P}_{12}\langle 0|[\hat{\mathbf{q}}, 2[\hat{\kappa}^{\omega_1}, \hat{V}^{\omega_2} + \hat{H}^{\omega_2}] + [\hat{\kappa}^{\omega_1}, [\hat{\kappa}^{\omega_2}, \hat{H}^{(0)}]] + \omega_2[\hat{\kappa}^{\omega_1}, \hat{\kappa}^{\omega_2}]]|0\rangle$

Here the operator $\hat{P}_{12}$ symmetrizes with respect to the frequencies ω_1 and ω_2

$$(84) \quad \hat{P}_{12}A(\omega_1, \omega_2) = \frac{1}{2}[A(\omega_1, \omega_2) + A(\omega_2, \omega_1)]$$

and the elements of the second-order perturbed Kohn–Sham matrix

$$(85) \quad \hat{H}^{\omega_1,\omega_2} = \sum_{pq} f_{pq}^{\omega_1,\omega_2} E_{pq}$$

are given by

$$(86) \quad f_{pq}^{\omega_1,\omega_2} = j_{pq}^{\omega_1,\omega_2} + v_{xc,pq}^{\omega_1,\omega_2}$$

with Coulomb and exchange–correlation parts

$$(87) \quad j_{pq}^{\omega_1,\omega_2} = \sum_{rs} g_{pqrs} D_{rs}^{\omega_1,\omega_2}$$

$$(88) \quad v_{xc,pq}^{\omega_1,\omega_2} = \langle \phi_p | \int \frac{\delta v_{xc}}{\delta \rho(\mathbf{r}')} \rho^{\omega_1,\omega_2}(\mathbf{r}') \, d\tau' | \phi_q \rangle$$
$$+ \hat{P}_{12} \langle \phi_p | \iint \frac{\delta^2 v_{xc}}{\delta \rho(\mathbf{r}') \delta \rho(\mathbf{r}'')} \rho^{\omega_1}(\mathbf{r}') \rho^{\omega_2}(\mathbf{r}'') \, d\tau' \, d\tau'' | \phi_q \rangle$$

The solution of Eq. (83) requires the separation of the second-order density matrix elements into first- and second-order parts

$$(89) \quad \rho^{\omega_1,\omega_2}(\mathbf{r}) = \sum_{pq} \phi_p^*(\mathbf{r}) \phi_q(\mathbf{r}) D_{pq}^{\omega_1,\omega_2}$$

$$(90) \quad D_{rs}^{\omega_1,\omega_2} = \bar{D}_{rs}^{\omega_1,\omega_2} + \bar{\bar{D}}_{rs}^{\omega_1,\omega_2}$$

$$(91) \quad \bar{D}_{rs}^{\omega_1,\omega_2} = \hat{P}_{12} \langle 0|[\hat{\kappa}^{\omega_1}, [\hat{\kappa}^{\omega_2}, E_{rs}]]|0 \rangle$$

$$(92) \quad \bar{\bar{D}}_{rs}^{\omega_1,\omega_2} = \langle 0|[\hat{\kappa}^{\omega_1,\omega_2}, E_{rs}]|0 \rangle$$

such that the second-order Hamiltonian may be written as the sum of two operators that depend on the first- and second-order parameters, respectively,

$$(93) \quad \hat{H}^{\omega_1,\omega_2} = \hat{\bar{H}}^{\omega_1,\omega_2} + \hat{\bar{\bar{H}}}^{\omega_1,\omega_2}$$

As in the linear case, the equation for the second-order parameters Eq. (83) may then written in matrix form

$$(94) \quad [\mathbf{E} - (\omega_1 + \omega_2)\mathbf{S}]\kappa^{\omega_1,\omega_2} = \mathbf{V}^{\omega_1,\omega_2}$$

where all first-order parameters have been collected on the right-hand side.

$$(95) \quad \mathbf{V}^{\omega_1,\omega_2} = 2\hat{P}_{12}\{\langle 0|[\hat{q}, [\hat{\kappa}^{\omega_1}, [\hat{\kappa}^{\omega_2}, \hat{H}^{(0)}]]]|0\rangle + \langle 0|[\hat{q}, [\hat{\kappa}^{\omega_1}, \omega_2\hat{\kappa}^{\omega_2}]]|0\rangle$$
$$+ 2\langle 0|[\hat{q}, [\kappa^{\omega_1}, \hat{H}^{\omega_2} + V^{\omega_2}]]|0\rangle + \langle 0|[\hat{q}, \hat{\bar{H}}^{\omega_1,\omega_2}]|0\rangle\}$$

The quadratic response function may now be calculated as

$$\langle\langle \hat{A}; \hat{V}, \hat{V}\rangle\rangle_{\omega_1,\omega_2} = \langle 0|[\hat{\kappa}^{\omega_1,\omega_2}, \hat{A}]|0\rangle + \hat{P}_{12}\langle 0|[\hat{\kappa}^{\omega_1}, [\hat{\kappa}^{\omega_2}, \hat{A}]]|0\rangle \quad (96)$$

where the first-order responses $\hat{\kappa}^{\omega_1}$ and $\hat{\kappa}^{\omega_2}$ are obtained from Eq. (70) and the second-order response $\hat{\kappa}^{\omega_1,\omega_2}$ from Eq. (83). We note that the first term in Eq. (96) resembles the evaluation of the linear response function,

$$\langle 0|[\kappa^{\omega_1,\omega_2}, \hat{A}]|0\rangle = -\mathbf{A}^\dagger \kappa^{\omega_1,\omega_2} = -\mathbf{A}^\dagger [\mathbf{E} - (\omega_1+\omega_2)\mathbf{S}]^{-1}\mathbf{V}^{\omega_1,\omega_2} \quad (97)$$

An alternative approach is to solve the adjoint linear response equation for $\mathbf{A}$ at frequency $\omega_1 + \omega_2$, as done in our implementation.

In short, for a given property $\hat{A}$ and two periodic perturbations $\hat{B}$ and $\hat{C}$ with associated frequencies ω_b and ω_c respectively, the evaluation of the quadratic response equation $\langle\langle \hat{A}; \hat{B}, \hat{C}\rangle\rangle_{\omega_b,\omega_c}$ is carried out by first solving the three linear response equations

$$\kappa^{A^\dagger}[\mathbf{E} - (\omega_a+\omega_b)\mathbf{S}] = \mathbf{A}^\dagger \quad (98)$$

$$(\mathbf{E} - \omega_b \mathbf{S})\kappa^B = \mathbf{B} \quad (99)$$

$$(\mathbf{E} - \omega_c \mathbf{S})\kappa^C = \mathbf{C} \quad (100)$$

and then evaluating

$$\langle\langle \hat{A}; \hat{B}, \hat{C}\rangle\rangle_{\omega_b,\omega_c} = \kappa^{\hat{A}^\dagger}\mathbf{V}^{\omega_b,\omega_c} + \hat{P}_{bc}\langle 0|[\hat{\kappa}^{\omega_b}, [\hat{\kappa}^{\omega_c}, \hat{A}]]|0\rangle \quad (101)$$

The main complication associated with extending a Hartree–Fock quadratic response code to Kohn–Sham DFT is the evaluation of the exchange–correlation contribution to $\mathbf{V}^{\omega_b,\omega_c}$. We refer to [41] for detailed expressions and Section 3 for a discussion of its implementation.

2.6.3 Cubic response

For the cubic response we need to determine $\hat{\kappa}^{(3)}$. By expanding Eq. (68) (with Q replaced by $\hat{q}$) we obtained the following equation

$$\langle 0|[\hat{q}, [\hat{\kappa}^{(3)}, \hat{H}^{(0)}] + \hat{H}^{(3)}]|0\rangle$$
$$+ i\langle 0|[\hat{q}, \dot{\hat{\kappa}}^{(3)} + \frac{1}{2}[\hat{\kappa}^{(1)}, \dot{\hat{\kappa}}^{(2)}] + \frac{1}{2}[\hat{\kappa}^{(2)}, \dot{\hat{\kappa}}^{(1)}] + \frac{1}{6}[\hat{\kappa}^{(1)}, [\hat{\kappa}^{(1)}, \dot{\hat{\kappa}}^{(1)}]]]|0\rangle$$
$$= -\langle 0|[\hat{q}, [\hat{\kappa}^{(2)}, \hat{V}(t) + \hat{H}^{(1)}] + [\hat{\kappa}^{(1)}, \hat{H}^{(2)}] + \frac{1}{2}[\hat{\kappa}^{(1)}, [\hat{\kappa}^{(1)}, \hat{V}(t) + \hat{H}^{(1)}]]$$
$$+ \frac{1}{2}[\hat{\kappa}^{(1)}, [\hat{\kappa}^{(2)}, \hat{H}^{(0)}]] + \frac{1}{2}[\hat{\kappa}^{(2)}, [\hat{\kappa}^{(1)}, \hat{H}^{(0)}]]$$
$$+ \frac{1}{6}[\hat{\kappa}^{(1)}, [\hat{\kappa}^{(1)}, [\hat{\kappa}^{(1)}, \hat{H}^{(0)}]]]]|0\rangle$$

Transformation to the frequency domain yields an equation for $\hat{\kappa}^{\omega_1,\omega_2,\omega_3}$

(102) $\langle 0|[\hat{\mathbf{q}},[\hat{\kappa}^{\omega_1,\omega_2,\omega_3},\hat{H}^{(0)}]+\hat{H}^{\omega_1,\omega_2,\omega_3}]|0\rangle + (\omega_1+\omega_2+\omega_3)\langle 0|[\hat{\mathbf{q}},\hat{\kappa}^{\omega_1,\omega_2,\omega_3}]|0\rangle$

$= -\hat{P}_{123}\langle 0|[\hat{\mathbf{q}}, 3[\hat{\kappa}^{\omega_1,\omega_2},\hat{V}^{\omega_3}+\hat{H}^{\omega_3}]+3[\hat{\kappa}^{\omega_1},\hat{H}^{\omega_2,\omega_3}]$

$+3[\hat{\kappa}^{\omega_1},[\hat{\kappa}^{\omega_2},\hat{V}^{\omega_3}+\hat{H}^{\omega_3}]]+\frac{3}{2}[\hat{\kappa}^{\omega_1},[\hat{\kappa}^{\omega_2,\omega_3},\hat{H}^{(0)}]]$

$+\frac{3}{2}[\hat{\kappa}^{\omega_1,\omega_2},[\hat{\kappa}^{\omega_3},\hat{H}^{(0)}]]+[\hat{\kappa}^{\omega_1},[\hat{\kappa}^{\omega_2},[\hat{\kappa}^{\omega_3},\hat{H}^{(0)}]]]+3\omega_3[\hat{\kappa}^{\omega_1,\omega_2},\hat{\kappa}^{\omega_3}]$

$+3(\omega_2+\omega_3)[\hat{\kappa}^{\omega_1},\hat{\kappa}^{\omega_2,\omega_3}]+\omega_3[\hat{\kappa}^{\omega_1},[\hat{\kappa}^{\omega_2},\hat{\kappa}^{\omega_3}]]]|0\rangle$

Here the operator $\hat{P}_{123}$ symmetrizes with respect to the frequencies ω_1, ω_2, and ω_3

(103) $\hat{P}_{123}A(\omega_1,\omega_2,\omega_3) = \frac{1}{6}[A(\omega_1,\omega_2,\omega_3)+A(\omega_1,\omega_3,\omega_2)+A(\omega_2,\omega_1,\omega_3)$
$+A(\omega_2,\omega_3,\omega_1)+A(\omega_3,\omega_1,\omega_2)+A(\omega_3,\omega_2,\omega_1)]$

and the elements of the third-order perturbed Kohn–Sham matrix

(104) $\hat{H}^{\omega_1,\omega_2,\omega_3} = \sum_{pq} f^{\omega_1,\omega_2,\omega_3}_{pq} E_{pq} = \sum_{pq}(j^{\omega_1,\omega_2,\omega_3}_{pq}+v^{\omega_1,\omega_2,\omega_3}_{xc,pq})E_{pq}$

where Coulomb and exchange–correlation parts are given by

(105) $j^{\omega_1,\omega_2,\omega_3}_{pq} = \sum_{rs} g_{pqrs} D^{\omega_1,\omega_2,\omega_3}_{rs}$

(106) $v^{\omega_1,\omega_2,\omega_3}_{xc,pq} = \langle\phi_p|\int\frac{\delta v_{xc}}{\delta\rho(\mathbf{r}')}\rho^{\omega_1,\omega_2,\omega_3}(\mathbf{r}')d\tau'|\phi_q\rangle$

$+3\hat{P}_{123}\langle\phi_p|\iint\frac{\delta^2 v_{xc}}{\delta\rho(\mathbf{r}')\delta\rho(\mathbf{r}'')}\rho^{\omega_1}(\mathbf{r}')\rho^{\omega_2,\omega_2}(\mathbf{r}'')d\tau'd\tau''d\tau'|\phi_q\rangle$

$+\hat{P}_{123}\langle\phi_p|\iiint\frac{\delta^3 v_{xc}}{\delta\rho(\mathbf{r}')\delta\rho(\mathbf{r}'')\delta\rho(\mathbf{r}''')}\rho^{\omega_1}(\mathbf{r}')\rho^{\omega_2}(\mathbf{r}'')\rho^{\omega_2}(\mathbf{r}''')$
$\times d\tau'd\tau''d\tau'''|\phi_q\rangle$

In order to solve Eq. (102) we need to separate the contribution from the third-order parameters to the density matrix elements from the contributions from first- and second-order parameters

(107) $\rho^{\omega_1,\omega_2,\omega_3}(\mathbf{r}) = \sum_{pq}\phi_p^*(\mathbf{r})\phi_q(\mathbf{r})D^{\omega_1,\omega_2,\omega_3}_{pq}$

(108) $D^{\omega_1,\omega_2,\omega_3}_{rs} = \bar{D}^{\omega_1,\omega_2,\omega_3}_{rs}+\bar{\bar{D}}^{\omega_1,\omega_2,\omega_3}_{rs}$

$$\text{(109)} \quad \bar{D}_{rs}^{\omega_1,\omega_2,\omega_3} = \hat{P}_{123}\langle 0|\frac{3}{2}[\hat{\kappa}^{\omega_1},[\hat{\kappa}^{\omega_2,\omega_3},E_{rs}]] + \frac{3}{2}[\hat{\kappa}^{\omega_1,\omega_2},[\hat{\kappa}^{\omega_3},E_{rs}]]$$
$$+ [\hat{\kappa}^{\omega_1},[\hat{\kappa}^{\omega_2},[\hat{\kappa}^{\omega_3},E_{rs}]]]|0\rangle$$

$$\text{(110)} \quad \bar{\bar{D}}_{rs}^{\omega_1,\omega_2,\omega_3} = \langle 0|[\hat{\kappa}^{\omega_1,\omega_2,\omega_3},E_{rs}]|0\rangle$$

such that the third-order Hamiltonian may be written as the sum of two operators that depend on the lower- and third-order parameters, respectively,

$$\text{(111)} \quad \hat{H}^{\omega_1,\omega_2,\omega_3} = \hat{\bar{H}}^{\omega_1,\omega_2,\omega_3} + \hat{\bar{\bar{H}}}^{\omega_1,\omega_2,\omega_3}$$

As in the linear case, the equation for the third-order parameters Eq. (102) may then written in matrix form

$$\text{(112)} \quad [\mathbf{E} - (\omega_1 + \omega_2 + \omega_3)\mathbf{S}]\kappa^{\omega_1,\omega_2,\omega_3} = \mathbf{V}^{\omega_1,\omega_2,\omega_3}$$

where all first- and second-order parameters have been collected on the right-hand side

$$\text{(113)} \quad \mathbf{V}^{\omega_1,\omega_2,\omega_3} = \hat{P}_{123}\{\langle 0|[\hat{q},[\hat{\kappa}^{\omega_1},[\hat{\kappa}^{\omega_2},[\hat{\kappa}^{\omega_3},\hat{H}^{(0)}]]]]|0\rangle$$
$$+ \omega_3 \langle 0|[\hat{q},[\hat{\kappa}^{\omega_1},[\hat{\kappa}^{\omega_2},\hat{\kappa}^{\omega_3}]]]|0\rangle$$
$$+ \frac{3}{2}\langle 0|[\hat{q},[\hat{\kappa}^{\omega_1,\omega_2},[\hat{\kappa}^{\omega_3},\hat{H}^{(0)}]]]|0\rangle + 3\omega_3 \langle 0|[\hat{q},[\hat{\kappa}^{\omega_1,\omega_2},\hat{\kappa}^{\omega_3}]]|0\rangle$$
$$+ \frac{3}{2}\langle 0|[\hat{q},[\hat{\kappa}^{\omega_1},[\hat{\kappa}^{\omega_2,\omega_3},\hat{H}^{(0)}]]]|0\rangle + 3(\omega_2+\omega_3)\langle 0|[\hat{q},[\hat{\kappa}^{\omega_1},\hat{\kappa}^{\omega_2,\omega_3}]]|0\rangle$$
$$+ 3\langle 0|[\hat{q},[\hat{\kappa}^{\omega_1},[\hat{\kappa}^{\omega_2},\hat{V}^{\omega_3}+\hat{H}^{\omega_3}]]]|0\rangle + 3\langle 0|[\hat{q},[\hat{\kappa}^{\omega_1,\omega_2},\hat{V}^{\omega_3}+\hat{H}^{\omega_3}]]|0\rangle$$
$$+ 3\langle 0|[\hat{q},[\hat{\kappa}^{\omega_1},\hat{H}^{\omega_2,\omega_3}]]|0\rangle + \langle 0|[\hat{q},\hat{\bar{\bar{H}}}^{\omega_1,\omega_2,\omega_3}]|0\rangle.\}$$

The cubic response function may now be calculated as

$$\text{(114)} \quad \langle\langle \hat{A};\hat{V},\hat{V},\hat{V}\rangle\rangle_{\omega_1,\omega_2,\omega_3} = \langle 0|[\hat{\kappa}^{\omega_1,\omega_2,\omega_3},\hat{A}]|0\rangle + 3\hat{P}_{123}\langle 0|[\hat{\kappa}^{\omega_1},[\hat{\kappa}^{\omega_2,\omega_3},\hat{A}]]|0\rangle$$
$$+ 3\hat{P}_{123}\langle 0|[\hat{\kappa}^{\omega_1,\omega_2},[\hat{\kappa}^{\omega_3},\hat{A}]]|0\rangle + \hat{P}_{123}\langle 0|[\hat{\kappa}^{\omega_1},[\hat{\kappa}^{\omega_2},[\hat{\kappa}^{\omega_3},\hat{A}]]]|0\rangle$$

where the first-, second-, and third-order responses are obtained from Eqs. (70), (83), and (102), respectively. As for the quadratic case we note that the first term in Eq. (114) resembles the evaluation of the linear response function,

$$\text{(115)} \quad \langle 0|[\kappa^{\omega_1,\omega_2,\omega_3},\hat{A}]|0\rangle = -\mathbf{A}^\dagger \kappa^{\omega_1,\omega_2,\omega_3} = -\mathbf{A}^\dagger[\mathbf{E}-(\omega_1+\omega_2+\omega_3)\mathbf{S}]^{-1}\mathbf{V}^{\omega_1,\omega_2,\omega_3}$$

Similarly as for the quadratic case an alternative approach is to solve the adjoint linear response equation for $\mathbf{A}$ at frequency $\omega_1 + \omega_2 + \omega_3$, as done in our implementation.

Summing up, for a given property $\hat{A}$ and three periodic perturbations $\hat{B}$, $\hat{C}$ and $\hat{D}$ with associated frequencies ω_b, ω_c and ω_d respectively, the evaluation of the cubic response function $\langle\langle \hat{A}; \hat{B}, \hat{C}, \hat{D} \rangle\rangle_{\omega_b,\omega_c,\omega_d}$ is carried out by first solving the four first-order linear response equations

(116) $\quad \kappa^{A^\dagger}[\mathbf{E} - (\omega_b + \omega_c + \omega_d)\mathbf{S}] = \mathbf{A}^\dagger$

(117) $\quad (\mathbf{E} - \omega_b \mathbf{S})\kappa^B = \mathbf{B}$

(118) $\quad (\mathbf{E} - \omega_c \mathbf{S})\kappa^C = \mathbf{C}$

(119) $\quad (\mathbf{E} - \omega_d \mathbf{S})\kappa^D = \mathbf{D}$

the three second-order linear response equations

(120) $\quad (\mathbf{E} - (\omega_b + \omega_c)\mathbf{S})\kappa^{\omega_b,\omega_c} = \mathbf{V}^{\omega_b,\omega_c}$

(121) $\quad (\mathbf{E} - (\omega_b + \omega_d)\mathbf{S})\kappa^{\omega_b,\omega_d} = \mathbf{V}^{\omega_b,\omega_d}$

(122) $\quad (\mathbf{E} - (\omega_c + \omega_d)\mathbf{S})\kappa^{\omega_c,\omega_d} = \mathbf{V}^{\omega_c,\omega_d}$

and finally evaluating

(123) $\quad \langle\langle \hat{A}; \hat{B}, \hat{C}, \hat{D} \rangle\rangle_{\omega_b,\omega_c,\omega_d} = \kappa^{\hat{A}^\dagger} \mathbf{V}^{\omega_b,\omega_c,\omega_d} + 3\hat{P}_{bcd}\langle 0|[\hat{\kappa}^{\omega_b},[\hat{\kappa}^{\omega_c,\omega_d},\hat{A}]]|0\rangle$
$\quad\quad + 3\hat{P}_{bcd}\langle 0|[\hat{\kappa}^{\omega_b,\omega_c},[\hat{\kappa}^{\omega_d},\hat{A}]]|0\rangle + \hat{P}_{bcd}\langle 0|[\hat{\kappa}^{\omega_b},[\hat{\kappa}^{\omega_c},[\hat{\kappa}^{\omega_d},\hat{A}]]]|0\rangle$

Our implementation is based on the SCF [55] and MCSCF [56] cubic response code in DALTON [51]. The main complication associated with the extension to Kohn–Sham DFT is the evaluation of the exchange–correlation contributions to $\mathbf{V}^{\omega_b,\omega_c,\omega_d}$. We refer to [42] for detailed expressions.

2.7 Residues

By comparison with sum-over-state expressions for the exact case it is possible to identify transition moments and excited state properties from different residues of the response functions. This is in particular valuable for DFT where it is difficult to straightforwardly extend the theory to excited states and where we have no explicit representation of the excited state wave function.

2.7.1 Linear response

The matrices $\mathbf{E}$ and $\mathbf{S}$ in the expression for the linear response function in Eq. (80) have a common set of eigenvectors $\mathbf{X}_k$

(124) $\quad \mathbf{X}_k^\dagger \mathbf{E} \mathbf{X}_k = \omega_k$

(125) $\quad \mathbf{X}_k^\dagger \mathbf{S} \mathbf{X}_k = \pm 1$

Kohn–Sham Time-Dependent Density Functional Theory

Thus, the linear response function has poles where the absolute value of the frequency is equal to an excitation energy of the system. From the corresponding residues

$$\lim_{\omega \to \omega_f} (\omega - \omega_f) \langle\langle \hat{A}; \hat{B} \rangle\rangle_\omega = \mathbf{A}^\dagger \mathbf{X}_f \mathbf{X}_f^\dagger \mathbf{B} \tag{126}$$

and

$$\lim_{\omega \to -\omega_f} (\omega + \omega_f) \langle\langle \hat{A}; \hat{B} \rangle\rangle_\omega = \mathbf{B}^\dagger \mathbf{X}_f \mathbf{X}_f^\dagger \mathbf{A} \tag{127}$$

we may identify the matrix elements as

$$T_f^{A*} = \langle 0|\hat{A}|f\rangle = \mathbf{A}^\dagger \mathbf{X}_f \tag{128}$$

$$T_f^{B} = \langle f|\hat{B}|0\rangle = \mathbf{X}_f^\dagger \mathbf{B} \tag{129}$$

in agreement with the exact case where

$$\lim_{\omega \to \omega_f} (\omega - \omega_f) \langle\langle \hat{A}; \hat{B} \rangle\rangle_\omega = \langle 0|\hat{A}|f\rangle \langle f|\hat{B}|0\rangle \tag{130}$$

$$\lim_{\omega \to -\omega_f} (\omega + \omega_f) \langle\langle \hat{A}; \hat{B} \rangle\rangle_\omega = \langle 0|\hat{B}|f\rangle \langle f|\hat{A}|0\rangle \tag{131}$$

For future use we note also that

$$\lim_{\omega \to \omega_f} (\omega + \omega_f) \kappa^{A^\dagger} = \mathbf{B}^\dagger \mathbf{X}_f \mathbf{X}_f^\dagger = T_f^{A*} \mathbf{X}_f^\dagger \tag{132}$$

$$\lim_{\omega \to \omega_f} (\omega - \omega_f) \kappa^{B} = \mathbf{X}_f \mathbf{X}_f^\dagger \mathbf{B} = \mathbf{X}_f T_f^{B} \tag{133}$$

2.7.2 Quadratic response

The quadratic response functions has poles where the absolute value of the frequency parameters or their sum matches an excitation energy. We may also consider double residues where two poles matches at the same time. The single residue of most interest is

$$\lim_{\omega_b + \omega_c \to \omega_f} (\omega_b + \omega_c - \omega_f) \langle\langle \hat{A}; \hat{B}, \hat{C} \rangle\rangle_{\omega_b, \omega_c} = -T_f^A * T_f^{BC} \tag{134}$$

$$T_f^{BC} = \mathbf{X}_f^\dagger \mathbf{V}^{\omega_b, \omega_c} \tag{135}$$

or alternatively

$$\lim_{\omega_c \to \omega_f} (\omega_c - \omega_f) \langle\langle \hat{A}; \hat{B}, \hat{C} \rangle\rangle_{-\omega_b, \omega_c,} = -T_f^{AB} T_f^C \tag{136}$$

$$T_f^{AB} = -\kappa^A \mathbf{X}_f^{-\omega_b} - \frac{1}{2}[\langle 0|[\hat{\kappa}^{-\omega_b}, [\hat{X}_f, \hat{A}]]|0\rangle + \langle 0|[\hat{X}_f, [\hat{\kappa}^{-\omega_b}, \hat{A}]]|0\rangle] \tag{137}$$

where $\mathbf{X}_f^{\omega_b}$ is defined to satisfy

$$(138) \quad \lim_{\omega_c \to \omega_f} (\omega_c - \omega_f) \mathbf{V}^{\omega_b, \omega_c} = \mathbf{X}_f^{\omega_b} T_f^C$$

Even though Eqs. (135) and (137) have a different structure they give numerically identical results. This follows from the permutation symmetry of the response function. In this case there is no computational advantage to use one expression before the other, we have to solve the same set of response equations in both cases.

When $\omega_b = 0$ we can be interpret T_f^{AB} as the induced transition amplitude of an operator $\hat{A}$ between the ground state (0) and an excited state (f) due to a perturbation B. With A as a dipole operator and spin–orbit coupling introduced as a perturbation through B we can calculate singlet–triplet transition moments (phosphorescence) [57]. For $\omega_b = \frac{1}{2}\omega_f$ and dipole operators T_f^{AB} is the two-photon absorption amplitude.

From the double residue

$$(139) \quad \lim_{\omega_b \to -\omega_e} (\omega_b + \omega_e) \lim_{\omega_c \to \omega_f} (\omega_c - \omega_f) \langle\langle \hat{A}; \hat{B}, \hat{C} \rangle\rangle_{-\omega_b, \omega_c} = -T_{ef}^A T_e^{B\dagger} T_f^C$$

we can identify the transition moment T_{ef}^A between excited states (e) and (f) in accordance with the exact case

$$(140) \quad T_{ef}^A = \langle e|\hat{A}|f\rangle - \delta_{ef}\langle 0|\hat{A}|0\rangle$$

as

$$(141) \quad T_{ef}^A = -\kappa^A \mathbf{X}_{ef} - \frac{1}{2}[\langle 0|[\hat{X}_e^\dagger, [\hat{X}_f, \hat{A}]]|0\rangle + \langle 0|[\hat{X}_e, [\hat{X}_f^\dagger, \hat{A}]]|0\rangle]$$

where $\mathbf{X}_{ef}$ is defined to satisfy

$$(142) \quad \lim_{\omega_b \to -\omega_e} (\omega_b + \omega_e) \lim_{\omega_c \to \omega_f} (\omega_c - \omega_f) \mathbf{V}^{\omega_b, \omega_c} = \mathbf{X}_{ef} T_e^{B\dagger} T_f^C$$

When the two excited states are the same we get excited state first order properties

$$(143) \quad T_{ff}^A = \langle f|\hat{A}|f\rangle - \langle 0|\hat{A}|0\rangle$$

2.7.3 Cubic response

The cubic response functions has poles where the frequency parameters or their sum matches an excitation energy. As for the quadratic response function we may also consider double residues where two poles matches at the same time (triple residues turn out not to give anything new).

The single residue related to third-order transition moments is

$$(144) \quad \lim_{\omega_0 \to \omega_f} (\omega_0 - \omega_f) \langle\langle \hat{A}; \hat{B}, \hat{C}, \hat{D} \rangle\rangle_{-\omega_b, \omega_c, \omega_d} = -T_f^{A\dagger} T_f^{BCD}$$

$$(145) \quad T_f^{BCD} = \mathbf{X}_f^\dagger \mathbf{V}^{\omega_b, \omega_c, \omega_d}$$

or alternatively

(146) $$\lim_{\omega_d \to \omega_f} (\omega_d - \omega_f) \langle\langle \hat{A}; \hat{B}, \hat{C}, \hat{D} \rangle\rangle_{-\omega_b,-\omega_c,\omega_d} = -T_f^{ABC} T_f^D$$

(147) $$T_f^{ABC} = \kappa^{\hat{A}\dagger} \mathbf{X}_f^{\omega_b,\omega_c} + 3\hat{P}_{bcx}\langle 0|[\hat{\kappa}^{\omega_b},[\hat{\kappa}^{\omega_c},\hat{A}]]|0\rangle$$
$$+ 3\hat{P}_{bcx}\langle 0|[\hat{\kappa}^{\omega_b,\omega_c},[\hat{X}_f,\hat{A}]]|0\rangle + \hat{P}_{bcx}\langle 0|[\hat{\kappa}^{\omega_b},[\hat{\kappa}^{\omega_c},[\hat{X}_f,\hat{A}]]]|0\rangle$$

where $\mathbf{X}_f^{\omega_b \omega_c}$ is defined to satisfy

(148) $$\lim_{\omega_d \to \omega_f} (\omega_d - \omega_f) \mathbf{V}^{\omega_b,\omega_c,\omega_d} = \mathbf{X}_f^{\omega_b \omega_c} T_f^D$$

and $\hat{\kappa}_f^{\omega_c}$ is obtained by solving the modified second-order equation

(149) $$(\mathbf{E} - (\omega_c + \omega_f)\mathbf{S})\kappa_f^{\omega_c} = \mathbf{X}_f^{\omega_c}$$

As for the corresponding residues for quadratic response function, even though Eqs. (145) and (147) have a different structure they give identical numerical results. This time, however, there is a computational advantage to use Eq. (145) over Eq. (147). In the first case we have to solve an eigenvalue equation for the excited state in question and also the usual set of first- and second-order linear response equations. In the second case we have to solve an additional set of modified second-order response equations for each excited state. In particular, if we are interested in several excited states we need not solve more response equations using Eq. (145) but two sets of extra equations for each excited state using Eq. (147). If all operators are the same we can solve one less response equation using Eq. (145) (2 versus 3) even if we are only interested in one single excited state.

For the case $\omega_b = \omega_c = 0$ we can interpret T_f^{ABC} as the induced transition amplitude of an operator $\hat{A}$ between the ground state (0) and an excited state (f) due to perturbations B and C. With one static perturbation we get induced two-photon absorption. For $\omega_b = \omega_c = \frac{1}{3}\omega_f$ and dipole operators, T_f^{ABC} is the three-photon absorption amplitude.

From the double residue

(150) $$\lim_{\omega_c \to -\omega_e} (\omega_c + \omega_e) \lim_{\omega_d \to \omega_f} (\omega_d - \omega_f) \langle\langle \hat{A}; \hat{B}, \hat{C}, \hat{D} \rangle\rangle_{\omega_b,\omega_c,\omega_d} = -T_{ef}^{AB} T_e^{C\dagger} T_f^D$$

we can identify the two-photon transition moment T_{ef}^{AB} between excited states (e) and (f). When the states are the same we obtain excited state dynamic second order properties,

(151) $$T_{ff}^{AB} = \alpha_{AB}^0(\omega_b) - \alpha_{AB}^f(\omega_b)$$

where in the case of A and B being dipole operators α_{AB}^0 and α_{AB}^f are the dipole polarizabilities of the ground and excited state respectively. Also here we can get an alternative expression by taking one of the residues at the A operator. We do get a different set of response equations to solve, but there is no computational advantage to this approach in this case.

For further details of DFT calculations of excited state polarizabilities and three-photon absorption see [58, 59].

2.8 The Quasi-energy Method

The quasi-energy formulation provides an alternative time-dependent variation principle for deriving dynamical response functions. It is arguably more attractive than the Ehrenfest method in that it provides a unified framework for treating variational and non-variational wave functions, by analogy with time-independent theory, to which it naturally reduces in the limit of static perturbations [60–62]. As an additional advantage over the Ehrenfest method, the permutational symmetries (with respect to the exchange of operators) becomes manifest in the quasi-energy method.

The concept of quasi-energy arises naturally in TDDFT—indeed, it turns out to be nothing but the action integral of Runge and Gross [1], where the integration limits are chosen to span a period of the perturbation, scaled by the inverse of the period length. In the following, we discuss the application of the quasi-energy method to the calculation of response functions in Kohn–Sham DFT. Since the Kohn–Sham quasi-energy method follows closely the corresponding Hartree–Fock method, we here consider only those aspects of Kohn–Sham theory that differ from Hartree–Fock theory.

In Hartree–Fock theory, the quasi-energy and its time average are defined as follows

$$Q(t) = \langle t|\hat{H} - i\frac{d}{dt}|t\rangle \qquad (152)$$

$$\{Q\}_T = \frac{1}{T}\int_{-T/2}^{T/2} Q(t)\,dt \qquad (153)$$

where T constitutes one period of the time-dependent perturbation. In Kohn–Sham theory, there is an additional contribution to the quasi-energy from the exchange–correlation functional,

$$Q_{xc}(t) = E_{xc}[\rho(t)] \qquad (154)$$

The perturbation is expanded in its (discrete) Fourier components, each of which is further expanded in field strengths $\epsilon_x^{\omega_k}$ coupled to a quantum-mechanical operator,

$$\hat{V}(t) = \sum_{kx} e^{-i\omega_k t} \epsilon_x^{\omega_k} \hat{V}_x^{\omega_k} \qquad (155)$$

Included in the summation is the operator $\hat{A}$, the response functions of which we are calculating. We associate a nonzero frequency ω but zero perturbation strength with this operator.

The periodic perturbation Eq. (155) induces a change in the quasi-energy, which is expanded in orders of the perturbation:

(156) $$Q(t) = Q^{(0)}(t) + Q^{(1)}(t) + Q^{(2)}(t) + \cdots$$

The response function of order n is then recovered by differentiating the time-averaged perturbed quasi-energy $Q^{(n+1)}(t)$ with respect to the frequency-dependent field strengths,

(157) $$\langle\langle \hat{A}; \hat{B} \rangle\rangle_\omega = \frac{d^2\{Q^{(2)}\}_T}{d\epsilon_A^{-\omega} d\epsilon_B^\omega}$$

(158) $$\langle\langle \hat{A}; \hat{B}, \hat{C} \rangle\rangle_{\omega_1,\omega_2} = \frac{d^3\{Q^{(3)}\}_T}{d\epsilon_A^{-\omega_1-\omega_2} d\epsilon_B^{\omega_1} d\epsilon_C^{\omega_2}}$$

(159) $$\langle\langle \hat{A}; \hat{B}, \hat{C}, \hat{D} \rangle\rangle_{\omega_1,\omega_2,\omega_3} = \frac{d^4\{Q^{(4)}\}_T}{d\epsilon_A^{-\omega_1-\omega_2-\omega_3} d\epsilon_B^{\omega_1} d\epsilon_C^{\omega_2} d\epsilon_C^{\omega_3}}$$

The frequency associated with $\hat{A}$ is set equal to minus the sum of the perturbing frequencies so that the time-averaged quasi-energy does not vanish.

To calculate the perturbed quasi-energy Eq. (156), we must first determine the perturbed orbital-rotation parameters $\hat{\kappa}(t)$, which are likewise expanded in orders of the perturbation,

(160) $$\hat{\kappa}^{(1)}(t) = \sum_{kx} e^{-i\omega_k t} \epsilon_x^{\omega_k} \hat{\kappa}_x^{\omega_k}$$

(161) $$\hat{\kappa}^{(2)}(t) = \frac{1}{2} \sum_{klxy} e^{-i(\omega_k+\omega_l)t} \epsilon_x^{\omega_k} \epsilon_y^{\omega_l} \hat{\kappa}_x^{\omega_k} \hat{\kappa}_y^{\omega_l}$$

(162) $$\hat{\kappa}^{(3)}(t) = \frac{1}{6} \sum_{klmxyz} e^{-i(\omega_k+\omega_l+\omega_m)t} \epsilon_x^{\omega_k} \epsilon_y^{\omega_l} c_z^{\omega_m} \hat{\kappa}_x^{\omega_k} \hat{\kappa}_y^{\omega_l} \hat{\kappa}_z^{\omega_m}$$

From the $2n+1$ rule, it follows that, to calculate the quasi-energy to order $2n+1$, we need only determine $\hat{\kappa}(t)$ to order n, the contributions from higher orders being zero. Thus, to calculate the linear and quadratic response functions, we need only determine $\hat{\kappa}^{(1)}$. Note that, even though the contribution of for example $\hat{\kappa}^{(2)}$ to $Q^{(2)}$ vanishes, its contribution to the exchange–correlation part of the quasi-energy $Q_{\text{xc}}^{(2)}$ is nonzero but cancelled by a similar contribution to $Q^{(2)} - Q_{\text{xc}}^{(2)}$. In our discussion of $Q_{\text{xc}}^{(2)}$ and $Q_{\text{xc}}^{(3)}$ we shall therefore ignore all contributions that do not depend on $\hat{\kappa}^{(1)}$, in accordance with the $2n+1$ rule. To calculate the cubic response functions, we need in addition to $\hat{\kappa}^{(1)}$ also $\hat{\kappa}^{(2)}$, but not $\hat{\kappa}^{(3)}$ or $\hat{\kappa}^{(4)}$. Accordingly we will only consider contribution from $\hat{\kappa}^{(1)}$ and $\hat{\kappa}^{(2)}$ to $Q_{\text{xc}}^{(4)}$.

2.8.1 Linear response

In linear response theory, the exchange–correlation energy is expanded to second order in $\hat{\kappa}^{(1)}$:

(163) $$\bar{E}_{\text{xc}}^{(2)} = \int \frac{\delta E_{\text{xc}}}{\delta \rho(\mathbf{r})} \bar{\rho}^{(2)}(\mathbf{r},t) d\tau + \frac{1}{2} \iint \frac{\delta^2 E_{\text{xc}}}{\delta \rho(\mathbf{r}) \delta \rho(\mathbf{r}')} \rho^{(1)}(\mathbf{r},t) \rho^{(1)}(\mathbf{r}',t) d\tau d\tau'$$

The perturbed densities are given by Eqs. (54) except that the contribution from $\hat{\kappa}^{(2)}$ to $\rho^{(2)}$ is ignored in accordance with the $2n+1$ rule, as indicated by the bar in $\bar{E}_{xc}^{(2)}$ and $\bar{\rho}^{(2)}$. Next, we expand $\hat{\kappa}^{(1)}(t)$ according to Eq. (160) and obtain

$$(164) \quad \bar{E}_{xc}^{(2)} = \frac{1}{2} \sum_{kl} e^{-i(\omega_k + \omega_l)t} \sum_{xy} \epsilon_x^{\omega_k} \epsilon_y^{\omega_l} \left[\int \frac{\delta E_{xc}}{\delta \rho(\mathbf{r})} \bar{\rho}_{xy}^{\omega_1,\omega_2}(\mathbf{r}) \, d\tau \right.$$
$$\left. + \iint \frac{\delta^2 E_{xc}}{\delta \rho(\mathbf{r}) \delta \rho(\mathbf{r}')} \rho^{\omega_1}(\mathbf{r}) \rho^{\omega_2}(\mathbf{r}') \, d\tau \, d\tau' \right]$$

where we have introduced

$$(165) \quad \rho_x^{\omega} = \langle 0 | [\hat{\kappa}_x^{\omega}, \hat{\rho}] | 0 \rangle$$

$$(166) \quad \bar{\rho}_{xy}^{\omega_1,\omega_2} = \hat{P}_{12} \langle 0 | [\hat{\kappa}_x^{\omega_1}, [\hat{\kappa}_y^{\omega_2}, \hat{\rho}]] | 0 \rangle$$

The first-order parameters $\kappa_x^{\omega_1}$ and $\kappa_y^{\omega_2}$ are determined by a variational condition on the second-order quasi-energy, the result of which is an equation identical to Eq. (70). Differentiating the time-averaged exchange–correlation quasi-energy with respect to the field strengths $\epsilon_a^{\omega_a}$, $\epsilon_b^{\omega_b}$, according to Eq. (157), we obtain

$$(167) \quad \langle\langle \hat{A}; \hat{B} \rangle\rangle_{\omega_b}^{xc} = \int \frac{\delta E_{xc}}{\delta \rho(\mathbf{r})} \rho^{\omega_a,\omega_b}(\mathbf{r}) \, d\tau + \iint \frac{\delta^2 E_{xc}}{\delta \rho(\mathbf{r}) \delta \rho(\mathbf{r}')} \rho^{\omega_a}(\mathbf{r}) \rho^{\omega_b}(\mathbf{r}') \, d\tau \, d\tau'$$

for the exchange–correlation energy contribution to the linear response function.

2.8.2 Quadratic response

The quadratic response function is obtained as the third derivative of the time-averaged quasi-energy. The program is then to expand the energy to third order in the first-order parameters:

$$(168) \quad \bar{E}_{xc}^{(3)} = \int \frac{\delta E_{xc}}{\delta \rho(\mathbf{r})} \bar{\rho}^{(3)}(\mathbf{r}, t) \, d\tau$$
$$+ \iint \frac{\delta^2 E_{xc}}{\delta \rho(\mathbf{r}) \delta \rho(\mathbf{r}')} \rho^{(1)}(\mathbf{r}, t) \bar{\rho}^{(2)}(\mathbf{r}', t) \, d\tau \, d\tau'$$
$$+ \frac{1}{6} \iiint \frac{\delta^3 E_{xc}}{\delta \rho(\mathbf{r}) \delta \rho(\mathbf{r}') \delta \rho(\mathbf{r}'')} \rho^{(1)}(\mathbf{r}, t) \rho^{(1)}(\mathbf{r}', t) \rho^{(1)}(\mathbf{r}'', t) \, d\tau \, d\tau' \, d\tau''$$

where

$$(169) \quad \bar{\rho}^{(3)}(\mathbf{r}, t) = \frac{1}{6} \langle 0 | [\hat{\kappa}^{(1)}(t), [\hat{\kappa}^{(1)}(t), [\hat{\kappa}^{(1)}(t), \hat{\rho}(\mathbf{r})]]] | 0 \rangle$$

Kohn–Sham Time-Dependent Density Functional Theory

Expanding the parameters as in the linear case Eq. (164), we obtain,

$$(170) \quad \bar{E}_{xc}^{(3)} = \frac{1}{6} \sum_{klm} e^{-i(\omega_k+\omega_l+\omega_m)t} \sum_{xyz} \epsilon_x^{\omega_k} \epsilon_y^{\omega_l} \epsilon_z^{\omega_m}$$

$$\times \left[\int \frac{\delta E_{xc}}{\delta \rho(\mathbf{r})} \bar{\rho}^{\omega_k,\omega_l,\omega_m}(\mathbf{r}) \, d\tau \right.$$

$$+ 2\hat{P}_{xyz} \iint \frac{\delta^2 E_{xc}}{\delta \rho(\mathbf{r}) \delta \rho(\mathbf{r}')} \rho^{\omega_k}(\mathbf{r}) \bar{\rho}^{\omega_l,\omega_m}(\mathbf{r}') \, d\tau \, d\tau'$$

$$\left. + \iiint \frac{\delta^3 E_{xc}}{\delta \rho(\mathbf{r}) \delta \rho(\mathbf{r}') \delta \rho(\mathbf{r}'')} \rho^{\omega_k}(\mathbf{r}) \rho^{\omega_l}(\mathbf{r}') \rho^{\omega_l}(\mathbf{r}'') \, d\tau \, d\tau' \, d\tau'' \right]$$

where the first- and second-order perturbed densities are given by Eqs. (165) and (166) and the third-order perturbed density by

$$(171) \quad \bar{\rho}_{xyz}^{\omega_1,\omega_2,\omega_3}(\mathbf{r}) = \hat{P}_{123} \langle 0 | [\hat{\kappa}_x^{\omega_1}, [\hat{\kappa}_y^{\omega_2}, [\hat{\kappa}_z^{\omega_3}, \hat{\rho}(\mathbf{r})]]] | 0 \rangle$$

As in the linear case, time averaging cancels the time-dependent phase factor if the sum of the frequencies is zero. The exchange–correlation energy contribution to the quadratic response function then becomes,

$$(172) \quad \langle\langle \hat{A}; \hat{B}, \hat{C} \rangle\rangle_{\omega_b,\omega_c}^{xc} = \frac{1}{6} \left[\int \frac{\delta E_{xc}}{\delta \rho(\mathbf{r})} \rho^{\omega_a,\omega_b,\omega_c}(\mathbf{r}) \, d\tau \right.$$

$$+ 3\hat{P}_{ABC} \iint \frac{\delta^2 E_{xc}}{\delta \rho(\mathbf{r}) \delta \rho(\mathbf{r}')} \rho^{\omega_a}(\mathbf{r}) \rho^{\omega_b \omega_c}(\mathbf{r}') \, d\tau \, d\tau'$$

$$\left. + \iiint \frac{\delta^3 E_{xc}}{\delta \rho(\mathbf{r}) \delta \rho(\mathbf{r}') \delta \rho(\mathbf{r}'')} \rho^{\omega_a}(\mathbf{r}) \rho^{\omega_b}(\mathbf{r}) \rho^{\omega_c}(\mathbf{r}) \, d\tau \, d\tau' \, d\tau'' \right]$$

2.8.3 Cubic response

The cubic response function is obtained as the fourth derivative of the time-averaged quasi-energy. Thus we expand the energy to fourth order in the first- and second-order parameters,

$$(173) \quad \bar{E}_{xc}^{(4)} = \int \frac{\delta E_{xc}}{\delta \rho(\mathbf{r})} \bar{\rho}^{(4)}(\mathbf{r},t) \, d\tau$$

$$+ \iint \frac{\delta^2 E_{xc}}{\delta \rho(\mathbf{r}) \delta \rho(\mathbf{r}')} (\frac{1}{2} \rho^{(2)}(\mathbf{r},t) \rho^{(2)}(\mathbf{r}',t) + \rho^{(1)}(\mathbf{r},t) \bar{\rho}^{(3)}(\mathbf{r}',t)) \, d\tau \, d\tau'$$

$$+ \frac{1}{2} \iiint \frac{\delta^3 E_{xc}}{\delta \rho(\mathbf{r}) \delta \rho(\mathbf{r}') \delta \rho(\mathbf{r}'')} \rho^{(1)}(\mathbf{r},t) \rho^{(1)}(\mathbf{r}',t) \rho^{(2)}(\mathbf{r}'',t) \, d\tau \, d\tau' \, d\tau''$$

$$+ \frac{1}{24} \iiiint \frac{\delta^4 E_{xc}}{\delta \rho(\mathbf{r}) \delta \rho(\mathbf{r}') \delta \rho(\mathbf{r}'') \delta \rho(\mathbf{r}''')}$$

$$\times \rho^{(1)}(\mathbf{r},t) \rho^{(1)}(\mathbf{r}',t) \rho^{(1)}(\mathbf{r}'',t) \rho^{(1)}(\mathbf{r}''',t) \, d\tau \, d\tau' \, d\tau'' \, d\tau'''$$

where

$$\bar{\rho}^{(4)}(\mathbf{r}, t) = \frac{1}{2} \langle 0|[\hat{\kappa}^{(2)}(t), [\hat{\kappa}^{(2)}(t), \hat{\rho}(\mathbf{r})]]|0\rangle \tag{174}$$

$$+ \frac{1}{6} \langle 0|[\hat{\kappa}^{(1)}(t), [\hat{\kappa}^{(1)}(t), [\hat{\kappa}^{(2)}(t), \hat{\rho}(\mathbf{r})]]]|0\rangle$$

$$+ \frac{1}{6} \langle 0|[\hat{\kappa}^{(1)}(t), [\hat{\kappa}^{(2)}(t), [\hat{\kappa}^{(1)}(t), \hat{\rho}(\mathbf{r})]]]|0\rangle$$

$$+ \frac{1}{6} \langle 0|[\hat{\kappa}^{(2)}(t), [\hat{\kappa}^{(1)}(t), [\hat{\kappa}^{(1)}(t), \hat{\rho}(\mathbf{r})]]]|0\rangle$$

$$+ \frac{1}{24} \langle 0|[\hat{\kappa}^{(1)}(t), [\hat{\kappa}^{(1)}(t), [\hat{\kappa}^{(1)}(t), [\hat{\kappa}^{(1)}(t), \hat{\rho}(\mathbf{r})]]]]|0\rangle$$

$$\bar{\rho}^{(3)}(\mathbf{r}, t) = \frac{1}{2} \langle 0|[\hat{\kappa}^{(1)}(t), [\hat{\kappa}^{(2)}(t), \hat{\rho}(\mathbf{r})]]|0\rangle \tag{175}$$

$$+ \frac{1}{2} \langle 0|[\hat{\kappa}^{(2)}(t), [\hat{\kappa}^{(1)}(t), \hat{\rho}(\mathbf{r})]]|0\rangle$$

$$+ \frac{1}{6} \langle 0|[\hat{\kappa}^{(1)}(t), [\hat{\kappa}^{(1)}(t), [\hat{\kappa}^{(1)}(t), \hat{\rho}(\mathbf{r})]]]|0\rangle$$

The bar indicates that, according to the $2n+1$ rule, we only consider contributions from the first- and second-order parameters. This also means that $\bar{\rho}^{(3)}(\mathbf{r}, t)$ differs from the quadratic case were only the first-order parameter were included. Expanding the parameters according to Eqs. (160) and (161), we obtain,

$$\bar{E}_{\text{xc}}^{(4)} = \frac{1}{24} \sum_{klmn} e^{-i(\omega_k+\omega_l+\omega_m+\omega_n)t} \sum_{xyzu} \epsilon_x^{\omega_k} \epsilon_y^{\omega_l} \epsilon_z^{\omega_m} \epsilon_u^{\omega_n} \tag{176}$$

$$\times \Bigg[\int \frac{\delta E_{\text{xc}}}{\delta \rho(\mathbf{r})} \bar{\rho}^{\omega_k,\omega_l,\omega_m,\omega_n}(\mathbf{r}) \, d\tau$$

$$+ 3\hat{P}_{xyzu} \iint \frac{\delta^2 E_{\text{xc}}}{\delta \rho(\mathbf{r})\delta \rho(\mathbf{r}')} \rho^{\omega_k,\omega_l}(\mathbf{r}) \rho^{\omega_m,\omega_n}(\mathbf{r}') \, d\tau \, d\tau'$$

$$+ 4\hat{P}_{xyzu} \iint \frac{\delta^2 E_{\text{xc}}}{\delta \rho(\mathbf{r})\delta \rho(\mathbf{r}')} \rho^{\omega_k}(\mathbf{r}) \bar{\rho}^{\omega_l,\omega_m,\omega_n}(\mathbf{r}') \, d\tau \, d\tau'$$

$$+ 6\hat{P}_{xyzu} \iiint \frac{\delta^3 E_{\text{xc}}}{\delta \rho(\mathbf{r})\delta \rho(\mathbf{r}')\delta \rho(\mathbf{r}'')} \rho^{\omega_k}(\mathbf{r}) \rho^{\omega_l}(\mathbf{r}') \rho^{\omega_m,\omega_n}(\mathbf{r}'') \, d\tau \, d\tau' \, d\tau''$$

$$+ \iiiint \frac{\delta^4 E_{\text{xc}}}{\delta \rho(\mathbf{r})\delta \rho(\mathbf{r}')\delta \rho(\mathbf{r}'')\delta \rho(\mathbf{r}''')}$$

$$\times \rho^{\omega_k}(\mathbf{r}) \rho^{\omega_l}(\mathbf{r}') \rho^{\omega_m}(\mathbf{r}'') \rho^{\omega_n}(\mathbf{r}''') \, d\tau \, d\tau' \, d\tau'' \, d\tau''' \Bigg]$$

where the perturbed densities are given by

(177)
$$\rho_x^{\omega_1}(\mathbf{r}) = \langle 0|[\hat{\kappa}_x^{\omega_1}, \hat{\rho}(\mathbf{r})]|0\rangle$$

$$\rho_{xy}^{\omega_1,\omega_2}(\mathbf{r}) = \langle 0|[\hat{\kappa}_{xy}^{\omega_1,\omega_2}, \hat{\rho}(\mathbf{r})]]|0\rangle + \hat{P}_{12}\langle 0|[\hat{\kappa}_x^{\omega_1}, [\hat{\kappa}_y^{\omega_2}, \hat{\rho}(\mathbf{r})]]|0\rangle$$

$$\bar{\rho}_{xyz}^{\omega_1,\omega_2,\omega_3}(\mathbf{r}) = \frac{3}{2}\hat{P}_{123}\langle 0|[\hat{\kappa}_x^{\omega_1}, [\hat{\kappa}_{yz}^{\omega_2,\omega_3}, \hat{\rho}(\mathbf{r})]]|0\rangle$$

$$+ \frac{3}{2}\hat{P}_{123}\langle 0|[\hat{\kappa}_{xy}^{\omega_1,\omega_2}, [\hat{\kappa}_z^{\omega_3}, \hat{\rho}(\mathbf{r})]]|0\rangle$$

$$+ \hat{P}_{123}\langle 0|[\hat{\kappa}_x^{\omega_1}, [\hat{\kappa}_y^{\omega_2}, [\hat{\kappa}_z^{\omega_3}, \hat{\rho}(\mathbf{r})]]]|0\rangle$$

$$\bar{\rho}_{xyzu}^{\omega_1,\omega_2,\omega_3,\omega_4}(\mathbf{r}) = 3\hat{P}_{1234}\langle 0|[\hat{\kappa}_{xy}^{\omega_1,\omega_2}, [\hat{\kappa}_{zu}^{\omega_3,\omega_4}, \hat{\rho}(\mathbf{r})]]|0\rangle$$

$$+ 2\hat{P}_{1234}\langle 0|[\hat{\kappa}_x^{\omega_1}, [\hat{\kappa}_y^{\omega_2}, [\hat{\kappa}_{zu}^{\omega_3,\omega_4}, \hat{\rho}(\mathbf{r})]]]|0\rangle$$

$$+ 2\hat{P}_{1234}\langle 0|[\hat{\kappa}_x^{\omega_1}, [\hat{\kappa}_{yz}^{\omega_2,\omega_3}, [\hat{\kappa}_u^{\omega_4}, \hat{\rho}(\mathbf{r})]]]|0\rangle$$

$$+ 2\hat{P}_{1234}\langle 0|[\hat{\kappa}_{xy}^{\omega_1,\omega_2}, [\hat{\kappa}_z^{\omega_3}, [\hat{\kappa}_u^{\omega_4}, \hat{\rho}(\mathbf{r})]]]|0\rangle$$

$$+ \hat{P}_{1234}\langle 0|[\hat{\kappa}_x^{\omega_1}, [\hat{\kappa}_y^{\omega_2}, [\hat{\kappa}_z^{\omega_3}, [\hat{\kappa}_u^{\omega_4}, \hat{\rho}(\mathbf{r})]]]]|0\rangle$$

Time averaging cancels the time-dependent phase factor if the sum of the frequencies is zero. The exchange–correlation energy contribution to the cubic response function then becomes:

(178)
$$\langle\langle \hat{A}; \hat{B}, \hat{C}, \hat{D} \rangle\rangle^{\text{xc}}_{\omega_b,\omega_c,\omega_d} = \frac{1}{24}\Bigg[\int \frac{\delta E_{\text{xc}}}{\delta\rho(\mathbf{r})}\bar{\rho}^{\omega_a,\omega_b,\omega_c,\omega_d}(\mathbf{r})\,d\tau$$

$$+ 3\hat{P}_{ABCD}\iint \frac{\delta^2 E_{\text{xc}}}{\delta\rho(\mathbf{r})\delta\rho(\mathbf{r}')}\rho^{\omega_a,\omega_b}(\mathbf{r})\rho^{\omega_c,\omega_d}(\mathbf{r}')\,d\tau\,d\tau'$$

$$+ 4\hat{P}_{ABCD}\iint \frac{\delta^2 E_{\text{xc}}}{\delta\rho(\mathbf{r})\delta\rho(\mathbf{r}')}\rho^{\omega_a}(\mathbf{r})\bar{\rho}^{\omega_b,\omega_c,\omega_d}(\mathbf{r}')\,d\tau\,d\tau'$$

$$+ 6\hat{P}_{ABCD}\iiint \frac{\delta^3 E_{\text{xc}}}{\delta\rho(\mathbf{r})\delta\rho(\mathbf{r}')\delta\rho(\mathbf{r}'')}\rho^{\omega_a}(\mathbf{r})\rho^{\omega_b}(\mathbf{r}')\rho^{\omega_c,\omega_d}(\mathbf{r}'')\,d\tau\,d\tau'\,d\tau''$$

$$+ \iiiint \frac{\delta^4 E_{\text{xc}}}{\delta\rho(\mathbf{r})\delta\rho(\mathbf{r}')\delta\rho(\mathbf{r}'')\delta\rho(\mathbf{r}''')}$$

$$\times \rho^{\omega_a}(\mathbf{r})\rho^{\omega_b}(\mathbf{r}')\rho^{\omega_c}(\mathbf{r}'')\rho^{\omega_d}(\mathbf{r}''')\,d\tau\,d\tau'\,d\tau''\,d\tau'''\Bigg]$$

3. IMPLEMENTATION

Numerical evaluations of Kohn–Sham matrix elements and exchange-correlation (xc) contributions to response vectors follow the same scheme. In contrast to the Coulomb and exact Hartree–Fock exchange contributions which are usually evaluated by summing analytically computed integrals between basis functions (c.f. Eq. (58)), one has to resort to an explicit numerical integration of the exchange-correlation contribution. Another difference in the evaluation of the xc contribution is that the functionals are non-linear with respect to the density and therefore time-dependent density functional theory requires knowledge of higher order derivatives of functionals. Evaluation of any of these contributions consists of three basic building blocks: Generation of the numerical grid, assembly of various prefactors depending on functional derivatives and expectation values of perturbed densities, and adding contributions to respective matrix elements. Generation of the numerical grid is a crucial component determining the accuracy of the xc evaluation. It is a well researched topic [63–65] and we are not going to dwell on it here. Instead, we will focus on the remaining parts, commenting occasionally on the efficiency aspects of appearing expressions. We only assume further that the grid consists of P points $\mathbf{r}_x$ with associated weights $w_x, x \in \{1 \ldots P\}$.

3.1 Implementation of Linear Transformations

Computing linear response of a molecular property from Eq. (80) requires knowledge of a response vector $\hat{\kappa}^\omega$ corresponding to the perturbation $\hat{V}^\omega$. This response vector can be found by solving a linear set of equations in Eq. (76). This set of equations is large and therefore is solved iteratively for $N^{\text{occ}} \times N^{\text{virt}}$ variables where N^{occ} is a number of occupied orbitals and N^{virt} is number of virtuals.

The xc contribution to the matrix being implicitly inverted is

$$(179) \quad f^\omega_{\text{xc},pq} = \int \left. \frac{\delta v_{\text{xc}}}{\delta \rho(\mathbf{r})} \right|_{\rho=\rho(\mathbf{r})} \rho^\omega(\mathbf{r})[k^\omega, \Omega_{pq}] d\mathbf{r}$$

$$(180) \quad \Omega_{pq} = \phi_p(\mathbf{r})\phi_q(\mathbf{r})$$

where $\rho^\omega(\mathbf{r})$ is given by Eq. (74). The integral in Eq. (179) is evaluated numerically by summing contributions from all the grid points with appropriate weights. At each grid point, we could compute the expectation value of the perturbed density $\rho^\omega(\mathbf{r})$, compute relevant derivatives of the exchange-correlation potential, and compute the contribution to pq matrix element of f^ω.

This is however not how the actual calculation is done. Generally, computing xc contributions is time-consuming because of usually large number of grid points and therefore one wants to minimize the amount of calculation performed at each grid point. Straightforward implementation of the scheme above will perform matrix-matrix multiplications at each grid point requiring N^3 of work at each grid point and scale like N^4 with the system size. A simple improvement would be to avoid constructing $\Omega_{pq} = \phi_p(\mathbf{r})\phi_q(\mathbf{r})$ matrices explicitly and instead multiply other terms by $\phi_p(\mathbf{r})$ from the left or right side when needed. This reduces the scaling by one order of magnitude to N^2 at each grid point and N^3. One can do better though: A constant amount of work per grid point (and a linear scaling in total) can be achieved by utilizing locality of basis set functions and the fact that the number of nonvanishing basis functions at given point in space does not change strongly as the system size increases. This requires however that the expression above is transformed to atomic-orbital (AO) basis. The transformation is done with help of the matrix of molecular orbital coefficients C.

The entire calculation looks as follows:
- a perturbed density $\tilde{\rho}^\omega$ in AO basis is created

$$(181) \quad \tilde{\rho}^\omega_{pq} = \{C[D^{MO}, \kappa^\omega]C^\dagger\}_{pq} = \sum_i^{occ} \sum_r C_{pi} \kappa^\omega_{ir} C_{qr} - C_{pr} \kappa^\omega_{ri} C_{qi}$$

- the xc contribution to the response vector transformation is being integrated numerically where only elements corresponding to basis functions $b_p(\mathbf{r})$ nonvanishing at given grid point $\mathbf{r}_x$ are actually computed,

$$(182) \quad e^\omega_{pq} = \sum_{x=1}^P w_x \left. \frac{\delta v_{xc}}{\delta \rho(\mathbf{r})} \right|_{\rho = \rho(\mathbf{r}_x)} \langle \tilde{\rho}^\omega(\mathbf{r}_x) \rangle b_p(\mathbf{r}_x) b_q(\mathbf{r}_x)$$

$$(183) \quad \langle \tilde{\rho}^\omega(\mathbf{r}_x) \rangle = \sum_{pq} b_p(\mathbf{r}_x) \tilde{\rho}^\omega_{pq} b_q(\mathbf{r}_x)$$

- finally, the response e^ω is transformed back to the MO basis and added to the Coulomb contribution:

$$(184) \quad f^{xc} = C^\dagger e^\omega C \Leftrightarrow f^\omega_{xc,pq} = \sum_{rs} C_{rp} e^\omega_{rs} C_{sq}$$

Observe, that for LDA-type functionals the first order variation of the potential can be expressed as the second order derivative of the functional with respect to the density,

$$(185) \quad E^{xc}[\rho] = \int F(\rho(\mathbf{r}))d\mathbf{r} \Rightarrow \frac{\delta v_{xc}}{\delta \rho(\mathbf{r})} = \frac{\partial^2 F}{\partial \rho^2}$$

It is critical to include in the sum in Eq. (182) only those grid points that contribute to the given matrix element e^ω_{pq}. The easiest way to achieve this is to divide grid points into spatial boxes and associate each box with a list of nonvanishing basis functions $b_p(\mathbf{r})$. The integration for a box needs then only evaluate the nonvanishing basis functions and matrix elements they contribute to.

Higher order transformations are performed in a similar fashion with the exception that more terms appears in the sums. For example, the quadratic response transformation is in general a sum of several matrices with appropriate prefactors (again, we omit gradient-dependent terms for brevity)

$$(186) \quad f^{xc} = \sum_x^P w_x \left(r\Omega^{BC} + r^B \Omega^C + r^C \Omega^B + r^{BC} \Omega \right)$$

where,

(187) $$\Omega^B = [\kappa^{\omega_B}, \Omega]$$

(188) $$\Omega^{BC} = \frac{1}{2}([\kappa^{\omega_B}, [\kappa^{\omega_C}, \Omega]] + [\kappa^{\omega_C}, [\kappa^{\omega_B}, \Omega]])$$

The coefficients r, r^B, r^C are expressed in terms of functional derivatives and expectation values of commutators $\rho^B = \langle \tilde{\rho}^{\omega_B}(\mathbf{r}) \rangle$,

(189) $$r = \frac{1}{2}\frac{\partial F}{\partial \rho}$$

(190) $$r^B = \rho^B \frac{\partial^2 F}{\partial \rho^2}$$

(191) $$r^{BC} = \rho^B \rho^C \frac{\partial^3 F}{\partial \rho^3} + \rho^{BC} \frac{\partial^2 F}{\partial \rho^2}$$

where we see third order derivatives of the exchange-correlation functional appearing. The quadratic response expression in Eq. (186) is formulated in terms of matrix commutators at a cost per grid point that scales as N^3. A speedup by one order of magnitude can be obtained by utilizing the structure of the $\Omega_{ij} = \phi_i \phi_j$ matrix and transforming the working formula to a form that uses only matrix–vector operations scaling as N^2 as suggested above for linear response. However, similarly to linear response, the expressions above can—and should—be evaluated in the atomic orbital basis to take advantage of the basis function locality. Such implementation gives manyfold speedup for all systems but the smallest ones.

3.2 Implementation of Functional Derivatives

As demonstrated above, time-dependent DFT may require higher derivatives of $E(\rho)$. The number of needed derivatives is multiplied additionally by the fact that in many practical applications an unrestricted formalism has to be used and α and β electron densities and gradients have to be treated as independent. Specifically, the derivatives need to be evaluated separately with respect to ρ_α, ρ_β and the corresponding density gradient components $[\nabla \rho_\alpha \cdot \nabla \rho_\alpha, \nabla \rho_\alpha \cdot \nabla \rho_\beta, \nabla \rho_\beta \cdot \nabla \rho_\beta]$.

For each functional, several set of functions needs to be implemented. Namely the functional itself, its first derivatives needed for the evaluation of the Kohn–Sham matrix, the second derivatives for the evaluation of the linear response transformations, the third derivatives for the quadratic response and finally the fourth derivatives are needed for the cubic response. Systematic handling of such a large number of derivatives is a daunting task: for the functionals that we have implemented, there are 5 different first order derivatives to compute, 12 second order derivatives and 23 third order derivatives. The number of fourth order derivatives is even higher. It is therefore not surprising that many have chosen to neglect the gradient dependence in the linear and higher response calculations [66] or compute the derivatives using finite-difference methods (as in [67]). We have however developed a systematic framework for analytical evaluation of functional derivatives, allowing also in many cases for automatic code generation.

4. SAMPLE APPLICATIONS

By choosing different operators for $\hat{A}$, $\hat{B}$, $\hat{C}$ and $\hat{D}$ in the expressions for the response functions derived in the previous section a wide range of different properties can be calculated. The most common example being the (hyper)polarizability were all operators are electric dipole moment operators. In this section we will present sample calculations of a few out of great many properties that are available from response functions.

Section 4.1.1 reviews second harmonic generation (SHG) for *para*-nitroaniline (PNA), Section 4.1.2 the polarizability and second hyperpolarizability of nitrogen and benzene, Section 4.1.3 the second hyperpolarizability of C_{60}, Section 4.2 the excited state polarizability of pyrimidine and *s*-tetrazine, Section 4.3 three-photon absorption, and finally, in Section 4.5 the electronic g-tensor and the hyperfine coupling tensor are reviewed as examples of open shell DFT response properties.

4.1 Hyperpolarizabilities

4.1.1 Para-nitroaniline

With an NH_2 donor and an NO_2 acceptor substituted on a phenyl ring, *para*-nitroaniline (PNA) shows an exceptionally strong charge-transfer character accompanied by a large polarizability and hyperpolarizability. It has served in the past as an important test system for experimental and theoretical investigations of hyperpolarizabilities, see for instance [69–73]. Of particular interest here is the parallel component of the SHG hyperpolarizability tensor $\beta_z^{SHG}(\omega) = \sum_k (\beta_{zkk} + \beta_{kzk} + \beta_{llz})$ where z is the molecular C_{2v} symmetry axis. Early Hartree–Fock and π-electron multiconfigurational self-consistent field (MCSCF) quadratic-response calculations [73] gave 5.7 and 8.2×10^{-30} esu, respectively, for this component at 1.17 eV, which is too low compared with an experimental value of 16.9×10^{-30} esu, as extrapolated from solvent measurements at the same frequency [69]. These discrepancies were first attributed to solvation but subsequent gas-phase measurements [70] yielded a similar value of 15.4×10^{-30} esu, however, seemingly using a different convention for defining β. Recently, Sałek et al. calculated the hyperpolarizability of PNA at the B3LYP level of theory, obtaining the values listed in Table 2, along with previous calculations and measurements of $\beta_z^{SHG}(\omega)$ in PNA.

Calculated excitation energies and oscillator strengths are listed in Table 1. The accurate description of the hyperpolarizability is strongly linked to the description of the intensive amino-to-nitro charge-transfer (CT) transition at 4.35 eV [68], which collects nearly all of the oscillator strength for the manifold of low-lying states. DFT represents a significant improvement on RPA, reducing the excitation energy, as

Table 1. The lowest excitation energies (eV) and oscillator strengths of PNA[a]. Reproduced from [41]

Sym	Root	4-31G B3LYP EE	4-31G B3LYP OS	4-31G LDA EE	4-31G LDA OS	6-31G* B3LYP EE	6-31G* B3LYP OS	6-31G* LDA EE	6-31G* LDA OS	Sadlej B3LYP EE	Sadlej B3LYP OS	Sadlej LDA EE	Sadlej LDA OS	cc-pVDZ B3LYP EE	cc-pVDZ B3LYP OS	cc-pVDZ LDA EE	cc-pVDZ LDA OS
1	1	3.80	0.3511	3.43	0.2901	4.13	0.3418	3.71	0.2712	3.84	0.3436	3.47	0.2825	4.13	0.3324	3.72	0.2622
	2	5.90	0.0478	5.56	0.0319	6.38	0.0077	5.52	0.0133	5.78	0.0185	5.47	0.0353	6.32	0.0020	5.38	0.0191
	3	6.42	0.0336	5.54	0.0788	6.46	0.0424	5.87	0.1047	6.45	0.0188	5.63	0.1194	6.62	0.0281	5.89	0.1183
2	1	4.57	0.0001	4.14	0.0017	4.74	0.0019	4.35	0.0000	4.51	0.0004	4.12	0.0005	4.71	0.0066	4.36	0.0022
	2	5.03	0.0459	4.60	0.0318	5.27	0.0574	4.72	0.0400	4.94	0.0474	4.47	0.0322	5.23	0.0547	4.66	0.0394
	3	5.97	0.0852	5.49	0.0026	6.18	0.0653	5.58	0.0208	5.94	0.0780	5.48	0.0261	6.12	0.0637	5.50	0.0198
3	1	4.24	0.0003	4.08	0.0002	4.44	0.0000	4.17	0.0000	4.20	0.0003	3.93	0.0002	4.42	0.0000	4.14	0.0000
	2	5.12	0.0029	4.62	0.0000	6.05	0.0000	4.50	0.0000	4.78	0.0017	4.32	0.0000	5.88	0.0000	4.36	0.0000
	3	6.11	0.0057	5.03	0.0027	6.50	0.0004	6.21	0.0003	5.71	0.0060	4.72	0.0020	5.95	0.0003	5.74	0.0002

[a] Experimental excitation energy [68] is 4.35 eV.

Table 2. The hyperpolarizability $\beta_z^{SHG}(\omega)$ average of PNA in units of 10^{-30} esu (1 au = 8.639418×10^{-33} esu), using the B-convention. *Reproduced from [74]*

ω(eV)	0	0.65	1.17	1.364	1.494
RHF[a]	4.09	4.50	5.68	6.50	7.24
MCSCF[a]	5.93	6.52	8.20	9.38	10.43
MP2[b]			12.0		
LDA/ALDA[c]			16.99		
LB94/ALDA[c]			21.16		
B3LYP[d]	6.72	7.94	12.33	16.28	21.16
B3LYP[e]	6.85	8.15	12.94	17.41	23.09
CCSD[f]	5.82		8.47		
CCSD[g]	7.50	8.24	11.52	13.83	16.12
CCSD[h]	7.38		11.37		
Exp.[i]		9.6±0.5	16.9±0.4	25±1	40±3
Exp.[j]			15.44±0.63		

[a] From [73].
[b] From [75] using RPA dispersion.
[c] From [76].
[d] From [77] using the aug-cc-pVDZ basis.
[e] From [77] using the Sadlej basis.
[f] From [74] using the cc-pVDZ basis.
[g] From [74] using the aug-cc-pVDZ basis.
[h] From [74] using a stripped down aug-cc-pVTZ basis.
[i] Experiment from [69] extrapolated from solvent measurements.
[j] Experiment from [70] in the gas phase.

expected from the increased correlation contribution to β from the σ electrons. Still, DFT results are clearly dependent on the functional, the local exchange functionals giving too low excitation energies (and too large β). The hybrid B3LYP functional gives the best result—that is, 4.13 eV in the cc-pVDZ basis—consistent with the nonlocal nature of the CT excitation.

At low frequencies, the agreement of the B3LYP values with experiment is reasonable. Thus, at 1.17 eV, the B3LYP model gives 12.9×10^{-30} esu, which is 16% lower than the experimental result. At higher frequencies, however, the B3LYP values are in poorer agreement with experiment, underestimating the experimental values by 30% at 1.364 eV and by more than 40% at 1.494 eV (see Table 2). However, in view of the large uncertainties in the experimental values and the fact that they were obtained by extrapolation from solvent measurements, these discrepancies may also arise from problems with the experimental measurements at high frequencies.

In Table 2, we have listed the values of $\beta_z^{SHG}(\omega)$ calculated using the CCSD model with different basis sets, including a stripped down aug-cc-pVTZ basis. At 1.17 eV, the CCSD model gives a value of 11.4×10^{-30} esu, somewhat lower than the B3LYP result of 12.9×10^{-30} esu and much lower than the experimental value of 15.4×10^{-30} esu. Since, at higher frequencies, the discrepancy between the B3LYP and CCSD models increases even more, it appears that the very large differences observed between the experimental measurements and the B3LYP values must arise from problems with the experimental measurements rather than with the calculations. In fact, from a comparison of the B3LYP and CCSD results, it appears that the B3LYP model gives an overestimation rather than an underestimation of the hyperpolarizabilities at high frequencies.

4.1.2 Nitrogen and benzene

Static and dynamic polarizabilities and hyperpolarizabilities of benzene and molecular nitrogen were computed and analyzed in [42] with respect to basis set and selected DFT method. The static values are reproduced in Tables 3 and 4.

The experimental value [78, 80] for the average polarizability α_{ave} is 11.76 a.u. for the nitrogen molecule and 69.51 a.u. for benzene. For the best basis set, i.e. the aug-cc-pVTZ basis set, we observe that all of the methods, including HF, are within 5% error. The HF method always slightly underestimates the polarizability α, due the general tendency of this method to overestimate excitation frequencies. The LDA approximation based on the uniform electron gas model, on the other hand, overestimates. The source of this behavior, which in this case is related to the tendency to underestimate the excitation frequencies, especially the HOMO-LUMO gap, is well understood: The LDA potential is simply not attractive enough in the outer region, due to the spurious problem of self-interaction [82]. As a correction to this problem the LB94 potential with the proper asymptotic limit has been proposed [83]. Using this potential and the LDA kernels, we obtain an excellent agreement with experiment for the nitrogen molecule. The α_{ave} value

Table 3. The basis set dependence of the average static polarizability α_{ave} and the second hyperpolarizability γ_{ave} of the nitrogen molecule. The HF, the DFT Potential/Kernel combinations and the CCSD methods were used. daug/taug-cc-pVTZ denotes doubly/triply augmented cc-pVTZ basis set. *Reproduced from [42]*

Potential/ Kernel	HF/ HF	LDA/ LDA	LB94/ LDA	B3LYP/ LDA	B3LYP/ B3LYP	PBE/ PBE	CCSD	Expt.
Polarizability α_{ave} in a.u.								
aug-cc-pVDZ	11.30	12.04	11.47	10.89	11.78	12.05	11.57	
aug-cc-pVTZ	11.54	12.23	11.76	11.11	12.00	12.24	11.65	
daug-cc-pVTZ	11.57	12.28	11.77	11.15	12.04	12.28	11.67	
taug-cc-pVTZ	11.57	12.28	11.77	11.15	12.04	12.28	11.67	
aug-cc-pVQZ	11.54	12.26	11.70	11.14	12.04	12.27	11.62	11.76[a]
Second hyperpolarizability γ_{ave} in a.u.								
aug-cc-pVDZ	456.0	923.2	634.5	640.5	773.9	925.0	684.4	
aug-cc-pVTZ	560.1	1020.8	706.0	737.2	879.9	1022.8	745.2	
daug-cc-pVTZ	697.1	1312.3	834.4	947.2	1132.4	1315.0	895.4	
taug-cc-pVTZ	710.2	1327.5	841.6	958.3	1144.3	1330.1	903.1	
aug-cc-pVQZ	654.3	1182.7	795.5	860.6	1020.2	1185.0	822.1	917 ± 5[b]

[a] From [78].
[b] From [79].

Table 4. The basis set dependence of the average static polarizability α_{ave} and the second hyperpolarizability γ_{ave} of benzene. The HF, the DFT Potential/Kernel combinations and the CCSD methods were used. *Reproduced from [42]*

Potential/ Kernel	HF/ HF	LDA/ LDA	LB94/ LDA	B3LYP/ LDA	B3LYP/ B3LYP	PBE/ PBE	CCSD	Expt.
Polarizability α_{ave} in a.u.								
cc-pVDZ	56.48	58.58	60.18	52.44	57.89	58.59		
cc-pVTZ	62.08	64.39	65.53	57.78	63.48	64.41		
cc-pVQZ	65.20	67.52	67.99	60.68	66.56	67.54		
4-31G+pd	63.40	66.44	65.70	59.73	64.36	66.47		
6-31G+sd	62.27	64.00	65.47	57.46	63.10	64.01		
6-31G+spd	67.06	69.53	69.11	62.56	68.57	69.56		
aug-cc-pVDZ	68.38	70.99	70.60	63.87	70.01	71.02	69.24	67.5[a]
aug-cc-pVTZ	68.65	71.10	70.53	64.05	70.21	71.13		69.5[b]
Second hyperpolarizability γ_{ave} in 10^3 a.u.								
4-31G	1.01	1.37	1.49	1.06	1.21	1.37		
6-31G	1.80	1.43	1.55	1.12	1.26	1.44		
cc-pVDZ	1.67	2.68	2.85	1.95	2.26	2.68		
cc-pVTZ	2.59	4.18	4.03	3.09	3.60	4.19		
cc-pVQZ	3.97	6.30	5.70	4.73	5.45	6.31		
4-31G+pd	15.11	22.61	14.45	17.78	20.34	22.66		
6-31G+sd	2.21	3.36	3.19	2.54	2.97	3.36		
6-31G+spd	10.02	14.91	10.01	11.44	13.40	14.94		
aug-cc-pVDZ	11.63	16.44	11.05	12.90	15.18	16.47	14.31	
aug-cc-pVTZ	13.28	17.92	11.75	14.27	16.78	17.96		16.4[c]

[a] From [78].
[b] From [80].
[c] From [81].

of benzene also decreased compared to LDA, but not significantly. We have also used the LDA, BLYP and B3LYP kernels in connection with the B3LYP potential. The B3LYP/LDA method, where the B3LYP potential and LDA kernels are combined, provides surprisingly low values; 11.11 a.u. for nitrogen and 64.1 for benzene. The B3LYP/BLYP method where the BLYP kernels were used in addition to the B3LYP potential underestimates the reference value even more. Finally, the pure B3LYP/B3LYP method where both the potential and the kernels are of the B3LYP type functional, slightly overestimates the reference value of 11.76 (69.51 for benzene) resulting in 12.00 (70.2), both being close to the reference value. The PBE functional shows results similar to LDA; the values of the polarizability are overestimated by a few percent. The CCSD result is only 0.11 a.u lower than experiment and is lower than most of the DFT methods. It may be concluded that the DFT methods tend to overestimate values of the polarizability α_{ave}.

The second hyperpolarizability γ_{ave} is a more difficult case. It is both more sensitive to the basis set and to the selected DFT method. Reported gas phase static experimental values are 917 ± 5 a.u. [79] and 16.4×10^3 a.u. [84] for γ of N_2 and benzene, respectively. When we compare the performance of the methods,

Table 5. The static and dynamic average polarizability α_{ave} and the second hyperpolarizability γ_{ave} of the nitrogen molecule in the aug-cc-pVTZ basis set. The HF, the DFT Potential/Kernel combinations and the CCSD methods were used. *Reproduced from [42]*

Potential/ Kernel	HF/ HF	LDA/ LDA	LB94/ LDA	BLYP/ BLYP	B3LYP/ LDA	B3LYP/ BLYP	B3LYP/ B3LYP	PBE/ PBE	CCSD	Expt.
Polarizability $\alpha(-\omega;\omega)$ in a.u.										
457.9 nm	11.81	12.58	12.06	12.66	11.34	11.19	12.32	12.59	11.93	12.06[a]
488.0 nm	11.77	12.53	12.03	12.62	11.31	11.16	12.28	12.54	11.89	12.03[a]
514.8 nm	11.75	12.50	12.00	12.59	11.29	11.14	12.25	12.51	11.87	12.00[a]
∞	11.53	12.23	11.76	12.31	11.11	10.96	12.00	12.24	11.65	11.76[a]
Second hyperpolarizability $\gamma(-3\omega;\omega,\omega,\omega)$ in a.u.										
457.9 nm	1072.9	2630.1	1485.5	2990.6	1479.9	1337.1	2081.3	2637.4	1529.1	
488.0 nm	981.6	2286.0	1338.9	2567.8	1343.3	1216.9	1839.5	2292.0	1382.9	
514.8 nm	920.8	2072.2	1243.0	2309.8	1253.3	1137.5	1685.6	2077.4	1287.1	
∞	560.1	1020.8	706.0	1086.0	737.2	675.1	879.9	1022.8	745.2	917±5[b]

[a] From [78].
[b] From [79].

qualitatively very similar but more pronounced trends are found as those for the polarizability α. HF underestimates by 20–25% compared to CCSD in the same diffuse basis set. The LDA model has a strong tendency to overestimate the reference values for the same reasons as for α. Overestimation is reduced by adding the LB94 correction to the potential. For the case of benzene, the reduction in the average γ goes below the HF value. The B3LYP/LDA, B3LYP/BLYP and B3LYP/B3LYP series follow the same trend as for α and pure B3LYP gives a balanced result also in the case of γ, approaching the experimental values. The PBE functional, consistent with the α results, performs similarly as the LDA functional which once again indicates that gradient corrections are not important for the description of properties. Among the studied potentials, the LB94 potential with the unique property of proper asymptotic behavior seems to perform very well for the polarizability α but tend to underestimate the second hyperpolarizability γ.

That the frequency dependent values can be obtained as readily as the frequency independent ones is an attractive feature of response theory as experimental measurements of (hyper)polarizabilities most often are carried out at non-zero frequencies.

The dispersion behavior of α and γ for a few selected frequencies can be found in Tables 5 and 6 for nitrogen and benzene, respectively. We have selected the ESHG process for benzene and the THG processes for nitrogen to study the frequency dependence of γ. The conclusions are consistent between the processes: When the CCSD dispersion is taken as a reference value, we observe that the DFT methods provide satisfactory dispersion for the polarizability α. However, for the hyperpolarizability γ there is a significant tendency to deviate from the CCSD dispersion curve. The only exception is the LB94/LDA method, where the deviation from the CCSD dispersion is found to be almost constant for the examined

Table 6. The static and dynamic average polarizability α_{ave} and the second hyperpolarizability γ_{ave} of benzene in the aug-cc-pVTZ basis set. HF and different DFT Potential/Kernel combinations were used. Reproduced from [42]

Potential/ Kernel	HF/ HF	LDA/ LDA	LB94/ LDA	BLYP/ BLYP	B3LYP/ LDA	B3LYP/ BLYP	B3LYP/ B3LYP	PBE/ PBE	Expt.
Polarizability $\alpha(-\omega;\omega)$ in a.u.									
620.0 nm	71.1	73.9	73.2	74.4	65.9	64.3	73.0	73.9	
670.0 nm	70.8	73.5	72.8	74.0	65.6	64.0	72.6	73.5	
693.4 nm	70.6	73.3	72.6	73.8	65.5	63.9	72.4	73.4	
∞	68.6	71.1	70.5	71.5	64.1	62.6	70.2	71.1	69.5[a]
Second hyperpolarizability $\gamma(-2\omega;\omega,\omega,0)$ in 10^3 a.u.									
620.0 nm	23.60	32.71	18.29	39.88	22.99	19.90	31.40	32.79	26.80±0.5[b]
670.0 nm	21.43	29.51	17.01	35.25	21.26	18.44	28.19	29.58	24.54±0.5[b]
693.4 nm	20.66	28.39	16.54	33.70	20.64	17.92	27.07	28.45	24.54±0.6[b]
∞	13.28	17.92	11.75	19.89	14.27	12.48	16.78	17.96	16.46[c]

[a] From [80].
[b] From [84].
[c] From [81].

range of frequencies. To sum up what we have observed for nitrogen and benzene: LDA/LDA and PBE/PBE give similar and generally too large values—the PBE gradient correction is not an improvement for properties. The same holds for the BLYP functional. LB94/LDA gives systematically lower values: the difference is not large for the polarizability but significant for the hyperpolarizability. This shows that an improved potential does not necessarily leads to better properties. The B3LYP potential with the same kernels provide balanced results in good agreement with experiment.

4.1.3 C_{60} fullerene

The chemistry and physics of fullerenes have constituted one of the most fast growing research fields during the last decade [90]. A summary of the early results for the second hyperpolarizability can be found in [91, 92]. There are a number of factors that make comparison of these results difficult, for instance the type of optical process, the phase of the samples, and the reference standard [91, 93]. The theoretical results, on the other hand, seem to be more consistent, especially among those from the first-principle calculations, such as *ab initio* Hartree-Fock and the density functional theory (DFT) methods [14, 89, 94, 95]. The recent applications of time-dependent DFT [14, 96] to NLO properties of the fullerenes has improved the situation considerably.

Two tailored basis sets were used in [42]: A 4-31G+pd basis set based upon a standard 4-31G basis set augmented with diffuse p(0.0780) and d(0.1870) exponents. The other basis set cc-pVDZ+spd is based upon the standard cc-pVDZ basis set, augmented with s(0.0469), p(0.0800) and d(0.3140) exponents. Experimental data for the polarizability α, which have been deduced from the measurements of the dielectric constant in thin films, are in the range of 579–595 a.u. [85–87]. Another

Table 7. The average static polarizability α_{ave} and the second hyperpolarizability γ_{ave} of the C_{60} fullerene, using 4-31G+pd and cc-pVDZ+spd taylored basis sets. Several DFT Potential/Kernel combinations were used. *Reproduced from [42]*

Potential/Kernel	LB94/LB94	B3LYP/LDA	B3LYP/BLYP	B3LYP/B3LYP	PBE/PBE	Expt.
Polarizability α_{ave} in a.u.						
4-31G+pd		496.2	495.9	535.7	544.7	
cc-pVDZ + spd	544.1	506.3	506.0	547.0	554.9	579–595[a]
Second hyperpolarizability γ_{ave} in 10^3 a.u.						
4-31G+pd		65.58	65.25	76.31	80.32	
cc-pVDZ + spd	87.02	96.84	96.38	118.27	119.23	93 ± 14[b]

[a] From [85–87].
[b] Experimental value from [88] corrected for dispersion as described in [89].

measurement by Antoine et al. [97] provides the static polarizability α value as 516 ± 54 a.u. In the calculated data in Table 7, we can observe similar trends as for the smaller nitrogen and benzene molecules. The results are within 15% of the thin film experiment and are well within the error bars of the experiment of [97]. The remaining deficiency can be attributed to the basis set and the state of the sample in the first case. However, we also obtain very good agreement with theoretical results of other authors. Our values for LB94 (544), B3LYP (547), and PBE (544) compare well with the analytical results of van Gisbergen et al; [98] LDA (557) and LB94 (544) and Iwata et al; [96] VWN(541), BLYP(545), and LB94(544), obtained using a numerical real-space method.

For the hyperpolarizability γ experimental data based on the measurement of χ^3 are not unambiguous, see discussion of Norman et al. [89]. The experimental value of $170 \pm 24 \times 10^3$ a.u. was given in [88]. The estimation of the dispersion by a two-state model brings this value down to $93 \pm 14 \times 10^3$ a.u. [89]. We see that the B3LYP/LDA/cc-pVDZ+spd basis set gives results within the error bars of the experiment, while B3LYP seems to overestimate somewhat and LB94 slightly underestimate. However, we need to keep in mind that we have seen the deficiency in the basis set already for the α values and that also the reliability of the estimation of the dispersion effect by a two-state model might be questioned.

When we compare to the theoretical work of other authors, the situation is more interesting for γ: van Gisbergen et al. [98] obtained γ values from finite-field differentiation of the first hyperpolarizability β using a Slater basis. They obtained an LB94 γ value of 65.5×10^3 a.u. which is considerably lower than what we obtain. Due to their use of LDA kernels this does not give identical results as the analytical calculation we carried out. The results of van Gisbergen et al. were criticized by Yabana and coworkers [96] for deficiencies in the basis set. Using a basis set free real space numerical method and the VWN, BLYP and LB94 potentials, they obtained a LB94 result of 94300 which is in excellent agreement with our, analytical calculations. This indicates that the problem with the results of van Gisbergen et al. is not due to the basis set, but that it is methodological. We note, however, that the LB94 results of Yabana et al. are not fully consistent since all the calculations were performed with pseudo potentials derived from LDA calculations.

4.2 Excited State Polarizabilities

Excited state properties of molecules are often important parameters in different models of interacting systems and chemical reactions. For example, excited state polarizabilities are key quantities in the description of electrochromic and solvatochromic shifts [99–103]. In gas phase there has been a series of experiments were excited state polarizabilities have been determined from Laser Stark spectroscopy by Hese and coworkers [104–106]. However, in the experiments most often not all the tensor components can be determined uniquely without extra information from either theory or other experiments.

Calculations of analytic excited state properties for correlated methods have been reported by several groups [107–118]. Excited state dynamic properties from cubic response theory were first obtained by Norman et al. at the SCF level [55] and by Jonsson et al. at the MCSCF [56] level, and in a subsequent study a polarizable continuum model was applied to account for solvation effects [119]. Hättig et al. presented a general theory for excited state response functions at the CC level using a quasi-energy formulation [120] which was subsequently implemented and applied at the CCSD level [121, 122]. The first TD DFT calculation of dynamic excited state polarizabilities, which we will shortly review here, was presented in [58] for pyrimidine and s-tetrazine utilizing the double residue of the cubic response function derived in Section 2.7.3.

For both S_0 and S_1 of pyrimidine we have used the experimental ground state geometry from [123]. For s-tetrazine S_0 we used the experimental geometry

Table 8. The static dipole polarizability for the ground state and the first excited singlet state of pyrimidine in a.u. Coupled cluster values from [121]. The polarizability anisotropy parameter is defined as $\gamma = (\alpha_{xx} + \alpha_{yy})/2 - \alpha_{zz}$. Reproduced from [58]

Component	SCF	CCS	CC2	CCSD	LDA	BLYP	B3LYP
				S_0 ground state			
α_{xx}	64.6	69.3	71.1	67.8	69.3	69.9	67.9
α_{yy}	66.5	71.5	74.1	70.2	72.4	72.8	70.6
α_{zz}	36.7	39.7	38.3	37.5	37.5	37.7	37.0
γ		30.7	34.3	31.5	33.3	33.6	32.3
				S_1 exited state			
α_{xx}		111.9	128.6	111.8	118.8	117.6	107.2
α_{yy}		76.0	75.8	71.4	73.0	73.6	71.8
α_{zz}		46.2	44.5	42.2	41.2	42.4	41.3
γ		47.8	57.7	49.4	54.7	53.2	48.2
$\Delta\alpha_{zz}$		6.5	6.2	4.7	3.7	4.7	4.3

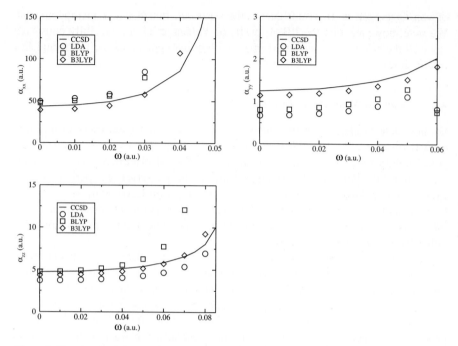

Figure 1. The change of the dynamic polarizability of pyrimidine upon excitation to the lowest excited singlet state. *Reproduced from [58]*

[124, 125] and for S_1 the optimized structure of Stanton and Gauss [115]. This is also the basis set and geometries used for pyrimidine and s-tetrazine in [121] making our results directly comparable to theirs.

The static polarizabilities for the ground and first singlet excited state of pyrimidine employing the LDA, BLYP and B3LYP functionals can be found in Table 8. For comparison we list also CCS, CC2, and CCSD values from [121]. It is interesting to note that the DFT errors (compared to CCSD) for the polarizabilities are much smaller than for the excitation energies and that all the DFT methods provide better results than CC2 which is not even better than CCS. The B3LYP ground and excited state polarizabilities are within 2% and 5%, respectively, of the CCSD values. Considering that the errors in excitation energies in many cases were much larger for LDA and BLYP it is somewhat surprising that these functionals still produces quite reasonable excited state polarizaiblities, even though the results are not as good as for B3LYP. We can conclude a general good agreement for the static ground and excited state polarizabilities of pyrimidine for all the DFT methods.

For the pyrimidine molecule we also calculated the frequency dependence of the S_1 state polarizability up to the first resonance. The change of the dynamic polarizability upon excitation is displayed in Fig. 1, for the α_{xx}, α_{yy}, and α_{zz} components, respectively. The dispersion of the ground state polarizabilities in the same frequency interval as for the excited state does not differ significantly from

Table 9. The static dipole polarizability for the ground state and the first excited singlet state of *s*-tetrazine in a.u. Numbers in parenthesis are for the S_1 geometry. The polarizability anisotropy parameter is defined as $\gamma = (\alpha_{yy} + \alpha_{zz})/2 - \alpha_{xx}$. Reproduced from [58]

Comp.	SCF[a]	CASSCF[a]	CASPT2[b]	CCSD[c]	CCSD resp.[d]	B3LYP[e]	Exp.[f]
				S_0 ground state			
α_{xx}	29.8	29.8	32.6	32.0	32.7	32.3	
α_{yy}	52.7	53.4	55.1	54.4	56.0	54.0	
α_{zz}	57.4	53.8	60.0	59.0	60.7	62.1	
γ	25.2	23.8	25.0	24.7	26.7	25.8	5.4
				S_1 exited state			
α_{xx}	28.7	38.2	32.4	31.3	31.9 (31.1)	31.1 (31.3)	
α_{yy}	70.8	74.4	67.2	74.0	80.1 (83.2)	75.8 (78.1)	
α_{zz}	63.7	66.6	66.0	63.9	61.5 (65.7)	71.0 (65.5)	
γ	38.4	42.3	34.2	37.6	41.1 (43.4)	42.3 (40.5)	45.2
$\Delta\alpha_{xx}$	−1.1	−1.6	−0.2	−0.7	−0.9 (−0.7)	−1.1 (−0.9)	−17.5

[a] Response calcaulation in a [4s4p2d/2s1p] basis, [126].
[b] Finite field calculation in the Sadlej POL basis, [127].
[c] Finite field calculation in the Sadlej POL basis, [115].
[d] Response calculation in the Sadlej POL basis, [121].
[e] B3LYP response calculation in Sadlej's POL basis, this work.
[f] Laser Stark spectroscopy, [105].

the static case and are therefore not presented here. As can be seen in the figures B3LYP generally compares well to the CCSD values almost all the way up to the first resonance, whereas LDA and BLYP tend to deviate from the CCSD results much earlier. We may note that the CCS values (not displayed here) are completely wrong for a large part of the frequency range and that the CC2 values generally are of the same quality as the B3LYP numbers.

The static polarizabilities for the ground and first singlet excited state of *s*-tetrazine for the B3LYP functional is presented in Table 9 compared to several other methods as well as experiment. As for pyrimidine the B3LYP ground state polarizabilities are within 2% of the CCSD values and there is in general good agreement between the different DFT methods. For the excited state there is good general agreement but larger differences than for pyrimidine. The largest deviation is observed for the $S_1\alpha_{zz}$ component were B3LYP is 15% larger than CCSD. It is also notable that the CCSD $S_1\alpha_{yy}$ component differs by 10% even though the only difference between the CCSD results from [115] and [121] is that the latter analytic results do not contain orbital relaxation. As noted before [115, 121, 127] the general good agreement between all the different theoretical methods warrants a reinvestigation of the experiment [105].

4.3 Three-photon Absorption

Though experimentally verified already in 1964 by Singh and Bradley [128], three-photon absorption (3PA) is far less examined than the two-photon absorption (TPA) analog. However, the increased attention directed toward non-linear optical processes [129–131] have lately been broadened to include higher order multi-photon excitations [132–134]. The simultaneous absorption of three photons inherits many of the profitable characteristics of the extensively studied TPA process, such as spatial confinement due to higher order dependence on the intensity, and increased penetrability due to that fundamental excitations can be reached by longer wavelengths. In addition, it also enables the spectroscopic access of states which are TPA forbidden.

The three-photon absorption cross section is related to the fifth order susceptibility which clearly forms a challenging computational task. Most often we are interested in the resonant absorption to a particular state, f, in which case the cross section can be expressed in terms of the third-order transition moment to this state, which in turn is much easier to compute. Hence, for 3PA it is sufficient to evaluate the third-order transition moment, T_f^{abc}, as

$$(192) \quad T_f^{abc} = \sum \mathcal{P}_{abc} \sum_{m,n} \frac{\langle 0|\mu_a|m\rangle \langle m|\mu_b|n\rangle \langle n|\mu_c|f\rangle}{(\omega_m - 2\omega_f/3)(\omega_n - \omega_f/3)}$$

which in turn can be identified from the single residue of the cubic response function as described in Section 2.7.3. In order to ascertain three-photon probabilities and cross-sections in an isotropic medium, the third order transition moments has to be orientationally averaged as devised by McClain [135]. For linearly polarized light the three-photon probabilities are given by

$$(193) \quad \delta^{3PA}{}_L = \frac{1}{35}(2\sum_{ijk} T_{iij}T_{kkj} + 3\sum_{ijk} T_{ijk}T_{ijk})$$

and the three-photon cross section, σ^{3PA}, is defined as

$$(194) \quad \sigma^{3PA} = \frac{4\pi^3 a_0^8 \alpha}{3c_0} \frac{\omega^3 g(\omega)}{\Gamma_f} \delta^{3PA}$$

where $g(\omega)$ relates to the spectral line profile and Γ_f to the lifetime broadening of the final state [136]. Provided that CGS units are used for a_0 and c_0 and atomic units for δ^{3PA}, ω and Γ_f, the final cross sections will given in units of $cm^6 \cdot s^2 \cdot photon^{-1}$.

4.3.1 Small molecules

Calculated excitation energies and three-photon probabilities for HF, NH_3, and H_2O can be found in Table 10. The oscillatory predictions in the sequence CCS-CC2-CCSD observed for the excitation energies is substantially more pronounced for δ^{3PA}, which may not be surprising considering the order of the property. The moderate contributions from triples, as estimated from the insignificant differences between excitation energies predicted by CCSD and CC3, strengthen the predictive credibility of the δ^{3PA} values obtained at the CCSD level. Even though the accuracy

Table 10. Excitation energies in eV and three-photon probabilities, δ^{3PA}, for linearly polarized light in 10^4 a.u. for HF, NH$_3$, and H$_2$O as calculated by response theory at CC, DFT and HF levels with aug-cc-pVTZ basis set. Reproduced from [59]

State	CCS		CC2		CCSD		CC3	B3LYP		BLYP		LDA		HF	
HF															
1Π	11.66	0.80	9.87	2.80	10.35	1.66	10.38	9.38	2.39	8.58	3.96	8.95	3.48	11.60	0.66
2Π	15.01	0.90	13.03	2.14	13.65	1.60	13.71	12.18	1.17	11.26	1.16	11.64	0.98	14.98	0.88
2Σ	15.18	0.55	13.94	3.10	14.32	1.10	14.36	13.22	0.56	12.50	2.56	12.82	0.87	15.14	0.50
NH$_3$															
A_2''	7.42	12.20	6.36	29.08	6.58	19.12	6.55	5.90	138	5.35	252	5.61	164	7.38	28.4
E''	8.88	9.72	7.83	47.42	8.13	25.70	8.12	7.16	117.5	6.49	220	6.82	180	8.86	23.0
H$_2$O															
1B_2	8.65	4.19	7.20	13.96	7.57	8.35	7.58	6.87	11.9	6.24	26.1	6.54	19.5	8.60	3.46
1B_1	10.32	0.61	8.86	3.72	9.33	2.06	9.35	8.28	2.96	7.50	5.51	7.87	4.72	10.27	0.49
1A_1	10.90	3.57	9.53	8.56	9.91	5.26	9.91	9.01	6.60	8.35	8.82	8.61	10.6	10.87	2.98
2B_2	11.78	5.24	10.35	54.66	10.79	26.69	10.82	9.80	17.9	8.98	39.8	9.31	26.5	11.75	3.45
2A_2	12.43	14.65	10.88	****	11.35	****	11.35	10.46	560	9.82	139	10.14	249	12.35	11.0

for excitation energies may not inarguably transfer to estimates of δ^{3PA}, we will consider them as tentative benchmark values for comparison with the results attained by Hartree-Fock and by the DFT functionals B3LYP, BLYP and LDA.

It is also shown in Table 10 that excitation energies are underestimated compared to CC3 and CCSD, by approximately 1 eV for B3LYP and around 2 eV for BLYP and LDA. Hartree-Fock on the the other hand overshoots around 1–2 eV.

All functionals, and HF to some extent, do predict similar trends, which overall are in reasonable accordance with the CC-result as long as the excitation energies do not exceed the threshold of resonance; i.e. 3/2 the energy of the first excited state. For the CC2 and CCSD calculations of the H_2O molecule, the $2A_2$ state matches exactly 3/2 of the energy of the lowest excited state $1B_2$ and thereby escapes the numerical precision of the program. Among the DFT methods and non-resonant states, expectedly B3LYP appears as the most well tempered method which rather closely follows the CCSD-results.

Fig. 2 displays the predictions by CCSD, B3LYP, HF for all non-resonant states. While the excitation energies predicted by CCSD neatly is bracketed between the HF and B3LYP results the ordering between the estimations for δ^{3PA} is more irregular. The HF results are uniformly the lowest, but CCSD and B3LYP interchangeably predict the largest value. A mean deviation between B3LYP and CCSD can be estimated to 40%, which can be compared with almost a factor 3 for HF and CCSD. The overall mutual agreement between CCSD and B3LYP seems to support the use of B3LYP for exploring δ^{3PA} for larger structures, optionally in conjunction with another low-scaling method such as HF.

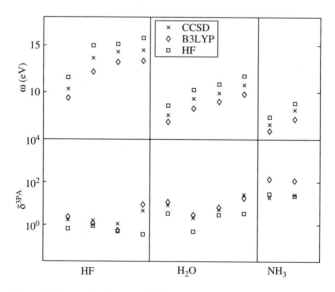

Figure 2. Comparison between CCSD, B3LYP and HF results obtained for the non-resonant states of HF, NH_3 and H_2O. *Reproduced from [59]*

4.3.2 Chromophores

In Table 11 we display the three-photon absorption probabilities, δ^{3PA}, for the first excited state for a series of modified *trans*-stilbene (TS) and dithienothiophene (DTT) molecules depicted in Fig. 3. Introducing electron accepting (A) and electron donating (D) groups to conjugated systems has the well known effect of localizing the otherwise de-localized highest occupied and lowest unoccupied orbitals (HOMO-LUMO) and thereby establishing an effective charge-transfer path across the molecule. Due to the increase of transition dipole moment guiding this transition and an overall alignment involving all transition dipole moments, this technique leads to enhancements of several orders of magnitude for TPA [137, 138]. The corresponding effect on 3PA of this process is less explored, but can be expected to benefit in the same manner as TPA. The qualitative mutual agreement between HF and DFT is comforting, though the intensities may differ with one order of magnitude. Except for reversing TS-DD and TS-AA, HF predicts the same ordering of the systems with respect to increasing δ^{3PA} as DFT. We note that the TS and DTT systems behave analogously upon substitution, though, the enhancement when attaching substituents is in general predicted to be more dramatic with DFT than for HF. We partly ascribe the quite large difference of 3PA cross sections between HF and DFT to the tendency to overshoot, respectively, (slightly) undershoot excitation energies, and partly to that the interstate transition moments seem to be systematically too small in the HF case.

A homologous (AA or DD) substitution will raise δ^{3PA} approximately by a factor between 2 and 18. Indisputably, AD substituted compounds give the largest responses and supersedes the non-substituted systems with at least one order of

Table 11. Excitation energies in eV, three-photon probabilities, δ^{3PA}, for linear polarized light in 10^6 a.u. and three-photon cross sections, σ^{3PA} in 10^{-82} cm^6 s^2 as calculated by response theory at HF and DFT levels with 6-31G basis set. *Reproduced from [59]*

Molecule	Exp.		HF			DFT		
	ω	ω	δ^{3PA}	σ^{3PA}	ω	δ^{3PA}	σ^{3PA}	
TS	≈ 4.0[a]	4.59	12.3	0.121	4.08	76.6	0.533	
TS-DD	3.32[b]	4.35	32.9	0.277	3.67	261	1.32	
TS-AA		4.22	77.8	0.598	3.34	138	0.525	
TS-AD	3.06[c]	4.05	289	1.96	2.78	5990	13.1	
DTT		3.17	806.08	2.63	2.66	11574	22.2	
DTT-DD(101)	2.88[d]2.67[e]	3.10	1382.17	4.20	2.50	31842.6	50.8	
DTT-AA	2.92[e]	3.01	2590.45	7.25	2.25	114951	134	
DTT-AD(102)	2.85[d]	2.97	4808.37	12.9	1.92	649761	472	

[a] From [139, 140].
[b] From [141].
[c] From [142].
[d] From [143].
[e] From [144].

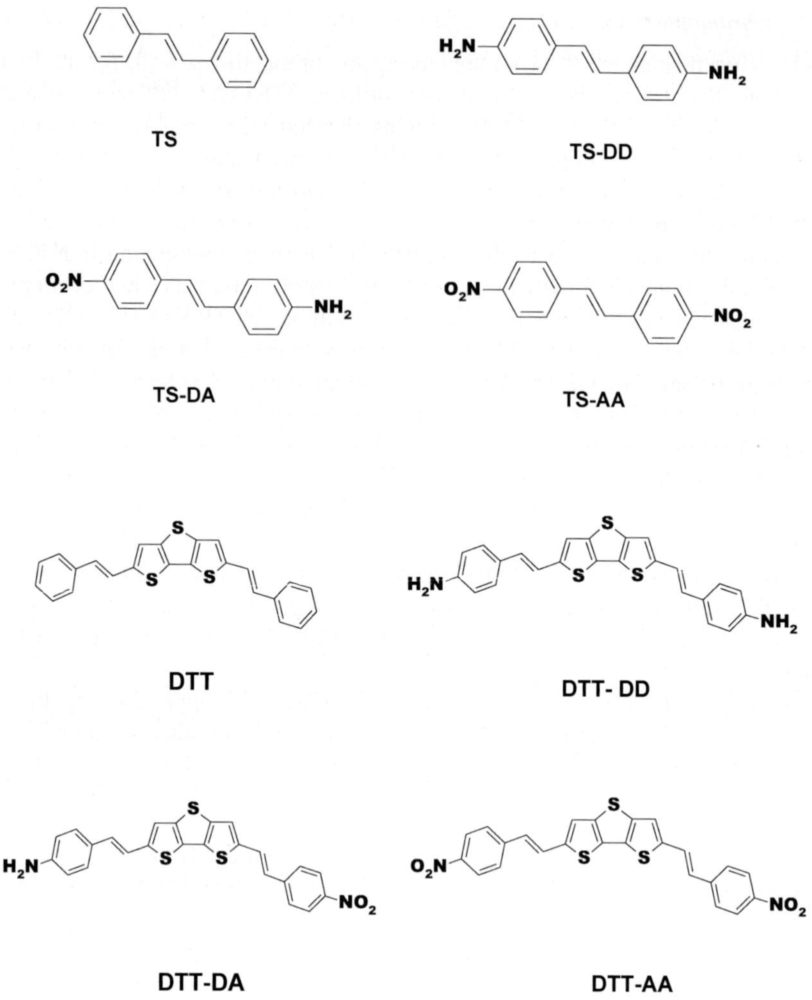

Figure 3. Molecular structures. *Reproduced from [59]*

magnitude and often close to two. As demonstrated by the substantial differences between the TS- and DTT-based systems, the electron richness of the basic building block, interpreted as the strength of the π-center, also strongly influences the σ^{3PA}.

4.4   Birefringences

That response theory encompasses a great number of of properties of various kind, was nicely illustrated in [145] and [146], which presented and analyzed results from the theory of this review applied to five different types of birefringences. These were the Kerr, Cotton–Mouton, Buckingham, Jones and Magneto-electric birefringences applied on benzene, respectively, hexafluorobenzene. The computational analysis was thus carried out with DFT using analytical (up to cubic) response theory and, for the purpose of comparison, with the Hartree-Fock SCF and CCSD wave function models. Different DFT functionals were employed in the study, and in some instances also different functionals for the energy and wave function determining step and for the subsequent response calculation. The general conclusion was that DFT proves to be a suitable approach for these rather exotic and, in all cases, demanding properties. The "standard" approach, where a single functional was applied all through the response calculations performed in general far better than that where a different functional is employed for the response part. The B3LYP functional was found to yield on average the best agreement with coupled cluster results, and reproduced in general with fairly good accuracy experimental data, which however, in some cases contain huge error bars. For a proper definition of these birefringences and for details of their evaluation, we refer to the original articles [145, 146].

4.5 EPR Spin Hamiltonian Parameters

As illustrating examples for the use of the open-shell time-dependent DFT, we review some examples of its recent applications concerning electron paramagnetic resonance (EPR) parameters. This choice find motivation in that EPR techniques are most important tools in the current arsenal of experimentalists for investigations of paramagnetic species, i.e. molecules with non-vanishing total electronic spin. Irrespective of EPR methodology chosen for a particular application, the general motivation is always the same, namely, to gather information about electronic structure and geometry of the compound in order to supply the experimentalist with microscopic interpretation and understanding of the data. However, extracting this information from EPR measurements represents often a formidable challenge in actual cases. Conventionally, one relates the so-called spin Hamiltonian parameters, which are the underlying quantities of EPR spectral analysis, to molecular structure via known empirical relationships. These relationships therefore form the keys for structural information [147, 148]. However, for molecules with complex electronic structure it often turns out difficult to predict geometrical structure with an acceptable and uniform precision via such empirical "*spin Hamiltonian parameter–molecular structure*" relationships. The development of density functional theory methods for evaluation of NMR and EPR spin Hamiltonian parameters serves as an alternative possibility to more rigorously relate geometrical and electronic structure with these parameters that has emerged in recent years [148, 149]. In fact, nowadays DFT calculations of EPR spin Hamiltonian parameters have become a significant aid for experimentalists to interpret their measurements. In the following part of this chapter we will shortly review some applications of spin-restricted open-shell density functional response theory for evaluation of two main EPR spin Hamiltonian parameters, namely the electronic g-tensor and the hyperfine coupling tensor.

4.5.1 Electronic g-tensor

The electronic g-tensor is a fundamental parameter in descriptions of the electronic Zeeman effect and one of the key elements in characterization of EPR spectra. It couples the external magnetic field, **B**, with the total spin angular momentum, **S**, of the molecule and is conventionally evaluated as the second derivative of the molecular energy:

$$(195) \quad \mathbf{g} = \frac{1}{\mu_B} \frac{\partial^2 E}{\partial \mathbf{S} \partial \mathbf{B}} \bigg|_{S=0, B=0}$$

where μ_B is the Bohr magneton. It is convenient to separate the electronic g-tensor into two parts, $\mathbf{g} = g_e \mathbf{1} + \Delta \mathbf{g}$, the free electron g-factor ($g_e = 2.0023$), and the so-called g-tensor shift, $\Delta \mathbf{g}$, which accounts for the effects of electronic interactions in the molecule. The electronic g-tensor shift, correct up to second order in the fine structure constant α, has contributions in first- and second order of perturbation theory which are based on the partitioning of the electronic Breit–Pauli Hamiltonian into a non-relativistic and a relativistic part. The first-order contributions are the relativistic mass-velocity correction to the electronic Zeeman effect and the so-called diamagnetic, or gauge-correction, terms, which correspond to the field-dependent parts of the one- and two-electron spin-orbit operators. The second-order paramagnetic contribution, which is often dominating, is the sum of cross terms between the orbital Zeeman operator and the canonical part of the one- and two-electron spin-orbit operators. This contribution can be evaluated in response theory employing the linear response function

$$(196) \quad \Delta g_{so}^{xy} = \frac{1}{S}[\langle\langle L_O^x; \hat{H}_{so(1e)}^y \rangle\rangle_0 + \langle\langle L_O^x; \hat{H}_{so(2e)}^y \rangle\rangle_0]$$

where $\mathbf{L}_O$ is the total orbital angular momentum operator with respect to a gauge origin O, and $\hat{H}_{so(1e)}$ and $\hat{H}_{so(2e)}$ are the one- and two-electron spin-orbit operators. S is the spin quantum number of the reference state (assuming maximum projection $M_S = S$) and originates from the Wigner-Eckart relations that have been applied in relating the spin Hamiltonian to the full Hamiltonian.

Computation of the spin-orbit contribution to the electronic g-tensor shift can in principle be carried out using linear density functional response theory, however, one needs to introduce an efficient approximation of the two-electron spin-orbit operator, which formally can not be described in density functional theory. One way to solve this problem is to introduce the atomic mean-field (AMFI) approximation of the spin-orbit operator, which is well known for its accurate description of the spin-orbit interaction in molecules containing heavy atoms. Another two-electron operator appears in the first order diamagnetic two-electron contribution to the g-tensor shift, but in most molecules the contribution of this operator is negligible and can be safely omitted from actual calculations. These approximations have effectively resolved the DFT dilemma of dealing with two-electron operators and have so allowed to take a practical approach to evaluate electronic g-tensors in DFT. Conventionally, DFT calculations of this kind are based on the unrestricted

Kohn–Sham formalism, which suffers from the spin contamination problem, and on the sum-over-states or coupled perturbed Kohn–Sham approaches. In recent articles devoted to computations of electronic g-tensors we advocated the use of an alternative approach, namely linear response theory based on the spin-restricted open-shell Kohn–Sham formalism, which is free from spin contamination problem (see Theory section). In the following we briefly review the applicability of this approach for some paramagnetic compounds.

DFT methods are well known for providing accurate electronic g-tensor values for various main group radicals, where unrestricted and spin-restricted DFT formalisms both give similar results [150] and which for most compounds agree with experimental data up to 500 ppm. In order to achieve such accuracy one can recommend to employ the AMFI approximation for spin-orbit operators and perform calculations using the BP86 exchange–correlation functional in a basis set that is sufficiently flexible in the valence region and at the same time is augmented by at least one set of d and p type polarization functions. The success achieved by DFT in predictions of electronic g-tensors of the main group radicals does unfortunately not extend into the domain of transition metal compounds, where both unrestricted and spin-restricted formalisms underestimate the experimental g-tensor components by 40–60% (see Table 12). The limited capability of current DFT in computation of electronic g-tensors of transition metal compounds can be traced back to an inaccurate description of excitation energies for such compounds. This in turn is caused by shortcomings in ordinary density functional response theory in describing a mix between single and double excitations encountered in open-shell molecules, as well as by deficiencies of currently available exchange-correlation

Table 12. Electronic g-tensors of transition metals compounds evaluated with various exchange correlation functionals.[a,b] *Reproduced from [150]*

Molecule	Δg_{ii}	LDA	BLYP	B3LYP	UBP[c]	UB3LYP[d]	Exp.[e]
TiF$_3$	$\Delta g_\parallel$	0.3	0.3	0.2	−1.7	−1.1	−11.1
	$\Delta g_\perp$	−47.0	−32.9	−49.1	−42.8	−41.8	−111.9
VOF$_4^{2-}$	$\Delta g_\parallel$	−37.0	−30.3	−39.5	−36.0	−34.1	−58.8
	$\Delta g_\perp$	−29.6	−21.8	−24.2	−28.0	−25.0	−51.1
VOCl$_4^{2-}$	$\Delta g_\parallel$	−20.6	−16.0	−28.3	−18.0	−20.5	−51.1
	$\Delta g_\perp$	−17.9	−14.4	−16.9	−20.0	−20.6	−32.2
CrOF$_4^-$	$\Delta g_\parallel$	−15.9	−12.4	−17.5	−19.0	−18.1	−33.5
	$\Delta g_\perp$	−27.9	−21.4	−21.9	−29.0	−27.0	−50.7
Cu(NO$_3$)$_2$	Δg_{11}	126.5	120.0	173.3	...	171.2	246.6
	Δg_{22}	31.6	30.9	44.4	...	46.0	49.9
	Δg_{33}	29.0	28.9	43.9	...	44.6	49.9

[a] Calculation performed at the geometries taken from [151].
[b] Electronic g-tensor shifts are given in the principal axis system. Values are in ppt.
[c] [152]. Unrestricted coupled perturbed Kohn–Sham calculations in gauge invariant atomic orbitals.
[d] [151]. Unrestricted coupled perturbed Kohn–Sham calculations.
[e] Experimental data taken from [151].

functionals. Apparently, one way to at least partially improve this situation is to introduce local hybrid exchange–correlation functionals which recently have been implemented by Kaupp et al. [153] for the unrestricted DFT formalism. However, the latter approach is not entirely satisfactory as it suffers from spin contamination. Their calculations accordingly show a significant deterioration of accuracy with increase of the spin-contamination. Further improvements will be given by implementation of local hybrid exchange-correlation functionals for the spin restricted DFT formalism combined with a response formalism which accounts for mixing of single and double excitations in open-shell molecules. Hopefully, this development will allow to resolve problems with the accuracy of DFT for electronic g-tensors of transition metal compounds and make the DFT formalism an uncontested approach for quantitative prediction of electronic g-tensors in large molecular systems of experimentalist interest.

The development of DFT computations of electronic g-tensors has mainly focused on improving the accuracy and applicability for isolated systems, while only little attention has been devoted to account for environmental effects. Most studies of solvent or matrix effects on electronic g-tensors have adopted the supermolecular approach, in which the solvent molecules are explicitly introduced into the model used in the calculations. Recently, we developed an electronic g-tensor formalism in which solvent effects are accounted for by the polarizable continuum model [154]. We applied this approach to investigate solvent effects on electronic g-tensors of di-t-butyl nitric oxide (N-I) and diphenyl nitric oxide (N-II). Calculations were

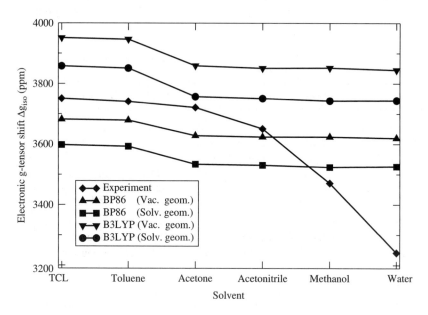

Figure 4. Experimental and theoretical isotropic g-tensor shifts of di-t-butyl nitric oxide for different solvents. *Reproduced from [154]*

carried out for both protic (methanol, water) and aprotic solvents (carbon tetrachloride, toluene, acetone, acetonitrile). In the case of aprotic solvents, the PCM model was capable of qualitative predictions of the solvent effects on the g-tensor (see Fig. 4); calculated isotropic g-shifts decrease with increase of dielectric constant of the solvent, while in the case of protic solvents the experimentally observed pattern could not be reproduced. This owes to the fact that the PCM model is only designed to handle long-range electrostatic effects, while in case of protic solvents hydrogen bonding is responsible for the major part of the solvent effect on the g-tensor. This limitation of the PCM model is well known and calls for a semi-continuum model that includes a the solvation shell of molecules with explicit treatment and which is augmented by PCM for the long-range effects. Such a semi-continuum PCM model gives accurate electronic g-tensors for the N-I and N-II molecules solvated in methanol and water and provides a computationally well defined and inexpensive way for modeling electronic g-tensors of large bioradicals.

4.5.2   Hyperfine coupling tensor

Another important EPR spin Hamiltonian parameter, featured in most observed EPR spectra, is the hyperfine coupling tensor for a magnetic nucleus K in a molecule, which describes the interaction of the electric spin angular momentum with the magnetic field created by the nucleus. According to this definition the hyperfine coupling tensor is evaluated as a second molecular energy derivative with respect to the nuclear spin, $\mathbf{I}_K$, and the total spin angular momentum:

$$(197) \quad \mathbf{A}_K = \frac{\partial^2 E}{\partial \mathbf{S} \partial \mathbf{I}_K}\bigg|_{\mathbf{S}=0, \mathbf{I}_K=0}$$

The hyperfine coupling tensor $\mathbf{A}_K$ is correct to second order $\mathcal{O}(\alpha^2)$ and consists of two contributions; the Fermi contact and spin-dipolar terms, which describe the interaction between the magnetic moments of the electrons with the magnetic moment of nucleus K via the Fermi contact and the classical dipolar mechanisms, respectively. The evaluation of the two contributions is in principle straightforward as it only involves computations of expectation values of the corresponding terms of the Breit–Pauli Hamiltonian. However, in practice these calculations are non-trivial due to the need for an accurate account of spin-polarization (especially close to the nuclei) and electron correlation simultaneously. DFT methods based on the unrestricted Kohn–Sham formalism account for spin polarization and are capable of treating larger molecular systems, but, as commented in the previous subsection, the unrestricted Kohn–Sham formalism suffers from spin-contamination, something that causes significant deterioration of the hyperfine coupling values as also observed for transition metal compounds in the work of qKaupp et al. [156]. An alternative way to calculate hyperfine coupling constants in DFT is to employ a spin-restricted Kohn–Sham formalism, which, however, does not account for spin polarization and therefore can not give reliable results when this effect is crucial. One way to overcome this problem of the spin-restricted Kohn–Sham

Table 13. Calculated isotropic HFC constants, in MHz, of transition metal compounds and their dependence on the exchange-correlation functionals. *Reproduced from [155]*

Molecule	Isotope	B3LYP	BHPW91	UB3LYP[a]	UBHPW91[a]	Exp.[b]
TiO	^{47}Ti	−248.6	−211.1	−252.8	−227.0	−241.0(60)
	^{17}O	−7.9	3.7	−4.9	1.6	...
VN	^{51}V	1317.4	1764.9	1388.9	1081.7	1311.8
	^{14}N	4.7	35.7	3.2	−7.2	...
VO	^{51}V	791.2	784.5	829.5	753.4	778.0(2)
	^{17}O	−2.5	1.8	1.1	8.0	0(4)
MnO	^{55}Mn	508.6	904.5	521.8	504.7	479.9(100)
	^{17}O	−9.4	−24.2	−8.0	−8.8	...
MnH	^{55}Mn	309.7	242.9	331.8	276.3	279.4(12)
	^{1}H	28.1	7.1	28.0	10.1	20.7(39)
TiF$_3$	^{47}Ti	−158.2	−175.2	−192.2	−149.4	−184.8(4)
	^{19}F	1.6	−13.8	−5.6	−24.3	8.3(4)
MnO$_3$	^{55}Mn	1476.2	1900.8	1735.5	1111.7	1613(6)
	^{17}O	−6.4	8.4	2.6	19.0	...

[a] [156]. Unrestricted Kohn–Sham formalism.
[b] Experimental data taken from [156].

formalism is to introduce spin-polarization in the property calculations through a restricted-unrestricted approach, as was described in the Theory part of this chapter.

The restricted-unrestricted approach (RU) in [155] has been applied to calculate the isotropic hyperfine coupling constants of a sequence of organic radicals and transition metal compounds. In the case of organic compounds, both spin-restricted and unrestricted approaches could accurately describe the isotropic hyperfine coupling constants which matched the accuracy achieved by coupled cluster methods. The situation is different for transition metal compounds for which the overall quality of the RU results is slightly better than the corresponding unrestricted results (see Table 13), independently on the exchange-correlation functional used in calculations.

The restricted-unrestricted approach not only improves results from the unrestricted approach, but also allows to rigorously describe the effect of spin polarization for the hyperfine coupling constants as well as to provide ways to analyze the behavior of spin polarization (response term in RU approach, see Theory part) in problematic cases. The RU approach therefore provides a higher degree of control over the calculation and its analysis compared to the unrestricted formalism. It can consequently be recommended for investigations of hyperfine coupling constants in various molecular systems.

4.6 Outlook

In this chapter we reviewed modern Kohn–Sham time-dependent density functional theory and its applications to linear and non-linear properties. As evident from a variety of application examples, DFT methods undoubtedly hold a prominent position among quantum chemistry approaches designed for calculations of molecular properties. Despite these achievements, density functional theory remains an evolving field of research in which new methods are developed and in which old ones are constantly improved. In view of the current state of DFT, in our opinion, two major methodological developments will have significant impact on the computation of molecular properties in the future: 1) implementation of efficient linear scaling techniques applicable to large molecules; 2) development of new DFT methods for evaluation of properties of molecular systems with arbitrary ground states, such as near-degenerate and low spin ground states. In recent years exceptional progress in development of linear scaling DFT response methods has been achieved, see e.g. [157–159], but currently existing codes are mainly experimental and remaining issues of algorithmic character must be resolved before they become practical tools for molecular properties.

The second area of density functional theory, which we suggest will be important in the future, is rather unexplored. For instance, only a few of the proposed methods capable of treating molecular systems, in which static electron correlation is important, have been implemented [160–162]. One can foresee development of DFT approaches for molecular systems which feature large static correlation, that are, for instance, commonly encountered in active sites of various proteins and in compounds with exceptional magnetic properties. Furthermore, research in this area will probably stimulate development of general DFT response methods, which are capable to correctly describe multiplet excitations from a given ground state, and consequently lead to the introduction of new exchange-correlation kernels beyond the single electron excitation formulation of current time-dependent DFT response theory [163]. Research along both above mentioned directions are pursued by various groups in the quantum chemistry community, something that probably will maintain time-dependent density functional theory as a leading approach for the evaluation of molecular properties.

ACKNOWLEDGEMENTS

The theory reviewed in this paper has recently been implemented in the 2.0 version of the Dalton program, available at http://www.kjemi.uio.no/software/dalton/dalton.html. The authors acknowledge long time collaboration with their coauthors of the Dalton program. In the area of density functional theory and applications they acknowledge in particular the collaboration with Professors Trygve Helgaker, Oslo, Kenneth Ruud, Tromsø, and Antonio Rizzo, Pisa.

REFERENCES

[1] Runge, E., Gross, E.K.U.: Phys. Rev. Lett. **52**, 997 (1984)
[2] Wendin, G.: Phys. Lett. **51A**, 291 (1975)
[3] Stott, M.J., Zaremba, E.: Phys. Rev. **A21**, 12 (1980)
[4] Stott, M.J., Zaremba, E.: Phys. Rev. **A22**, 2293 (1980)

[5] Zangwill, A., Soven, P.: Phys. Rev. **A21**, 1561 (1980)
[6] Kohn, W.: Phys. Rev. Lett. **56**, 2219 (1986)
[7] Amusia, M.Ya., Chernysheva, L.V., Gribakin, G.F., Tsemekhman, K.L.: J. Phys. **B23**, 393 (1990)
[8] Wendin, G., Wästberg, B.: Phys. Rev. **B48**, 14764 (1993)
[9] Lee, A.M., Colwell, S.M.: J. Chem. Phys. **101**, 9704 (1994)
[10] Bauernschmitt, R., Ahlrichs, R.: Chem. Phys. **256**, 454 (1996)
[11] Casida, M.E.: In: J.M. Seminario (ed.) Recent Developments and Applications of Modern Density Functional Theory pp. 391–440. Elsevier, Amsterdam (1996)
[12] Görling, A.: Phys. Rev. **A54**, 3912 (1996)
[13] Jamorski, C., Casida, M.E., Salahub, D.R.: J. Chem. Phys. **104**, 5134 (1996)
[14] van Gisbergen, S.J.A., Snijders, J.G., Baerends, E.J.: Phys. Rev. Lett. **78**, 3097 (1997)
[15] Champagne, B., Perpète, E.A., van Gisbergen, S.J.A., Baerends, E.J., Snijders, J.G., Soubra-Ghaoui, C., Robins, K.A., Kirtman, B.: J. Chem. Phys. **109**, 10489 (1998)
[16] van Gisbergen, S.J.A., Snijders, J.G., Baerends, E.J.: J. Chem. Phys. **109**, 10644 (1998)
[17] Nagy, A.: Phys. Rep. **298**, 1 (1998)
[18] Stratmann, R.E., Scuseria, G.E.: J. Chem. Phys. **109**, 8218 (1998)
[19] van Gisbergen, S.J.A., Snijders, J.G., Baerends, E.J.: Comp. Phys. Comm. **118**, 119 (1999)
[20] van Caillie, C., Amos, R.D.: Chem. Phys. **308**, 249 (1999)
[21] Görling, A., Heinze, H.H., Ruzankin, S.Ph., Staufer, M., Rösch, N.: J. Chem. Phys. **110**, 2785 (1999)
[22] Grimme, S., Waletzke, M.: J. Chem. Phys. **111**, 5645 (1999)
[23] Hessler, P., Park, J., Burke, K.: Phys. Rev. Lett. **82**, 378 (1999)
[24] Hessler, P., Park, J., Burke, K.: Phys. Rev. Lett. **83**, 5184 (1999)
[25] Hirata, S., Head-Gordon, M., Bartlett, R.J.: J. Chem. Phys. **111**, 10774 (1999)
[26] Yabana, K., Bertsch, G.F.: Int. J. Quant. Chem. **75**, 55 (1999)
[27] Yabana, K., Bertsch, G.F.: Phys. Rev. **A60**, 3809 (1999)
[28] Casida, M.E., Salahub, D.R.: J. Chem. Phys. **113**, 8918 (2000)
[29] Hainze, H.H., Görling, A., Rösch, N.: J. Chem. Phys. **113**, 2088 (2000)
[30] Kootstra, F., de Boeij, P.L., Snijders, J.G.: J. Chem. Phys. **112**, 6517 (2000)
[31] Larsen, H., Jørgensen, P., Olsen, J., Helgaker, T.: J. Chem. Phys. **113**, 8908 (2000)
[32] Lein, M., Gross E.K.U., Perdew, J.P.: Phys. Rev. **B61**, 13431 (2000)
[33] Tozer, D.J., Handy, N.C.: Phys. Chem. Chem. Phys. **2**, 2117 (2000)
[34] Furche, F.: J. Chem. Phys. **114**, 5982 (2001)
[35] Maitra, N.T., Burke, K.: Phys. Rev. **A63**, 042501 (2001)
[36] Grüning, M., Gritsenko, O.V., van Gisbergen, S.J.A., Baerends, E.J.: J. Chem. Phys. **116**, 9591 (2002)
[37] Heinze, H.H., Sala, F.D., Görling, A.: J. Chem. Phys. **116**, 9624 (2002)
[38] Autschbach, J., Ziegler, T.: J. Chem. Phys. **116**, 891 (2002)
[39] van Faassen, M., de Boeij, P.L., van Leeuwen, R., Beger, J.A., Snijders, J.G.: J. Chem. Phys. **118**, 1044 (2003)
[40] Tretiak, S., Chernyak, V.: J. Chem. Phys. **119**, 8809 (2003)
[41] Sałek, P., Vahtras, O., Helgaker, T., Ågren, H.: J. Chem. Phys. **117**, 9630 (2002)
[42] Jansik, B., Sałek, P., Jonsson, D., Vahtras, O., Ågren, H.: J. Chem. Phys. **122**, 054107 (2005)
[43] Wang, J., Becke, A.D., Smith, Jr, V.H.: J. Chem. Phys. **102**, 3477 (1995)
[44] Löwdin, P.-O.: Phys. Rev. **97**, 1490 (1955)
[45] Harriman, J.E.: Theoretical foundations of electron spin resonance. Academic, New York, 1978
[46] Kohn, W., Sham, L.J.: Phys. Rev. **140**, A1133 (1965)
[47] Parr, R.G., Yang, W.: Density-functional theory of atoms and molecules. Oxford University press, New York, 1989
[48] Okazaki, I., Sato, F., Yoshihiro, T., Ueno, T., Kashiwagi, H.: J. Mol. Struct. (Theochem) **451**, 109 (1998)
[49] Filatov, M., Shaik, S.: Chem. Phys. **288**, 689 (1998)
[50] Russo, T.V., Martin, R.L., Hay, P.J.: J. Chem. Phys. **101**, 7729 (1994)

[51] Helgaker, T., Jensen, H.J.A., Jørgensen, P., Olsen, J., Ruud, K., Ågren, H., Auer, A.A., Bak, K.L., Bakken, V., Christiansen, O., Coriani, S., Dahle, P., Dalskov, E.K., Enevoldsen, T., Fernandez, B., Hättig, C., Hald, K., Halkier, A., Heiberg, H., Hettema, H., Jonsson, D., Kirpekar, S., Kobayashi, R., Koch, H., Mikkelsen, K.V., Norman, P., Packer, M.J., Pedersen, T.B., Ruden, T.A., Sanchez, A., Saue, T., Sauer, S.P.A., Schimmelpfennig, B., Sylvester-Hvid, K.O., Taylor, P.R., Vahtras, O., DALTON, an ab initio electronic structure program, Release 1.2.See //www.kjemi.uio.no/software/dalton/dalton.html (2001)
[52] Hamilton, T.P., Pulay, P.: J. Chem. Phys. **84**, 5728 (1986)
[53] Fernandez, B., Jørgensen, P., Byberg, J., Olsen, J., Helgaker T., Jensen, H.J.A.: J. Chem. Phys. **97**, 3412 (1992)
[54] Olsen, J., Jørgensen, P.: J. Chem. Phys. 82, 3235 (1985)
[55] Norman, P., Jonsson, D., Vahtras, O., Ågren, H.: Chem. Phys. **203**, 23 (1996)
[56] Jonsson, D., Norman, P., Ågren, H.: J. Chem. Phys. **105**, 6401 (1996)
[57] Tunell, I., Rinkevicius, Z., Sałek, O.V.P., Helgaker, T., Ågren, H.: J. Chem. Phys. **119**, 11024 (2003)
[58] Jansik, B., Jonsson, D., Sałek P., Ågren, H.: J. Chem. Phys. **121**, 7595 (2004)
[59] Cronstrand, P., Jansik, B., Jonsson, D., Luo Y., Ågren, H.: J. Chem. Phys. **121**, 9239 (2004)
[60] Rice, J.E., Handy, N.C.: J. Chem. Phys. **97**, 4959 (1991)
[61] Christiansen, O., Jørgensen, P., Hättig, C.: Int. J. Quant. Chem. **68**, 1 (1998)
[62] Saue, T.: In Schwerdtfeger, P. (ed.) Relativistic Electronic Structure Theory. Part 1. Fundamentals, Chap. 7 p. 332. Elsevier, Amsterdam (2002)
[63] Becke, A.D.: J. Chem. Phys. **88**, 2547 (1988)
[64] Stratmann, R.E., Scuseria, G.E., Frisch, M.J.: Chem. Phys. Lett. **257**, 213 (1996)
[65] Lindh, R., Malmqvist, P.-Å., Gagliardi, L.: Theor. Chem. Acc. **106**, 178 (2001)
[66] van Gisbergen, S.J.A., Osinga, V.P., Gritsenko, O.V., van Leeuwen, R., Snijders, J.G., Baerends, E.J.: J. Chem. Phys. **105**, 3142 (1996)
[67] Adamo, C., Barone, V.: J. Chem. Phys. **110**, 6158 (1999)
[68] Bertinelli, F., Palmieri, P., Brillante, A., Taliani, C.: Chem. Phys. **25**, 333 (1977)
[69] Teng, C., Garito, A.: Phys. Rev. **B28**, 6766 (1983)
[70] Kaatz, P., Donley, E., Shelton, D.: J. Chem. Phys. **108**, 849 (1998)
[71] Karna, S., Prasad, P., Dupuis, M.: J. Chem. Phys. **94**, 1171 (1991)
[72] Ågren, H., Vahtras, O., Koch, H., Jørgensen, P., Helgaker, T.: J. Chem. Phys. **98**, 6417 (1993)
[73] Luo, Y., Ågren, H., Vahtras, O., Jørgensen, P.: Chem. Phys. **207**, 190 (1993)
[74] Sałek, P., Helgaker, T., Vahtras, O., Ågren, H., Jonsson, D., Gauss, J.: Mol. Phys. **103**, 439 (2005)
[75] Sim, F., Chin, S., Dupuis, M., Rice, J.E.: J. Chem. Phys. **97**, 1158 (1993)
[76] van Gisbergen, S.J.A., Snijders, J.G., Baerends, E.J.: J. Chem. Phys. **111**, 6652 (1999)
[77] Sałek, P., Vahtras, O., Helgaker, T., Ågren, H.: J. Chem. Phys. **117**, 9630 (2002)
[78] Alms, R.G., Burnham, A.K., Flygare, W.H.: J. Chem. Phys. **63**, 3321 (1975)
[79] Shelton, D.: Phys. Rev. **A42**, 2578 (1990)
[80] Bogaard, M.P., Buckingham, A.D., Corfield, M.G., Dunmur, D.A., White, A.H.: Chem. Phys. Lett. **12**, 558 (1972)
[81] Ward, J.F., Elliot, D.S.: J. Chem. Phys. **69**, 5438 (1978)
[82] van Gisbergen, S.J.A., Osinga, V.P., Gritsenko, O.V., van Leeuwen, R., Snijders, J.G., Baerends, E.J.: J. Chem. Phys. **105**, 3142 (1996)
[83] van Leeuwen, R., Baerends, E.J.: Phys. Rev. **A49**, 2421 (1994)
[84] Shelton, D.P.: J. Opt. Soc. Am. **B2**, 1880 (1985)
[85] Hebard, A.F., Haddon, R.C., Fleming, R.M., Korton, A.R.: Appl. Phys. Lett. **59**, 2109 (1991)
[86] Lambin, P.H., a. Lucas, A., Vigneron, J.P., Phys. Rev. **B46**, 1794 (1992)
[87] Ecklund, P.: Bull Am. Phys. Soc. **37**, 191 (1992)
[88] Meth, J.S., Vanherzeele, H., Wang, Y.: Chem. Phys. Lett. **197**, 26 (1992)
[89] Norman, P., Luo, Y., Jonsson, D., Ågren, H.: J. Chem. Phys. **106**, 8788 (1997)
[90] Cioslowski, J.: Electronic Structure Calculations on Fullerenes and Their Derivatives. Oxford University Press, London (1995)

[91] Geng, L., Wright, J.C.: Chem. Phys. Lett. **249**, 105 (1996)
[92] Kajzar, F., Taliani, C., Zamboni, R., Rossini, S., Danieli, R.: Synth. Metals **77**, 257 (1996)
[93] Kajzar, F., Taliani, C., Zamboni, R., Rossini, S., Danieli, R.: Fullerenes: Status and Perspectives In C. Taliani, G. Ruani and R. Zamboni (eds.), World Scientific, Singapore (1992)
[94] Guo, D., Mazumdar, S., Dixit, S.N.: Synth. Metals. **49**, 1 (1992)
[95] Ruud, K., Jonsson, D., Taylor, P.R.: J. Chem. Phys. **114**, 4331 (2001)
[96] Iwata, J.-I., Yabana, K., Bertsch, G.F.: J. Chem. Phys. **115**, 8773 (2001)
[97] Antoine, R., Dugourd, P., Rayane, D., Benichou, E., Chandezon, M.B.F., Guet, C.: J. Chem. Phys. **110**, 9771 (1999)
[98] van Gisbergen, S.J.A., Snijders, J.G., Baerends, E.J.: Phys. Rev. Lett. **78**, 3097 (1997)
[99] Morales, R.G.E.: J. Chem. Phys. **86**, 2550 (1982)
[100] Ghoneim, N., Suppan, P.: J. Chem. Soc., Faraday Trans. **86**, 2079 (1990)
[101] Sinha, H.K., Thomson, P.C.P., Yates, K.: Can. J. Chem. **68**, 1507 (1990)
[102] Chongwain, P.T., Iweibo, I.: Spectrochim. Acta **A47**, 713 (1991)
[103] Iweibo, I., Chongwain, P.T., Obi-Egbedi, N.O., Lesi, A.F.: Spectrochim. Acta **A47**, 705 (1991)
[104] Heitz, S., Weidauer, D., Hese, A.: Chem. Phys. Lett. **176**, 55 (1991)
[105] Heitz, S., Weidauer, D., Hese, A.: J. Chem. Phys. **95**, 7952 (1991)
[106] Heitz, S., Weidauer, D., Rosenow, B., Hese, A.: J. Chem. Phys. **96**, 976 (1992)
[107] Shepard, R., Lischka, H., Szalay, P.G., Kovar, T., Ernzerhof, M.: J. Chem. Phys. **96**, 2085 (1992)
[108] Foresman, J.B., Head-Gordon, M., Pople, J.A., Frisch, M.J.: J. Phys. Chem. **96**, 135 (1992)
[109] Stanton, J.F., Gauss, J.: Theor. Chim. Acta **91**, 267 (1995)
[110] Celani, P., Werner, H.-J.: J. Chem. Phys. **119**, 5044 (2003)
[111] Van Caille, C., Amos, R.D.: Chem. Phys. Lett. **308**, 249 (1999)
[112] Van Caille, C., Amos, R.D.: Chem. Phys. Lett. **317**, 159 (2000)
[113] Amos, R.D.: Chem. Phys. Lett. **364**, 612 (2002)
[114] Furche, F., Ahlrichs, R.: J. Chem. Phys. **117**, 7433 (2002)
[115] Stanton, J.F., Gauss, J.: J. Chem. Phys. **104**, 9859 (1996)
[116] Stanton, J.F.: J. Chem. Phys. **99**, 8840 (1993)
[117] Stanton, J.F., Gauss, J.: J. Chem. Phys. **100**, 4695 (1994)
[118] Stanton, J.F., Gauss, J.: J. Chem. Phys. **103**, 8931 (1995)
[119] Jonsson, D., Norman, P., Ågren, H., Sylvester-Hvid, K.O., Mikkelsen, K.V.: J. Chem. Phys. **109**, 6351 (1998)
[120] Hättig, C., Christiansen, O., Coriani, S., Jørgensen, P.: J. Chem. Phys. **109**, 9219 (1998)
[121] Hättig, C., Christiansen, O., Coriani, S., Jørgensen, P.: J. Chem. Phys. **109**, 9237 (1998)
[122] Christiansen, O., Hättig, C., Jorgensen, P.: Spectrochim. Acta **A55**, 509 (1999)
[123] Cradock, S., Liescheski, P.B., Rankin, D.W.H., Robertson, H.E.: J. Am. Chem. Soc. **110**, 2758 (1988)
[124] Job, V.A., Innes, K.K.: J. Mol. Spectrosc. **71**, 299 (1978)
[125] Innes, K.K., Brumbaugh, D.V., Franks, L.A.: Chem. Phys. **59**, 439 (1981)
[126] Jonsson, D., Norman, P., Ågren, H.: Chem. Phys. **224**, 201 (1997)
[127] Schütz, M., Hutter, J., Lüthi, H.P.: J. Chem. Phys. **103**, 7048 (1995)
[128] Singh, S., Bradley, L.T.: Phys. Rev. Lett. **12**, 612 (1964)
[129] Reinhardt, B.A.: Photonics Science News **4**, 21 (1999)
[130] Reinhardt, B., Brott, L., Clarson, S., Dillard, A., Bhatt, J., Kannan, R., Yuan, L., He, G., Prasad, P.: Chem. Mater. **10**, 1863 (1998)
[131] Albota, M., Beljonne, D., Brédas, J., Ehrlich, J., Fu, J., Heikal, A., Hess, S., Kogej, T., Levin, M., Marder, S., McCord-Maughon, D., Perry, J., Röckel, H., Rumi, M., Subramaniam, G., Webb, W., Wu, X., Xu, C.: Science, **281**, 1653 (1998)
[132] Guang, S., Przemyslaw, P., Markowicz, P., Lin, T., Prasad, P.: Nature **415**, 767 (2002)
[133] Wang, D., Zhan, C., Chen, Y., Li, Y.J., Lu, Z., Nie, Y.: Chem. Phys. **369**, 621 (2003)
[134] Zhan, C., Li, D., Zhang, D., Li, Y., Wang, D., Wang, T., Lu, Z., Zhao, L., Nie, Y., Zhu, D.: Chem. Phys. **353**, 138 (2002)
[135] McClain, W.: J. Chem. Phys. **55**, 2789 (1971)

[136] Sutherland, R.: Handbook of nonlinear optics Marcel Dekker Inc., New York (1996)
[137] Norman, P., Luo, Y., Ågren, H.: J. Chem. Phys. **111**, 7758 (1999)
[138] Cronstrand, P., Luo, Y., Ågren, H.: J. Chem. Phys. **117**, 11102 (2002)
[139] Hohlneicher, G., Dick, B.: J. Photochem. **27**, 215 (1984)
[140] Gudipati, M., Mauds, M., Daverkausen, J., Hohlneicher, G.: Chem. Phys. **192**, 37 (1995)
[141] Robinson, J.: Handbook of Spectroscopy, Vol. II, CRC press, Florida (1974)
[142] Rumi, M., Ehrlich, J., Heikal, A., Perry, J., Barlow, S., Hu, Z., McCord-Maughon, D., Parker, T., Röckel, H., Thayumanavan, S., Marder, S., Beljonne, D., Brédas, J.: J. Am. Chem. Soc. **122**, 9500 (2000)
[143] Kim, O.K., Lee, K.-S., Woo, H.Y., Kim, K.-S., He, G.S., Swiatkiewicz, J., Prasad, P.: Chem. Mater. **12**, 284 (2000)
[144] Ventelon, L., Moreaux, L., Mertz, J., Blanchard-Desce, M.: Chem. Commun. **1999**, 2055 (1999)
[145] Rizzo, A., Cappelli, C., Jansík, B., Jonsson, D., Sałek, P., Coriani, S., Ågren, H.: J. Chem. Phys. **121**, 8814 (2004)
[146] Rizzo, A., Cappelli, C., Jansík, B., Jonsson, D., Sałek, P., Coriani, S., Ågren, H.: J. Chem. Phys. **122**, 234314 (2005)
[147] Neese, F., Salomon, E.: In M. Drillon and J. Miller (eds.), Magnetoscience – From Molecules to Materials, vol. 4, New York (Wiley, 2002)
[148] Neese, F.: Current Opinion in Chemical Biology **7**, 125 (2003)
[149] Kaupp, M., Bühl, M., Malkin, V.G.: Calculation of NMR and EPR parameters. Theory and applications, Wiley–VCH, Weinheim, 2004).
[150] Rinkevicius, Z., Telyatnyk, L., Sałek, P., Vahtras, O., Ågren, H.: J. Chem. Phys. **119**, 10489 (2003)
[151] Neese, F.: J. Chem. Phys. **115**, 11080 (2001)
[152] Patchkovskii, S., Ziegler, T.: J. Phys. Chem. **A105**, 5490 (2001)
[153] Arbuznikov, A.V., Kaupp, M.: J. Phys. Chem. **A391**, 16 (2004)
[154] Rinkevicius, Z., Telyatnyk, L., Ruud, K., Vahtras, O.: J. Chem. Phys. **121**, 5051 (2004)
[155] Rinkevicius, Z., Telyatnyk, L., Vahtras, O., Ågren, H.: J. Chem. Phys. **121**, 7614 (2004)
[156] Munzarova, M.L., Kaupp, M.: J. Phys. Chem. **A103**, 9966 (1999)
[157] Watson, M.A., Sałek, P., Macak, P., Jaszuński, M., Helgaker, T.: Chemistry – A European Journal **10**, 4627 (2004)
[158] Watson, P.M.M., Sałek, P., Helgaker, T.: J. Chem. Phys. **121**, 2915 (2004)
[159] Rubensson, E., Sałek, P.: J. Comput. Chem. **26**, 1628 (2005)
[160] Sala, F.D., Görling, A.: J. Chem. Phys. **118**, 10439 (2003)
[161] Vitale, V., Sala, F.D., Görling, A.: J. Chem. Phys. **122**, 244102 (2005)
[162] Filatov, M., Shaik, S.: Chem. Phys. **332**, 409 (2000)
[163] Casida, M.E.: J. Chem. Phys. **122**, 054111 (2005)

CHAPTER 6

NON-LINEAR PULSE PROPAGATION IN MANY-PHOTON ACTIVE ISOTROPIC MEDIA

A. BAEV, S. POLYUTOV, I. MINKOV, F. GEL'MUKHANOV
AND H. ÅGREN

Theoretical Chemistry, Roslagstullsbacken 15, Royal Institute of Technology, S-106 91 Stockholm, Sweden

Abstract: It is an experimental fact that light propagation in a medium is sensitively dependent on the shape and intensity of the optical pulse as well as on the electronic and vibrational structure of the basic molecular units. We review in this paper results of systematic studies of this problem for isotropic media. Our theoretical approach is based on numerical solutions of the density matrix and Maxwell's equations and a quantum mechanical account of the complexity of the many-level electron-nuclear medium. This allows to accommodate a variety of non-linear effects which accomplish the propagation of strong light pulses. Particular attention is paid to the understanding of the role of coherent and sequential excitations of electron-nuclear degrees of freedoms. We highlight the combination of quantum chemistry with classical pulse propagation which allows to estimate the optical transmission from cross sections of multi-photon absorption processes and from considerations of propagation effects, saturation and pulse effects. It is shown that in the non-linear regime it is often necessary to account simultaneously for coherent one-step and incoherent step-wise multi-photon absorption, as well as for off-resonant excitations even when resonance conditions prevail. The dynamic theory of non-linear propagation of a few interacting intense light pulses has been successfully applied to study, for example, frequency-upconversion cavity-less lasing in a chromophore solution, namely in an organic stilbenechromophore 4-[N-(2-hydroxyethyl)-N-(methyl)amino phenyl]-4'-(6-hydroxyhexyl sulphonyl) dissolved in dimethyl sulphoxide. Furthermore, the theory has been used to explain observed differences between spectral shapes of one- and two-photon absorption in the di-phenyl-amino-nitro-stilbene molecule. The present simulations evidence that the reason for this effect is the competition between two-step and coherent two-photon absorption processes

1. INTRODUCTION

The field of non-linear optics is to a large extent driven and motivated by the anticipation of large technological dividends. The use of lasers in modern technology is now commonplace, ranging in application from high-density data storage on optical disks [1, 2] to improved surgical techniques in ophthalmology, neurosurgery, dermatology and biology imaging [3, 4]. Non-linear optics has also the potential to revolutionize future telecommunication and computer technologies [5–7]. A new extension of non-linear optics and non-linear spectroscopy is biophotonics, which involves a fusion of photonics and biology [4]. However, it is a common view that for future use more sophisticated understanding of the basic mechanisms underlying non-linear phenomena will be required. The scientific problems of this kind motivate us to explore non-linear propagation of light pulses through complex media, where details of the quantum structure of the molecular units become essential. Ideas and theories can be tested through wide parameter ranges, and quantitative support can be given to complement the qualitative understanding obtained through the use of general arguments and simple models. It is our belief that the techniques of modern computer simulations of the coupled matter and Maxwell's equations will transform the arsenal of theoretical tools to qualitatively higher levels, paving the way to the crucial understanding of non-linear phenomena.

Non-linear optics has developed into an extensive branch of science presenting many models which mimic certain aspects or special effects. However, when light propagates through a real system one faces the problem of having different physical effects operating simultaneously, in particular so in the non-linear regime. In such cases a unified theoretical approach is needed which takes into account the complexity of the many-level electron-nuclear medium as well as a variety of non-linear effects which regulate the propagation of strong light pulses. The main goal of our review is to demonstrate the usefulness of such a unified theoretical tool in some fundamental problems of non-linear optics, like many-photon absorption and upconverted lasing. Understanding the formation of many-photon spectra is essential in order to tailor structure-property relations for non-linear materials and so to improve their use in technical applications. For example, one salient feature of two-photon spectra is that they in general are very different from their one-photon counterparts. The character of many-photon absorption in condensed phases is often complicated by the competition between one-step and sequential absorption channels.

We begin in the next section (Section 2) by describing a general theoretical tool for the solution of the Maxwell's and matter density matrix equations. We pay attention to the importance of two mechanisms of light-matter interactions, namely those given by the coherent and incoherent, or step-like, channels. The shape of the wings of the spectral line is of crucial importance in the competition between these different excitation channels (Section 2.4). We then review the application of the general theory to studies of the role of saturation effects and pulse shapes in many-photon absorption, and bidirectional propagation of stimulated emission (Section 3.4). The role of the vibrational degrees of freedom is discussed in Section 4 devoted to the spectral shape of one- and two-photon absorption. Findings and conclusions are highlighted in Section 6.

2. GENERAL THEORY OF NON-LINEAR PULSE PROPAGATION

2.1 Non-Linear Polarization

When light propagates through matter it induces some displacement of the charge distribution inside the molecules. Such an influence can be rather easily understood with use of the forced harmonic oscillator model in which atoms, constituting the material, are seen as charge distributions pushed away from their equilibrium state when exposed to the electric field. The contribution from the magnetic field part of the light is much weaker and is usually neglected giving rise to the so-called electric-dipole approximation. These induced microscopic displacements of charge distributions or induced electric dipoles, which in the linear approximation oscillate with the frequency of an applied electric field, add up to the macroscopic polarization. The latter is proportional to the field applied: $\mathbf{P}(t, \mathbf{r}) = \varepsilon_0 \chi^{(1)} \mathbf{E}(t, \mathbf{r})$, where the tensor $\chi^{(1)}$ is the first order susceptibility—an intrinsic characteristic of the given material. The linear coupling model holds while the amplitude of the electric field is small compared to the intra-atomic field (10^9 V/cm for the hydrogen atom). For larger amplitudes the linear motion of the displaced charges will be distorted and non-linear terms will be important. Provided that these new terms are still small compared to the linear term one can write down a general expansion for the polarization in a power series in $\mathbf{E}(t, \mathbf{r})$ [5, 8]:

(1) $$\mathbf{P}(\mathbf{r}, t) = \mathbf{P}_L(\mathbf{r}, t) + \mathbf{P}_{NL}(\mathbf{r}, t) = \int_{-\infty}^{+\infty} dt \int d\mathbf{r}\ \boldsymbol{\mathcal{P}}(\mathbf{k}\omega) e^{-i(\omega t - \mathbf{k}\cdot\mathbf{r})}$$

(2) $$\boldsymbol{\mathcal{P}}(\mathbf{k}\omega) = \varepsilon_0 \sum_{n=1}^{\infty} \frac{1}{n!} \int_{-\infty}^{+\infty} d\omega_1 \ldots d\omega_n \int_{-\infty}^{+\infty} d\mathbf{k}_1 \ldots d\mathbf{k}_n$$
$$\times \chi^{(n)}_{j_1\ldots j_n}(-\mathbf{k}\omega; \mathbf{k}_1\omega_1, \ldots, \mathbf{k}_n\omega_n) \mathcal{E}_{j_1}(\mathbf{k}_1\omega_1) \ldots \mathcal{E}_{j_n}(\mathbf{k}_n\omega_n)$$
$$\times \delta\Big(-\mathbf{k} + \sum_{\alpha=1}^{n} \mathbf{k}_\alpha\Big) \delta\Big(-\omega + \sum_{\alpha=1}^{n} \omega_\alpha\Big)$$

Here $\chi^{(n)}_{j_1\ldots j_n}(-\mathbf{k}\omega; \mathbf{k}_1\omega_1, \ldots, \mathbf{k}_n\omega_n)$ is the nth order susceptibility. The δ-functions show the momentum and energy conservation (phase matching):

$$\mathbf{k} = \sum_{\alpha=1}^{n} \mathbf{k}_\alpha, \quad \omega = \sum_{\alpha=1}^{n} \omega_\alpha$$

The conventional formula for $\boldsymbol{\mathcal{P}}(\mathbf{k}\omega)$ follows directly from Eq. (2) when the fields are monochromatic: $\mathcal{E}(\mathbf{k}_j\omega_j) = \delta(\mathbf{k}_j - \mathbf{k}_j^0)\delta(\omega_j - \omega_j^0)\mathcal{E}(\mathbf{k}_j^0\omega_j^0)$. It is necessary to note that the expansion (2) can break down for very high intensities when various saturation effects come into play.

The induced polarization displays some important properties. Firstly, as mentioned above, the induced electric dipoles oscillate with the frequency of the perturbing field. As far as any oscillating dipole emits radiation, with the frequency of the oscillation, the optical field that induced the polarization will, in turn, be modified. Moreover, the polarization will contain terms that oscillate at double, triple and so on frequencies and will even contain a non-oscillating, direct current, component in addition to the linear component oscillating at the input frequency. In a material with a center of symmetry (isotropic media such as achiral glasses, liquids and gases) the even order terms in the expansion of the polarization are absent for symmetry reasons. The lowest order non-linearity is then the cubic one. This term is responsible for all four-wave mixing processes such as third-harmonic generation and the quadratic electrooptic effect, and for intensity dependent refractive index effects such as self-focusing. One of the most important manifestations of the cubic non-linearity, from both spectroscopic and technological points of view, is two-photon absorption.

Higher-order non-linearities are not so widely studied as the second- and third-order terms. The effects are usually very small, but if the process is resonantly enhanced they can easily be detected. It should be mentioned here that the susceptibility is in general complex and frequency dependent because of the finite response time of the medium.

2.2 Maxwell's Equations

The polarization expansion (2) over a power series in the electric field is the constitutive relation that describes how matter responds to the applied optical field. The interaction works both ways and the polarized medium will modify existing fields and create new ones. This is governed by Maxwell's equations. When the intensity of the applied field is high, which means that the number of photons interacting with matter is large, it is justified to use the classical representation of the electromagnetic field. For homogeneous, non-conductive and non- magnetic media (permeability, μ, is equal to that of free space, μ_0), Maxwell's equations (SI system of units) read [6]:

(3) $\quad \nabla \times \mathbf{H}(\mathbf{r}, t) = \dfrac{\partial \mathbf{D}(\mathbf{r}, t)}{\partial t}, \quad \nabla \times \mathbf{E}(\mathbf{r}, t) = -\mu_0 \dfrac{\partial \mathbf{H}(\mathbf{r}, t)}{\partial t}$

(4) $\quad \nabla \cdot \mathbf{H}(\mathbf{r}, t) = 0, \quad \nabla \cdot \mathbf{D}(\mathbf{r}, t) = \varrho$

Here $\mathbf{D}(\mathbf{r}, t)$ is the displacement vector, ϱ is free charge density, ε_0 is the free space permittivity:

(5) $\quad \mathbf{D}(\mathbf{r}, t) = \varepsilon_0 \mathbf{E}(\mathbf{r}, t) + \mathbf{P}(\mathbf{r}, t), \quad \varepsilon_0 \mu_0 = \dfrac{1}{c^2}$

It follows, after all necessary transformations, that the field $\mathbf{E}(\mathbf{r}, t)$ satisfies the wave equation:

$$(6) \quad \nabla \times [\nabla \times \mathbf{E}(\mathbf{r}, t)] + \frac{1}{c^2} \frac{\partial^2 \mathbf{E}(\mathbf{r}, t)}{\partial t^2} = -\frac{1}{\varepsilon_0 c^2} \frac{\partial^2 \mathbf{P}(\mathbf{r}, t)}{\partial t^2}$$

Provided that there are no free charges in the medium under consideration, Eq. (6) can be rewritten as follows:

$$(7) \quad -\Delta \mathbf{E}(\mathbf{r}, t) - \frac{1}{\varepsilon_0} \nabla(\nabla \cdot \mathbf{P}(\mathbf{r}, t)) + \frac{1}{c^2} \frac{\partial^2 \mathbf{E}(\mathbf{r}, t)}{\partial t^2} = -\frac{1}{\varepsilon_0 c^2} \frac{\partial^2 \mathbf{P}(\mathbf{r}, t)}{\partial t^2}$$

A traveling wave representation of the electric field and polarization enables us to make the following factorization:

$$(8) \quad \mathbf{E}(\mathbf{r}, t) = \frac{1}{2} \boldsymbol{\mathcal{E}}(\mathbf{r}, t) e^{-\imath[\omega t - \mathbf{k} \cdot \mathbf{r} + \varphi(\mathbf{r}, t)]} + c.c.$$

$$(9) \quad \mathbf{P}(\mathbf{r}, t) = \frac{1}{2} \boldsymbol{\mathcal{P}}(\mathbf{r}, t) e^{-\imath[\omega t - \mathbf{k} \cdot \mathbf{r} + \varphi(\mathbf{r}, t)]} + c.c.$$

where $\boldsymbol{\mathcal{E}}(\mathbf{r}, t)$, $\boldsymbol{\mathcal{P}}(\mathbf{r}, t)$ and $\varphi(\mathbf{r}, t)$ are slowly varying functions of position-vector and time. This Slowly-Varying Envelope Approximation (SVEA) is justified when the electric field and the polarization amplitudes, as well as the phase $\varphi(\mathbf{r}, t)$, do not change appreciably in an optical frequency period. SVEA breaks down in the case of ultrashort ($\tau \sim 1$ fs) pulses when the inverse pulse duration becomes comparable with the carrier frequency. Another assumption, which is commonly used in non-linear optics of gas and liquid phases, is that the susceptibility itself is a slowly varying function of $\mathbf{r}$ in the wavelength scale. In this case one can neglect the second term at the left-hand side of the equation (7). This is, however, not true in the general case and one of the examples where this assumption does not hold is photonic crystals—artificially created periodic structures with variable dielectric constant $\varepsilon(\mathbf{r})$ [7]. The spatial scale of modulation of the dielectric constant for these crystals is comparable with the wavelengths of incoming light (at least in the optical region).

An extraction of the fast variables in equation (7) results in the following paraxial wave equation, connecting the slowly varying amplitudes $\boldsymbol{\mathcal{E}}(\mathbf{r}, t)$ and $\boldsymbol{\mathcal{P}}(\mathbf{r}, t)$:

$$(10) \quad \left(\frac{\partial}{\partial z} + \frac{1}{c} \frac{\partial}{\partial t} - \frac{\imath}{2k} \Delta_\perp \right) \boldsymbol{\mathcal{E}} = \frac{\imath k}{\varepsilon_0} \boldsymbol{\mathcal{P}}$$

The cross Laplacian, $\Delta_\perp$, is important for narrow light beams, for systems with self-focusing. It is worthwhile to stress here that most of the currently cherished approaches applied for solving Eq. (10), are based on the power series expansion (2) of the polarization over the laser field amplitude and on an account of only the coherent contributions to the corresponding polarizabilities. However, this approximation breaks down even for fairly short laser pulses, which is the case in many types of experiments carried out, when various saturation limits come into play. Moreover, large homogeneous broadenings in solutions, which appear to be typical experimental media, changes the correlation between coherent and incoherent processes quite drastically [9, 10]. Thus, the series (2) diverges when the intensity of the light is rather high, for example when the intensity is higher than the saturation intensity.

An alternative to the power series expansion (2) of the polarization approach is to explicitly account for the coupling of multi-photon processes in a strong field. The formalism is based on a Fourier expansion of the density matrix (see Section 2.3) and allows to deal with photon-matter interaction events of any order, depending on the resonant conditions introduced [11]. The corresponding theory is generalized to the case of few interacting strong fields [12] to be able to model processes like mirrorless laser generation (upconverted stimulated emission induced by multi-photon absorption [11, 13]). This theory is based on a strict solution of the density matrix equations of a many-level system without using an expansion of the density matrix over powers of the light intensity and without using the rotatory wave approximation. In this way the saturation effects as well as the coherent and incoherent processes are accounted for explicitly.

2.3 Density Matrix Equations

The non-linear susceptibilities of the expansion (2) can be evaluated by means of the density matrix formalism. Recall that the non-linear polarization of a non-polar medium is equal to the induced dipole moment of a unit volume which, in turn, can be easily represented by standard methodology of quantum mechanics as an expectation value of the dipole moment. After some straightforward derivations we get:

$$(11) \quad \mathbf{P} = \langle \mathbf{d}(t) \rangle = \mathrm{Tr}\left[\rho(t)\mathbf{d}(t)\right] = \sum_{\beta\alpha} \mathbf{d}_{\beta\alpha}(t)\rho_{\alpha\beta}(t)$$

where $\mathbf{d}_{\beta\alpha}(t) = \mathbf{d}_{\beta\alpha}\exp(\imath\omega_{\beta\alpha}t)$ is the transition dipole moment, $\omega_{\beta\alpha} = (E_\beta - E_\alpha)/\hbar$ is the frequency of the quantum transition $\beta \to \alpha$, and $\rho_{\alpha\beta}(t)$ is the density matrix of the medium which obeys the following equation in the interaction representation:

$$(12) \quad \hat{L}\rho(t) = \frac{\imath}{\hbar}[\rho(t), V(t)], \quad \mathrm{Tr}[\rho(t)] = N, \quad \hat{L} = \frac{\partial}{\partial t} + \mathbf{v}\cdot\nabla + \hat{\Gamma}$$

$$V(t) = e^{\imath H_0 t/\hbar} V e^{-\imath H_0 t/\hbar}, \quad V = -\mathbf{E}(\mathbf{r}, t)\cdot\mathbf{d}(t)$$

Here, V is the interaction of the electric field with the molecule, N is the concentration of the absorbing molecules, $\mathbf{v}$ is the thermal velocity of molecules, and the term $\mathbf{v}\cdot\nabla$ is responsible for the Doppler effect. The relaxation matrix, $\hat{\Gamma}$, contains the rates of various radiative and non-radiative transitions.

The physical meaning of the density matrix elements arises from its definition: The diagonal elements are related to level populations and the off-diagonal elements are related to the polarization of the medium. Thus, the kinetic equations for populations, $\rho_{\alpha\alpha}$, and off-diagonal elements, $\rho_{\alpha\beta}$, of the density matrix read

$$(13) \quad \left(\frac{\partial}{\partial t} + \mathbf{v}\cdot\nabla + \Gamma_{\alpha\beta}\right)\rho_{\alpha\beta} = \delta_{\alpha,\beta}\sum_{\gamma>\alpha}\Gamma_\gamma^\alpha \rho_{\gamma\gamma} + \frac{\imath}{\hbar}\sum_\gamma(\rho_{\alpha\gamma}V_{\gamma\beta} - V_{\alpha\gamma}\rho_{\gamma\beta})$$

It is often convenient to rewrite the field-system interaction via the Rabi frequencies $G_{\alpha\beta} = \boldsymbol{\mathcal{E}}\cdot\mathbf{d}_{\alpha\beta}/2\hbar$:

$$(14) \quad \frac{1}{\hbar}V_{\alpha\beta} = e^{\imath\omega_{\alpha\beta}t}G_{\alpha\beta}\left(e^{\imath\mathbf{k}\cdot\mathbf{r}-\imath(\omega t+\varphi)} + e^{-\imath\mathbf{k}\cdot\mathbf{r}+\imath(\omega t+\varphi)}\right)$$

Let us now switch to the case of a many-mode field [12]. We consider the propagation of N electromagnetic fields, $\boldsymbol{\mathcal{E}}_j\exp(-\imath\omega_j t + \imath\mathbf{k}_j\cdot\mathbf{r} - \imath\varphi_j)$, through a non-linear many-level medium. These fields with frequencies, ω_j, wave vectors, $\mathbf{k}_j$, and phases, φ_j, form the total electromagnetic wave:

$$(15) \quad \mathbf{E} = \frac{1}{2}\sum_{j=1}^N \boldsymbol{\mathcal{E}}_j\, e^{-\imath\omega_j t + \imath\mathbf{k}_j\cdot\mathbf{r} - \imath\varphi_j} + \text{c.c.}$$

We will then assume that the initial phases φ_j are constant, $\dot\varphi_j = 0$. We want to consider here non-linear interaction of waves with frequencies and wave vectors that do not match:

$$(16) \quad \sum_j n_j\omega_j \neq 0, \quad \sum_j n_j\mathbf{k}_j \neq 0$$

Many applied and scientific problems satisfy these conditions. The opposite limiting case which will not be touched here is complete phase matching.

Let us now seek for a solution of the density matrix equations (13) making use of the Fourier transform:

$$(17) \quad \rho_{\alpha\beta} = e^{\imath\omega_{\alpha\beta}t}\sum_n r_{\alpha\beta}^n\, e^{\imath(n\circ\mathbf{k}\cdot\mathbf{r} - n\circ\omega t - n\circ\varphi)}$$

Here we introduce the N-dimensional vector n and the N-dimensional scalar product

$$(18) \quad n \equiv (n_1, n_2, \cdots, n_N), \quad n\circ\omega \equiv \sum_{j=1}^N n_j\omega_j, \quad n\circ\mathbf{k} \equiv \sum_{j=1}^N n_j\mathbf{k}_j$$

with $n_j = 0, \pm 1, \pm 2, \cdots \pm\infty$ as the number of photons of the field $\boldsymbol{\mathcal{E}}_j$; $\omega = (\omega_1, \omega_2, \cdots, \omega_N)$, and $\mathbf{k} = (\mathbf{k}_1, \mathbf{k}_2, \cdots, \mathbf{k}_N)$. We refer to $r_{\alpha\beta}^n$ as the density matrix in the photon occupation number representation. In fact, this is a mixed representation because $r_{\alpha\beta}^n$ depends also on the quantum states α, β of a molecule.

To clarify the physical meaning of the Fourier expansion (Eq. 17) we consider the case of two fields $n = (2, -1)$. In this case $r_{\alpha\beta}^n = r_{\alpha\beta}^{(2,-1)}$ describes absorption of two photons ($n_1 = 2$) with the frequency ω_1 and emission of one photon ($n_2 = -1$) with the frequency ω_2. The substitution of the Fourier expansion (17) into the density matrix equation (13) using the conditions (16) gives the following equations:

$$(19) \quad \left(\frac{\partial}{\partial t} - \imath(n \circ (\omega - \mathbf{k} \cdot \mathbf{v}) - \omega_{\alpha\beta}) + \Gamma_{\alpha\beta}\right) r^n_{\alpha\beta} = \delta_{\alpha\beta} \sum_{\gamma > \alpha} \Gamma^\alpha_\gamma r^n_{\gamma\gamma}$$

$$+ \imath \sum_{j=1}^{N} \sum_{\gamma} \left[\left(r^{n-1_j}_{\alpha\gamma} + r^{n+1_j}_{\alpha\gamma}\right) G^{(j)}_{\gamma\beta} - G^{(j)}_{\alpha\gamma}\left(r^{n-1_j}_{\gamma\beta} + r^{n+1_j}_{\gamma\beta}\right)\right]$$

The origin of the upper indices of the density matrix, $r^{n\pm 1_j}_{\alpha\gamma}$, in the field term

$$(20) \quad n \pm 1_j \equiv (n_1, n_2, \cdots, n_j \pm 1, \cdots, n_N)$$

is the absorption or emission of one photon of the jth mode due to the molecule-field interaction.

A few remarks are necessary, concerning the general properties of the density matrix. Keeping in mind Eq. (17) and that the density matrix is Hermitian ($\rho_{\alpha\beta} = \rho^*_{\beta\alpha}$) we obtain the following symmetry relations:

$$(21) \quad r^n_{\alpha\beta}(\omega, \mathbf{k}) = \left(r^{-n}_{\beta\alpha}(\omega, \mathbf{k})\right)^*, \quad r^n_{\alpha\beta}(\omega, \mathbf{k}) = \left(r^n_{\alpha\beta}(-\omega, -\mathbf{k})\right)^*$$

The first equation indicates that the population is real

$$(22) \quad r^0_{\alpha\alpha} = r^{0*}_{\alpha\alpha}$$

The particle conservation law $\left(\sum_\alpha \rho_{\alpha\alpha} = N, \partial \sum_\alpha \rho_{\alpha\alpha}/\partial t = 0\right)$ and Eq. (17) results in

$$(23) \quad \sum_\alpha r^0_{\alpha\alpha} = \mathcal{N}$$

$$\sum_\alpha r^n_{\alpha\alpha} = 0, \quad \text{except} \quad n = 0, \quad \text{or} \quad n \circ (\mathbf{k} \cdot \mathbf{v} - \omega) = 0$$

The upper equation indicates that only $r^0_{\alpha\alpha}$ has the meaning of a population of the αth level. Meanwhile $r^n_{\alpha\alpha}(n \neq 0)$ describes fast oscillations of population, $\exp(-\imath n \circ \omega t)$, near the population $r^0_{\alpha\alpha}$ (see Eq. 17).

Thus, Eq. (19) together with the wave equations (10) solve the problem of propagation of a strong multi-mode field through a non-linear many-level medium without restrictions on the mode intensities.

2.4 Relaxation Hierarchy; Role of Collisions; Scattering Duration Time

Let us imagine that because of some perturbation a molecule is excited from its ground state to an upper level. After a while this molecule will decay to the ground state with the released energy converted to a photon. A recorded emission spectrum will show that the emission line is broadened. Apparently, a longer lifetime would give a narrower line according to Heisenberg's uncertainty principle, $\Delta E \Delta t \sim \hbar$. The Weisskopf-Wigner theory states that for isolated molecules the spectral line broadening for transitions between electronic states β and α, with finite decay rates Γ_β and Γ_α, is equal to half the sum of these decay rates [14]. This approach, which neglects interaction between a solute (molecular chromophore) and solvent, is justified in the optical region only for rarefied gases. The situation changes drastically for molecules in solutions, where the concentration of buffer (solvent) particles is much higher than in the gas phase and where the collisions broaden the spectral lines. Homogeneous line broadening is essentiallly a dynamical process caused by collisions between solute and solvent, which are sufficiently fast to affect different chromophores in the same way. The spectral shape of a solute in a solvent experiences also static or inhomogeneous broadening due to the fact that the resonant frequency of solute molecule is different in different local environments [15]. The question which broadening mechanism is the major one is far from clear (compare [16] and [17]). In our study we include the broadening phenomenologically and due to this the effective broadening of the spectral line includes both mechanisms of the broadening. For simplicity we will refer below to the broadenings caused by dynamical or static interaction with solvent as to collisional broadening.

We study here organic molecules that strongly change the permanent dipole moment under electronic excitation. Due to this, the interaction potential $U_J(R)$

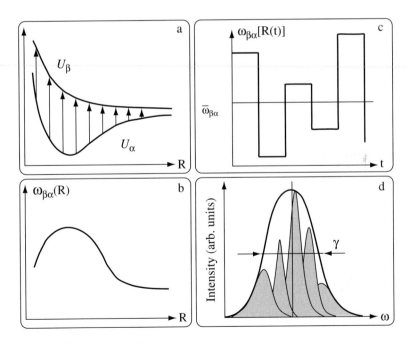

Figure 1. Illustration of collisional broadening

between chromophore and solvent molecule is different for different electronic states (Fig. 1a). This means that the resonant frequency changes during collisions, $\omega_{\beta\alpha}(R) = U_\beta(R) - U_\alpha(R) \neq const$ (Fig. 1a and b). Let us consider first the case of a rarefied gas. During the mean time between collisions the molecules do not interact and the transition frequency is constant $\omega_{\beta\alpha}(\infty)$. When the molecules approach the van der Waals or Weisskopf radius they interact during a short time (~ 10 fs) and change the frequency. Such a sharp random change of the resonant frequency is depicted in Fig. 1c. The mean time between collisions decreases with increase of molecular concentration. This makes the mean time between collisions to be comparable with the duration of collision. Due to this circumstance, the modulation of $\omega_{\beta\alpha}(R[t])$ in the time domain is not sharp. One can see directly from Fig. 1c and d the origin of the collisional broadening γ. The rate γ is the amplitude of deviation $\omega_{\beta\alpha}(R[t])$ from the expectation value $\bar{\omega}_{ij}$: $\gamma = \sqrt{(\omega_{\beta\alpha}(R[t])) - \bar{\omega}_{ij})^2}$. The asymmetry of R-dependence of $\omega_{\beta\alpha}(R)$ results in blue or red shift of spectral line.

The collisional broadening can be expressed by a simple formula [23]

$$(24) \quad \gamma_{\beta\alpha} = N\bar{v}\sigma_{\beta\alpha}$$

where $\sigma_{\beta\alpha}$ is the cross section of dephasing collisions, N is the concentration of buffer particles and $\bar{v}$ is the thermal velocity. The total spectral line broadening now reads:

$$(25) \quad \Gamma_{\alpha\beta} = \frac{\Gamma_\alpha + \Gamma_\beta}{2} + \gamma_{\alpha\beta}$$

Motivated by numerous experiments, we study the interaction of light with a solution. In solutions, the dephasing rate, $\gamma_{\alpha\beta}$, makes the rate of relaxation for the polarization large compared to the decay rate of the population: $\Gamma_{\alpha\beta} \gg \Gamma_{\alpha\alpha}$. We will consider the quite common case of a pump pulse with a duration, τ, longer than the time of decay of polarization and $\omega_{\alpha\beta}^{-1}$

$$(26) \quad \tau \gg \Gamma_{\alpha\beta}^{-1}, \frac{1}{\omega}, \frac{1}{|\omega_{\alpha\beta}|}, \frac{1}{\sqrt{\omega_{\alpha\beta}^2 + \Gamma_{\alpha\beta}^2}}$$

This allows to neglect the time derivatives in all equations, except for $r_{\alpha\alpha}$ in Eq. (19).

We have to stress here that there are many works devoted to the experimental measurements of γ available from the literature. A commonly used electronic dephasing rate for large conjugated organic molecules in solution is $\gamma \approx 0.1$ eV [17] near the resonance. Photon-echo measurements give $\gamma \approx 0.01$ eV for the same compounds [18, 19], while according to another experimental technique [20, 21] $\gamma \approx 0.02$–0.06 eV. Probably, the photon-echo measurements give more accurate value of γ. Apparently the dephasing rate is sensitive to local environment. All in all, we could not find the dephasing rates for molecules of our interest and, due to this we performed simulations for different values of γ (0.01–0.1 eV).

Furthermore, as it is seen from numerous experiments, the far red wing of the absorption profile of chromophores in solutions does not appear to follow a Lorentzian decay, as implemented in the conventional expression for the frequency dependent linear absorption cross section, but rather some kind of fast exponential, Urbach-like, decay [22, 23].

To model this non-Lorentzian decay we introduce a homogeneous broadening which depends on the wave length of the exciting light, or, equivalently, on the detuning from resonance:

(27) $\quad \Gamma_{\alpha\beta}(\Omega) = \Gamma_{\alpha\beta}^{(0)} + \gamma(\Omega)$

where we extracted the natural broadening (spontaneous decay)

(28) $\quad \Gamma_{\alpha\beta}^{(0)} = \frac{1}{2}(\Gamma_{\alpha\alpha} + \Gamma_{\beta\beta})$

which does not depend on the detuning from resonance $\Omega = \omega - \omega_{\alpha\beta}$ and collisional broadening which depends on the detuning:

(29) $\quad \gamma(\Omega) = \gamma(0) e^{-(\lambda - \lambda_{\alpha\beta})/a}, \quad \lambda > \lambda_{\alpha\beta}$

where a has dimensionality of the length and can be treated as fitting parameter. Here,

(30) $\quad \lambda = \frac{2\pi}{k} = \frac{2\pi c}{n\omega}, \quad \lambda_{\alpha\beta} = \frac{2\pi c}{n\omega_{\alpha\beta}}, \quad \lambda - \lambda_{\alpha\beta} = 2\pi c \left(\frac{\omega_{\alpha\beta} - \omega}{n\omega\omega_{\alpha\beta}} \right)$

where n is the refraction index. Let us note that Eq. (29) is only valid for $\lambda > \lambda_{\alpha\beta}$, the Ω-dependence for $\lambda < \lambda_{\alpha\beta}$ is weaker. One can see from Eqs. (27) and (29) that $\Gamma_{\alpha\beta}$ decreases in the red wing:

(31) $\quad \Gamma_{\alpha\beta}(\Omega) \to \Gamma_{\alpha\beta}^{(0)}, \quad |\Omega| \gg \gamma(0)$

which can be easily understood in terms of scattering duration time. The process of photoabsorption from the ground state is necessarily followed by the decay to the same ground state, either radiative or non-radiative. In case of pure radiative decay, the entire process can be considered as elastic photon scattering. The scattering duration time is inversely proportional to the photon detuning from the molecular resonance, $\tau_{scat} \sim 1/\Omega$ [24]. When the detuning is large, the scattering time can be shorter than the time of collisions, and hence, the natural broadening dominates.

Taking into account that the linear photoabsorption cross section reads

(32) $\quad \sigma(\Omega) = \frac{\zeta \Gamma_{ij}(\Omega)}{\Omega^2 + \Gamma_{ij}^2(\Omega)}$

where $\zeta = \sigma(0)\Gamma_{ij}(0)$ is expressed through the resonant values of the cross section and the dephasing rate.

We can extract the parameter a from experimental measurements. For example, for Rhodamine B in ethanol the parameter a is found [25] equal to 80 nm, assuming exponential decay from the profile peak maximum at 542 nm to the far red wing at 600 nm and $\gamma(0)$ set to 0.1 eV.

3. SAMPLE APPLICATIONS: AMPLIFIED SPONTANEOUS EMISSION

Three-photon active (3PA) materials have been studied extensively over the last few years owing to their potential applications in the fields of telecommunications and biophotonics [26–28, 30]. Two major advantages of these materials—longer excitation wavelengths and much better spatial confinement—make them attractive in comparison with two-photon absorption (2PA) based materials [29]. One of the most important applications of 3PA materials is three-photon pumped frequency-upconversion cavity-less lasing [26, 30]. Short infra-red (IR) pulses induce the ASE process via 3-photon absorption followed by fast non-radiative decay to a long-lived state which collects population. Conventional experiments with a pulsed longitudinal pump [27, 28, 30] show that stimulated emission occurs in both forward and backward directions with respect to the pump pulse.

Our dynamic theory of non-linear propagation of a few interacting intense light pulses has been successfully applied to study this frequency-upconversion cavity-less lasing in a chromophore solution [11, 13, 30], namely in an organic stilbenechromophore 4-[N-(2-hydroxyethyl)-N-(methyl)amino phenyl]-4'-(6-hydroxyhexyl sulphonyl) (abbreviated as APSS) dissolved in dimethyl sulphoxide (DMSO). The influence of the solvent was modeled by the collisional dephasing rate which is assumed to be the same, 0.01 eV, for all transitions. For details of the electronic structure calculations we refer the reader to [11]. The experimental data, such as concentration of molecules in the solvent, the length of the active medium (cuvette length), input laser intensity and pulse duration, are taken from [30].

3.1 Theory

3.1.1 Formulation of the problem

We explore the propagation along the z−axis of a strong pump, $\mathcal{E}\exp(-\iota\omega t+\iota kz)$, and an ASE field through a non-linear many-level medium. The total electromagnetic field, $\mathbf{E}=\mathbf{E}_{pump}+\mathbf{E}_{ASE}$, consists of pump and probe (ASE) contributions. The ASE field, $\mathbf{E}_{ASE}$, consists of two components

$$(33) \quad \mathbf{E}_{ASE} = \frac{\mathcal{E}^R}{2}e^{-\iota\omega_R t+\iota k_R z-\iota\varphi_R} + \frac{\mathcal{E}^L}{2}e^{-\iota\omega_L t-\iota k_L z-\iota\varphi_L} + c.c.$$

propagating, respectively, to the right and to the left (R-component and L-component below) along the z−axis which is parallel to the propagation direction of the pump field. The R-component propagates in the same direction as the pump field. The strengths, $\mathcal{E}^\mu$, the frequencies, ω_μ, and the wave vectors, $\mathbf{k}_\mu$, of the ASE fields are indexed by $\mu = R, L$. Three-photon absorption of the pump field with the forthcoming non-radiative decay to a long-lived state results in a population inversion leading to the ASE (see Fig. 2). In turn, the ASE field affects the population distribution when it becomes strong. This effect makes the propagation of the ASE field non-linear.

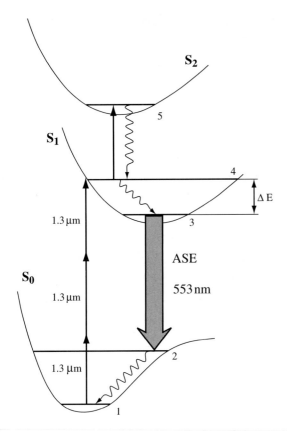

Figure 2. Energy level diagram

3.1.2 Induced polarization

The polarization oscillating with the frequencies of the pump and the ASE fields has the following structure:

$$(34) \quad \mathbf{P} = \mathrm{Tr}(\mathbf{d}\rho) = \boldsymbol{\mathcal{P}} e^{-\imath \omega t + \imath k z} + \boldsymbol{\mathcal{P}}^R e^{-\imath \omega_R t + \imath k_R z - \imath \varphi_R} + \boldsymbol{\mathcal{P}}^L e^{-\imath \omega_L t - \imath k_L z - \imath \varphi_L} + c.c.$$

where the trace is taken over the energy levels. According to (33) we select polarizations $\boldsymbol{\mathcal{P}}$ and $\boldsymbol{\mathcal{P}}^\mu$ which oscillate with the frequencies of the pump and ASE fields, respectively. To know the polarization we need the transition dipole moments, $\mathbf{d}_{\beta\alpha}(t) = \mathbf{d}_{\beta\alpha} \exp(\imath \omega_{\alpha\beta} t)$, and the density matrix, $\rho_{\alpha\beta}$, of the medium.

3.1.3 Paraxial Maxwell's equation

The substitution of the total field $\mathbf{E}$ and the polarization $\mathbf{P}$ in Maxwell's equations and a selection of contributions with the frequencies ω, ω_R and ω_L, and momenta $\mathbf{k}$, $\mathbf{k}_R$ and $-\mathbf{k}_L$ results in the following paraxial wave equations for the amplitudes of the pump field and the amplitudes of the R- and L-component of the ASE field (SVEA is applied, see Section 2.2):

$$(35) \quad \left(\frac{\partial}{\partial z} + \frac{1}{c}\frac{\partial}{\partial t} - \frac{\imath}{2k}\Delta_\perp\right)\mathcal{E} = \frac{\imath k}{\varepsilon_0}\mathcal{P}$$

$$\left(\frac{\partial}{\partial z} + \frac{1}{c}\frac{\partial}{\partial t} - \frac{\imath}{2k_R}\Delta_\perp\right)\mathcal{E}^R = \frac{\imath k_R}{\varepsilon_0}\mathcal{P}^R$$

$$\left(-\frac{\partial}{\partial z} + \frac{1}{c}\frac{\partial}{\partial t} - \frac{\imath}{2k_L}\Delta_\perp\right)\mathcal{E}^L = \frac{\imath k_L}{\varepsilon_0}\mathcal{P}^L$$

where the SI system of units is used. The phase, φ_μ, of the ASE fields is assumed to be constant.

3.1.4 Density matrix

Let us assume that ASE field is weak and neglect the Doppler effect which is small in liquids. This allows to seek a solution of the density matrix 19 as the sum of contributions of strong pump and weak ASE fields

$$(36) \quad \rho_{\alpha\beta} = e^{\imath\omega_{\alpha\beta}t}\left[\sum_n r_{\alpha\beta}^{(n)} e^{-\imath(n\omega t + \varphi)} + \eta_{\alpha\beta} e^{\imath(\omega_\mu t - \varphi_\mu)} + \eta^*_{\beta\alpha} e^{-\imath(\omega_\mu t - \varphi_\mu)}\right]$$

Making use of Eq. (19) we derived the density matrix related to the ASE field [11, 12]:

$$(37) \quad \eta^\mu_{\alpha\beta} \approx \frac{\imath\sum_\gamma \left(r_{\alpha\gamma} G^\mu_{\gamma\beta} - G^\mu_{\alpha\gamma} r_{\gamma\beta}\right)}{\Gamma_{\alpha\beta} - \imath(\omega_\mu - \omega_{\alpha\beta})}, \quad \mu = L, R$$

where $G^\mu_{\gamma\beta} = \mathcal{E}^\mu \cdot \mathbf{d}_{\alpha\beta}/2\hbar$ are the Rabi frequencies of the ASE field and $r_{\alpha\gamma}$ are the elements of the density matrix induced by the pump field (see [11, 12] for details). Apparently, the populations $r_{\alpha\alpha}$, created by the pump field contribute mainly to the density matrix (37). In our case, the main mechanism initiating redistribution of populations is three-photon absorption. To ensure this in the modeling we applied special resonant conditions: $\omega = \omega_{10}/3$, where ω is the frequency of the pumping field and ω_{10} is frequency of the vertical transition from the ground to the first excited state of the molecule. The off-diagonal elements $r_{\alpha\beta}$ are very small because of their off-resonant character, though they are included explicitly. We neglected the time derivative in Eq. (37) because of the large value of the dephasing rate in solutions (see [9–13]).

In order to relate the wave equations (34) with the solution of the density matrix equations (19), we use the following definitions:

$$(38) \quad \mathcal{P} = \sum_{\beta\alpha}\mathbf{d}_{\beta\alpha} r^{(1)}_{\alpha\beta}, \quad \mathcal{P}^\mu = \sum_{\beta\alpha}\mathbf{d}_{\beta\alpha} \eta^\mu_{\alpha\beta}$$

3.1.5 Final wave equations for the intensities

Substitution of the polarizations $\mathcal{P}$ and $\mathcal{P}^\mu$ in the wave equations (34) yields finally the following paraxial equations for the pump and ASE fields:

(39)
$$\left(\frac{\partial}{\partial z}+\frac{1}{c}\frac{\partial}{\partial t}-\frac{\imath}{2k}\Delta_\perp\right)\mathcal{E}=\frac{\imath k}{\varepsilon_0}\sum_{\beta\alpha}\mathbf{d}_{\beta\alpha}r^{(1)}_{\alpha\beta}$$

$$\left(\frac{\partial}{\partial z}+\frac{1}{c}\frac{\partial}{\partial t}-\frac{\imath}{2k_R}\Delta_\perp\right)\mathcal{E}^R=\frac{\imath k_R}{\varepsilon_0}\sum_{\beta\alpha}\mathbf{d}_{\beta\alpha}\eta^R_{\alpha\beta}$$

$$\left(-\frac{\partial}{\partial z}+\frac{1}{c}\frac{\partial}{\partial t}-\frac{\imath}{2k_L}\Delta_\perp\right)\mathcal{E}^L=\frac{\imath k_L}{\varepsilon_0}\sum_{\beta\alpha}\mathbf{d}_{\beta\alpha}\eta^L_{\alpha\beta}$$

Often the transverse inhomogeneity of the fields is small. In such cases, one can directly write down the equations for the intensities $I=c\varepsilon_0|\mathcal{E}|^2/2$ and $I^\mu=c\varepsilon_0|\mathcal{E}^\mu|^2/2$:

(40)
$$\left(\frac{\partial}{\partial z}+\frac{1}{c}\frac{\partial}{\partial t}\right)I=2\hbar\omega\,\Im m\sum_{\beta\alpha}\tilde{G}_{\beta\alpha}r^{(1)}_{\alpha\beta}$$

$$\left(\frac{\partial}{\partial z}+\frac{1}{c}\frac{\partial}{\partial t}\right)I^R=g^R I^R$$

$$\left(-\frac{\partial}{\partial z}+\frac{1}{c}\frac{\partial}{\partial t}\right)I^L=g^L I^L$$

where we introduce the gain of ASE fields

(41) $\quad g^\mu=\dfrac{2\omega_\mu}{c\varepsilon_0}\Im m\sum_{\beta\alpha}\mathbf{d}_{\beta\alpha}\tilde{\eta}^\mu_{\alpha\beta},\quad \tilde{\eta}^\mu_{\alpha\beta}=\eta^\mu_{\alpha\beta}/\mathcal{E}^\mu$

When the photons are linearly polarized, the Rabi frequencies $\tilde{G}_{\alpha\beta}=(\boldsymbol{\mathcal{E}}^*\cdot\mathbf{d}_{\alpha\beta})/2\hbar$ coincide with the above defined Rabi frequencies. We assume real transition dipole moments and amplitudes of the electromagnetic fields and use the following expressions for the field amplitudes, $\mathcal{E}=\sqrt{2I/c\epsilon_0}$, $\mathcal{E}^\mu=\sqrt{2I^\mu/c\epsilon_0}$. Here $\boldsymbol{\mathcal{E}}=\mathbf{e}\mathcal{E}$, $\boldsymbol{\mathcal{E}}^\mu=\mathbf{e}^\mu\mathcal{E}^\mu$; $\mathbf{e}$ and $\mathbf{e}^\mu$ are the polarization vectors of strong pump and ASE fields.

It is worthwhile to stress that the L- and R-components of the ASE field have different frequencies in general [27]. The experimental data [30] show that $\omega_R=\omega_L$ for the studied APSS molecule. Using this fact in our simulations, we find that $\tilde{\eta}^R_{\alpha\beta}=\tilde{\eta}^L_{\alpha\beta}$, and, hence $g^R=g^L=g$.

3.1.6 Averaging over orientations

As the molecules in solution have random orientations, we have to average the right-hand side of the field equations over all molecular orientations: $\mathcal{P}^\mu\to\langle\mathcal{P}^\mu\rangle$ or $\eta^\mu_{\alpha\beta}\to\langle\eta^\mu_{\alpha\beta}\rangle$. We perform the orientational averaging over the angle $0\le\varphi\le 2\pi$ between the molecular axis and the polarization vector, $\parallel x$, numerically [11]. We note that our averaging procedure ignores rotations of molecules, consistent with the fact that they are very slow in comparison to the inverse broadening of the spectral transitions.

3.1.7 Influence of the ASE field on populations

The strict equations for populations take into account the change of populations by the ASE field:

$$(42) \quad \left(\frac{\partial}{\partial t} + \Gamma_{\alpha\alpha}\right) r_{\alpha\alpha} = \sum_{\beta > \alpha} \Gamma_\beta^\alpha r_{\beta\beta} + W_\alpha + W_\alpha^R + W_\alpha^L$$

The pump and ASE fields change the population of the αth level with probabilities per unit time

$$(43) \quad W_\alpha = 2\Im m \sum_\beta \left[G_{\alpha\beta} r_{\beta\alpha}^{(1,0)} - r_{\alpha\beta}^{(1,0)} G_{\beta\alpha} \right]$$

$$W_\alpha^\mu = 2\Im m \sum_\beta \left[G_{\alpha\beta}^\mu \eta_{\beta\alpha}^\mu - \eta_{\alpha\beta}^\mu G_{\beta\alpha}^\mu \right]$$

respectively. The pump and ASE fields are assumed linearly polarized with the same direction of polarization vector. Equation (42) is strict: The only approximation we used is the density matrix, $\eta_{\alpha\beta}^\mu$ (37), induced by the ASE field.

When the system is long, the ASE field becomes high enough to change the populations, in accordance with Eq. (42). It leads to a decrease of the population inversion, $r_{33} - r_{22}$ and, hence, to a cease of the amplification.

3.2 Numerical Simulations

3.2.1 Organization of the code

Let us describe the scheme of the solution of the density matrix and field equations. The code consists of three main blocks. The pump field block appears first in the code. It is split up in two main parts: the first one solves the algebraic density matrix equations for all the components of the Fourier series expansion (this is the most time consuming step because it involves matrices of large dimension), the second one deals with the wave equation at a given time-point through the whole active medium with the parameters obtained from the first part. We used the following initial and boundary conditions for the pump intensity $I(t, z)$: $I(0, z) = I_0 \exp(-(z/c\tau + 4)^2)$ and $I(t, z) = I_0 \exp(-(t/\tau - 4)^2)$, which are consistent with the initial pulse shape, $I(t, z) = I_0 \exp[-(t - z/c - t_0)^2/\tau^2]$.

The second block is the solution of the wave equation for ASE intensity developing out of uniform spontaneous noise (see Section 3.2.3). The final, third, block of the code gives the solution to differential equations for populations with pump and ASE intensities evaluated in the previous two blocks. All three blocks are looped with respect to time while the space dependence is vectorized. The orientational

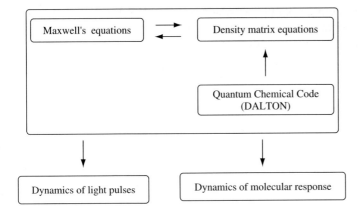

Figure 3. The flow-diagram of our simulations

averaging is performed according to the procedure described in Section 3.1.6. The code is written in such a way that it accepts any number of levels, specified by the matrices of transition and permanent dipole moments, transition frequencies and relaxation constants. Another important parameter related to numerical simulations is the number of Fourier components in the expansion (17). Our tests, performed before the main calculations, have shown that for n = 4 the numerical solution to the algebraical density matrix problem is quite stable, i.e. for any input values of the pump field intensity and of the populations the values of the off-diagonal elements contributing to the polarization ($r^{(1)}_{\alpha\beta}$ and $\eta_{\alpha\beta}$) do not change if we increase this number. The energy conservation law has been checked numerically at all instances. The flow-diagram of our simulations is shown in Fig. 3.

3.2.2 Five-state model

An analysis of the calculated transition dipole moments of the APSS molecule shows that only transitions between the singlet S_0, S_1 and S_2 states dominate in the non-linear optical process of interest (see Fig. 2) and we therefore restricted our simulations to comprise these three states. To explain the observed delay between the ASE and pump pulses, we need to take into account the vibrational structure. We model this structure for the levels S_1 and S_0 by including two pairs of vibrational levels, (4, 3) and (2, 1), respectively, making use of the one-mode approximation. The vibrational levels 3 and 1 are the lowest vibrational states for the electronic states S_1 and S_0, respectively (Fig. 2). The vibrational state 4 models the group of vibrational states nearby the vertical photoabsorption transition $S_0 \rightarrow S_1$. Apparently, the vertical transition has larger probability. The lasing transition takes place from the vibrational level 3 to the vibrational level 2 which corresponds to the vertical decay $S_1 \rightarrow S_0$ (Fig. 2). The transition matrix element is a product of the electronic matrix element and the Franck-Condon amplitude between vibrational states ν and ν':

$$d_{S_i\nu,S_j\nu'} = d_{S_i,S_j}\langle\nu|\nu'\rangle \qquad (44)$$

Our model gives the following picture which is in agreement with the discussed experiment [30]: the lasing level 3 is populated mainly due to two distinct channels – $1 \to 3$ and $1 \to 4 \to 3$. The low intensive adiabatic transition $1 \to 3$ populates instantaneously the lasing level 3. Therefore, this channel results in an ASE pulse without delay relative to the pump pulse. However, a delay time between the ASE and pump pulses takes place for the strong vertical channel $1 \to 4 \to 3$. Indeed, the instantaneously populated level 4 is quenched non-radiatively into the lasing level 3. An inversion, $r_{33} - r_{22}$, between the lasing levels 3 and 2 is created during this non-radiative decay.

Unfortunately, we lack information on FC factors for the $S_0 \to S_1$ and $S_1 \to S_2$ excitations, and we had therefore to introduce phenomenological FC amplitudes, aiming to get a reasonable fit with the experimental data on the three-photon absorption coefficient and the energy conversion coefficient: $\langle 1|4 \rangle = \langle 3|2 \rangle = 0.99994$, $\langle 4|2 \rangle = \langle 1|3 \rangle = \sqrt{1 - \langle 1|4 \rangle^2} = 0.0100$. These FC amplitudes satisfy the completeness condition for the two-level approximation: $\langle 3|\nu \rangle^2 + \langle 4|\nu \rangle^2 = 1$, $\langle 2|\nu \rangle^2 + \langle 1|\nu \rangle^2 = 1$. All FC amplitudes involving upper level 5 are assumed to be the same: $\langle 5|\nu \rangle = 1\sqrt{2}$.

3.2.3 Spontaneous noise and initial conditions

The ASE process is initiated by the spontaneously generated noise photons (SPs) with intensity

$$(45) \quad I_{sp} = \hbar \omega \frac{\Delta \omega}{\lambda^2} \frac{\Omega}{8\pi}$$

Due to amplification, only SPs from the spectral interval $\Delta \omega \approx \Gamma/\sqrt{gL}$ participate in the lasing. The solid angle $\Omega = \Delta S/L^2$ in which SPs are emitted can be estimated through the waist radius, $w_0 \approx 30\,\mu m$, of the focal spot of the pump pulse: $\Delta S \approx \pi w_0^2$. This gives $I_{sp} \approx 10^{-4}\,W/cm^2$. In the simulations, we used the following initial and boundary conditions for the ASE intensity $I(t,z): I_\mu(0,z) = I_\mu(t,0) = I_{sp} \approx 10^{-4}\,W/cm^2$.

As is well known [31], the ASE intensity stops to be sensitive to I_{sp} when the active medium is long and that the ASE intensity exceeds the saturation intensity, I_{sat}. This is easy to see from the equation $d\kappa_\mu/dz = g\kappa_\mu/(1+\kappa_\mu)$ for the saturation parameter $\kappa_\mu(z) = I_\mu(z)/I_{sat}$. The saturation parameter is related to the resonant saturation parameter as $\kappa_\mu = \kappa_\mu^r \Gamma^2/[(\omega_\mu - \omega_{32})^2 + \Gamma^2]$. The intensity grows exponentially, $\kappa_\mu(z) = \kappa_\mu(0)\exp(gz)$, until the point z_0, where $\kappa_\mu = 1$. The ASE intensity looses memory about the SP noise at $z > z_0$: $\kappa_\mu(z) \approx 1 + g(z - z_0)$.

3.3 Pump Pulse Propagation

The profile of the pump pulse shown on Fig. 4 demonstrates the decrease of the intensity I at the expense of non-linear photoabsorption during propagation. One can see that the pump pulse moves without essential change of the shape.

Non-Linear Pulse Propagation in Many-Photon Active Isotropic Media 229

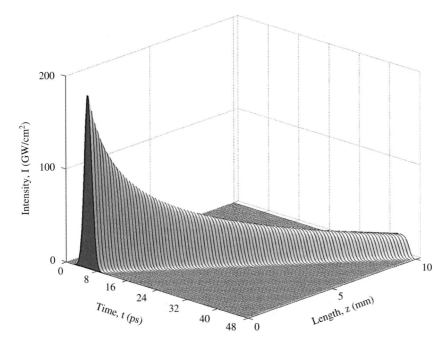

Figure 4. Propagation of the pump pulse through a non-linear medium of APSS molecules in solution. Initial intensity is $190\,\text{GW/cm}^2$, the pulse duration (FWHM) is 2 ps, the wavelength is $1.3\,\mu\text{m}$, the sample length is 10 mm. Input data are collected in Table 1a

3.3.1 Effective three-photon absorption cross section

In the quoted experimental paper [30] the effective three-photon absorption cross section $\sigma^{(3)}$ is obtained making use of the formula

$$(46) \quad \gamma = \frac{1}{2L} \frac{I_0^2/I^2 - 1}{I_0^2}, \quad \sigma^{(3)} = \frac{\gamma}{N},$$

where L is the length of the non-linear medium, I_0 and I are the peak intensities of the pump pulse before and after the absorber, respectively. This formula is obtained from the equation for the light intensity $dI(z)/dz = -\gamma I^3(z)$, assuming that $\sigma^{(3)}$ does not depend on the intensity of the pump pulse. This equation assumes that the three-photon absorption is the major process. This assumption ceases to be valid when the one-photon off-resonant absorption becomes important (see discussion in Section 3.3.3). Clearly, the cross section $\sigma^{(3)}$ defined by Eq. (46) does not coincide with the strict three-photon absorption cross section when the intensity of the pump pulse is large and, hence, $\sigma^{(3)}$ depends on the intensity.

We calculate γ making use of Eq. (46) with the purpose to compare the theoretical value with the measured one [30] (see Table 1. Here the headings a, b, c...f correspond to different sets of the initial simulation parameters. As one can see, the output parameters, such as γ and $\Delta\tau$, change as some of the input parameters vary.)

Table 1. Basic results of simulations, where γ is the 3PA coefficient, $\Delta \tau$ is the delay between the pump and ASE pulses, η is the efficiency of nonlinear conversion of the pump energy to the ASE energy. $I_0 = 190\,\text{GW/cm}^2$. (1ph—one-photon absorption dominates)

Input	Simulations						
	a	b	c	d	e	f	
Γ, eV	0.01	0.01	0.01	0.01	0.01	0.1	
Γ_{44}^{-1}, ps	30	100	10	30	30	30	
$\langle 1\|4 \rangle$	0.99994	0.99994	0.99994	0.8367	0.9487	0.99994	
$\langle 1\|3 \rangle$	0.01	0.01	0.01	0.5477	0.3162	0.01	
Output	Simulations						Exp. [30]
γ, $10^{-4}\,\text{cm}^3/\text{GW}^2$	8.8	8.8	8.8	6.3	8.2	1ph	0.5
$\Delta \tau$, ps	2.6	1.0	2.3	0	0.5	25	12
ASE FWHM, ps	22	70	9.9	3	12	40	40
η, %	4	3	4.6	2.5	3.1	0.01	2.1

It is worth noting that the effective cross section $\sigma^{(3)}$ depends in general also on the intensity of the ASE pulse, I_μ. However, the ASE pulse is delayed relative to the pump pulse (in the studied experimental situation) and, therefore, does not significantly influence the propagation of the pump pulse (see also Section 3.4.2). The experiment, as well as our simulations, show that the pump pulse runs ahead of the ASE pulse. It means that $\sigma^{(3)}$ is not noticeably influenced by the ASE pulse.

3.3.2 Ab initio computations of three-photon absorption cross sections

The microscopic origin of the three-photon absorption can be traced to the fourth order hyperpolarizability. Despite the high order of this property, it is still attainable due to two decisive steps of simplification. First, at resonance, i.e. under excitation with a frequency equal to one-third to the excitation energy of the final state, $\omega_f/3$, a fraction of the terms will dominate the summation completely, which leads to a formulation where the fourth order hyperpolarizability can be expressed as a product of third order transition dipole moments:

$$(47) \quad T_{abc} = \sum \mathcal{P}_{a,b,c} \sum_{n,m} \frac{\langle 0|\mu_a|m\rangle\langle m|\mu_b|n\rangle\langle n|\mu_c|f\rangle}{(\omega_m - 2\omega_f/3)(\omega_n - \omega_f/3)}$$

where $\sum \mathcal{P}_{a,b,c}$ denotes permutations with respect to the Cartesian indexes a, b, and c. This contains apparently an infinite summation, but can be identified as the first residue of the second order hyperpolarizability. In the response formalism it corresponds to the single residue of the cubic response function, as implemented in the DALTON package [32, 33].

Even though the response approach represents a significant simplification, it is still computationally demanding for three-photon absorption, and the third order transition moments are therefore obtained at the RPA level, i.e. cubic response

Table 2. Excitation energies of the final state, $\hbar\omega_e$, the third order transition moment, T_{zzz} and the three-photon probability, δ^L_{3P}, calculated at SCF level with the basis sets 6-31G and 6-31G*. Geometry is optimized by DFT/B3LYP with 6-31G*

6-31G			6-31G*		
$\hbar\omega_e$, eV	T_{zzz}, a.u.	δ^L_{3P}, a.u.	$\hbar\omega_e$, eV	T_{zzz}, a.u.	δ^L_{3P}, a.u.
4.18	31693.08	0.196×10^9	4.10	29380.22	0.168×10^9
5.46	6519.29	0.899×10^7	5.33	5734.72	0.328×10^7

theory applied to a singe-determinant SCF reference state [34]. These calculations are carried out for 6-31G and 6-31G* basis sets. The geometry of the APSS molecule was optimized with a B3LYP functional and a 6-31G* basis set using the GAUSSIAN package. The results of our calculations are collected in Table 2.

We recalculated the three-photon absorption coefficient, γ, making use of the formulas which relate the third order transition dipole moment, T_{abc}, three-photon transition probability, δ_{3p} and γ:

$$(48) \quad \gamma = \frac{N\pi\hbar^{-1}}{2c^2\varepsilon_0^3\lambda} \cdot \frac{\delta^L_{3p}}{\Gamma_{\alpha\beta}}, \quad \delta^L_{3P} = \frac{1}{35}(2\delta_F + 3\delta_G)$$

$$\delta_F = \sum_{ijk} T_{iij}T_{kkj}, \quad \delta_G = \sum_{ijk} T_{ijk}T_{ijk}$$

where λ is the wavelength of the incoming photon. The three-photon transition probability for linearly polarized light, δ^L_{3p}, is averaged over molecular orientations. To estimate the quantity (47), we used experimental data from [30]. The three-photon absorption coefficient is found to be equal to 1.06×10^{-29} m^3/W^2 or 0.11×10^{-4} cm^3/GW^2 for the 6-31G basis set and 0.91×10^{-29} m^3/W^2 or 0.091×10^{-4} cm^3/GW^2 for the 6-31G* basis set and for $\Gamma_{\alpha\beta} = 0.1$ eV. This latter theoretical value is 5 times lower than the one evaluated from the experiment. We did the same estimation also for $\Gamma_{\alpha\beta} = 0.01$ eV. The results are the following: $\gamma = 1.06 \times 10^{-4}$ cm^3/GW^2 for the 6-31G basis set and 0.91×10^{-4} cm^3/GW^2 for the 6-31G* basis set, which is remarkably close to the experimental value. However, at the SCF level the energies are usually overestimated which could give an underestimated value of the three-photon absorption coefficient.

3.3.3 One-photon versus three-photon absorption

We are faced with the important problem to elucidate the competition between resonant (here three-photon) absorption and off-resonant one-photon absorption for pumping of the excited state. From previous work on two-photon excitation [9, 10, 35] we learnt that the balance is most delicately dependent on factors like pulse lengths, excited state lifetimes and saturation. The transition dipole moment, $d_{S_1S_0} \approx -9.25$ D, between the ground and the first excited electronic states of the studied molecule is the largest one and oriented mainly along the molecular x axis.

Due to this, one can use the two-level approximation both for one- and three-photon absorption. The length of one-photon absorption is

$$(49) \quad l_{1p} = \frac{1}{N\sigma_{1p}}, \quad \sigma_{1p} = \frac{kd_{S_1 S_0}^2 \cos^2 \vartheta}{\hbar \varepsilon_0} \frac{\Gamma}{\Gamma_s^2 + (\omega - \omega_{41})^2}$$

$$\Gamma_s = \Gamma \sqrt{1 + \kappa_r \cos^2 \vartheta}$$

Here, ϑ is the angle between $\mathbf{e}$ and $\mathbf{d}_{S_1 S_0}$, and $\kappa_r = I/I_s$ is the resonant saturation parameter with the saturation intensity $I_s = \varepsilon_0 c \hbar^2 \Gamma T_{44}/2d_{S_1 S_0}^2$ ($I_s \approx 7.9 \times 10^5$ W/cm^2 for $\Gamma = 0.01$ eV). The length of the three-photon absorption reads:

$$(50) \quad l_{3p} = \frac{1}{\gamma I^2}, \quad \delta_{3p}^L = \frac{1}{7} \left(\frac{54 d_{S_1 S_0}^3}{\omega_{41}^2} \right)^2$$

The ratio

$$(51) \quad \frac{l_{3p}}{l_{1p}} \approx 1.23 \times 10^{-2} \left(\frac{190 \, \text{GW/cm}^2}{I} \right)^2 \left(\frac{\Gamma}{0.01 \, \text{eV}} \right)^2 \left(\frac{9.25 D}{d_{S_1 S_0}} \right)^4$$

indicates that the role of the one-photon pumping becomes important for large Γ. In this estimation we neglected the "one-photon" saturation parameter, κ_r, which is equal to $\kappa_r = 2.4 \times 10^5$ for $I = 190$ GW/cm^2, $\Gamma = 0.01$ eV. In this case, $\Gamma_s \approx \Gamma \sqrt{1 + \kappa_r/3} \approx 2.84$ eV becomes comparable with $|\omega - \omega_{41}| = 2\omega_{41}/3 = 2.31$ eV. This means that the saturation does not essentially change the estimation (51).

We see that the one- and three-photon pumping of the excited state compete with each other and this competition is very sensitive to the homogeneous broadening Γ. However, the results of experimental measurements of Γ are rather vague. A commonly used electronic dephasing rate for large conjugated organic molecules in solution is $\Gamma = 0.1$ eV [17]. The photon-echo measurements give $\Gamma \approx 0.01$ eV for the same compounds [18, 19], while according to another experimental technique [20, 21] $\Gamma \approx 0.02 - 0.06$ eV.

Let us now estimate l_{1p} and l_{3p} for $I = 190$ GW/cm^2, $N = 3.6 \times 10^{19}$ cm^{-3}, $\Gamma_{44} = 2.2 \times 10^{-5}$ eV, $d_{S_1 S_0} = -9.25$ D, and $\cos^2 \vartheta = 1/3$ neglecting the saturation. Eqs. (48) and (50) give $l_{1p} = 1.1$ cm and $l_{3p} = 0.014$ cm for $\Gamma = 0.01$ eV, and $l_{1p} = 0.11$ cm and $l_{3p} = 0.14$ cm for $\Gamma = 0.1$ eV. Thus, the three-photon pumping dominates when $\Gamma = 0.01$ eV, while when $\Gamma = 0.1$ the one-photon off-resonant population of the excited state becomes the major one.

Our conclusions based on estimation (51) need support from strict simulations, which, however, are not easily obtained because we cannot select one-photon or three-photon pumping using strict equations. This means that we can use only some fingerprints of one- or three-photon excitations. One of these fingerprints is the transmission of the pump field

$$(52) \quad T = \frac{J(L)}{J(0)}, \quad J(z) = \int I(t, z) dt$$

which is a ratio of integral intensities. The transmission shown on Fig. 5 demonstrates qualitatively different dependences on the pump intensity for $\Gamma = 0.01\,\text{eV}$ and $\Gamma = 0.1\,\text{eV}$. The decrease of T with increase of I for $\Gamma = 0.01\,\text{eV}$ agrees with the trend of the transmission for the three-photon absorption, $T = 1/\sqrt{1+2L\gamma J^2(0)}$ (see Eq. 46). When $\Gamma = 0.1\,\text{eV}$, the transmission increases with I. This is in agreement with the one-photon transmission: $T = \exp(-L/l_{1p})$. Indeed, according to Eq. (48) the length of the one-photon absorption increases for large intensities due to the saturation. So, one can conclude that when the intensity is quite large the three-photon absorption dominates for $\Gamma = 0.01\,\text{eV}$, while the one-photon pumping is more important when $\Gamma = 0.1\,\text{eV}$. Apparently, the one-photon absorption dominates for quite small intensities because of small three-photon absorption, γI^2.

Another fingerprint is the unrealistically large three-photon absorption, $\gamma \sim 1\,\text{cm}^3/\text{GW}^2$ for $\Gamma = 0.1\,\text{eV}$. This value was obtained for different sets of parameters making use of strict simulations and Eq. (46) based on the assumption that the three-photon absorption dominates. So the large magnitude of γ compared with the experimental value $0.5 \times 10^{-4}\,\text{cm}^3/\text{GW}^2$ indicates that the one-photon absorption channel is the dominating one for $\Gamma = 0.1\,\text{eV}$.

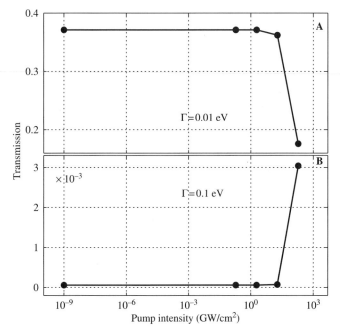

Figure 5. Transmission (52) versus input pump intensity. **A:** $\Gamma = 0.01\,\text{eV}$, parameters from Table 1a. **B:** $\Gamma = 0.1\,\text{eV}$, parameters from Table 1f.

3.3.4 Role of saturation on three-photon absorption

As shown in our previous works [9, 10] saturation effects are of great importance for long-pulse ($\tau > 100$ fs) multi-photon absorption processes. They lead to certain dependencies for the absorption cross section on the input intensity of the incoming light. A sufficiently accurate estimation of the saturation intensities can be obtained making use of the following expressions:

$$(53) \quad I_{41}^{(1)} = \frac{c\varepsilon_0 \hbar^2 \Gamma_{44}}{2|d_{41}|^2 \Gamma_{41}} \left[\frac{1}{(\omega - \omega_{41})^2} + \frac{1}{(\omega + \omega_{41})^2} \right]^{-1}$$

$$I_{41}^{(3)} = 2c\varepsilon_0 \hbar^2 \left[\frac{\Gamma_{44} \Gamma_{41}}{\hbar^4 T_{xxx}^2} \right]^{1/3}$$

where $I_{41}^{(1)}$ and $I_{41}^{(3)}$ are the saturation intensities for off-resonant, $1 \to 4$, one-photon and a resonant, $1 \to 4$, three-photon transitions, respectively, and T_{xxx} is the third order transition dipole moment (47). For the input intensities less than the three-photon saturation intensity a measured three-photon absorption cross section estimated according to Eq. (46) should not depend on the length of the active medium. For intensities higher than the one-photon saturation intensity, the measured three-photon absorption cross section drops down increasing the intensity. The reason is that the incoherent, step-wise, contribution to the total cross section is saturated and the coherent three-photon absorption process becomes dominant (the saturation intensity for the three-photon absorption cross section is much higher than for the one-photon process). If, due to some reasons, the step-wise processes are eliminated then the measured total three-photon absorption cross section will be fully represented by the pure coherent contribution. One reason for such an elimination would be a shortening of the pulse duration, i.e. when the contribution of step-wise processes is suppressed with the factor of $\Gamma_{ii} \tau$ [9].

To check whether the parameters of our system satisfy coherent three-photon "purity", we carried out two calculations with pump pulse duration FWHM = 2 ps. The first calculation based on input data from Table 1a corresponds to the suppression factors $\Gamma_{ii} \tau \sim 10^{-3} - 10^{-1}$ (depending on Γ_{ii}). In the second calculation the suppression factors were decreased by three orders of magnitude by decreasing all Γ_{ii} in 10^3 times. Both calculations gave almost the same three-photon absorption coefficient, γ (46) (the relative difference $\Delta\gamma/\gamma \approx 7\%$). This is a direct evidence of negligibly small step-wise absorption processes compared to the coherent three-photon absorption.

Applying the estimations (52) to our model we found the saturation intensities to be equal to 35 GW/cm^2 for off-resonant one-photon transitions $1 \to 4$, 0.035 MW/cm^2 for resonant one-photon transitions $3 \to 2$ and 169 GW/cm^2 for resonant three-photon transitions $1 \to 4$. For comparison, the saturation intensity for off-resonant one-photon transitions $4 \to 5$ is equal to 1.02×10^4 GW/cm^2. These numbers indicate that we face strong saturation effects for both one- and three-photon processes. Indeed, the calculated three-photon absorption coefficient was found to be equal to 15.3×10^{-4} cm^3/GW^2 for the initial intensity of 70 GW/cm^2

and to $8.8 \times 10^{-4}\,\text{cm}^3/\text{GW}^2$ for the initial intensity of $190\,\text{GW}/\text{cm}^2$. We should note here that for smaller intensities the contribution of one-photon absorption could be significant, and, therefore, the estimation according to Eq. (46) does not hold.

3.4 Propagation of the ASE Pulses

3.4.1 Inversion, gain and threshold

Clearly, lasing is possible when the pump intensity exceeds the ASE threshold: $I > I_{th}$. Let us estimate I_{th}. In the case of ASE, the gain, $g = N(\sigma_{amp} - \sigma_{abs}) \sim I - I_{th}$, is given by the difference between amplification and absorption cross sections and can be written for the studied system as

$$(54) \quad g \approx \frac{k_\mu}{\varepsilon_0 \hbar}\left[(r_{33} - r_{22})\frac{d_{32}^2}{\Gamma} - (r_{11} - r_{44})\frac{d_{41}^2 \Gamma}{\Omega_\mu^2 + \Gamma^2}\right]$$

Here $\Omega_\mu = \omega_\mu - \omega_{41}$, $\Gamma_{41} \approx \Gamma_{32} \approx \Gamma$. In Eq. (54) we used the four-level approximation $(1, 2, 3, 4)$, the resonant condition for a lasing photon, $\omega_\mu = \omega_{32}$ (Fig. 2), and took into account only the strongest photoabsorption channel $1 \to 4$. We also did not consider inhomogeneous broadening of molecular transitions caused by variations of the molecular environment. The solution of the equations for r_{33} and r_{44} and the threshold condition $g = 0$ give the following estimation of the threshold intensity:

$$(55) \quad I_{th} \approx \left[\frac{\hbar\omega\Gamma_{44}}{N\sigma^{(3)}} \cdot \frac{r_{33}}{\tau\Gamma_4^3}\right]^{1/3}$$

where $r_{33} \approx N/[1 + (1 + \Omega_\mu^2/\Gamma^2)d_{32}^2/d_{41}^2] \approx 7 \times 10^{-5} N$ for $d_{32}/d_{41} = \langle 3|2\rangle/\langle 4|1\rangle = 1$. Making use of this number and of the experimental value $\gamma = 0.5 \times 10^{-4}\,\text{cm}^3/\text{GW}^2$ [30], we find that the threshold intensity to be equal to $I_{th} \approx 20\,\text{GW}/\text{cm}^2$ which is one order of magnitude smaller than the incident pump intensity $I \approx 190\,\text{GW}/\text{cm}^2$ [30].

3.4.2 Delay of the ASE pulse relative to the pump pulse

The ASE dynamics is characterized by two important parameters, the delay between the maxima of the pump and ASE pulses, $\Delta\tau$, and the temporal width of the ASE pulse, Δ (full width at half maximum (FWHM)). The fulfillment of the ASE threshold condition, $g = 0$, takes some time, which implies that the delay time $\Delta\tau$ is non-zero in the general case. The first impression which arises is to relate $\Delta\tau$ to the lifetime of level 4, $1/\Gamma_{44}$, because the lasing level 3 is populated during this particular time. According to the experiment [30] the delay between pump and right-propagating ASE pulses is equal to 10–15 ps. This value is shorter than $1/\Gamma_{44} = 30$ ps. Our simulations show the same trend (see Table 1). From Table 1 we see that $\Delta\tau$ varies in the range 0–25 ps and is sensitive to Γ, $\Gamma_{\alpha\alpha}$, and to the FC factors. The conclusion is that the delay time is a rather complicated function of different parameters and that $\Delta\tau$ is not related directly to $1/\Gamma_{44}$ (see [11] for details).

3.4.3 Temporal width of the ASE pulse

Our simulations gives temporal width $\Delta \approx 22$ ps (Table 1a) for the right-propagating ASE pulse which is close to the experimental value of 40 ps. We see both from the experiment and from our simulations that the width of the ASE pulse is much shorter than the averaged fluorescence lifetime of $1/\Gamma_{33} = 720$ ps. Such a temporal narrowing or "gain-narrowing" is a characteristic of stimulated emission.

3.4.4 Formation of the ASE pulses

We will try to understand the formation of the right- and left-propagating ASE pulses following a scheme outlined in [13]. We assume an infinitely "narrow" source of photons, moving to the right along the z-axis with the speed of light. "Narrow" means that the lasing threshold is suddenly overcome at a point inside the medium at a certain time and then suppressed at the same time because the population inversion is eliminated by the emitted photons. This source is the gain. So, at each point of the medium a bunch of photons is emitted in equal numbers to the right and to the left (Fig. 6).

This number depends on the value of the gain. We suppose the emission is discrete, with time lapses Δt. The photons emitted to the right at time instant $t = 0$ will then move further with the gain. At time Δt a new portion of photons will be emitted. The gain will be at the point $c\Delta t$ as well as the photons emitted at the preceding moment. So, the R-component of the ASE would be infinitely narrow in this case and would be growing with negative gradient as the gain is being depleted. In contrast, the left-propagating ASE component would be broad. The reason is the following. At time $t = 0$ a bunch of photons is emitted to the left. At time Δt the gain coordinate is $c\Delta t$ and the previously emitted left-propagating photons are

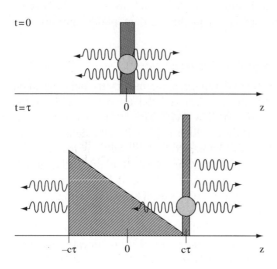

Figure 6. Illustration of narrow gain propagation—formation of the L- and R-components of the ASE

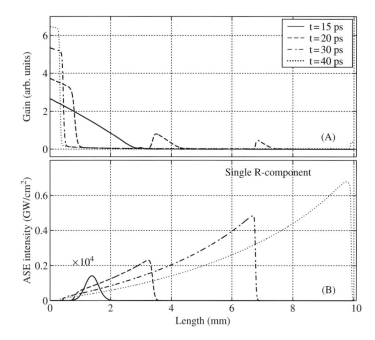

Figure 7. Snapshots of the (**A**) gain and (**B**) single R-component at different time instants

located at $-c\Delta t$. In a case of only two emission acts (for instance the gain is then negative) the temporal width of the left-propagating ASE component is $2\Delta t$. The pulse would look like a saw jag if the gain were decreasing. Let us stress once again that such an asymmetry of the right- and left-propagating components of the ASE owes to the right-propagating longitudinal pump.

Apparently, the picture of formation of the ASE pulses given above is a naive one. In reality, the gain is continuously decreasing through the medium if observed at a certain instant of time (see for example Figs. 7 and 8 at t = 15 ps). The reason for this is the population of the lasing level 3 through the exponential decay of level 4. Such character of the gain is maintained until the intensity of the ASE components is smaller than the saturation intensity of the lasing transition. When the ASE intensity becomes comparable with the saturation intensity, the ASE starts to change the populations and, hence, the gain (Figs. 7 and 8). The broad dips of the gain are formed by the leading edges of the R- and L-components. The leading edges with high intensity cut off the gain due to an abrupt decrease of the population inversion. Because of the continuous gain the R-component of the ASE becomes broad, with a steep front and a long, slowly decreasing tail. The forward pulse is growing chasing the leading edge of the decreasing gain. This leading edge of the gain looks like our model "narrow" source of photons—it supports the growth of the leading edge of the forward ASE pulse. The source amplitude decreases because

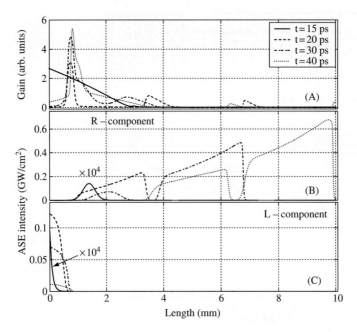

Figure 8. Snapshots of the (**A**) gain and (**B, C**) R- and L-components at different time instants

the pump pulse, creating the population inversion, is being absorbed when it runs through the medium.

As mentioned earlier, the formation of the L-component of the ASE takes place at a very short distance from the entrance of the cell, about 1 mm. As we can see in Fig. 8, the peaks of the gain supplying photons to R- and L-components are located at the same distance from the left edge of the cell at different time instants. The right-hand slope of the gain, spanning to the right, is "eaten away" by the forward ASE pulse because its intensity at this short distance of 1 mm reaches the saturation intensity of the lasing transition. It is necessary to note that the output energies of the R- and L-components of the ASE differ substantially due to the asymmetry with respect to the pump.

3.4.5 Efficiency and maximum of conversion

The net conversion coefficient, $\eta = \eta'/(1 - T)$, is a very important parameter defining how much of the absorbed energy of a pump pulse is converted to ASE radiation. Here, η' is the ratio of the energy of the overall output ASE (both forward and backward) to the energy of the incident pump pulse and T is the transmission of the pump pulse. To make comparison between the R- and L-components of the ASE we evaluated the partial coefficients, η'_μ and η_μ[13]. The overall coefficients can be found as a sum of the partial ones.

The results of our calculations of partial η_μ are visualized in Fig. 9. We used six different values of the concentration for comparison. As one can see, for each

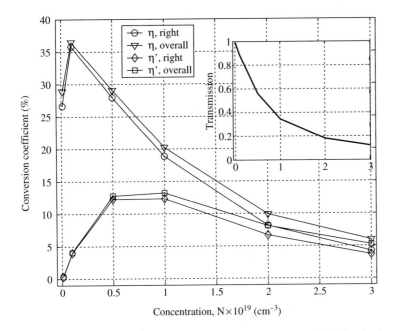

Figure 9. Conversion coefficient versus concentration. "Right" means simulations for the one-component case and "overall" means simulations for both forward and backward components

value of the concentration the net conversion coefficient, η_R, is larger in the case when the left-propagation is not accounted for. The cause behind this observation is that a part of the absorbed energy is consumed by the left-propagating ASE pulse in the opposite case.

Note that the difference between η_R for the single R-component case and the two-component case grows with decrease of the molecular concentration. The reason for this is found in the temporal oscillations of the ASE R-component at the end of the active medium when both R- and L-components are accounted for. A numerical check showed that the oscillations cannot be attributed to the accuracy of the simulations. The period and amplitude of the oscillations strongly depend on the concentration, as demonstrated in Fig. 10(A).

As the concentration decreases, the period of oscillations increases and the peaks become wider. A similar behaviour is observed for the L-component as demonstrated in Fig. 10(B). However, as we can see in the figure, the amplitude of oscillations is larger for the L-component which certainly reflects the asymmetry of the ASE propagation with respect to the pump.

Based on our observations we concluded that oscillations built up in a certain point of the medium are blurred as the ASE pulse propagates to the right with an increase of intensity. The closer to the left edge the pulse emerges, the longer distance it travels through. Apparently, for higher concentrations of the absorbing molecules the formation of the ASE pulse is faster in space and time according

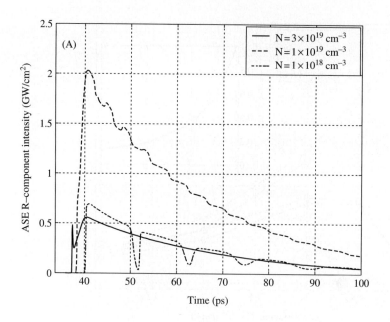

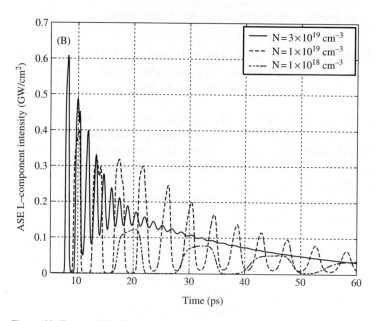

Figure 10. Temporal distributions: **A**—of R-component intensity at the end of the cell at different concentrations; **B**—of L-component intensity at the entrance to the cell at different concentrations

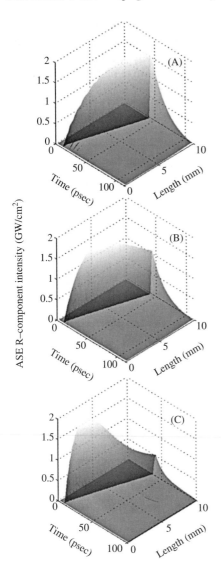

Figure 11. 3D dependencies of R-component intensity under different concentrations: **(A)** 1×10^{19} cm^{-3}, **(B)** 2×10^{19} cm^{-3}, **(C)** 3×10^{19} cm^{-3}

to Eq. (39). The degree of delay of the leading edge of the R- and L-components depending on concentration seen in Fig. 10 proves this statement.

Let us now pay attention to the fact that the efficiency of the conversion is a non-linear function of concentration of the active molecules (see Fig. 9). We know that the peak intensity of the ASE in an optically thin medium is determined by the interruption of the population inversion which occurs when the ASE intensity approaches some critical value. This value refers to the resonant saturation intensity of the lasing transition and is independent of the concentration. In our case, the saturation intensity of the $3 \to 2$ transition is approximately equal to $3\,\text{MW}/\text{cm}^2$. As the ASE pulse propagates, it gains intensity until some maximal value is reached. This maximal value is determined by the pump depletion, i.e. when the pump pulse intensity is less than the lasing threshold intensity the ASE ceases to grow and experiences reabsorption as it propagates through the medium. If the pump pulse is almost depleted near the entrance to the cell, which is the case for higher concentrations, the right-propagating ASE pulse starts to be reabsorbed far from the right edge of the cell. In the case of lower concentrations, the ASE pulse is formed further from the left edge of the cell, which means a shorter reabsorption path for the R-component. At given length of the gain medium (10 mm in our case), the output energy of the forward ASE pulse is maximal for a certain value of the concentration. It then drops down as the concentration decreases, because the ASE starts to be formed so close to the right edge of the cell that it does not have time to reach the maximum. Fig. 11 shows that the maximum level of the ASE R-component intensity is approximately the same for all values of the concentration. But for the lowest value this maximum is reached at the very end of the medium. Thus, a decrease of the concentration shifts the maximum of the intensity towards the end of the cell.

4. SPECTRAL PROFILES OF TWO-PHOTON ABSORPTION: COHERENT VERSUS TWO-STEP TWO-PHOTON ABSORPTION

One salient feature of two-photon spectra is that they in general are very different from their one-photon counterparts. Two-photon absorption of polyatomic molecules is strongly influenced by vibrational interaction, providing a general broadening of the spectra with fine structure. As for other types of electronic transitions this merely reflects the fact that the potential surface changes going from the ground to the final state of the optical excitation. In the conventional theory of photoexcitation, using the sudden approximation, the spectral profile is defined by the Franck-Condon factors between the ground and the excited state. This means that the two-photon profile has to copy the profile of the one-photon absorption as a consequence of the Franck-Condon principle. However, experiments, e.g. [36, 37], indicate a strong violation of this statement. Our explanation of this is based on two notions: First of all, considering the full spectrum, one-photon as well as coherent TPA absorption are formed by resonance transitions to a set of excited electronic states, each with its particular cross section for the two processes. Secondly, the conventional coherent TPA absorption (one-step absorption) is accompanied by a two-step absorption, which has the same order of magnitude as the one-step TPA and which in general both broadens and distorts the profile of the particular band. A third reason, which in general is less important and not operating for the example reviewed here, is that the spectral shapes of one- and two-photon

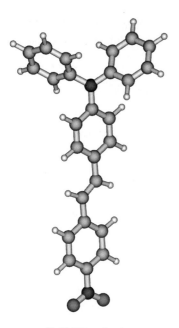

Figure 12. N-101 molecule

absorption can be different also due to vibronic coupling between the states. As illustration we review results of simulations [38] of the two-photon absorption of the the di-phenyl-amino-nitro-stilbene molecule, called N-101 (Fig. 12), experimentally studied recently in [36].

The effective TPA cross section $\sigma^{(2)} = \sigma_1^{(2)} + \sigma_2^{(2)}$ is the sum of one-step (coherent):

$$\sigma_1^{(2)} = \frac{k}{c\varepsilon_0^2 \hbar^3} \sum_i \langle S_{i0} \rangle \sum_{\nu_i} \frac{\Gamma_{i0}(\omega) \langle 0_0 | \nu_i \rangle^2}{(2\omega - \omega_{i\nu_i,00_0})^2 + \Gamma_{i0}^2(\omega)} \tag{56}$$

$$\langle S_{i0} \rangle = \frac{1}{15} \sum_{nm} (s_{nn} s_{mm} + s_{nm} s_{nm} + s_{nm} s_{mn})$$

$$s_{mn} = \sum_{\alpha} \frac{d_{i\alpha}^{(m)} d_{\alpha 0}^{(n)}}{\omega - \omega_{\alpha 0}} + \frac{d_{i0}^{(m)} \Delta d_{ii}^{(n)}}{\omega - \omega_{i0}}, \quad \Delta \mathbf{d}_{ii} = \mathbf{d}_{ii} - \mathbf{d}_{00}$$

and two-step contributions:

$$\sigma_2^{(2)} = \frac{k}{c\varepsilon_0^2 \hbar^3} \sum_{i,\alpha\nu_\alpha} \left[d_{i0}^2 d_{\alpha i}^2 \Lambda_{i\alpha}^{i0} P_{i0}(\omega) \frac{R_{ii}}{\Gamma_{ii}} + \right. \tag{57}$$

$$\left. + \sum_\beta d_{\beta 0}^2 d_{i\alpha}^2 \Lambda_{i\alpha}^{\beta 0} P_{\beta 0}(\omega) \frac{R_{\beta\beta}}{\Gamma_{\beta\beta}} \frac{\Gamma^{\beta i}}{\Gamma_{ii}} \right] \frac{\Gamma_{\alpha i}(\omega) \langle 0_i | \nu_\alpha \rangle^2}{(\omega - \omega_{\alpha\nu_\alpha,i0_i})^2 + \Gamma_{\alpha i}^2(\omega)}$$

Here, $\omega_{\alpha\nu_\alpha,i\nu_i}$ is the frequency of electron-vibrational transition $|i\nu_i\rangle \to |\alpha\nu_\alpha\rangle$; $\langle \nu_i | \nu_\alpha \rangle$ is the Franck-Condon (FC) amplitude between vibrational states ν_i and ν_α of electronic states i and α; $R_{jj} = (\Gamma_{jj}/I(z,t))e^{-\Gamma_{jj}t}\int_{-\infty}^{t} e^{\Gamma_{jj}\tau} I(z,\tau)d\tau$; the probability of non-resonant absorption $P_{i0}(\omega)$ and the anisotropy factor $\Lambda_{i\alpha}^{\beta 0}$ are

$$(58) \qquad P_{i0}(\omega) = \Gamma_{i0}(\omega)\left[\frac{1}{(\omega-\omega_{i0}^V)^2+\Gamma_{i0}^2(\omega)} + \frac{1}{(\omega+\omega_{i0}^V)^2}\right]$$

$$(59) \qquad \Lambda_{i\alpha}^{\beta 0} = \frac{1+2\cos^2(\mathbf{d}_{\beta 0},\mathbf{d}_{i\alpha})}{15}.$$

One can see that the vibrational profile of the one-photon absorption

$$(60) \qquad \sigma = \frac{k}{3\hbar\varepsilon_0}\sum_i \frac{d_{i0}^2 \Gamma_{i0}\langle 0_0|\nu_i\rangle^2}{(\omega-\omega_{i\nu_i,00})^2+\Gamma_{i0}^2}$$

copies the vibrational profile of the one-step TPA (55) for the same final electronic state. The vibrational profile of the two-step TPA (56) differs qualitatively from the one-step TPA profile because of the difference in the FC factors in Eqs. (56) and (55).

Our simulations are based on response theory in the framework of the Hartree-Fock (HF) method, outlined in [32]. In the simulations we take into account the first five excited electronic states. The important parameter is the energy of the vertical transition from the ground to the first excited electronic state, ω_{10}^V. In the calculations of the TPA cross sections, we used the intermediate value $\omega_{10}^V = 3.01$ eV as motivated in [38]. Another fitting parameter we used is the ratio $\Gamma_{10}(\omega_{10}/2)/\Gamma_{11} \approx 10^{-3}$. We extracted this value from the experimental profile of the one-step absorption measured in a broad energy region [25] (see also discussion in Section 2.4). The width of the spectral line near the resonance is assumed be equal to $\Gamma_{ij}(\omega_{res}) = 0.01$ eV. The many-dimensional FC amplitudes are calculated using the harmonic approximation. The gradients of the excited state potentials are obtained making use of a code for analytical derivation of the excited state gradients, implemented in the DALTON suite of programs.

We assume that the pulse is long ($R_{jj} \approx 1$) and the molecule has time to relax to the lowest vibrational level of the electronic level i. The case of short pulse differs qualitatively as illustrated in Fig. 13. First of all we simulated the spectra of one-photon (Fig. 14) and coherent two-photon absorption, (Fig. 15). One can see that the one-photon spectrum consists of two vibrational bands related to transitions to the first ($0 \to 1$) and the fourth ($0 \to 4$) electronic states. This is in agreement with experiment [36]. The coherent TPA cross section is formed mainly due to the TPA transition to the first electronic state $0 \to 1$ (see Fig. 15). One can see clearly that this band copies strictly the $0 \to 1$ band in the one-photon absorption (Fig. 14). One-step TPA cannot explain the long wavelength band in the experimental spectrum [36]. As one can

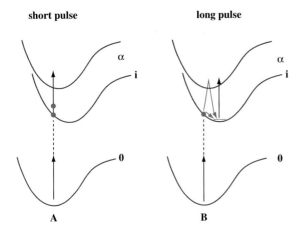

Figure 13. Scheme of transitions for short and long pulses

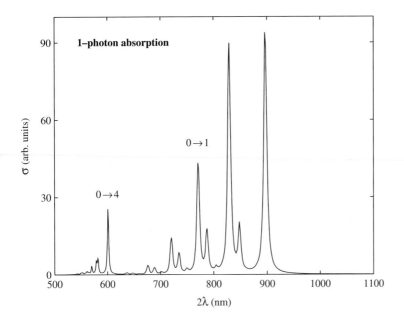

Figure 14. The cross section of the one-photon absorption

see from our simulations, this band appears when the two-step TPA process is taken into account. The origin of this band is the electron-vibrational transitions $0 \to 1, 1 \to 2$ and $0 \to 1, 1 \to 3$ to the second and the third electronic states. The central part of the TPA profile is influenced by the two-step TPA transitions from the first excited state to the electronic states 2, 3, 4 and 5. The TPA transitions

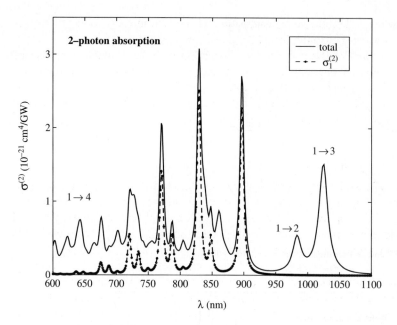

Figure 15. The cross section of the two-photon absorption

to higher electronic states give negligible contribution. Thus, the two-step TPA absorption explains the red shift of the TPA spectral profile comparing to the one-photon absorption.

5. FUTURE PROSPECTS OF *AB-INITIO* SIMULATIONS OF PULSE PROPAGATION THROUGH NONLINEAR MEDIA

The simulation technique which we presented is still restricted in a few aspects. First of all, by the size of a chromophore molecule. At present, sufficiently accurate (less than a few tenths of an eV in a relative energy scale) ab initio electronic structure calculations can be performed on molecules containing about a hundred of heavy (non hydrogen) atoms. A typical electronic structure calculation of a molecule with 50 atoms, using DFT and a double zeta basis set on a 2.8 GHz Pentium Xeon processor, takes roughly a day. This barrier is being slowly lifted with the increase of the computational power of the modern computers. Even more important is the development of different algorithms, allowing linear scaling of certain steps of the computational procedure. These advances will make possible the theoretical investigation of even larger, possibly biologically important, chromophores.

Another aspect of the chromophore size is the amount of vibrational frequencies to be accounted for. As this number increases quite fast (3N)—the computational time becomes unrealistically large. Thus, smart approximations need to be made in order to account only for the important vibrational degrees of freedom. More theoretical and code development work should be done in order to easily manage the electron-nuclear coupling inside the chromophores.

We achieved good agreement of our simulations with the available experimental data. This proves the possibility of reversing the roles—theoretical design of a molecule with valuable properties followed by synthesis and eventual experimental confirmation. Still, the time required and the complexity of the simulations prevent us from leading the experiment in a desired direction.

6. SUMMARY

Owing to successful synthesis of different multi-photon absorption based materials there is much ongoing research activity that aims to capitalize on the potential of such materials for optical applications. Although multiphoton materials in general are possible to design for a particular application, their characterization often poses an arduous or time-consuming undertaking in the laboratory. It is therefore relevant to make a versatile modeling toolbox accessible for laboratory simulations of prospective materials under different experimental conditions and that can aid in the design and the characterization. This review presents the standpoint that it is essential to combine into such a toolbox quantum chemistry methods for predicting basic electronic properties and cross sections, with classical Maxwell's theory, in order to investigate the properties of the materials with respect to the interaction with electromagnetic fields of various wavelengths and strengths. An important asset of the toolbox is that it transcends the power series approach in which the non-linear polarization is expanded over powers of the electric field, accounting for the coupling of multi-photon processes, and that it goes beyond the so-called rotatory wave approximation meaning that off-resonant in addition to resonant effects can be considered. In fact, as reviewed in this paper, in the non-linear regime it is often necessary to account simultaneously for coherent one-step and incoherent step-wise multi-photon absorption, as well as for off-resonant excitations even when resonance conditions prevail. This dynamic theory of non-linear propagation of a few interacting intense light pulses has been successfully applied to studies of varying phenomena involving propagation of light, some of them being reviewed here.

We expect that our dynamical theory based on the density matrix formalism can be useful in modeling other fundamental and applied problems of nonlinear optics like spectral-hole burning, self-focusing, white light generation, the dynamics of the spectral profiles of amplified stimulated emission etc. The scope is certainly wide for future researh in these areas.

ACKNOWLEDGEMENTS

This work was supported by the Swedish Research Council (VR). The authors acknowledge a grant from the photonics project, run jointly by the Swedish Materiel Administration (FMV) and the Swedish Defense Research Agency (FOI). Fruitful discussions with Prof. K. Kamada are acknowledged.

REFERENCES

[1] Parthenopoulos, D.A., Rentzepis, P.M.: 3-dimensional optical storage memory, Science **245**, 843–845 (1989)
[2] Dvornikov, A.S., Rentzepis, P.M.: Novel organic ROM materials for optical 3D memory devices, Opt. Comm. **136**, 1–6 (1997)
[3] Denk, W., Strickler, J.H., Webb, W.W.: 2-photon laser scanning fluorescence microscopy, Science **248**, 73–76 (1990)
[4] Prasad, P.N.: Introduction to Biophotonics. Wiley, New Jersey, (2003)
[5] Mukamel, S.: Principles of Nonlinear Optical Spectroscopy. Oxford University Press, New York, (1995)
[6] Scully, M.O., Zubairy, M.S.: Quantum optics. Cambridge University Press, Cambridge, (1997)
[7] Sakoda, K.: Optical properties of photonic crystals. Springer Verlag, Berlin (2001)
[8] Newell, A.C., Moloney, J.V.: Nonlinear Optics. Addison-Wesley Publ. Co., New York, (1992)
[9] Gel'mukhanov, F., Baev, A., Macak, P., Luo, Y., Ågren, H.: Dynamics of two-photon absorption by molecules and solutions, J. Opt. Soc. Am. B **19**, 937–945 (2002)
[10] Baev, A., Gel'mukhanov, F., Macak, P., Luo, Y., Ågren, H.: General theory for pulse propagation in two-photon active media, J. Chem. Phys. **117**, 6214–6220 (2002)
[11] Baev, A., Gel'mukhanov, F., Rubio-Pons, O., Cronstrand, P., Ågren, H.: Upconverted lasing based on many-photon absorption: An all dynamic description, J. Opt. Soc. Am. B **21**, 384–396 (2004)
[12] Baev, A., Gel'mukhanov, F., Kimberg, V., Ågren, H.: Nonlinear propagation of strong multi-mode fields, J. Phys. B: At. Mol. Opt. Phys. **36**, 3761–3774 (2003)
[13] Baev, A., Kimberg, V., Polyutov, S., Gel'mukhanov, F., Ågren, H.: Bidirectional description of amplified spontaneous emission induced by three-photon absorption, J. Opt. Soc. Am. B **22**, 385–393 (2005)
[14] Weisskopf, V., Wigner, E.: Z. Phys. **63**, 54 (1930)
[15] Marcus, R.A.: On the theory of shifts and broadening of electronic spectra of polar solutes in polar media, J. Chem. Phys. **43**, 1261–1274 (1965)
[16] Matyushov, D.V., Ladanyi, B.M.: Nonlinear effects in dipole solvation. II. Optical spectra and electron transfer activation, J. Chem. Phys. **107**, 1375–1387 (1997)
[17] Rumi, M., Ehrlich, J.E., Heikal, A.A., Perry, J.W., Barlow, S., Hu, Z., McCord-Maughon, D., Parker, T.C., Röckel, H., Thayumanavan, S., Marder, S.R., Beljonne, D., Brédas, J.-L.: Structure-property relationships for two-photon absorbing chromophores: Bis-donor diphenylpolyene and bis(styryl)benzene derivatives, J. Am. Chem. Soc. **122**, 9500–9510 (2000)
[18] Becker, P.C., Fragnito, H.L., Bigot, J.-Y., Brito Cruz, C.H., Fork, R.L., Shank, C.V.: Femtosecond photon-echoes from molecules in solution, Phys. Rev. Lett. **63**, 505–507 (1989)
[19] Bigot, J.-Y., Portella, M.T., Schoenlein, R.W., Bardeen, C.J., Migus, A., Shank, C.V.: Non-Markovian dephasing of molecules in solution measured with 3-pulses femtosecond photon echoes, Phys. Rev. Lett. **66**, 1138–1141 (1991)
[20] Lawless, M.K., Mathies, R.A.: Excited state structure and electronic dephasing time of nile blue from absolute resonance Raman intensities, J. Chem. Phys. **96**, 8037–8045 (1992)
[21] Kummrow, A., Lau, A., Lenz, K.: Time-resolved study of ultrafast dephasing processes in solution, Phys. Rev. A **55**, 2310–2320 (1997)
[22] Knox, R.S.: Theory of Excitons. Academic Press, New York (1963)
[23] Allard, N., Kielkopf, J.: The effect of neutral nonresonant collisions on atomic spectral lines, Rev. Mod. Phys. **54**, 1103–1182 (1982)
[24] Gel'mukhanov, F., Sałek, P., Privalov, T., Ågren, H.: Duration of X-ray Raman scattering, Phys. Rev. A **59**, 380–389 (1999)
[25] Polyutov S., Minkov I., Gel'mukhanov F., Kamada K., Baev A., and Ågren H., Spectral profiles of two-photon absorption: Coherent versus two-step two-photon absorption, Mater. Res. Soc. Symp. Proc. v.846, Warrendale, PA, 2005, DD1.2
[26] He, G.S., Prasad, P.N.: Three-photon absorbing materials: Characterization and applications, in Nonlinear optical transmission and multiphoton processes in organics, Todd Yeates, A., Kevin D. Belfield, Francois Kajzar, and Christopher M. Lawson, (eds) Proc SPIE **5211**, 1–12 (2003)

[27] He, G.S., Dai, J., Lin, T.-C., Markowicz, P.P., Prasad, P.N.: Ultrashort 1.5 μm laser excited upconverted stimulated emission based on simultaneous three-photon absorption. Opt. Lett. **28**, 719–721 (2003)

[28] He, G.S., Helgeson, R., Lin, T.-C., Zheng, Q., Wudl, F., Prasad, P.N.: One-, two-, and three-photon pumped lasing in a novel liquid dye salt system. IEEE J. Quantum Electron., **39**, 1003–1008 (2003)

[29] Reinhardt, B.A.: Two-photon technology: new materials and evolving applications, Photonics Science News **4**, 21–33 (1999)

[30] He, G.S., Markowicz, P.P., Lin, T.-C., Prasad, P.N.: Observation of stimulated emission by direct three-photon excitation, Nature **415**, 767–770 (2002)

[31] Pantell, R.H., Puthoff, H.E.: Fundamentals of quantum electronics. Wiley, New York (1969)

[32] DALTON, a molecular electronic structure program, Release 2.0 (2005), see, http://www.kjemi.uio.no/software/dalton/dalton.html

[33] Olsen, J., Jørgensen, P.: Time dependent response theory with applications in to self consistence field (SCF) and multiconfigurational self consistent field (MCSCF) wave functions, I.F.A. PRINT, Aarhus Universitet, 1994

[34] Norman, P., Jonsson, D., Vahtras, O., Ågren, H.: Non-linear electric and magnetic properties obtained from cubic response functions in the random phase approximation, Chem. Phys. **203**, 23–42 (1996)

[35] Mitkova Dushkina, N., Ullrich, B.: Intensity dependence of two-photon absorption in CdS measured by photoluminescence excited by femtosecond laser pulses, Opt. Eng. **41**, 2365–2368 (2002)

[36] Lin, T.-C., He, G.S., Prasad, P.N., Tan, L.-S.: Degenerate nonlinear absorption and optical power limiting properties of asymmetrically substituted stilbenoid chromophores, J. Mater. Chem. **14**, 982–991 (2004)

[37] He, G.S., Lin, T.-C., Dai, J., Prasad, P.N.: Degenerate two-photo-absorption spectral studies of highly two-photon active organic chromophores, J. Phys. Chem. **120**, 5275–5284 (2004)

[38] Polyutov, S., Minkov, I., Gel'mukhanov, F., Ågren, H.: Interplay of one- and two-step channels in electro-vibrational two-photon absorption, J. Phys. Chem. A 109, 9507 (2005)

CHAPTER 7

COLLECTIVE AND COOPERATIVE PHENOMENA IN MOLECULAR FUNCTIONAL MATERIALS

ANNA PAINELLI[1] AND FRANCESCA TERENZIANI[2]

[1] *Dip. Chimica GIAF, Parma University & INSTM-UdR Parma, Viale delle Scienze 17/A, 43100 Parma, Italy, anna.painelli@unipr.it*
[2] *Synthèse et ElectroSynthèse Organiques (CNRS, UMR 6510), Institut de Chimie, Université de Rennes 1, Campus Scientifique de Beaulieu, Bât 10A, F-35042 Rennes Cedex, France, francesca.terenziani@univ-rennes1.fr*

Abstract: We discuss cooperative and collective behavior resulting from classical electrostatic intermolecular interactions in molecular materials with negligible intermolecular overlap. With reference to materials based on push-pull chromophores, we discuss the merits of several approximation schemes for the calculation of linear and non-linear optical susceptibilities. Collective and cooperative behavior is recognized in important deviations of the material properties from the oriented gas approximation scheme, and/or from the exciton model. Extreme collective and cooperative behavior in attractive clusters is discussed, where bistable behavior and the phenomenon of multielectron-transfer appear

Keywords: Molecular materials, molecular functional materials, intermolecular interactions, non-linear optical properties, push-pull chromophores, polarizability, optical spectra, excitons

1. INTRODUCTION

Molecular functional materials are the Holy Grail of modern materials science, with promising applications in the fields of molecular electronics [1] and photonics [2, 3]. Indeed molecular materials (mm) already found an appealing and rewarding application in organic-light emitting devices, that first devised in the 1980's, are nowadays present in the market [4]. From a different perspective, fundamental biological processes often involve supramolecular arrangements of functional molecules, as nicely demonstrated by the light harvesting complexes central in the photosynthetic process [5]. A thorough understanding of these processes will unveil some fundamental mechanism of life, offering at the same time important clues for the optimal engineering of molecular devices.

Molecular materials of interest for applications, molecular functional or intelligent materials, are materials that respond in a qualitatively different way to different inputs: non-linearity is the qualifying property for functional behavior. Delocalized electrons are an obvious source of non-linearity [6]: interesting classes of materials then involve, to cite a few, charge-transfer (CT) complexes and salts, materials based on π-conjugated molecules or polymers, inorganic complexes, and so on.

In this work we will limit attention to mm made up of large π-conjugated molecules of interest for NLO applications. Intermolecular distances larger than the sum of Van der Waals radii point to negligible overlap of molecular orbitals on different molecules, so that electrons are basically localized on the molecular units. Non-linear, functional behavior is, in these materials, a consequence of the large electronic delocalization *within* each molecular unit, related to the presence of a large π-conjugated backbone. Non-trivial collective and cooperative behavior in materials of this kind was foreseen in the early 60's by McConnell and coworkers who, in a seminal paper titled *Collective Electronic States in Molecular Crystals*, wrote: *Based on rough calculations, we believe it not unlikely that such strong interactions* (charge resonance coupling, responsible for collective behavior) *are to be found in molecular crystals of large dye-like molecules, and perhaps even in molecular aggregates that play a role in photosynthesis or other biological phenomena* [7]. In this contribution we develop this suggestion and demonstrate that collective and cooperative behavior in molecular functional materials appears as a consequence of the strong polarizability and hyperpolarizabilities of large π-conjugated chromophores.

Molecular crystals are the simplest example of mm. According to textbook descriptions [8], in molecular crystals the intermolecular (electrostatic and/or Van der Waals) forces are pretty weak if compared with the strong chemical forces acting within each molecule. The molecular units then keep their identity in the crystal and their properties are only marginally affected by the surrounding [9]. This description is the base of the oriented gas model (OGM) for molecular crystals [10], where the properties of the materials are calculated as the sum of the properties of the isolated molecules, of course accounting for the specific mutual orientation of the molecules in the solid.

Deviations from the additive behavior predicted by OGM are usually referred to as collective and/or cooperative effects. Of course OGM is expected to fail in so called CT molecular crystals, i.e in the special class of molecular crystals where, due to the finite overlap of frontier orbitals on adjacent molecules, CT degrees of freedom become relevant [11]. In these systems intermolecular electronic delocalization is observed, as demonstrated by the appearance of an intense CT absorption band lying at lower energy than the localized molecular excitations. Of course the CT band, and many related spectroscopic phenomena, can only be described by relaxing the OGM approximation [11, 12]. Collective and cooperative phenomena in CT materials are very interesting in several respects, but here we concentrate on more subtle phenomena occurring in mm with negligible intermolecular delocalization, i.e. in materials where CT states lye at much higher energies than local molecular excitations, as it occurs when intermolecular distances are larger than Van der Waals distances. The low-energy physics of these materials is dominated by local, Frenkel-like excitations [13]. Collective and cooperative behavior in these materials cannot be ascribed to intermolecular delocalization, but results from classical electrostatic intermolecular interactions.

Deviations from OGM were recognized early on spectroscopic properties of molecular crystals: Davydov shifts and splittings of absorption bands in molecular crystals are clear deviations from OGM and were rationalized based on the excitonic model (EM) [10, 14, 15, 16, 17]. This same model proved extremely successful to describe the complex and technologically relevant spectroscopy of molecular aggregates, i.e. of clusters of molecules that spontaneously self-assemble in solution or in condensed phases [18]. Much as it occurs in molecular crystals, due to intermolecular electrostatic interactions the local bound electron-hole pair created upon photoexcitation travels in the lattice and the corresponding wave function describes an extended *delocalized* object called an exciton. We explicitly remark that the Frenkel picture of the exciton, as a bound electron-hole pair, both residing on the same molecule, survives, or better is the basis for the excitonic picture. The delocalization of the exciton refers to the fact that the relevant wave function describes a Frenkel exciton (a bound e-h pair) that travels in the lattice, and this is of course possible even when electrons and/or holes are, separately, totally localized. In other terms, the EM describes localized charges, but delocalized excitations.

The simple additivity of the OGM is lost in the EM: excitons are collective objects whose wave function extends over the lattice, so that they cannot be described in any local picture. In the following we will reserve the term *collective* to indicate all phenomena that are intrinsically related to delocalized states.

EM, allowing for the exciton motion, played a central role in understanding energy transport in mm, at least as long as Forster mechanism dominates over the Dexter exchange process [19]. Instead, not allowing for charge motion, EM cannot describe charge transport. As a matter of fact charge transport requires a finite overlap between orbitals on different molecules and is therefore beyond the scope of this contribution. However, we underline that non-additive behavior was early recognized in the context of charge transport. A charge carrier inside a mm in fact polarizes the surrounding molecules and the resulting electric field screens the charge. Several schemes of different complexity were proposed to calculate the induced polarization as sum of contributions from properly defined subunits, but, as it was recognized early, the problem is intrinsically non-linear and hence non-additive [16, 17]. In fact each subunit, being it a molecule, a molecular fragment or even an atom, is a polarizable object that responds to the presence of a charge by generating a local electric field that sums up to the field generated by all other subunits and by the added charge, leading to a complex self-consistent problem that is the basis for non-linearity [20, 21, 22, 23, 24, 25].

The problem of the polarization of the material in response to an added charge shares the same physics with another long-standing problem [26]: the calculation of linear and non-linear optical responses of mm. In fact the molecules are polarized by any electric field, being it an externally applied field or generated by a charge carrier. Linear and non-linear optical responses of mm are technologically relevant and a lot of theoretical and experimental effort was spent to optimize molecular responses [27, 28, 29, 30, 31]. Quite reliable structure-properties relationships have been devised at the molecular level, but supramolecular interactions can affect heavily the responses at the material level [32, 33], and reliable relationships between the supramolecular structure and the material responses are still lacking. In diluted samples the simple additive behavior of OGM possibly applies, even if local field corrections must be properly accounted for [34]. However for medium-large concentrations, and particularly for molecules with large (hyper-)polarizabilities (i.e. exactly for those molecules that are more interesting for applications), the readjustment of the molecular charge distribution in response to the interaction with the applied field and with the field generated by all surrounding molecules lead to very important non-linear effects that, depending on the lattice geometry, can either amplify or reduce the response, as it will be discussed in the following.

Apart from some very special cases that will be discussed in Section 7, in mm the gs describes a collection of molecules each one in its local gs. Non-additivity of static NLO responses, that are gs properties, cannot be ascribed to delocalization and hence is not a signature of collective behavior. Non-additivity of NLO responses can be understood in these terms: the response of each molecule is truly local, but it is affected in a self-consistent way by the interaction with the surrounding molecules. Non-additivity of static NLO responses thus results from *cooperative* behavior.

Whereas the distinction between collective and cooperative effects can appear artificial, it is obvious that, since optical responses are gs properties, their non-additivity cannot be ascribed to the delocalized nature of excited states. On the other hand, static responses can be calculated from sum-over-state (SOS) expressions involving excited state energies and transition dipole moments [35]. And in fact the exciton model has been recently used by several authors to calculate and/or discuss linear and non-linear optical responses of mm [36, 37, 38, 39, 40, 41, 42]. But the excitonic model hardly accounts for cooperativity and one may ask if there is any link between collective effects related to the delocalized nature of exciton states and cooperative effects in the gs, related to the self-consistent dependence of the local molecular gs on the surrounding molecules.

The so called supermolecule approach offers a powerful and interesting way to investigate the properties of mm by performing quantum chemical calculations on clusters of molecules of finite size (the 'supermolecule'). The approach was pioneered by the Wagniere group in the 80's [36]. Adopting a Pariser-Parr-Pople model for the supermolecule, the authors were able to appreciate the multifaced role of intermolecular interactions in mm: on one side electrostatic intermolecular forces between polarizable molecules lead to a variation of the molecular charge distribution; on the other hand, the interactions between transition dipole moments and/or permanent dipole moments lead to anomalous excitation spectra that hardly reconcile with the prediction of either OGM or EM. Di Bella and coworkers [37] presented an extensive discussion of NLO responses of a dimer of paranitroaniline (a prototypical push-pull chromophore) in different geometries. Based on a ZINDO approximation they observe the deformation of the molecular charge distribution as a result of intermolecular interactions and discuss its role in the definition of the second-order NLO response. More recently, similar studies were presented by several authors, [42, 43] that, studying several dimers of different push-pull chromophores, underlined again the non-trivial role of intermolecular interactions. Other studies on dimers of push-pull chromophores where presented based on ab initio approaches [44].

The supermolecule approach has the obvious advantage of a complete description of intermolecular interactions, at least at the level of the model adopted to describe the molecular structure. However the approach is computationally demanding, so that, depending on the size of the molecule, only small clusters and/or not too refined models (specifically not too large basis sets) can be treated. Apart from these technical limitations, that will be overcome with the increase of computational power, extracting relevant information from the analysis of numerical data is a demanding and subtle task. In this respect we notice that Hamada [45] presented very interesting results of ab initio calculations of second order NLO response of dimers of 2-methyl-4-nitroaniline. He observed important deviations from the OGM model, with the response strongly reduced as a consequence of intermolecular interactions. He ascribed this phenomenon to the reduction of the molecular polarity in the dimer as a consequence of the molecular polarizability. This is for sure an important observation, but, as it will be discussed in detail in Section 6, the effects of intermolecular interactions on NLO responses are more subtle.

Recently an extensive review has been published covering the calculation of NLO properties in the solid state [44]. We refer the interested reader to this work for an extensive coverage of previous literature devoted to intermolecular interactions and their effects on optical responses of mm. In this work we will discuss models for collective and cooperative effects as occurring in mm with particular emphasis on the relation between the description of excited states and linear and non-linear static optical responses. We will mention a few seminal papers where the concepts of collective and cooperative behavior appeared. The proposed references then follow a very personal and unavoidably incomplete view of the very rich literature in the field.

The paper is organized as follows. Section 2 shortly introduces the exciton model and its approximations. Section 3 reviews calculations of ground state properties (mainly the polarization and polarizability) paying special attention to the mean-field approximation. Push-pull chromophores, the special family of polar and polarizable molecules studied in this contribution, are presented in Section 4, with a brief discussion of their properties in solution and of relevant models. In Section 5 we present a model for interacting push-pull chromophores that will be the basis for the discussion of collective and cooperative effects in relevant materials. Static susceptibilities of clusters of push-pull chromophores are discussed in Section 6, focusing attention on cooperative effects in the ground state. Excited state properties are addressed in Section 7, with special emphasis to systems where intermolecular interactions lead to extreme consequences. Section 8 finally summarizes main results.

2. THE EXCITONIC PICTURE: A MODEL FOR HARDLY POLARIZABLE MOLECULES

EM applies to clusters of non-overlapping and weakly interacting molecules describing either a molecular crystal or an aggregate. Intermolecular interactions are treated perturbatively, so that the zeroth order basis for the cluster is the direct product of the local basis relevant to each molecular sites. For the sake of simplicity we shall refer to the case when just a ground state (gs), $|g_i\rangle$, and a single excited state, $|e_i\rangle$, are defined on each molecular site. The zeroth order energies are the sum of local energies, and, assigning energy 0 to the gs, states with n local excitations have energy $n\hbar\omega_0$, where ω_0 is the excitation energy of the isolated molecule. The first effect of intermolecular interactions is a perturbative correction of the excitation energy: $\omega = \omega_0 + D$, where D, the Davydov shift, is the difference between the energy of interaction of the excited molecule and the unexcited molecule with all other molecules in the cluster [10, 46].

States with n local excitations are degenerate, and any tiny perturbation induced by intermolecular interactions mixes them effectively. EM accounts for the mixing of degenerate states so that n is a conserved quantity. The relevant Hamiltonian is [46]:

$$(1) \quad H_J = \frac{1}{2}\sum_{i,j} J_{ij}(b_i^\dagger b_j + b_j b_i^\dagger)$$

where i, j indeces run on the molecular sites, $b_i^\dagger$ creates a local excitation on the i-th site and J_{ij} is the matrix element that transfers the excitation between sites i and j. In the simple dipolar approximation for electrostatic forces, J_{ij} measures the interaction between the transition dipole moments on the i and j molecules. It dyes off quickly with the intermolecular distance so that often only the nearest-neighbor exciton hopping is retained. In any case, states with n local excitations are mixed by the Hamiltonian in Eq. (1) to form delocalized n-exciton states. Of course, due to delocalization, the additivity of the OGM is lost and collective phenomena appears.

The simplest example is that of a one-dimensional cluster of N equivalent molecules with nearest neighbor J interaction, periodic boundary conditions and intermolecular spacing r. The N states with a single local excitation are mixed to give a band of 1-exciton states with energies $E_k = 2J\cos(kr)$, and wave vector $k = 0, \pm\pi/Nr, \pm 2\pi/Nr, \ldots, \pi/r$, as sketched in Fig. 1.

Only the $k = 0$ state is optically allowed, and for attractive/repulsive J it lies at the bottom/top of the band. In attractive $J < 0$ lattices the absorption spectrum is red-shifted with respect to ω_0 by $2J$ (J-aggregates), whereas a blue-shift by the same amount is expected for repulsive interactions ($J > 0$, H-aggregates). The most impressive collective phenomenon in this simple model is recognized in the fluorescence behavior. In H-aggregates the fast relaxation of excited states to the bottom of the 1-exciton band leads to an optically forbidden state so that fluorescence is strongly suppressed in the material. In J-aggregates instead fluorescence is

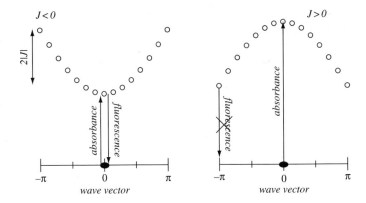

Figure 1. One-exciton band (energy vs wave vector) for a 15-site lattice described by the Hamiltonian in Eq. (1), with only nearest-neighbor exciton hopping, J, and constant intermolecular spacing, $r = 1$. Left and right panels refer to negative and positive J, respectively. The arrows mark one-photon absorption and emission processes. The emission process for $J > 0$ is forbidden

allowed and the corresponding transition dipole moment has contributions from all molecular transition dipole moments leading to a collective emission often referred to as super-radiance [18].

The same J-interaction appearing in Eq. (1) enters the Hamiltonian with the following term [46]:

$$H'_J = \frac{1}{2}\sum_{i,j} J_{ij}(b_i^\dagger b_j^\dagger + b_i b_j) \quad (2)$$

This term breaks down the excitonic approximation mixing up states whose exciton number differs by two units. This so called non-Heitler-London term has negligible effects for systems with $J \ll 2\omega_0$ [47]. So the excitonic approximation is expected to work well for clusters of molecules with large excitation energies and not too large transition dipole moments, i.e. for hardly polarizable molecules.

EM was quite extensively and successfully applied to model optical spectra of molecular crystals and aggregates. Extensions were discussed [18] to account for disorder, whose effects are particularly important in aggregates, and to include the coupling between electronic degrees of freedom and molecular vibrations [48], needed to properly describe the absorption and emission bandshapes. However, as it was already recognized in original papers [7, 46], other terms enter the excitonic Hamiltonian. Electrostatic interactions between local excitations can in fact be introduced as:

$$H_{ee} = \frac{1}{2}\sum_{i,j} U_{ij} b_i^\dagger b_i b_j^\dagger b_j \quad (3)$$

where U_{ij} measures the difference between the electrostatic interactions of the two i, j molecules in the excited and in the gs. This term is small for clusters of non-polar molecules, but for polar molecules it can be pretty large. In the dipolar approximation it is proportional to the interaction between the two molecular mesomeric dipole moments, where the mesomeric dipole moment measures the difference between the molecular dipole moments in the excited and ground state. Exciton-exciton interactions have been discussed in models for aggregates and crystals of polar chromophores [49], as well as in related models for CT crystals with a mixed stack motif and a largely neutral gs [50]. Important effects of exciton-exciton interactions are recognized in the appearance of bound biexciton states [49] or even in the formation of exciton-strings [50].

The Hamiltonian in Eq. (3) conserves the exciton number and, by itself, does not break the excitonic approximation. However large exciton-exciton interactions may eventually lead to the breakdown of EM. Consider in fact the case of large attractive interactions between adjacent excitons. The energy of the two-exciton state with two nearby excitons is lowered with respect to 2ω then making the non-Heitler-London term in Eq. (2) much more effective. But for clusters of polar molecules an additional term appears that directly breaks the excitonic approximation [46]:

$$(4) \qquad H_w = \sum_{i,j} W_{ij} (b_i^\dagger + b_i) b_j^\dagger b_j$$

where, in the dipole approximation, W_{ij} measures the interaction between the transition dipole moment and the mesomeric dipole moment on sites i and j. This term mixes up states whose excitation number differs by just one unit and its effects are negligible when $W_{ij} \ll \omega_0$. This requirement is more stringent that that enforced by the neglect of the non-Heitler London term (cf the discussion following Eq. (2)), and is particularly stringent for materials based on polar molecules whose mesomeric dipole moment are larger than transition dipole moments, and hence $W_{ij} > J_{ij}$.

The short discussion in this Section demonstrates that EM is hardly applicable to describe mm based on largely polarizable molecules: in these materials in fact transition dipole moments are large and excitation frequencies are low. The situation is even worse for mm based on polar and polarizable molecules, where U- and W-like interactions appear. Push-pull chromophores are a very interesting family of polar and largely (hyper)polarizable molecules, characterized by an intense optical transition in the visible region, whose large solvatochromism is related to a large mesomeric dipole moment [51, 52]. Materials based on push-pull chromophores are extremely attractive for second order NLO applications [28, 29, 33, 53, 54, 55], but they are also actively investigated in the related field of molecular electronics [56, 57]. According to previous discussion these materials cannot be described within the standard excitonic picture and large deviations from EM are expected.

3. POLARIZABLE MOLECULES: THE MEAN-FIELD APPROXIMATION FOR GS PROPERTIES

Static linear and non-linear polarizabilities measure the successive derivatives of the gs dipole moment vs an applied static electric field. More precisely for extended systems the (hyper)polarizabilities are defined as intensive properties, i.e. as derivatives of the polarization, P, the gs dipole moment per unit volume. But the proper definition of P, and hence of the polarizabilities, is a tricky problem in extended systems, where several choices are possible for the definition of the basic unit cell, leading to different values of P. In a different perspective, periodic boundary conditions, that are needed to preserve the symmetry of the extended system, are difficult to reconcile with finite P, or, for what matters, with a uniform static electric field. A very elegant and powerful solution of this problem was recently proposed by Resta [58], that defined P in systems with periodic boundary conditions as the phase of the gs wave function. The general formulation of P as a phase bypasses the unit cell problem in extended insulators and opens the way to the calculation of corresponding polarizability and hyperpolarizabilities [26, 59, 60]. In extended mm with strictly non-overlapping molecules, however, the P problem is irrelevant. In this case in fact the definition of the unit cell as a non-overlapping unit is unambiguous: the charge redistribution is purely intramolecular and conventional molecular approaches to the polarization are adequate [26].

The neglect of intermolecular overlap solves the unit-cell problem, but, in spite of that, the calculation of linear and non-linear polarizabilities of mm or of molecular aggregates is not at all easy. Deviations of the polarizabilities from the additive OGM behavior were recognized and discussed early [61]. The first improvement on OGM approximation accounts for local field corrections: basically the response of the material is again calculated as the sum of the responses of the molecular units, but accounting for the fact that each molecule experiences a local field that is different from the external field. Several approximation schemes have been proposed for the non-trivial calculation of local-field corrections [44] ranging from simple Lorentz field models to detailed calculations of local fields in the point-dipole approximation [20]. Indeed the definition of proper local fields is a subtle problem since the dipole approximation for extended molecular units is questionable: several approaches have then been proposed to describe each molecule as a collection of local dipoles that are by themselves polarizable [20]. Describing the lattice as a collection of polar and polarizable dipoles lead unavoidably to a complex self-consistent problem [21, 22, 23] whose solution becomes more and more demanding as the model adopted to describe the molecular units becomes more sophisticated and hence more realistic. An important consideration is that, particularly for polar molecules, large permanent local fields are found in the material that can profoundly alter the gs polarity of the molecular units and hence all their properties [23].

This observation paves the way to the work by Tsiper and Soos that proposed a mean-field approximation for the calculation of the linear polarizability of molecular crystals and films [25, 62, 63, 64]. The approach is based again on the neglect of intermolecular overlap. A quantum chemical model is adopted for each molecular unit, and its interaction with the surrounding molecules is described by introducing into the molecular Hamiltonian an electrostatic potential that is calculated based on the geometry of the cluster and on the charge distribution on the molecular units. The problem is clearly self-consistent, since the charge distribution on each molecule, and hence the generated potential, depend on the potential itself, but at each iteration the QM problem to be solved just describes a single molecule. Of course at each iteration one must also calculate the electrostatic potential, but this is easily done through standard Ewald summation schemes [62]. Once convergence is reached, the total polarization P of the lattice can be calculated and its derivatives with respect to an applied electric field yield the polarizabilities. The method is pretty general, and the quality of the results depend on the quality of the quantum chemical model adopted to describe the molecular units and of the model adopted to describe the charge distribution. So far results have been published for the linear polarizability of crystals and films of non-polar molecules (including polyacenes and PTCDA), described at a INDO level and adopting a Lödwin description of atomic charges [25, 62, 63, 64]. The intramolecular charge fluxes are well described within this approach and they add to the atomic polarizabilities giving the largest contribution to the linear optical response of the material (the dielectric constant). Particularly important effects from charge reorganization are observed in materials based on PTCDA, a quadrupolar molecule [62, 63]. Even larger effects are expected in materials based on polar and polarizable molecules where local fields are very large and can largely affect the molecular polarity and polarizability [22, 23].

4. PUSH-PULL CHROMOPHORES: AN INTERESTING FAMILY OF POLAR AND (HYPER-)POLARIZABLE MOLECULES

Push-pull chromophores are molecules made up by an electron donor (D) and an acceptor (A) group connected by a π-conjugated bridge. The neutral (DA) and zwitterionic (i.e. charge separated, D^+A^-) states have similar energies and both contribute to the gs. These molecules are actively investigated in several, apparently unrelated fields. Push-pull chromophores are the molecules of choice for second-order NLO applications [28, 29, 33, 53, 54], are typical solvation probes [52], and are useful model systems for electron transfer [65]. All these applications exploit the presence of a low-lying excited-state characterized by a different electronic distribution from the gs. Good solvation probes have an electronic absorption and/or emission band well separated from the other transitions, with good intensity (i.e., a sizable transition dipole moment), and whose position strongly depends on the solvent polarity. This last requirement is easily fulfilled if the mesomeric dipole moment is large [52]. A large mesomeric dipole moment implies a large charge redistribution upon excitation, so that the absorption process basically models a photoinduced electron transfer, whereas the emission process models a spontaneous electron transfer [65]. On the other hand, large transition and mesomeric dipole moments guarantee for large NLO responses [66]. Push-pull chromophores are then a very interesting class of polar and highly (hyper-)polarizable molecules, and we adopt the versatile acronym of pp chromophores for them that also reads as an acronym of polar-polarizable chromophores.

In the last years we devoted considerable effort to understand the properties of pp chromophores in different environments. The basic idea is that pp chromophores respond largely and non-linearly to electric fields, being them externally applied field, or internal fields generated by the interaction with the surrounding. In this section we will briefly review our work on NLO responses of pp chromophores and on their spectroscopic behavior in solution. This work set the stage for building up models for mm based on pp chromophores, where intermolecular electrostatic interactions dominate the physics of the material.

Our choice is to use extremely simple models where a few basic interactions are accounted for. Our aim is not the detailed and accurate modeling of all the material properties, but rather to understand the basic physics governing the behavior of pp chromophores in different environments, and sorting out relevant interactions and parameters. Simple models are instrumental in this respect, moreover, being amenable to exact, or at least, non-perturbative solutions, they offer stringent reliability tests for common approximation schemes.

We describe the electronic structure of pp chromophores based on an old, but extremely powerful model, originally proposed by Mulliken [67] to describe DA complexes in solution. The model is based on the assumption that the low-energy physics of pp chromophores is dominated by the resonance between the neutral and the charge separated (zwitterionic) structures. Two basis states, $|DA\rangle$ and $|D^+A^-\rangle$, separated by an energy $2z$ and mixed by a matrix element $-\sqrt{2}t$, completely define the electronic Hamiltonian. The solution of this problem is trivial and was already discussed by several authors (see, e.g. [68] and reference therein). For future reference we explicitly write the ground and excited states:

(5) $$|g\rangle = \sqrt{1-\rho}|DA\rangle + \sqrt{\rho}|D^+A^-\rangle$$
$$|e\rangle = \sqrt{\rho}|DA\rangle - \sqrt{1-\rho}|D^+A^-\rangle$$

where $\rho = (1 - z/\sqrt{z^2 + 2t^2})/2$ measures the weight of $|D^+A^-\rangle$ in the gs and is therefore a measure of the molecular polarity. Following Mulliken [67], we recognize that in the adopted basis the dipole moment operator is strongly dominated by μ_0, the dipole moment relevant to $|D^+A^-\rangle$, so that all quantities of interest for spectroscopy can be derived as follows [68]:

(6) $$\mu_G = \langle G|\hat{\mu}|G\rangle = \mu_0 \rho$$

(7) $$\mu_E = \langle E|\hat{\mu}|E\rangle = \mu_0 (1-\rho)$$

(8) $$\mu_{CT} = \langle G|\hat{\mu}|E\rangle = \mu_0 \sqrt{\rho(1-\rho)}$$

(9) $$\hbar\omega_{CT} = \mathcal{E}_E - \mathcal{E}_G = \sqrt{2}t/\sqrt{\rho(1-\rho)}$$

On this basis close expressions for static NLO responses were written, that proved particularly useful since they relate linear and non-linear susceptibilities to easily

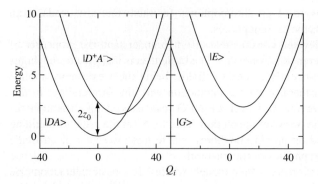

Figure 2. Potential energy surfaces for the basis states (left panel) and for the adiabatic ground and excited states (right panel)

accessible spectroscopic observables [66, 68, 69]. In this respect, the use of simple models in guiding the synthesis of molecules with desired properties proved extremely powerful [30, 31, 70, 71].

To properly describe optical spectra of pp chromophores in solution the Mulliken model has to be extended to account for the coupling of electronic and vibrational degrees of freedom, i.e. for the different molecular geometries associated with the two basis states. In the framework of the Holstein model [72, 73], we assign the two basis states two harmonic potential energy surfaces (PES) with exactly the same curvatures, but displaced along the coupled coordinates, Q_i, as sketched in Fig. 2, left panel, as to account for the different geometry associated with the two basis states. The resulting model describes linear electron-phonon (e-ph) coupling, with the z parameter (half the energy gap between the two basis states) linearly depending on vibrational coordinates (cf Fig. 2). The model is easily diagonalized in the adiabatic approximation yielding to the anharmonic PES for the ground and excited states sketched in the right panels of Fig. 2 [74, 75, 76, 77]. Anharmonicity is a clear signature of non-linear behavior and results, in our model for linear e-ph coupling, from the strongly non-linear response of the electronic system [78, 79].

Common spectroscopic techniques test small portions of the ground and/or excited state PES either around the gs minimum (IR and non-resonant Raman spectra, electronic absorption spectra.) or in the proximity of the excited state minimum (steady-state fluorescence). These spectra are then satisfactorily described in the *best harmonic approximation*, a local harmonic approach that approximates the PES with parabolas whose curvatures match the exact curvatures calculated at the specific position of interest [78]. Anharmonicity in this approach manifests itself with the dependence of harmonic frequencies and relaxation energies on the actual nuclear configuration [79]. Along these lines we predicted softened (hardened) vibrational frequencies for the ground (excited) state [74], amplified and ρ-dependent infrared and Raman intensities [68, 74], different Frank-Condon factors for absorption and steady-state emissions leading to narrower emission than absorption bands [75, 80].

The non-linearity of the electronic response to the vibrational perturbation, evidenced by the anharmonicity of the PES in the right panel of Fig. 2, shows up most directly in NLO responses. The large static electric fields involved in NLO susceptibilities force the systems to explore large regions around the equilibrium: anharmonic corrections to static NLO responses are very large and increase with the order on linearity, as extensively discussed in [77, 78] where a detailed test of the adiabatic approximation was also presented. Interesting consequences of e-ph coupling, or better of the non-linear response of the electronic system to vibrational perturbation, are recognized in NLO responses at optical frequencies. In particular new purely vibrational channels to dynamic NLO responses, such as SHG or TPA where demonstrated as due to e-ph coupling leading to very interesting effects on bandshapes and intensities [81, 82]. Some of the effects predicted for TPA spectra have found at least qualitative confirmation in recent experimental data [83], demonstrating that, in spite of its crudeness, the proposed model catches the basic physics of pp chromophores. However a more extensive model validation requires the introduction of environmental interactions. As a first step in this direction we considered pp chromophores in solution: this comparatively simple system offers a good test for models of environmental interactions, also allowing for an extensive validation against the very large body of available experimental data.

The solvent surrounding a polar molecule polarizes itself generating a reaction field at the chromophore position [51, 52]. The electronic polarization of the solvent is very fast and, as such, only results in a renormalization of the electronic states [84]. The slow orientational component of the solvent polarization, as occurring in polar solvents instead plays essentially the same role as internal vibrations, the main difference being that the relevant solvation coordinate is a very slow, actually overdamped coordinate [74, 85]. The similarity of polar solvation and vibrational coupling is not accidental: in pp chromophores molecular vibrations induce a flux of electronic charge back and forth between the D and A sites and hence plays exactly the same role as an electric field [86]. Much as with the reaction field, the amplitude of the oscillations self-consistently depends on the molecular polarity.

A pp chromophore dissolved in a polar solvent readjusts its polarity in response to the polarity of the surrounding, and its properties are affected accordingly. But other more subtle phenomena can be predicted, including amplification of NLO responses [74, 87], and inhomogeneous broadening effects as observed e.g in vibrational and resonant-Raman spectra [74]. Particularly interesting results were obtained modeling time-resolved absorption and fluorescence spectra of pp chromophores in polar solvents [88], were we were able to reproduce quantitatively the time evolution of the emission frequency and bandshapes as experimentally observed for several dyes.

The described approach to spectroscopic properties of pp chromophores in solution is semiempirical in nature: it is based on a specific model for the electronic structure, molecular vibration, and solute-solvent interactions that allows to calculate a large variety of (low-energy) spectroscopic properties, including steady-state and time resolved electronic and vibrational spectra, in terms of few microscopic parameters to be extracted from a judiciously chosen subset of experimental data. The approach has been validated and its predictive capabilities have been demonstrated via an extensive and detailed comparison with experimental data on several chromophores [74, 75, 88, 89]. The resulting picture is both robust and simple enough to offer a good starting point to describe the more complex behavior of dense clusters of pp chromophore where intermolecular interactions play a role.

5. INTERACTING POLAR AND POLARIZABLE MOLECULES: A TOY MODEL

Based on the discussion in the previous Section we define the following Hamiltonian for a cluster of non-overlapping pp chromophores:

$$(10) \quad H = \sum_i \left(2z\hat{\rho}_i - \sqrt{2}t\hat{\sigma}_{x,i}\right) + \sum_{i>j} V_{ij}\hat{\rho}_i\hat{\rho}_j$$

where i runs on the N molecular sites. The first term above describes each chromophore in terms of the two-state Mulliken model introduced in the previous section, and the last term accounts for electrostatic interchromophore interactions. We have defined $\hat{\rho}_i = (1 - \hat{\sigma}_{z,i})/2$ as the operator measuring the amount of CT from D to A in the i-th molecules, and $\hat{\sigma}_{z/x,i}$ as the z/x Pauli matrices defined on the two basis states for the i-th chromophores. Since the charge distribution on the i-th chromophore is fully defined by the expectation value of $\hat{\rho}_i$, the operator representing the electrostatic interaction between i and j chromophores is proportional to $\hat{\rho}_i\hat{\rho}_j$ through a proportionality constant, V_{ij}, measuring the interaction energy between the two fully zwitterionic (D^+A^-) molecules. Of course several models are possible for V_{ij}, including dipolar or multipolar approximations, possibly accounting for screening effects.

The natural basis set for the above Hamiltonian is given by the 2^N functions obtained from the direct product of the two basis states, $|DA\rangle$ and $|D^+A^-\rangle$ on each molecular site. The corresponding Hamiltonian matrix is easily diagonalized for clusters of finite dimension. Specifically, by exploiting translational symmetry, we obtained exact solutions for systems with up to 16 sites and periodic boundary conditions.

The model can be extended to account for e-ph coupling, but dealing with the coupled electronic and bosonic problem is computationally very demanding. Preliminary results have been obtained only for dimers [90]. As long as the adiabatic approximation holds, i.e. as long as electronic degrees of freedom (including intermolecular ones) are faster than vibrational degrees of freedom, the coupling between electrons and molecular vibrations can be accounted for via a self-consistent renormalization of molecular parameters [89] so that in that limit the basic physics of interacting chromophores is described by the Hamiltonian in Eq. (10).

The Hamiltonian in Eq. (10) is very simple but describes a fairly rich physics. Indeed it encompasses both the mf and the excitonic models for interacting chromophores. To demonstrate this important point we rewrite the same Hamiltonian on a different basis. Specifically, we define a new basis by rotating the two original basis states on each chromophore, $|DA\rangle$ and $|D^+A^-\rangle$, by an arbitrary angle. This amounts to use on each site a couple of orthogonal states $|g_i\rangle$ and $|e_i\rangle$ defined, according to Eq. (5), as a linear combination of the two original states, with an arbitrary mixing coefficient ρ, ranging from 0 to 1: the extreme values 0, 1 correspond to rotation angles of 0 and $\pi/2$, respectively. We define the vacuum state, $|0\rangle$, as the state where all molecules are in the $|g_i\rangle$ state: the complete basis is then obtained by repeated applications of the hard-boson creation operator $b_i^\dagger$ that, applied to $|g_i\rangle$ switches it to $|e_i\rangle$. With these definitions the Hamiltonian in Eq. (10) can be rewritten as the sum of three terms [91]:

$$H = H_{mf} + H_{ex} + H_{uex} \tag{11}$$

with

$$H_{mf} = \sum_i \left[2(1-2\rho)(z+M\rho) + 4\sqrt{2}t\sqrt{\rho(1-\rho)} \right] \hat{n}_i \tag{12}$$

$$+ \sum_i \left[2\sqrt{\rho(1-\rho)}(z+M\rho) - \sqrt{2}t(1-2\rho) \right] (\hat{b}_i^\dagger + \hat{b}_i)$$

$$H_{ex} = \sum_{i,j>i} V_{ij} \left[\rho(1-\rho)(\hat{b}_i^\dagger \hat{b}_j + \hat{b}_j \hat{b}_i^\dagger) + (1-2\rho)^2 \hat{n}_i \hat{n}_j \right] \tag{13}$$

$$H_{uex} = \sum_{i,j>i} V_{ij} \rho(1-\rho)(\hat{b}_i \hat{b}_j + \hat{b}_j^\dagger \hat{b}_i^\dagger) \tag{14}$$

$$+ \sum_{i,j\neq i} V_{ij}(1-2\rho)\sqrt{\rho(1-\rho)}(\hat{b}_i^\dagger + \hat{b}_i)\hat{n}_j$$

where $\hat{n}_i = \hat{b}_i^\dagger \hat{b}_i$ counts the excitations on the ith site, and $M = \sum_{i,j>i} V_{ij}/N$ is the Madelung energy. The transformation from the Hamiltonian in Eq. (10) to Eq. (11) is general, but the above expressions have been written for the special case where all molecular sites are equivalent; inequivalent sites can be treated analogously, but the notation becomes cumbersome.

The first term (H_{mf}) above describes an effective Hamiltonian for non-interacting chromophores, and then defines the mean-field approximation. In the mf approximation in fact the gs describes a collection of molecules each one in the local $|g_i\rangle$ states, so that it corresponds to the vacuum state for a specific choice of ρ. With this choice H_{exc} and H_{uex} are irrelevant in the definition of the gs. The requirement that $|0\rangle$ corresponds to the lowest eigenstate of H_{mf} fixes the optimal mf value of ρ as to impose the vanishing of the term in $(b_i^\dagger + b_i)$ in H_{mf}. This amount to fix ρ at the value relevant to a molecule feeling the electrostatic potential generated by all surrounding molecules in their local gs [91]. The self-consistent mf problem can be easily solved along the lines originally proposed by Soos and coworkers working on a similar Hamiltonian in a different context [92].

The mf solution for the gs fully accounts for the molecular polarizability: each molecular (polarizable) dipole readjusts itself in response to the local field created at its locations by all other (polarizable) dipoles. The feedback loop generated by interacting pp chromophores is the key to understand cooperative effects on gs properties, as it will be discussed in Section 6. However, the mf gs describes the molecular units as totally uncorrelated and therefore cannot account for collective behavior. In spite of that the mf gs is a good starting point to define the excitonic approximation. In particular, once ρ is fixed to the mf value the mf Hamiltonian simplifies in:

$$(15) \quad H_{mf} = \hbar\omega_{CT} \sum_i \hat{n}_i$$

where $\hbar\omega_{CT}$ is the local molecular excitation energy, whose dependence on ρ is defined by Eq. (9). This Hamiltonian, assigning energy $\hbar\omega_{CT}$ to each local excitation defines the zeroth order functions for the excitonic approximations. Indeed, the local excitation, $|g_i\rangle \to |e_i\rangle$ is fully defined by the mf solution. All relevant spectroscopic quantities are defined as functions of the mf ρ by Eqs. (6)–(9).

H_{ex} in Eq. (13) collects all terms in the Hamiltonian conserving the number of excitation and therefore defines the excitonic Hamiltonian. The first term in 13 accounts for exciton hopping and corresponds to H_J in Eq. (1), with $J_{ij} = V_{ij}\rho(1-\rho)$ describing the electrostatic interaction between transition dipole moments (cf Eq. (8)). The second term in (13) accounts for exciton-exciton interaction and corresponds to H_{e-e} in Eq. (3), with $U_{ij} = V_{ij}(1-2\rho)^2$ measuring the electrostatic interaction between mesomeric dipole moments (cf. Eq. (6), (7)). The sum of the mf Hamiltonian in Eq. (15) and of the excitonic term in Eq. (13), $H_{mf} + H_{exc}$, defines the best excitonic approximation to the general Hamiltonian in Eq. (10). In fact, relying on the mf description of the gs and of local excitations, it is based on the best uncorrelated description for the molecules in the cluster.

The third portion of the total Hamiltonian, H_{uex} in (14), collects all terms beyond the excitonic approximation, being responsible for the mixing of states with different number of excitons. The first term in this ultraexcitonic Hamiltonian corresponds to the non-Heitler London term in Eq. (2), as originating from the interaction between transition dipole moments. The second term describes the interaction between transition and mesomeric dipole moments and corresponds to the term in Eq. (4) with $W_{kl} = V_{kl}(1-2\rho)\sqrt{\rho(1-\rho)}$ (cf. Eqs. (8)–(7)).

All the terms in our transformed Hamiltonian were already derived by Agranovich [46] in the discussion of clusters of two-level molecules interacting via electrostatic forces. The main novelty of the proposed approach is that it is based on local (molecular) ground and excited states as obtained in the mf approximation and therefore it explicitly accounts for the dependence of the local ground and excited state on the supramolecular interactions. In the standard approach instead the term in $(b_i^\dagger + b_i)$ in Eq. (12) is disregarded [46], and the local wave functions do not depend on the supramolecular geometry. The strength of our approach is in its ability to follow the evolution of the properties of the supramolecular systems when the supramolecular structure is changed, then accounting at the same time for collective and cooperative behavior.

6. INTERACTING POLAR AND POLARIZABLE MOLECULES: COOPERATIVE EFFECTS ON OPTICAL SUSCEPTIBILITIES

In this Section we discuss optical susceptibilities of some representative clusters. Specifically, we consider one-dimensional arrays of equivalent molecules (periodic boundary conditions are imposed) with the three geometries sketched in Fig. 3, where the arrows represents the dipolar pp chromophores. In cluster A all chromophores are oriented in the same direction, perpendicularly to the stack axis. This geometry then corresponds to repulsive intermolecular interactions. Clusters B and C instead describe attractive lattices. In cluster B the antiparallel orientation of chromophores leads to a structure with two molecules per unit cell, an inversion center lying in the middle of each pair of molecules.

Several choices are possible for the definition of V_{ij}, the interaction energy between two zwitterionic molecules at i, j sites. The simplest approximation models $|D^+A^-\rangle$ molecules as point dipoles. The dipolar approximation is however poor for pp dyes, that are fairly elongated molecules. We therefore adopt a different model that, while retaining the simplicity of the dipolar approximation, accounts for the finite size of the chromophore. Therefore we model the zwitterionic molecule as a rigid rod of length l with the positive and negative charges located at the two ends. This defines the basic unit of electrostatic energy: $v = e^2/l$. In the following we discuss unscreened interactions and introduce the dimensionless inverse intermolecular distance, $w = l/r$, where r is the distance between the chromophores.

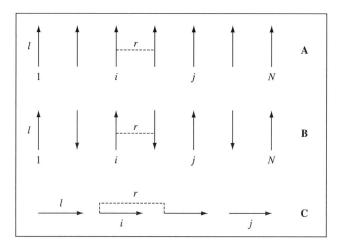

Figure 3. The three one-dimensional lattices considered in this paper. Each molecule is sketched by an arrow, representing its permanent dipole moment

For C-clusters the condition $w < 1$ applies. For any cluster V_{ij} interactions can be defined in terms of v and w parameters [91].

Fig. 4 shows the evolution of the gs polarity (ρ) with the inverse intermolecular distance for the three lattices with $N = 16$, and $v = 1$ (here and hereafter all energies are measured in units with $\sqrt{2}t = 1$). For pp chromophores $\sqrt{2}t \sim 1$ eV, so that $v = 1$ corresponds to molecular lengths of the order of 14 Å. For each cluster geometry, results are reported for $z = 1$ and $z = -1$ (top and bottom panels, respectively), corresponding to neutral and zwitterionic isolated molecules ($w = 0$), respectively. For the repulsive A cluster the gs polarity decreases with increasing w, i.e. with decreasing intermolecular distance. Of course the opposite behavior, with ρ increasing with increasing w, is observed in the attractive B and C clusters. In this respect, for A clusters the $z = -1$ case is particularly interesting: the isolated zwitterionic chromophores can be driven towards a neutral gs by simply packing them together at intermolecular distances smaller than about 2/3 of the molecular length (say 10 Å for molecules 14 Å long). For attractive B and C clusters neutral isolated chromophores ($z = 1$) can be turned zwitterionic by packing them at intermolecular distances of about 1/2 and 4/3 of the molecular length, respectively.

The possibility to completely reverse the nature of the molecular gs (from neutral to zwitterionic or viceversa) by simply putting molecules close together, and/or by changing their intermolecular packing or distance is an important consequence of cooperativity. This offers a new and powerful tool to tune the molecular and hence the material properties via a careful supramolecular design. On the back-side

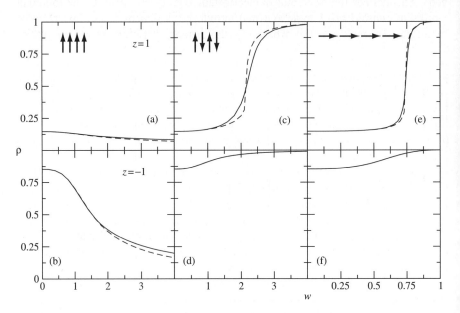

Figure 4. Ground state polarity vs the inverse intermolecular distance for A, B and C, lattices with $v=1$ and $z = \pm 1$. Dashed lines: mf results for $N = \infty$; continuous lines: exact results for $N = 16$.

however it makes the prediction of the material properties a difficult task that requires a careful modeling and understanding of supramolecular interactions.

The mf approximation offers simple clues to cooperativity: within mf each molecule is affected by the electric field created by the surrounding molecules. Cooperative behavior then results from the self-consistent dependence of the molecular polarity from the polarity of the surrounding molecules. The mf ρ, shown as dashed lines in Fig. 4, compares well with exact results obtained from the complete diagonalization of the Hamiltonian in (10) (full lines in Fig. 4). More interestingly the mf approach rationalizes the qualitatively different shape and slope of the $\rho(w)$ curves for attractive and repulsive clusters. In fact, in the mf picture, the molecular polarity is obtained by solving the two state molecular problem defined in Section 4, but with an effective z that self-consistently depends on the molecular polarity: $z \to z + M\rho$. The slope of the ρ curve vs M is then:

$$(16) \quad \frac{\partial \rho}{\partial M} = \frac{d\rho}{dz}\rho\left(1 - M\frac{d\rho}{dz}\right)^{-1}$$

Since $d\rho/dz < 0$, for repulsive lattices ($M > 0$) the $\rho(M)$ slope varies smoothly with M, and hence we observe fairly smooth $\rho(w)$ curves. For attractive clusters ($M < 0$) the $\partial \rho / \partial M$ slope can eventually change sign when M increases in magnitude, going through a divergence at $M = -2$. This explains the sharp curves calculated for B and C clusters (cf panels (c) and (e) in Fig. 4), and also leads to the interesting prediction of bistable behavior in attractive lattices as it will be discussed in the next Section.

Exact static optical susceptibilities are conveniently calculated as successive derivatives on an applied electric field of the gs polarization, defined, in linear aggregates, as the dipole moment per unit length. The linear polarizability (α), the first and second hyperpolarizabilities (β and γ, respectively) obtained for 16-sites clusters are shown as full lines in Fig. 5. Left, middle and right panels refer to A, B and C clusters, respectively, for the parameters that, in Fig. 4 drive the system through the neutral-zwitterionic interface. Susceptibilities show a strong and non-trivial dependence on the intermolecular distance, and, to understand the physical origin of this complex and interesting behavior we shall discuss several approximated results.

OGM, as discussed in Section 1, represents the simplest approach to NLO responses: the response of the cluster is defined as the sum of the molecular responses, properly accounting for the chromophore orientation in the cluster. This approach is for sure very poor for clusters of pp chromophores, since it disregards the dependence of the molecular polarity, and hence of all molecular properties, on intermolecular interactions. A slightly more refined approach is a mf-based OGM. The mf gs describes a collection of uncorrelated molecules, each one in the local (mf) gs. Accordingly, the mf-OGM approach calculates the susceptibilities of the cluster as the sum of the susceptibilities of the (oriented) molecules in their local mf gs. The mf-OGM approach for sure improves over the crude OGM, since it accounts for the polarizability of the molecular units and hence for the dependence of the

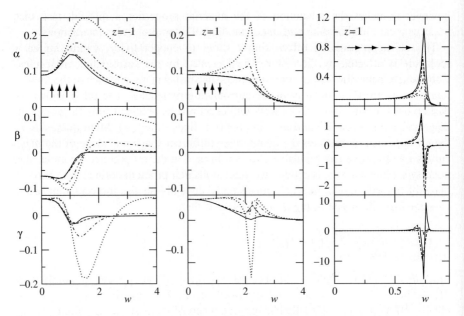

Figure 5. Linear polarizability (α), first and second hyperpolarizabilities (β and γ) vs the inverse intermolecular distance for A, B and C clusters (left, middle and right panels, respectively) with $N = 16$, $v=1$ and $z = \pm 1$. Dotted lines: mf-oriented gas approximation; dot-dashed line: excitonic approximation; dashed lines: mf approximation; continuous lines: exact results. The β response for B-clusters vanishes by symmetry

molecular polarity (and hence of the molecular properties) on the supramolecular structure. However the mf-OGM results, reported in Fig. 5 as dotted lines, grossly deviate from exact susceptibility, a surprising result, particularly if contrasted with the good quality of the mf approximation in the calculation of ρ (cf. Fig. 4). Indeed the mf-OGM fails since it does not account for local field corrections. Deviation of mf-OGM results from exact susceptibilities increase with the order on non-linearity and are particularly important for clusters with $\rho \sim 0.5$, where molecules are maximally polarizable. This suggests that the most important contributions to local field corrections arise from the molecular polarizability rather than from the molecular polarity (we notice that the effect of the local electric field generated by the molecules in their gs is already accounted for by the variation of ρ with the supramolecular arrangement and is therefore already included in the mf-OGM result). Another important observation is that intermolecular interactions strongly tune the molecular polarity, hence affecting the molecular and material responses, but this does not exhaust the role of intermolecular interactions. Depending on the cluster geometry and on model parameters, mf-OGM can either over- or underestimate NLO responses: the role of supramolecular interactions is far from trivial and a detailed modeling is needed to devise reliable relationships between the supramolecular structure and the properties of mm.

Linear and non-linear susceptibilities calculated within the mf approach, but relaxing the OGM approximation, lead to much better results, as shown by the dashed lines in Fig. 5. These results are calculated from the field derivatives of the total mf polarization, i.e. from the field derivatives of the sum of molecular dipole moments, rather than summing up derivatives of the molecular dipole moments as done in mf-OGM. Apart from very narrow regions around the neutral-zwitterionic interface in attractive lattices where the mf approximation itself is rather poor (cf results for ρ in Fig. 4, and the discussion in the next Section), the mf approximation to linear and non-linear susceptibilities is fairly good, suggesting that cooperativity dominates the gs properties of the material. This observation gives confidence on the reliability of mf approaches to static optical responses, as that proposed by Tsiper and Soos [25, 62, 63, 64] for the calculation of the linear polarizability of clusters of non-polar molecules, and suggests its extension to clusters of polar molecules, and to the calculation of non-linear susceptibilities.

Sum-over-state (SOS) expressions [35] relate static susceptibilities to optical excitations. Of course (hyper)polarizabilities obtained as field derivatives of P coincide with SOS results, provided that the calculations refer to the same Hamiltonian. In this respect it is particularly interesting to discuss the merit of the excitonic approximation in the calculation of susceptibilities. Our best excitonic approximation to the Hamiltonian for interacting molecules in Eq. (10) defines the vacuum states for the excitonic model as the mf gs, i.e. adopts as gs the best uncorrelated solution of the total Hamiltonian. Therefore exactly the same OGM and mf-OGM susceptibilities discussed above can be obtained also in the EM approach from the derivatives or the relevant gs polarization. As a matter of fact mf-OGM results can also be obtained from a SOS calculation, summing over the excited states of the cluster described by the mf Hamiltonian in Eq. (15), or, equivalently, by summing up the SOS susceptibilities calculated for the chromophores in their mf gs. Instead, susceptibilities obtained by summing over the excited states calculated in the excitonic approximation (i.e. by diagonalizing $H_{mf} + H_{ex}$) are different, and are reported as dot-dashed lines in Fig. 5. The excitonic susceptibilities improve over mf-OGM, but badly deviates from exact results. So, accounting for exciton delocalization, as described in the excitonic picture, partly accounts for the cooperative effects in the gs, but EM estimates of linear or non-linear polarizabilities are not reliable. We notice in particular that the failure of the excitonic model for polarizabilities is already apparent in regions where the mf approximation to the same quantities is very good. Our best excitonic model starts with the mf gs and then fully accounts for the molecular polarizability in the definition of the local ground and excited states. However, not allowing for the mixing of states with different number of excitations, it does not allow for the reorganization of the molecular polarity in response to excitations on nearby sites. So we expect ultraexcitonic corrections to be particularly large in the definition of excited states with a large number of excitations and particularly so for polar and polarizable molecules. Similarly, large ultraexcitonic contributions to linear and non-linear polarizabilities are understood since the electric field itself mixes up states with a different number of excitations and this mixing breaks down the excitonic approximation. In other terms, reliable estimates of (hyper)polarizabilities of molecular clusters cannot be obtained within the EM approximation that, not allowing for the mixing of states with a different number of excitations, does not account for the readjustment of the charge distribution on the molecular units in response to the applied fields: (hyper)polarizabilities cannot be reliably described by models that do not properly account for the molecular (hyper)polarizabilities.

7. EXCITED STATES IN CLUSTERS OF POLAR AND POLARIZABLE MOLECULES: EXTREME COLLECTIVE AND COOPERATIVE BEHAVIOR

To better appreciate the merits and limits of EM when applied to clusters of pp chromophores we now discuss in some detail excited states. Fig. 6 shows the complete set of excitation energies obtained in the exciton approximation (i.e. diagonalizing $H_{mf} + H_{exc}$). Specifically, for each eigenstate we plot the energy (setting to zero the gs energy) vs the exciton number, n. In order to avoid overcrowded figures we report data for clusters with $N=6$. The same parameters are chosen as in Fig. 5, but data are reported only for two w values for each cluster, corresponding to interactions of weak and medium strength (upper and lower panels, respectively): the insets show the relevant $\rho(w)$ curves, with the vertical dotted lines marking the w values of the corresponding parent panels.

Since n is conserved in EM, in all panels of Fig. 6 the excitonic states group in bands with the same n. The energy spread of the excitonic bands is determined by the exciton hopping, while the exciton-exciton interactions are responsible for (de)stabilization of multiexcitonic states. Within each band, states can be labeled by their wave-vector, k; for example, for clusters A and B, the $k=0$ one-exciton state is at the top of the one-exciton band, while for C clusters it is at the bottom (see states marked by error bars). As a matter of fact, A and C clusters correspond to H- and J-aggregates, respectively (cf Fig. 1). B clusters are somehow different: intermolecular interactions are attractive (as much as in J-aggregates), but there are two molecules per unit cell and hence two $k=0$ states, and out of them only that one lying at the top of the one-exciton band (corresponding to the antisymmetric state with respect to the inversion center lying between each couple of chromophores) is allowed by one photon absorption.

Ultraexcitonic terms in the Hamiltonian mix states with a different number of excitons: exact excitation energies, reported in Fig. 7, are not perfectly aligned in bands. For weak interaction (upper panels) deviations from EM are small, and the exciton number is almost conserved, confirming the validity of EM to describe low-lying excitations for clusters of weakly interacting molecules. But for larger interactions (lower panels) the failure of the excitonic approximation becomes apparent. We underline that increasing intermolecular interactions (increasing w) has a twofold effect: on one side it increases the absolute value of ultraexcitonic mixing in Hamiltonian 14, on the other side, for the chosen parameters, it drives the

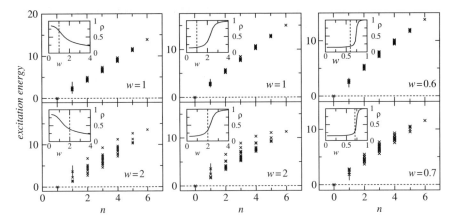

Figure 6. Excitation energy calculated in the excitonic approximation against the number of excitons n, for A (left panels), B (middle panels) and C clusters (right panels) with $N=6$, $v=1$, for two different w. The z parameter is fixed to -1 for the A cluster, to 1 for the B and C clusters. States on the zero energy axis correspond to the gs. Error bars measure the squared transition dipole moment from the gs to the relevant states. Insets show the $\rho(w)$ mf curve for the relevant parameters, with the dotted vertical line marking the w value for which results are reported in the parent panel

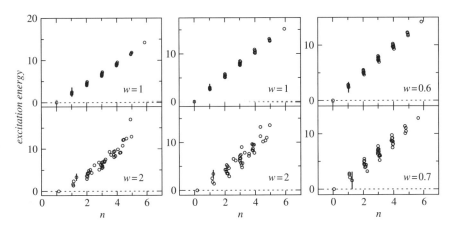

Figure 7. The same excitation spectrum as in Fig. 6, but obtained from the diagonalization of the complete Hamiltonian

molecules towards states with $\rho \to 0.5$, where the molecules are more polarizable and hence the EM approximation more dangerous.

Error bars in Figs. 6 and 7 measure for each excited state, the squared transition dipole moment from the gs and are therefore proportional to the intensity of the relevant transition (notice that in all panels a single state has appreciable intensity). In EM, the absorption of one photon always populates the zero-wavevector one-exciton state, creating just a single exciton. While this stays approximately true for weak intermolecular interactions (upper panels), EM is completely spoiled for stronger interactions. In fact data in the lower panels of Fig. 7 show that the absorption of one photon populates a state having an exciton number sizeably different from 1. So ultra-excitonic terms in the Hamiltonian breaks the $\Delta n = 1$ selection rule for optical transitions. This interesting effect becomes even more important in attractive lattices where the charge crossover becomes discontinuous, as it will be shortly discussed below.

While discussing $\rho(w)$ curves in the previous Section, we anticipated that within mf bistable behavior is predicted for attractive (B or C) clusters for specific choices of model parameters. This is shown in Fig. 8 where S-shaped $\rho(w)$ curves obtained in mf for a C-cluster clearly indicate the presence of a bistability region, i.e. a region where for the same w two states with different polarity are both stable (the states lying in the portion of the $\rho(w)$ curve with negative slope correspond to unstable states). Bistability signals the presence of a discontinuous neutral-zwitterionic interface [91], that closely resembles the well known neutral-ionic phase transition observed in charge-transfer crystals with a mixed (D-A) stack motif [11, 92]. Bistability is hardly recognized in the exact $\rho(w)$ curves, since only stable states are found by exact diagonalization, neither metastable nor unstable states are addressed: exact diagonalization always leads to single-valued $\rho(w)$ curves. Hints

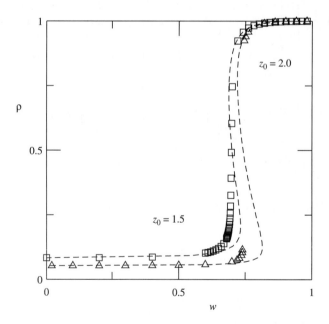

Figure 8. Exact ground state polarity vs the inverse intermolecular distance for a 16-site C cluster with $v = 2$ and $z = 1.5$ (squares) and 2 (circles). Dashed lines show $N = \infty$ mf results

of discontinuity are however easily appreciated in the exact curves in Fig. 8 (squares and triangles).

The appearance of bistability regions in clusters of pp chromophores is a very interesting result. It represents the extreme consequence of cooperativity, and has no counterpart for isolated molecules. Moreover, if we are able to prepare our material in the close proximity of the bistability region, any tiny variation of external conditions can easily switch the material between two phases with macroscopically different properties, a quite appealing phenomenon in view of the possibility to produce molecular-based switches.

The mf ρ in Fig. 8 show sizeable deviations from the exact result in the proximity of the discontinuous crossover: this marks the appearance of correlation of the electronic motion on different molecules [91, 93]. But before addressing this important point we concentrate attention on the excitation spectrum near the discontinuous charge crossover. In Fig. 9 the n-dependence of excitation energies is reported for a 10-site C cluster with parameters corresponding to a discontinuous neutral-zwitterionic crossover (as in Fig. 8, $z = 1.5$). Results are reported for $w = 0.69$, corresponding to a neutral gs, just before the abrupt transition to the zwitterionic gs. For the sake of clarity, only $k = 0$ eigenstates are reported, with circles and crosses referring to exact and excitonic results, respectively. Error bars in the figures are proportional to the squared transition dipole moment from the gs.

Data in Fig. 9 shows that excitonic bands relevant to states with different n overlap in energy: in these conditions ultra-excitonic terms mix up degenerate or quasi-degenerate states and are therefore responsible for large deviations from EM. The exciton number is not even approximately conserved for exact results in Fig.9, and, what is even more impressive, the energy of excited-states shows a

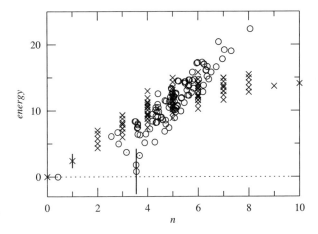

Figure 9. The (zero-wavevector) spectrum of a 10 site C-cluster with $v = 2$, $z = 1.5$ and $w = 0.69$. Excitation energies of zero-wavevector states are reported vs n. Circles and crosses refer to exact and excitonic eigenstates, respectively. Error bars are proportional to the squared transition dipole moment from the ground-state

non-monotonic dependence on n: a state with $n \sim 3.5$ has lower energy that several states with lower n. As a matter of fact, the low-energy $n \sim 3.5$ state is the state with the largest oscillator strength. This means that the absorption of one photon creates more that 3 excitations at a time, switching more than three molecules from the local ground to the local excited state. This result completely spoils the excitonic picture where a single excitation is created upon photon absorption.

To better understand this result we go back to the original basis built from the $|DA\rangle$ and $|D^+A^-\rangle$ states and calculate the number of $|D^+A^-\rangle$ molecules that are created upon photoexcitation, or, equivalently, the number of electrons that are transferred from D to A moieties: $\Delta = N(\langle E_1|\hat{\rho}|E_1\rangle - \langle G|\hat{\rho}|G\rangle)$ (negative Δ means that electrons are transferred from A^- to D^+). In EM the absorption of a photon creates a single excitation, switching a molecule from the local ground to the local excited state. Then in this approximation $\Delta = 1 - 2\rho$, and the number of transferred electrons upon photoexcitation is always smaller than 1. The upper panels in Fig. 10 shows the evolution with w of Δ, calculated for a 16-site C lattice near to a continuous and a discontinuous neutral to zwitterionic interface (left and right panels, respectively, cf. bottom panels, where the relevant $\rho(w)$ curves are shown). The dotted lines mark the extreme limits of the excitonic approximation for Δ, i.e. $|\Delta| < 1$: no more than a single electron is transferred upon photoexcitation according to EM. This simple result is spoiled by ultraexcitonic coupling: deviations are minor near a continuous interface (left panels), but become important near a discontinuous interface: for the parameters in Fig. 10, up to 6 electrons are transferred upon absorption of a single photon.

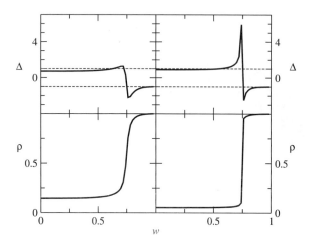

Figure 10. Upper panels: number of zwitterionic species created upon photoexcitation (Δ) vs the inverse interchromophore distance. Dotted lines delimitate the Δ-region allowed according to the exciton model. Lower panels: corresponding evolution of the gs polarity. Results are obtained for a 16-site C cluster. Left panels: $v = 1$, $z = 1$; right panels: $v = 2$, $z = 2$

To investigate in greater detail this phenomenon we define the following l-th order correlation function:

$$(17) \quad f_l = \sum_{i=1}^{N} \langle \hat{\rho}_i \hat{\rho}_{i+1} \cdots \rho_{i+l-1} \rangle - \sum_{i=1}^{N} \langle \hat{\rho}_i \rangle^l$$

This function exactly vanishes for uncorrelated states, i.e. states that can be defined as products of local molecular states (as in the mf and EM assumption). Positive (negative) f_l indicates instead an increased (decreased) probability of finding l nearby zwitterionic molecules with respect to the uncorrelated state at the same average polarity. Fig. 11 shows the l dependence of f_l calculated for the most optically active state (left panel) and for the gs (right panel) of the same cluster as in Fig. 9. For the active excited state, a sizeable weight is found of wave functions with several (say 2 to 6) nearby fully zwitterionic molecules (we will call these wave functions 'I-droplets'). This demonstrates that the phenomenon of photoinduced multielectron transfer corresponds to a concerted motion of electrons on several nearby molecular sites [93].

But also the gs deviates from the uncorrelated limits, with $f_l \neq 0$, in Fig. 11, panel b), even if, quite predictably, deviations are much smaller than for excited states. Finite f_l near the discontinuous neutral-zwitterionic crossover show that the gs cannot be described as the product of local molecular states and therefore does not coincide with the mf or excitonic vacuum state: the very same gs is collective in nature and I-droplet states contribute to the gs. This is the key to understand multielectron transfer: wave functions with I-droplet character have, in C-clusters, very large permanent dipole moments, so that their finite amplitude in the gs is the origin of sizeable transition dipole moments towards states characterized by a high I-droplet character.

Discontinuous charge crossovers can also be observed for B clusters, that, much as C clusters have attractive intermolecular interactions. Multielectron transfer instead is not observed for B geometry. B and C lattices in fact have different symmetry

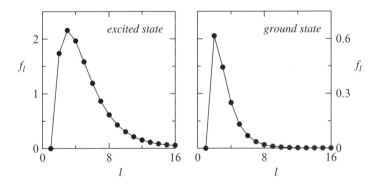

Figure 11. The l dependence of the f_l correlation function for a 16-site C-cluster with $v = 2$, $z = 1.5$ and $w = 0.69$. Left and right panels refer to the excited state with the largest transition dipole moment, and to the gs, respectively. Lines are drawn as guide for eyes

and hence a qualitatively different spectroscopic behavior. In particular, lattice C has one molecule per unit cell and all $k = 0$ state are optically allowed, even if most of them have negligible intensity. The two molecules per cell in lattice B are exchanged by reflection, so that only $k = 0$ antisymmetric states are accessible by one-photon absorption from the (totally-symmetric) gs. The special low-energy state with large n is a dark state in B-clusters, but corresponds to the state with the largest oscillator strength in C-clusters.

The phenomenon of multielectron transfer is an extreme manifestation of cooperative and collective behavior. Collective behavior is recognized in the nature of the state reached upon photoexcitation with large correlation of electrons on several nearby molecules. But the gs itself near the discontinuous charge crossover is collective in nature, even if at a lower extent than excited states. The cooperative nature of multielectron transfer is recognized in the fact that a single-electron operator, like the transition dipole moment, is responsible for the motion of several electrons. Cooperative and collective phenomena are strictly connected near a discontinuous charge crossover where intermolecular interactions have extremely large effects on the material behavior: cooperativity and collectivity have in our model the very same origin. Indeed electrostatic intermolecular interactions dominate the physics of the material at the discontinuous charge crossover: they are responsible for the appearance of bistability, the extreme manifestation of cooperative behavior, as well as for excitonic and ultraexcitonic mixing of states with the same and different exciton number, and hence for nontrivial collective behavior.

8. CONCLUSIONS

Molecular functional materials, and specifically mm for NLO applications, are based on large π-conjugated molecules whose large (hyper)polarizability is responsible for the qualifying properties of the material itself, but, at the same time, makes our understanding of the material properties quite difficult. Non-trivial collective and cooperative behavior in materials based on large conjugated dyes were predicted 40 years ago [7] and have been discussed at different level of sophistication in different approaches ranging from the crude OGM [16, 17], to the mean field approximation for the calculation of polarization and polarizability [25], to the excitonic model for optical spectra [18]. Sophisticated quantum chemical calculations on supermolecular structures also shed some light on the complex role of intermolecular interactions in mm [44], but these calculations are computationally too demanding to allow for a wide statistics. Moreover the underlying models are very complex and extracting a coherent picture from the resulting data is a very hard task.

We follow a different, complementary strategy based on the definition of extremely simple (toy) models for the material where only few interactions are accounted for. Toy models are easily diagonalized on large systems, and an enormous amount of calculations can be performed testing different parameters regimes at an irrelevant fraction of the computational power required by quantum chemical calculations. This opens the way to appreciate the role of different interactions and to test approximation schemes, leading, in favorable cases to a clear physical picture of the phenomena of interest. Of course toy models lack the accuracy of more refined models, and, being semiempirical in nature, must rely on an extensive validation against experimental data.

In this contribution we discuss mm based on pp chromophores, a very interesting class of molecules for applications in molecular photonics and electronics. Push-pull chromophores are both polar and polarizable and this makes the role of intermolecular interactions particularly important. The toy model we propose for clusters of pp chromophores neglects intermolecular overlap, just accounting for classical electrostatic intermolecular interactions, and describes each pp chromophore based on a two state model. The two-state model for pp chromophores has been discussed and validated via an extensive comparison with the spectroscopic properties of several dyes in solution [74, 75, 90]. The emerging picture is safe and led to the definition of a reliable set of molecular parameters for selected dyes. This analysis then offers valuable information to be inserted into models for clusters of interacting chromophores, in a the bottom-up modeling strategy that was nicely exemplified in Ref. [90].

Whereas in this paper we focus attention on mm based on pp chromophores, some fundamental concepts emerge from the discussion with wider applicability. In Section 5 we defined the best EM for the working Hamiltonian as the EM that has the same gs, and the same local excitations as obtained in the mf approximation. The relation between the mf and the excitonic approach is indeed fundamental. In EM excitations are created on top of an uncorrelated gs: the mf solution then gives the best gs for the excitonic calculation. The definition of the local states for the excitonic problem as the eigenstates of the local mf Hamiltonian unambiguously defines the excitonic and ultraexcitonic Hamiltonian for any supramolecular arrangement, based on the adopted model for the isolated molecule. Uncovering a direct link from the molecular to the supramolecular description gives an important contribution to our understanding of mm, and is a fundamental step to devise approaches to guide the chemical synthesis of functional mm from the molecular to the supramolecular level. The mf description of mm can be set up for different kinds of molecules described at different level of sophistication and is therefore of general applicability.

The reliability of mf approximation for the calculation of static linear and non-linear susceptibilities is another important result that is expected to hold for mm with negligible intermolecular overlap and not too near to charge instabilities (i.e. to phase transitions or to precursor of phase transitions in finite size systems). This is related to the uncorrelated nature of the gs in these materials, that is fairly well captured within mf. On the opposite, EM does not properly account for the molecular polarizability: it can possibly describe low-lying excited states in clusters of weakly interacting and hardly polarizable molecules, but it is for sure inadequate to calculate linear and non-linear susceptibilities for mm of interest for applications.

As for clusters of pp chromophores we have shown that classical electrostatic intermolecular interactions lead to the appearance of new phenomena that widely extend the scope of applications of mm. The possibility to tune the polarity of the molecule by simply affecting the geometry of the cluster adds new value to the supramolecular engineering of mm. To fully exploit the potential of mm, we must however extend our knowledge of structure properties relationships from the molecular to the supramolecular level. The discontinuous charge crossover and the related bistable regime open the possibility of molecular-based switches that can be driven by applying pressure or stresses to the sample. The observation of multielectron transfer suggests on one side the possibility of photoinduced transformations that were indeed observed [94] in related materials like CT salts with a mixed stack motif. On the other side, the possibility to move several electrons upon absorption of a single photon opens new perspectives for efficient photoconversion devices.

ACKNOWLEDGEMENTS

We gratefully thank Aldo Brillante, Alberto Girlando and Zoltán G. Soos for discussions and correspondence. Support from the Italian MIUR is acknowledge through project FIRB-RBNE01P4JF and by INSTM through PRISMA 2002. The work of F.T. was in part done while PhD student in Parma with a grant supported by INSTM. F.T. acknowledges support by a Marie Curie Intra-European Fellowship within the 6th European Community Framework Programme.

REFERENCES

[1] Heath, J.R., Ratner, M.A.: Physics Today, May issue, **43**, and references therein (2003)
[2] Prasad, P.N., Reinhardt, B.A.: Chem. Mater. **2**, 660 (1990)
[3] Hide, F., Diaz-Garcia, M.A., Schwartz, B.J., Heeger, A.J.: Acc. Chem. Res. **30**, 430 (1997)
[4] Friend, R.H., Gymer, R.W., Holmes, A.B., Burroughes, J.H., Marks, R.N., Taliani, C., Bradley, D.D.C., Dos Santos, D.A., Brédas, J.L., Lögdlund, M., Salaneck, W.R.: Nature, **397**, 121 and references therein (1999)
[5] Pulleritis, T., Sundström, V.: Acc. Chem. Res. **29**, 381 (1996)
[6] Skotheim, T.A., Elsenbaumer, R.L., Reynolds, J.R.: (eds) Handbook of Conducting Polymers, Marcel Dekker, New York (1998)
[7] Krugler, J.I., Montgomery, C.G., McConnell, H.M.: J. Chem. Phys. **41**, 2421 (1964)
[8] See, e.g., Burns, G., Solid State Physics, Chap. 6, Academic, San Diego (1990)
[9] Dykstra, C.E., Adv. Chem. Phys. **126**, 1 (2003)
[10] Craig, D.P., Walmsley, S.H.: Excitons in Molecular Crystals, Benjamin (1968)
[11] Soos, Z.G., Klein, D.J.: In Molecular Association: Including Molecular Complexes, Vol. 1, p. 1, R. Foster (ed.) Academic, New York (1975)
[12] Painelli, A., Girlando, A.: J. Chem. Phys. **84**, 5665 (1986)
[13] Frenkel, J.I.: Phys. Rev. **37**, 17 (1931)
[14] Davydov, A.S.: Theory of Molecular Excitons, Plenum Press, New York and references therein (1971)
[15] Craig, D.P.: J. Chem. Soc. 2302 (1955)
[16] Pope, M., Swenberg, C.E.: Electronic Processes in Organic Crystals and Polymers, Oxford Univeristy Press, New York (1999)
[17] Silinsh, E.A., Capek, V.: Organic Molecular crystals, AIP Press (1994)

[18] Knoester, J.: In Organic Nanostructures: Science and Applications, Proceedings of the International School of Physics Enrico Fermi, CXLIX Course, Agranovich, M., La Rocca, G.C., (eds) IOS Press, Amsterdam, and references therein (2002)
[19] Scholes, G.D.: Annu. Rev. Phys. Chem. **54**, 57 and references therein (2003)
[20] Malagoli, M., Munn, R.W.: J. Chem. Phys. **107**, 7926 and reference therein (1997)
[21] Reis, H., Papadopoulos, M.G., Munn, R.W.: J. Chem. Phys. **109**, 6828 (1998)
[22] Reis, H., Papadopoulos, M.G., Calaminici, P., Jug, K., Koster, A.M.: Chem. Phys. **261**, 359 (2000)
[23] Reis, H., Papadopoulos, M.G., Theodorou, D.N.: J. Chem. Phys. **114**, 876 (2001)
[24] Mazur, G., Petelenz, P.: Chem. Phys. Lett. **324**, 161 (2000)
[25] Soos, Z.G., Tsiper, E.V., Pascal Jr., R.A.: Chem. Phys. Lett. **342**, 652 (2001)
[26] Soos, Z.G., Tsiper, E.V., Painelli, A., Lumin. J.: **110**, 332 (2004)
[27] Dick, B.: (ed.) Chem. Phys. **245**, Special issue on Molecular Nonlinear Optics: Materials, Phenomena and Devices, (1999)
[28] Kanis, D.R., Ratner, M.A., Marks, T.J.: Chem. Rev. **94**, 195 (1994)
[29] Marder, S.R., et al.: Nature **388**, 845 (1997)
[30] Blanchard-Desce, M., Alain, V., Bedworth, P.V., Marder, S.R., Fort, A., Runser, C., Barzoukas, M., Lebus, S., Wortmann, R.: Chem. Eur. J. **3**, 1091 (1997)
[31] Barzoukas, M., Blanchard-Desce, M.: J. Chem. Phys. **113**, 3951 (2000)
[32] Science **295**, special issue on Supramolecular Chemistry and Self-Assembly, edited by Service, R.F., Szuromi, P., Upperbrink, J (2002)
[33] Dalton, L.R.: J. Phys.: Condens. Matter **15**, R897 and references therein (2003)
[34] Bloembergen, N.: Nonlinear Optics, Addison-Wesley, Reading MA (1992)
[35] Orr, B.J., Ward, J.F.: Mol. Phys. **20**, 513 (1971)
[36] Dirk, C.W., Twieg, R.J., Wagniere, G.: J. Am. Chem. Soc. **108**, 5387 (1986); Wagniere, G.H., Hutter, J.B.: J. Opt. Soc. Am. B **6**, 693 (1989)
[37] Di Bella, S., Ratner, M.A., Marks, T.J.: J. Am. Chem. Soc. **114**, 5842 (1992)
[38] Spano, F.C.: J. Chem. Phys. **96**, 8109 (1992)
[39] Abe, S.: Chem. Phys. **264**, 355 (2001)
[40] Belkin, M.A., Shen, Y.R., Flytzanis, C.: Chem. Phys. Lett. **363**, 479 (2002)
[41] Tretiak, S., Mukamel, S.: Chem. Rev. **102**, 3171 and references therein (2002)
[42] Datta, A., Pati, S.K.: J. Chem. Phys. **118**, 8420 (2003)
[43] Guillaume, M., Botek, E., Champagne, B., Castet, F., Ducasse, L.: Int. J. Quantum Chem. **90**, 1378 (2002)
[44] Champagne, B., Bishop, D.M.: Adv. Chem. Phys. **126**, 41 (2003)
[45] Hamada, T.: J. Phys. Chem. **100**, 8777 (1996)
[46] Agranovich, V.M., Galanin, M.D.: Electronic Excitations Energy Transfer in Condensed Matter, North-Holland, Amsterdam (1982)
[47] Bakalis, L.D., Knoester, J.: J. Chem. Phys. **106**, 6964 (1997)
[48] Spano, F.C., Siddiqui, S.: Chem. Phys. Lett. **314**, 481 (1999)
[49] Spano, F.C., Agranovich, V., Mukamel, S.: J. Chem. Phys. **95**, 1400 (1991); Spano, F.C., Manas, E.S.: J. Chem. Phys. **103**, 5939 (1995)
[50] Ezaki, H., Tokihiro, T., Hanamura, E.: Phys. Rev. B **50**, 10506 (1994); Kuwata-Gonokami, M., Peyghambarian, N., Meissner, K., Fluegel, B., Sato, Y., Ema, K., Shimano, R., Mazumdar, S., Guo, F., Tokihiro, T., Ezaki, H., Hanamura, E.: Nature **367**, 47 (1994)
[51] Liptay, W.: Angew. Chem. **8**, 177 (1969)
[52] Reichardt, C.: Chem. Rev. **94**, 2319 (1994)
[53] Marder, S.R., Cheng, L.T., Tiemann, B.G., Friedli, A.C., Blanchard-Desce, M., Perry, J.W., Skindhoej, J.: Science **263**, 511 (1994)
[54] Verbiest, T., Houbrechts, S., Kauranen, M., Clays, K., Persoons, A., Mater, J.: Chem. **7**, 2175 (1997)
[55] Rekai, E.D., Baudin, J.-B., Jullien, L., Ledoux, I., Zyss, J., Blanchard-Desce, M.: Chem. Eur. J. **7**, 4395 (2001)
[56] Metzger, R.M.: Chem. Rev. **103**, 3802 and references therein (2003)

[57] Ashwell, G.J., Chwialkowska, A., High, L.R.H.: J. Mater. Chem. **14**, 2389, and references therein (2004)
[58] Resta, R.: J. Phys.: Condens. Matter **14**, R625 (2002)
[59] Nunes, R.W., Gonze, X.: Phys. Rev. B **63**, 155107 and references therein (2001)
[60] Soos, Z.G., Bewick, S.A., Peri, A., Painelli, A.: J. Chem. Phys. **120**, 6712 and references therein (2004)
[61] Ye, P., Shen, Y.R.: Phys. Rev. B **28**, 4288 , and references therein (1983)
[62] Tsiper, E.V., Soos, Z.G.: Phys. Rev. B **64**, 195124 (2001)
[63] Tsiper, E.V., Soos, Z.G., Gao, W., Kahn, A.: Chem. Phys. Lett. **360**, 47 (2002)
[64] Tsiper, E.V., Soos, Z.G.: Phys.Rev. B **68**, 085301 (2003)
[65] Myers Kelley, A.B.: J. Phys. Chem. **103**, 6891 (1999)
[66] Oudar, J.L., Chemla, D.S.: J. Chem. Phys. **66**, 2664 (1977)
[67] Mulliken, R.S.: J. Am. Chem. Soc. **74**, 811 (1952)
[68] see e.g. Painelli, A.: Chem. Phys. Lett. **285**, 352 and references therein (1998)
[69] Barzoukas, M., Runser, C., Fort, A., Blanchard-Desce, M.: Chem. Phys. Lett. **257**, 531 (1996)
[70] Kogej, T., Beljonne, D., Meyers, F., Perry, J.W., Marder, S.R., Brédas, J.L.: Chem. Phys. Lett. **298**, 1 (1998)
[71] Abbotto, A., Beverina, L., Bozio, R., Bradamante, S., Ferrante, C., Pagani, G.A., Signorini, R.: Adv. Mater. **12**, 1963 (2000)
[72] Holstein, T.: Ann. Phys. **8**, 325 (1959)
[73] Schatz, G.C., Ratner, M.A.: Quantum mechanics in Chemistry, Prentice-Hall, Englewood Cliffs, NJ (1993)
[74] Painelli, A., Terenziani, F.: J. Phys. Chem. A. **104**, 11041 (2000); Terenziani, F., Painelli, A., Comoretto, D.: J. Phys. Chem. A **104**, 11049 (2000)
[75] Boldrini, B., Cavalli, E., Painelli, A., Terenziani, F.: J. Phys. Chem. A **106**, 6286 (2002)
[76] Painelli, A., Del Freo, L., Terenziani, F.: In: Molecular Low Dimensional Materials for Advanced Applications, Graja, A., Agranovich, V.M., Kajzar, F.: (eds) Kluwer Academic Publisher, Dordrecht, The Netherlands, p. 113 (2002)
[77] Del Freo, L., Terenziani, F., Painelli, A.: J. Chem. Phys. **116**, 755 (2002)
[78] Del Freo, L., Painelli, A.: Chem. Phys. Lett. **338**, 208 (2001)
[79] Painelli, A., Del Freo, L., Terenziani, F.: Synth. Met. **121**, 1465 (2001)
[80] Painelli, A., Terenziani, F.: Chem. Phys. Lett. **312**, 211 (1999)
[81] Painelli, A., Del Freo, L., Terenziani, F.: Chem. Phys. Lett. **346**, 470 (2001)
[82] Painelli, A., Del Freo, L., Terenziani, F.: In Nonlinear Optical Responses of Molecules, Solids and Liquids: Methods and Applications, p. 39, Papadopoulos, M.G., (ed.) Research Signpost, India (2001)
[83] Zojer, E., Wenseleers, W., Halik, M., Grasso, C., Barlow, S., Perry, J.W., Marder, S.R., Brédas, J.L., Chem Phys Chem **5**, 982 (2004)
[84] Gehlen, J.N., Chandler, D., Kim, H.J., Hynes, J.T.: J. Phys. Chem. **96**, 1748 (1992)
[85] Mukamel, S.: Principles of Nonlinear Optical Spectroscopy Oxford University Press, New York (1995)
[86] Painelli, A., Terenziani, F.: Synth. Metals, **124**, 171 (2001)
[87] Painelli, A., Terenziani, F.: Synth. Metals, **109**, 229 (2000)
[88] Terenziani, F., Painelli, A.: Chem. Phys. **295**, 35 (2003)
[89] Terenziani, F., Painelli, A., Girlando, A., Metzger, R.M.: J. Phys. Chem. B **108**, 10743 (2004)
[90] Terenziani, F., Painelli, A.: J. Lumin. **112**, 474 (2005)
[91] Terenziani, F., Painelli, A.: Phys. Rev. B **68**, 165405 (2003)
[92] Soos, Z.G., Keller, H.J., Moroni, W., Nothe, D.: Ann. N.Y. Acad. Sci. **313**, 442 (1978)
[93] Painelli, A., Terenziani, F.: J. Am. Chem. Soc. **125**, 5624 (2003)
[94] Iwai, S., Tanaka, S., Fujinuma, K., Kishida, H., Okamoto, H., Tokura, H.: Phys. Rev. Lett. **88**, 057402 (2002)

CHAPTER 8

MULTICONFIGURATIONAL SELF-CONSISTENT FIELD-MOLECULAR MECHANICS RESPONSE METHODS

KURT V. MIKKELSEN
Danish Center for Scientific Computing, Department of Chemistry, University of Copenhagen, DK-2100 Copenhagen Ø, Denmark

Abstract: The fundamental aspects of response theory for the multiconfigurational self-consistent field electronic structure method coupled to molecular mechanics force fields are outlined. An overview of the theoretical developments presented in the work by Poulsen et al. is given. Poulsen et al. have developed multiconfigurational self-consistent field molecular mechanics (MCSCF/MM) response methods to include third order molecular properties and these approaches are discussed

Keywords: Multiconfigurational self-consistent field molecular mechanics response methods, MCSCF/MM energies, MCSCF/MM response properties to third order

1. INTRODUCTION

The present contribution concerns an outline of the response theory for the multiconfigurational self-consistent field electronic structure method coupled to molecular mechanics force fields and it gives an overview of the theoretical developments presented in the work by Poulsen et al. [7, 8, 9]. The multiconfigurational self-consistent field molecular mechanics (MCSCF/MM) response method has been developed to include third order molecular properties [7, 8, 9]. This contribution contains a section that describes the establishment of the energy functional for the situation where a multiconfigurational self-consistent field electronic structure method is coupled to a classical molecular mechanics field. The second section provides the necessary background for forming the fundamental equations within response theory. The third and fourth sections present the linear and quadratic, respectively, response equations for the MCSCF/MM response method. The fifth section gives a short overview of the results obtained by the MCSCF/MM response method. A conclusion is given in the final section.

2. ENERGY FUNCTIONAL

The starting point for the MCSCF/MM response method is the Hamiltonian for the total system containing both the quantum mechanical and the classical system and it is given by the sum of three terms: the Hamiltonian of the quantum mechanical system in vacuum ($\hat{H}_{QM}$), the Hamiltonian, represented as a force field, for the classical system ($\hat{H}_{MM}$) and the interactions between the quantum mechanical and the classical system ($\hat{H}_{QM/MM}$). This is written as

$$(1) \quad \hat{H} = \hat{H}_{QM} + \hat{H}_{QM/MM} + \hat{H}_{MM}$$

The interaction operator between the classical and quantum mechanical subsystems is given by three terms: (i) the electrostatic interactions ($\hat{H}^{el}$), (ii) the polarization interactions ($\hat{H}^{pol}$) and (iii) the van der Waals interactions ($\hat{H}^{vdw}$)

$$(2) \quad \hat{H}_{QM/MM} = \hat{H}^{el} + \hat{H}^{vdw} + \hat{H}^{pol}$$

In the following equations, the indices i and m run over all the electrons and nuclei, respectively, in the quantum mechanical subsystem and s runs over all the sites in the classical subsystem. The coordinates corresponding to particles and sites are denoted $\bar{r}_i$, $\bar{R}_m$, and $\bar{R}_s$, respectively.

The electrostatic term is given by

$$(3) \quad \hat{H}^{el} = -\sum_{s=1}^{S}\sum_{i=1}^{N} \frac{q_s}{|\bar{r}_i - \bar{R}_s|} + \sum_{s=1}^{S}\sum_{m=1}^{M} \frac{q_s Z_m}{|\bar{R}_s - \bar{R}_m|}$$

The polarization interaction term between the quantum and classical subsystems is given by

$$(4) \quad \hat{H}^{pol} = \frac{1}{2}\sum_{i=1}^{N}\sum_{a=1}^{A} \frac{\bar{\mu}_a^{ind} \cdot (\bar{R}_a - \bar{r}_i)}{|\bar{R}_a - \bar{r}_i|^3} - \frac{1}{2}\sum_{m=1}^{M}\sum_{a=1}^{A} \frac{Z_m \bar{\mu}_a^{ind} \cdot (\bar{R}_a - \bar{R}_m)}{|\bar{R}_a - \bar{R}_m|^3}$$

and the index a runs over polarization sites. The induced dipole moment ($\bar{\mu}_a^{ind}$) is proportional to the total electric field and it is given by

$$(5) \quad \bar{\mu}_a^{ind} = \alpha \left(\vec{E}^e(\bar{R}_a) + \vec{E}^m(\bar{R}_a) + \vec{E}^s(\bar{R}_a) + \vec{E}^{ind}(\bar{R}_a) \right)$$

where the isotropic polarizability at the polarization site is given by α. The electric fields are due to the electric field associated with (i) the electrons in the quantum mechanical subsystem, (ii) the nuclei in the quantum mechanical subsystem with the charges Z_m, (iii) the charges, q_s, in the classical subsystem and (iv) the induced dipole moments within the classical subsystem.

The van der Waal term $\hat{H}^{vdw}$ is written as

$$(6) \quad \hat{H}^{vdw} = \sum_{a=1}^{A} \sum_{m:center} 4\epsilon_{ma} \left[\left(\frac{\sigma_{ma}}{|\bar{R}_m - \bar{R}_a|} \right)^{12} - \left(\frac{\sigma_{ma}}{|\bar{R}_m - \bar{R}_a|} \right)^{6} \right]$$

$$= \sum_{a=1}^{A} \sum_{m:center} \left[\frac{A_{ma}}{|\bar{R}_m - \bar{R}_a|^{12}} - \frac{B_{ma}}{|\bar{R}_m - \bar{R}_a|^{6}} \right]$$

with $A_{ma} = 4\epsilon_{ma}\sigma_{ma}^{12}$ and $B_{ma} = 4\epsilon_{ma}\sigma_{ma}^{6}$. Note that the indices i and m run over all the electrons and nuclei within the quantum mechanical subsystem and the index s runs over all the interaction sites within the classical subsystem.

The total QM/MM energy is given by

$$(7) \quad E_{QM/MM} = E^{el} + E^{pol} + E^{vdw}$$

$$= \tilde{O}_{mm'}^{As'} - \sum_{s=1}^{S} \langle N_s \rangle - \frac{1}{2}\alpha \sum_{a=1}^{A} \langle Rr_a \rangle \{\langle Rr_a \rangle + \bar{O}_{mm'}^{aS'}\}$$

$$+ \underbrace{E^{vdw} + E_{S,M}^{el,nuc}}_{\text{independ. of elec.}}$$

with the following definition for $\bar{O}_{mm'}^{aS'}$

$$(8) \quad \bar{O}_{mm'}^{aS'} = 2\sum_{m=1}^{M} \frac{Z_m(\bar{R}_a - \bar{R}_m)}{|\bar{R}_a - \bar{R}_m|^{3}} + \sum_{s' \notin a} \frac{q_{s'}(\bar{R}_a - \bar{R}_{s'})}{|\bar{R}_a - \bar{R}_{s'}|^{3}}$$

$$+ \sum_{a' \neq a} \left\{ \frac{3(\bar{\mu}_{a'}^{ind} \cdot (\bar{R}_a - \bar{R}_{a'}))(\bar{R}_a - \bar{R}_{a'})}{|\bar{R}_a - \bar{R}_{a'}|^{5}} - \frac{\bar{\mu}_{a'}^{ind}}{|\bar{R}_a - \bar{R}_{a'}|^{3}} \right\}$$

and for $\tilde{O}_{mm'}^{As'}$

$$(9) \quad \tilde{O}_{mm'}^{As'} = \frac{1}{2}\alpha \sum_{m=1}^{M} \sum_{a=1}^{A} -Z_m \frac{(\bar{R}_a - \bar{R}_m)}{|\bar{R}_a - \bar{R}_m|^{3}}$$

$$\cdot \left[\sum_{m'=1}^{M} Z_{m'} \frac{(\bar{R}_a - \bar{R}_{m'})}{|\bar{R}_a - \bar{R}_{m'}|^{3}} + \sum_{s' \notin a} q_{s'} \frac{(\bar{R}_a - \bar{R}_{s'})}{|\bar{R}_a - \bar{R}_{s'}|^{3}} \right.$$

$$\left. + \sum_{a' \neq a} \left\{ \frac{3(\bar{\mu}_{a'}^{ind} \cdot (\bar{R}_a - \bar{R}_{a'}))(\bar{R}_a - \bar{R}_{a'})}{|\bar{R}_a - \bar{R}_{a'}|^{5}} - \frac{\bar{\mu}_{a'}^{ind}}{|\bar{R}_a - \bar{R}_{a'}|^{3}} \right\} \right]$$

Furthermore, the terms involving Rr_a are defined by

$$(10) \quad Rr_a = \sum_{pq} t^a_{pq} E_{pq}$$

and

$$(11) \quad \langle Rr_a \rangle = \frac{\langle 0|Rr_a|0\rangle}{\langle 0|0\rangle} = \sum_{pq} D_{pq} t^a_{pq}$$

where

$$(12) \quad t^a_{pq} = \langle \phi_p | \frac{\bar{r}_i - \bar{R}_a}{|\bar{r}_i - \bar{R}_a|^3} | \phi_q \rangle$$

and the terms involving N_s are given by

$$(13) \quad N_s = \sum_{pq} n^s_{pq} E_{pq}$$

$$(14) \quad \langle N_s \rangle = \sum_{pq} D_{pq} n^s_{pq}$$

where

$$(15) \quad n^s_{pq} = \langle \phi_p | \frac{q_s}{|\bar{R}_s - \bar{r}_i|} | \phi_q \rangle$$

Finally, the term describing the nuclear part of the electrostatic interaction is given by

$$(16) \quad E^{el,nuc}_{S,M} = \sum_{s=1}^{S} \sum_{m=1}^{M} \frac{q_s Z_m}{|\bar{R}_m - \bar{R}_s|}$$

3. THE MCSCF WAVEFUNCTION

The total electronic free energy for the QM/MM model is given by

$$(17) \quad \mathcal{F}_{QM/MM}(\lambda) = \mathcal{E}_{vac}(\lambda) + \mathcal{E}_{QM/MM}(\lambda)$$

where electronic wave function is presented by the parameters λ. In the case of a MCSCF electronic wave function we have that

$$(18) \quad |0\rangle = exp\left[\sum_{r>s} \kappa_{sr}(E_{sr} - E_{rs})\right] \sum_i c_i |\Theta_i\rangle$$

where the λ parameters are represented by a combined set of orbital $\{\kappa\}$ and configurational $\{c_i\}$ parameters. The function $|\Theta_i\rangle$ denotes the set of configuration state functions (CSFs).

Utilizing the above expression for the electronic wave function, the energy functional is expanded to second order in the non-redundant electronic parameters λ, $\lambda^{(k)}$ where k indicates the current iteration.

$$\text{(19)} \quad \Delta \mathcal{E}^{(2)}(\lambda - \lambda^{(k)}; \lambda^{(k)}) = \mathcal{E}^{(2)}(\lambda - \lambda^{(k)}; \lambda^{(k)}) - \mathcal{E}(\lambda^{(k)})$$

$$= g^T(\lambda - \lambda^{(k)}) + \frac{1}{2}(\lambda - \lambda^{(k)})^T H (\lambda - \lambda^{(k)})$$

The QM/MM energy in Eq. (7) gives rise to the following gradient contribution

$$\text{(20)} \quad \frac{\partial E_{QM/MM}}{\partial \lambda_i} = -\sum_{s=1}^{S} \frac{\partial \langle N_s \rangle}{\partial \lambda_i} - \alpha \sum_{a=1}^{A} \frac{\partial \langle Rr_a \rangle}{\partial \lambda_i} \left\{ \langle Rr_a \rangle + \frac{1}{2} \bar{O}_{mm'}^{aS'} \right\}$$

and it is convenient to introduce the following operator

$$\text{(21)} \quad T^g = \sum_{s=1}^{S}(-N_s) - \alpha \sum_{a=1}^{A} \left\{ \langle Rr_a \rangle + \frac{1}{2} \bar{O}_{mm'}^{aS'} \right\} Rr_a$$

Thereby, the configuration part of the gradient is given by

$$\text{(22)} \quad \frac{\partial E_{QM/MM}}{\partial c_\mu} = \sum_{s=1}^{S} \left(-\frac{\partial \langle N_s \rangle}{\partial c_\mu} \right) - \alpha \sum_{a=1}^{A} \frac{\partial \langle Rr_a \rangle}{\partial c_\mu} \left\{ \langle Rr_a \rangle + \frac{1}{2} \bar{O}_{mm'}^{aS'} \right\}$$

$$= 2\left[\langle \mu | T^g | 0 \rangle - \langle 0 | T^g | 0 \rangle c_\mu \right]$$

and the orbital part of the gradient is given by

$$\text{(23)} \quad \frac{\partial E_{QM/MM}}{\partial \kappa_{pq}} = \sum_{s=1}^{S} \left(-\frac{\partial \langle N_s \rangle}{\partial \kappa_{pq}} \right) - \alpha \sum_{a=1}^{A} \frac{\partial \langle Rr_a \rangle}{\partial \kappa_{pq}} \left\{ \langle Rr_a \rangle + \frac{1}{2} \bar{O}_{mm'}^{aS'} \right\}$$

$$= 2\langle 0 | [E_{pq}, T^g] | 0 \rangle$$

The Hessian contribution to the energy functional, $\sigma_i^{(k)}$, is determined by a linear transformation algorithm utilizing trial vectors $\boldsymbol{b}^{(k)}$

$$\text{(24)} \quad \sigma_i^{(k)} = \sum_j \frac{\partial^2 E}{\partial \lambda_j \partial \lambda_i} b_i^{(k)}$$

The following relates to a CSF trial vector,

$$(25) \quad \sigma_j^{c,QM/MM} = \sum_\nu \frac{\partial^2 E_{QM/MM}}{\partial c_\nu \partial \lambda_j} b^{(c)}$$

$$= -\sum_{s=1}^S \sum_\nu \frac{\partial^2 \langle N_s \rangle}{\partial c_\nu \partial \lambda_j} b_\nu^{(c)}$$

$$- \alpha \sum_{a=1}^A \left\{ \langle Rr_a \rangle + \frac{1}{2} \bar{O}_{mm'}^{aS'} \right\} \sum_\nu \frac{\partial^2 \langle Rr_a \rangle}{\partial c_\nu \partial \lambda_j} b_\nu^{(c)}$$

$$- \alpha \sum_{a=1}^A \frac{\partial \langle Rr_a \rangle}{\partial \lambda_j} \sum_\nu \frac{\partial \langle Rr_a \rangle}{\partial c_\nu} b_\nu^{(c)}$$

This gives the following compact expressions

$$(26) \quad \sigma_\mu^{c,QM/MM} = 2\left[\langle \mu | T^g | B \rangle - \langle 0 | T^g | 0 \rangle b_\mu^{(c)}\right]$$
$$+ 2\left[\langle \mu | T^{xc} | 0 \rangle - \langle 0 | T^{xc} | 0 \rangle c_\mu\right]$$

and

$$(27) \quad \sigma_{pq}^{c,QM/MM} = 2\left[\langle 0 | [E_{pq}, T^g] | B \rangle + \langle 0 | [E_{pq}, T^{xc}] | B \rangle\right]$$

having the operator, T^{xc} defined as

$$(28) \quad T^{xc} = -2\alpha \sum_{a=1}^A \langle 0 | Rr_a | B, \rangle Rr_a$$

and

$$(29) \quad |B\rangle = \sum_\mu b_\mu^{(c)} |\Theta_\mu\rangle$$

An orbital trial vector leads to the following results

$$(30) \quad \sigma_j^{o,QM/MM} = \sum_{pq} \frac{\partial^2 E_{QM/MM}}{\partial \lambda_j \partial \kappa_{pq}} b_{pq}^{(o)}$$

$$= -\sum_{s=1}^S \sum_{pq} \frac{\partial^2 \langle N_s \rangle}{\partial \lambda_j \partial \kappa_{pq}} b_{pq}^{(o)}$$

$$- \alpha \sum_{a=1}^A \left\{ \langle Rr_a \rangle + \frac{1}{2} \bar{O}_{mm'}^{aS'} \right\} \sum_{pq} \frac{\partial^2 \langle Rr_a \rangle}{\partial \lambda_j \partial \kappa_{pq}} b_{pq}^{(o)}$$

$$- \alpha \sum_{a=1}^A \frac{\partial \langle Rr_a \rangle}{\partial \lambda_j} \sum_{pq} \frac{\partial \langle Rr_a \rangle}{\partial \kappa_{pq}} b_{pq}^{(o)}$$

with the effective expressions given as

(31) $$\sigma_j^{o,QM/MM} = 2\langle\mu|T^{yo}|0\rangle + 2\left[\langle\mu|T^{xo}|0\rangle - \langle 0|T^{xo}|0\rangle c_\mu\right]$$

and

(32) $$\sigma_{pq}^{o,QM/MM} = 2\langle 0|[E_{pq}, T^{yo}]|0\rangle + 2\langle 0|[E_{pq}, T^{xo}]|0\rangle$$
$$+ \sum_t \left(\langle 0|[E_{tq}, T^g]|0\rangle b_{pt} - \langle 0|[E_{tp}, T^g]|0\rangle b_{qt}\right)$$

Here, the effective operators T^{yo} and T^{xo} are defined as

(33) $$T^{yo} = \sum_{s=1}^{S}(-V^s) - \alpha \sum_{a=1}^{A}\left\{\langle Rr_a\rangle + \frac{1}{2}\bar{O}_{mm'}^{aS'}\right\}Q^a$$

and

(34) $$T^{xo} = -\alpha \sum_{a=1}^{A}\langle 0|Q^a|0\rangle Rr_a$$

The terms V^s and Q^a are defined as

(35) $$V^s = \sum_{pq} V_{pq}^s E_{pq}$$

with

(36) $$V_{pq}^s = \sum_r \left[\kappa_{pr} n_{rq}^s - n_{pr}^s \kappa_{rq}\right]$$

and

(37) $$Q^a = \sum_{pq} Q_{pq}^a E_{pq}$$

with

(38) $$Q_{pq}^a = \sum_r \left[\kappa_{pr} t_{rq}^a - t_{pr}^a \kappa_{rq}\right]$$

The procedure for optimization of the MCSCF/MM wave function is similar to that seen for the vacuum [2, 3] and reaction field [4, 5] approaches within MCSCF wave functions.

4. RESPONSE EQUATIONS

In order to determine time-dependent molecular properties utilizing the MCSCF/MM approach it is necessary to consider the time evolution of the appropriate operators and this is done by applying the Ehrenfest's equation for the evolution of an expectation value of an operator, X

$$(39) \quad \frac{d}{dt}\langle X \rangle = \langle \frac{\partial X}{\partial t} \rangle - i\langle [X, H] \rangle$$

and the Hamiltonian for the quantum mechanical subsystem is given by

$$(40) \quad H = H_0 + W_{QM/MM} + V(t)$$

Here, a state $|0\rangle$ is perturbed by an external field represented by the perturbation operator $V(t)$ that describes the interactions between the quantum subsystem and the external field. The Hamiltonian H_0 describes the isolated quantum mechanical subsystem and $W_{QM/MM}$ denotes the interaction operator describing the interactions between the quantum mechanical and the classical mechanical subsystems.

The expectation values are determined from the time-dependent wave function $|0^t\rangle$

$$(41) \quad \langle \ldots \rangle = \langle {}^t 0 | \ldots | 0^t \rangle$$

The external interactions between the quantum mechanical subsystem and the external field are expressed through the interaction operator in the frequency domain

$$(42) \quad V(t) = \int_{-\infty}^{\infty} d\omega V^\omega \exp[-(i\omega + \eta)t]$$

Here η is a positive infinitesimal number that ensures the proper boundary conditions $V(t \to -\infty) = 0$. The term V^ω is the Fourier transform of $V(t)$.

We obtain the reference state, $|0\rangle$ as a solution to the following Hamiltonian

$$(43) \quad (H_0 + W_{QM/MM})|0\rangle = E_0 |0\rangle$$

where [6]

$$(44) \quad W_{QM/MM} = E^{vdw} + E_{S,M}^{el,nuc} + \tilde{O}_{mm'}^{As'} + T^g$$

The operator T^g is defined as

$$(45) \quad T^g = \sum_{s=1}^{S}(-N_s) - \alpha \sum_{a=1}^{A}\{\langle Rr_a \rangle + \frac{1}{2}\bar{O}_{mm'}^{aS'}\}Rr_a$$

In order to determine the time-evolution of expectation values it is crucial to know the time-evolution of time-transformed operators $T^{\dagger,t}$. The time-transformed operators are defined as

(46) $\quad T^{\dagger,t} = \begin{pmatrix} q^t \\ R^t \\ q^{\dagger,t} \\ R^{\dagger,t} \end{pmatrix}$

where $q_k^t = exp[i\kappa(t)]q_k exp[-i\kappa(t)]$ and $R_n^t = exp[i\kappa(t)]exp[iS(t)]R_n exp[-iS(t)]exp[-i\kappa(t)]$. We have the following for the orbital excitation operators

(47) $\quad q_k^\dagger = E_{pq} \equiv a_{p\alpha}^\dagger a_{q\alpha} + a_{p\beta}^\dagger a_{q\beta} \quad p > q$

We represent the state transfer operators as

(48) $\quad R_n^\dagger = |n><0|$

The next step utilizes the time transformed operators and the Ehrenfest's equation where the object of the game is to establish building blocks for how the time transformed operators are changed due to the external perturbation.

(49) $\quad \dfrac{d}{dt}\langle T^{\dagger,t}\rangle = \langle \dfrac{\partial T^{\dagger,t}}{\partial t}\rangle - i\langle [T^{\dagger,t}, H_0]\rangle$
$\qquad \qquad - i\langle [T^{\dagger,t}, V(t)]\rangle - i\langle [T^{\dagger,t}, W_{QM/MM}]\rangle$

Our present interest is on the added contributions due to the QM/MM term

(50) $\quad -i\langle [T^{\dagger,t}, W_{QM/MM}]\rangle = -i\langle [T^{\dagger,t}, T^g]\rangle$
$\qquad \qquad = G_{QM/MM}^{(a)} + G_{QM/MM}^{(b)} + G_{QM/MM}^{(c)}$

where

(51) $\quad G_{QM/MM}^{(a)} = -i(-\alpha) \sum\limits_{a,pq,p'q'} t_{pq}^a \langle {}^t 0|[T^{\dagger,t}, E_{pq}]|0^t\rangle t_{p'q'}^a \langle {}^t 0|E_{p'q'}|0^t\rangle$

(52) $\quad G_{QM/MM}^{(b)} = -i(-\dfrac{1}{2}\alpha) \sum\limits_{a,pq} \bar{O}_{mm'}^{aS'} \langle {}^t 0|[T^{\dagger,t}, E_{pq}]|0^t\rangle t_{pq}^a$

(53) $\quad G_{QM/MM}^{(c)} = -i(-\sum\limits_{s,pq} \langle {}^t 0|[T^{\dagger,t}, E_{pq}]|0^t\rangle n_{pq}^s)$

The two terms denoted $G_{QM/MM}^{(a)}$ and $G_{QM/MM}^{(b)}$ represent the polarization interactions between the two subsystems, and the term $G_{QM/MM}^{(c)}$ gives the electrostatic interactions between the quantum mechanical and the classical subsystems.

The incorporation of the time evolution of the electronic wave function is taken care of by the following parameterization

(54) $|0'\rangle = exp[i\kappa(t)]exp[iS(t)]|0\rangle$

The two unitary operators $exp[i\kappa(t)]$ and $exp[iS(t)]$ ensure that one is able to perform unitary transformations in the orbital and configuration space, respectively.

4.1 Linear Response Equations

Here, the implication of the three terms on the structure of the linear response equations will be considered and for the first term $G^{(a)}_{QM/MM}$ we will illustrate the modifications related to the operator $q_k^t = exp[i\kappa(t)]q_k exp[-i\kappa(t)]$

(55) $G^{(a)}_{QM/MM} = -i(-\alpha) \sum_{a,pq,p'q'} t^a_{pq} t^a_{p'q'} \langle 0|exp[-iS(t)]exp[-i\kappa(t)]$

$\times [exp[i\kappa(t)]q_k exp[-i\kappa(t)], E_{pq}] exp[i\kappa(t)]exp[iS(t)]|0\rangle$

$\times \langle 0|exp[-i\kappa(t)]exp[-iS(t)]E_{p'q'} exp[i\kappa(t)]exp[iS(t)]|0\rangle$

At this point it is convenient to introduce the following states

(56) $|0^R\rangle = -\sum_n S_n R_n^\dagger |0\rangle = -\sum_n S_n |n\rangle$

$\langle 0^L| = \sum_n \langle 0|(S'_n R_n^\dagger) = \sum_n S'_n \langle n|$

and in the case of linear response theory we consider terms linear in $S(t)$ and $\kappa(t)$, and therefore we write $G^{(a)}_{QM/MM}$ as

(57) $G^{(a)}_{QM/MM} = -\alpha \sum_{a,pq,p'q'} t^a_{p'q'} \{(\langle 0|[q_k, E_{pq}]|0^R\rangle + \langle 0^L|[q_k, E_{pq}]|0\rangle) t^a_{pq}$

$+ Q^a_{pq} \langle 0|[q_k, E_{pq}]|0\rangle\} \langle 0|E_{p'q'}|0\rangle$

$- \alpha \sum_{a,pq,p'q'} t^a_{pq} \{(\langle 0|E_{p'q'}|0^R\rangle + \langle 0^L|E_{p'q'}|0\rangle) t^a_{p'q'}$

$+ Q^a_{p'q'} \langle 0|E_{p'q'}|0\rangle\} \langle 0|[q_k, E_{pq}]|0\rangle$

where we have used the one-index transformed integrals defined as

(58) $Q^a_{pq} = \sum_r [\kappa_{pr} t^a_{rq} - t^a_{pr} \kappa_{rq}]$

(59) $Q^a = \sum_{pq} Q^a_{pq} E_{pq}$

where $Q^a_{pq}(Q^a)$ is the index tranformed integral.

Similarly we find for the two other terms $G^{(b)}_{QM/MM}$ and $G^{(c)}_{QM/MM}$ when inserting the operator q_k^t

(60) $$G^{(b)}_{QM/MM} = -\frac{1}{2}\alpha \sum_{a,pq} \bar{O}^{aS'}_{mm'}\{(\langle 0|[q_k, E_{pq}]|0^R\rangle + \langle 0^L|[q_k, E_{pq}]|0\rangle)t^a_{pq} + Q^a_{pq}\langle 0|[q_k, E_{pq}]|0\rangle\}$$

and

(61) $$G^{(c)}_{QM/MM} = -\sum_{s,pq}\{(\langle 0|[q_k, E_{pq}]|0^R\rangle + \langle 0^L|[q_k, E_{pq}]|0\rangle)n^s_{pq} + V^s_{pq}\langle 0|[q_k, E_{pq}]|0\rangle\}$$

For a compact and effective representation we use the following effective operators, T^g (Eq. 45), T^{xc}, T^{xo}, and T^{yo}. They are given by

(62) $$T^{xc} = -\alpha \sum_{a=1}^{A} \{\langle 0|Rr_a|0^R\rangle + \langle 0^L|Rr_a|0\rangle\} Rr_a$$

(63) $$T^{xo} = -\alpha \sum_{a=1}^{A} \langle 0|Q^a|0\rangle Rr_a$$

(64) $$T^{yo} = \sum_{s=1}^{S}(-V^s) - \alpha \sum_{a=1}^{A}\left\{\langle Rr_a\rangle + \frac{1}{2}\bar{O}^{aS'}_{mm'}\right\} Q^a$$

Having these effective operators we are able to rewrite the three terms $G^{(a)}_{QM/MM}$, $G^{(b)}_{QM/MM}$, and $G^{(c)}_{QM/MM}$ and we obtain

(65) $$W^{[2]}(q_k) = -i\langle[q_k, W_{QM/MM}]\rangle = -\langle 0^L|[q_k, T^g]|0\rangle - \langle 0|[q_k, T^g]|0^R\rangle$$
$$-\langle 0|[q_k, T^{xc}]|0\rangle$$
$$-\langle 0|[q_k, T^{yo} + T^{xo}]|0\rangle$$

The same can be done for the three other time-transformed operators $q_k^{\dagger,t}$, R_n^t, and $R_n^{\dagger,t}$. For the time-transformed operators involving the state transfer operators R_n^t we obtain

(66) $$W^{[2]}(R_n) = -i\langle[R_n, W_{QM/MM}]\rangle = -\langle n|T^g|0^R\rangle - \langle 0|T^g|0\rangle S_n(t)$$
$$-\langle n|T^{xc}|0\rangle$$
$$-\langle n|T^{yo} + T^{xo}|0\rangle$$

Finally, we are able to write the MCSCF/MM contributions to the linear response function as

$$(67) \quad W^{[2]}_{j(k)} N^1 = - \begin{pmatrix} \langle 0|[q_j, W]|0^{1R}\rangle + \langle 0^{1L}|[q_j, W]|0\rangle \\ \langle j|W|0^{1R}\rangle \\ \langle 0|[q_j^\dagger, W]|0^{1R}\rangle + \langle 0^{1L}|[q_j^\dagger, W]|0\rangle \\ \langle 0^{1L}|W|j\rangle \end{pmatrix}$$

$$- \begin{pmatrix} \langle 0|[q_j, W(^1\kappa)+2A^1]|0\rangle \\ \langle j|W(^1\kappa)+2A^1|0\rangle \\ \langle 0|[q_j^\dagger, W(^1\kappa)+2A^1]|0\rangle \\ -\langle 0|W(^1\kappa)+2A^1|j\rangle \end{pmatrix} - \langle 0|W|0\rangle \begin{pmatrix} 0 \\ {}^1S_j \\ 0 \\ {}^1S'_j \end{pmatrix}$$

and here a new set of effective operators has been defined

$$(68) \quad T = -\alpha \sum_{a=1}^{A} \langle 0|Rr_a|0\rangle Rr_a$$

$$(69) \quad W = T - \frac{1}{2}\alpha \sum_{a=1}^{A} \bar{O}^{aS'}_{mm'} - \sum_{s=1}^{S} N_s = T^g$$

$$(70) \quad 2A^1 = -\alpha \sum_{a=1}^{A} \left\{ \langle 0|Rr_a(^1\kappa)|0\rangle + \langle 0^{1L}|Rr_a|0\rangle + \langle 0|Rr_a|0^{1R}\rangle \right\} Rr_a$$

4.2 Quadratic Response Equations

This subsection presents the modification of the response equations when considering the quadratic response equations for calculating third order time-dependent molecular properties.

As in the case of the linear response a set of convenient and effective operators is introduced and the operators are given by

$$(71) \quad T = -\alpha \sum_{a=1}^{A} \langle 0|Rr_a|0\rangle Rr_a$$

$$(72) \quad W = T - \frac{1}{2}\alpha \sum_{a=1}^{A} \bar{O}^{aS'}_{mm'} - \sum_{s=1}^{S} N_s$$

$$(73) \quad 2A^1 = -\alpha \sum_{a=1}^{A} \left\{ \langle 0|Rr_a(^1\kappa)|0\rangle + \langle 0^{1L}|Rr_a|0\rangle + \langle 0|Rr_a|0^{1R}\rangle \right\} Rr_a$$

and

$$(74) \quad 2A^{12} = -\alpha \sum_{a=1}^{A} \{ \langle 0|Rr_a(^1\kappa, {}^2\kappa)|0\rangle$$
$$+ 2(\langle 0^{1L}|Rr_a(^2\kappa)|0\rangle + \langle 0|Rr_a(^2\kappa)|0^{1R}\rangle)$$
$$+ \langle 0^{1L}|Rr_a|0^{2R}\rangle + \langle 0^{2L}|Rr_a|0^{1R}\rangle \} Rr_a$$

Having these definitions we are able to write the QM/MM contributions to the quadratic response equations as

(75) $\left(W^{[3]}_{jl_1l_2} + W^{[3]}_{jl_2l_1}\right)^1 N_{l_1}{}^2 N_{l_2}$

$$= \frac{1}{2}P(1,2)\begin{pmatrix}\langle 0|[q_j, W(^1\kappa, {}^2\kappa)+2A^{12}+4A^1({}^2\kappa)]|0\rangle \\ \langle j|W(^1\kappa, {}^2\kappa)+2A^{12}+4A^1({}^2\kappa)|0\rangle \\ \langle 0|[q_j^\dagger, W(^1\kappa, {}^2\kappa)+2A^{12}+4A^1({}^2\kappa)]|0\rangle \\ -\langle 0|W(^1\kappa, {}^2\kappa)+2A^{12}+4A^1({}^2\kappa)|j\rangle\end{pmatrix}$$

$$+ P(1,2)\begin{pmatrix}\langle 0|[q_j, W(^2\kappa)+2A^2]|0^{1R}\rangle + \langle 0^{1L}|[q_j, W(^2\kappa)+2A^2]|0\rangle \\ \langle j|W(^2\kappa)+2A^2|0^{1R}\rangle \\ \langle 0|[q_j^\dagger, W(^2\kappa)+2A^2]|0^{1R}\rangle + \langle 0^{1L}|[q_j^\dagger, W(^2\kappa)+2A^2]|0\rangle \\ -\langle 0^{1L}|W(^2\kappa)+2A^2|j\rangle\end{pmatrix}$$

$$+ \frac{1}{2}P(1,2)\begin{pmatrix}\langle 0^{1L}|[q_j, W]|0^{2R}\rangle + \langle 0^{2L}|[q_j, W]|0^{1R}\rangle \\ 0 \\ \langle 0^{1L}|[q_j^\dagger, W]|0^{2R}\rangle + \langle 0^{2L}|[q_j^\dagger, W]|0^{1R}\rangle \\ 0\end{pmatrix}$$

$$+ 2P(1,2)\langle 0|A^2|0\rangle \begin{pmatrix}0 \\ {}^1S_j \\ 0 \\ {}^1S_j'\end{pmatrix}$$

$$+ P(1,2)\,{}^1S_n^{\,2}S_n' \begin{pmatrix}\langle 0|[q_j, T]|0\rangle \\ \langle j|T|0\rangle \\ \langle 0|[q_j^\dagger, T]|0\rangle \\ -\langle 0|T|j\rangle\end{pmatrix}$$

The physical significance of the terms involving the T, A^i, A^{ij} operators is related to the solvent polarization interactions. The interactions between the effective charges of the solvent and the solute's electronic charge distribution is denoted W.

The similarities between the coupling to a molecular mechanics field or to a dielectric medium are obvious when investigating the derivation of Eq. (75) and the derivation presented in [10]. It is also clear that Eq. (75) has the same structure as the response equations for the molecule in vacuum [1] and the only changes due to the QM/MM interaction are represented by the effective QM/MM operators. Therefore, the implementation of this QM/MM response method into an existing response program requires changes of existing subroutines for calculating W [3] in Eq. (75). The implementation of these MCSCF/MM response equations enables investigation of linear and nonlinear molecular properties within the MCSCF/MM model. The actual use of computational resouces of the MCSCF/MM calculations is comparable to that of an ordinary MCSCF vacuum calculations. The extra work requires an extra storage and retrieval of one-electron integrals.

5. APPLICATIONS OF THE MCSCF/MM RESPONSE METHOD

This section contains a short overview of the results of the applications presented in the work by Poulsen et al., [7, 8, 9]. This work represents the utilization of the multiconfigurational self-consistent field molecular mechanics (MCSCF/MM) response method to calculate molecular properties up to the level of third order molecular properties [7, 8, 9]. Having a MCSCF/MM response method enables investigations of solvation energies and potential energy surfaces with higher accuracy compared to uncorrelated electronic structure methods or density functional methods. The advantages of the MCSCF/MM approach are related to its flexibility for investigating general systems and states such as excited and ionized states that are not described well by single determinant wave functions or semi-empirical density functional methods. Initially, the method has been utilized for the solvation of water and how solvent effects affect the molecular properties of water. The QM/MM system is given by a sample of 128 H_2O molecules and one of these is selected as the quantum mechanical subsystem and the other 127 H_2O molecules represent the classical subsystem.

The calculations of ground state energies have shown that the method provides good agreement with the experimental values of the solvation enthalpy and the solvation energy of water. Furthermore, we have seen that the polarization term from Eq. (2) contributes ~ 20-25% of the total QM/MM energy for the ground state. This is a term that is often neglected but from our applications it plays a crucial part in understanding solvent effects and solvation. In the case of excited and ionized states, we find that the polarization term is of larger importance having a contribution that is about five times larger than that seen in the ground state. For these cases, the magnitude of the polarization term is about the same as the magnitude for the electrostatic term. Generally, we find that the contribution to the total energy due to the QM/MM interactions is much larger in the excited and ionized state than in the ground state. Finally, the values for the solvent binding energy shift obtained using the QM/MM approach are in excellent agreement with experiment. For the induced dipole moment we have obtained using the MCSCF/MM model a value of around 0.8 D which is in good agreement with the experimental data.

The MCSCF/MM calculations of electronic excitation processes considered the transitions to the first and second electronically excited states of H_2O. Compared to the dielectric continuum model, we have observed that the excitation energies obtained using the QM/MM model are uniformly equal to or larger than those obtained using the continuum model. The solvent shifts of the excitation energies are in good agreement with experimental results. Additionally, we observed that the polarization terms contribute significantly. The transition moments are enhanced when going from a gas-phase representation to a QM/MM model and the calculations of transition moments clearly illutrates how the solvent perturbs the electronic transitions substantially and especially forbidden transitions. This aspect is not treated correctly by the continuum models due to the lack of intermolecular descriptions of short-range interactions.

The polarizabilities calculated using the MCSCF/MM response method show that the polarizabilities increase with frequency in the same way as the corresponding gas-phase results and in general the MCSCF/MM values are about 10% larger than the corresponding HF/MM values. For the anisotropies, we observe that the MCSCF/MM results and the dielectric continuum results are similar in magnitude.

The MCSCF/MM quadratic response functions give the hyperpolarizability tensor and the two-photon absorption cross sections. The latter is given by the residues of the quadratic response functions. Generally, all the individual tensor components are shifted substantially compared to the results from a corresponding vacuum calculation. The MCSCF/MM qudratic response calculations lead to a sign change in the average value of the hyperpolarizability which is also observed experimentally. Quadratic response calculation within the MCSCF/MM approach without the polarization interactions gives significantly smaller values for the average hyperpolarizability which illustrates the importance of including the polarization terms in the QM/MM model. The two-photon absorption cross section are also strongly perturbed by the presence of the solvent and in some cases enhance the transitions significantly in the case of forbidden transitions. The results from the MCSCF/MM model compare very well with the available experimental data on two-photon cross sections of liquid water.

6. SUMMARY

The main purpose of this presentation has been to give an overview of the theoretical developments concerning MCSCF/MM response theory. The overview has presented the necessary contributions arising from the coupling to the classical molecular mechanics field for performing calculations of zeroth, first, second and third order molecular properties within the multiconfigurational response framework. The focus has been on methodology development of MCSCF/MM response theory. The MSCSF/MM method is a rather promising method not only for studying ground state solvent effects, but also excited and ionized states, calculations of frequency-dependent linear and nonlinear polarizabilities, transition moments, and vertical excitation energies. The MCSCF/MM response method is a recent molecular response method for obtaining frequency-dependent molecular properties for a solute perturbed by solvent interactions. This is achieved by treating the solute on a quantum mechanical level and the solvent is described by a molecular mechanics force field. The coupling between the two parts is included directly in the optimized wave function.

ACKNOWLEDGEMENTS

K.V.M. thanks Statens Naturvidenskabelige Forskningsråd, Statens Tekniske Videnskabelige Forskningsråd, the Danish Center for Scientific Computing and the EU-network NANOQUANT for support.

REFERENCES

[1] Hettema, H., Jensen, H.J.Aa., Jørgensen, P., Olsen, J.: J. Chem. Phys., **97**, 1174–1190 (1992)
[2] Jensen, H.J.Aa., Ågren, H.: Chem. Phys. Lett., **110**, 140 (1984)
[3] Jensen, H.J.Aa., Ågren, H.: Chem. Phys., **104**, 229 (1986)
[4] Mikkelsen, K.V., Ågren, H., Jensen, H.J.Aa., Helgaker, T.: J. Chem. Phys., **89**, 3086–3095 (1988)
[5] Mikkelsen, K.V., Cesar, A., Ågren, H., Jensen, H.J.Aa.: J. Chem. Phys., **103**, 9010–9023 (1995)
[6] Mikkelsen, K.V., Jørgensen, P., Jensen, H.J.Aa.: J. Chem. Phys., **100**, 6597–6607 (1994)
[7] Poulsen, T.D., Kongsted, J., Osted, A., Ogilby, P.R., Mikkelsen, K.V.: J. Chem. Phys., **115**, 2393–2400 (2001a)
[8] Poulsen, T.D., Ogilby, P.R., Mikkelsen, K.V.: J. Chem. Phys., **115**, 7843–7851 (2001b)
[9] Poulsen, T.D., Ogilby, P.R., Mikkelsen, K.V.: J. Chem. Phys., **116**, 3730–3738 (2002)
[10] Sylvester-Hvid, K.O., Mikkelsen, K.V., Jonsson, D., Norman, P., Ågren, H.: J. Chem. Phys., **109**, 5576–5584 (1998)

CHAPTER 9

SOLVATOCHROMISM AND NONLINEAR OPTICAL PROPERTIES OF DONOR-ACCEPTOR π-CONJUGATED MOLECULES

WOJCIECH BARTKOWIAK

Institute of Physical and Theoretical Chemistry, Wrocław University of Technology, Wybrzeże Wyspiańskiego 27, 50-370 Wrocław, Poland, e-mail: wojciech.bartkowiak@pwr.wroc.pl; bartkowiak@kchk.ch.pwr.wroc.pl

Abstract: We review the theoretical approaches based on the perturbation theory, namely few-states approximations. These approaches are extensively used in the description of the solvent effects on nonlinear response of molecular systems. The connection between the nonlinear optical response and solvatochromic behavior of the donor-acceptor π-conjugated molecules is considered. The general relations between molecular (hyper)polarizabilities and two-photon absorption for the positively and negatively solvatochromic compounds are presented

Keywords: solvatochromism, solvent effect, intramolecular charge-transfer (CT), hyperpolarizabilities, two-photon absorption, nonlinear optical properties

1. INTRODUCTION

In this contribution we review the connection between the nonlinear optical (NLO) response and the solvatochromic behavior of the important class of organic molecules, namely donor-acceptor π-conjugated compounds (D-π-A). In these compounds, also called push-pull chromophores, an electron-donating group D is conjugated to an electron-acceptor substituent A through a system of the localized π-bonds (Scheme 1). It is well established that such compounds exhibit the following properties [1–7]:
- large ground state dipole moment,
- an intense low-lying ($\pi \rightarrow \pi^*$) transition in the UV-Vis spectral region which is assigned to the intramolecular charge-transfer (CT) occurring along the molecule axis,

- large values of the molecular first-order electronic hyperpolarizabilities (β) in comparison with other non-centrosymmetric organic molecules.

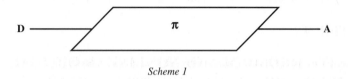

Scheme 1

It should be noticed that the lowest CT excited state is found to dominate both linear and nonlinear optical responses of the D-π-A chromophores [3–7]. The term 'CT excited state' denotes the photoinduced Franck-Condon electronic excited state which differs significantly from the ground electronic state. This difference is mainly connected with a large change of the permanent dipole moment of the D-π-A molecule during the electronic excitation. The perturbative approach may be used to studying the response of the chromophores of the D-π-A type to external electric field. The simplification of the perturbative expressions for molecular (hyper)polarizabilities gives an insight into the relation between the spectroscopic parameters and nonlinear optical properties of considered molecular systems [8–13]. The approximate sum-over-states expression for the two-level system was proposed by Oudar and Chemla [8]. On the basis of the few-states approximations one can express the NLO properties in terms of the transition moments (or oscillator strengths), the excitation energies, and the dipole moments differences. The few-states approximations may be very useful in the design of chromophores with large NLO response [3–13].

The theoretical and experimental works clearly shows that the environment plays a remarkable role in the considerations of the first- (β) and the second-order hyperpolarizabilities (γ) of the D-π-A type chromophores [14–45]. Not only the magnitudes of β and γ, but also the signs of these properties can be affected by change of the solvent polarity. On the other hand, the linear polarizability (α) is less affected by the solvent than the higher order polarizabilities, namely β and γ.

The development of the theoretical models and computational methods, which can accurately account for the solvent effects on the NLO properties of molecular systems is one of the largest challenges in the contemporary quantum chemistry. The important conceptional and computational difficulty arises from the fact that the nature of the solute-solvent interaction in the condensed phase is extremely complicated. In modern quantum chemistry the solute-solvent interactions are modeled by a number of different approaches that can be divided into three main groups: supermolecular approximations, discrete simulations and continuum models.

Most of the proposed models are based on the classical electrostatic description. For the detailed study of these models we refer readers to monographs, reviews and original papers [27, 30, 31, 33, 40, 46–58]. Numerous theoretical works have also been conducted to understand the relation between the micro- and macroscopic optical properties [50, 51, 59–63]. The methodology developed on this field plays a key role in a direct comparison between the calculated and experimental values for α, β, and γ as well as for the macroscopic susceptibilities.

This article is devoted to the methodology of predicting the direction of the changes of molecular (hyper)polarizabilities values as a function of the solvent polarity. Since the environmental effect on the two-photon absorption (TPA) is still poorly understood, we will consider the two-level approximation to describe the influence of the solvent effects on TPA from the ground to the CT excited state of the D-π-A type chromophores. Only electronic contributions will be taken into account. In contrast to the TPA process, the substantial progress in theoretical description of the solvent influence on the vibrational (hyper)polarizabilities has been observed recently [64–67].

2. SOLVATOCHROMISM PHENOMENON

The term 'solvatochromism' is used to describe the change of position, intensity and shape of the UV-Vis absorption band of the chromophore in solvents of different polarity [1, 2]. This phenomenon can be explained on the basis of the theory of intermolecular solute-solvent interactions in the ground $|g\rangle$ and the Franck-Condon excited state $|e\rangle$. We will consider only the effect of the solute-solvent interaction on the electronic absorption and nonlinear optical response of a dilute solution of the solute. This way we avoid the explicit discussion of the solute-solute interaction, which significantly obscures the picture of the solvatochromism phenomenon.

It should be noticed that, in many theoretical works, the term 'solvent polarity' is defined by the values of the relative electric permittivity, ε_r, also called dielectric constant. However, such a definition is by no means precise. The existence of hydrogen bonds (H-bonds) between solute and solvent molecules is one of the important limitations of the use of the continuum models based on the theory of dielectrics. In modern physical chemistry of solutions in order to quantitatively describe the solvatochromism phenomenon various empirical scales of the polarity are used. The exhaustive reviews on this topic have been presented by Reichardt [1, 2].

According to the work of Li, Cramer and Truhlar, the solvatochromic shift of the maximum absorption band ($\Delta\omega$) can be divided into the following contributions [68]:

(1) $\quad \Delta\omega = \Delta\omega_E + \Delta\omega_D + \Delta\omega_H$

where $\Delta\omega_E$ is the pure electrostatic contribution, $\Delta\omega_D$ denotes the dispersion contribution and $\Delta\omega_H$ is connected with the short-range specific interaction between the solvent and solute, e.g., hydrogen bonding. The deeper insight into the nature of the above contributions is crucial for the better understanding of the relations between NLO response and the solvatochromism. Hence, the more detailed analysis will be presented here.

2.1 The Electrostatic Contribution

The electronic transition from the ground state to the CT excited state leads to the substantial change of the electronic density of solute molecule. Hence, a large change in the permanent dipole moment is observed during the excitation process. Due to the difference in the solute-solvent electrostatic interaction energy between the ground and the CT excited state of the solute the strong solvatochromic shift occurs going from the gas phase to the polar solvents. The basic fundamental contributions to a modern understanding of the influence of the solute-solvent electrostatic interactions ($\Delta\omega_E$) on the electronic absorption are often based on the classical Onsager's reaction field theory. The excellent monographs and reviews have to be addressed here [1, 2, 69–73]. In all of these theories solvent is usually treated as an isotropic dielectric medium characterized by its relative electric permittivity (ε_r) and refractive index (n). Starting from this assumption, such theories predict the linear dependence of solvatochromic shift on dipole moment difference between the ground and the CT excited state ($\Delta\mu_{gCT}$). In particular, the combination of the perturbation theory and the reaction field method has lead Amos and Burrows to establishing the relation between the solvent effects and the transition energy from the ground to the CT excited state of the form [71]:

(2) $$\hbar\Delta\omega_{gCT} = A\Delta\mu_{gCT}(\mu_{CT} + \mu_g) + B\Delta\mu_{gCT}\mu_g$$

where

$$A = \frac{1}{a^3}\frac{n^2-1}{2n^2+1}, \quad B = \frac{2}{a^3}\frac{\varepsilon_r-1}{2\varepsilon_r+1} - \frac{n^2-1}{2n^2+1}$$

and $\hbar\Delta\omega_{gCT} = \hbar\Delta\omega_{gCT}^{gas} - \hbar\Delta\omega_{gCT}^{sol}$ is the change in the transition energy, μ_g and μ_{CT} denote the ground state and the CT excited state solute dipole moment, respectively and $\Delta\mu_{gCT} = \mu_g - \mu_{CT}$. The parameter a stands for a radius of a spherical cavity occupied by the solute molecule. On the basis of the above equation it is possible to simply explain the direction of the shifts of the absorption bands with the increase of the solvent polarity (negative and positive solvatochromism) [1, 2, 70–72]. The positive solvatochromism (red or bathochromic shift) is exhibited by the molecules with the larger polarization for the CT excited state (zwitterionic form) than for the ground state (neutral form), $\mu_g < \mu_{CT}$ (see Scheme 2). For such molecules, highly polar solvents cause better stabilization for the excited than the ground state. Hence, the excitation energy is significantly decreased.

In the case of the negative solvatochromism (blue or hypsochromic shift), the ground state is better stabilized by the polar solvent than the CT excited state,

Positive solvatochromism $\quad S_0: D-\pi-A \xrightarrow{h\nu} S_1: {}^+D-\pi-A^-$

Negative solvatochromism $\quad S_0: {}^+D-\pi-A^- \xrightarrow{h\nu} S_1: D-\pi-A$

Scheme 2

$\mu_g > \mu_{CT}$. It leads to the enhancement of the excitation energy of the solute. For the schematic representation of the two types of excitations see Scheme 2.

2.2 The Dispersion Contribution

The dispersion energy plays a role both in polar and in nonpolar solvents. The quantum chemical and classical description of the influence of the dispersion effect, $\Delta\omega_D$, on the excitation energy of molecules in the condensed phase is much more complicated than that of the electrostatic solute-solvent interactions, $\Delta\omega_E$. However, it is well established that the dispersion effects lead to the red shift (positive solvatochromism) of the electronic absorption band [1, 2, 68, 69, 73, 74]. It arises from the fact that the excited state exhibits larger linear polarizability than the ground state. Hence, the dispersion contribution to the interaction energy between the excited solute molecules with the environment is substantialy larger than in the case of the dispersion interactions of the solute in the ground state. It is worth noticing that in the case of the intramolecular CT transition the influence of the dispersion contribution, $\Delta\omega_D$, on the position of the electronic absorption band in the polar solvents is less important than the impact of the electrostatic effects, $\Delta\omega_E$ [1, 2].

2.3 The H-bonds Contribution

The H-bond formation may affect the energies of various excited states in different ways [1, 2, 68]. It is well established that the specific H-bond interactions strongly influence the n → π* transition of carbonyl compounds [1, 2, 68, 69, 75, 76]. In this case, the H-bonding interactions ($\Delta\omega_H$) stabilize the ground state better than the less-dipolar excited state. Hence, the H-bonds donating solvents exhibit the solvent-induced blue shifts (negative solvatochromism). On the other hand, the presence of H-bonds can also strongly influence the intensity of the CT absorption band (π → π*) as it is observed in the case of the p-nitroaniline (PNA) molecule [77].

The theoretical and experimental investigations of the specific H-bonding effects on the nonlinear optical response of the donor-acceptor chromophores were also carried out. The widely analyzed the prototypical PNA molecule can form H-bonds involving the NH groups and electron donor atoms of solvent. Huyskens et al. have shown that the formation of specific solute-solvent interactions such as H-bonds always increases hyperpolarizability β [35]. The statistical study of a large number of H-bonding solvents has lead Huyskens et al. to the following expression for β:

(3) $\quad \beta = \beta_{gas} + \gamma^0 a\sqrt{\mu_S/V_S} + (1-\gamma^0) b\sqrt{\mu_S}$

where a and b are constants for a given solute molecule, μ_S is the gaseous dipole moment, and V_S denotes the solvent molar volume. γ^0 stands for the period of the time during which the solute is not involved in H-bonding with the solvent molecules. Luo et al. have considered the possibility of forming H-bonding between PNA and different solvent molecules (acetone, methanol, and acetonitrile) using the polarizable continuum model (PCM) combined with the supermolecular approach [42]. The significant influence of the H-bonds on the electronic structure of PNA has been shown. The substantial effects of the H-bonds on the spectroscopic as well as NLO properties of aminobenzodifuranone (ABF) derivative exhibiting the largest positive solvatochromic shift in comparison with the other known chromophores have been illustrated by Bartkowiak and Lipkowski [78]. It has been shown that the H-bonds formation involving NH groups of the ABF molecule and donor atoms of a hexamethylphosphoramide (HMPA) solvent substantially increases the value of β in comparison with the gas phase.

3. SOLVATOCHROMISM AND MOLECULAR (HYPER)POLARIZABILITIES

A substantial change in the NLO response of the D-π-A chromophores as a function of the solvent polarity may be understood on the basis of the simple two-state model combined with the solvatochromic phenomenon. The origin of these changes is related to the solute-solvent interaction in the ground and CT excited state. Considering the D-π-A molecules exhibiting the monotonic behavior of the NLO properties as a function of the solvent polarity, one can draw the following conclusions: Positively solvatochromic molecules, i.e. exhibiting red shift of the CT absorption band, are characterized by the increase of α, β, and γ values while increasing the solvent polarity [15–20, 26–38, 40–43]. For the negative solvatochromism manifesting itself in the blue shift of the CT absorption band, the solvent dependence of α, β, and γ shows opposite trends [14, 38, 39, 79]. On the other hand, there is a group of the D-π-A type molecules exhibiting the so-called reversible solvatochromism. This type of solvatochromism is observed in particular for relatively long push-pull polyene molecules and various merocyanine dyes [1, 22–25, 28, 45, 80, 81]. Their ground state geometry is strongly affected by the presence of solvent. The ground state of such molecules can be viewed as a combination of two valence-bond (VB) forms, namely neutral and zwitterionic. Both of these forms differ in the extent of the charge separation (Scheme 3).

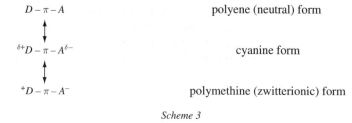

Scheme 3

The relative contribution of these VB forms depends on the strength of the donor and acceptor substituents, the structure of π-conjugated bridge, as well as on the solvent polarity [1, 14, 21–25, 28, 38, 45, 80, 81]. Hence, the solvatochromic behavior of such systems is determined by the VB form dominating in the ground state for the given solvent.

Marder et al. [11, 21–25, 80, 82–84], in their pioneering works, have shown that structural parameters such as bond length alternation (BLA) and bond order alternation (BOA) strongly influence the values of α, β, and γ of organic compounds with delocalized π-electron systems. The bond length alternation (BLA) is usually defined as a difference between the average single and double bond distances in the π-conjugated pathway. Hence, this structural parameter can be applied to the quantitative description of evolution of the ground state geometry from the neutral polyene structure (D-π-A) to a highly polar zwitterionic form (D^+-π-A^-). The intermediate structure of the cyanine-type is composed of the equal contributions from neutral and charge-separated resonance forms (Scheme 3). During the past years, the relationship between molecular structure (BLA) and molecular linear and nonlinear polarizabilities for several donor-acceptor organic molecules has been established [11, 21–25, 80, 82–86]. The molecular (hyper)polarizabilities exhibit minima and maxima as a function of BLA.

Barzoukas, Blanchard-Desce and Thompson et al. have developed simple two-form formalism based on VB-CT approach in order to derive analytical relations between the ground state polarization and NLO properties for donor-acceptor π-conjugated molecules [87–91]. This approach is known as a valence-bond state model (VBSM). It should be mentioned that in the two-form formalism the validity of the two-state models for α, β, and γ is assumed. In the two-form approach, a MIX parameter, characterizing the mixing between the limiting neutral and zwitterionic resonance forms is introduced. Since the MIX parameter is proportional to the change in the dipole moment between the ground and CT excited states, it can be extracted from the experimental data similarly as the remaining parameters included in the model. Recently, the two-form model has been applied in order to obtain the qualitative insight into mutual relations between hyperpolarizabilities as well as two-photon cross section and BLA for the donor-acceptor π-conjugated organic molecules [92]. This topic is presented by Bartkowiak and Zaleśny in one of the chapters of the present book.

Lu, Goddard III, Perry et al. have independently proposed the valence-bond charge-transfer (VB-CT) model to predict (hyper)polarizabilities of the donor-acceptor π-conjugated molecules [93–94]. The VB-CT model accounts for the dependence of the (hyper)polarizabilities on the charge transfer energy. The (hyper)polarizabilities are related to the BLA parameter which from the other side is proportional to the fraction of the wave function having CT characters. The VBSM and VB-CT models have been succesfully applied to predict the solvent and BLA effects on the NLO properties for the large group of the donor-acceptor π-conjugated molecules. The usefulness and popularity of the above simple models comes from the fact that for numerous systems not only qualitative but also quantitative picture of the NLO response has been achieved.

It should be noticed that an important contribution to understand solvatochromism and NLO response of molecules of the D-π-A type has been given by Painelli et al. [95–100]. These authors have developed a simple non-perturbative model for the description of the NLO response and low-energy spectral properties of numerous donor-acceptor systems. A polar molecule in solution is modeled in terms of the two electronic states linearly coupled to molecular vibrations and to so-called solvation coordinate. This coordinate describes orientational degrees of freedom of the surrounding solvent.

In the present contribution we will discuss the direction of the changes of the NLO response and the solvatochromic behavior as a function of solvent polarity of the D-π-A chromophores. The best starting point for these considerations seems to be the simple two-state model for the first-order hyperpolarizability (β) [8]. To avoid the extreme complexity of the sum-over-states (SOS) expression [101], Oudar and Chemla proposed the relation between the dominant component of β along the molecular axis (let it be the x-axis) and the spectroscopic parameters of the low-lying CT transition [8]. The use of the two-level approximation in the static case ($\omega = 0.0$) has lead to the following expression for the static xxx component of the first-order hyperpolarizability tensor:

$$(4) \quad \beta_{xxx}(0) = 6 \frac{\langle g| r_x |CT\rangle^2 \Delta\mu_{gCT,x}}{(\omega_{gCT})^2}$$

where $\hbar\omega_{gCT}$ is the transition energy from the ground to the CT excited state, and $\langle g| r_x |CT\rangle$ denotes the transition dipole moment. The atomic units and the T convention are adopted in the above equation [102].

Alternatively, noticing that $\langle g| r_x |CT\rangle^2$ is proportional to the oscillator strength, f, the expression for $\beta_{xxx}(0)$ can be rewritten as

$$(5) \quad \beta_{xxx}(0) = 9 \frac{f \Delta\mu_{gCT,x}}{(\omega_{gCT})^3}$$

The above expressions become slightly modified when the frequency dependent β_{xxx} is evaluated. In order to calculate the first-order hyperpolarizability for a second-harmonic generation (SHG) process ($\beta_{xxx}(2\omega)$), the denominator in Eqs. (4) and (5) should be modified by a factor reflecting the frequency dependence (the dispersion factor) of the form:

$$(6) \quad (1 - \omega^2/\omega_{gCT}^2)(1 - 4\omega^2/\omega_{gCT}^2)$$

where ω is the polarization frequency (below the electric resonance).

Thus, the equation for the off-resonance $\beta_{xxx}(2\omega)$ can be expressed as

$$(6a) \quad \beta_{xxx}(2\omega) = \frac{\beta_{xxx}(0)}{(1 - \omega^2/\omega_{gCT}^2)(1 - 4\omega^2/\omega_{gCT}^2)}$$

The Oudar-Chemla equation has been tested for many donor-acceptor π-conjugated organic molecules. The two-state model works quite well in most cases. However, it should not be left unmentioned that the values of β_{xxx} are usually overestimated in comparison to the more advanced quantum chemical calculations [3, 31, 38–40]. Hence, the two-state model should be treated only as a rather rough approximation to the SOS method. On the other hand, for the most applications, the relative values of the first-order hyperpolarizabilies are of the higher importance. The two-state model allows to establish the structure-NLO properties relationship in terms of relatively simple spectroscopic parameters.

The two-state model is also applied to determine the first-order hyperpolarizabilities based on the experimental measurements of the spectroscopic quantities [103]. The ground state and the CT excited state dipole moments, excitation energy, as well as transition dipole moment (or oscillator strength) can be determined through the solvatochromic effect measurements. In particular, the first-order hyperpolarizability can be obtained in such a way by employing Eq. (4) or Eq. (6a).

The influence of the solute-solvent interactions on β_{xxx} is easily understandable in terms of the two-state model. This approach clearly shows that β_{xxx} strongly depends on the excitation energy, $\hbar\omega_{gCT}$. According to Eq. (5), β_{xxx} is inversely proportional to the third power of the excitation energy.

The excitation energies ($\hbar\omega_{gCT}$) in Eqs. (4), (5), and (6a) correspond to the experimental UV-Vis absorption band peaks. As it was mentioned previously, the solvatochromic effect manifests itself as a shift of the positions of the low-energy CT bands in electronic absorption spectra of dissolved solute molecules. The shift arises from the interactions with solvents of various polarity. Hence, from the theoretical point of view, the leading contribution to the solvent effect on β_{xxx} should be approximately described by a different change of the CT transition energy. This relation has been confirmed theoretically and experimentally [15, 16, 19, 20, 27, 29, 31, 34, 38–40, 42]. On the other hand, the influence of the solvent on β_{xxx} is also reflected in a substantial changes of the values of $\langle g|r_x|CT\rangle^2$ (or f) and $\Delta\mu_{gCT}$. In the case of the positively solvatochromic compounds the increase of β_{xxx} is followed by the red shift of the CT absorption band. It translates into the decreasing value of $\hbar\omega_{gCT}$ and increasing value of the product $f\Delta\mu_{gCT}$ going from the gas-phase to the polar solvents. These two contributions can have a comparable influence on the value of β_{xxx}. In [38] the example of 4-nitro-4'-aminostilbene (ANS) was considered. In that paper, the product of $f\Delta\mu_{gCT}$ causes the increase of the β_{xxx} value by about 46%, while the transition energy ($\hbar\omega_{gCT}$) change—by about 53% on going from the gas phase to aqueous solutions. Thus, this example clearly shows that the methods based exclusively on the transition energy shifts (e.g. the solvatochromic methods) may lead to the erroneous results. The presence of the dispersion factor in Eq. (6a) indicates that the inclusion of the dispersion effects lead to the larger enhancement $\beta_{xxx}(2\omega)$ in comparison to the static case, $\beta_{xxx}(0)$, when the solvent effects are taken into account. There are much more theoertical and experimental studies in the literature devoted to the positive than to the negative solvatochromism phenomenon of the chromophores of the D-π-A type. The chromophores exhibiting positive solvatochromism show significantly smaller sensitivity on the specific interactions. This makes them more suitable for the description of the solvent polarity parameters in terms of unspecific electrostatic interactions [1, 2, 104].

In general, in the case of the chromophores exhibiting large negative solvatochromism, the presence of polar environment leads to the substantial decrease of β_{xxx} values [14, 38, 39, 79]. The important examples are heterocyclic betaine dyes. This class of compounds exhibits the zwitterionic character even in solvents of relatively small polarity. The well known representative of the betaine dyes is the so-called Reichardt's betaine dye [1, 2]. This type of molecules has large negative value of the vector part of the first-order hyperpolarizability, β_μ. β_μ is evaluated using the electric field induced second-harmonic generation (EFISH) technique [47, 49, 102]. For the one-dimensional polar donor-acceptor molecules β_μ is proportional to β_{xxx}. According to the two-level model, the negative sign of β_μ is due to the sign of $\Delta\mu_{gCT}$. There is a dramatic decrease in the polarity of the betaine dyes upon excitation from the ground state to the CT excited state. This change of $\Delta\mu_{gCT}$ is reflected in a considerable negative solvatochromic shift of the intramolecular CT absorption band. Such an effect leads to a significant decrease of the values of β_{xxx} in the polar solution in comparison to the gas phase. It comes from the fact that the increasing $\Delta\mu_{gCT}$ is partially neutralized by diminished intensity of the CT absorption band. The quantum chemical calculations confirm these findings [38, 39]. A similar trends are observed for the negatively solvatochromic merocyanine dyes in the polar solvents [14, 45]. The solvent effects

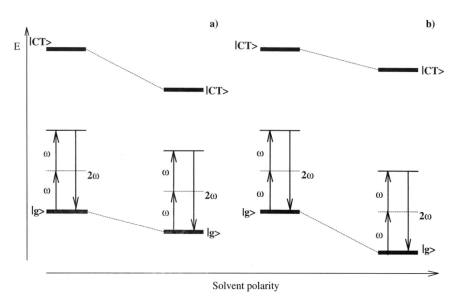

Figure 1. Two-level picture corresponds to the second-harmonic generation (SHG) process in the gas phase and in polar solvent: a) positively solvatochromic molecule, b) negatively solvatochromic molecule

on the first-order hyperpolarizabilities, $\beta_{xxx}(2\omega)$ within the two-level picture for the positively as well as negatively solvatochromic molecules are schematically presented on Figure 1.

A similar reasoning can be applied to linear polarizabilities, α, and second-order hyperpolarizabilities, γ, for the donor-acceptor π-conjugated molecules. The two-state models for α and γ are given in Appendix A. Similarly as in the case of β, the lowest CT excited state gives also a significant contribution to the values of α and γ for the analyzed molecules [11, 14, 17–19, 38, 40]. Also, the behaviour of α and γ as a function of the solvent polarity seems to reflect the tendencies for β. However, the α values appear to be less sensitive to the solvent effects than β and γ [14, 19, 26, 31, 36, 38]. Recently, for the case of static vector components of β and scalar part of γ, it has been shown that the calculated ratio of β^{sol}/β^{gas} is approximately equal to the $\gamma^{sol}/\gamma^{gas}$ ratio for a variety of the donor-acceptor chromophores exhibiting the positive solvatochromism [40]. An explanation of these observations can be proposed on the basis on the two-state models for α and γ.

4. SOLVATOCHROMISM AND TPA CROSS SECTION

The two-photon absorption phenomenon of molecular systems is connected with a process of a simultaneous absorption of two-photon. The electronic contribution to the two-photon absorption is usually studied on the basis of the Monson and McClain formalism [105, 106]. The two-photon absorption cross section from a ground $|g\rangle$ to a final state $|f\rangle$ (δ_{gf}) for molecules in isotropic media is defined as

$$(7) \quad \langle \delta_{gf} \rangle = \langle |S_{gf}(\vec{\mu}_1, \vec{\mu}_2)|^2 \rangle = \frac{1}{30} \sum_{ij} [S_{ii}S_{jj}^* F + S_{ij}S_{ij}^* G + S_{ij}S_{ji}^* H]$$

where $F = F(\vec{\mu}_1, \vec{\mu}_2)$, $G = G(\vec{\mu}_1, \vec{\mu}_2)$, and $H = H(\vec{\mu}_1, \vec{\mu}_2)$ are the polarization variables and S_{ij} denotes the two-photon transition moment. The equation for the two-photon matrix elements,

$$(8) \quad S_{ij}^{gf} = \sum_k \left(\frac{\langle g|\vec{\mu}_1 \vec{r}|k\rangle \langle k|\vec{\mu}_2 \vec{r}|f\rangle}{\omega_{kg} - \omega_1} + \frac{\langle g|\vec{\mu}_2 \vec{r}|k\rangle \langle k|\vec{\mu}_1 \vec{r}|f\rangle}{\omega_{kg} - \omega_2} \right)$$

can be obtained directly from the time-dependent perturbation theory, choosing the interaction of a molecule with an electromagnetic field as a perturbation [106]. Labels i, j in Eq. (8) refer to the Cartesian coordinates and $\vec{\mu}_1$, $\vec{\mu}_2$ denote polarization of photons with energies $\hbar\omega_1$, $\hbar\omega_2$. Summation in Eq. (8) runs over all the intermediate states $|k\rangle$ including the ground $|g\rangle$ and the final state $|f\rangle$. If we consider the simultaneous absorption of the two photons from one monochromatic laser beam, the energies of both photons are equal, thus one can write: $\hbar\omega_1 = \hbar\omega_2 = \frac{1}{2}\hbar\omega_{gf}$. In the case of both photons polarized linearly with a parallel polarization, Eq. (7) can be rewritten in the simplified form:

$$(9) \quad \langle \delta_{gf} \rangle = \frac{1}{15} \sum_{ij} [S_{ii}S_{jj}^* + 2S_{ij}S_{ij}^*]$$

Alternatively, the TPA cross section can also be shown to be proportional to the imaginary part of the averaged second-order hyperpolarizability $<\gamma>$, defined at the absorption frequency of $\omega(\delta_{gf} \propto Im\gamma(-\omega; \omega, -\omega, \omega))$. It has been shown by Luo et al. that for TPA into a particular excited state, both approaches are equivalent [107].

Direct comparison between $<\delta_{gf}>$ (in atomic units) estimated theoretically from Eq. (9) and from the experimental data requires the knowledge of the line shape function $g(\omega)$ for investigated molecule. The experimental TPA cross section (in the conventional cm^4 s/photon units) is connected with the $<\delta_{gf}>$ via the following equation [108, 109]:

$$(10) \quad \sigma_{gf}^{(2)} = \frac{8\pi^3 \alpha^2 \hbar^3}{e^4} \frac{\omega^2 g(\omega)}{\Gamma_f/2} \langle \delta_{gf} \rangle$$

Here α is a fine-structure constant, Γ_f stands for the lifetime broadening of final state and $\hbar\omega$ denotes the photon energy.

As it was mentioned in the Introduction, in contrast to the extensive investigations on the solvent influence on the molecular (hyper)polarizabilites, there are only few theoretical works on the TPA cross section (δ) of the donor-acceptor π-conjugated organic molecules in which TPA from the ground to the CT excited state is considered. Below, we give short review of these works.

Luo et al. have investigated the influence of the solvent polarity on the NLO properties of a simple donor-acceptor polyene molecule (Scheme 4) at the *ab initio* level of theory applying the continuum model of solvent [107]. It has been shown that the evolution of the TPA cross section with respect to the bond length alternation (BLA) closely follows that of the static first-order hyperpolarizability, β. The TPA cross section is strongly dependent on the geometrical changes. Moreover, these authors have noticed that the solvent effect on the TPA cross section (for the CT excited state) in the typical donor-acceptor polyene molecule exhibiting the positive solvatochromism is smaller than the influence of the solvent on the values of β.

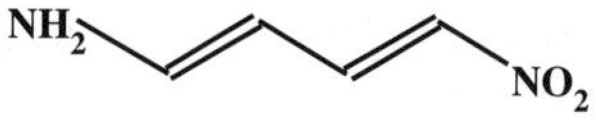

Scheme 4

Das and Dudis have found that the second-order transition moments (S_{ij}) of the PNA molecule are strongly influenced by the presence of the solvent (Scheme 5) [110]. It has been also demonstrated that the diagonal polarizability components $\alpha_{ij} = \{\alpha_{xx}, \alpha_{yy}, \alpha_{zz}\}$ are weakly affected by the solvents in comparison with S_{ij}. In that work, α_{ij} and S_{ij} have been evaluated by using the modified sum-over-states (MSOS) approach combined with the semicontinuum model for the solvent.

Zaleśny, Bartkowiak and co-workers have considered the influence of the solvent polarity on the hyperpolarizabilities and TPA cross section of the simplest

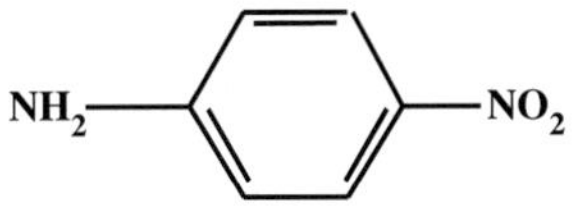

Scheme 5

pyridinium-N-phenolate betaine dye [4-(1-pyridinium-1-yl)phenolate] (Scheme 6) [111]. The molecule investigated in that study, also known as Betaine-30, is a less substituted derivative of Reichrd's dye [2,6-diphenyl-4-(2,4,6-triphenyl-N-pyridinium-1-yl)phenolate]. Reichardt has found that Betaine-30 exhibits the largest blue shift of the longest-wavelength absorption band in comparison with other known solvatochromic compounds. According to the results of calculations, the extremely decreased values of the TPA cross section and molecular hyperpolarizabilities have been observed. The quantum chemical calculations have been carried out in the all-valence INDO-like approximation. The reaction field contribution of the solvent has been evaluated using the non-cavity quantum-mechanical Langevin dipoles/Monte Carlo (QM/LD/MC) technique. It should be noticed that the heterocyclic betaine dyes are not typical donor-acceptor π-conjugated compounds. The CT electronic transition in these molecules arises owing to the interaction of the π orbitals of the donor and acceptor moieties through the σ bond (π-σ-π bond coupling). Hence, the structure of the molecule can be schematically described as D-π-σ-π-A.

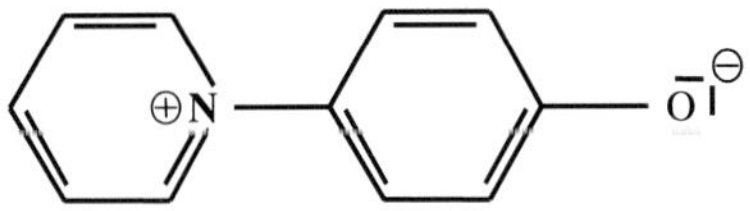

Scheme 6

The drastical enhancement of the TPA cross section in the presence of the solvent for the two-photon polymerization initiator [4-trans-[p-(N,N-Di-n-butylamino)-p-stilbenyl vinyl]piridine (DBASVP) has been illustrated in a recent work by Wong et al. [112]. The DBASVP is the typical D-π-A molecule exhibiting the positive solvatochromism (scheme 7). Hence, the lowest excited state of the DBASVP molecule has been found to be a CT state, which completely dominates the linear absorption spectrum. Wong et al. have combined the time-dependent density functional theory and the polarized continuum model (PCM) to evaluate the solvatochromic shift, TPA cross-section, and oxidation potential of the DBASVP molecule in different solutions.

Recently, the general considerations related to the solvent effects on the TPA cross section and molecular (hyper)polarizabilities have been presented by Bartkowiak et al. [113]. On the basis of the full quantum chemical calculations as well as discussion within the simple two-state models it has been shown that the solvent

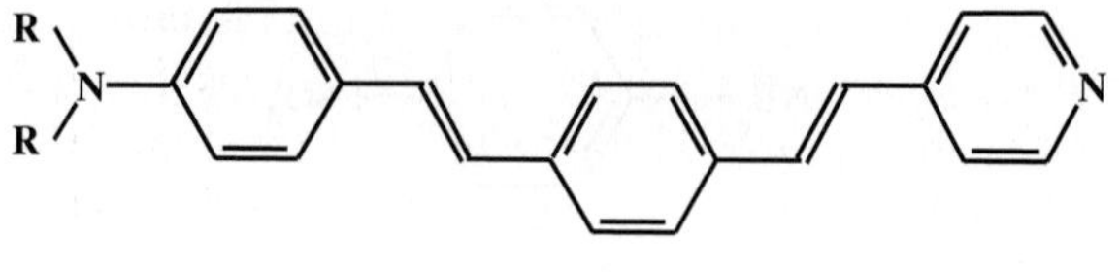

R=CH$_2$CH$_2$CH$_2$CH$_3$

Scheme 7

dependence in the case of the TPA cross-section is significantly larger than for the molecular (hyper)polarizabilities. This general conclusion is restricted to the positively solvatochromic D-π-A compounds exhibiting a monotonic behavior with respect to the polarity of the solvents. In general, there is an important discrepancy between works of Luo et al. [107] and Bartkowiak et al. [113]. As was mentioned above, Luo et al. have shown that the first-order hyperpolarizability is more solvent sensitive than the TPA cross section for the positively solvatochromic donor-acceptor polyene.

It has been shown in a few theoretical studies that the TPA cross section δ of the donor-acceptor π-conjugated molecules can be correctly described, similarly as the molecular (hyper)polarizabilities by a simple two-state model, involving only the ground and the CT excited state [111–114]. Hence, the solvent effect on δ can be discussed within this approximation. It leads to the better understanding of the theoretical results cited in the above papers. Including only the ground and the CT excited state in the Eq. (8), one obtains the two-state approximate equation for dominant component along the molecular axis (in this case chosen as x axis):

$$(11) \quad S_{xx}^{gCT} = \frac{4 \langle g | r_x | CT \rangle \Delta \mu_{gCT,x}}{\omega_{gCT}}$$

In order to obtain the expression of $\langle \delta_{gCT} \rangle$, one can combine Eq. (11) with Eq. (9). Finally, the expression for TPA cross section becomes:

$$(12) \quad \langle \delta_{gCT} \rangle \approx \langle \delta_{gCT}^{xx} \rangle = \frac{16}{5} S_{xx}^2 = \frac{16}{5} \frac{\langle g | r_x | CT \rangle^2 (\Delta \mu_{gCT,x})^2}{\omega_{gCT}^2}$$

It should be noted that the meaning of the spectroscopic parameters in above equations is the same as in the case of Eq. (4) for β_{xxx}. Alternatively, the two-state expression for $\langle \delta_{gCT} \rangle$, similarly as in the case of β_{xxx}, can be written (in atomic units) as

$$(13) \quad \langle \delta_{gCT}^{xx} \rangle = \frac{24 f (\Delta \mu_{gCT,x})^2}{5 \omega_{gCT}^3}$$

A direct comparison between Eq. (5) and Eq. (13) shows that the TPA cross section is a quadratic function of $\Delta \mu_{gCT,x}$ ($\langle \delta_{gCT}^{xx} \rangle \sim f (\Delta \mu_{gCT,x})^2$) while β_{xxx} being a linear one $(\beta_{xxx} \sim f\Delta\mu_{gCT,x})$. In other words, the TPA cross section is directly connected to the static first-order hyperpolarizability through the relation: $\langle \delta_{gCT}^{xx} \rangle \sim \beta_{xxx}\Delta\mu_{gCT,x}$. This important observation indicates that the influence of the solvent effects on the TPA cross section of the positively solvatochromic D-π-A molecules should be larger than that for β_{xxx} [113]. It should be noticed that the structure of Eq. (13) shows that sign of $<\delta_{gCT}>$ is always positive (in opposite to β_μ for the negatively solvatochromic dyes). This result has been confirmed by the direct quantum chemical calculations including the solvent effect [111]. Moreover, the simple two-state relations presented in this section allow for the systematic analysis of the theoretical data obtained in the works cited above [110–113]. On the basis of the limited available theoretical results, it is possible to formulate a hypothesis that the directions of changes in the values of the TPA cross section as a function of the solvent polarity are similar to the molecular (hyper)polarizabilities (see also Figure 2). Unfortunately, there are no systematic results of experimental measurements of the TPA cross section in different solvents for the donor-acceptor π-conjugated molecules. This is important limitation for the discussion presented in this review. Moreover, it should be remembered that, according to Eq. (10), the direct comparison between experimental and theoretical data requires the knowledge of the line shape function $g(\omega)$ for any investigated molecule. This quantity depends on the many parameters such as the given solvent in which measurements are carried out. It is very difficult to establish of $g(\omega)$ on the theoretical way [115]. It can be a much more important solvent effect that the pure dielectric response of the medium.

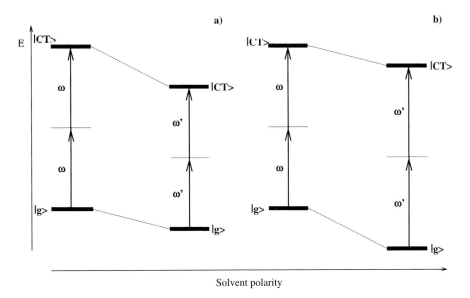

Figure 2. Transition diagrams involved in the two-level description of TPA in the gas phase and in polar solvent: a) positively solvatochromic molecule, b) negatively solvatochromic molecule

In general, the investigations based on the quantum chemical calculations as well as on the other literature data show that the following order of magnitude holds for the TPA cross section and the molecular (hyper)polarizabilities:

(14) $\quad <\delta>^{sol}/<\delta>^{gas} > \beta^{sol}/\beta^{gas} \approx \gamma^{sol}/\gamma^{gas} > \alpha^{sol}/\alpha^{gas}$

This relation is restricted for (a) the positively solvatochromic donor-acceptor π-conjugated molecules, (b) the two-photon absorption from the ground state to the CT excited state, and (c) the static molecular (hyper)polarizabilities. The above ordering was confirmed for the extended variety of the donor-acceptor π-conjugated molecules exhibiting the monotonic behavior as a function of the solvent polarity [40, 113, 116]. However, this trend dose not hold in some cases (see [40]).

In the case of the negatively solvatochromic compounds, according to our best knowledge there is probably only one theoretical work for TPA cross section in the gas phase and in polar solvent [111]. Hence, a generalization is impossible at the moment.

5. CONCLUDING REMARKS

In this review, we focus our interest on the donor-acceptor π-conjugated molecules. This type of molecules is promising for applications in various nonlinear optical devices because of the significant values of the molecular hyperpolarizabilities. The high NLO responses of the donor-acceptor π-conjugated molecules are related to the intramolecular charge transfer (CT) excited state. The optimization of materials for nonlinear optical devices requires understanding of NLO processes as a function of electronic and geometrical molecular structure. On the other hand, the environmental effect on the molecule in a crystal, film, solution etc., may lead to the substantially change their nonlinear optical response in comparison with the gas phase. It is an additional fact, which should be taken into account in design of new nonlinear optical materials.

The understanding and reliable prediction of the influence of the solute-solvent interactions on the nonlinear optical properties of molecular systems is a significant issue for a width range of theoretical and experimental areas of studies. In this review, it was shown that the simple two-state approximations combined with the solvatochromic methods are an effective tools in prediction the direction of the changes of molecular nonlinear responses as a function of solvent polarity. This methodology based on the description of the solvent effects at the molecular level should be treated as a supporting for the most sophisticated quantum chemical approaches.

APPENDIX A. TWO-LEVEL MODELS FOR α AND γ

Including only the ground and the CT excited state in the sum-over states (SOS) expressions for the static α and γ, one obtains the two-state approximate equations for dominating components along the molecular axis (in this case x axis) [101, 102, 117]:

(A1) $\quad \alpha_{xx} = 2 \dfrac{\langle g|r_x|CT\rangle^2}{\omega_{gCT}}$

(A2) $\quad \gamma_{xxxx} = 24 \dfrac{\langle g|r_x|CT\rangle^2 (\Delta\mu_{gCT,x})^2 - \langle g|r_x|CT\rangle^4}{(\omega_{gCT})^3}$

The meaning of the spectroscopic parameters in above expressions is the same as in the case of β_{xxx} (see Eq. (4) in the main text).

ACKNOWLEDGEMENTS

This work was sponsored by the Polish Committee for Scientific Research (grant No T09A 05926).

REFERENCES

[1] Reichardt, C.: Solvents and Solvent Effects in Organic Chemistry, VCH, Weinheim (1988)
[2] Reichardt, C.: Chem. Rev. **94**, 2319–2358 (1994)
[3] Kanis, D.R., Ratner, M.A., Marks, T.J.: Chem. Rev. **94**, 195–242 (1994)
[4] Verbiest, T., Houbrechts, S., Kauranen, M., Clays, K., Persoons, A.: J. Mater. Chem., **7**, 2175–2189 (1997)
[5] Zyss, J., Chemla, D.S. (eds): Nonlinear Optical Properties of Organic Molecules and Crystals, Academic, Orlando (1987)
[6] Nalwa, H.S., Miyata, S.: Nonlinear Optics of Organic Molecules and Polymers, CRC Press, New York (1997)
[7] Bosshard, C., Sutter, K., Prêtre, P., Hulliger, J., Flörsheimer, M., Kaatz, P., Günter, P.: Organic Nonlinear Optical Materials, Gordon & Breach, Basel (1995)
[8] Oudar, J.L., Chemla, D.S.: J. Chem. Phys. **66**, 2664–2668 (1977)
[9] Kuzyk, M.G., Dirk, C.W.: Phys. Rev. A **41**, 5098–5109 (1990)
[10] Dirk, C.W., Cheng, L.-T., Kuzyk, M.G.: Int. J. Quantum Chem. **43**, 27–36 (1992)
[11] Meyers, F., Marder, S.R., Pierce, B.M., Brédas, J.L.: J. Am. Chem. Soc. **116**, 10703–10714 (1994)
[12] Constrand, P., Luo, Y., Ågren, H.: Chem. Phys. Lett. **352**, 262–269 (2002)
[13] Constrand, P., Norman, P., Luo, Y., Ågren, H.: J. Chem. Phys. **121**, 2020–2029 (2004)
[14] Levine, B.F., Bethea, C.G., Wasserman, E., Leenders, L.: J. Chem. Phys. **68**, 5042–5045 (1978)
[15] Teng, C.C., Garito, A.F.: Phys. Rev. Lett. **50**, 350–352 (1983)
[16] Teng, C.C., Garito, A.F.: Phys. Rev. B **28**, 6766–6773 (1983)
[17] Cheng, L.-T., Tam, W., Stevenson, S.H., Meredith, G.R., Rikken G., Marder, S.R.: J. Phys. Chem. **95**, 10631–10642 (1991)
[18] Cheng, L.-T., Tam, W., Marder, S.R., Stiegman, A.E., Rikken, G., Spangler, C.W.: J. Phys. Chem. **95**, 10643–10652 (1991)
[19] Maslianitsin, I.A., Shigorin, V.D., Shipulo, G.P.: Chem. Phys. Lett. **194**, 355–358 (1992)
[20] Sen, R., Majumdar D., Bhattacharyya, S.P.: Chem. Phys. Lett. **190**, 443–446 (1992)
[21] Marder, S.R., Bretan, D.N., Cheng, L.T.: Science **252**, 103–106 (1991)
[22] Marder, S.R., Perry, J.W., Bourhill, G., Gorman, C.B., Tiemann, B.G., Mansour, K.: Science **261**, 186–189 (1993)
[23] Marder, S.R., Gorman, C.B., Meyers, F., Perry, J.W., Bourhill, G., Brédas, J.L., Pierce, B.M.: Science **265**, 632–635 (1994)

[24] Bourhill, G., Brédas, J.L., Cheng, L.-T., Marder, S.R., Meyers, F., Perry, J.W., Tiemann, B.G.: J. Am. Chem. Soc. **116**, 2619–2620 (1994)
[25] Runser, C., Fort, A., Barzoukas, M., Combellas, C., Suba, C., Thiébault, A., Graff, R., Kintzinger, J.P.: Chem. Phys. **193**, 309–319 (1995)
[26] Lu, D., Marten, B., Cao, Y., Ringnalda, M.N., Friesner, R.A., Goddard, W.A., III: Chem. Phys. Lett. **242**, 543–547 (1995)
[27] Di Bella, S., Marks, T.J., Ratner, M.A.: J. Am. Chem. Soc. **116**, 4440–4448 (1994)
[28] Albert, I.D.L., Marks, T.J., Ratner, M.A.: J. Phys. Chem. **100**, 9714–9725 (1996)
[29] Stähelin, M., Burland, D.M., Rice, J.E.: Chem. Phys. Lett. **191**, 245–250 (1992)
[30] Yu, J., Zerner, M.C.: J. Chem. Phys. **100**, 7487–7494 (1994)
[31] Mikkelsen, K.V., Luo, Y., Ågren, H., Jørgensen, P.: J. Chem. Phys. **100**, 8240–8250 (1994)
[32] Larsson, P.-E., Kristensen, L.M., Mikkelsen, K.V.: Chem. Phys. **75**, 449–456 (1999)
[33] Dehu, C., Meyers, F., Hendrickx, E., Clays, K., Persoons, A., Marder, S.R., Brédas, J.L.: J. Am. Chem. Soc. **117**, 10127–10128 (1995)
[34] Woodford, J.N., Pauley, M.A., Wang, C.H.: J. Phys. Chem. A **101**, 1989–1992 (1997)
[35] Huyskens, F.L, Huyskens, P.L., Persoons, A.P.: J. Chem. Phys. **108**, 8161–8171 (1998)
[36] Bartkowiak, W., Lipiński, J.: Computers & Chemistry **1**, 31–37 (1998)
[37] Bartkowiak, W., Lipiński, J.: Chem. Phys. Lett. **292**, 92–96 (1998)
[38] Lipiński, J., Bartkowiak, W.: Chem. Phys. **245**, 263–276 (1999)
[39] Bartkowiak, W., Lipiński, J.: J. Phys. Chem. A **102**, 5236–5240 (1998)
[40] Bartkowiak, W., Zaleśny, R., Niewodniczański, W., Leszczynski, J.: J. Phys. Chem. A **105**, 10702–10710 (2001)
[41] Zhu, W., Wu, G.-S., Jiang, Y.: Int. J. Quantum Chem. **86**, 347–355 (2002)
[42] Wang, C.-K., Wang, Y.-H., Su, Luo, Y.: J. Chem. Phys. **119**, 4409–4412 (2003)
[43] Cammi, R., Frediani, L., Mennucci, B., Ruud, K.: J. Chem. Phys. **119**, 5818–5827 (2004)
[44] Ray, P.C.: Chem. Phys. Lett. **394**, 354–360 (2004)
[45] Ray, P.C.: Chem. Phys. Lett. **395**, 269–273 (2004)
[46] Tomasi, J., Cammi, R., Mennucci, B.: Int. J. Quantum Chem. **75**, 783–803 (1999)
[47] Kurtz, H.A., Dudis D.S.: In: Lipkowitz, K.B., Boyd, D.B., (eds) Review in Computational Chemistry, vol. 12, VCH, New York (1998)
[48] Champagne B., Kirtman, B.: In: Nalwa, H.S. (eds) Handbook of Advanced Electronic and Photonic Materials and Devices, vol. 9, Academic, New York (2001)
[49] Bishop, D.M., Norman, P.: In: Nalwa, H.S. (eds) Handbook of Advanced Electronic and Photonic Materials and Devices, vol. 9, Academic, New York (2001)
[50] Reis, H., Papadopoulos, M.G.: In: Papadopoulos, M.G. (eds) On the Non-linear Optical Response of Molecules, Solids and Liquids: Methods and Applications, Research Signpost Trivadrum, India (2003)
[51] Cammi, R., Mennucci, B., Tomasi: In: Papadopoulos, M.G. (eds) On the Non-linear Optical Response of Molecules, Solids and Liquids: Methods and Applications, Research Signpost Trivadrum, India (2003)
[52] Mikkelsen, K.V.: In: Papadopoulos, M.G. (eds) On the Non-linear Optical Response of Molecules, Solids and Liquids: Methods and Applications, Research Signpost Trivadrum, India (2003)
[53] Mikkelsen, K.V., Luo, Y., Ågren H., Jørgensen, P.: J. Phys. Chem. **102**, 9362–9367 (1995)
[54] Sylvester-Hvid, K.O., Mikkelsen, K.V, Jonsson, D., Norman, P., Ågren, H.: J. Chem. Phys. **109**, 5576–5584 (1998)
[55] Jørgensen, S., Ratner, M.A., Mikkelsen, K.V.: J. Chem. Phys. **115**, 8185–8192 (2001)
[56] Kongsted, J., Osted, A., Mikkelsen, K.V., Christiansen, O.: J. Chem. Phys. **119**, 10519–10535 (2003)
[57] Maroulis, G.: J. Chem. Phys. **113**, 1813–1820 (2000)
[58] Bartkowiak, W., Zaleśny, R., Kowal, M., Leszczynski, J.: Chem. Phys. Lett. **362**, 224–228 (2002)
[59] Wortmann, R., Bishop, D.M.: J. Chem. Phys. **108**, 1001–1007 (1998)
[60] Munn, R.W., Luo, Y., Macák, P., Ågren, H.: J. Chem. Phys. **114**, 3105–3108 (2001)
[61] Macák, P., Norman, P., Luo, Y., Ågren, H.: J. Chem. Phys. **112**, 1868–1875 (2000)

[62] Cammi, R., Mennucci, B., Tomasi, J.: J. Phys. Chem. A **102**, 870–875 (1998)
[63] Cammi, R., Mennucci, B., Tomasi, J.: J. Phys. Chem. A **104**, 4690–4698 (2000)
[64] Zuliani, P., Del Zoppo M., Castiglioni C., Zerbi, G., Andraud, C., Botin, T., Collet, A.: J. Phys. Chem. **99**, 16242–16247 (1995)
[65] Cammi, R., Mennucci, B., Tomasi, J.: J. Am. Chem. Soc. **120**, 8834–8847 (1998)
[66] Bartkowiak, W., Misiaszek, T.: Chem. Phys. **261**, 353–357 (2000)
[67] Reis, H., Papadopoulos, M.G., Avramopoulos, A.: J. Phys. Chem. A **107**, 3907–3917 (2003)
[68] Li, J., Cramer, C.J., Truhlar, D.G.: Int. J. Quantum. Chem. **77**, 264–280 (2000)
[69] Cramer, C.J., Truhlar, D.G.: Chem. Rev. **99**, 2161–2200 (1999)
[70] Suppan, P.: J. Photochem. Photobiol. A **50**, 293–330 (1990)
[71] Amos, A.T., Burrows, B.L.: In: Advances in Quantum Chemistry, Academic, New York (1973)
[72] Liptay, N.: In: Lim, E.C. (eds) Excited States, Academic, New York (1974)
[73] Tomasi, J., Persico, M.: Chem. Rev. **94**, 2027–2094 (1994)
[74] Grozema, F.C., van Duijnen, P.T.: J. Phys. Chem. A **102**, 7984–7989 (1998)
[75] Gao, J.: J. Am. Chem. Soc. **116**, 9324–9328 (1994)
[76] Fox, T., Rösch, N.: Chem. Phys. Lett. **191**, 33–37 (1992)
[77] Kovalenko, S.A., Schanz, R., Farztdinov, V.M., Henning, H., Ernsting, N.P.: Chem. Phys. Lett. **323**, 312–322 (2000)
[78] Bartkowiak, W., Lipkowski, P.: J. Mol. Mod., in press.
[79] Abe, J., Shirai, Y.: J. Phys. Chem. B **101**, 576–582 (1997)
[80] Bublitz, G.U., Ortiz, R., Runser, C., Fort, A., Barzoukas, M., Marder, S.R., Boxer, S.G.: J. Am. Chem. Soc. **119**, 2311–2312 (1997)
[81] Jacques, P.: J. Phys. Chem. **90**, 5535–5539 (1986)
[82] Mayers, F., Brédas, J.L., Pierce, B.M., Marder, S.R.: Nonlinear Opt. **14**, 61–71 (1995)
[83] Kogej, T., Mayers, F., Marder, S.R., Silbey, R., Brédas, J.L.: Synth. Met. **85**, 1141–1142 (1997)
[84] Marder, S.R., Perry, J.W., Tiemann, B.G., Gorman, C.B., Gilmour, S., Biddle, S.L., Bourhill, G.: J. Am. Chem. Soc. **115**, 2524–2526 (1993)
[85] Kirtman, B., Champagne, B., Bishop, D.M.: J. Am. Chem. Soc. **122**, 8007–8012 (2000)
[86] Balakina, M.Y., Li, J., Geskin, V.M., Marder, S.R., Brédas, J.L.: J. Chem. Phys. **113**, 9599–9609 (2000)
[87] Barzoukas, M., Runser, C., Fort, A., Blanchard-Desce, M.: Chem. Phys. Lett. **257**, 531–537 (1996)
[88] Blanchard-Desce, M., Barzoukas, M.: J. Opt Soc. Am. B **15**, 302–3007 (1998)
[89] Thompson, W.H., Blanchard-Desce, M., Hynes, J.T.: J. Phys. Chem. A **102**, 7712–7722 (1998)
[90] Thompson, W.H., Blanchard-Desce, M., Alain,V., Muller, J., Fort, A., Barzoukas, M., Hynes, J.T.: J. Phys. Chem. A **103**, 3766–3771 (1999)
[91] Laage, D., Thompson, W.H., Blanchard-Desce, M., Hynes, J.T.: J. Phys. Chem. A **107**, 6032–6046 (2003)
[92] Bartkowiak, W., Zaleśny, R., Leszczynski.: J. Chem. Phys. **287**, 103–112 (2003)
[93] Lu, D., Chen, G., Perry, J.W., Goddard, W.A., III: J. Am. Chem. Soc. **116**, 10679–10685 (1994)
[94] Chen, G., Lu, D., Goddard, W.A., III: J. Chem. Phys. **101**, 5860–5864 (1994)
[95] Painelli, A., Terenziani, F.: Chem. Phys. Lett. **312**, 211–220 (1999)
[96] Painelli, A.: Chem. Phys. **245**, 185–197 (1999)
[97] Painelli, A., Terenziani, F.: Synth. Met. **116**, 135–138 (2001)
[98] Painelli, A., Terenziani, F.: Synth. Met. **124**, 171–173 (2001)
[99] Painelli, A., Terenziani, F.: Synth. Met. **109**, 229–233 (2000)
[100] Boldrini, B., Cavalli, E., Painelli, A., Terenziani, F.: J. Phys. Chem. A **106**, 6286–6294 (2002)
[101] Orr, B.J., Ward, J.F.: Mol. Phys. **20**, 513–526 (1971)
[102] Willetts, A., Rice, J.E., Burland, D.M., Shelton, D.P.: J. Chem. Phys. **97**, 7590–7599 (1992)
[103] Sworakowski, J., Lipiński, J., Ziółek, Ł., Palewska, K., Nešpůrek, S.: J. Phys. Chem. A **100**, 12288–12294 (1996)
[104] Effenberger, F., Wűrtner, F.: Angew. Chem. Int. Ed. Engl. **32**, 719–721 (1993)
[105] Manson, P.R., McClain, W.M.: J. Phys. Chem. **53**, 29–37 (1970)

[106] McClain, W.M., Harris, R.A.: In: Lim, E.M. (eds) Excited States, vol. 3, Academic, New York (1977)
[107] Luo, Y., Norman, P., Macak, and Ågren, H.: J. Phys. Chem. A **104**, 4718–4722 (2000)
[108] Bishop, D.M., Luis, J.M., Kirtman, B.: J. Chem. Phys. **116**, 9729–9739 (2002)
[109] Norman, P., Cronstrand, P., Ericsson, J. Chem. Phys. **285**, 207–220 (2002)
[110] Das, G.P., Dudis, D.S.: J. Phys. Chem. A **104**, 4767–4771 (2000)
[111] Zaleśny, R., Bartkowiak, W., Styrcz, S., Leszczynski, J.: J. Phys. Chem. A **106**, 4032–4037 (2002)
[112] Wang, C.-K., Zhao, K., Su, Y., Ren, Y., Zhao, X., Luo, Y.: J. Chem. Phys. **119**, 1208–1213 (2003)
[113] Bartkowiak, W., Zaleśny, R., Skwara, B.: J. of Computational Methods in Sciences and Engineering **4**, 551–558 (2004)
[114] Wang, C.-K., Macak, P., Luo, Y., Ågren, H.: J. Chem. Phys. **114**, 9813–9820 (2001)
[115] Das, G.P., Yeates, A.T., Dudis, D.S.: Chem. Phys. Lett. **361**, 71–78 (2002)
[116] Bartkowiak, W.: Unpublished results
[117] Combellas, C., Mathey, G., Thiébault, A., Kajzar, F.: Nonlinear Opt. **12**, 251–256 (1995)

CHAPTER 10

SYMMETRY BASED APPROACH TO THE EVALUATION OF SECOND ORDER NLO PROPERTIES OF CARBON NANOTUBES

L. DE DOMINICIS AND R. FANTONI
Advanced Technologies Department, ENEA Frascati Via E. Fermi 45, 00044 Frascati – Italy

Abstract: This chapter presents a review of the direct implantation of symmetry into calculations of carbon nanotubes (CNTs) second order nonlinear optical properties (NLO). Emphasis is given to potentiality of the method to estimate quantitatively the magnitude of first hyperpolarizability for several CNTs topologies. The main advantage of performing calculations with symmetrized eigenfunctions, relies on the direct identification of the state-to-state transitions contributing to the hyperpolarizability. An estimated value of $\beta \sim 10^{-30}$ esu for the non-resonant hyperpolarizability of chiral CNTs is obtained

Keywords: Nonlinear spectroscopy; hyperpolarizability; carbon nanotubes; irreducible representations

1. INTRODUCTION

The discovery of Carbon Nanotubes (CNTs) dates back to 1991 [1]. Since then CNTs, for their excellent electrical and mechanical properties, have been an hot topic in several scientific and technological fields ranging from chemistry to mechanics [2]. Most of the remarkable properties of CNTs are a direct consequence of their peculiar topology. From a topological point of view, a single wall CNT is a graphite sheet rolled up into a cylinder with nanometer size diameter. The wrapping procedure leads to considerable changes in the topological space of the formed structure (CNT) with respect to the original graphite sheet. First of all, due to the large aspect ratio (length/diameter $\sim 10^4$), a CNT can be considered a quasi one-dimensional (1D) structure. In addition, the wrapping procedure maps the symmetry group of graphite into a new class of symmetry operations leaving the CNT structure invariant. In particular, according to the direction of the wrapping axis with respect the basic vectors of the graphite unit cell, the CNT may display a chiral structure [3]. The wrapping procedure, at the basis of CNT formation from a graphite sheet, then singles out a transformation from a 2D achiral topology to a quasi-1D topological space with, eventually, a chiral structure.

The drastic change of the topological space accessible to the delocalized π electrons coming from the C=C bond, is expected to strongly influence the properties of their wavefunction and consequently the CNT optical response. The reduced dimensionality of the topological space leads to a quantum confinement of the π electronic wavefunction in the radial and circumferential directions with subsequent quantization of the angular momentum component along the tube axis (band index m). As a consequence, plane wave motion occurs only along the nanotube axis corresponding to a large number of closely spaced allowed wavevector k.

In addition, as far as electron-electron interaction is neglected, the π electrons are subject to a potential with the full spatial symmetry of the CNT topology. The electronic wavefunctions can then be classified according to their transformation properties under the symmetry operations leaving the CNT invariant. As far as dipole approximation holds, both the linear and nonlinear optical response of a CNT are governed by matrix elements of the dipole operator between two electronic wavefunctions. Being the selection rules of dipole matrix elements essentially determined by the wavefunctions symmetry, it turns out that the optical properties of CNT are essentially rooted in their topology.

The most striking demonstration of the deep influence of direct implantation of symmetry into CNTs linear optical properties calculation, is reported in the works of Bozović [4] and Damnjanović [5]. On the other side, the nonlinear optical properties (NLO) of CNTs have attracted a certain interest in the last decade. The first attempt of modeling third order CNT nonlinearity has to be ascribed to Xie [6], Wan [7], Margulis [8] and Jiang [9]. The works of Slepyan [10, 11] shed light on the high harmonic generation processes both from an individual CNT and a CNTs rope. On the experimental side, the third order NLO of CNTs have been demonstrated by several works. Optical limiting properties of CNT was observed by Vivien [12], Jin [13]. Third harmonic generation has been studied by Stanciu [14] and Koronov [15] at femtosecond time regime, while Wang [16] probed Kerr type nonlinear processes. Degenerate four wave mixing experiments by Liu [16] and Botti [17] allowed to estimate in $\gamma \sim 10^{-11}$ esu the second hyperpolarizability of CNT at nanosecond time scale. Less attention has been paid to second order NLO of CNTs. Experimental evidence of second harmonic generation (SHG) is reported by De Dominicis [18] and Koronov [15] at nanosecond and femtosecond time regime respectively, while theoretical investigations on the role played by CNT topology and quantum confinement in affecting SHG are still at an early stage [19].

The issue of modeling the second order NLO of CNTs is strictly related to the advances in CNTs assembling technology. In fact, as far as fully oriented non-centrosymmetric assembling of CNTs are being developed [20], samples characterized by a non-vanishing second order susceptibility becomes available, allowing the study of second order NLO of CNTs.

In the present chapter an approximate method for quantitative estimation of CNT first hyperpolarizability is presented. The method fully takes advantage of the algorithm developed by Damnjanovic [21] to calculate CNT eigenstates by means of the modified projector technique with tight binding approach. The direct incorporation of symmetry into calculations allows to identify the state-to-state transitions contributing to β both for electron with k at the origin and inside the irreducible domain (ID) $[0, \pi/a]$, where a is the translational periodicity of the CNT. Under appropriate approximations, whose validity can be only checked *a posteriori*, the developed code runs on commercially available PC. Within this limits, values of β up to 10^{-30} esu (1 esu = $3.7 \cdot 10^{-21}$ $C^3 J^{-2} m^3$) are obtained, thus positioning CNTs among the most efficient second order nonlinear optical materials.

The chapter is organized as follows. In Section 2 the symmetry properties of CNTs and the selection rules for electronic transitions are described. In Section 3 the form of β tensor is determined, together with the several state-to-state contributions coming from electrons inside the ID. In Section 4 the principles for estimation of the magnitude of β are reported together with the results of a calculation for several CNT topologies.

2. CARBON NANOTUBES SYMMETRY PROPERTIES AND ELECTRONIC TRANSITIONS SELECTION RULES

CNTs topology is completely determined [2] by two integers (n_1, n_2). The particular cases (n, n) and $(n, 0)$ give rise to armchair (**A**) and zigzag (**Z**) CNTs respectively. **A** and **Z** CNTs are characterized by invariance under mirror reflection and then can be grouped as achiral CNTs. All other possible combinations of (n_1, n_2) have a chiral topology. The symmetry properties of a CNT are gathered in the symmetry group **G** giving all the transformations which leave the CNT invariant. Both for chiral and achiral CNTs **G** is a line group (symmetry group of an object translationally periodic along a line) given by [3] (the Hermann-Mauguin international notation is used)

(1) $$G_c = Lq_p 22$$
$$G_{ZA} = L2N_N/mcm$$

where N is the greatest common divisor of n_1 and n_2. The q parameter is given by

(2) $$q = 2(n_1^2 + n_2^2 + n_1 n_2)/NR$$

with $R = 3$ or $R = 1$ whether $(n_1 - n_2)/3N$ is an integer or not. The parameter p is expressed in terms of n_1, n_2 and N by numerical functions [3].

Between the elements of **G** there are screw axes $(C_q^r | Na/q)^t$, consisting of a rotation of $2\pi rt/q$ around the tube axis followed by a translation of Nta/q along the tube axis direction. Here t is an integer and the helicity parameter r and the translational period a given by

$$(3) \quad r = \frac{q}{N} Fr \left[\frac{N}{qR} \left(3 - 2 \frac{n_1 - n_2}{n_1} \right) + \frac{n}{n_1} \left(\frac{n_1 - n_2}{N} \right)^{\varphi\left(\frac{n_1}{N}\right) - 1} \right]$$

$$a = \sqrt{\frac{3q}{2RN}} a_0$$

where $F[x]$ is the fractional part of x, $\varphi(n)$ the Euler function and $a_0 = 2.461\text{Å}$. Other elements are rotations C_N^s of $2\pi s/N$ around tube axis ($s = 0, 1, \ldots N\text{-}1$) and rotations (U, U') of π around a direction perpendicular to the tube axis (Fig. 1). For armchair and zigzag CNT we have additional vertical (σ_v) and horizontal (σ_h) mirror planes and glide planes (σ_v', σ_h').

Let us define the orbit of a carbon atom in a CNT as the set of atoms generated by the symmetry transformations from any arbitrary chosen initial atoms. Within this scheme, a CNT is a mono-orbit system in which the atom C_{tsu} is obtained by the action of

$$(4) \quad Z_{Tr} = \left(C_q^r | na/q \right)^t C_n^s U^u$$

on the initial one C_{000}. The subgroup Z_{Tr} is called the transversal of the CNT symmetry group G and is obtained by neglecting all the site symmetry transformations (set of transformations for which the initial atom is a fixed point). It is interesting to note that for chiral CNT $Z_{Tr} = G$.

All the achiral CNTs have a point of inversion symmetry located on the tube axis. For achiral CNTs with even n this can be easily demonstrated. In fact, for even n, C_2 is a symmetry transformation, which together with σ_h acts as the inversion transformation I on CNT structure.

This consideration has a very important consequence on the central issue of this work. In fact, the presence of an inversion centre makes vanishing the third

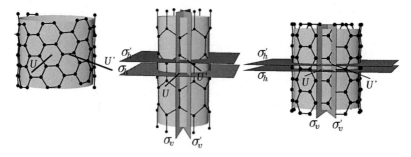

Figure 1. Symmetry of the single wall CNT (8,6), (6,0) and (6,6). The horizontal rotational axes U and U' are symmetries of all the tubes, while the mirror planes (σ_v, σ_h), the glide plane σ_v' and the roto-reflectional plane σ_h' are symmetries of the achiral tubes only (from [3])

rank tensor responsible for the second order nonlinear optical response of achiral CNTs. In the following, attention will then be restricted only to chiral CNTs, which, lacking inversion symmetry, have a non-vanishing second order nonlinear optical response. Within this context, it is important to mention that Slepyan [6] demonstrated theoretically, following a power expansion method of the current density, the absence of even high harmonics generation in chiral CNT illuminated by an intense laser pulse.

As far as electron-electron interaction is neglected, the Hamiltonian H a π electron in a CNT commutes with all the element of G making, according to the basic theory of group representation [22], the electronic eigenfunctions a set of basis functions for the irreducible representations of G. In fact, the basis functions $\{\phi_i\}_{i=1...l}$ of an irreducible representation of dimension l are characterized by the property

$$(5) \quad g\phi_i = \sum_{k=1}^{l} g_{ik}\phi_k \quad \forall g \in G$$

where g_{ik} are the matrix elements of the irreducible representation.

It is easy to verify that if $[H, G] = 0$, the l degenerate eigenfunctions $\{\phi_i\}_{i=1...l}$ of H which belong to the same eigenvalue ε obey to the transformation law in Eq. (5). In fact we have

$$(6) \quad H\phi_j = \varepsilon\phi_j \quad j = 1\ldots l \rightarrow gH\phi_j = H(g\phi_j) = \varepsilon(g\phi_j)$$

Then $g\phi_j$ is still an eigenfunction of H with eigenvalue ε. But the $\{\phi_i\}_{i=1...l}$ are linearly independent and a linear combination of degenerate eigenfunctions is still an eigenfunction with the same eigenvalue, then we must have

$$(7) \quad g\phi_j = \sum_{k=1}^{l} g_{jk}\phi_k \quad \forall g \in G$$

The demonstration that the coefficients in the summation in Eq. (7) are with the matrix elements of the irreducible representations of G is out of the scope of the present treatment and can be found in [22].

It is then demonstrated that the irreducible representations of the symmetry group G decompose the electronic state space of the CNT into invariant subspaces in which the eigenfunctions of H act as bases.

The irreducible representations (irreps) of the line groups in (1) have been extensively studied by Bozovic [23] and Damnjanovic [24].

The irreps are given as a function of the parameters $k \in [0, \pi/a]$ and $m = (-q/2, q/2]$. The $k = 0$ condition means the suppression of all the translations making the symmetry operations to collapse into a point group. This point group P is called isogonal and is given, for chiral and achiral CNT, by

$$(8) \quad P_c = D_q$$
$$P_{ZA} = D_{(2N)h}$$

Within this scheme the parameter k, being associated with translational symmetries, plays the role of a quasi-linear momentum, while m is the component of a quasi-angular momentum along the tube axis. The two quantum numbers behave differently when an optical excitation induces a state-to-state transition. On one hand, k is a conserved quantity, being the parametrization of the CNT translational symmetries which form a subgroup of G. On the other hand, m is not a conserved physical quantity in electromagnetic interaction being associated to the isogonal rotations which are not a subgroup of G, as dictated by the non-symmorphic (P not a subgroup of G) nature of G both for achiral and chiral CNTs.

An electronic state can be then labelled as $|k, m, \Pi\rangle$, with the quantum numbers k and m, plus an additional quantum number Π gathering all the other possible symmetries (U, σ_v, σ_h).

As far as dipole approximation holds, the probability of an electronic transition due to the interaction of CNT with an optical field, is governed by the position operator matrix elements taken between the initial (i) and final (f) states

$$(9) \quad \langle k_f, m_f, \Pi_f | \vec{r} | k_i, m_i, \Pi_i \rangle$$

General selection rules governing the process can be obtained by observing that the decomposition onto irreducible representations of the position operator both for chiral and achiral CNTs contains only contributions with $k = 0$ and $m = 0, 1$. In fact, the matrix elements in Eq. (9) can be expressed in the Wigner-Eckart form as [5]

$$(10) \quad \langle k_f, m_f, \Pi_f | \vec{r} | k_i, m_i, \Pi_i \rangle$$
$$= \langle k_f, m_f, \Pi_f | 0, m, \Pi; k_i, m_i, \Pi_i \rangle \cdot \vec{r} \left(k_f, m_f, \Pi_f \| 0, m, \Pi \| k_i, m_i, \Pi_i \right)$$

where $\vec{r}\left(k_f, m_f, \Pi_f \| 0, m, \Pi \| k_i, m_i, \Pi_i\right)$ and $\langle k_f, m_f, \Pi_f | 0, m, \Pi; k_i, m_i, \Pi_i \rangle$ are the reduced matrix element and the Clebsch-Gordan coefficient, respectively. The Clebsh-Gordan coefficients, being independent on $\vec{r}$, are subject to quite general selection rules given by [5]

$$(11) \quad k_f = k_i$$
$$m_f - m_i = m$$
$$\Pi_f = \Pi \Pi_i$$

In order to better understand the importance of the selection rules in Eq. (11) it must be outlined that the dispersion relations $\varepsilon_m(k)$, giving the electron energy as a function of k, have a band structure labelled by the quantum number m. The selection rules in Eq. (11) determine that in a CNT, under the action of an optical field, the electrons are subject to interband direct transitions with $\Delta m = 0, 1$.

Moreover, the meaning of the third selection rules in Eq. (11) is strictly related to symmetry labelling of the $|k, m, \Pi\rangle$ states. The symmetry properties of the

Evaluation of Second Order NLO Properties of Carbon Nanotubes   325

Table 1. Irreps of CNT electronic states at the origin and inside the ID

	Chiral	Achiral
$k = 0$	$_0E_m \ _0A_0^\pm \ _0A_{q/2}^\pm$	$_0E_m^\pm \ _0A_0^\pm \ _0B_0^\pm \ _0A_n^\pm \ _0B_n^\pm$
$k \in (0, \pi/a)$	$_kE_m$	$_kG_m \ _kE_0^{a/b} \ _kE_n^{a/b}$

electronic states of a CNT, as dictated by the irreps of G, are listed in Table 1 as a function of CNT topology [24] and of quantum numbers k and m. The symmetry of electronic states for achiral topology is reported only for the sake of completeness. The dimension of a representation gives the degeneracy of the energy level $\varepsilon_m(k)$. The A and B states are one-dimensional, the states E two dimensional while G states span a four dimensional representation. For chiral CNTs, the A states have the superscript $+$ or $-$ according to the parity of the eigenfunction under the U symmetry operation. For achiral CNTs the a or b superscript indicates parity with respect to σ_v, while parity under σ_h is indicated by $+$ or $-$ superscript. It is worthwhile to note that achiral CNT have a well defined parity under σ_v only at the origin of the irreducible domain.

Within the formalism of the irreducible representation the position operator is decomposed in $_0E_1 + _0A_0^-$ for chiral CNTs and $_0E_1^+ + _0A_0^-$ for achiral topology.

Armed with this formalism is possible to determine the allowed optical transitions in dipole approximation as determined by the matrix element in Eq. (9).

Let start to analyze the element for chiral CNT. A x,y,z reference frame intrinsic to the CNT is introduce with the z axis along the tube axis. For chiral CNT, U is the only additional symmetry operation, which is equivalent to the inversion of z axis. An electronic wavefunction labelled with "+" corresponds to a distribution invariant under U, and is then characterized by $\Pi = +1$. On the other hand a wavefunction with "−" change its sign under U and then has $\Pi = -1$. Because of z has $\Pi = -1$, it turns out that, the third selection rules in Eq. (11) asserts that for chiral CNT we have the following non-vanishing matrix elements

(12) $\langle _0A_0^\pm |z| _0A_0^\mp \rangle \langle _0A_{q/2}^\pm |z| _0A_{q/2}^\mp \rangle \langle _kE_m |z| _kE_m^* \rangle$

Where the asterisk indicates an electron in conduction band. Note that in Eq. (12) $\Delta m = 0$ because the irreducible representation of z is characterized by $m = 0$.

When analyzing the matrix element in Eq. (9) with x and y, it must be considered that the two operators have a $m = 1$ irreps, which, in view of the second selection rules in Eq. (11), results in the following non-vanishing elements

(13) $\langle _0A_0^\pm |x| _0E_1 \rangle$ $\langle _0A_0^\pm |y| _0E_1 \rangle$

$\langle _kE_m |x| _kE_{m+1}^* \rangle$ $\langle _kE_m |y| _kE_{m+1}^* \rangle$

$\langle _0A_{q/2} |x| _0E_{q/2-1} \rangle$ $\langle _0A_{q/2} |y| _0E_{q/2-1} \rangle$

3. CNT HYPERPOLARIZABILITY

At microscopic scale, to second order in the incident field, the induced dipole moment μ on a carbon nanotube is given by

$$\mu_i = \alpha_{ij} E_j + \beta_{ijk} E_j E_k + \ldots \quad (14)$$

where i, j, k run over the axes of an intrinsic CNT reference frame, α_{ij} is the polarizability tensor and β_{ijk} the first hyperpolarizability. The efficiency of a CNT as SHG emitter is then related to the magnitude of the first hyperpolarizability. The SHG hyperpolarizability can be written in dipole approximation as [25]

$$\beta_{ijk}(2\omega;\omega,\omega) = e^3 \sum_{lmn} \left[\frac{\langle l|i|m\rangle \langle n|j|m\rangle \langle n|k|l\rangle}{(\omega_{nl} - 2\omega - i\gamma_{nl})(\omega_{ml} - \omega - i\gamma_{ml})} + \ldots \right] \quad (15)$$

where l, m, n run over the electronic eigenstates, e is the electronic charge and the terms summarized by dots are obtained by permuting the first term. The parameters γ_{nl} are related to the state lifetime τ by the relation

$$\gamma_{nl} = \frac{1}{\tau_n} + \frac{1}{\tau_l} \quad (16)$$

The form of the β tensor for chiral CNT can be established quite generally from consideration based on the symmetry group G_C. As described previously, chiral CNT are invariant under U symmetry transformation. If the x axis is taken along U, the transformation corresponds to $x \to x$, $y \to -y$, $z \to -z$. It follows that, under U, we have $\beta_{zzz} \to -\beta_{zzz}$, $\beta_{zzy} \to -\beta_{zzy}$, $\beta_{yyz} \to -\beta_{yyz}$.

These transformations rules are in accordance with the invariance of the CNT under U if and only if

$$\beta_{zzz} = \beta_{zzx} = \beta_{xzz} = \beta_{zzy} = \beta_{yzz} = \beta_{zxx} = \beta_{xzx} = \beta_{zyy} = \beta_{yzy} = 0 \quad (17)$$

The first hyperpolarizability tensor for chiral CNT has then the form

$$\beta(2\omega;\omega,\omega) = \begin{pmatrix} 0 & 0 & 0 & xyz & xzy & 0 & 0 & 0 & 0 \\ 0 & 0 & 0 & 0 & 0 & yzx & yxz & 0 & 0 \\ 0 & 0 & 0 & 0 & 0 & 0 & 0 & zxy & zyx \end{pmatrix} \quad (18)$$

It is interesting to note that for chiral CNT the symmetry properties of the hyperpolarizability tensor are governed by the isogonal point group. In fact the hyperpolarizability tensor of a system with D_q (q even) symmetry point group transforms like [26] the tensor in Eq. (18). This finding enforces the statement that CNTs with achiral topology have a vanishing hyperpolarizability. In fact, the isogonal point group of achiral CNTs is $D_{(2N)h}$, which is characterized [26] by $\beta = 0$. In the zero temperature approximation, if, in Eq. (15), the state l is taken in the valence band, the summation over m and n runs over conduction band states. The selection rules in Eqs. (12) and (13) allow to identify the state to state transition contributing to summation in Eq. (15). In fact, the necessary condition for a non-vanishing CNT electronic hyperpolarizability is the existence of three electronic eigensates satisfying, in view of the selection rules in Eqs. (12) and (13) and the tensorial form in Eq. (18), the condition

(19) $\langle k, m, \Pi | x | k, m \pm 1, \Pi' \rangle \langle k, m, \Pi' | y | k, m \pm 1, \Pi'' \rangle$
$\langle k, m, \Pi'' | z | k, m, \Pi \rangle \neq 0$

where the state $|k, m, \Pi\rangle$ singles out an irreps of the symmetry group G_C.

The contributions $\beta_{xyz}^{(0)}$ from $k=0$ and $\beta_{xyz}^{(k,m)}$ for k inside the ID for a given m, as determined from the allowed transitions reported in Eq. (12) and (13), are shown in Figs. 2 and 3.

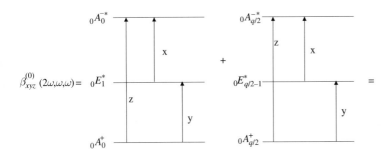

Figure 2. Diagrams of state to state transitions contribution to hyperpolarizability for $k=0$

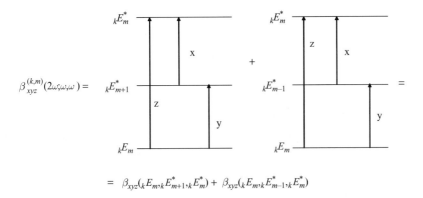

Figure 3. Diagrams of state to state transitions contribution to hyperpolarizability for k inside the ID

The hyperpolarizability of a chiral CNT is obtained by integrating $\beta_{xyz}^{(k,m)}$, after summation over band index m, for k inside the ID and adding the $k=0$ contribution $\beta_{xyz}^{(0)}$.

$$(20) \quad \beta_{xyz}(2\omega;\omega,\omega) = \beta_{xyz}^{(0)} + \lim_{\varepsilon \to 0} \sum_{m=-q/2+1}^{q/2} \left(\int_{0+\varepsilon}^{(\pi/a)-\varepsilon} \beta_{xyz}(_kE_m, {}_kE_{m+1}^*, {}_kE_m^*)dk \right.$$

$$\left. + \int_{0+\varepsilon}^{(\pi/a)-\varepsilon} \beta_{xyz}(_kE_m, {}_kE_{m-1}^*, {}_kE_m^*)dk \right)$$

As far as ω and 2ω do not match an allowed transition among the ones involved in Figs. 2 and 3, the contributions to β in Eq. (15) are all out of resonance and their magnitude is of the same order. On contrary, as ω or 2ω approaches the energy of a state-to-state transition contributing to β, the associated term becomes resonant and dominates with respect the others. In both cases, for a quantitative determination of the magnitude of the β tensor an exact knowledge of the band structure $\varepsilon_m(k)$, as dictated by CNT topology, is required. In the off-resonant case the band structure, in principle, considerably complicates the integration over k. On contrary, for resonant interaction, the integration over k collapses into a discrete summation, but the quantitative determination depends now also on the knowledge of the γ_{nl} parameters Eq. (16).

4. QUANTITATIVE CALCULATION OF β

The quantitative determination of the magnitude of the β_{xyz} tensor component is based on the explicit knowledge of the spatial dependence of the $|k,m,\Pi\rangle$ electronic wavefunctions. As pointed out in Section 2 the $|k,m,\Pi\rangle$ functions transform as the basis functions of the irreps of the symmetry group G and can be obtained with the modified projector technique method [23] in tight binding approximation. The so obtained symmetrized wavefunctions preserve the transformation properties as dictated by the irreps of the symmetry group G.

For example, the wavefunction $|_0A_0^+\rangle$, corresponding to the one dimensional irreducible representation of G and obtained with the projector technique in tight binding approximation, results to be fully symmetric under the symmetry transformations of G.

In the following, in order to simplify calculations and notation, the attention is restricted to CNTs with $N=1$ and hence $s=0$. Despite the extension of calculations to CNTs with $N \neq 1$ is straightforward, it must be noted that almost all the CNTs chiral topologies are characterized by $N=1$. For such a class of CNTs the whole topology is generated by the set of transformation $(C_q^r|a/q)^t U^u$. Being a CNT a mono-orbit system, as pointed out in Section 2, the position vector $\vec{R}_{tu}$ of a carbon atom is obtained with the transformation

$$(21) \quad \vec{R}_{tu} = (C_q^r|a/q)^t U^u \vec{R}_{00}$$

where $\vec{R}_{00}$ is the position vector of the atom chosen as origin, given in cylindrical coordinates by

(22) $\quad \vec{R}_{00} = (\rho_0, \phi_0, z_0) = \left(D/2, 2\pi \dfrac{n_1+n_2}{qR}, \dfrac{n_1-n_2}{\sqrt{6qR}} a_0 \right)$

$$D = \dfrac{1}{\pi}\sqrt{Rq/2} a_0$$

It turns out that

(23) $\quad \vec{R}_{tu} = \left(\rho_0, (-1)^u \phi_0 + 2\pi \dfrac{rt}{q}, (-1)^u z_0 + t\dfrac{a}{q} \right)$

If $|tu\rangle$ indicates an orbital centred on the atoms at $\vec{R}_{tu}$, in a similar fashion we have $|tu\rangle = (C_q^r|a/q)^t U^u |00\rangle$, where $|00\rangle$ is the orbital centred on the atom chosen as origin. Armed with this formalism, it is possible to write down (Table 2) the symmetry adapted electronic eigenfunctions in tight binding approximation, also called generalized Bloch eigenfunctions, as calculated by Damnjanovic [21, 24] with the modified projector technique method.

Where

(24) $\quad \psi_m^k(t) = \dfrac{ka + 2\pi mr}{q} t \qquad h_m^k = Arg(h_m^1(k))$

$\quad h_m^1(k) = \sum_t H_{t1} e^{i\psi_m^k(t)} \qquad H_{t1} = \langle 00| H |t1 \rangle$

As the expressions in Table 2 show, the generalized Bloch functions are constructed by taking a linear combination of atomic orbitals centred on the carbon atom sites and modulated with a phase factor. The so obtained functions are the projection of a tight binding constructed eigenfunction on the irreducible bases of the group G_c. For example, the symmetry transformation U of G_c corresponds to the $0 \leftrightarrow 1$ exchange in the generalized Bloch functions in Table 2.

It is easy to verify that while the one dimensional representation A have a well defined parity under U, the two dimensional representation E have not a defined symmetry under U, in agreement with the general statement in Section 2. From

Table 2. Irreducible representation and generalized Bloch functions for the electronic states of a chiral carbon nanotubes. G_C is the dimension of the symmetry group

Irreducible Representation	Generalized Bloch functions
$_0A_m^\Pi (m=0, q/2)$	$\|0m\Pi\rangle = \dfrac{1}{\sqrt{G_C}} \sum_t e^{-i\psi_m^0(t)} (\|t0\rangle + \Pi \|t1\rangle)$
$_kE_m$	$\|km\rangle = \dfrac{1}{\sqrt{G_C}} \sum_t e^{-i\psi_m^k(t)} \left(\|t0\rangle \pm e^{ih_m^k} \|t1\rangle \right)$
$_kE_m^*$	$\|km\rangle = \dfrac{1}{\sqrt{G_C}} \sum_t e^{-i\psi_m^k(t)} \left(\|t1\rangle \pm e^{ih_m^k} \|t0\rangle \right)$

a theoretical point of view, summations over the index t in Table 2 are extended from $-\infty$ to $+\infty$ but for calculations the summation can be truncated at an integer t_{cutoff} for which boundary effects are negligible. The integer t_{cutoff} is related to a characteristic CNT length L_{char} to which all calculated values of hyperpolarizability in the following will be referred. At first approximation we can assume that $L_{char} \approx 10^3(a/q)$. This means that nearly 10^3 group elements characterize the tube of length L_{char}. In this case, for a carbon nanotube with topology $(10, 1)$ L_{char} is nearly equal to 200 nm.

The introduction of the generalized Bloch states reported in Table 2 allows to obtain an explicit form for the matrix element in Eqs. (12) and (13) and then ultimately of $\beta_{xyz}^{(0)}$ and $\beta_{xyz}^{(k,m)}$. Quantitative calculation of $|\beta_{xyz}(2\omega;\omega,\omega)|$ is then possible once that the summation is truncated at some $t \leq t_{cutoff}$ and only nearest neighbours integrals are taking into account.

As pointed out, before the quantitative calculation of β requires the exact knowledge of CNT energy band structures $\varepsilon_m(k)$. For the a non-resonant CNT-laser interaction, when both ω and 2ω do not match the energy difference $\varepsilon_m(k) - \varepsilon_{m'}(k)$ of a direct transition between two different bands, at a first approximation the k, m dependence of denominators can be neglected. In such a limit, the term in Eq. (15) have the following approximate expression

$$(25) \quad \frac{\langle l|i|m\rangle \langle n|j|m\rangle \langle m|k|l\rangle}{(\omega_{nl}-2\omega)(\omega_{ml}-\omega)} \approx \frac{\langle l|i|m\rangle \langle n|j|m\rangle \langle m|k|l\rangle}{2\omega^3}$$

In this case a further simplification occurs, because the summation and integration variable are separable allowing to obtain the various contribution to β in Eq. (20) in a closed form.

As explicative example, the form of $\beta_{xyz}^{(k,m)}$ is here reported for $t = 0$

$$(26) \quad \beta_{xyz}^{(k,m)} \approx \{\cos\psi_m^k(-n_2)A(-n_2,z) + \cos\psi_m^k(n_1)A(n_1,z)$$
$$+ \cos\psi_m^k(n_1-n_2)A(n_1-n_2,z)\} \cdot$$
$$\cdot \{\cos\psi_m^k(-n_2)A(-n_2,x) + \cos\psi_m^k(n_1)A(n_1,x)$$
$$+ \cos\psi_m^k(n_1-n_2)A(n_1-n_2,x)\} \cdot$$
$$\cdot \{\cos\psi_m^k(-n_2)A(-n_2,y) + \cos\psi_m^k(n_1)A(n_1,y)$$
$$+ \cos\psi_m^k(n_1-n_2)A(n_1-n_2,y)\}$$

where

$$(27) \quad A(n_1 z) = \langle n_1 0|z|00\rangle + \langle 01|z|n_1 0\rangle + \langle n_1 1|z|00\rangle + \langle 01|z|n_1 1\rangle$$

The problem of the numerical estimation of β is now reduced to the choice of an atomic wavefunction $|tu\rangle$.

Evaluation of Second Order NLO Properties of Carbon Nanotubes

At first approximation, the localized states $|tu\rangle$ are assumed as one p orbital pointing along the normal of the CNT surface and centred on the atom at position $\vec{R}_{tu}$

$$(28) \quad \langle \vec{r}|tu\rangle = \phi_{tu}^p(\vec{r}) = \vec{c}_{tu} \cdot (\vec{r} - \vec{R}_{tu}) \exp\left[-\frac{Z|\vec{r} - \vec{R}_{tu}|}{2a_B} \right]$$

where $\vec{c}_{tu}$ is a versor normal to CNT surface at atomic position, $Z = 3.65$ is the effective nuclear charge as calculated following the Slate method and a_B the Bohr radius. With the replacement $|tu\rangle \to \phi_{tu}^p(\vec{r})$ in the wavefunctions in Table 2, quantitative calculations of optical properties of chiral CNT are possible once that the summation is truncated at some t. In this case the dimension of the symmetry group G_C has to be taken equal to the number of atomic sites included in summation.

The approximation to p-orbitals, together with the restriction of interaction to nearest neighbours, limits the validity of the following calculations to thick

Table 3. For each investigated topology the CNT diameter, the contribution to β_{xyz} from electrons with $k=0$ and the calculated value of β_{xyz}, are listed

CNT topology	Diameter (Å)	$\left|\beta_{xyz}^{(0)}\right|$ (esu)	$\left|\beta_{xyz}\right|$ (esu)
(7, 5)	8.183	$3.61 \cdot 10^{-36}$	$1.1 \cdot 10^{-30}$
(8, 3)	7.719	$5.28 \cdot 10^{-35}$	$1.3 \cdot 10^{-30}$
(8, 5)	8.9	$2.17 \cdot 10^{-36}$	$1.01 \cdot 10^{-30}$
(9, 1)	7.477	$2.00 \cdot 10^{-31}$	$4.1 \cdot 10^{-30}$
(10, 7)	11.59	$1.19 \cdot 10^{-37}$	$3.3 \cdot 10^{-30}$
(10, 3)	9.24	$1.8 \cdot 10^{-35}$	$0.74 \cdot 10^{-30}$
(11, 2)	9.5	$3.2 \cdot 10^{-37}$	$5.2 \cdot 10^{-30}$
(11, 4)	10.5	$6.39 \cdot 10^{-36}$	$1.21 \cdot 10^{-30}$
(11, 5)	11.11	$2.5 \cdot 10^{-34}$	$2.9 \cdot 10^{-30}$
(11, 8)	12.95	$3.00 \cdot 10^{-40}$	$1.58 \cdot 10^{-30}$
(12, 7)	13.04	$6.7 \cdot 10^{-41}$	$0.63 \cdot 10^{-30}$
(12, 11)	15.06	$2.3 \cdot 10^{-33}$	$0.6 \cdot 10^{-30}$
(13, 2)	11.05	$7.48 \cdot 10^{-39}$	$4.2 \cdot 10^{-30}$
(13, 4)	12.06	$4.6 \cdot 10^{-36}$	$3.04 \cdot 10^{-30}$
(13, 5)	12.16	$7.03 \cdot 10^{-40}$	$2.16 \cdot 10^{-30}$
(13, 6)	13.18	$1.76 \cdot 10^{-34}$	$2.9 \cdot 10^{-30}$
(13, 7)	13.77	$8.79 \cdot 10^{-35}$	$2.19 \cdot 10^{-30}$
(13, 8)	14.38	$7.5 \cdot 10^{-42}$	$0.63 \cdot 10^{-30}$
(13, 10)	15.65	$1.55 \cdot 10^{-41}$	$2.6 \cdot 10^{-30}$
(13, 12)	16.97	$2.11 \cdot 10^{-33}$	$0.81 \cdot 10^{-30}$
(14, 1)	11.38	$2.3 \cdot 10^{-31}$	$1.36 \cdot 10^{-30}$
(14, 3)	12.31	$4.88 \cdot 10^{-39}$	$1.9 \cdot 10^{-30}$
(14, 5)	13.37	$2.02 \cdot 10^{-36}$	$4.45 \cdot 10^{-30}$
(14, 11)	17.01	$2.9 \cdot 10^{-43}$	$0.98 \cdot 10^{-30}$
(14, 13)	18.33	$1.73 \cdot 10^{-33}$	$0.82 \cdot 10^{-30}$
(15, 1)	12.16	$2.3 \cdot 10^{-31}$	$0.72 \cdot 10^{-30}$
(15, 14)	19.68	$1.58 \cdot 10^{-33}$	$0.56 \cdot 10^{-30}$
(16, 7)	16.005	$1.58 \cdot 10^{-44}$	$2.47 \cdot 10^{-30}$

nanotubes (diameter greater than 7 Å) being the approximations plausible for them only.

The results of non-resonant hyperpolarizability quantitative estimation for several CNTs topologies and with $\omega = 1$ eV, are reported in Table 3. In the calculations all series have been truncated at $t = 1$.

Within the model considered and in the limit of the described approximations, the code estimates a value $\beta \sim 10^{-30}$ esu for most of the chiral CNT topologies. No regular dependence of β on CNT diameter, ranging from 7.7 to 19.68 Å for the investigated topologies, has been found in simulations.

The contribution from electrons at the origin of the ID is, for most of the simulated topologies, order of magnitudes less than the contribution coming from inside the ID. For $(n_1, n_1 - 1)$ and $(n_1, 1)$ topologies, the $k = 0$ contribution is only three and one order of magnitude, respectively, less than that from inside the ID. It is interesting to note that these topologies, having a structure close to the armchair and zigzag CNTs, are characterized by a low degree of chirality. This result is in agreement with the findings in Section 2, where it was asserted that hyperpolarizability of achiral CNTs is most determined by the transformations properties of the $k = 0$ irreps.

It must be noted that no significant differences have been found between calculations performed for $t = 0$ and $t = 1$. This is a direct consequence of restriction to nearest neighbours interaction and of the mono-orbit nature of the CNT topology. In fact, within this scheme, the single carbon atom in the CNT constitutes an elementary cell for the whole structure and hence the properties of the CNT, at first approximation, are a replica of the properties of the elementary cell as determined by its interaction with nearest neighbours. The extension of sums to $t > 1$ is expected to refine the estimations and also to account for the reduced effect of quantum confinement at increasing CNT diameters. Nevertheless, the extension of summation to $t > 1$ stretches the limits of commercial PC computational resources and extends considerably the computational time.

Conversely, if also non-nearest neighbours interactions are taken into account, the truncation to $t << t_{cutoff}$ is not more valid and parallel calculation methods are required to run the code.

If the exciting laser field interacts with the CNT resonantly, the magnitude of the hyperpolarizabilty tensor is also affected by the lifetimes of the involved electronic levels. Lifetime of electronic states in a CNT is mostly determined by electron-electron scattering events. As far as electrons in a CNT are treated as a Fermi liquid (despite there are experimental evidence of Luttinger liquid behaviour [27]) the lifetime of an electronic state with an energy E above the Fermi level E_F, is given by

$$(29) \quad \tau \propto \frac{1}{(E - E_F)^2}$$

Lifetime increases for weakly excited state due to phase space limitations for scattering events near the Fermi level. Hertel [28] demonstrated experimentally that

τ varies from 235 fs for $E - E_F = 0.03$ eV to 18 fs for $E - E_F > 1.5$ eV. It results that, for excitation wavelengths in the near infrared ($\omega \sim 1$ eV), the hyperpolarizability can be enhanced up to one order of magnitude under resonant interaction regime. Hyperpolarizabilities of the order of 10^{-29} esu are then in principle possible for selected chiral CNT topologies. This value is slightly lower then the estimated hyperpolarizability of chromophores and Sol-Gel, usually accounted as the most efficient second order nonlinear optical materials [29].

5. CONCLUSIONS

We have presented the description of an approximate method to calculate the electronic hyperpolarizability for chiral CNTs topologies. The method extends to NLO calculations the formalism developed by Damnjanovic [5] to model the influence of topology on CNT linear optical response. The direct incorporation of CNT symmetry properties into the set of basis function used for calculation, allows to identify the state-to-state transitions which give a contribution to β. Despite the heavy assumption of neglecting the detailed bands structure of the CNT and the restriction of calculations to few atoms, the estimated values are typical of systems with delocalized π electrons. This result essentially reflects the profound meaning of direct incorporation of symmetry into calculations. The values of resonant hyperpolarizability obtained with the developed method allows to set CNTs next to the most efficient second order nonlinear optical materials. Nevertheless, due to the high optical damage threshold [13], CNTs have to be considered very promising second order optical materials for application in the high laser power regime (P $\sim 10^{13}$ W/cm^2).

Being the selection rules selecting the various contributions, rooted in the CNT topology as dictated by the couples of integers (n_1, n_2), the method establishes, in principle, a connection between NLO properties and growing mechanisms of CNTs. In fact, for a given exciting laser frequency ω, an assembling of CNTs with a well-specified topology (n_1, n_2) tailoring the band structure at resonance with an allowed contribution to β, is expected to strongly enhance the quantum yield of SHG from the sample.

ACKNOWLEDGEMENTS

The author are deeply indebted with dr. S. Botti for suggesting the present theoretical investigation with her experiments on catalyst free carbon nanotube laser synthesis.

Thank are due also to dr. L. Asylian for participating to the iniatial part of this work.

REFERENCES

[1] Iijima, S.: Helical microtubules of graphitic carbon, Nature, **354**, 56–58 (1991)
[2] Saito, R., Dresselhaus, G., Dresselhaus, M.S.: Physical properties of carbon nanotubes. Imperial College, London (1998)

[3] Damnjanović, M., Milošević, I., Vuković, T., Sredanović, R.: Full symmetry, optical activity, and potentials of single wall and multiwall nanotubes. Phys. Rev. B., **60**, 2728–2738 (1999)
[4] Božović, I., Božović, N., Damnjanović, M.: Optical dichroism in carbon nanotubes. Phys. Rev. B. **62**, 6971–6974 (2000)
[5] Damnjanović, M., Milošević, I., Vuković, T., Marinković, T.: Wigner-Eckart theorem in the inductive spaces and applications to optical transitions in nanotubes. J. Phys. A, **37**, 4059–4068 (2004)
[6] Xie, R.H., Jiang, J.: Nonlinear optical properties of armachair nanotube. Appl. Phys. Lett., **71**, 1029–1031 (1997)
[7] Wan, X., Dong, J., Xing, D.Y.: Optical properties of carbon nanotubes. Phys. Rev. B., **59**, 6756–6759 (1999)
[8] Margulis, A.V., Sizikova, T.A.: Theoretical study of third order nonlinear optical response of semiconductor carbon nanotube. Physica B, **245**, 173–189 (1998)
[9] Slepyan, G.Y., Maksimenko, S.A., Kalosha, V.P., Hermann, J., Campbell, E.E.B., Hertel, IV.: Higly efficient high harmonic generation by metallic carbon nanotubes. Phys. Rev. A., **60**, R777–R780 (1999)
[10] Slepyan, G.Y., Maksimenko, S.A., Kalosha, V.P., Gusakov, A.V., Hermann, J.: High order harmonic generation by conduction electrons in carbon nanotubes ropes. Phys. Rev. A., **63**, 53808 1–53808 10 (2001)
[11] Vivien, L., Anglaret, E., Riehl, D., Bacou, F., Journet, C., Goze, C., Andrieux, M., Brunet, M., Lafonta, F., Bernier, P., Hache, F.: Single wall carbon nanotubes for optical limiting. Chem. Phys. Lett., **307**, 317–319 (1999)
[12] Jin, Z.X., Sun, X., Xu, G.Q., Goh, S.H., Ji, W.: Nonlinear optical properties of some polymer/multi walled carbon nanotubes composites. Chem. Phys. Lett., **318**, 505–510 (2000)
[13] Stanciu, C., Ehlich, R., Petrov, V., Steinkellner, O., Herrmann, J., Hertel, I.V., Slepyan, G.Y., Khrutchinski, A.A., Maksimenko, S.A., Rotermund, F., Campbell, E.E.B., Rohmund, F.: Experimental and theoretical study of third harmonic generation in carbon nanotubes. Appl. Phys. Lett., **81**, 4064–4066 (2002)
[14] Koronov, S.O., Akimov, D.A., Ivanov, A.A., Alfimov, M.V., Botti, S., Ciardi, R., De Dominicis, L., Asilyan, L., Podshivalov, A.A., Sidorov Biryukov, D.A., Fantoni, R., Zheltikov, A.M.: Femtosecond optical harmonic generation as a non-linear spectroscopic probe for carbon nanotubes. J. Raman Spectrosc., **34**, 1018–1024 (2003)
[15] Wang, S., Huang, W., Yang, H., Gong, Q., Shi, Z., Zhou, X., Qiang, D., Gu, Z.: Large and ultrafast third order optical nonlinearity of single wall carbon nanotubes at 820 nm. Chem. Phys. Lett., **320**, 411–414 (2000)
[16] Liu, X., Si, J., Chang, B., Xu, G., Yang, Q., Pan, Z., Xie, S., Ye, P.: Third order optical nonlinearity of carbon nanotubes. Appl. Phys. Lett., **74**, 164–166 (1999)
[17] Botti, S., Ciardi, R., De Dominicis, L., Asilyan, L., Fantoni, R., Marolo, T.: DFWM measurements of third order susceptibility of single wall carbon nanotubes grown without catalyst. Chem. Phys. Lett., **378**, 117–121 (2003)
[18] De Dominicis, L., Botti, S., Asilyan, L., Ciardi, R., Fantoni, R., Terranova, M.L., Fiori, A., Orlanducci, S., Appolloni, R.: Second and third harmonic generation in single walled carbon nanotubes at nanosecond time scale. Appl. Phys. Lett., **85**, 1418–1420 (2004)
[19] De Dominicis, L., Fantoni, R., Botti, S., Asilyan, L.S., Ciardi, R., Fiori, A., Appolloni, R.: Symmetry point group description of second harmonic generation in carbon nanotubes. Las. Phys. Lett **1**, 172–175 (2004)
[20] Kocabas, C., Meitl, M.A., Gaur, A., Shim, M., Rogers, J.A.: Aligned arrays of single wall carbon nanotubes generated from random networks by orientationally selected laser ablation. Nanoletters, **4**, 2421–2426 (2004)
[21] Damnjanović, M., Vuković, T., Milosević, I.: Modified group projectors: tight-binding method. J. Phys. A., **33**, 6561–6571 (2000)
[22] Cornwell, J.F.: Group theory and electronic energy bands in solids. In: Wolhlfarth, E.P. (ed.) Selected topics in solid state physics, vol.X, p. 52, North Holland, London (1969)

[23] Bozović, I.B., Vujičić, M., Herbut, F.: Irreducible representation of the symmetry group of polymer molecules I. J. Phys. A, **11**, 2133–212147 (1978)
[24] Damnjanović, M., Vuković, T., Milosević, I.: Carbon nanotubes band assignation, topology, Bloch states and selection rules. Phys. Rev. B, **65**, 45418–45422 (2002)
[25] Boyd, R.W.: Nonlinear Optics p. 139, Academic, New York (1992)
[26] Butcher, P.N. Cotter, D.: The elements of nonlinear optics Cambridge University Press (1990)
[27] Bockrath, M., Cobden, D.H., Lu, J., Rinzler, A.G., Smalley, R.E., Balents, L., McEuen, P.L.: Luttinger liquid behaviour in carbon nanotube. Nature, **397**, 598–601 (1999)
[28] Hertel, T., Moos, G.: Influence of excited electron lifetimes on the electronic structure of carbon nanotubes. Chem. Phys. Lett., **320**, 359–364 (2000)
[29] Innocenzi, P., Brusatin, G., Abbotto, A., Beverina, L., Pagani, G.A., Casalboni, M., Sarcinelli, F., Pizzoferrato, R.: Entrapping of push-pull zwitterionic chromophores in hybrid matrices for photonic applications. Journal of Sol-Gel Science and Technology, **26**, 967–970 (2003)

CHAPTER 11

ATOMISTIC MOLECULAR MODELING OF ELECTRIC FIELD POLING OF NONLINEAR OPTICAL POLYMERS

MEGAN R. LEAHY-HOPPA, JOSEPH A. FRENCH,
PAUL D. CUNNINGHAM, AND L. MICHAEL HAYDEN
Department of Physics, University of Maryland, Baltimore County, Baltimore Maryland, USA

Abstract: The orientation of the nonlinear optical chromophore in a guest-host polymer system under the application of an external electric field plays an important role in the electro-optic activity in the material. The process of electric field poling of nonlinear optical chromophores in polymer systems has been studied through both Monte Carlo simulations and atomistic molecular modeling simulations. We review the progress of simulations in this area as well as describe our efforts and progress in understanding the process of electric field poling at an atomistic level of theory

Keywords: Atomistic molecular modeling; nonlinear optical polymers

1. INTRODUCTION

Molecular modeling is a valuable tool which can aide in the design and modification of a wide variety of molecules from drugs to proteins through investigations of the structure and properties of the materials. Industries from the pharmaceutical industry to the fuel industry use molecular modeling to design new drugs, better fuel additives, and improve polymers for packaging. Molecular modeling can be used as a tool to investigate nonlinear optical polymers in an effort to identify the structures which give rise to experimentally observed features and properties in the material. The studies reported here investigate the behavior of the polymers under the application of an external electric field. Electric field poling of nonlinear optical polymers is a widely used experimental technique. Modeling of this process, as described in this chapter, will be compared with experimental results from literature.

Nonlinear optical (NLO) polymers have been widely studied for their use in electro-optic (EO) applications. The EO coefficient is one parameter that can be optimized to improve the performance of NLO materials and is directly proportional to the applied electric poling field. Both atomistic and statistical modeling methods have been previously applied in an effort to understand the poling process on a microscopic level in these types of materials [1–4].

Although there are only a few groups which have studied the electric field poling of NLO chromophores [1–4], molecular modeling has been used for a wide variety of research over the past two decades. Theodorou and Suter [5–7] developed an atomistic model to study the molecular structure of amorphous glassy polymers in order to enable the predictions of structural and thermodynamic properties of the bulk polymer. They used their model to further investigate mechanical properties of amorphous polypropylene. Suter and others [8–10] followed this work by investigating chain dynamics, conformational changes, and packing effects in the bulk of a polycarbonate polymer glass. Mattice and co-workers [11, 12] further investigated structural and conformational properties of glassy polymers by modeling amorphous polybutadiene. Rigby and Roe [13–16] used alkane-like chains to explore many properties of the bulk liquid and glass of these long-chain molecules. The dependence of the glass transition temperature on different properties of the chains, the short range ordering of the chains and the orientational correlations between sub-chains, density, pressure, and temperature dependence of chain conformational distributions, and the distribution of free volume in the system are among the properties Rigby and Roe investigated in the liquid and glass phases of the alkane-like chains. Greenfield and Theodorou [17] and Misra and Mattice [18] also studied free volume distribution in both liquid and glassy polypropylene and atactic polybutadienes respectively. Time dependent properties such as local chain dynamics near and above the glass transition temperature [19, 20] and local dynamics of polyisoprene chains [21], bulk amorphous polybutadienes [22], and polyisoprene chains [23] have also been studied. Diffusion properties of small molecules in polymers have also been the topic of research for several groups [24–26]. Several studies have focused on the development of methods for using molecular modeling to determine glass transition temperatures of polymers [27–29].

More recently, Robinson and co-workers [2, 4, 30] have used Monte Carlo simulations in order to investigate the electric field poling and its effects in organic nonlinear optical materials. They investigated the role of intermolecular forces between the chromophore molecules. In addition, concentration effects have been studied. Kim and Hayden [1] used atomistic molecular modeling to examine the electric field poling in a guest-host NLO polymer as well as static conformational properties of the polymer and dopant. Makowska-Janusik and co-workers [3] followed Kim and Hayden's work using atomistic molecular modeling and investigated guest-host NLO systems with various chromophore dopants.

Both classical and quantum mechanical methods are employed for various levels of molecular modeling. While quantum mechanical methods can be used with small molecules to yield precise results, classical force-field methods provide faster yet approximate solutions for large molecules or molecular systems that could not be studied using quantum methods. Although these two types of methods differ fundamentally in their calculations of the energy of the system, the two methods are complementary. Classical methods are often parameterized using data from quantum simulations. For studying large molecular systems, classical molecular dynamics must be employed to investigate these materials since the systems are too large for quantum mechanical calculations.

Molecular dynamics employs Newtonian mechanics to model the time evolution of the system. The positions, velocities, and accelerations of each atom in the system are calculated from the force-field potential. Newtonian mechanics describes the relationship between the potential felt by each atom, the forces on each atom, and, therefore, the accelerations, velocities, and positions of each atom at each time step of the simulation. From the time evolution of the system, we can calculate many properties of the system. In this chapter, we describe the history, methods, and results of the work on the electric field poling of nonlinear optical polymeric guest-host systems.

Kim and Hayden [1] were first to employ fully atomistic molecular modeling to study the static and dynamic properties of a guest-host NLO polymeric system. Their system consisted of the polymer host poly(methyl-methacrylate) (PMMA) and the chromophore guest N,N-dimethyl-p-nitroaniline (DPNA) with 3% mass fraction of DPNA in the system. They studied static and dynamic properties of the system in both poled and unpoled states at two different densities corresponding to systems above and below the glass transition temperature, T_g. They also investigated the locations of the chromophores with respect to the polymer backbone and torsional angles of the polymer chain for the static case. In the dynamic case, they investigated the orientation of the dipole moments of the NLO chromophores during a poling simulation. They compared the calculated values of $<\cos\theta>$ and $<\cos^3\theta>$ obtained from the simulation trajectories to the theoretical values predicted by the Langevin functions, $L_1(p)$ and $L_3(p)$, which correspond to the orientational distribution for an ensemble of non-interacting dipoles in the presence of an external electric field. They found that the calculated values of $<\cos\theta>$ and $<\cos^3\theta>$ agreed with the theoretical prediction above T_g but not below T_g. Additionally, for the sub-T_g system, the chromophores oriented themselves with the electric field noticeably slower than those in the systems poled above T_g.

Makowska-Janusik, et al. [3] extended Kim and Hayden's work, modeling the electric field poling and the cooling process for three chromophores of differing shapes. Both Kim and Hayden and Makowska-Janusik investigate electric field poling in guest-host systems in which a polymer host is doped with an NLO guest chromophore.

Robinson and Dalton [2] employed both equilibrium statistical mechanics using Piekara's [31] mean field approximation and Monte Carlo simulations to study the effects of the poling field on NLO chromophores. Their investigation also included studies of the effects of the shape of the chromophore and the number density of chromophores in the system on the EO coefficient. One significant finding of this study is the roll-off of the EO coefficient with increased chromophore loading density. They found that the EO coefficient increases with number density to a peak value after which the EO coefficient steadily decreases with increased chromophore loading [2]. This phenomenon was attributed to competition between the chromophore-chromophore inter-dipolar energy and the electrostatic energy of the applied field. The resulting "effective field" felt by the chromophores is described in terms of a vectoral combination of the applied field and the local dipolar fields due to near-by chromophores. Robinson and Dalton demonstrate consistent results between calculations using the two different models. The Monte Carlo simulations provide more detailed information about the systems than the equilibrium statistical mechanics using Piekara's approximation [2]. One of the goals of atomistic modeling is to provide even more detailed structural information of the poling process.

2. THEORY

Electro-optic activity in guest-host materials can be quantified in terms of the electro-optic coefficient, r_{33}, which can be expressed as

$$(1) \qquad r_{33} = \left| \frac{2Nf\beta \langle \cos^3 \theta \rangle}{n^4} \right|,$$

where N is the chromophore number density, f is a product of local field factors, $\beta(\omega; 0, \omega)$ is the molecular hyperpolarizability, and n is the index of refraction. The order parameter $<\cos^3 \theta>$ reveals the level of ordering that has been achieved in the material under the application of the electric field. This quantity plays a significant role in understanding the materials under investigation [2].

In order to understand the overall orientation of the chromophores within guest-host systems under differing external electric fields, we examine the orientational alignment of the dipole moment of the chromophores with respect to the direction of the external electric field. In the non-interacting rigid gas model, the intermolecular electrostatic interactions are ignored and one can describe a general order parameter of

$$(2) \qquad \langle \cos^n \theta \rangle = L_n(p)$$

where $L_n(p)$ is the nth order Langevin function [31] with θ the angle between the dipole moment of the chromophore and the external electric field vector and p given by $\mu E/kT$. Here, μ is the dipole moment of the chromophore, E is the strength of the applied electric field, k is Boltzmann's constant, and T is the temperature of the simulation in Kelvin. Given the above assumptions, one can then express the order parameter as

$$(3) \qquad L_1(p) = \coth(p) - 1/p = \langle \cos \theta \rangle,$$

$$L_3(p) = \left(1 + 6/p^2\right) L_1(p) - 2/p = \langle \cos^3(\theta) \rangle.$$

This model predicts a linear increase in the electro-optic coefficient with an increase in the chromophore concentration. For small dipole moment molecules and for extremely dilute concentrations of chromophores, this theory reasonably predicts the order parameter achieved experimentally and via atomistic molecular modeling. But even for small dipole molecules we do not find a linear increase in the electro-optic coefficient with increasing the chromophore concentration, but instead find that the electro-optic increases sub-linearly (see Figure 7 later). This roll-off of the electro-optic coefficient has been seen both experimentally and through Monte Carlo simulations [2]. The roll-off can be attributed to electrostatic interchromophore interactions, which must be included in the model in order to accurately predict the order parameter.

When the intermolecular electrostatic energy cannot be neglected, we can use the methodology of London [30, 32] to obtain an expression accounting for the "effective field" the chromophores experience

(4) $\quad <\cos^3 \theta> = L_3(p) \left[1 - L_1^2(W/kT) \right],$

where W is the chromophore-chromophore interaction energy and is expressed as

(5) $\quad W = (1/R^6) \left[(2\mu^4/3kT) + 2\mu^2\alpha + 3I\alpha^2/4 \right].$

Here, R is the average distance between chromophores, α is the polarizability, and I is the ionization potential of the chromophores [30, 33]. The multiplicative factor which scales the Langevin function in Equation (4) is highly dependent on the interchromophore separation distance. Using simulation parameters for typical NLO dopant molecules, Figure 1 shows a graph of the bracketed quantity in Equation (4) as a function of interchromophore separation distance for three different dipole moments. Although the polarizability of a molecule is dependent on the dipole moment, we illustrate the trend of the functional form of this bracketed quantity with a change in dipole moment in Figure 1. Since the classical force-field being used does not take into account the polarizability, we only vary the quantity which is taken into account by the force-field. Since we can vary the dipole moment of a chromophore by changing the charges assigned to the atoms in the chromophore, we investigate the effect of the change in the dipole moment of the chromophore in the bracketed quantity in Equation (5) and hence its effect on the overall order predicted by Equation (4).

In Equation (5) we observe the combination of three types of electrostatic interactions. The first term in brackets (μ^4/R^6) denotes a dipole-dipole interaction. The second term ($\mu^2\alpha/R^6$) denotes a dipole interaction with an induced dipole. The third term ($I\alpha^2/R^6$) denotes an induced dipole interaction with another induced dipole. Given the simulation parameters from Figure 1 and a dipole moment of 15 D, the dipole-induced dipole interactions are more than one order of magnitude smaller than the dipole-dipole interactions and the induced dipole-induced dipole interactions are smaller by a factor of about three. From these estimations, it can easily be seen that for molecules with large dipole moments, it is important to include the dipole-dipole interactions and perhaps induced dipole-induced dipole electrostatic interactions.

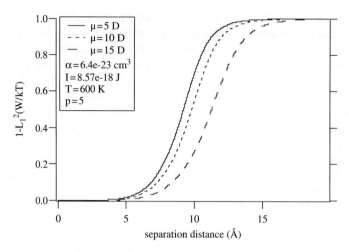

Figure 1. [$1-L_1^2(W/kT)$] as a function of interchromophore separation distance in Angstroms. Note the narrow region over which the factor dramatically decreases the predicted order parameter

3. METHODS/TECHNIQUES

Both Monte Carlo statistical mechanical simulations and atomistic molecular dynamics simulations have been employed to study the phenomenon of the electric field poling of NLO materials. The Monte Carlo approach has been used by the groups of Robinson and Dalton [2, 30, 33, 34]. These workers have investigated the role of the shapes and concentrations of the chromophores, the role of intermolecular interactions, and the competition between the intermolecular interactions and the poling process. The atomistic molecular dynamics of the poling of NLO chromophores has been studied by Kim and Hayden [1] as well as Makowska-Janusik, et al. [3]. Both of these groups have investigated the poling of small chromophores with low dipole moments at dilute concentrations. Kim and Hayden have additionally investigated the static conformational properties of both poled and unpoled systems. Makowska-Janusik, et al., have extended the work of Kim and Hayden to include the cooling process under the continued application of the electric field.

3.1 Monte Carlo Statistical Mechanics

Robinson and Dalton use Monte Carlo statistical mechanics to explore concentration and shape dependencies of the chromophores. Monte Carlo methods provide valuable information about the distribution of a collection of chromophores but are not able to provide atomistic information about the systems. The Monte Carlo simulations performed by Robinson and Dalton employ an array of point dipoles on a periodic lattice with the given parameters for the shape of the chromophores and the chromophore spacing adjustable to achieve the desired chromophore concentration. The model system consisted of 1000 chromophores on a body-centered cubic lattice with 3 dimensions (10 × 10 × 10). The lattice spacing was controlled to vary the chromophore concentration. These studies have provided valuable information regarding the optimal loading concentration for a chromophore of a given shape and dipole moment. In performing their simulations, dipole i was chosen to rotate to a new angle through a step of step size δ ($\cos\theta_i$) and δ (ϕ_i) to the new angles $\cos\theta_i$ and ϕ_i. The change in energy from the move, ΔE_i, was calculated and the move was accepted if the criteria $\Delta E < 0$ was met. In other words, if the move resulted in a decrease in energy for the system, the move was accepted. If the criteria was not met, e.g. $\Delta E > 0$ and the system experienced an increase in energy due to the move, the probability of that move, $\exp(-\Delta E_i/kT)$, was compared with a random number uniformly generated between 0 and 1. If the probability of the move was greater than the random number generated, the move was accepted, otherwise, the move was rejected. This implies that if the move produces a small change in energy, the move is most likely to be accepted. This procedure assures a Boltzmann distribution in $\cos\theta_i$ and ϕ_i [2, 35].

3.2 Molecular Modeling

Kim and Hayden and Makowska-Janusik, et al., both employ classical molecular dynamics to model the behavior of the chromophores under the application of an electric field. Although the method is classical, the force-fields are parameterized using data from both empirical data sets and ab-initio calculations.

In our current work, in addition to that of Kim and Hayden [1], the molecular dynamics were performed using the DISCOVER (98.0) [36] molecular modeling package employing the consistent valence force-field (CVFF) [37, 38]. The force-field parameters used can be found in Kim and Hayden's reference. Makowska-Janusik, et al., used the GROMACS [39–41] molecular dynamics package and a CVFF force-field with different parameters [3]. The potential energy was calculated using the terms

(6) $\quad V = V_{bond} + V_{angle} + V_{torsion} + V_{oop} + V_{nonbond} + \left(V_{field}\right)$

where V_{bond} is the potential energy of bond stretching, V_{angle} is the potential energy of the angle bending interactions, $V_{torsion}$ is the potential energy of the torsional interactions, V_{oop} is the potential energy of the out of plane interactions, and $V_{nonbond}$ encompasses both non-bonded van der Waals and Coulombic interactions. The potential energy of the external electric field, V_{field}, is calculated using the summation over all the particles of the dot product of the position vector of each particle with the product of the charge of the particle with the applied field,

(7) $\quad V_{field} = \sum_{i=1}^{n} \mathbf{r}_i \cdot (q_i \mathbf{E})$,

where i is the particle index. The inclusion of the energy of the external field in the potential energy calculation is limited to the poling stage of the calculation.

Group-based cutoff distances were used in the non-bond energy term calculations in order to avoid artificial splitting of dipoles. Neutral charge groups were created using the partial charges assigned by the force-field for the polymer host and by the semi-empirical method MOPAC [42] for the chromophore. The cutoff distances varied as a function of the total size of the amorphous cell. For van der Waals interactions, the cut-off distance used was 9.5 Å and for the Coulomb interactions, the cut-off distance is set equal to half the length of one side of the amorphous cell, which in our case is cubic. We found that in most cases the force-field assigned partial charges underestimate the dipole moment of the chromophore significantly. In order to more reasonably represent the dipole moment of the chromophores, and therefore the interactions between chromophores, we find it necessary to replace the partial charges assigned by the force-field with that from a MOPAC calculation with the AM1 Hamiltonian. Semi-empirical and quantum mechanical calculations of the dipole moment of chromophores as well as other material parameters, such as the first hyperpolarizabiilty β, are widely accepted values of these parameters [33, 43–45].

The amorphous cells were generated using a Monte Carlo method implemented in Cerius^2 [46] at the given densities with three-dimensional periodic boundary conditions. In order to avoid large energy jumps between sequential configurations in the molecular dynamics simulations, the systems were first equilibrated. We used a similar equilibration procedure as Kim and Hayden. Experimentally, chromophore reorientation occurs on time scales larger than nanoseconds, which is the currently realizable computational time scale. In order to expedite the orientation process to computational timescales, the temperature and applied electric field are increased [47] while maintaining a constant ratio in the parameter p. The temperature is chosen to be in the rubbery region for the material which is determined through constant pressure and temperature dynamics studies for each material under investigation. Additionally, for polymers, it has been experimentally shown that as pressure increases, the glass transition temperature, T_g, of the system increases by approximately 20 °C per 1000 atmospheres of pressure [48–51]. Because of this rise in T_g with an increase in pressure, the density of the system is also chosen to keep the pressure in the system at atmospheric pressure.

The NPT studies involve performing molecular dynamics on the system while allowing the volume and therefore density of the system to vary at a given temperature and pressure. We begin with a system at a high temperature, e.g., 700 K, and perform molecular dynamics to allow the system to reach an equilibrium configuration. Monitoring the density of the system throughout the simulation, we can find the equilibrium density of the system. Then, the system is cooled in an increment of 50 K and molecular dynamics are performed until the system once again reaches equilibrium. The equilibrium density is again recorded and the process continues in 50 K increments until the system has been cooled to about 200 K. This process allows us to plot the density of the system as a function of temperature. Examining this plot, we observe a break in the curve which defines the simulated glass transition temperature. Figure 2 shows an example of this plot for a chromophore which we call ezFTC (Figure 3) as the guest in the polymer host poly(methyl methacrylate), PMMA, in a 9% by weight concentration.

The NPT study allows us to determine the appropriate temperature and density to ensure the simulations are occurring in the rubbery region of the material. It should be noted that the simulated and experimental glass transition temperatures for many materials are different. The simulated T_g is usually found to be higher than that determined experimentally [3, 28]. This high glass transition temperature results from the high cooling rate of the system, 50 degrees in 1.5 ns or 3×10^{10} degrees/sec. The procedure was adapted from a study by Soldera [28] in which the simulated glass transition temperature of two tacticities of PMMA was investigated. For syndiotactic PMMA the simulated T_g was found to be 211 °C or 484 K and for isotactic PMMA the simulated T_g was found to be 157 °C or 430 K. For comparison, the experimental T_g of PMMA is around 100 °C. Although there is a difference between the two values, as the experimental T_g is the important quantity

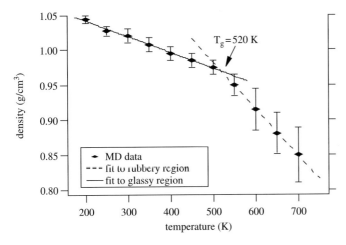

Figure 2. NPT study of ezFTC in PMMA (9% by weight). The crossing point for the two fit functions defines the simulated glass transition temperature of the composite. The two fit lines are linear fits to the data using least squares analysis

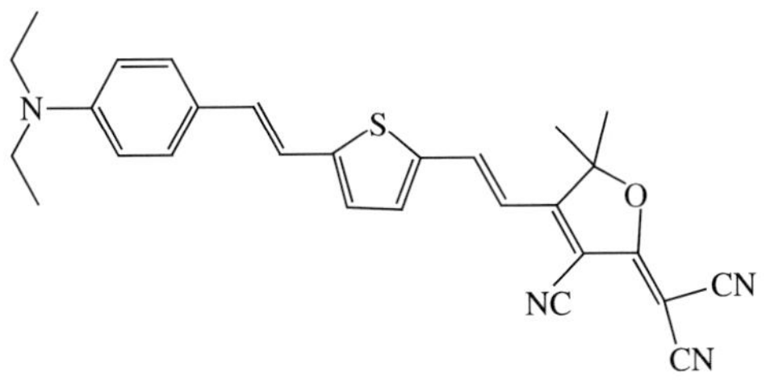

Figure 3. Structure of ezFTC

for determining optimal poling conditions in the laboratory, the simulated T_g is the quantity that is important for determining optimal simulation poling conditions.

Once the simulation parameters have been determined, the system equilibration process can begin. Thirty to fifty amorphous cells are constructed from which those with the lowest energy are selected. All cells undergo energy minimization until convergence is reached using a conjugate gradient method with a convergence criteria of $0.1\,\text{kcal}\,\text{mol}^{-1}\,\text{Å}^{-1}$. We then choose the desired number of configurations with the lowest energy in order to make certain that all starting configurations are energetically stable and will not experience a sudden large jump or drop in energy during the next phase of the equilibration process. For example, if the systems are in a high-energy configuration, there will be a large difference in energy between successive steps in the molecular dynamics and the simulation will terminate. The second step is the relaxation by NVT molecular dynamics (MD) for 100–200 ps at the poling temperature, recording the configuration every 100 fs. The lowest potential energy configuration of the MD is further minimized for 500 iterations with the same convergence criteria as the initial energy minimization. The last configuration of this second minimization is the beginning configuration for the poling stage of the simulation. The poling stage of the simulation consists of 1–2 ns of NVT MD, recording the configuration every 1 ps, with the applied electric field turned on for the entire length of the simulation. The temperature of all the MD simulations is controlled with the velocity scaling method [52] with a temperature window of 10 K around the simulation temperature. The NVT MD was performed using a 1 fs time step using the velocity Verlet method [53]. The procedure and parameters employed are similar to those of Kim and Hayden [1]. Calculations were executed on an SGI Challenge-XL server with 0.248 s/step of CPU time for the MIPS R10000 processor and a XEON 2 GHz based LINUX server.

4. RESULTS

The Monte Carlo simulation work of Robinson and Dalton has shown good agreement with experimental work. The electro-optic coefficient of a material experimentally exhibits what we will call a roll-off with increased chromophore concentration past a peak value. The EO coefficient increases linearly with increasing chromophore concentration at low concentrations for all chromophores and at all concentrations for chromophores with small dipole moments. For chromophores with larger dipole moments, the roll-off of the electro-optic coefficient is dependent on the shape of the chromophore and the dipole moment. For more spherical chromophores, the EO coefficient slightly deviates from the predicted linear value at higher concentrations. The actual EO coefficient is lower than that predicted by the non-interacting rigid gas model, but at high concentrations, the EO coefficient is still increasing with increasing chromophore concentration, just at a slower rate. For ellipsoidal-shaped chromophores, the electro-optic coefficient increases linearly at low chromophore concentrations but after a certain concentration, determined by the dipole moment of the chromophore, the EO coefficient actually decreases in value. This is the roll-off of the electro-optic coefficient referred to earlier in this paper. Robinson and Dalton's Monte Carlo simulations have accurately modeled the experimental roll-off of the electro-optic coefficients adjusting for the shape of the chromophore (spherical or ellipsoidal) and the value of the dipole moment. Although this finding is extremely valuable, it is limited in its predictive capabilities for future development of chromophores. The only adjustable parameters with respect to the chromophores in the systems are the shape, dipole moment, and concentration of the chromophores. Detailed structural information cannot be obtained through these types of simulations. For more detailed structural information, we must turn to atomistic molecular modeling.

Both Kim and Hayden [1] and Makowska-Janusik, et al., [3] have used a dilute concentration of a small, nearly spherically shaped chromophore with a small dipole moment, N,N-dimethyl-p-nitroaniline (DPNA) in a guest-host configuration with the polymer host poly(methyl methacrylate) (PMMA) (Figure 4). The order parameter obtained from their simulations, $<\cos^3 \theta>$, which is directly proportional to the electro-optic coefficient, matched that predicted by the non-interacting rigid gas model in both studies. Given the dilute concentration of only 3% by weight of chromophore and the small dipole moment of the chromophore ($< 4D$) it is expected that the chromophores in this system would follow the non-interacting rigid gas model. Makowska–Janusik, et al., expanded this poling work to include not only DPNA but two other chromophores, 4-(dimethylamino)-4'-nitrostilbene (DMANS) and N,N'-di-n-propyl-2,4-dinitro-1,5-diaminobenzene (DPDNDAB). DMANS is a more rod-like chromophore than DPNA and DPDNDAB is more spherical (see Figure 4).

In the work of Mankowska-Janusik, et al., [3] for DPNA, the orientational alignment of the chromophores was in agreement with the prediction from the non-interacting rigid gas model. For the DPDNDAB chromophore, which is larger, only at extremely high poling fields did the order in the system approximate that predicted by the non-interacting rigid gas model. At lower fields, the order in the system is significantly less than that predicted by the non-interacting rigid gas model. For DMANS, even at high poling fields, the orientation achieved in the system is not equivalent to the predicted values. Because the DMANS chromophore is a more elongated chromophore, there is a greater chance of entanglement with the polymer matrix. The entanglement with the polymer would inhibit the mobility of the chromophores, therefore leading to a lower degree of orientation with the poling field than with the smaller chromophores. The DMANS chromophore did not align with the poling field as well as the smaller chromophores, pointing to the fact that this chromophore had decreased mobility in the polymer host.

Recently we have been working toward the extension of the earlier work to include larger chromophores representative of those we use experimentally, in addition to the integration of potential terms which specifically include the electrostatic interactions between the chromophores within the system. All the systems we have investigated are guest-host systems with the polymer host PMMA. Similarly to Kim and Hayden, we found agreement between the poled order exhibited by small

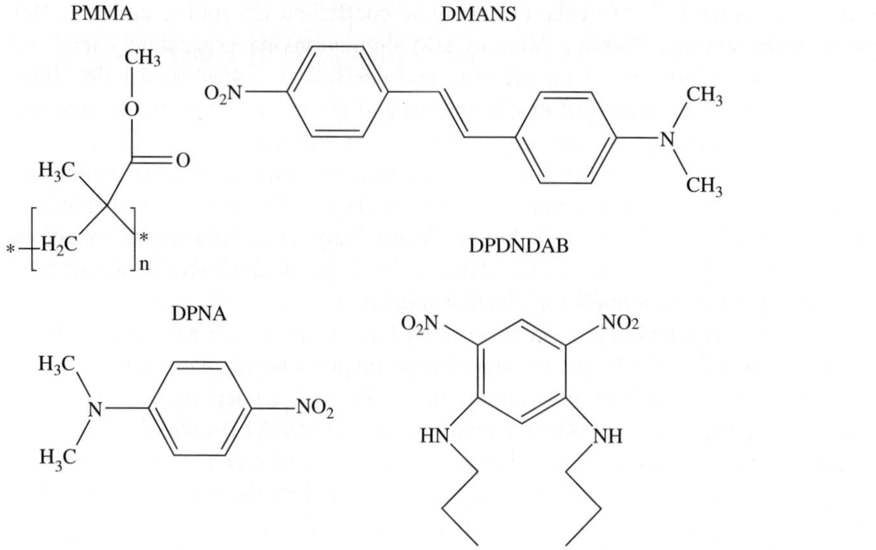

Figure 4. Diagram of the three chromophores DPNA, DMANS, DPDNDAB, and the polymer PMMA

chromophores and the order predicted by the non-interacting rigid gas model. The orientational order in the system is strongly influenced by the strength of the poling field allowing any level of ordering in the system by manipulating the strength of the poling field while keeping all other simulation parameters constant, i.e., temperature, density. Figure 5 shows the effect of changing the poling field by a factor of 5 while keeping the other simulation parameters constant in the system.

The DPNA chromophore has a small dipole moment, less than 4 D, as calculated using the partial charges on the atoms assigned by the force-field. It would be difficult to see the concentration effect of the roll-off of the electro-optic coefficient with that small of a dipole moment.

In order to increase the dipole moment of the chromophore, we substituted the partial charges associated with the chromophore from the semi-empirical calculation MOPAC in the place of the force-field charges. Using the MOPAC charges, DPNA has a dipole moment around 8 D. Although this dipole moment is more than two times larger than that obtained using the force-field charges, it is still too small to see chromophore-chromophore interactions at typical concentrations. We created a chromophore which we named dinitrovinylmethylpyridine (DNVMP) (Figure 6) which is approximately the same length and shape as DPNA, but has a dipole moment of 13 D using the MOPAC charges. According to Robinson and Dalton's Monte Carlo simulations, an elliptically shaped chromophore with a dipole moment around 13 D should exhibit a roll-off of the electro-optic coefficient with a peak value occurring around a concentration of 4.5×10^{20} molecules/cm^3.

Figure 7 shows the results of a concentration study of DNVMP in PMMA. At the lower concentrations, the linear trend in the data agree with the prediction of the

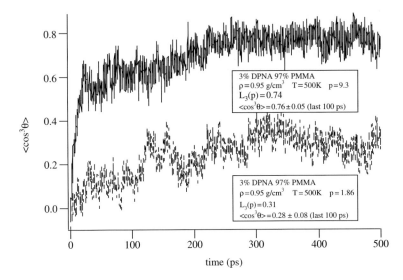

Figure 5. The strength of the poling field directly impacts the level of order achievable in the system. We can artificially induce a higher level of order in the system by fixing the simulation parameters and adjusting the value of the applied electric field

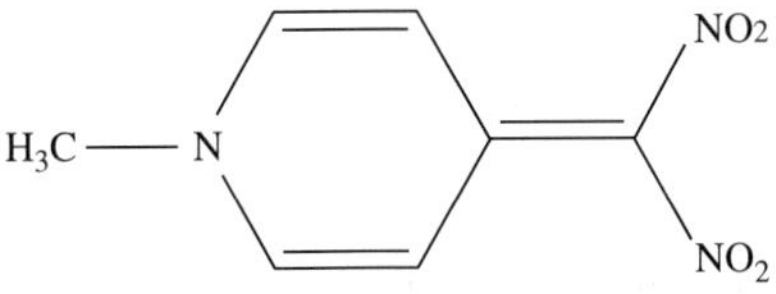

Figure 6. Structure of DNVMP

non-interacting rigid gas model. In the highest concentration, however, the order observed in the system is significantly less than that predicted by the non-interacting rigid gas model. The deviation from the straight line fit to the first points can easily be seen.

Recall that not only does the dipole moment of the chromophore play a role in the poling of the systems but also the shape of the chromophore. The DNVMP chromophore has a large dipole moment but is a small chromophore. It is more spherical in shape than elliptical. We initially used the DNVMP chromophore since we know that small chromophores like DPNA are able to be poled well with reasonable simulation parameters, temperature and applied field values. Moving to a chromophore with larger dipole moment, but not changing the shape, minimized the number of parameters we changed between studies, therefore limiting the number of factors which may have caused differences between the poling behavior of the DPNA system and the DNVMP system. In Figure 7, although we do see a deviation from the linear relationship of the electro-optic coefficient with chromophore

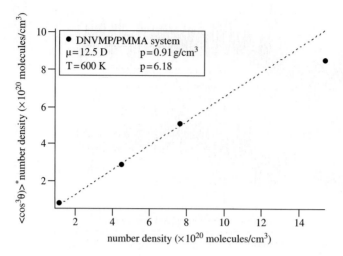

Figure 7. Concentration study of DNVMP in PMMA. Data points are the solid circles. Line is a linear fit to the first three points of data. The most concentrated system experiences a deviation from the linear curve. Here the strength of the poling field is 1.2 kV/μm

concentration, we do not see as large a roll-off as is expected with an elliptical chromophore with the same dipole moment. This leads us to conclude that the DNVMP chromophore, although it has a large dipole moment, is too spherical in shape to see the sharp roll-off as is seen in systems with chromophores with large dipole moments which are elliptical in shape. It is evident that in order to model the roll-off of the electro-optic coefficient with the increase in chromophore concentration, we need to not only increase the size of the dipole moment of the chromophore, but also to make the shape of the chromophore elliptical.

The shift to the elliptical chromophore is a large jump from the simulations with DNVMP and DPNA since they both are small chromophores. Determining the simulation parameters for which a dilute solution of the larger dopant gives the simulated order parameter equal to those predicted by the non-interacting rigid gas model is non-trivial. The chromophore we chose to use for this stage in the simulations we call ezFTC (Figure 3). First we performed an NPT study (Figure 2) to determine the temperature and density of the polymer composite of ezFTC and PMMA which corresponds to atmospheric pressure. The break between the two fit functions for the glassy and rubbery regions defines the simulated glass transition temperature of the system. We use the curve to guide our choice of starting parameters so that we ensure that poling occurs in the rubbery region.

We chose the temperature of 600 K for our simulations and initially choose a density of $0.9 \, g/cm^3$ so that we were performing the simulations in the rubbery region of the polymer. Examination of the poling results (Figure 8) reveals that at this density and temperature, and a dilute concentration (5% by weight), the system does not behave as the non-interacting rigid gas model predicts. The predicted order for this system has a value of 0.60. The system's calculated order parameter

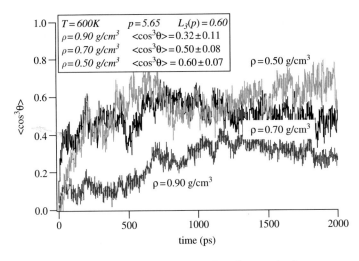

Figure 8. Poling of ezFTC in PMMA as a function of system density

was 0.32 ± 0.11. The ezFTC chromophores are much larger than the DPNA and DNVMP chromophores which are each approximately 8 Å in length whereas the ezFTC chromophore is about 22 Å long. We have attributed the lack of alignment in the ezFTC system at 0.90 g/cm^3 to steric hindrance. The large chromophores need more free volume in the system in order to be able to freely rotate. The chromophores need to be able to push the polymer out of the way in order to have enough room to rotate to align with the applied field. By lowering the density, we begin to see alignment which agrees with the predicted value from the non-interacting rigid gas model. The comparison between the systems with densities of $0.70 \text{ g/cm}^3 (<\cos^3 \theta> = 0.50 \pm 0.08)$ and $0.50 \text{ g/cm}^3 (<\cos^3 \theta> = 0.60 \pm 0.07)$ does not show an extremely large difference in the ordering of the chromophores in the systems, however, the systems at 0.50 g/cm^3 ordered as predicted by the non-interacting rigid gas model. All three density system calculations consisted of 16 amorphous cells which were equilibrated using the technique described earlier and were poled for 2 ns. The average order parameter, $<\cos^3 \theta>$, has been time averaged over the last 1 ns of simulation. Based on this study of the ordering of the systems at different densities, we chose the lowest density to perform a concentration study for the ezFTC chromophore in PMMA.

In the concentration study, several different concentrations of the polymer and chromophore were created and poled in order to see the effect of increased concentration on the orientational order in the systems. Due to the low density in the systems, the range of concentrations we were able to study is only from 0.45×10^{20} to 5.57×10^{20} molecules/cm^3.

It is important to understand the magnitude of all contributions to the energy of the system before choosing a poling field. Although the chromphores will align more with a strong poling field, if the poling field interaction energy with the dipole moment of the chromophores is much greater than inter-chromophore interaction energies, there will be no way to investigate the inter-chromophore effects in the poling process. Since these inter-chromophore effects play a large role in the experimental poling efficiency, it is essential to consider these interactions in the modeling of the poling process. For example, for the ezFTC chromophore system, with a dipole moment of 14 D, a polarizability of 6.43×10^{-23} cm^3, and an ionization potential of 1.29×10^{-18} J, Figure 9 shows the values of the poling field interactions with the dipoles (solid horizontal lines), and the three inter-molecular interactions between, two dipoles, a dipole and an induced dipole, and two induced dipoles (dashed lines). Given an inter-chromophore interaction distance of 15 Å, the poling field interaction with the dipole moment for the $p = 5.65$ case, where the poling field strength is 1 kV/μm, is five times as large as the largest of the three inter-molecular interactions, the dipole-dipole interaction energy. At this p and an interchromophore separation distance of 15 Å, we would not expect to see any inter-chromophore effects due to the overwhelming poling field interaction energy. In our ezFTC simulations, the inter-chromophore separation distances are 6 to 9 Å. According to Figure 9, intermolecular interactions between dipoles are on the order of or greater than the interaction energy between the poling field and the chromophores. From Figure 10, however, it is clear that intermolecular interactions are not driving the ordering of the system as the concentration of the system is increased. In addition, visual inspection of the locations of the chromophores within the simulation boxes reveals a phase separation between the chromophores and the polymer at the higher concentrations. The concentrations

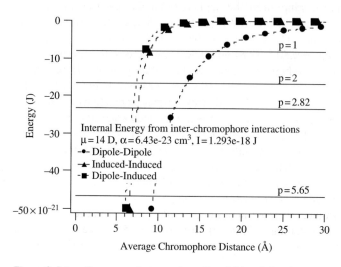

Figure 9. Interaction energy between the poling field and the chromophore (solid lines) for various values of the parameter p. Interaction energy between two dipoles, two induced dipoles, and a dipole and an induced dipole (dashed lines). The vertical axis is the interaction energy for each of the intermolecular interactions as well as for the poling field interaction with the chromophore

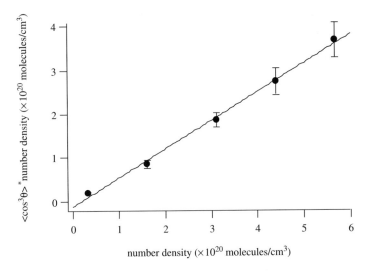

Figure 10. Concentration effect of ezFTC chromophores in PMMA for the parameters corresponding to a value $p = 5.65$, $E = 1\,\text{kV}/\mu\text{m}$, $T = 600\,\text{K}$. The linear increase in the electro-optic coefficient with an increase in concentration corresponds to the poling field-dipole interaction energy overwhelming the inter-chromophore interactions. Solid line is a linear fit to the data (circles)

we are using in this simulation are significantly higher than what is used experimentally for all except the lowest two concentrations shown on the graph on Figure 10. With a similar chromophore, experimental concentrations of the system reach a maximum electro-optic activity around 25% by weight. For comparison, the second point on the graph in Figure 10 has a concentration of 25%. The linear increase in the electro-optic activity in the material with increasing concentration shown in Figure 10 may be a result of the phase separation in the systems with the close packing of the chromophores having a minimum energy configuration in the poled state.

The inter-chromophore interactions between the dipoles competes with the interaction between the poling field and the chromophores. These inter-chromophore interactions, if present, should reduce the ordering in the systems with chromophores which are in close proximity to their neighbors. In the case of the ezFTC system at the density under investigation, $0.50\,\text{g}/\text{cm}^3$, if chromophore-chromophore interactions were observable we would expect a significant reduction in ordering at concentrations greater than $5 \times 10^{20}\,\text{molecules}/\text{cm}^3$. However, we do not see any reduction of order (Figure 10) for increasing chromophore concentration. This might be explained by recalling (Figure 9) that for $p = 5.65$, the poling field interaction is significantly stronger than the inter-chromophore interactions at this concentration.

In our current model, the non-bond interactions between atoms take two interactions into account, the Coulombic interaction between atoms and the interaction between the poling field and the atoms. The Coulombic interaction is inversely proportional to the distance between atoms and directly proportional to the charges associated with those atoms,

(8) $$U_{Coulomb} \sim \frac{qq}{r}$$

This can be readily compared with the dipole-dipole interaction,

(9) $$U_{dipole-dipole} \sim \frac{\mu^2}{r^3},$$

which is directly proportional to the dipole moment squared and inversely proportional to the distance between dipoles cubed. The dipole moment, however, is simply a combination of the charges and locations of the atoms in the chromophores. It can be seen, then, that the Coulomb interactions include (strictly) the dipole-dipole interactions since the dipole-dipole interactions are a combination of charges and distances. Therefore, without the addition of any extra terms in the potential force-field, the largest of the inter-chromophore interactions should be observable as long as the poling field interaction does not overpower the inter-chromophore interaction. Finding the correct balance between the poling field strength, which affects poling efficiency because of steric hindrance, and the interactions between the chromophores is not a trivial task.

5. DISCUSSION

Atomistic molecular modeling studies of the electric field poling of nonlinear optical polymers have been studied by at least two groups over the past few years. Methods for determining the simulation parameters corresponding to appropriate experimental conditions are ongoing. It is important to keep the simulations rooted in experiments as their applicability is limited if the simulation parameters keep the simulations far from reality.

In performing molecular dynamics, the first important quantity to determine is the simulation glass transition temperature of the system. To do this, constant temperature and pressure molecular dynamics is performed at a temperature much higher than the "expected" glass transition temperature of the material, 700 K for example, and repeated to "slowly" cool the system to below the transition temperature, 200 K for example. The linear fits to the beginning and end of the data map out two regions in the material, the glassy region and the rubbery region. The crossing point of the two fit functions defines the simulation glass transition temperature. Recall that this temperature will be higher than the corresponding glass transition temperature of the material due to the high cooling rate. A temperature change of 50 degrees per 1.5 ns yields a cooling rate of about 3×10^{10} degrees per second. This high cooling rate is unavoidable with the current computational time scales.

Once the temperature and density of the simulation at atmospheric pressure have been determined from the NPT study, the strength of the poling field should be chosen so that inter-molecular forces such as the dipole-dipole interactions are not overwhelmed. This requires some knowledge of the concentrations at which the simulations will be performed, however, as a rough calculation, the chromophore concentration can be approximated as $1/\sqrt[3]{N}$ where N is the number density of the chromophores in the system. Using this information to approximate the inter-chromophore interaction distance, the choice of poling field can be made.

Before beginning the concentration study, a simulation with a single chromophore dopant should be performed to verify ordering equivalent to the thermodynamic prediction of the rigid gas model. If the orientational order of this system does not correspond to the non-interacting rigid gas model prediction, the density of the amorphous cells should be adjusted such that the orientational order in the system is in agreement with the non-interacting rigid gas model, i.e., $<\cos^3\theta> = L_3(p)$.

Although no one has yet reproduced the experimental roll-off of the electro-optic coefficient with increased chromophore concentration using the atomistic molecular modeling methods to date, significant progress has been made. We are currently working on integrating the three explicit terms for the dipole-dipole interaction, the dipole-induced dipole interaction, and the induced dipole-induced dipole interaction into the simulation. The results of this work will be published elsewhere.

ACKNOWLEDGEMENTS

This work was partially supported by the National Science Foundation (ECS-0139457) and the STC program of the National Science Foundation (DMR 0120967).

REFERENCES

[1] Kim, W.-K., Hayden, L.M.: Fully atomistic modeling of an electric field poled guest-host nonlinear optical polymer, J. Chem. Phys. **111**, 5212–5222 (1999)

[2] Robinson, B.H., Dalton, L.R.: Monte carlo statistical mechanical simulations of the competition of intermolecular electrostatic and poling-field interactions in defining macroscopic electro-optic activity for organic chromophore/polymer materials, J. Phys. Chem. A **104**(20), 4785–4795 (2000)

[3] Makowska-Janusik, M., Reis, H., Papadopoulos, M.G., Economou, I.G., Zacharopoulos, N.: Molecular dynamics simulations of electric field poled nonlinear optical chromophores incorporated in a polymer matrix, J. Phys. Chem. B **108**(2), 588–596 (2004)

[4] Nielsen, R.D., Rommel, H.L., Robinson, B.H.: Simulation of the loading parameter in organic nonlinear optical materials, J. Phys. Chem. B **108**(25), 8659–8667 (2004)

[5] Theodorou, D.N., Suter, U.W.: Detailed molecular structure of a vinyl polymer glass, Macromolecules **18**, 1467 (1985)

[6] Theodorou, D.N., Suter, U.W.: Local structure and the mechanism of response to elastic deformation in a glassy polymer, Macromolecules **19**, 379–387 (1986)

[7] Theodorou, D.N., Suter, U.W.: Atomistic modeling of mechanical properties of polymeric glasses, Macromolecules **19**, 139–154 (1986)

[8] Hutnik, M., Argon, A.S., Suter, U.W.: Quasi-static modeling of chain dynamics in the amorphous glassy polycarbonate of 4, 4,′-isopropyulidenediphenol, Macromolecules **24**(22), 5970 (1991)

[9] Hutnik, M., Argon, A.S., Suter, U.W.: Conformational characteristics of the polycarbonate of 4, 4′-isopropylidenediphenol, Macromolecules **24**, 5956–5961 (1991)

[10] Hutnik, M., Gentile, F.T., Ludovice, P.J., Suter, U.W., Argon, A.S.: An atomistic model of the amorphous glassy polycarbonate of 4, 4′-isopropylidenediphenol, Macromolecules **24**, 5962–5969 (1991)

[11] Kim, E.G., Misra, S., Mattice, W.L.: Atomistic models of amorphous polybutadienes. 2. Poly(1.4-trans-butadiene), poly(1,2-butadiene), and a random copolymer of 1,4-trans-butadiene, 1,4-cis-butadiene, and 1,2-butadiene, Macromolecules **26**, 3424–3431 (1993)

[12] Li, Y., Mattice, W.L.: Atom-based modeling of amorphous 1,4-cis-polybutadiene, Macromolecules **25**, 4942–4947 (1992)

[13] Rigby, D., Roe, R.J.: Molecular dynamics simulation of polymer liquid and glass. 4. Free volume distribution, Macromolecules **23**(26), 5312–5319 (1990)

[14] Rigby, D., Roe, R.-J.: Molecular dynamics simulation of polymer liquid and glass. I. Glass transition, J. Chem. Phys. **87**(12), 7285–7292 (1987)

[15] Rigby, D., Roe, R.J.: Molecular dynamics simulation of polymer liquid and glasss. Ii. Short range order and orientation correlation, J. Chem. Phys. **89**(8), 5280–5290 (1988)

[16] Rigby, D., Roe, R.J.: Molecular dynamics simulation of polymer liquid and glass. 3. Chain conformation, Macromolecules **22**, 2259–2264 (1989)

[17] Greenfield, M.L., Theodorou, D.N.: Geometric analysis of diffusion pathways in glassy and melt atactic polypropylene, Macromolecules **26**, 5461–5472 (1993)

[18] Misra, S., Mattice, W.L.: Atomistic models of amorphous polybutadienes. 3. Static free volume, Macromolecules **26**, 7274 (1993)

[19] Fan, C.F., Cagin, T., Shi, W., Smith, K.A.: Local chain dynamics of a model polycarbonate near glass transition temperature: A molecular dynamics simulation, Macromol. Th. Sim. **6**, 83–102 (1997)

[20] Takeuchi, H., Roe, R.-J.: Molecular dynamics simulation of local chain motion in bulk amorphous polymers. I. Cynamics above the glass transition, J. Chem. Phys. **94**, 7446 (1991)

[21] Adolf, D.B., Ediger, M.D.: Brownian dynamics simulations of local motions in polyisoprene, Macromolecules **24**, 5834 (1991)

[22] Kim, E.-G., Mattice, W.L.: Local chain dynamics of bulk amorphous polybutadienes: A molecular dynamics study, J. Chem. Phys. **101**, 6242–6250 (1994)

[23] Moe, N.E., Ediger, M.D.: Computer simulations of polyisoprene local dynamics in vacuum, solution, and the melt: Are conformational transitions always important? Macromolecules **29**(16), 5484–5492 (1996)

[24] Muller-Plathe, F., Rogers, S.C., Gunstern, W.F.v.: Diffusion coefficients of penetrant gases in polyisobutylene can be calculated correctly by molecular dynamics simulations, Macromolecules **25**, 6722–6724 (1992)

[25] Pant, P.V.K., Boyd, R.H.: Molecular dynamics simulation of diffusion of small penetrants in polymers, Macromolecules **26**, 679–686 (1993)

[26] Gusev, A.A., Suter, U.W., Moll, D.J.: Relationship between jelium transport and molecular motions in a glassy polcarbonate, Macromolecules **28**, 2582–2584 (1995)

[27] Soldera, A.: Energetic analysis of the two pmma chain tacticities and pma through molecular dynamics simulations, Polymer **43**, 4269–4275 (2002)

[28] Soldera, A.: Comparison between the glass transition temperatures of the two pmma tacticities: A molecular dynamics simulation point of view, Macromol. Symp. **133**, 21–23 (1998)

[29] Paul, W.: Molecular dynamics simulations of the glass transition in polymer melts, Polymer **45**,3901–3905 (2004)

[30] Dalton, L.R., Harper, A.W., Robinson, B.H.: The role of London forces in defining noncentrosymmetric order of high dipole moment-high hyperpolarizability chromophores in electrically poled polymeric thin films, Proc. Nat. Acad. Sci. USA **94**, 4842–4847 (1997)

[31] Piekara, A.: A theory of electric polarization, electro-optical kerr effect and electric saturation in liquids and solutions, Proc. Royal Soc. London. Series A, Math. Phys. Sci. **172**(950), 360–383 (1939)

[32] London, F.: The general theory of molecular forces, Trans. Farad. Soc. **33**, 8–26 (1937)

[33] Harper, A.W., Sun, S., Dalton, L.R., Garner, S.M., Chen, A., Kalluri, S., Steier, W.H., Robinson, B.H.: Translating microscopic optical nonlinearity into macroscopic optical nonlinearity: The role of chromophore-chromophore electrostatic interactions, J. Opt. Soc. of Am. B **15**(1), 329–337 (1998)

[34] Robinson, B.H., Dalton, L.R., Harper, A.W., Ren, A., Wang, F., Zhang, C., Todorova, G., Lee, M., Aniszfeld, R., Garner, S., Chen, A., Steier, W.H., Houbrecht, S., Persoons, A., Ledoux, I., Zyss, J., Jen, A.K.Y.: The molecular and supramolecular engineering of polymeric electro-optic materials, Chem. Phys. **245**, 35 (1999)

[35] Allen, M.P., Tildesley, D.J.: Computer simulation of liquids. Clarendon Press, Oxford (1987)

[36] Molecular Simulation, Inc, Cerius^2, San Diego, CA.

[37] Hagler, A.T., Huler, E., Lifson, S.: Energy functions for peptides and proteins. I. Derivation of a consistent force field including the hydrogen bond for amide crystals, J. Am. Chem. Soc. **96**(17), 5319–5327 (1974)

[38] Kitson, D.H., Hagler, A.T.: Theoretical studies of the structure and molecular dynamics of a peptide crystal, Biochem. **27**(14), 5246–5257 (1988)

[39] Berendsen, H.J.C., Spoel, D.v.d., Drunen, R.v.: Gromacs: A message-passing parallel molecular dynamics implementation, Comp. Phys. Comm. **91**(1–3), 43–56 (1995)

[40] Lindahl, E., Hess, B., Spoel, D.J.v.d.: Gromacs 3.0: A package for molecular simulation and trajectory analysis, J. Mol. Model. **7**(8), 306–317 (2001)

[41] Spoel, D.J.v.d., Buuren, A.R.v., Apol, E., Tieleman, P.J., Sijbers, A.L.T.M., Hess, B., Feenstra, K.A., Lindahl, E., Drunen, R.v., Berendsen, H.J.C.: Gromacs-user manual, Department of Biophysical Chemistry, University of Groningen, Groningen, Germany, 2002

[42] Stewart, J.J.P.: Mopac, a semi-empirical molecular orbital program, Quantum Chemical Program Exchange No. 455, 1983

[43] Hayden, L.M., Sinyukov, A.M., Leahy, M.R., French, J., Lindahl, P., Herman, W., Twieg, R.J., He, M.: New materials for optical rectification and electrooptic sampling of ultrashort pulses in the terahertz regime, J. Polym. Sci.:Part B: Polym. Phys. **41**, 2492–2500 (2003)

[44] Harris, K.D., Ayachitula, R., Strutz, S.J., Hayden, L.M., Twieg, R.J.: Dual use chromophores for photorefractive and irreversible photochromic applications, Appl. Opt. **40**(17), 2895–2901 (2000)

[45] Sinyukov A.M., Hayden, L.M.: Efficient electro-optic polymers for thz systems, J. Phys. Chem. B **108**, 8515–8522 (2004)

[46] Molecular Simulation, Inc, Amorphous builder, San Diego, CA.

[47] Young, J.A., Farmer, B.L., Hinkley, J.A.: Molecular modeling of the poling of piezoelectric polyimides, Polym. **40**(10), 2787–2795 (1999)

[48] Hayden, L.M., Brower, S.C., Strutz, S.J.: Pressure dependence of the depoling temperature in nonlinear optical polymers, Macromolecules **30**(9), 2734–2737 (1997)

[49] O'Reilly, J.M.: The effect of pressure on glass temperature and dielectric relaxation time of polyvinyl acetate, J. Polym. Sci. **57**, 429 (1962)

[50] Olabisi, O., Simha, R.: Pressure-volume-temperature studies of amorphous and crystallizable polymers. I. Experimental, Macromolecules **8**(2), 206–210 (1975)

[51] Quach, A., Simha, R.: Pressure-volume-temperature properties and transitions of amorphous polymers; polystyrene and poly (orthomethylstyrene), J. Appl. Phys. **42**(12), 4592 (1971)

[52] Woodcock, L.V.: Isothermal molecular dynamics calculations for liquid salts, Chem. Phys. Lett. **10**, 257 (1971)

[53] Swope, W.C., Anderson, H.C., Berens, P.H., Wilson, K.R.: A computer simulation method for the calculation of equilibrium constants for the formation of physical clusters of molecules: Application to small water clusters, J. Chem. Phys. **76**, 637 (1982)

CHAPTER 12

NONLINEAR OPTICAL PROPERTIES OF CHIRAL LIQUIDS
Electric-dipolar pseudoscalars in nonlinear optics

PEER FISCHER[1] AND BENOÎT CHAMPAGNE[2]
[1] *The Rowland Institute at Harvard, Harvard University, Cambridge, MA 02142, USA*
[2] *Laboratoire de Chimie Théorique Appliquée Facultés Universitaires Notre-Dame de la Paix (FUNDP), Rue de Bruxelles, 61, B-5000 Namur, Belgium*

Abstract: We give an overview of linear and nonlinear optical processes that can be specific to chiral molecules in isotropic media. Specifically, we discuss the pseudoscalars that underlie nonlinear optical activity and chiral frequency conversion processes in fluids. We show that nonlinear optical techniques open entirely new ways of exploring chirality: Sum-frequency-generation (SFG) at second-order and BioCARS at fourth-order arise in the electric-dipole approximation and do not require circularly polarized light to detect chiral molecules in solution. Here the frequency conversion in itself is a measure of chirality. This is in contrast to natural optical activity phenomena which are based on the interference of radiation from induced oscillating electric and magnetic dipoles, and which are observed as a differential response to right and left circularly polarized light. We give examples from our SFG experiments in optically active solutions and show how the application of an additional static electric field to sum-frequency generation allows the absolute configuration of the chiral solute to be determined via an electric-dipolar process. Results from *ab initio* calculations of the SFG pseudoscalar are presented for a number of chiral molecules

Keywords: chiral molecules; optical activity; pseudoscalars; liquids; second-order nonlinear optics; sum-frequency generation; SFG; electric field induced SFG; nonlinear optical activity

1. INTRODUCTION

The term "chirality", first introduced by Lord Kelvin in his Baltimore lectures, describes any system or structure that possesses a sufficiently low symmetry such that it is distinct from its mirror image [1, 2]:

"I call any geometrical figure, or group of points, chiral, and say that it has chirality if its image in a plane mirror, ideally realized, cannot be brought to coincide with itself."

Chirality is found at all physical length scales: from interactions mediated by the weak force to the anatomy of living organisms. It plays a particularly important role in biochemistry, as most biological molecules, including proteins, DNA, and their building blocks – the amino-acids[1] and sugars – are chiral. Surprisingly, all living organisms contain almost only 'left-handed' amino-acids and 'right-handed' sugars. This exclusive homochirality (having all molecules of one type of the same handedness) has the important consequence that the pharmaceutical activity of many biological molecules is often directly related to their chirality. Besides sugars, which organisms can only metabolize in the D-form, there are many examples of the different physiological action of the two mirror-image forms (enantiomers) of a chiral molecule. Despite clear differences in biological action, the enantiomers of a chiral molecule are, however, exactly alike in all chemical and physical properties except those that involve a left-right difference. This makes the observation and detection of chirality a challenging task. Only under a chiral influence, such as another chiral molecule or circularly polarized light, can interactions distinguish between the mirror-image forms of a chiral molecule. Optical methods are often the only practical physical means to probe molecular chirality, and a liquid's ability to rotate the plane of polarization of a linearly polarized light beam traversing it (in the absence of a static magnetic field), is the classical distinguishing characteristic of a chiral liquid, i.e. one that is "optically active". Conventional optical activity phenomena, such as optical rotation and circular dichroism, are based on the interference of induced oscillating electric- and magnetic (and electric-quadrupole) moments, and arise from a differential response to left and right circularly polarized light.

In the presence of intense electromagnetic radiation from a laser, the induced moments may show a nonlinear dependence on the incident field strength. Nonlinear optical activity as well as new chiroptical phenomena may now be observed.

Nonlinear optical activity phenomena arise at third-order and include intensity dependent contributions to optical rotation and circular dichroism, as well as a coherent form of Raman optical activity. The third-order observables are – like their linear analogs – pseudoscalars (scalars which change sign under parity) and require electric-dipole as well as magnetic-dipole transitions. Nonlinear optical activity is circular differential.

Sum-frequency generation (SFG) at second-order and the nonlinear Raman spectroscopy BioCARS at fourth-order can also probe chiral molecules. They have no analog in linear optics. We show that both are only symmetry allowed in a fluid, if the fluid is chiral. However, in contrast to optical activity phenomena, these processes arise entirely from induced electric-dipoles (without magnetic or quadrupolar transitions) and they are not circular differential. All laser beams can be linearly polarized and no polarization modulation is required as the detection of a sum-frequency (*viz.* five-wave mixing) photon is in itself a measure of the solution's chirality. Since an achiral solvent can not contribute to the signal, these techniques are sensitive, background-free probes of molecular chirality. The SFG

---

[1] All amino acids (except for glycine) have a chiral carbon atom.

pseudoscalar corresponds to the isotropic part of the second-order electric-dipolar first hyperpolarizability, which is the focus of this chapter. We discuss our recent SFG experiments, and *ab initio* computations, as well as a new chiral electro-optic effect that makes it possible to determine the sign of the second-order susceptibility and hence the absolute configuration (handedness) of a chiral solute.

Nonlinear optical phenomena are generally discussed in terms of an induced polarization $\vec{P}(t)$ written as a power series in the applied electric field $\vec{E}$ [3],

$$(1) \quad \vec{P}(t) = \varepsilon_0 \left(\overleftrightarrow{\chi}^{(1)} \vec{E} + \overleftrightarrow{\chi}^{(2)} \vec{E}\vec{E} + \overleftrightarrow{\chi}^{(3)} \vec{E}\vec{E}\vec{E} + \overleftrightarrow{\chi}^{(4)} \vec{E}\vec{E}\vec{E}\vec{E} + \ldots \right)$$

Most linear optical phenomena such as refraction, absorption and Rayleigh scattering are described by the first term in Eq. (1) where $\overleftrightarrow{\chi}^{(1)}$ is the linear susceptibility tensor. The higher order terms and susceptibilities are responsible for nonlinear optical effects. The second-order susceptibility tensor $\overleftrightarrow{\chi}^{(2)}$ underlies SFG, whereas and BioCARS arises within $\overleftrightarrow{\chi}^{(4)}$. As we are concerned with optical effects of randomly oriented molecules in fluids, we need to consider unweighted orientational averages of the susceptibility tensors in Eq. (1). We will show that the symmetries of the corresponding isotropic components $\chi^{(2)}$ and $\chi^{(4)}$ correspond to time-even pseudoscalars: the hallmark of chiral observables [2].

In order to describe linear and nonlinear optical activity, it becomes necessary to consider susceptibilities other than the electric-dipole susceptibilities in Eq. (1). We will only briefly discuss such nonlocal terms.

This chapter is organized as follows: In Section 2 we discuss the general symmetry requirements of chiroptical processes in isotropic media. In particular, we consider linear and nonlinear optical activity and we describe how frequency conversion at second-order (and at fourth-order) is specific to chiral molecules in fluids. In Section 3 we discuss our work on SFG in optically active solutions, and the computation of the SFG pseudoscalar is described in Section 4. Results from recent computations are given in Section 5. Conclusions are drawn in Section 6.

Sections 2 and 3 are in part based on a recent review article by Fischer and Hache [4]. The article gives an overview of chiral nonlinear optical spectroscopies in solution and at interfaces, and includes a discussion of second-harmonic generation from chiral surfaces [4].

2. PSEUDOSCALARS AND SYMMETRY

Parity, or space inversion, is the symmetry operation that interconverts a chiral molecule into its mirror image. All coordinates (x, y, z) are replaced everywhere by $(-x, -y, -z)$ under space inversion [2]. A chirality specific response in liquids and gases requires that the isotropic component of the susceptibility is odd under parity. Further, since the isotropic part of any tensor is necessarily a scalar, it follows that pseudoscalars – independent of the choice of coordinate axes and of opposite sign for enantiomers – underlie chiral observables in fluids. The isotropic medium may be a gas or a liquid. Typically, however, the experiments are conducted in liquids whereas, so far, the computations are for molecules in the gas phase. We also require that the pseudoscalars are even with respect to time-reversal symmetry, as we do not consider the application of a static magnetic field and since we assume that the liquid is stationary [2, 5].

In summary, we seek susceptibilities that are
- time-even,
- parity-odd, and
- have a non-vanishing isotropic part.

Susceptibilities at all orders can satisfy these requirements, but only those at even-order can do so within the electric-dipole approximation. We now discuss pseudoscalars at the different orders, but we shall concentrate on those that arise within the electric-dipole approximation.

2.1 Linear Optical Activity

In a liquid, the electric-dipole polarization linear in the field is given by

$$\vec{P} = \varepsilon_0 \chi^{(1)} \vec{E} + \ldots, \quad (2)$$

where the achiral scalar $\chi^{(1)}$ is related to the molecular polarizability via

$$\chi^{(1)} = \frac{N}{\varepsilon_0} \frac{1}{3} (\alpha_{xx} + \alpha_{yy} + \alpha_{zz}) \equiv \frac{N}{\varepsilon_0} \bar{\alpha}. \quad (3)$$

N is the number density. In order to describe natural optical activity it becomes necessary to go beyond the electric-dipole approximation. Apart from the polarization $\vec{P} = N \langle \vec{\mu}_{ind} \rangle$, we also need to consider the magnetization $\vec{M} = N \langle \vec{m}_{ind} \rangle$ and the quadrupole density $\overleftrightarrow{Q} = N \langle \overleftrightarrow{\Theta}_{ind} \rangle$. $\vec{\mu}_{ind}$, $\vec{m}_{ind}$, and $\overleftrightarrow{\Theta}_{ind}$ are respectively the induced molecular electric dipole, magnetic dipole and electric quadrupole moments. For a chiral molecule we need to consider [2, 6, 7]

$$\vec{\mu}_{ind} = \overleftrightarrow{\alpha} \vec{E} + \omega^{-1} \overleftrightarrow{G'} \dot{\vec{B}} + \ldots \quad (4)$$

$$\vec{m}_{ind} = -\omega^{-1} \overleftrightarrow{G'} \dot{\vec{E}} + \ldots$$

where the dot denotes a derivative with respect to time, and where $\overleftrightarrow{G'}$ is the optical rotation tensor, which is a function of the rotational strength. We have not considered $\overleftrightarrow{\Theta}_{ind}$ and any quadrupolar contributions to Eq. (4) as they average to zero in an isotropic medium. However, the electric quadrupole induced by the electric field as well as the electric dipole induced by the electric field gradient of the electromagnetic wave need to be considered should the molecules be oriented (anisotropic) [8]. We also note that in condensed media, such as liquids, the field at the molecule will be different from the applied optical electric field, $\vec{E}$, due to

Nonlinear Optical Properties of Chiral Liquids

induced dipole-dipole interactions of the surrounding molecules. $\vec{E}$ in microscopic expressions, such as Eq. (4), should be replaced with a 'local field', which is in the Lorentz model approximately given by $\left(\frac{\varepsilon+2}{3}\right)\vec{E}$.

Considering an effective polarization that combines the polarization, magnetization and quadrupole density [9, 10],

$$(5) \quad \vec{P}_{\text{eff}} = \vec{P} + \frac{i}{\omega}\vec{\nabla} \times \vec{M} - \vec{\nabla} \cdot \overset{\leftrightarrow}{Q}$$

we can write for an optically active fluid [4]:

$$(6) \quad \overset{\leftrightarrow}{P}^{\pm}_{\text{eff}} = \varepsilon_0 \left(\frac{N\overline{\alpha}}{\varepsilon_0} \pm \frac{2N\overline{G'}n_0}{\varepsilon_0 c}\right)\vec{E}^{\pm}$$

where the upper sign corresponds to right-circularly polarized and the lower to left-circularly polarized light, and where n_0 is the linear refractive index. The effective polarization is of the form $\vec{P}_{\text{eff}} = \varepsilon_0 \chi_{\text{eff}} \vec{E}$, and since the refractive index of a non-magnetic dielectric is, in general, given by

$$(7) \quad n = (1 + \chi_{\text{eff}})^{1/2}$$

we obtain the refractive index of an optically active liquid for right (+) and left (−) circularly polarized light:

$$(8) \quad n^{(\pm)} \approx n_0 \pm g_0, \quad \text{where} \quad g_0 \equiv \frac{N\overline{G'}}{\varepsilon_0 c}$$

The optical rotation θ in radians developed over a path length l is a function of the wavelength λ and the circular birefringence, and is given by [2]:

$$(9) \quad \theta = \frac{\pi l}{\lambda}\left(n^{(-)} - n^{(+)}\right) \approx -\frac{2\pi l}{\lambda}\frac{N}{\varepsilon_0 c}\overline{G'}$$

The linear pseudoscalar $\overline{G'}$ is the isotropic part of the optical rotation tensor, and $\overline{G'} \equiv (G'_{xx} + G'_{yy} + G'_{zz})/3$. Time-dependent perturbation may be used to obtain a sum-over-states expression for $\overline{G'}$ away from resonance [2, 7]:

$$(10) \quad \overline{G'} = -\frac{2}{3\hbar}\sum_{j \neq g}\frac{\omega}{\omega_{jg}^2 - \omega^2}\text{Im}\left[\langle g|\vec{\mu}|j\rangle \cdot \langle j|\vec{m}|g\rangle\right]$$

where ω_{jg} is the Bohr angular frequency in the basis set for which g is the ground state, and all the other symbols have their usual meaning. The electric dipole moment is odd under parity whereas the magnetic transition dipole is parity-even. Time-reversal inverts the direction of momenta and spins but leaves charge invariant. It follows that the transition electric-dipole moment is symmetric under time-reversal and that the magnetic moment is time-antisymmetric. However, the imaginary part of the time-odd magnetic moment is time-even. Hence, $\overline{G'}$ is a time-even pseudoscalar. We will now show that the nonlinear pseudoscalar at second-order arises entirely in the electric-dipole approximation.

2.2 The Sum-Frequency Generation Pseudoscalar

We can write the polarization of a sum-frequency generation process $\omega_3 = \omega_1 + \omega_2$ for two incident monochromatic waves at ω_1 and ω_2 as

(11) $\quad \vec{P}(\omega_3) = \varepsilon_0 \overleftrightarrow{\chi}^{(2)} \vec{E}(\omega_1) \vec{E}(\omega_2)$

In an isotropic medium, such as a liquid, the polarization is given by the vector cross product of the electric fields

(12) $\quad \vec{P}(\omega_3) = \varepsilon_0 \chi^{(2)} \vec{E}(\omega_1) \times \vec{E}(\omega_2)$

and the isotropic part of the second order susceptibility takes the form

(13) $\quad \chi^{(2)} = \dfrac{N}{2\varepsilon_0} \dfrac{1}{6} (\beta_{xyz} - \beta_{xzy} + \beta_{yzx} - \beta_{yxz} + \beta_{zxy} - \beta_{zyx}) \equiv \dfrac{N}{2\varepsilon_0} \overline{\beta}$

The term in parentheses vanishes for any molecule that possesses reflection planes, a center of inversion, or rotation-reflection axes, and $\overline{\beta}$ is thus only non-zero for a chiral molecule. It is of opposite sign for the enantiomers of a chiral molecule. In Rayleigh-Schrödinger perturbation theory the isotropic component of the first electric-dipolar hyperpolarizability at $\omega_3 = \omega_1 + \omega_2$ may be written as

(14) $\quad \overline{\beta} = \dfrac{(\omega_2 - \omega_1)}{6\hbar^2} \sum_{j,k} \vec{\mu}_{gk} \cdot (\vec{\mu}_{kj} \times \vec{\mu}_{jg})$

$$\left\{ \dfrac{1}{(\tilde{\omega}_{jk} - \omega_3)(\tilde{\omega}_{jg} - \omega_2)(\tilde{\omega}_{jg} - \omega_1)} \right.$$

$$+ \dfrac{1}{(\tilde{\omega}^*_{kj} + \omega_3)(\tilde{\omega}^*_{kg} + \omega_2)(\tilde{\omega}^*_{kg} + \omega_1)}$$

$$+ \dfrac{1}{(\tilde{\omega}_{kg} - \omega_3)(\tilde{\omega}_{jg} - \omega_2)(\tilde{\omega}_{jg} - \omega_1)}$$

$$\left. + \dfrac{1}{(\tilde{\omega}^*_{jg} + \omega_3)(\tilde{\omega}^*_{kg} + \omega_2)(\tilde{\omega}^*_{kg} + \omega_1)} \right\}$$

where the summation is over all excited states j, k. By allowing the transition frequency to be the complex quantity defined by $\tilde{\omega}_{jk} = \omega_{jk} - \left(\frac{i}{2}\right)\Gamma_k$, where ω_{jk} is the real transition frequency and Γ_k is the population decay rate of the upper level k, the theory is appropriate for near-resonant frequencies. The electric-dipole transition moments are defined as $\vec{\mu}_{gj} = \langle g| \vec{\mu} |j\rangle$. It is seen in Eq. (14) that $\overline{\beta}$ vanishes for a two-state system, and for degenerate incident frequencies, i.e. SHG with $\omega_1 = \omega_2$. Even though diagonal (single state) terms often play an important role in resonance phenomena, all diagonal contributions ($j = k$) to $\overline{\beta}$ vanish. Furthermore, $\overline{\beta}$ has no static limit and, as we shall see, its dispersion is consequently much more dramatic than that of a regular nonzero tensor component of the first hyperpolarizability. In practice, $\overline{\beta}$ needs to be near resonance for there to be an appreciable sum-frequency response.

We note, that when deducing the expression for the Pockels effect ($\omega = \omega + 0$) from the fully dynamic sum-over-states expression for $\overline{\beta}$, care has to be taken that both the optical frequency and the associated complex damping terms are set to zero [11, 12]. It is then seen that the Pockels effect vanishes in any liquid [11, 12].

We have discussed the symmetries of linear optical activity and sum-frequency generation. The former is an odd-order process that requires a nonlocal response tensor in order to be specific to chiral molecules in solution, whereas the latter is an even-order response where the dominant electric-dipolar susceptibility is a probe of chirality. These observations can be extended to pseudoscalars at third- and fourth-order.

2.3 Pseudoscalars at Order n

We can deduce the symmetry of a response tensor by considering the operators that enter the numerator of its quantum mechanical expression. For example, the product of three electric-dipole transition moment operators in Eq. (14) render SFG a parity-odd and time-even process. It follows that a third-order process requires nonlocal magnetic-dipole contributions in order to be parity-odd and that a local fourth-order process is parity-odd within the electric-dipole approximation. Some pseudoscalars that arise at order n are tabulated below.

Both the third-order and the fourth-order susceptibilities have non-vanishing isotropic parts [13]. The pseudoscalar $\hat{m}\hat{m}\hat{\mu}$ underlies the magnetochiral effect in linear optics; it is not circular differential and has recently been observed [14–16]. In nonlinear optics a $\hat{m}\hat{m}\hat{\mu}$ effect is predicted to give rise to inverse magnetochiral

Table 1. Some chirality specific susceptibilities in fluids and the operators that enter the numerator of the corresponding quantum mechanical expressions

Order $\chi^{(n)}$	Operators	Pseudoscalar	Examples of associated optical phenomena
(1)	$i\hat{m}\hat{\mu}$	$g_0, \overline{G'}$	optical activity (optical rotation, circular dichroism)
(2)	$\hat{\mu}\hat{\mu}\hat{\mu}$ $\hat{m}\hat{m}\hat{\mu}$ $i\hat{\Theta}\hat{m}\hat{\mu}$	$\chi^{(2)}, \overline{\beta}$	three-wave mixing: sum- and difference frequency generation magnetochiral effect, inverse magnetochiral birefringence, etc.
(3)	$i\hat{m}\hat{\mu}\hat{\mu}\hat{\mu}$ $\hat{\Theta}\hat{\mu}\hat{\mu}\hat{\mu}$	g_2	four-wave mixing: nonlinear optical activity (nonlinear optical rotation, nonlinear circular dichroism) coherent Raman optical activity
(4)	$\hat{\mu}\hat{\mu}\hat{\mu}\hat{\mu}\hat{\mu}$	$\chi^{(4)}$	five-wave mixing: e.g. BioCARS

birefringence, the radiation induced generation of a static magnetization. Other effects exist that have pseudoscalars associated with them which are not listed in the Table.

2.3.1 Nonlinear optical activity

In the case of degenerate four-wave mixing, i.e. $\omega = \omega + \omega - \omega$, a nonlocal $\chi^{(3)}$ may support nonlinear optical activity and thus intensity dependent contributions to optical rotation and circular dichroism [4, 13, 17–19]. In analogy to Eq. (8) we can include nonlinear optical activity phenomena by writing [4].

$$(15) \quad n^{\pm} \approx (n_0 + n_2 I) \pm (g_0 + g_2 I)$$

where I is the intensity of the light beam, n_2 is the usual nonlinear index of refraction, and where we have introduced a nonlinear optical activity index, g_2. Nonlinear optical rotation has been observed experimentally in the gas phase by Cameron and Tabisz [20], and nonlinear circular dichroism has recently been observed by Mesnil and Hache in solutions of the ruthenium(II) tris(bipyridyl) salt [21, 22].

In the case of non-degenerate frequencies, the nonlocal third-order effects may give rise to chiral pump-probe spectroscopies. The only observation of a coherent Raman optical activity process to date is also due to a third-order pseudoscalar. Spiegel and Schneider have observed Raman optical activity in coherent anti-Stokes Raman scattering in a liquid of (+)-trans-pinane and report chiral signals that are $\sim 10^{-3}$ of the conventional electric-dipolar CARS intensity [23].

2.3.2 BioCARS

The nonlinear polarization of a fourth-order process in a liquid is for two incident fields at ω_1 and ω_2 given by [24, 25]

$$(16) \quad \vec{P}(3\omega_1 - \omega_2) = \varepsilon_0 \chi^{(4)}(3\omega_1 - \omega_2)\left(\vec{E}(\omega_1) \times \vec{E}(\omega_2)\right)\left(\vec{E}(\omega_1) \cdot \vec{E}(\omega_1)\right)$$

There is one experimental report of such a process in liquids to date. Using two noncollinear frequency degenerate beams ($\omega_1 = \omega_2$) Shkurinov et al. report the observation of weak signals from aqueous solutions of arabinose that they attribute to the electric-dipolar pseudoscalar at fourth-order [4, 26].

A new coherent chiral Raman spectroscopy that arises when $3\omega_1 - \omega_2$ in Eq. (16) is resonant with the angular frequency of a vibration has been proposed by Koroteev [27]. Known as BioCARS, such a spectroscopy could exclusively probe chiral vibrations that are simultaneously Raman and hyper-Raman active [27, 28]. BioCARS has not yet been observed.

We now concentrate on our experimental observations and computations of $\bar{\beta}$ and $\chi^{(2)}$.

3. SUM-FREQUENCY GENERATION IN LIQUIDS

Two optical fields of different frequency may interact coherently in a chiral liquid to generate light at their sum- (or difference) frequency, as first predicted by Giordmaine [29]. From Eq. (12) it follows that the electric fields at ω_1, ω_2 and ω_3 need to span the X, Y, and Z directions of a Cartesian frame. Hence, a non-collinear beam geometry is required where two beams are polarized parallel – and one beam is polarized perpendicular to the plane defined by the input beams. This would suggest that the two incident beams make a right angle. However, momentum conservation favours collinear beams. The optimum angle to observe SFG in an optically active solution is thus a balance of these two requirements. Figure 1 shows a schematic of the experimental arrangement.

For a solution that contains only the R- and S-enantiomers of a chiral molecule, we can write the isotropic part of the electric-dipolar second-order susceptibility as

$$(17) \quad \chi^{(2)} = \frac{1000 N_A}{\varepsilon_0} ([R] - [S]) \overline{\beta}_R$$

where $\overline{\beta}_R$ is the $\overline{\beta}$ of the R-enantiomer, N_A is Avogadro's number, and where the square brackets denote a concentration in mol/l. We have used the identity $\overline{\beta}_R = -\overline{\beta}_S$. It is seen that $\chi^{(2)}$ is zero for a racemic solution (where $[R] = [S]$). Since the intensity at the sum-frequency is proportional to the square of the isotropic component of the second-order susceptibility, $I_{SFG} \propto |\chi^{(2)}|^2$, SFG can in general not distinguish between optical isomers.

This is illustrated in Figure 2 where we have performed an enantiomeric excess titration of 1,1'-bi-2-naphthol (BN) in tetrahydrofuran. The solution is initially R-(+)-BN and through the addition of S-(−)-BN the solution is taken through its racemic mid-point to one that has an enantiomeric excess of S-(−)-BN, whilst the

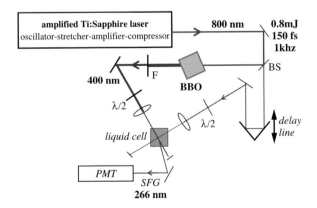

Figure 1. Experimental arrangement for the observation of SFG from a chiral liquid. The optical elements shown include a beam splitter (BS), a filter (F), a photomultiplier tube (PMT), and a nonlinear crystal (BBO)

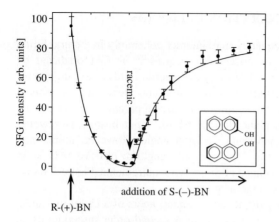

Figure 2. Titration starting with R-(+)-BN to which S-(−)-BN is added until the solution is racemic. At this point some of the racemic solution is removed, and more S-(−)-BN is added. The SFG signal is proportional to the square of the (fractional) concentration difference of the R-(+) and S-(−) enantiomers, which is plotted by the solid line [30]. Inset: structure of R-(+)-BN

total concentration ($[R] + [S]$) is kept at 0.4 M. I_{SFG} obeys the expected quadratic dependence on the mole fraction of the enantiomeric difference.

Within the noise of the experiment, no signal is recorded for the racemic mixture. SFG is thus effectively background free: any achiral signals (solvent, higher-order multipolar contributions) are either weak or can be eliminated by choosing appropriate beam polarizations [30, 31]. Table 2 lists the polarization combinations that allow chiral bulk SFG signals to be discerned from SFG signals that have an achiral origin.

Not only is chiral SFG background-free, but it is entirely electric-dipolar. This is significant, as magnetic-dipole (and electric-quadrupole) transitions are typically much weaker than electric-dipole transitions. Nevertheless, the absolute strength of the SFG signals is low. In principle, this is not a problem, as it is possible to detect low light levels with single photon counting methods, especially as there is little or no background in SFG from chiral liquids. However, in practice at least one of the three frequencies needs to be near or on (electronic or vibrational) resonance

Table 2. Polarization combinations of the three fields and the origin of the SFG signal. For S-polarized light the electric field vector is orthogonal to the plane defined by the two incident beams, and for P-polarized light it is parallel to the plane [31]

Polarizations	Chiral SFG	Achiral SFG
PPP	—	yes
SPP, PSP, PPS	yes	—
PSS, SPS, SSP	—	yes
SSS	—	—

Table 3. Comparison between linear optical activity and nonlinear optical sum-frequency generation from a chiral liquid [4]

	Optical activity	Sum-frequency generation
Pseudoscalar	$\overline{G'} \propto \omega \mathrm{Im}[\vec{\mu}_{gj} \cdot \vec{m}_{jg}]$ electric- and magnetic-dipolar	$\overline{\beta} \propto (\omega_1 - \omega_2)\vec{\mu}_{gk} \cdot [\vec{\mu}_{kj} \times \vec{\mu}_{jg}]$ electric-dipolar
Signal	$\sim \overline{G'}$ different response to cp light	$\sim |\overline{\beta}|^2$ intensity at sum-frequency
Chiral probe	circularly polarized light	x, y, z components of three linearly polarized light beams
Signal contains	chiral and achiral response	only chiral response, no background

for there to be a measurable SFG signal. The concomitant linear absorption near resonance further limits the conversion efficiency, which is already low as the SFG process can not be phase-matched in liquids. Table 3 summarizes differences between linear optical activity and chiral SFG.

3.1 Electric-field Induced SFG

As the sum-frequency signal is proportional to the square of the enantiomeric concentration difference, SFG can in general not distinguish between the enantiomers of a chiral solute. However, Buckingham and Fischer have shown that the application of a static electric field to SFG makes it possible to determine the sign of the pseudoscalar $\chi^{(2)}$ and thus the absolute configuration of the chiral solute [32]. The static field does not change the phase matching conditions of the sum-frequency process, but it gives rise to an electric-field induced contribution to the signal. The combined sum-frequency polarization along x is given by

$$(18) \quad P_x(\omega_3) = \varepsilon_0 \left[\underbrace{\chi^{(2)} E_y(\omega_1) E_z(\omega_2)}_{\text{chiral}} + \underbrace{\chi^{(3)} E_y(\omega_1) E_y(\omega_2) E_x(0)}_{\text{achiral}} \right]$$

where we assume that the ω_1 beam travels along the z-direction and has its electric field vector oscillating along y, and the ω_2 beam to be plane polarized in the yz-plane (see Figure 3). The beat between chirality-sensitive SFG (a second-order process) and achiral electric-field induced sum-frequency generation (a third order process) yields a contribution to the intensity ($I_{\mathrm{SFG}} \propto |\vec{P}(\omega_3)|^2$) that changes sign with the enantiomer. The cross-term linear in the static electric field is [33]

$$(19) \quad I_{\mathrm{SFG}}(E) \propto \mathrm{Re}[\chi^{(2)}(\chi^{(3)})^*]E(0)I(\omega_1)I(\omega_2)$$

The effect can therefore be used to determine the absolute sign of the isotropic part of the sum-frequency hyperpolarizability. The effect has recently been observed in solutions of 1,1'-bi-2-naphthol [33]. Figure 3 shows that the $I_{\mathrm{SFG}}(E)$ signals depend linearly on the strength of the static electric field and that they change sign with the enantiomer. *Ab initio* computations can be used to relate the sign of the pseudoscalar to the absolute configuration of the enantiomers [33].

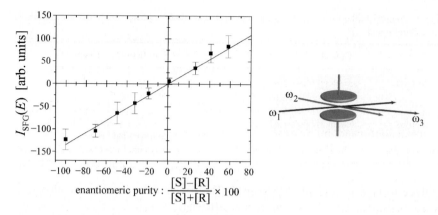

Figure 3. Intensity of SFG linear in the static field measured as a function of the fractional concentration difference in R-(+)-BN and S(−)-BN. Beam geometry for SFG(E) [33]

4.   COMPUTATIONAL METHODS

Hyperpolarizabilities can be calculated in a number of different ways. The quantum chemical calculations may be based on a perturbation approach that directly evaluates sum-over-states (SOS) expressions such as Eq. (14), or on differentiation of the energy or induced moments for which (electric field) perturbed wavefunctions and/or electron densities are explicitly calculated. These techniques may be implemented at different levels of approximation ranging from semi-empirical to density functional methods that account for electron correlation through approximations to the exact exchange-correlation functionals to high-level *ab initio* calculations which systematically include electron correlation effects.

Additional approximations are typically made. The application of external electric fields not only perturbs the electron densities, but it can also modify the nuclear configuration. However, a global calculation that evaluates hyperpolarizabilities with a sum over rovibronic states [34] is often computationally too demanding for polyatomic molecules. Instead, one can use a two-step procedure which sequentially, rather than simultaneously, treats the effects of the applied electric fields upon the motion of the electrons and nuclei. In the so-called clamped-nucleus approximation [35], the effects of the electric field on the electron distribution are determined first, giving the electronic contributions to the hyperpolarizability. In a second step, the field-induced relaxation on the potential energy surface is considered. It gives rise to vibrational hyperpolarizabilities, which can be decomposed into a nuclear relaxation and a curvature part or into a pure vibrational contribution and a zero-point vibrationally-averaged (ZPVA) correction [36, 37]. Although important for many non-linear optical processes [36], the pure vibrational contribution to second-order sum-frequency generation is negligible for wavelengths in the UV/visible. Although preliminary investigations have shown that the ZPVA correction to the first hyperpolarizability of *p*-nitroaniline is less than 20% of its electronic counterpart [38], less is known about the ZPVA correction and its importance for compounds like binaphtol or helicenes. We have therefore limited our theoretical investigations of the SFG pseudoscalar to the purely electronic contributions to the hyperpolarizability.

In the condensed phase the field at the molecule will be different from the applied macroscopic field due to induced dipole-dipole (and higher order multipolar) interactions with the surrounding molecules, as briefly mentioned in Section 2.1 [39]. In addition, the molecules' properties are changed due to the interaction with the surrounding medium. Several computational schemes have been proposed to address these effects. They are essentially based on the extension of the Onsager reaction field cavity model and give effective hyperpolarizabilities, i.e. molecular hyperpolarizabilities induced by the external fields that include the modifications due to the surrounding molecules as well as local (cavity) field effects [40–42]. These condensed-phase effects have, however, not yet been included in the SFG hyperpolarizability calculations, which are therefore strictly gas-phase calculations.

4.1 CIS-SOS

A sum-over-states calculation of the (electronic) hyperpolarizability and hence the pseudoscalar is particularly attractive as it makes use of transition moments and frequencies that can be directly related to spectroscopic quantities. The use of phenomenological damping terms ensures that the SOS expression in Eq. (14) can also be used near resonance. This is important, as the pseudoscalar has no static limit and in practice resonance enhancement is required so that a measurable SFG signal can be obtained.

Accurate SOS calculations of the first hyperpolarizabilities in turn require accurate ground and excited state wavefunctions and derived properties, such as excitation energies, dipole transition moments, and dipole moments. The excited state wavefunctions and energies are typically calculated at the Configuration Interaction Singles (CIS) level of approximation, and a large number of such studies on conjugated compounds employing semi-empirical Hamiltonians have been published [43]. *Ab initio* SOS/CIS studies are less frequent [44]. They can be applied to any kind of system, but suffer from two drawbacks: the truncation of the sum (also present in semi-empirical calculations) and an overestimation of the excitation energies. Limited computational resources often make it necessary that the SOS is truncated after inclusion of only a few excited states. Related to truncated SOS/CIS expressions is the loss of size consistency. Although the absence of intruder states cannot be guaranteed, a study of the evolution of the SOS hyperpolarizability as function of the number of excited states enables one to assess its convergence (for the particular basis) [45]. This is illustrated in Fig. 4 for a CIS/6-311++G** SOS calculation of $\overline{\beta}$ for R-monofluoro-oxirane (structure is shown in Fig. 5).

The overestimation of the excitation energies is in general due to a neglect of electron correlation effects. Although several highly-correlated schemes have been elaborated [46–51], it is often sufficient to simply downshift all the excitation

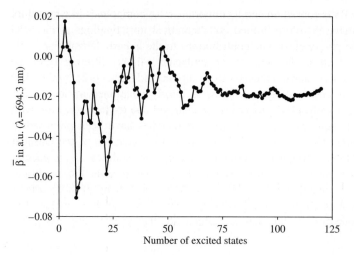

Figure 4. CIS/6-311++G** calculation of $\overline{\beta}(-3\omega, 2\omega, \omega)$ as a function of the number of excited states included in the SOS expression for R-monofluoro-oxirane ($\lambda = 694.3$ nm, $\Gamma = 0$, the scissor operator: 2.27 eV)

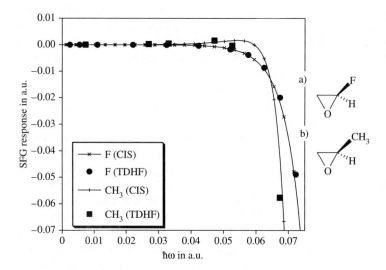

Figure 5. Dispersion of R-monofluoro-oxirane (a) and R-propylene oxide (b) determined at the CIS and TDHF levels of approximation. A damping factor of 1000 cm^{-1} has been used in the CIS calculations. Both methods employ the 6-311++G** basis set

energies by a fixed amount using a so-called 'scissor operator'. Typically, the excited state energies are reduced such that the calculation agrees with the experimental UV absorption spectrum or such that the lowest-lying states agree with the energies obtained from highly-correlated calculations. The correction is particularly important for calculations that concern the SFG hyperpolarizability/pseudoscalar near resonance [30].

4.2 TDHF

Apart from SOS approaches, hyperpolarizabilities can be evaluated by differentiating the energy or the dipole moment with respect to the applied electric field(s):

$$(20) \quad \beta = \left(\frac{\partial^2 \mu(E)}{\partial E^2}\right)_{E=0} = -\left(\frac{\partial^3 W(E)}{\partial E^3}\right)_{E=0}$$

In principle, the differentiation is either done numerically in the so-called finite-field methods, or in an analytical scheme, or a combination of both. Numerical finite-field calculations are limited to derivatives with respect to static fields. Since SFG is an optical process that involves dynamic oscillating fields, it becomes necessary to use an analytical approach, such as the time-dependent Hartree Fock (TDHF) method.

TDHF [52, 53] is one of the most widely-employed *ab initio* techniques to evaluate nonlinear-optical response tensors. The TDHF approach is size consistent but cannot account for the finite lifetime of the excited states. The matrices of the TDHF equation are expanded in a Taylor series of the perturbation due to the static and/or dynamic electric fields and are solved for each order [52, 53]. The so-obtained successive field-derivatives of the density matrix are then inserted into the expressions for the hyperpolarizability,

$$(21) \quad \beta_{xyz}(-3\omega; 2\omega, \omega) = -\text{Tr}\left[\boldsymbol{\mu}_x \mathbf{D}^{yz}(2\omega, \omega)\right]$$

where $\boldsymbol{\mu}_x$ is the dipole moment matrix in the x direction and $\mathbf{D}^{yz}(2\omega, \omega)$ is the second-order derivative of the density matrix with respect to electric fields at 2ω and ω along y and z, respectively. The original TDHF scheme [52, 53] has recently been generalized to evaluate the SFG response (Eq. (21)) using the $2n+1$ rule [54]. The latter consists in evaluating Eq. (21), which apparently depends on second-order response wavefunctions, using only first-order quantities. Extensions of the TDHF approach have been developed that can account for the effects of electron correlation [46, 48, 55–58].

In order to compare *ab initio* SOS/CIS with TDHF calculations we have evaluated the SFG response of two model chiral compounds, R-(+)-propylene oxide and R-mono-fluoro-oxirane. The computed excited state energies have all been down-shifted by 2.675 eV for R-(+)-propylene oxide (and 2.27 eV for R-mono-fluoro-oxirane) [59]. Figure 5 shows that both methods calculate very similar magnitudes and dispersions of the SFG pseudoscalar.

Apart from calculations at the *ab initio* level, the TDHF scheme has also been used with semi-empirical Hamiltonians. This makes it possible to calculate larger molecules while partially accounting for electron correlation effects. Indeed, when estimating the dynamic first hyperpolarizability of reference push-pull π-conjugated compounds, the semi-empirical TDHF scheme performs generally better than the *ab initio* TDHF approach in comparison with high-level *ab initio* methods [60].

4.3   DFT

In addition to the traditional wavefunction CIS-SOS and TDHF methods, density functional theory (DFT) can be used to calculate frequency dependent hyperpolarizabilities [61–67]. For small to medium-sized molecules DFT methods calculate hyperpolarizabilities that are in rather good agreement with *post* Hartree-Fock methods. However, the conventional exchange-correlation functionals are local functions of the density (and its derivatives) and cannot properly describe the ultra-non-local effects associated with the hyperpolarizabilities of large (conjugated) molecules [68]. Newer DFT methods such as those based on current-density-functional theory [69] or exact exchange [70] require further development before they can be used to accurately model dynamic first hyperpolarizabilities of push-pull π-conjugated systems.

5.   CALCULATIONS OF $\overline{\beta}$

The isotropic part of the first hyperpolarizability, $\overline{\beta}$, may be calculated using a number of methods – ranging from a simple single-centre chiral molecular orbital approach to *ab initio* calculations at varying levels of approximation (see Section 4). We have computed $\overline{\beta}$ for a number of chiral molecules in order to establish the typical strength and frequency dispersion of the pseudoscalar [30, 54, 59, 71]. $\overline{\beta}$ is a measure of the signal strength in SFG experiments and a convenient measure that facilitates direct comparison with achiral second-order nonlinear optical processes.

For a given class of molecules it is interesting to explore whether the presence of certain functional groups can enhance $\overline{\beta}$. We survey calculations based on the AM1 semi-empirical Hamiltonian and the TDHF scheme that explore how molecular structure, here for helical molecules such as helicenes and heliphenes, influences the magnitude of $\overline{\beta}$ [54, 72, 73].

5.1   Strength of $|\overline{\beta}|$

The simplest possible chiral molecule consists of a central atom bonded to three different, non-coplanar substituents. The bonding can be modeled by four sp³ molecular orbitals ψ_g, ψ_ℓ, ψ_m, and ψ_n formed from the s, p_x, p_y, and p_z atomic orbitals located on the central atom. The molecular ground state is chosen to be chiral and, following [71], is predominately of s character. A particular set is given by:

$$(22) \quad (\Psi_g, \Psi_\ell, \Psi_m, \Psi_n) = \begin{pmatrix} 0.93 & 0.10 & 0.20 & 0.30 \\ 0.093 & -0.99 & 0.02 & 0.03 \\ 0.26 & 0 & -0.95 & -0.16 \\ 0.26 & 0 & 0.22 & -0.94 \end{pmatrix} \begin{pmatrix} s \\ p_x \\ p_y \\ p_z \end{pmatrix}$$

In this phenomenological model the transition energies to the states ψ_ℓ, ψ_m, and ψ_n are respectively taken to be 45,000, 50,000, and 55,000 cm^{-1}. The transition dipole moment $\langle s|\mu_x|p_x\rangle = \langle s|\mu_y|p_y\rangle = \langle s|\mu_z|p_z\rangle$ is taken to be 1.0 debye (D) (= 0.3935 a.u.). From Eq. (22) it follows that the permanent dipole moment has the components $\mu_x = 0.19$ D, $\mu_y = 0.37$ D, and $\mu_z = 0.56$ D. Substituting these quantities into Eq. (14) for $\lambda = 694$ nm and $\Gamma = 1000$ cm^{-1}, we compute that the sum-frequency pseudoscalar $\overline{\beta}(-3\omega; 2\omega, \omega) = 0.0087$ a.u. [1.0 atomic unit (a.u.) of first hyperpolarizability = 3.2063×10^{-53}C^{3}m^{3}J$^{-2} = 8.641 \times 10^{-33}$ esu], which should be compared to an individual tensor component of the first hyperpolarizability, such as $\beta_{xyz}(-3\omega; 2\omega, \omega) = 1.38$ a.u. [59].

Similarly small magnitudes of $\overline{\beta}(-3\omega; 2\omega, \omega)$ are obtained in *ab initio* calculations on the chiral molecules R-(+)-propylene oxide and R-monofluorooxirane (their structures are shown in Fig. 5). For $\hbar\omega < 0.04$ a.u., $|\overline{\beta}(-3\omega; 2\omega, \omega)|$ is less than 10^{-3} a.u.. Closer to resonance, at $\hbar\omega = 0.07$ (0.08) a.u. for propylene oxide (monofluoro-oxirane), *i.e.* less than 0.01 a.u. below the first excited state, the magnitude of $\overline{\beta}(-3\omega; 2\omega, \omega)$ is still < 0.3 a.u.. The finding that $|\overline{\beta}|$ is several orders of magnitude smaller compared to the strength of an individual hyperpolarizability tensor component [71] calls into question the first experimental report on SFG from optically active liquids that suggested that $\overline{\beta}(-3\omega; 2\omega, \omega)$ of arabinose may be 14 a.u. (at $\omega = 2\pi c/(694\,\text{nm})$) [74, 75].

5.2 Dispersion of $|\overline{\beta}|$

The $|\overline{\beta}|$ calculations (*vide supra*) suggest that the SFG pseudoscalar is, even near resonance, much weaker than a regular nonzero tensor component of the first hyperpolarizability [30, 59, 71]. Since $\overline{\beta}$ has no static limit, its dispersion is, however, much more dramatic. We have computed the dispersion of R-(+) and S-(−)-1, 1′-bi-2-naphtol (BN) to illustrate this point. Figure 6 shows the frequency dependence of $\overline{\beta}(-3\omega; 2\omega, \omega)$ for R-(+)−1, 1′-bi-2-naphtol calculated *ab initio* at the TDHF/6-31G* level of approximation. For $\hbar\omega = 1.5$ eV ($\lambda = 828$ nm), $\overline{\beta}(-3\omega; 2\omega, \omega)$ amounts to 0.4862 a.u., which is still rather small considering that it is so close to resonance – at approximately 1.61 eV the 2ω beam is resonant with the lowest lying absorption peak of BN. Similar calculations (TDHF/6-31G) estimate that the corresponding pseudoscalars for [4]-helicene and [5]-helicene are of comparable strength [54].

Even though the nonlinearity of BN is predominantly electric-dipolar, recent experiments show that nonlocal higher-order multipolar (magnetic and electric quadrupolar) contributions to SFG from BN are measurable [30]. However, the polarization of all three fields in bulk SFG experiments may always be chosen such that only the chiral electric-dipolar signals are observed.

Figure 7 shows the enhancement by many orders of magnitude that the SFG signal (which is proportional to the square of $\overline{\beta}$) from BN experiences over a relatively small wavelength range [30]. Shown is the dispersion of $\overline{\beta}$ relative to the (achiral) vector component of the first hyperpolarizability.

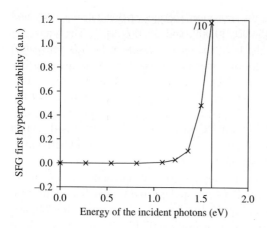

Figure 6. $\overline{\beta}(-3\omega; 2\omega, \omega)$ of R-(+)-1,1'-bi-2-naphtol as a function of the incident photon energy

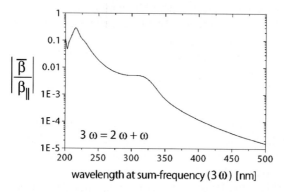

Figure 7. Dispersion of the chirality specific (chiral) pseudoscalar relative to for the SFG process in optically active 1,1'-bi-2-naphthol. Configuration Interaction Singles SOS calculation with the cc-pVDZ basis, damping = 3000 cm^{-1}

5.3 Molecular Structure and $|\overline{\beta}|$

Among the many classes of chiral molecules, helical systems are particularly fascinating. Their structure is relevant to proposed mechanisms of handedness induction in relation to chiral amplification [76]. Helicenes ([N]-H) are helical molecules formed from N-ortho-fused benzene rings (Fig. 8) which display considerable rotatory power [77]. Helicenes are presently the subject of intense synthesis efforts that try to functionalize these molecules in order to attain enhanced electric, magnetic, and optical properties [78, 79]. Phenylenes ([N]-P), or heliphenes, constitute another class of helical aromatic compounds for which syntheses have recently been reported [80, 81]. They are made up of N benzene rings fused together with N − 1 cyclobutadiene rings (Fig. 8).

The magnitude of $\overline{\beta}(-3\omega; 2\omega, \omega)$ has been investigated for [N]-H and [N]-P at the TDHF/AM1 level as a function of the number of units [72]. Figure 9 shows that, with the exception of the smallest members of the series, which are approximately planar, $\overline{\beta}(-3\omega; 2\omega, \omega)$ increases approximately linearly with N. Indeed, when $N = 3$ (i.e. anthracene in the helicene series) the molecules are planar and achiral, and therefore $\overline{\beta}$ is necessarily zero. For $N > 3$ the [N]-H and [N]-P series display similar chiral SFG responses, indicating that the strength of $\overline{\beta}(-3\omega; 2\omega, \omega)$ depends primarily on the size of the helix (the number of turns).

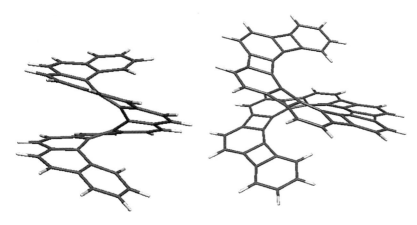

Figure 8. Representation of [10]-H (left) and [10]-P (right)

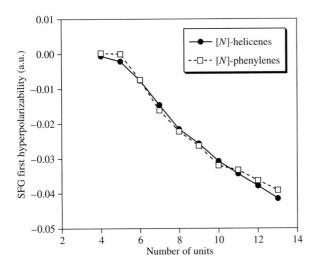

Figure 9. $\overline{\beta}(-3\omega; 2\omega, \omega)$ for [N]-H and [N]-P as a function of N ($\hbar\omega = 0.5$ a.u.)

Table 4. $\bar{\beta}(-3\omega; 2\omega, \omega)$ of substituted tetrathia-[7]-helicenes as a function of the number and position of the D/A NH_2/NO_2 substituents. All quantities are in atomic units (a.u.) and have been obtained at the TDHF/AM1 level with $\lambda = 1907$ nm. Representation of tetrathia-[7]-helicene and numbering of the substituent positions

M/7	M'/8	N/2	N'/13	$\bar{\beta}$(SFG)
H	H	H	H	−0.38
H	H	NO_2	NH_2	−0.05
NH_2	H	NO_2	H	0.52
NH_2	H	H	NO_2	−0.27
NO_2	H	H	NH_2	−1.97
NO_2	H	NH_2	H	−2.36
NO_2	NH_2	H	H	−0.58
NO_2	NO_2	H	H	−3.51
NH_2	NH_2	H	H	−0.87
H	H	NO_2	NO_2	0.67
H	H	NH_2	NH_2	−1.30
NO_2	NO_2	NH_2	NH_2	−8.43
NH_2	NH_2	NO_2	NO_2	1.35
NH_2	NO_2	NH_2	NO_2	−0.68
NH_2	NO_2	NO_2	NH_2	0.41

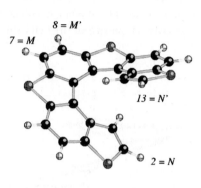

It is interesting to explore how substitution of the helical molecules affects $\bar{\beta}(-3\omega; 2\omega, \omega)$. It was found that hexahelicene substituted with an amino/nitro (NH_2/NO_2) donor/acceptor pair at the terminal positions hardly affects the strength of $\bar{\beta}(-3\omega; 2\omega, \omega)$ [72]. The same was found for octahelicene, even though its electric-field induced second-harmonic generation (EFISH) second-order response is strongly affected upon such a substitution. However, a similar investigation of substitution effects for tetrathia-[7]-helicene [73] demonstrates that here an appropriately chosen position of the donor/acceptor pair can enhance $\bar{\beta}(-3\omega; 2\omega, \omega)$ by up to two orders of magnitude Table 4. In particular, strong acceptor groups on the central sites (M, M') lead to large SFG responses while donor groups maximize the SFG response when placed at the ends of the helix (N, N'), or vice versa. Nevertheless, much work remains to be done in order to systematically unravel what functional groups maximize the SFG pseudoscalar. Comparison of such calculations with experiments will certainly be helpful. Apart from the strength of $|\bar{\beta}|$ it is also important to consider the solubility of the compounds.

6. CONCLUSIONS

Symmetry arguments show that parity-odd, time-even molecular properties which have a non-vanishing isotropic part underlie chirality specific experiments in liquids. In linear optics it is the isotropic part of the optical rotation tensor, G', that gives rise to optical rotation and vibrational optical activity. Pseudoscalars can also arise in nonlinear optics. Similar to the optical rotation tensor, the odd-order susceptibilities require magnetic-dipole (electric-quadrupole) transitions to be chirally sensitive.

Nonlocal third-order susceptibilities can give rise to nonlinear optical rotation and nonlinear circular dichroism.

Interestingly, pseudoscalars formed at even orders ($\chi^{(2)}, \chi^{(4)}$) arise within the electric dipole approximation and have no analog in linear optics. They give rise to frequency conversion processes only if the liquid is optically active and it is thus the generation of a photon which is the measure of a liquid's chirality.

We focus on sum-frequency generation (SFG) and present results from our experiments. We demonstrate how electric-field induced sum-frequency generation can be used to determine the handedness of the chiral solute.

The molecular and spectral properties of the SFG pseudoscalar $\overline{\beta}$ are discussed. We survey a number of different computational methods that may be used to calculate $\overline{\beta}$. Examples are given for small chiral molecules that possess a single chiral center as well as larger, conjugated molecules that exhibit a helical structure due to steric interference. The calculations show that the magnitude of $\overline{\beta}$ is, even near resonance, small compared with a regular nonzero tensor component of the first hyperpolarizability. Calculations on helicenes suggest that the chiral SFG response can be enhanced by appropriately placed functional groups. General structure-property relationships have not yet been deduced.

ACKNOWLEDGEMENTS

P.F. is grateful for a fellowship from the Rowland Institute at Harvard. B.C thanks the Belgian National Fund for Scientific Research for his Research Director position.

REFERENCES

[1] Kelvin, L.: Baltimore Lectures. C.J. Clay & Sons, London (1904)
[2] Barron, L.D.: Molecular Light Scattering and Optical Activity. Cambridge University Press, Cambridge (2004)
[3] Butcher, P.N., Cotter, D.: The Elements of Nonlinear Optics. Cambridge University Press, Cambridge (1990)
[4] Fischer, P., Hache, F.: Chirality **17**, 421 (2005)
[5] Barron, L.D., Buckingham, A.D.: Accounts Of Chemical Research. **34**, 781 (2001)
[6] Condon, E.U.: Reviews of Modern Physics. **9**, 432 (1937)
[7] Buckingham, A.D.: In: J.O. Hirschfelder Advances in Chemical Physics. Interscience, New York, vol. 12, p. 107 (1967)
[8] Buckingham, A.D., Dunn, M.B.: Journal of The Chemical Society A -Inorganic Physical Theoretical, 1988 (1971)
[9] Rosenfeld, L.: Theory of Electrons. Dover, New York (1965)
[10] Pershan, P.S.: Phys. Rev. **130**, 919 (1963)
[11] Buckingham, A.D., Fischer, P.: Phys. Rev. A **61**, 035801 (2000)
[12] Buckingham, A.D., Fischer, P.: Phys. Rev. A **63**, 047802 (2001)
[13] Wagnière, G.: J. Chem. Phys. **77**, 2786 (1982)
[14] Wagnière, G.H.: Chem. Phys. **245**, 165 (1999)
[15] Rikken, G.L.J.A., Raupach, E.: Nature. **390**, 493 (1997)
[16] Kleindienst, P., Wagnière, G.H.: Chem. Phys. Lett. **288**, 89 (1998)

[17] Akhmanov, S.A., Zharikov, V.I.: JETP Lett. **6**, 137 (1967)
[18] Atkins, P.W., Barron, L.D.: Proc. Roy. Soc. A **304**, 303 (1968)
[19] Akhmanov, S.A., Lyakhov, G.A., Makarov, V.A. et al.: Opt. Acta **29**, 1359 (1982)
[20] Cameron, R., Tabisz, G.C.: Mol. Phys. **90**, 159 (1996)
[21] Mesnil, H., Hache, F.: Phys. Rev. Lett. **85**, 4257 (2000)
[22] Hache, F., Mesnil, H., Schanne-Klein, M.-C.: Phys. Rev. B **60**, 6405 (1999)
[23] Spiegel, H., Schneider, F.W.: In: Spectroscopy of Biological Molecules – State of the Art. A. Bertoluzza, C. Fagnano and P. Monti (eds) Wiley, p. 317 (1989)
[24] Koroteev, N.I.: In: Frontiers in Nonlinear Optics. The Serguei Akhmanov memorial volume, H. Walther, N.I. Koroteev and M. Scully (eds) Inst. of Physics Publishing, Bristol, p. 228 (1993)
[25] Romero, L.C.D., Meech, S.R., Andrews, D.L.: Journal Of Physics B-Atomic Molecular And Optical Physics. **30**, 5609 (1997)
[26] Shkurinov, A.P., Dubrovskii, A.V., Koroteev, N.I.: Physical Review Letters. **70**, 1085 (1993)
[27] Koroteev, N.I.: Biospectroscopy. **1**, 341 (1995)
[28] Volkov, S.N., Konovalov, N.I., Koroteev, N.I. et al.: Quantum Electronics. **25**, 62 (1995)
[29] Giordmaine, J.A.: Physical Review. **138**, 1599 (1965)
[30] Fischer, P., Wise, F.W., Albrecht, A.C.: Journal Of Physical Chemistry. A **107**, 8232 (2003)
[31] Fischer, P., Beckwitt, K., Wise, F.W. et al.: Chemical Physics Letters. **352**, 463 (2002)
[32] Buckingham, A.D., Fischer, P.: Chemical Physics Letters. **297**, 239 (1998)
[33] Fischer, P., Buckingham, A.D., Beckwitt, K. et al.: Phys. Rev. Lett. **91**, 173901 (2003)
[34] Bishop, D.M.: Rev. Mod. Phys. **62**, 343 (1990)
[35] Bishop, D.M., Kirtman, B., Champagne, B.: J. Chem. Phys. **107**, 5780 (1997)
[36] Kirtman, B., Champagne, B.: Int. Rev. Phys. Chem. **16**, 389 (1997)
[37] Bishop, D.M.: Adv. Chem. Phys. **104**, 1 (1998)
[38] Quinet, O., Champagne, B., Kirtman, B.: J. Mol. Struct. (Theochem) **633**, 199 (2003)
[39] Bishop, D.M.: Int. Rev. Phys. Chem. **13**, 21 (1994)
[40] Wortmann, R., Bishop, D.M.: J. Chem. Phys. **108**, 1001 (1998)
[41] Macak, P., Norman, P., Luo, Y., et al.: J. Chem. Phys. **112**, 1868 (2000)
[42] Tomasi, J., Cammi, R., Mennucci, B., et al.: Phys. Chem. Chem. Phys. **4**, 5697 (2002)
[43] Docherty, V.J., Pugh, D., Morley, J.O.: J. Chem. Soc. Faraday Trans. II **81**, 1179 (1985)
[44] Spassova, M., Monev, V., Kanev, I. et al.: In: Quantum Systems in Chemistry and Physics, A. Hernandez-Laguna (eds) Kluwer, Dordrecht, vol. 1: Basic Problems and Model Systems, p. 101 (2000)
[45] Kanis, D.R., Ratner, M.A., Marks, T.J.: Chem. Rev. **94**, 195 (1994)
[46] Stanton, J.F., Bartlett, R.J.: J. Chem. Phys. **98**, 7029 (1993)
[47] Morley, J.O.: J. Phys. Chem. **99**, 10166 (1995)
[48] Christiansen, O., Jørgensen, P., Hättig, C.: Int. J. Quantum Chem. **68**, 1 (1998)
[49] Grimme, S., Waletzke, M.: Phys. Chem. Chem. Phys. **2**, 2075 (2000)
[50] Kohn, A., Hättig, C.: J. Chem. Phys. **119**, 5021 (2003)
[51] Grimme, S., Izgorodina, E.I.: Chem. Phys. **305**, 223 (2004)
[52] Sekino, H., Bartlett, R.J.: J. Chem. Phys. **85**, 976 (1986)
[53] Karna, S.P., Dupuis, M.: J. Comp. Chem. **12**, 487 (1991)
[54] Quinet, O., Champagne, B.: Int. J. Quantum Chem. **85**, 463 (2001)
[55] Hättig, C., Hess, B.A.: J. Chem. Phys. **105**, 9948 (1996)
[56] Aiga, F., Itoh, R.: Chem. Phys. Lett. **251**, 372 (1996)
[57] Olsen, J., Jørgensen, P.: J. Chem. Phys. **82**, 3235 (1985)
[58] Luo, Y., Ågren, H., Jørgensen, P. et al.: Adv. Quantum Chem. **26**, 165 (1995)
[59] Champagne, B., Fischer, P., Buckingham, A.D.: Chemical Physics Letters **331**, 83 (2000)
[60] Botek, E., Champagne, B.: Appl. Phys. B **74**, 627 (2002)
[61] Zangwill, A.: J. Chem. Phys. **78**, 5926 (1983)
[62] van Gisbergen, S.J.A., Snijders, J.G., Baerends, E.J.: J. Chem. Phys. **109**, 10644 (1998)
[63] Aiga, F., Tada, T., Yoshimura, R.: J. Chem. Phys. **111**, 2878 (1999)
[64] Iwata, J.I., Yabana, K., Bertsch, G.F.: J. Chem. Phys. **115**, 8773 (2001)

[65] Heinze, H.H., Della Sala, F., Görling, A.: J. Chem. Phys. **116**, 9624 (2002)
[66] Salek, P., Vahtras, O., Helgaker, T. et al.: J. Chem. Phys. **117**, 9630 (2002)
[67] Tretiak, S., Chernyak, V.: J. Chem. Phys. **119**, 8809 (2003)
[68] Champagne, B., Perpete, E.A., Jacquemin, D. et al.: J. Phys. Chem. A **104**, 4755 (2000)
[69] van Faassen, M., de Boeij, P.L., van Leeuwen, R. et al.: J. Chem. Phys. **118**, 1044 (2003)
[70] Bulat, F., Toro-Labbe, A., Champagne, B. et al.: J. Chem. Phys. **123**, 014319 (2005)
[71] Fischer, P., Wiersma, D.S., Righini, R. et al.: Physical Review Letters **85**, 4253 (2000)
[72] Botek, E., Champagne, B., Turki, M. et al.: Journal Of Chemical Physics **120**, 2042 (2004)
[73] Champagne, B., Andre, J.M., Botek, E. et al.: ChemPhysChem **5**, 1438 (2004)
[74] Rentzepis, P.M., Giordmaine, J.A., Wecht, K.W.: Physical Review Letters **16**, 792 (1966)
[75] Giordmaine, J.A., Rentzepis, P.M.: J. Chim. Phys. **64**, 215 (1967)
[76] Jiang, H., Dolain, C., Léger, J.M. et al.: J. Am. Chem. Soc. **126**, 1034 (2004)
[77] Martin, R.H.: Angew. Chem., Int. Ed. Engl. **13**, 649 (1974)
[78] Verbiest, T., Van Elshocht, S., Kauranen, M. et al.: Science **282**, 913 (1998)
[79] Verbiest, T., Sioncke, S., Persoons, A. et al.: Angew. Chem. Int. Ed. **114**, 4038 (2002)
[80] Schmidt-Radde, R.H., Vollhardt, K.P.C.: J. Am. Chem. Soc. **114**, 9713 (1992)
[81] Han, S., Bond, A.D., Disch, R.L. et al.: Angew. Chem. Int. Ed. **41**, 3227 (2002)

CHAPTER 13

RECENT PROGRESS IN MOLECULAR DESIGN OF IONIC SECOND-ORDER NONLINEAR OPTICAL MATERIALS

PARESH CHANDRA RAY

Jackson State University, Department of Chemistry, 1400 J.R. Lynch Street, Jackson, MS, USA

Abstract: This chapter deals with recent and important developments in the field of the molecular design of ionic organic materials with and without metals for second-order nonlinear optics. The first section discusses 1) the origin of optical nonlinearity, 2) the relationship between microscopic and macroscopic polarizabilities and 3) the importance of ionic chromophores as second-order nonlinear optical (NLO) materials. The second section reviews 4) the current experimental and theoretical developments in the design of dipolar and octupolar ionic chromophores for second-order nonlinear optics and 5) the progress on zwitterionic second-order NLO materials. The third section presents 6) possible device applications based on ionic chromophores

Keywords: Nonlinear optical materials; Ionic octupolar; Zwitterionic; Ab initio; Stilbazolium; Aggregates

1. BACKGROUND

Photonics is playing an ever-increasing role in our modern information society. Photon is gradually replacing the electron, the elementary particle in electronics. Several books and reviews have appeared dealing with the theory of nonlinear optics and the structural characteristics and applications of nonlinear optical molecules and materials [1–18]. The earliest nonlinear optical (NLO) effect discovered was the electro-optic (EO) effect. The linear EO coefficient r_{ijk} defines the Pockel effect, discovered in 1906, while the quadratic (nonlinear) EO coefficient s_{ijk} relates to the Kerr effect, discovered 31 years later (1875). Truly, all-optical NLO effects were not discovered until the discovery of lasers. Second harmonic generation (SHG) was first observed in a single crystal of quartz by Franken et al. [1] in 1961. They frequency doubled the output of a ruby laser (694.3 nm) into the ultraviolet (347.15 nm) with a conversion efficiency of only about 10^{-4}% in their best experiments, but the ground had been broken. The early discoveries often originated in two- or multi-photon spectroscopic studies being conducted at that time. Parametric amplification was observed in lithium niobate ($LiNbO_3$) by two-wave mixing in temperature-tuned single crystals [2]. Rentzepis and Pao [3] made the first observation of SHG in an organic material, benzpyrene, in 1964. Heilmeir examined hexamethylenetetramine single crystal SHG in the same year [4]. Two other organic materials followed rapidly: hippuric acid and benzil [6]. Benzil was the first material that proved relatively easy to grow into large single crystals. Over the last three decades the study of nonlinear optical process in organic and polymer systems has enjoyed rapid and sustained growth. One indication of the growth is the increase in the number of articles published in refereed society journals. The four years period 1980–1983 saw the publication of 124 such articles. In the next four years period 1984–1987, the production of articles increased to 736 (nearly six times). From 1988–1992, the number of articles increased to more than 4000. In the last decade, academia, industry and government laboratories have been working in this field to replace electronics by photonics and as a result, the number of publications reached more than 50,000.

The rapid growth of the field is mainly due to the technological promise of these materials. Materials with high NLO activities are useful as EO switching elements for telecommunication and optical information processing. For communication, the electron as a carrier in a metallic conductor has been replaced by the photon in an optical fiber. Multi-wavelength optical communication increases the capacity on an optical network by orders of magnitudes over electronic communication. Traditionally, the materials used to measure second-order NLO behavior were inorganic crystals, such as lithium niobate ($LiNbO_3$) and potassium dihydrogen phosphate (KDP). The optical nonlinearity in these materials is to a large fraction caused by the nuclear displacement in an applied electric field, and to a smaller fraction by the movement of the electrons. This limits the bandwidth of the modulator. Organic materials have a number of advantages over inorganic materials for NLO applications, such as (i) their dielectric constants and refractive indices are much smaller, (ii) polarizabilities are purely electronic and therefore faster and (iii) molecules are compatible with the polymer matrix. The ease of modification of organic molecular structures makes it possible to synthesize tailor-made molecules and to fine-tune the properties for the desired application. In the case of second-order nonlinear optical processes, the macroscopic nonlinearity of the material (bulk susceptibility) is derived from the microscopic molecular nonlinearity and the geometrical arrangement of the NLO-chromophores.

So, optimizing the nonlinearity of a material begins at the molecular structural level, which requires a detailed understanding of the origin of an induced nonlinear polarization and its relationship to the molecular electronic structure. For this reason, much attention has been paid to the theoretical calculations of the nonlinear optical response by ab initio and semi-empirical methods, providing the chemist with the information that indicates which synthetic strategy should be followed [16–24].

Recent Progress in Molecular Design

In addition, the experimental data obtained from the NLO characterization of the chromophores can be used to verify and improve computational concepts. The theory of NLO has been described thoroughly by Chemla and Zyss [7, 8], Prasad and Williams [9] which will be shortly summarized.

2. THEORY OF NONLINEAR OPTICS

NLO is concerned with how the electromagnetic field of light waves interacts with the electromagnetic fields of matter and other light waves. The interaction of an electromagnetic field with matter induces a polarization in that matter. In linear optics, there is an instantaneous displacement (polarization) of the electron density of an atom by the electric field E of the light wave. The displacement of the electron density away from the nucleus results in a charge separation (an induced dipole) with a moment μ. For a weak field, the displacement of the charge from the equilibrium position (polarization) is proportional to the strength of the applied field.

(1)   Polarization(p) = αE

Thus the plot of polarization as a function of the applied field is a straight line whose slope is the linear polarizabilty, α, of the molecule or atom. If the field oscillates with a frequency, then the induced polarization will have the same frequency and phase, if the response is instantaneous. Most applications of experiments with NLO are carried out on bulk or macroscopic materials and in this case, the linear polarization can be defined as

(2)   P = χE

where χ is the linear susceptibility of a collection of molecules (on which the parameters of the dielectric constant and refractive index have a bearing).

When a molecule is subjected to laser light (that is—very high intensity electric field), its polarizability can change and be driven beyond the normal regime. Therefore, the polarization, which is a function of the applied field and leads to nonlinear effect, can be expressed as,

(3) $P = \alpha E + \beta E^2 + \gamma E^3 + \ldots\ldots\ldots$

where β is the first molecular hyperpolarizability (second-order effect) and γ is the second molecular hyperpolarizability (third-order effect). Typical values of β's are in the order of 10^{-30} esu (esu unit mean that the dimensions are in CGS units and the charge is in electrostatic units, thus "β in esu" means β in the units of $cm^3 esu^3/erg^2$), γ's are in the order of 10^{-36} esu and α's are in the order of 10^{-24} esu. With increasing field strengths, nonlinear effects become more important due to the higher powers of the field E. Since α is much greater than β and γ, NLO effects were not commonly observed before the invention of lasers. For the electric field of Q-switched YAG laser light, $\sim 10^4$ stat volts/cm, the contribution to P from βE^2

is 10^{-4} (D). These polarizations are infinitesimal on the scale of chemical thinking. Yet, these small polarizations are responsible for the exotic effects described throughout this chapter. For a macroscopic system this equation can be written as

(4) $$P = \chi^{(1)}E + \chi^{(2)}E^2 + \chi^{(3)}E^3 + \ldots\ldots$$

$\chi^{(1)}$ (a second-rank tensor) is the first-order susceptibility, $\chi^{(2)}$ (third-rank tensor) is the second-order nonlinear susceptibility, describing the action of two electric field vectors into a polarization, and $\chi^{(3)}$ (fourth-rank tensor) is the nonlinear susceptibility describing third-order processes. When equation (2) is written as

(5) $$P = \chi_{eff} E$$

The nonlinear index of refraction is related to the applied electric field, E, through the electric-field-dependent susceptibility, χ_{eff}, of a material,

(6) $$n^2 = \varepsilon = 1 + 4\pi\chi_{eff}$$

with ε being the dielectric constant of the material at optical frequencies. The induced polarization therefore results in a modulation of the refractive index of a material. If one of the acting fields is a static dc-field (E_0) and the other an optical field, $E(\omega)$, the polarization at the fundamental frequency (ω) will depend on the amplitude of the applied electric field. This is known as the linear EO effect (or Pockels effect), with its magnitude being proportional to $\chi(2)$. This is the NLO phenomenon responsible for optical switching [15, 16]. Other second-order NLO are sum- and difference-frequency generation (where two different fundamental wavelengths (ω_1, ω_2) interact with each other) and SHG in which the two interacting fields are of the same frequency. The manifestation of SHG can clearly be seen by substituting a sinusoidal field into the linear and first nonlinear term of the following equation:

(7) $$P = \chi^{(1)}E_0 \sin(\omega t) + \chi^{(2)}E_0^2 \sin^2(\omega t) + \ldots\ldots\ldots$$

Since $\sin^2(\omega t) = 1/2 - 1/2\cos(2\omega t)$,

(8) $$P = 1/2\chi^{(2)}E_0^2 + \chi^{(1)}E_0 \sin(\omega t) - 1/2\chi^{(2)}E_0^2 \cos(2\omega t) + \ldots\ldots$$

Equation (8) shows the presence of new frequency components in addition to having the fundamental frequency ω: a frequency-independent one (optical rectification), and the 2ω contribution.

3. SYMMETRY CONSIDERATIONS

The response of centrosymmetric molecules to an external field is given by $P(-E) = -P(E)$. This relation expresses the requirement that the induced polarization of centrosymmetric molecules is opposite and of equal magnitude when

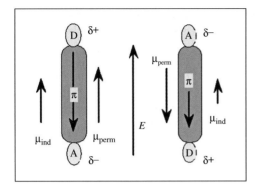

Figure 1. Electric-field-induced polarization from donor to acceptor and vice versa in a D-A molecule

the field is reversed. In order for the equation (3) to satisfy this condition, all coefficients of even powers of E ($\beta, \delta \ldots$) have to be equal to zero. Hence, only non-centrosymmetric molecules have a non-zero β value, since then $P(-E) \neq -P(E)$. This is illustrated in the Figure 1, where the polarization in para-disubstituted benzene is larger in the direction from the donor (D) to the acceptor (A) than in the opposite way, since electrons are more easily moved from an electron-rich (donor) to an electron-poor (acceptor) environment.

The requirement of non-centrosymmetry is not restricted to the molecular level, but also applies to the macroscopic nonlinear susceptibility, $\chi^{(2)}$, which means that the NLO molecules have to be organized in a non-centrosymmetric alignment. The first measurements of the macroscopic second-order susceptibility, $\chi^{(2)}$, have been performed on crystals without centrosymmetry [5]. However, many organic molecules crystallize in a centrosymmetric way. Other condensed oriented phases such as Langmuir-Blodgett (LB) films and poled polymers therefore seem to be the most promising bulk systems for NLO applications.

4. IMPORTANCE OF ORGANIC IONIC NLO MATERIALS

In the recent years, an intense worldwide effort has been focused on the research, design and development of new materials with large optical non-linearities due to their potential applications in various optical devices [7–16]. The EO effect is a second-order NLO effect. Not all organic materials display second-order NLO properties. At the molecular level, they need to be non-centrosymmetric. A large number of organic π-conjugated molecules have been investigated [7–16, 25–30] in the last twenty years. The outcome of the results has helped to establish certain guidelines for molecular design to get good second order NLO materials. However, roughly more than 80% of all π-conjugated organic molecules crystallize in centro-symmetric space groups, therefore producing materials with no second order bulk susceptibility $\chi^{(2)}$. To overcome this problem, ionic organic chromophores are considered to be an important class of materials for applications in second-order NLO. Due to their ionic interaction, they conquer the dipole attraction and in fact, ionic compounds exhibit excellent bulk SHG efficiency [7–16, 32–34]. The largest SHG efficiency to date is reported for a stilbazolium salt [32–34]. The main advantage of using ionic compounds for second-order NLO are (i) the alignment of the ionic chromophore into a polar structure can be controlled by changing the counter ion [32–34], and (ii) the dipolar interaction that provides a strong driving force for centerosymmetric crystallization in neutral compounds is countered by the columbic interaction, thus favoring non-centrosymmetric space groups with good SHG efficiency.

The field of ionic organics for NLO has been hampered by the lack of a widely applicable, simple and fast screening procedure for NLO applications. The technique that used to be used to determine how good an organic molecule was for second-order NLO, was the Electric-Filed–Induced Second Harmonic Generation (EFISHG) technique. Since a second-order effect can only be observed from a non-centrosymmetric bulk arrangement, an external electric dc field was applied over a solution of neutral candidate molecules with a dipole. Neutral dipolar molecules were the only candidates that were studied for prototype of an organic EO modulator. However, non-centrosymmetry does not automatically imply a dipolar molecule, or, more generally, vectorial properties. Molecules without dipole moment can also exhibit high second-order NLO properties. However, they cannot be oriented in an electric field, due to the absence of dipole moment and therefore EFISGH cannot be used for the measurement of their β values. Similarly, although ionic species have maximum SHG response in bulk, their β values cannot be measured by EFISHG technique, since ionic species will migrate, rather than rotate, under the influence of an electric field. The β values of these ionic chromophores had not been evaluated until the hyper-Rayleigh scattering (HRS) method was established. Second–order nonlinear light scattering [35–36] or the HRS technique was discovered soon after the availability of the pulsed ruby laser but only after the advent of reliable, electro-optically Q-switched Nd^{3+}-YAG laser, Clays et al. [36] reinvented this technique to measure the hyperpolarizabilities of molecules in solution. The technique is both experimentally and theoretically much simpler and more widely applicable, and has quickly became the technique of choice for the determination of the first hyperpolarizabilities of a wealth of newly designed and synthesized chromophores [36–64].

In this chapter, we will give an overview of recent major advances in the design of ionic NLO materials. An understanding of the electronic origin of molecular NLO response is of fundamental scientific interest as well as a crucial component in the development of state-of-the-art NLO materials. Quantum chemical chromophore structure-NLO response analysis permits researchers to identify the electronic structure signature characteristic of enhanced macroscopic NLO response, and ultimately to design molecular structure with potentially optimal NLO susceptibilities. Many theoretical papers [65–80] have addressed the NLO response of ionic organic molecules. Here, we will also discuss the overview of the recent literature

5. STILBAZOLIUM BASED DIPOLAR CHARGED ORGANIC COMPOUNDS

5.1 Effect of Counter Ions on SHG Properties

Ionic organic compounds have been known in the last two decades as an interesting class of NLO materials, especially for the development of single crystals [7–10, 32, 34] and LB films [7–9, 33, 81] with high NLO properties. Most attention has been devoted in this series on 4-N,N-dimethylamino-4'-N'-methyl–stilbazolium tosylate (DAST) and this is due to the fact that the largest SHG efficiency reported till now is from DAST salt and single crystals of different structures that can be grown simply by changing the counter ions. Marder et al. have shown [26] (data in Table 1) that the variation of the counter ions in the ionic stilbazolium salts lead to a material with the highest SHG efficiency reported till to date and it is about 1000 times that of the urea reference. The measurement has been performed using 1907 nm fundamental wavelength.

Their data suggest that dipolar ionic compounds show higher tendency to crystallize in non-centrosymmetric fashion than the corresponding dipolar covalent

Table 1. Powder SHG efficiencies for compounds of the form RCH=CHC5H4NCH3+X- (Reprinted with permission from (S.R. Marder, J.W. Perry and P.W. Schaefer, (1989) *Science*, 245, 626). Copyright [1989] AAAS

R	X$^-$			
	$CF_3SO_3^-$	BF_4^-	$p\text{-}CH_3C_6H_4SO_3^-$	Cl^-
$4\text{-}CH_3OC_6H_4\text{-}$	50	0	120	60
	54	0	100	270
$4\text{-}CH_3OC_6H_4\text{-}CH=CH\text{-}$	0.0	1.0	28	48
	0.0	2.2	50	4.3
$4\text{-}CH_3SC_6H_4\text{-}$	0.0	0.0		0.0
	0.0	0.0	1.0	0.0
$2,4\text{-}(CH_3)_2C_6H_3\text{-}$	40	5.5	0.0	0.4
	67	2.9	0.08	0.7
$C_{10}H_8\text{-}(\text{pyrenyl})$	0.8		37	
	1.1		14	
$4\text{-}(CH_2CH_2CH_2CH_2N)C_6H_4\text{-}$	0.5	5.2	0.2	1.1
	0.06	0.05	0.03	0.0
$4\text{-}BrC_6H_4\text{-}$	0.0	0.0	1.7	22
	0.0	0.02	5.0	100
$4\text{-}(CH_3)_2NC_6H_4\text{-}$	0.0	75	1000	0.0
	0.0		15	0.0
$4\text{-}(CH_3)_2NC_6H_4\text{-}CH=CH\text{-}$	~500	350	115	0.0
	5.0	4.2	5	0.0

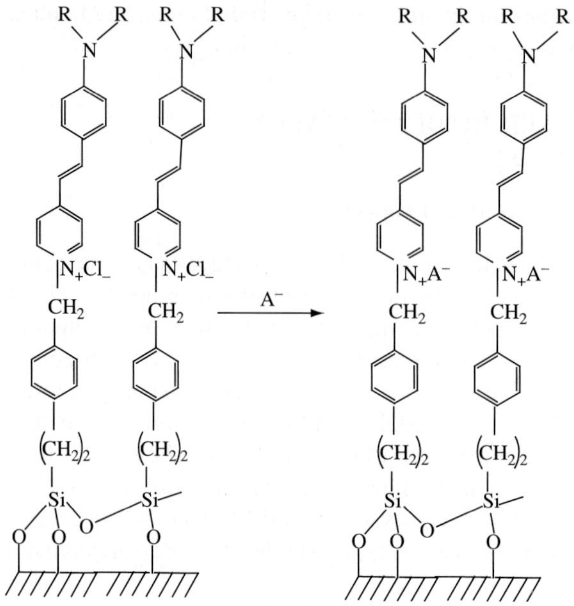

Figure 2. Ion Exchange scheme in LB films containing stilbazolium chromophores. (Reprinted with permission from (S. Di-Bella, I. Fragala, M.A. Ratner and T.J. Marks, (1995) *Chem. Mater.*, 7, 400). Copyright (1995) American Chemical Soceity.)

compounds. It also has been demonstrated that the counter ion can play a significant role in crystallographic packing architecture. Roscoe et al. [82] has demonstrated that the counter ions of these stilbazolium salts can be exchanged using LB films using the following scheme. Polarized SHG measurement at 1064 nm indicates that the second–order NLO response from counter ion varies from 34% to 44% as moved from I^- to ethyl orange.

These results indicate that the local environment of the chromophore or chromophore cation-anion interactions play an important role in the bulk SHG properties of stilbazolium derivatives. The relative NLO contributions due to packing architecture and the role of anion identity in bulk SHG response cannot be well understood from the above experimental data. To overcome this problem, Di Bella et al. [80] have investigated theoretical studies on the dependence of NLO response in chromophoric ion pairs by changing the nature and packing arrangement of the constituent species, as shown in Figure 2. They investigated theoretically the effect of anion environment on the second-order NLO properties of stilbazolium ion using (intermediate neglect of differential overlap package of Zerner) ZINDO/SOS (Sum-Over-States) method. Recently we have shown [59] that the calculated dimmer hyperpolarizabilities depend strongly on relative molecular orientation. We have found that this increment in hyperpolarizability is due to increase in the oscillator strength and the change in the dipole moment. We have used DFT(density functional theory)/3-21G/SCRF(self-consistent reaction field) scheme to calculate the structure of dimers of compound at different configuration (eclipsed, slipped and tilted) as shown in Figure 3. This suggests a powerful new technique of arranging donor and acceptor groups in a manner that significantly enhances the NLO properties of a system. Figure 4a shows how the β value varies with the interplanar separation between two monomers. At interplanar distance

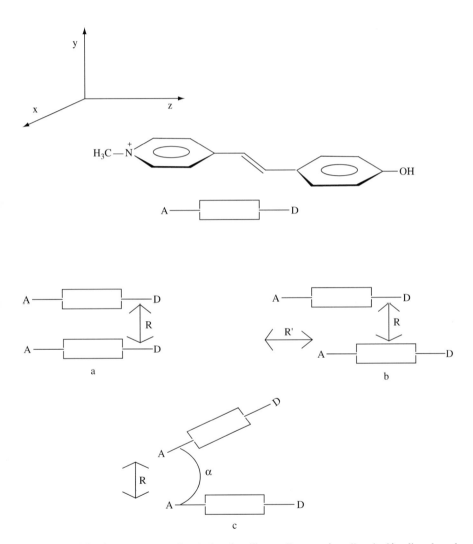

Figure 3. Molecular geometry of substituted stilbene dimmer a) eclipsed, b) slipped and c) tilted conformation. (Figure has been adapted from reference 59) (Reprinted with permission from (Z. Sainudeen and P.C. Ray, (2005) *Inter. J. Quan. Chem*, 1054, 348. Copyright (2005) Wiley InterScience.)

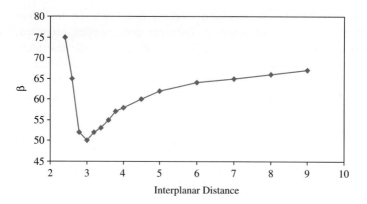

Figure 4a. Plot of β vs. interplanar distance (Å). (Reprinted with permission from (Z. Sainudeen and P.C. Ray, (2005) *Inter. J. Quan. Chem*, 1054, 348. Copyright (2005) Wiley InterScience.)

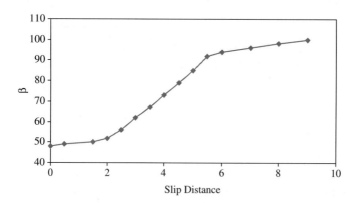

Figure 4b. Plot of β vs. slip distance (Å). (Reprinted with permission from (Z. Sainudeen and P.C. Ray, (2005) *Inter. J. Quan. Chem*, 1054, 348. Copyright (2005) Wiley InterScience.)

near 3 Å, the van der Waals interaction in typical crystal, the plot reaches a flat minimum and then increases asymptotically. In case of slipped conformer (b), different behavior has been noted (shown in Figure 4b). In this arrangement, β remains unchanged till 2 Å and then increased monotonically with more separation. An interesting result (Figure 4c) has been found in case of tilted conformer (c). After a slit initial increase for smaller angle, β values fall of for larger angles and finally reached to zero at 180°, due to fully coplanar centrosymmetric structure.

The greatest increases in βzzz values are associated with the displacement (R′) of the anion from the pyridinium acceptor. To understand the origin of this increment,

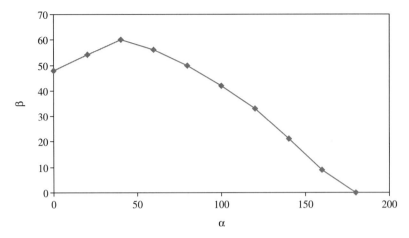

Figure 4c. Plot of β vs. tilt angle (in degree). (Reprinted with permission from (Z. Sainudeen and P.C. Ray, (2005) *Inter. J. Quan. Chem.*, 1054, 348. Copyright (2005) Wiley InterScience.)

they calculated the change in dipole moment between ground and charge-transfer excited state ($\Delta\mu_{10}$). According to the two-state model [82],

$$(9) \quad \beta^{\text{two state}} = \underbrace{\frac{3\mu_{eg}^2 \Delta\mu_{eg}}{E_{eg}^2}}_{\text{static factor}} \underbrace{\frac{\omega_{eg}^2}{(1-4\omega^2/\omega_{eg}^2)(\omega_{eg}^2 - \omega^2)}}_{\text{dispersion factor}} \cdots\cdots$$

where μ_{eg} is the transition dipole moment between the ground state $|g\rangle$ and the charge-transfer excited state $|e\rangle$, $\Delta\mu_{eg}$ is the difference in dipole moment and E_{10} is the transition energy. Their calculations indicate that $\Delta\mu_{10}$ values increase with the displacement of R' due to the reduced ground state dipole moment, μ_z along the CT axis and E_{eg} decreases due to the reduced energy gap between the relevant MOs (molecular orbital). Largest βzzz has been observed at R' ≥ 4.25 A⁰. These results indicate that the large hyperpolarizability and hence $\chi^{(2)}_{zzz}$ enhancement can be understood on the basis of plausible anion-cation packing configurations.

5.2 First Hyperpolarizabilities of Stilbazolium Ions

The molecular hyperpolarizability of charged ionic compounds was not directly accessible for measurement until the HRS method was reestablished [36]. Therefore only recently research has been focused on the effect of donor or acceptor substitution and elongation of the conjugation path length to demonstrate the engineering guidelines for enhancing molecular optical nonlinearities.

5.2.1 Effect of donor-acceptor strength

Duan et al. [67–69] investigated first hyperpolarizability of stilbazolium cation with different donors as shown in Figure 5 and their measured values are shown in Table 2.

The experimental β^0 values for stilbazolium cations are obtained using two-state model and from the experimental β values obtained from the literature [67–69], which was measured at a fundamental wavelength of 1064 nm light from Nd:YAG laser. The trends in static hyperpolarizabilities with the change of substituents follow the Hammett Parameter Constant (σ_p) as we have seen before for neutral molecules.

To compare the experimental measurements with the theoretical findings [77], we have performed the geometry optimization as well as hyperpolarizability calculations with solvent dielectric constant for methanol ($\varepsilon = 33$) using the SCRF [83] approaches as implemented in Gaussian Package [84]. β^0's calculated by DFT/6-31G**/SCRF methods are listed above. They are in quite good agreement with experimental values. The little discrepancy between experimental and theoretical values can be due to neglect of other effects in computation such

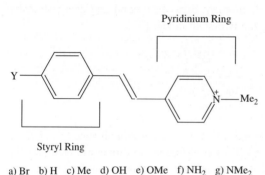

a) Br   b) H   c) Me   d) OH   e) OMe   f) NH_2 g) NMe_2

Figure 5. Donor-acceptor substituted stilbazolium ions used for this calculation

Table 2. Experimental (Exp) and theoretical static first hyperpolarizability (in 10^{-30} esu) values of substituted stibazolium cation

Compound No.	Y	$\beta^0_{ab\text{-initio}}$	$\beta^0_{ab\text{-initio}}$ (solvent)	β^0_{Exp}
5a	Br	79	128	160
5b	H	63	92	100
5c	Me	91	114	130
5d	OH	107	150	160
5e	OMe	128	164	150
5f	NH_2	190	270	
5g	NMe_2	218	295	370

as hydrogen bonding between solute and solvent specifically when the solvent is methanol. Another possibility can be due to resonance contributions to measured β's. This is definitely true for compound 5f, where the molecule absorbs quite significantly at 532 nm. Since the measurement has been done at 1064 nm and molecules have absorption at 532 nm, the resonance contribution to the HRS signal cannot be neglected. To understand the origin of the very high β's for stilbazolium cations, we calculated change in dipole moment between ground and first excited state ($\Delta\mu_{eg}$), oscillator strengths (f) and transition energies (E_{eg}) values for N, N-dimethyl amino substituted stilbazolium cation and corresponding nitro stilbene using ZINDO package employing correction vector (CV) method [18, 23–24]. We also included solvent effects through a SCRF [83]. According to the two-state model, the calculated static β^0's is proportional to the product of f, $\Delta\mu_{eg}$, and E_{eg}^{-3}. Calculated f and E_{eg} values are about the same for both compounds, but the $\Delta\mu_{eg}$ is about 3.02 times higher for stilbazolium cation than stilbene compound, and this could be the main factor for about 4.2 times higher β^0 for the stilbazolium cation.

5.2.2 Influence of the conjugation length

Elongation of the conjugation pathway is one of the primary design steps for increasing β values of neutral organic molecules and several studies have been performed in this direction [7–16]. Only very recently, investigation has been started for first hyperpolarizability of ionic molecules in that direction [37–41]. Clays et al. [37] have reported highly unusual effects on the influence of the conjugation length in first nonlinearity of extended conjugated stilbazolium ions as shown in Figure 6.

They have extended the conjugation from n = 1 to 5. Their experimental values show that the β value is at maximum at n = 3 for ionic N-aryl substituted stilbazolium chromophores. β value increases very sharply from n = 1 to 2 to 3 (100 to 1640 to 2045) and then decreases with the increment of chain length. Hence it is apparent that the design criteria for ionic NLO chromophore can diverge dramatically from those of purely organic molecules. To understand this unusual behavior, we performed quantum mechanical calculation for the same molecules using the BL3YP/6-31G**/SCRF scheme and chloroform solvent to calculate the structure of

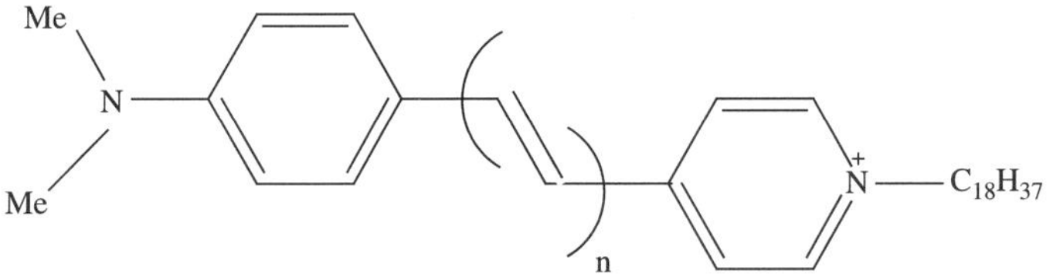

Figure 6. Structure of all-trans extended conjugated stilbazolium ions used in their investigation. (Reprinted from Chemical Physics Letter, P.C. Ray, (2004) *Chem. Phys. Lett.*, 394, 354, Copyright (2004), with permission from Elsevier)

all the hemicyanine dyes with the increment of number of double bonds from 2 to 7. The optimized geometry of the isolated all-trans molecule is almost linear. Frequency analysis indicates that all the structures are minima on the potential energy surface. To compare the theoretical β values with experimental data, we have computed the first hyperpolarizabilities for all the compounds with elongation of double bonds from n = 1 to 7, using the ZINDO/CV/SCRF method. Since molecules with n = 3 and above have strong absorption at 650 nm as reported by Clays et al. [37], we have used photon energy corresponding to 1907 nm excitation source for ZINDO calculation, though the experiment has been performed at 1300 nm. It is interesting to note that the experimental β values increase very sharply from n = 1 to 2 to 3 (100 to 1640 to 2045) and then decrease with the increment of chain length as shown in the Figure 7.

However, the theoretical values increase with the elongation of the chain length untill n = 7. To understand the origin of high β_{vec} hemicyanine dyes and the difference in experimental and theoretical findings, we have calculated λ_{max} and the change in dipole moment between ground and charge-transfer excited state ($\Delta\mu_{10}$) as shown in the Table 3 below.

Both $\Delta\mu_{10}$ and λ_{max} increase with n for all the compounds till n = 7. So the unusual effect observed by experimental studies cannot be explained using the

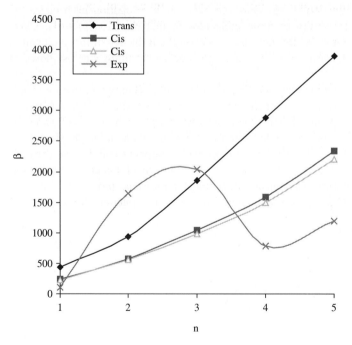

Figure 7. Plot of β vs. n for trans, cis and experimental values for extended conjugated stilbazolium. (Reprinted from Chemical Physics Letter, P.C. Ray, (2004) *Chem. Phys. Lett.*, 394, 354, Copyright (2004), with permission from Elsevier)

Table 3. Theoretical dynamic first hyperpolarizabilities (in 10^{-30} esu), λ max (nm) and $\Delta\mu_{10}$ values of the compound 1a and corresponding experimental* values for different conjugation lengths from n = 1 to 7. (Reprinted from Chemical Physics Letter, P.C. Ray, (2004) *Chem. Phys. Lett.*, 394, 354, Copyright (2004), with permission from Elsevier)

n	1	2	3	4	5	6	7
λ_{max} (exp)	496	524	546	556	570	–	–
λ_{max} (theo)	462	501	527	541	554	565	573
$\Delta\mu_{10}$ (theo)	18.8	24.6	28.5	31.2	36.6	40.7	44
β (theo)	360	740	1320	2090	2980	3690	4200
β (exp)	100	1640	2045	780	1200	–	–

(*) Experimental values have been taken from, *J. Opt. Soc. Am. B.* 17, (2000), 256.

two-state model. This can be due to several other factors as we discussed in our recent publication [77] and these are:

i) Cis-Trans isomerization in highly conjugated structures;
ii) Aggregation of ionic compounds in nonaqueous solvents like chloroform;
iii) Degradation of compounds during the measurements.

In the next session, we will discuss thoroughly each of the above-mentioned aspects.

To perform the first hyperpolarizability calculation for cis configuration, we have adapted two different strategies, cis conformation through single bonds and through double bonds as shown in Figure 8.

We have optimized the geometry using ab initio methods as we did for compound 5a-g in this chapter and then calculated their β using ZINDO/CV/SCRF methods. The plot of experimental and theoretical β in cis and trans form with n is shown in the Figure 7. First hyperpolarizabilities of cis conformers are about (50–60)% of trans conformers with equivalent chain lengths and the trends in β values with chain length are the same for both conformers. Even in the cis form, the theoretical β values are much higher than the experimentally measured values, whereas the

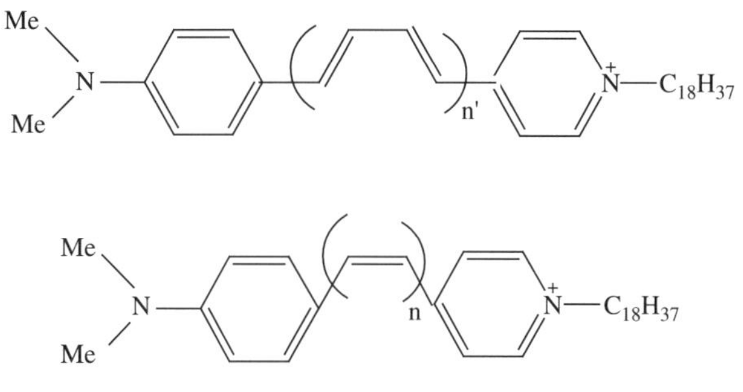

Figure 8. Different cis forms of extended conjugated stilbazolium ions have used for this. (Reprinted from Chemical Physics Letter, P.C. Ray, (2004) *Chem. Phys. Lett.*, 394, 354, Copyright (2004), with permission from Elsevier)

experimental values for n=2 and n=3 are much higher than the theoretical finding. Again, the fluorescence intensity is maximum for n=3 and the fluorescence intensity pattern follows closely the pattern of β values for n=1 to n=5. So it is hard to believe that the cis-trans isomerization is alone responsible for the observed discrepancies. A detailed non-resonant Raman spectroscopic investigation is necessary to resolve this question.

It is known in the literature [45, 63, 73, 86] that the HRS intensity increases with aggregation. We have shown [63] previously that the HRS intensity increases tremendously as melamine forms supra molecular aggregates with cyanuric acid. Clays et al. [45] also noted the same phenomena for bacteriorhodopsin trimer. Recently, Gross et al. has shown that [73] the aggregation is very common for ionic and zwitterionic molecules and it can affect the measured β values tremendously. The larger dipole will also favor dimerization for longer chromophores and it can form a centrosymmetric dimer in the solution that can alter β values as discussed by Gross et al. [73]. However, it is not clear why the β values are so high only for n=2 and n=3. Clays et al. [37] have measured the absorption spectra before and after the HRS measurement and since the values are the same, we can neglect the degradation of the compounds during measurements.

It is known that resonance effect can enhance the β values by an appreciable amount. Also due to the noncoherent nature, one cannot separate the multi-photon emission signal from the HRS signal. Though Clays et al. have used femtosecond HRS technique to separate the HRS signal from multi-photon fluorescence signal, the fluorescence intensity pattern follows closely the pattern of β values for n=1 to n=5. Since all the molecules have strong absorption at second harmonic light of the fundamental, one cannot neglect the resonance contribution to the observed experimental values. Only measurement at 1907 nm fundamental light can resolve the situation. Recently Zyss et al. [87] have measured the first hyperpolarizability of conjugated stilbazolium dyes (see Figure 9) using EFISHG technique at 1907 nm.

They have used donor groups and acceptor groups similar to the one used by Clays et al. [37]. We have calculated static first hyperpolarizabilities of the same compound using the TD-DFT (Time-dependent density functional theory), SOS-FF (sum-over-states finite field) method and compared our theoretical values of $\mu\beta^{(0)}$ with the experimental findings as shown in Figure 10. Interestingly, the EFISHG results indicate that the $\mu\beta^{(0)}$ value increases with the increment of conjugation and the trends of their experimental findings match very well with our theoretical results. Since the measurement has been done using 1907 nm, we can neglect the resonance contribution to their experimental values. Due to the coherent nature, EFISHG

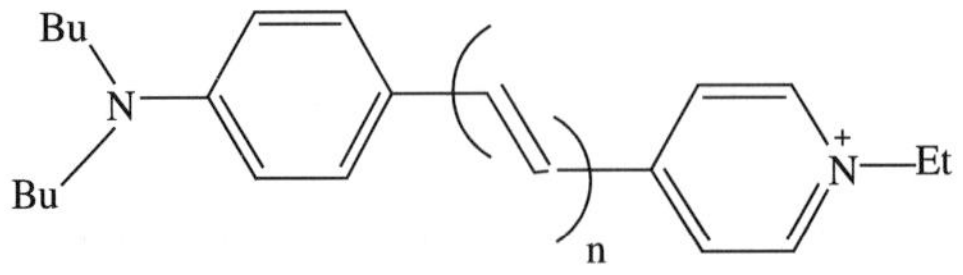

Figure 9. All trans extended conjugated stilbazolium ions used in their measurements

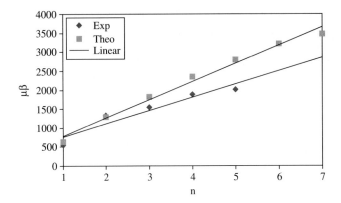

Figure 10. Experimental and theoretical μβ values for compounds shown in Figure 9

signal cannot mix with fluorescence signal. Therefore, multi-photon emission and resonance effect are possible main contributions to the observed experimental trend.

Coe et al. has synthesized and measured first hyperpolarizabilities of a series of pyridinium and stylbazolium salts [38–40]. Since he has discussed his work in detail in his chapter, I will not focus on that here.

6. IONIC ORGANOMETALLIC IONIC NLO CHROMOPHORES

Loucif et al. [88] have investigated the influence of complexation of ferrocenyl derivatives on the second-order hyperpolarizabilities β. They performed their β value measurements using dc EFISHG technique at 1.34 and 1.9 μm. Significantly increased β values were observed for these new bimetallic ferrocenyl derivatives. Their best β value (123.5×10^{-30} esu at 1.34 μm) is comparable to the highest reported values for organometallic complexes. The nature of the second metal ion has a weak influence on the β value, in consequence to the change in geometry of the associated complex. Laidlaw et al. [30] have reported the first hyperpolarizability, β, of a bimetallic complex ion, $[(CN)_5Ru-\mu-CN-Ru(NH_3)_5]^-$ and a novel organometallic analog, $[(\mu5-C_5H_5)Ru(PPh_3)_2-\mu-CN-Ru(NH_3)_5]^{3+}$. Their measured β values at the wavelength of 1064 nm using the HRS technique was greater than 10^{-27} esu which are among the largest reported for the solution species.

Pal et al. [89] have reported on substituted ferrocenyl compounds, where one of the cyclopentadienyl rings is linked to an aromatic Schiff base, that were synthesized and analyzed for their second-order nonlinearity (β). Their results indicate that the metal to ligand charge transfer (MLCT) transition dominates their second-order response. These compounds form charge transfer (CT) complexes with acceptors such as I_2, p-chloranil (CA), 2,3-dichloro-5,6-dicyano-1,4-benzoquinone (DDQ), tetracyanoethylene (TCNE), and 7,7,8,8-tetracyanoquinodimethane (TCNQ). The CT complexes exhibit much higher second-order response. Bisferrocenyl complexes where two ferrocene moieties are linked through the same aromatic Schiff base spacer were also synthesized and characterized by Pal et al. [89]. The β values of the bisferrocenyl complexes and their CT counterparts are much higher than the corresponding monoferrocene complexes.

Wostyn et al. [61] have reported the molecular nonlinear optical polarizability of lanthanate complexes containing stilbazolium ions. Their experimental results indicate that the hyperpolarizability is independent of the nature of the lanthanide, though the complex anion size is a function of the size of the ligand on the lanthanide cation. Andreu et al. [55] have synthesized a new chiral cyanine dye, 4′-[2-(methoxymethyl)pyrrolidinyl]-1-methylstilbazolium iodide (MPMS+I). They have reported that MPMS+I exhibits phase-matched SHG with the efficiency of up to 80 times that of urea.

Recently Coe et al. [39–40] have reported highly unusual effects on the influence of the conjugation length on the first nonlinearity of ionic organometallic chromophores based on stilbazolium ions. Their experimental values show that the first hyperpolarizability is maximum for n=2 in ionic organometallic complex with Ruthenium (II) amine donor and N-methyl pyridinum as an acceptor. To understand the origin of the unusual effect, we have computed and reported [77] the first hyperpolarizabilities for all the compounds with elongation of double bonds from n=1 to 5, using the ZINDO/CV/SCRF method in the acetonitrile solvent. For Ruthenium complexes (as shown in Figure 11), we have taken the mainframe structure from the crystal structure available in the literature [39–40].

Since these molecules have strong MLCT absorption around 550 nm as reported by Benjamin et al. [39–40] we have used photon energy corresponding to 1907 nm as the excitation source for ZINDO calculation though the experiment has been performed at 1064 nm. In case of ruthenium complexes, we observed unusual effects

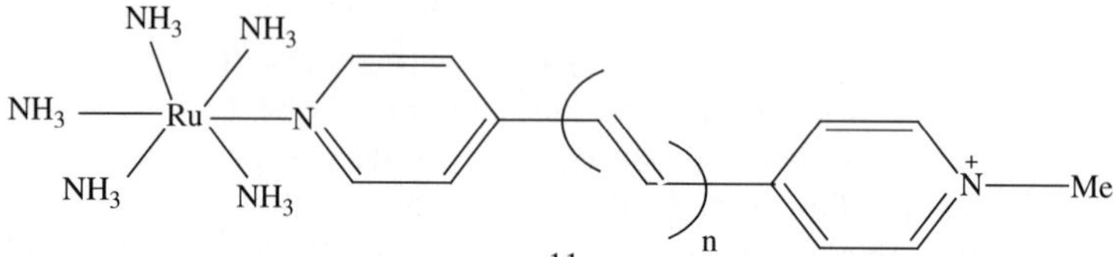

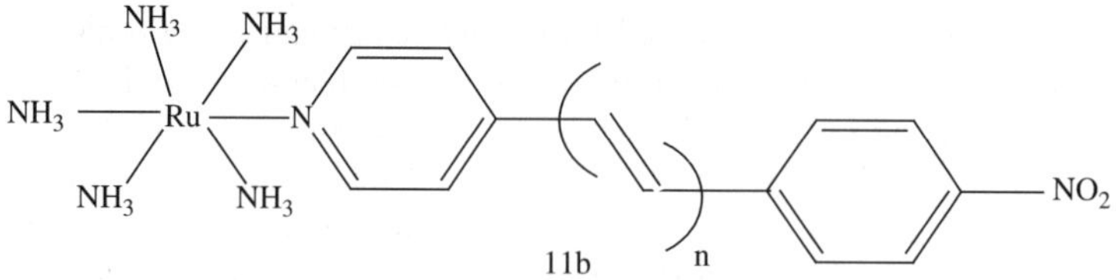

Figure 11. Extended conjugated, all trans. ruthenium complex of stilbazolium cations have used in our calculation. (Reprinted from Chemical Physics Letter, P.C. Ray, (2004) *Chem. Phys. Lett.*, 394, 354, Copyright (2004), with permission from Elsevier)

of π-conjugation extension on the first nonlinearity. Table 4 show the experimental and theoretical βs with n for the compound 11b. It is interesting to note that the trend of our theoretical result [77] matches very well with the experimental finding [39–40]. To understand this unusual effect, we have calculated λ_{max} for the MLCT transition and change in dipole moment between ground and charge-transfer excited state ($\Delta\mu_{10}$) that are listed in Table 4. One can note that the trends in β and $\Delta\mu_{10}$ with n are same for both compounds and also λ_{max} decreases with n. So the two-state model can interpret our results. Since both the pyridinium rings are primarily acceptors at both ends, it is not really surprising to see the above results. To find out if the above statement is correct or not, we have computed the first

Table 4. Theoretical dynamic first hyperpolarizabilities (in 10^{-30} esu), λ_{max} (nm) and $\Delta\mu_{10}$ values of the compound 1b and the corresponding experimental* values for different conjugation lengths from n = 0 to 4 (Reprinted from Chemical Physics Letter, P.C. Ray, (2004) *Chem. Phys. Lett.*, 394, 354, Copyright (2004), with permission from Elsevier)

n	0	1	2	3	4
λ_{max} (exp)	590	595	584	568	–
λ_{max} (theo)	568	574	560	545	530
$\Delta\mu_{10}$ (theo)	17.6	24.5	27.6	18.6	16.2
β (theo)	240	440	760	345	220
β (exp)	750	828	2593	1308	–

(*) Experimental values have been taken from, *J. Am. Chem. Soc.*, 126, (2004), 2004.

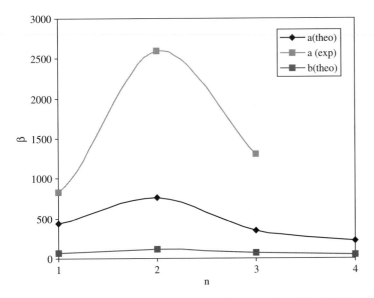

Figure 12. β vs. n for compound 11a and 11b. (Reprinted from Chemical Physics Letter, P.C. Ray, (2004) *Chem. Phys. Lett.*, 394, 354, Copyright (2004), with permission from Elsevier)

hyperpolarizabilities of compound 11b in the acetonitrile solvent. The difference between compound 11a and 11b is the presence of $-NO_2$ group as an acceptor instead of pyridinium ring. Plots of experimental β's [16] for the compound 11a, theoretical β's for the compound 11a and 11b with n is shown in Figure 12. One can note that though the pyridinium ring has been substituted by $-NO_2$ group in compound c, the trends in β values with n is almost the same. The only difference is that the βs are lower for the compound 11a, than the corresponding compound 11b. This can be due to the fact that the pyridinium ring is much stronger acceptor than $-NO_2$.

7. FIRST HYPERPOLARIZABILITIES OF RETINAL SCHIFF BASES

Hendrickx et al. [41–42] have reported the first hyperpolarizabilities of retinal, retinal Schiff base and retinal protonated Schiff base at 1064 nm excitation wavelength. Retinal protonated Schiff base is responsible for the linear and NLO properties of bacteriorhodopsin protein. Their measured hyperpolarizabilities are 3600×10^{-30} esu for retinal protonated Schiff base and 470×10^{-30} esu for retinal Schiff base. They also investigated theoretical understanding of the first hyperpolarizabilities of retinal derivatives. Results are shown in Table 5.

Schmalzlin et al. [90] have used the HRS technique to determine the molecular first hyperpolarizabilities β values of retinal Schiff base in its protonated and unprotonated form. Results of their HRS measurements performed at 1064, 1300 and 1500 nm were reported. The derived hyperpolarizabilities are self-consistent with the two-state model for all three wavelengths, but they are an order of magnitude lower than those reported by Hendrickx et al.

Table 5. Experimental and theoretical hyperpolarizabilities (in 10^{-30} esu) of various retinal derivatives. (Reprinted with permission from (E. Hendrickx, K. Clays, A. Persoons, C. Dehu and J.L. Bredas, (1995) J. Am. Chem. Soc., 117, 3547). Copyright (1995) American Chemical Society

Compound	$\hbar\omega_{eg}$(eV) MeOH (Exp)[a]	$\hbar\omega_{eg}$(eV) INDO (Theretical)	β 1064 nm MeOH (Exp)	β_ν^0 INDO/SOS (Theoretical)	β_ν^0 two-state INDO/SOS (Theoretical)	β^0 (Exp)
Vitamin A acetate	3.80	3.78	140	2.4	22.9	80
Retinoic acid	3.53	3.52	310	38.0	114.4	160
Retinal Schiff base	3.40	3.60	470	17.9	70.0	220
Retinal	3.26	3.50	730	41.5	120.5	300
RPSB	2.79	2.42	3600	214.6	388.6	900

[a] The measurements are performed in methanol. The columns refer to the first optical transition energy, $\hbar\omega_{eg}$ the experimental β(2ω; ω = 1064 nm) values, the calculated static β_ν^0 values (obtained over 40 states and using the two-state model), the experimental extrapolated static β^0 values.

8. FIRST HYPERPOLARIZABILITIES OF IONIC OCTUPOLAR MOLECULES

Zyss et al. [16, 30, 51] recognized in early nineties that this inherent conflict between dipole minimization and molecular hyperpolarizability is not essential. It could be ultimately resolved by enlarging the pool of candidate molecules to encompass non-centrosymmetric systems known as octopuses with symmetry ensured cancellation of their dipole moment as well as of any other physical property behaving like a vector under symmetry operations. A common way to design second-order NLO-active octupolar molecules is to develop non-centrosymmetrically substituted trigonal or tetrahedral π-conjugated systems that display efficient CT from the periphery to the center of the molecule. The presence of a 3-fold symmetry axis in octupolar 1,3,5 substituted aromatic ring systems can lead to better transparency characteristics [16, 30, 48–54] and the lack of a molecular dipole can enhance the prospects of non-centrosymmetric crystal packing [16]. Crystal violet (tris(p-(dimethylamino)phenyl)methyl ion or CV) is one of the intensely studied as the prototype octupolar molecules for NLO. It is cationic chromophore (as shown in Figure 13), which exhibited considerable NLO properties, which is comparable to traditional dipolar NLO materials.

Crystal violet, a trigonal conjugated cationic dye with electronic CT from peripheral dimethylamino donor groups to an electron deficient sp^2 hybridized central carbon atom. A resonantly enhanced β value of 580×10^{-30} esu has been reported at 1.064 nm, in acetone solution ($\lambda_{max} = 590$ nm) by Zyss et al. [16] The 1st hyperpolarizability, β, of crystal violet dye was measured at 1450 and 1500 nm by Rao et al. [91]. The resonance-free β value, $\beta^{(0)}$ for this octupole is comparable with that of the dipolar dye Disperse Red 1 but with the nonlinearity-transparency trade-off worse for the octupole. Symmetric cyanine dyes of the kind Me_2N^+:CH-(CH:CH)n-NMe$_2$ normally exhibit no first hyperpolarizabilities but have relatively long absorption

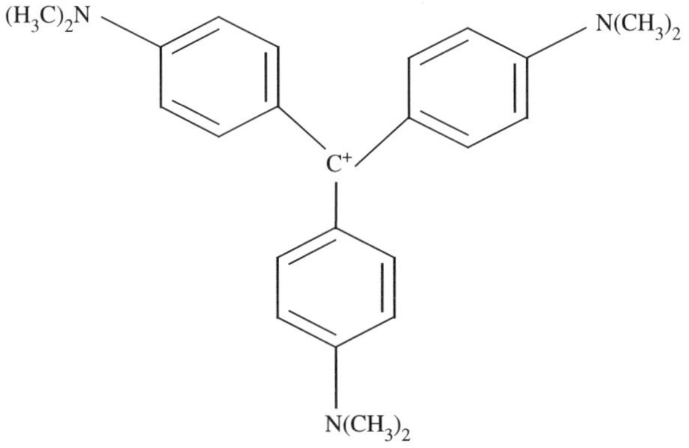

Figure 13. Structure of crystal violet

wavelengths. Therefore they are not suitable NLO chromophores for second-order applications. However, if these chromophores are converted into octupolar molecules either by coupling an extra aldiminium group onto the dye molecules or by grafting three vinamidinium units onto a benzene ring, NLO chromophores with considerable first hyperpolarizabilities are reported by Stadler et al. [52]. To prove this, the β values at 1064 nm, of both kinds of octupoles in solution have been determined via the HRS technique. We have already shown [53] experimentally that the symmetrically substituted triazines have larger first hyperpolarizabilities than their corresponding benzene analogues. Cho et al. [48] have reported 1,3,5 tricyano-2,4,6-tris(vinyl) benzene derivatives with very large second-order NLO properties. Recently [79] we have reported first hyperpolarizabilities of the ionic octupolar NLO systems based on 1,3,5 substituted aromatic rings as shown in Figure 14.

Due to their ionic interaction, they conquer the dipole attraction and in fact, ionic dipolar compounds exhibit excellent bulk SHG efficiency [7–10]. Dynamic hyperpolarizabilities of 1,3,5 substituted benzenes and triazines cations calculated by ZINDO/CV/SCRF method are listed in Table 6. Experimental data [92] for the corresponding neutral octupolar molecules (as shown in Figure 15) are also listed in Table 6.

The comparison the magnitudes of the molecular first-order hyperpolarizabilities obtained from the HRS studies [35–36] does not require the computation of β projected onto dipole moment (μ), since the orientation averaged value is the

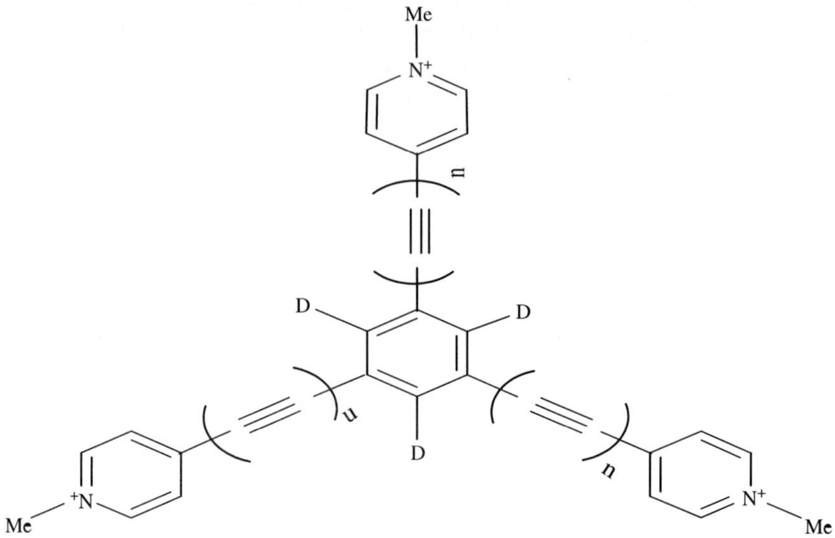

Figure 14. Structure of ionic extended conjugated octupolar compounds used in this investigation. (Reprinted from Chemical Physics Letter, P.C. Ray and J. Leszczynski, (2004) *Chem. Phys. Lett.*, 399, 162, Copyright (2004), with permission from Elsevier)

Table 6. Theoretical (ZINDO/CV/SCRF) dynamic first hyperpolarizabilities (in 10^{-30} esu) for compound 1 and 2 with different donor-acceptors. The conversion factor for β values are, 1×10^{-30} esu $= 371.1 \times 10^{-53}$ C^3 m^3 J^{-2} = 115.74 aus. (Reprinted from Chemical Physics Letter, P.C. Ray and J. Leszczynski, (2004) *Chem. Phys. Lett.*, 399, 162, Copyright (2004), with permission from Elsevier

Donor	$<\beta_{zzz}^{ionic}>$	$<\beta_{zzz}^{2a}>$	$<\beta_{zzz}^{2b}>$	$<\beta_{800}^{2b}>$	$<\beta_{exp}^{2b*}>$
NMe$_2$	210	75	40	48	—
NH$_2$	180	60	32	38	—
OMe	130	40	20	24	34
Me	80	26	11	14	18
F	89	28	13	16	19

Experimental data has been taken from G. Hendrickx, I Asselberghs, K. Clays and A. Persoons, *J. Org. Chem*, 69 (2004) 5077.

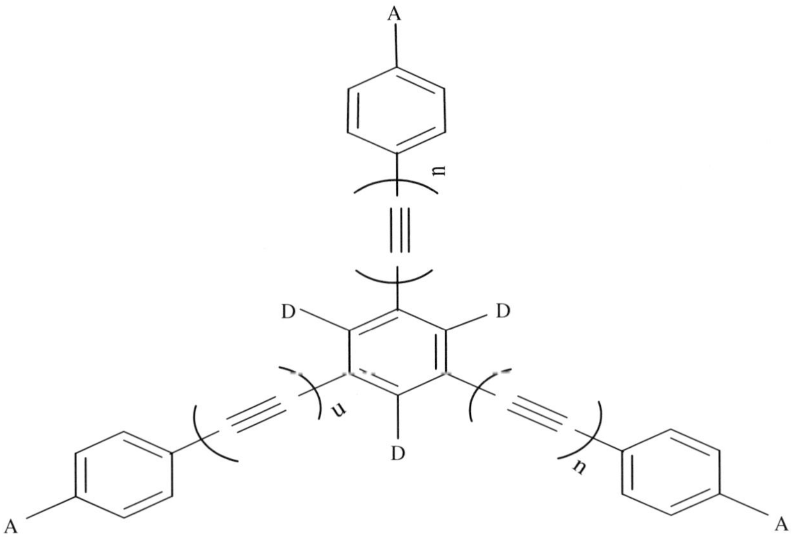

Figure 15. Structure of neutral extended conjugated octupolar compounds used in this calculation; a) A = -NO2; b) A = -H. (Reprinted from Chemical Physics Letter, P.C. Ray and J. Leszczynski, (2004) *Chem. Phys. Lett.*, 399, 162, Copyright (2004), with permission from Elsevier)

relevant parameter and is evaluated directly. In the HRS experiment [32], one measures average $<\beta^2>$ for any molecule, where,

(10) $\quad <\beta_{HRS}^2> = <\beta_{ZZZ}^2> + <\beta_{XZZ}^2>\ldots\ldots\ldots$

For a octupolar molecule with C$_3$ or D$_3$ or D$_{3h}$ symmetry,

(11) $\quad <\beta_{ZZZ}^2> = 24/105\ \beta_{ZZZ}^2 \ldots\ldots\ldots\ldots$

and

(12) $<\beta^2_{XZZ}> = 16/105\beta^2_{ZZZ}$..........

Though in the HRS expression, an isotropic average is made for molecule's all the β tensor components, indicating that the HRS is sensitive to all such contributions. But for octupolar molecules, only $<\beta_{zzz}>$ tensor contributes to the total β.

The trends in hyperpolarizabilities with the change of substituents follow the Hammett Parameter Constant (σ_p) as we can see from Figure 16.

It is interesting to note that β_{zzz}'s for ionic octupolar chromophores are 2–3 times higher than that of the corresponding neutral molecules. Three-level model can approximate the quadratic hyperpolarizability of a D3h symmetric molecule. The energy gap between the ground and two-fold degenerate excited state monotonically decreases and the transition dipole matrix elements monotonically increase with the increase of Hammett parameter constant for donor and as we move from neutral to ionic system for the same donor groups. Though the trends in experimental and theoretical β values for the compound 15b is same, the experimental β_{zzz}'s are always higher than our theoretical values and it can be due to two factors: i) the experimental values were measured at a incident wavelength of 800 nm light and calculation has been performed at 1907 nm. To compare with the experimental results perfectly, we also calculated β's using 800 nm excitation source as listed in Table 1. One can note that the β values calculated using 800 nm are higher than the corresponding β values calculated using 1907 nm and it is obvious according to the three-state model. ii) Since these molecules have some absorption at 400 nm, the experimental signal can have a significant contribution from two-photon fluorescence as discussed by several publications recently.

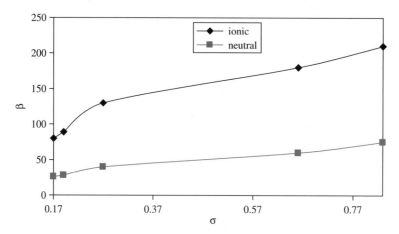

Figure 16. Plot of β_{zzz} vs. σ for compound 14 and 15. (Reprinted from Chemical Physics Letter, P.C. Ray and J. Leszczynski, (2004) *Chem. Phys. Lett.*, 399, 162, Copyright (2004), with permission from Elsevier)

Elongation of the conjugation pathway is one of the primary design steps for increasing β values of neutral and ionic dipolar organic molecules and several studies have been performed in this direction [7–16]. But to the best of our knowledge, there are no reports on the relevant studies for the octupolar molecules in this direction. We have used DFT/6-31G**/PCM (Polarizable Continuum Model) scheme using $CHCl_3$ solvent to optimized the structure of compound 14 and 15a with NMe_2 donor and ZINDO/CV/SCRF scheme to calculate their dynamic β values with increment of number of double bonds from 2 to 5. The optimized geometry of all the molecules is characterized by D_3 symmetry. Harmonic vibrational frequency analysis indicates that all the structures are minima on the potential energy surface. Figure 17 shows how the first hyperpolarizabilities increase with the elongation of chain lengths.

Our data indicate a significant increase in the first hyperpolarizability on elongation on the conjugation pathway. β increases 1.6 times as one increase the number of conjugated double bonds from 1 to 2. β values can be fitted with the function of $β_{zzz} \propto N^m$, where n is the number of conjugated π bonds. A linear dependence of $\log(β_{zzz})$ vs. log (n) is also observed for compounds 14 and 15, as shown in Figure 18. The exponent m provides insight into the chain length dependence of $β_{zzz}$. Though the trends in $β_{zzz}$'s for both series are same, always for a given number of π bonds (n values), the static first hyperpolarizabilities for ionic octupoles are much higher than corresponding neutral octuples and the slope is higher for ionic octupolar chromophores.

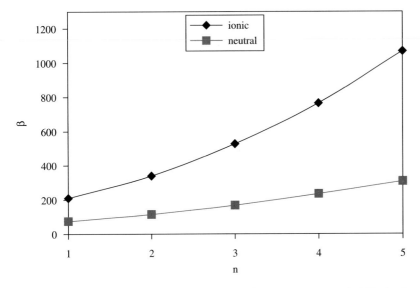

Figure 17. Plot of $β_{zzz}$ vs. n for compound 14 and 15a. (Reprinted from Chemical Physics Letter, P.C. Ray and J. Leszczynski, (2004) *Chem. Phys. Lett.*, 399, 162, Copyright (2004), with permission from Elsevier)

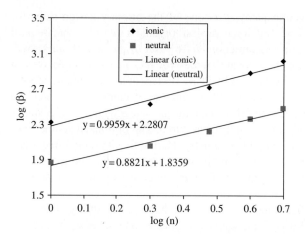

Figure 18. Plot of $\log(\beta_{zzz})$ vs. log (n) for compound 14 and 15. (Reprinted from Chemical Physics Letter, P.C. Ray and J. Leszczynski, (2004) *Chem. Phys. Lett.*, 399, 162, Copyright (2004), with permission from Elsevier)

9. ZWITTERIONIC DYES BASED ON PYRIDINIUM OR STILBAZOLIUM RING

Zyss et al. [93] have reported the synthesis and bulk SHG properties of a new class of stilbazolium derivatives in which the anionic and cationic parts are liked together by a n-alkyl chain. These molecules are inner salts in the zwitterionic families as shown in Figure 19. These zwitterionic molecules have a large dipole moment, which is independent of the intermolecular CT.

Among them, the most active compounds in SHG were obtained from sulfonato-propyl derivatives as shown in Table 7.

D = Donating group
X^- = Anion
R = Alkyl chain

D = Donating group
n = 2,3,4

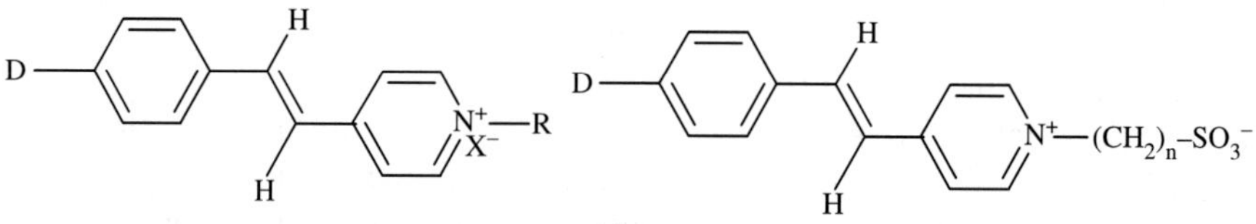

Figure 19. Structure of zwitterionic stilbazolium derivatives: 1) 4-substituted 4′-(N-alkyl)-stilbazolium, 2) 4-substituted 4′-(N-sulfonatoalkyl)-stilbazolium. (Reprinted with permission from (C. Serbutoviez, J.F. Nicould, I. Ledoux and J. Zyss, (1994) *Chem., Mater.*, 6, 1358). Copyright (1994) American Chemical Soceity.)

Table 7. Powder SHG values obtained from zwitterionic 4-substituted 4'-(N-sulfonatopropyl)-stilbazolium derivatives, where + denotes a signal comparable to or a few times greater than that of urea. ++, +++, and ++++ refer to 1,2, or 3 orders of magnitude greater signals respectively and − denotes no eye detectable SHG signal (Reprinted with permission from (C. Serbutoviez, J.F. Nicould, I. Ledoux and J. Zyss, (1994) Chem. Mater., 6, 1358). Copyright (1994) American Chemical Soceity.)

Compound 2	λ (SHG)	n = 2	n = 3	n = 4
			10, MTSPS	
D = CH$_3$S	1.06 μm	++	++++	−
			phase I: 2.4 × NPP	
			phase II: 1.3 × NPP	
			phase III: inactive	
	1.34 μm	++	++++	−
	1.06 μm	/	**11, BSPS**	/
D = nC$_4$H$_9$			++	
	1.34 μm	/	/	/
D = H			**12, SPS**	
	1.06 μm	−	++	−
			phase I: inactive	
			phase II: 15 × urea	
			phase III: inactive	
		−		−
	1.34 μm		++	

4-Methylthio-4'-(3-sulfonatopropyl) stilbazolium derivatives exhibits two highly SHG-active monohydrated crystalline phase and their measured SHG values were 2.4 and 1.3 times of NPP at 1.06 μm excitation.

9.1 Molecular Hyperpolarizabilities

Betaine analogues that are zwitterionic in the ground state, containing two oppositely charged heteroaromatic rings linked directly or through a vinyl unit have been considered as second order NLO chromophore theoretically and experimentally by Abe et al. [70–72] Lambert et al. [94] have studied a new family of zwitterionic NLO chromophores in which a polyene bridge is capped by phenyl rings substituted by NR_3^+ in one end and at the other end by BR_3^-. This molecular design leads to (i) a very high degree of ground state polarization (high dipole moment) and (ii) higher transparency in the visible wavelength region with respect to push-pull diphenylpolyenes, presumably due to a shorter conjugation length, from which the saturated groups can be considered to be at least partially excluded. Victor et al. [95] presented a quantum-chemical analysis of the molecular structure and second-and third-order polarizabilities in a series of promising NLO chromophores (as shown in Figure 20), the zwitterionic ammonio/borato diphenylpolyenes, $R_3N^+Ph(C=C)_nPhB^-R_3$, whose experimental results have been reported by Lambert et al. [94] The origin of the remarkable NLO response of these zwitterionic molecules was elucidated with the help of two complementary theoretical frameworks.

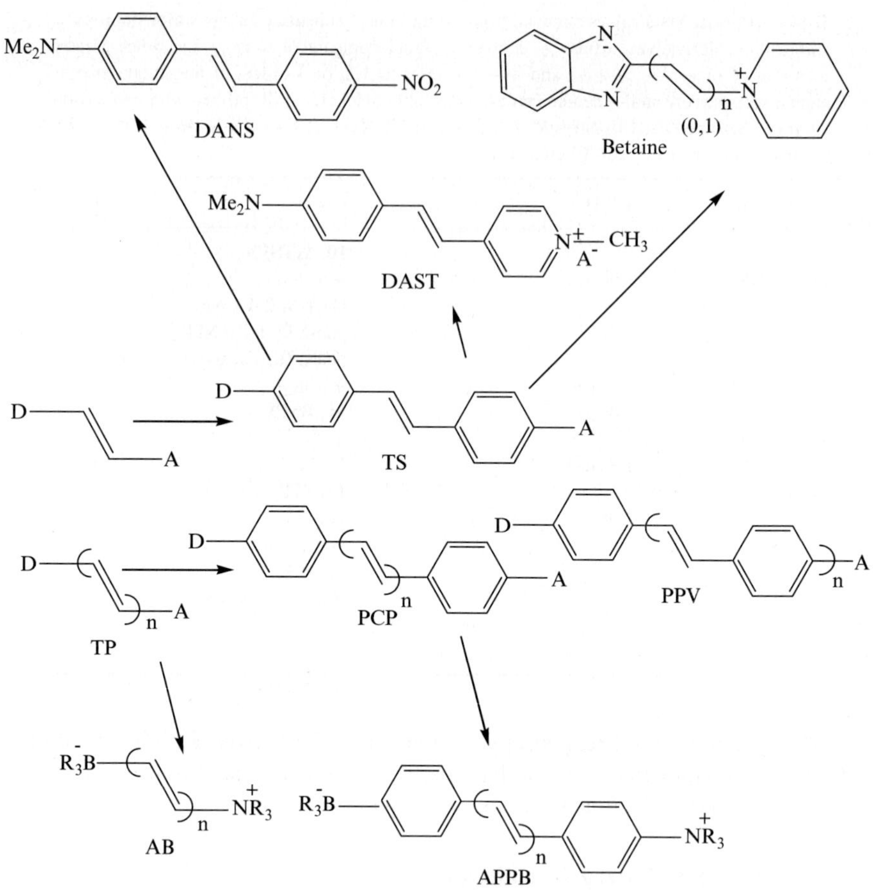

Figure 20. Structures of different aliphatic and aromatic zwitterionic molecules. (Reprinted with permission from (V.M. Geskin, C. Lambert and J.L. Bredas, (2003) *J. Am. Chem. Soc.*, 125 15651). Copyright (2003) American Chemical Soceity.)

Real-space finite-field results directly point to the most NLO-active segments in the molecules: these are primarily phenylene groups. The sum-over-states analysis highlights the essential role of a single excitation channel in the zwitterionic series. It consists of the ground state to lowest excited state channel, corresponding to the HOMO → LUMO excitation. This transition involves an inter-phenylene electron density transfer; again underlining the critical role of the phenylene rings in generating the NLO response. Bartkowiak et al. have repoted the solvent effect on the first hyperpolarizabilities [96] and two photon absorption cross sections [97] of zwitterionic betaine dyes. Szablewski et al. [86] have reported the synthesis and NLO properties of a series of novel zwitterionic chromophores. Their measured second-order nonlinearity using the HRS technique for one series of compound predicted $\mu\beta^{(0)} = 9500 \times 10^{-48}$ esu, one of the highest values reported till now in

Recent Progress in Molecular Design 411

the literature. Zwitterionic merocyanine dyes are very interesting especially for EO applications due to the large nonlinear optical susceptibilities and the good alignment of the chromophores in the co-crystal. Mericyanine dyes can switch from zwitterionic to quinoid character in ground state (Figure 21) with solvent polarity as shown below.

The merocyanines achieve this zwitterionic state in polar solvents largely because there is an increase in aromaticity to be gained from the charge separation. In order to

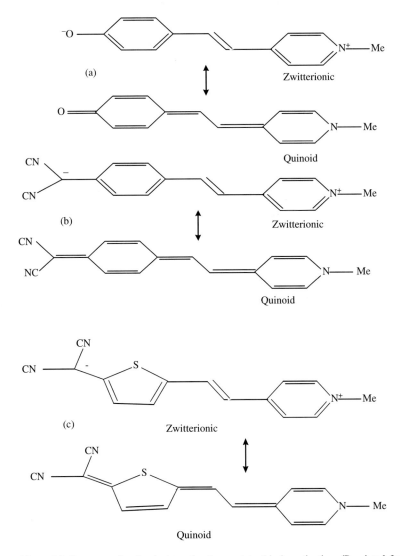

Figure 21. Structure of zwitterionic molecules used in this investigation. (Reprinted from Chemical Physics Letter, P.C. Ray, (2004) *Chem. Phys. Lett.*, 395, 269, Copyright (2004), with permission from Elsevier)

understand the effect of solvent-solute interactions on the NLO properties of the zwitterionic molecules, we have performed [78] a systematic study of the first hyperpolarizabilities of a series of merocyanine dyes (Figure 21) using TD-DFT calculations with fairly extensive basis set (6-31G**). TD-DFT performs the calculations of frequency-dependent response properties like electronic excitations and frequency dependent hyperpolarizabilities [21]. To understand how the solvent polarity affects the structure and NLO properties of the zwitterionic dyes, we have used SCRF approach with PCM [19], as implemented in Gaussian 03 [84].

To investigate how this structural change in different solvents affects β values, we have calculated the first hyperpolarizabilities of all the molecules using ZINDO/CV/SCRF method using different solvent parameters and 1907 nm as excitation source. Computed β values are plotted with the solvent parameter as shown in Figure 22.

Our calculation shows a remarkable solvent effect on the first hyperpolarizabilities of zwitterionic molecules. First order NLO responses are low and positive in the gas phase and then increases slowly with solvent polarity. Then it started to decrease with the solvent polarity. The β values remain negative in all the polar solvents and pauses highest values at a moderate ε (6–8) and then again decrease slowly with the increase in ε. This behavior is mainly due to the change of the structure from quinoid to zwitterionic form. Once it is in mostly zwitterionic form, β values decrease with the increase of solvent polarity. Recently similar behavior has been observed experimentally by Cross et al [73] and Abbotto et al.

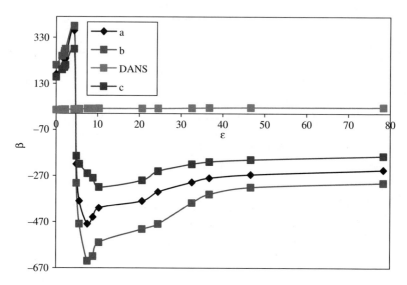

Figure 22. Plot of β vs. n for zwitterionic and neutral (DANS) dyes. (Reprinted from Chemical Physics Letter, P.C. Ray, (2004) *Chem. Phys. Lett.*, 395, 269, Copyright (2004), with permission from Elsevier)

Recent Progress in Molecular Design 413

[66, 74] for zwitterionic molecules. Changing the solvent from low to high dielectric causes not only an increase in magnitude of β but also a change in sign, therefore passing through zero at intermediate dielectric. There are clearly important consequences from this, in choice of solvent and molecular environment when evaluating NLO molecules. An excellent NLO response in solution might vanish when the active chromophore is dispersed in a matrix with suitable ε. The same chromophore can prove to perform very well under appropriately chosen solvent. The commonly established procedure for the NLO compound to report β values in one solvent may in certain cases be insufficient to draw definite conclusions on the overall chromophore performance and the prospect for different design strategies. To compare this result with the corresponding neutral and ionic molecules, we also computed β values for DANS in different solvents and their values plotted in Figure 22. First of all, the first nonlinearity is much lower for neutral molecules than the corresponding zwitterionic molecules. For neutral molecule, first hyperpolarizability is always positive and increases as we move from gas phase to solvent phase and keeps increasing with the solvent polarity. This phenomenon for neutral molecule is very common and the same trend has been reported in the literature [7–16].

To understand the origin of the remarkable solvent effects in zwitterionic molecules, we calculated the change in dipole moment between ground and charge-transfer excited state ($\Delta\mu_{10}$). Figure 23 shows the plot of the $\Delta\mu_{10}$ vs. ε for different solvents. It is interesting to note that the trend in the variation of $\Delta\mu_{10}$ and β with ε is quite similar, which follows two-state model perfectly. This confirms

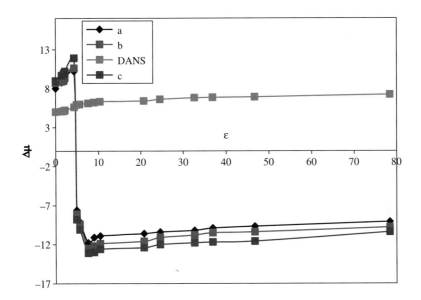

Figure 23. Plot of β vs. ε for zwitterionic and neutral dyes. (Reprinted from Chemical Physics Letter, P.C. Ray, (2004) *Chem. Phys. Lett.*, 395, 269, Copyright (2004), with permission from Elsevier)

that the $\Delta\mu_{10}$ could be the main factor for the remarkable solvent effect on the first hyperpolarizabilities of the zwitterionic molecules. We have also calculated the $\Delta\mu_{10}$ for DANS molecule in different solvent and plotted in the same Figure 23. For DANS molecule, $\Delta\mu_{10}$ is always positive and increases slowly with the variation of ε. As a result, β is positive for DANS molecule and increase slowly with solvent parameter.

10. DEVICE APPLICATIONS

DAST crystals are very interesting especially for EO applications due to the exceptionally large NLO susceptibilities and good alignment of the chromophores in the crystals. Pan et al. [98–102], have found that DAST crystal exhibits very pronounced NLO effect, with EO coefficients, e.g. $r_{111} = 47\,\text{pm/V}$ at 1535 nm. Their results indicate that DAST is a very interesting material for EO applications in the near infrared. Their measured EO coefficients in the spectral range of 700 to 1535 nm are shown in Table 8.

Meier et al. [98] have shown that DAST crystal is also a very interesting material for phase-matched parametric oscillation around the telecommunication wavelength at $\lambda = 1318$ and 1542 nm. Their results (as shown in Table 9) indicate that the second-order NLO coefficient d111 = 1010 pm/V at 1318 nm for DAST crystal.

They have shown that there exist phase-matching configurations for frequency doubling with large effective NLO coefficients at telecommunication wavelength, which is interesting for applications such as the cascading of the second-order nonlinearities and optical parametric oscillation. Thakur et al. [102] have reported (as shown in Figure 24) high EO modulation using single-crystal film of stilbazolium salts with light propagating perpendicular to the film.

Table 8. Linear EO coefficients r_{ijk} (pm/V), reduced half-wave voltages v_σ^{ijk} (kV), dielectric constants e_f, and linear polarization optical coefficients, f_{ijk} (m^2/C) of DAST (Reprinted with permission from (F. Pan, G. Knopfle, Ch. Bosshard, S. Follonier, R. Spreiter, M.S. Wong and P. Gunter, (1996) *Appl. Phys. Lett.*, 69, 13). Copyright (1996), American Institute of Physics)

	$\lambda = 1535$ nm	$\lambda = 1313$ nm	$\lambda = 800$ nm	$\lambda = 720$ nm
r_{111}	47 ± 8	53 ± 6	77 ± 8	92 ± 9
v_σ^{111}	3.5	2.4	0.79	0.53
r_{221}	21 ± 4	25 ± 3	42 ± 4	60 ± 6
v_σ^{221}	17	12	4.0	2.3
r_{333}	<0.1	<0.1	<0.1	0.8 ± 0.2
r_{311}	<0.1	<0.1	<0.1	0.7 ± 0.2
r_{223}	<0.1	<0.1	<0.1	<0.1
e_j @ 10^3–10^5 KHz:		$e_1 = 5.2\pm0.4$,	$e_2 = 4.1\pm0.4$,	$e_3 = 3.0\pm0.3$
f_{ijk} @ $\lambda = 800$ nm:		$f[1] = 2.1\pm0.3$,	$f_{221} = 1.5\pm0.3$,	$f_{113} = 0.85\pm0.13$

DAST exhibited large EO coefficients with low dielectric constants, e.g. $\varepsilon = 5.2$.

CHAPTER 14

CHARACTERIZATION TECHNIQUES OF NONLINEAR OPTICAL MATERIALS
An introduction to experimental nonlinear optical techniques

INGE ASSELBERGHS, JAVIER PÉREZ-MORENO AND KOEN CLAYS
Department of Chemistry, University of Leuven, Celestijnenlaan 200D, Leuven, Belgium

Abstract: Different techniques to characterize the strength of the second- and third-order nonlinear optical response are presented, with particular emphasis on the relationship between the macroscopic measurable quantities and the intrinsic molecular nonlinear properties

Keywords: Characterization techniques, Second-order nonlinear optical response, Third-order nonlinear optical response, Molecular hyperpolarizabilities, Nonlinear susceptibilities, Intensity-dependent refractive index

1. INTRODUCTION

Nonlinear optics (NLO) is the study of the interaction of intense light beams with matter, where the response of the media is nonlinear in the applied optical field. Nonlinear optics was born only four decades ago, when the first observation of second-harmonic generation was reported [1]. Despite its late start, nonlinear optics has undergone a very rapid growth, a growth that has benefited both the scientific and the technological world. Scientifically, with the development of nonlinear optics, the understanding of the essential interaction of light and matter has improved. Technologically, nonlinear optical processes are the basic tools of optoelectronic and photonic applications. In essence, optoelectronics and photonic applications are the equivalent to electronic applications, where the role of the electrons to store, manage and transmit information is played by photons.

Some of the relevant applications of nonlinear optics are currently used in laser technology and fiber communications, such as optical frequency conversion, optical parametric oscillation and amplification, the linear electrooptic effect (Pockels effect) and optical phase conjugation. Other applications, such as optical bistability and optical solitons are candidates to become the basic elements of optical computing and long-distance fiber communications.

The use of optical frequency conversion and optical parametric oscillation allows the generation of new frequencies from a source frequency [2, 3]. Second- and third-harmonic generation are particular cases of optical frequency conversion, where respectively, the original frequency is doubled or tripled. Optical frequency conversion and parametric oscillation devices are commonly used in laser technology [4–6].

The linear electrooptic effect is the change in the index of refraction of a medium due to the presence of a dc or low-frequency electric field, in such a manner that the change in the index of refraction depends linearly in the strength of the low-frequency electric field. The linear electrooptic effect is the mechanism behind optical intensity modulators that are used in optical switching and fiber-optics communications, where the optical signal is modulated at high frequencies (out to 110 GHz) [7–9].

Optical phase conjugation allows the reversal of the phase of the electromagnetic wave associated with a beam of light. It has applications in real-time adaptive optics where it is used to correct aberrations, in image processing and in optical computing [10–12].

The realization of stable optical solitons will have a significant impact in long distance optical communications and information technology. A stable optical soliton is able to propagate through an optical fiber without pulse broadening or dispersion (a feature that cannot be avoided in the linear regime) which minimizes information losses [13–15].

Optical bistability is a nonlinear process where for the same optical input intensity in a material, there are two possible output intensities and it is a consequence of nonlinear saturable absorption. Optical bistability allows the design of optical logical circuits and will play a fundamental role in the development of optical communication and optical computing [16–18].

The implementation of nonlinear optical applications depends strongly on the discovery and development of new nonlinear materials. Although in principle, all materials might show nonlinear optical effects, the response is generally too small to be of any use. Furthermore, in order to observe even-order nonlinear optical effects, the material has to be noncentrosymmetric at both the molecular and the bulk level. Therefore, the design of optoelectronic and photonic devices relies heavily in the development of new nonlinear optical materials.

Following the description of Prasad and Williams [19], nonlinear materials are classified in two categories: molecular materials and bulk materials.

Molecular materials are basically molecular units that are bonded chemically and that interact in the bulk through van der Waals interactions. In these materials, the optical nonlinearity is produced as a consequence of the molecular structure. Examples of this type of materials are organic crystals and polymers. Because the nonlinear response of the material is due to the nonlinear response of the molecular sites, it is possible to define microscopic nonlinear coefficients. The microscopic nonlinear coefficients are called the molecular hyperpolarizabilities and are the molecular equivalent of the bulk nonlinear susceptibilities. In this manner, the bulk susceptibility is related to the molecular susceptibilities by simple addition principles.

Since in molecular materials the nonlinear effect occurs fundamentally at the molecular level, it is possible to assemble new molecules by substituting groups of donors or acceptors to optimize the nonlinear optical response. The optimization of optical nonlinearities at the molecular level constitutes a field that incorporates elements of chemistry, physics, polymer science and material science, known as organic molecular engineering.

In contrast, the nonlinearities in bulk materials are due to the response of electrons not associated with individual sites, as it occurs in metals or semiconductors. In these materials, the nonlinear response is caused by effects of band structure or other mechanisms that are determined by the electronic response of the bulk medium. The first nonlinear materials that were applied successfully in the fabrication of passive and active photonic devices were in fact ferroelectric inorganic crystals, such as the potassium dihydrogen phosphate (KDP) crystal or the lithium niobate ($LiNbO_3$) [20–22]. In the present, potassium dihydrogen phosphate crystal is broadly used as a laser frequency doubler, while the lithium niobate is the main material for optical electrooptic modulators that operate in the near-infrared spectral range. Another ferroelectric inorganic crystal, barium titanate ($BaTiO_3$), is currently used in phase-conjugation applications [23].

New bulk materials have arisen recently, such as the quantum well structures derived from GaAs and II-VI semiconductors. The introduction of these new inorganic semiconductor materials has recovered the interest in inorganic materials. However, the engineering of efficient multiple well structures is not a trivial task. The design of such structures is very expensive especially in comparison with organic materials. Also, while in organic materials, optimization is achieved fundamentally at the molecular level, the nonlinear optical optimization of quantum well structures needs more elaborated approaches (such as band theory). For these reasons, inorganic semiconductor materials are expected to complement organic materials rather than compete with them in the design of photonic applications.

With respect to the traditional inorganic materials, organic materials are relatively new, and yet, they have become the leading systems for fabrication of optoelectronic and photonic devices. The main advantage of organic materials is the availability of optimization at the molecular level. Compared to inorganic materials they are relatively cheap to produce. Also, organic materials are easy to grow, manipulate and incorporate into different structures. Finally, another advantage of organic materials is their low dielectric constants, especially in comparison with inorganic crystals. Low dielectric constants are required in order to design efficient electrooptic devices, leading to a large operating bandwidth modulation (up to $> 10\,GHz$).

The origin of the nonlinearities at the microscopic level of organic structures is the existence of the delocalized π electrons of the molecules, which are loosely bound to the positive nuclear sites and have orbitals that can extend over the entire molecule. The existence of this unique chemical π bonding results in the largest nonabsorptive optical nonlinearities. Since in general, absorption and heat dissipation limits the response time, organic materials are preferred because in the nonabsorptive regime, the delocalized π electrons provide the fastest type of response.

The rest of this chapter is organized as follows: A brief section introduces the basic definitions needed to describe the nonlinear optical response, with some remarks about the different available notations. Next, a general overview of the experimental techniques used to characterize the nonlinear response is developed. The overview starts with second-order nonlinear characterization, where the Second-Harmonic Generation, Electric Field Induced Second-Harmonic Generation and Hyper-Rayleigh Scattering techniques are described. Also, general experimental techniques that are also applied to characterize the third-order nonlinear response such as the Maker Fringe technique or the Wedge-shaped technique are introduced. Finally, the basic elements of some characterization techniques of the third-order nonlinear response are presented: Third-Harmonic Generation, Electric-Field-Induced Second-Harmonic Generation, Degenerate Four-Wave Mixing and Optical Phase Conjugation, Optical Kerr-Gate, Self-focusing methods and Nonlinear Fabry-Perot methods.

2. NOTATION AND DEFINITIONS

2.1 The Electromagnetic Description

The generation of new frequency components from the original frequencies of the incident radiation field can be understood by the use of the Maxwell equations. In Gaussian units, the Maxwell equations can be written as [24]:

(1)
$$\nabla \times \vec{E} = -\frac{1}{c}\frac{\partial \vec{B}}{\partial t}$$

$$\nabla \times \vec{B} = \frac{1}{c}\frac{\partial \vec{E}}{\partial t} + \frac{4\pi}{c}\vec{J}$$

$$\nabla \cdot \vec{E} = 4\pi\rho$$

$$\nabla \cdot \vec{B} = 0$$

where $\vec{E} \equiv \vec{E}(\vec{r}, t)$ is the electric field, $\vec{B} \equiv \vec{B}(\vec{r}, t)$ is the magnetic field, $\vec{J} \equiv \vec{J}(\vec{r}, t)$ is the current density, $\rho \equiv \rho(\vec{r}, t)$ is the charge density and c is the speed of light.

The current density and the charge density are usually expanded in a series of multipoles:

(2)
$$\vec{J} = \vec{J}_0 + \frac{\partial \vec{P}}{\partial t} + c\nabla \times \vec{M} + \frac{\partial}{\partial t}(\nabla \cdot \vec{Q}) + \ldots$$

$$\rho = \rho_0 - \nabla \cdot \vec{P} - \nabla(\nabla \cdot \vec{Q}) + \ldots$$

where $\vec{P}$ is the electric dipole polarization, $\vec{Q}$ the electric quadrupole polarization and $\vec{M}$ the dipole magnetization. For many applications of nonlinear optics, the dipole magnetization and higher-order multipoles can be ignored. This is known as the electric dipole approximation. In this regime, the Maxwell equations become:

$$(3) \quad \nabla \times \vec{E} = -\frac{1}{c}\frac{\partial \vec{B}}{\partial t}$$

$$\nabla \times \vec{B} = \frac{1}{c}\frac{\partial}{\partial t}(\vec{E}+4\pi\vec{P}) + \frac{4\pi}{c}\vec{J_0}$$

$$\nabla \cdot (\vec{E}+4\pi\vec{P}) = 0$$

$$\nabla \cdot \vec{B} = 0$$

In the nonlinear regime, the electric dipole polarization is expanded in a Taylor series in terms of the total applied electric field. This approximation assumes that the electric field is small. The electric dipole polarization is then written as:

$$(4) \quad \vec{P} = \vec{P}^{(1)} + \vec{P}^{(2)} + \vec{P}^{(3)} + \ldots + \vec{P}^{(n)} + \ldots$$

where $\vec{P}^{(1)}$ is linear in the electric field, $\vec{P}^{(2)}$ is quadratic in the electric field, $\vec{P}^{(3)}$ is cubic in the electric field and so on. In general, the total applied electric field is the sum of different fields at different frequencies, so it is convenient to expand the field in terms of Fourier components.

Due to the fact that the study of nonlinear optics has been approached from many disciplines and has been growing very quickly, different notations have been developed to describe the nonlinear optical phenomena. This can sometimes be a source of confusion. Here we list the four most used conventions as reported by Shi and Garito [25]. The conventions differ in the way of defining the nonlinear polarization and the interacting fields in the frequency domain.

CONVENTION I

This convention defines the nth-order nonlinear susceptibility, $\chi^{(n)}$ in terms of the nth-order polarization and the electric fields as:

$$(5) \quad P_i^{\omega_{n+1}} = \sum_{jk\ldots m} \chi_{ijk\ldots m}^{(n)}(-\omega_{n+1}; \omega_1, \ldots, \omega_n) E_j^{\omega_1} E_k^{\omega_2} \ldots E_m^{\omega_n}$$

where $P_i^{\omega_{n+1}}$ is the i-component of the nth-order polarization field at frequency ω_{n+1}, $E_j^{\omega_1}$ is j-component of the electric field amplitude at frequency ω_1, and the $\chi_{ijk\ldots l}^{(n)}$ terms are the components the nth-order electric susceptibility of the medium, $\chi^{(n)}$. $\chi^{(n)}$ is a $(n+1)$-order tensor that determines completely the optical (linear and nonlinear) properties of the medium. Conservation of energy requires that $\omega_{n+1} = \omega_1 + \omega_2 + \ldots + \omega_n$ at all orders.

In this convention, the time-dependence of the electric and polarization fields are expanded in Fourier series as:

$$(6) \quad \vec{E}(t) = \frac{1}{2} \sum_{\omega \geq 0} [\vec{E}^{\omega} \exp(-i\omega t) + \vec{E}^{-\omega} \exp(i\omega t)], \text{ with } \left(\vec{E}^{\omega}\right)^{*} = \vec{E}^{-\omega}$$

and

$$(7) \quad \vec{P}(t) = \frac{1}{2} \sum_{\omega \geq 0} [\vec{P}^{\omega} \exp(-i\omega t) + \vec{P}^{-\omega} \exp(i\omega t)], \text{ with } \left(\vec{P}^{\omega}\right)^{*} = \vec{P}^{-\omega}$$

This notation is probably the simplest but it suffers from the fact that different susceptibilities do not converge to the same value in the zero-frequency limit. This is solved in Convention II by introducing frequency dependent numerical factors.

CONVENTION II

Convention II uses the following definitions:

$$(8) \quad P_i^{\omega_{n+1}} = \sum_{jk...m} K(-\omega_{n+1}; \omega_1, \ldots, \omega_n) \chi_{ijk...m}^{(n)}(-\omega_{n+1}; \omega_1, \ldots, \omega_n)$$

$$E_j^{\omega_1} E_k^{\omega_2} \ldots E_m^{\omega_n}$$

$$(9) \quad \vec{E}(t) = \frac{1}{2} \sum_{\omega \geq 0} [\vec{E}^{\omega} \exp(-i\omega t) + \vec{E}^{-\omega} \exp(i\omega t)], \text{ with } \left(\vec{E}^{\omega}\right)^{*} = \vec{E}^{-\omega}$$

and

$$(10) \quad \vec{P}(t) = \frac{1}{2} \sum_{\omega \geq 0} [\vec{P}^{\omega} \exp(-i\omega t) + \vec{P}^{-\omega} \exp(i\omega t)], \text{ with } \left(\vec{P}^{\omega}\right)^{*} = \vec{P}^{-\omega}$$

The numerical factors $K(-\omega_{n+1}; \omega_1, \ldots, \omega_n)$ are defined through:

$$(11) \quad K(-\omega_{n+1}; \omega_1, \ldots, \omega_n) = \frac{D}{2^{q-p}}$$

where q is the number of nonzero input frequencies and p is equal to zero if $\omega_{n+1} = 0$ and unity otherwise, while D is the number of distinguishable orderings in the input frequencies.

CONVENTION III

Convention III defines the nonlinear susceptibility in the same manner as convention II:

$$(12) \quad P_i^{\omega_{n+1}} = \sum_{jk...m} K(-\omega_{n+1}; \omega_1, \ldots, \omega_n) \chi_{ijk...m}^{(n)}(-\omega_{n+1}; \omega_1, \ldots, \omega_n)$$

$$\times E_j^{\omega_1} E_k^{\omega_2} \ldots E_m^{\omega_n}$$

However, the electric field and polarization are defined as:

(13) $$\vec{E}(t) = \sum_{\omega \geq 0}[\vec{E}^{\omega}\exp(-i\omega t) + \vec{E}^{-\omega}\exp(i\omega t)], \text{ with } \left(\vec{E}^{\omega}\right)^* = \vec{E}^{-\omega}$$

and

(14) $$\vec{P}(t) = \sum_{\omega \geq 0}[\vec{P}^{\omega}\exp(-i\omega t) + \vec{P}^{-\omega}\exp(i\omega t)], \text{ with } \left(\vec{P}^{\omega}\right)^* = \vec{P}^{-\omega}$$

CONVENTION IV

In convention IV the nonlinear susceptibility is defined in terms of the Taylor expansion:

(15) $$\chi^{(n)}_{ijk...m}(-\omega_{n+1}; \omega_1, \ldots, \omega_n) = \frac{\partial^n P_i^{\omega_{n+1}}}{\partial E_j^{\omega_1} \partial E_k^{\omega_2} \ldots \partial E_m^{\omega_n}}$$

which makes it very suitable for finite field calculations

The time-dependence of the electric and polarization fields are expanded in the same manner as in convention I:

(16) $$\vec{E}(t) = \frac{1}{2}\sum_{\omega \geq 0}[\vec{E}^{\omega}\exp(-i\omega t) + \vec{E}^{-\omega}\exp(i\omega t)], \text{ with } \left(\vec{E}^{\omega}\right)^* = \vec{E}^{-\omega}$$

and

(17) $$\vec{P}(t) = \frac{1}{2}\sum_{\omega \geq 0}[\vec{P}^{\omega}\exp(-i\omega t) + \vec{P}^{-\omega}\exp(i\omega t)], \text{ with } \left(\vec{P}^{\omega}\right)^* = \vec{P}^{-\omega}$$

Shi and Garito recommend conventions I and II, because they are the ones most used to report values in organic nonlinear optics. As pointed out previously, convention IV is widely used in finite field calculations, and experimentally it is also used to report values on gas phase atoms or molecules.

From the different definitions it is possible to relate the nonlinear susceptibilities from the first three conventions:

(18) $$\chi^{(n)\text{I}}_{ijk...m}(-\omega_{n+1}; \omega_1, \omega_2, \ldots, \omega_n) \\ = K(-\omega_{n+1}; \omega_1, \omega_2, \ldots, \omega_n)\chi^{(n)\text{II}}_{ijk...m}(-\omega_{n+1}; \omega_1, \omega_2, \ldots, \omega_n)$$

(19) $$\chi^{(n)\text{I}}_{ijk...m}(-\omega_{n+1}; \omega_1, \omega_2, \ldots, \omega_n) \\ = \frac{1}{2^{q-p}}\chi^{(n)\text{III}}_{ijk...m}(-\omega_{n+1}; \omega_1, \omega_2, \ldots, \omega_n)$$

and

(20) $$\chi^{(n)\text{II}}_{ijk...m}(-\omega_{n+1}; \omega_1, \omega_2, \ldots, \omega_n) \\ = \frac{2^{p-q}}{K(-\omega_{n+1}; \omega_1, \omega_2, \ldots, \omega_n)}\chi^{(n)\text{III}}_{ijk...m}(-\omega_{n+1}; \omega_1, \omega_2, \ldots, \omega_n)$$

with q and p defined in Eq. (11).

Convention IV is mostly used in the static limit, where the following relationship holds:

$$\chi_{ijk...m}^{(n)\text{II}}(0; 0, 0, \ldots, 0) = \frac{1}{n!} \chi_{ijk...m}^{(n)\text{IV}}(0; 0, 0, \ldots, 0) \tag{21}$$

As an example, we can consider the values of $\chi_{xxxx}^{(3)}(-2\omega; \omega, \omega, 0)$ which can be measured using Electric-Field-Induced Second-Harmonic Generation techniques. The values of $\chi_{xxxx}^{(3)}(-2\omega; \omega, \omega, 0)$ in the different conventions will obey the following relationships

$$\chi_{xxxx}^{(n)\text{I}}(-2\omega; \omega, \omega, 0) = \frac{3}{2}\chi_{xxxx}^{(n)\text{II}}(-2\omega; \omega, \omega, 0) = \frac{1}{2}\chi_{xxxx}^{(n)\text{III}}(-2\omega; \omega, \omega, 0) \tag{22}$$

This simple example warns about the danger of neglecting the differences between author conventions. Less used conventions as well as more details on the relationships between them can be found in [25].

Another source of confusion might be the units used to report nonlinear optical values. In order to compare the results of two different experiments or confront a theoretical prediction with experimental data, it is necessary to make sure the same units are used. Commonly, in nonlinear optics, CGS (esu) units or MKS units are used. Table 1 lists some fundamental quantities of nonlinear optics, as well as the corresponding units in both CGS (esu) and MKS, together with the conversion factors between the two systems of units [24, 25].

Thus, in order to compare values of the nonlinear electric susceptibilities, $\chi^{(n)}$, one has to make sure that the same system of units is used, and check out the conventions that have been used to describe the nonlinear optical response.

Because the electric susceptibilities fully determine the optical properties of the medium, one of the essential tasks of nonlinear optics is the characterization of the

Table 1. Fundamental quantities of nonlinear optics and its units in *CGS* (*esu*) and *MKS*. The conversion factor between both systems of units, Q, defined as $N_{CGS} = Q \cdot N_{MKS}$, where N_{CGS} is the quantity in *CGS* units, while N_{MKS} is the quantity in *MKS* units, is also provided

Quantity	CGS (esu)	Conversion Factor, Q	MKS
length, l	cm	10^{-2}	m
mass, m	g	10^{-3}	kg
force, F	dyn	10^{-5}	N
energy, E	erg	10^{-7}	J
charge, q	statcoulomb	3.336×10^{-10}	C
potential, V	statvolt	2.998×10^{-2}	V
electric field, E	statvolt/cm	2.998×10^{-4}	V/m
polarization, P	statvolt/cm	3.336×10^{-6}	Cm^{-2}
$\chi^{(1)}$	(none)	1.257×10^{1}	(none)
$\chi^{(2)}$	statvolt^{-1} cm	4.188×10^{-4}	mV^{-1}
$\chi^{(3)}$	statvolt^{-2} cm^{2}	1.397×10^{-8}	m^{2} V^{-2}

different tensorial components of the nonlinear electric susceptibilities. However, since in general the magnitude of the response decreases as the order of the nonlinearity increases, it is usually enough to characterize second- and third-order nonlinear susceptibilities, $\chi^{(2)}$ and $\chi^{(3)}$, respectively.

In organic materials, it is convenient to define microscopic nonlinear coefficients that relate the molecular dipole moment with the electric field applied to the molecule. Including the possibility of a permanent molecular dipole moment, μ_0, the molecular dipole moment is related to the electric field components in the frequency domain by:

$$(23) \quad \mu_i^\omega = \mu_i^0 \delta_{0,\omega} + \sum_j \alpha_{ij}^{(1)}(-\omega; \omega) \cdot E_j^\omega + \sum_{jk} \beta_{ijk}^{(2)}(-\omega; \omega_1, \omega_2) \cdot E_j^{\omega_1} E_k^{\omega_2}$$

$$+ \sum_{jkl} \gamma_{ijkl}^{(3)}(-\omega; \omega_1, \omega_2, \omega_3) \cdot E_j^{\omega_1} E_k^{\omega_2} E_l^{\omega_3} + \ldots$$

where $\alpha^{(1)}$, $\beta^{(2)}$ and $\gamma^{(3)}$ are the tensors corresponding to the first-, second-, and third-order molecular susceptibilities. The nonlinear molecular susceptibilities, $\beta^{(2)}$ and $\gamma^{(3)}$ are also called the first and second hyperpolarizabilities, respectively.

It has to be noted that the field applied to the molecule is not necessarily the electric field associated with the applied optical beam in the experimental set up. To account for this, one uses local field factors that vary depending on the particular characteristics of the molecular ensemble.

3. CHARACTERIZATION TECHNIQUES OF SECOND-ORDER SUSCEPTIBILITIES

3.1 Material Requirements: Symmetry Conditions in Second-order Nonlinear Optics

Neumann's principle states that under any symmetry operation on the system, the sign and the amplitude of the physical property should remain unchanged. This has a severe consequence for second-order effects: only non-centrosymmetric systems are allowed. A system is centrosymmetric when its physical properties remind unchanged under the inversion symmetry transformation ($x \to -x$, $y \to -y$, $z \to -z$).

If we consider a centrosymmetric system we can examine the influence of the inversion symmetry on the polarization of a general second-order nonlinear process:

$$(24) \quad P_i^{(2)}(\omega_\sigma) = \chi_{ijk}^{(2)} E_j^{\omega_1} E_k^{\omega_2}$$

Upon inversion, the electric field and the polarization transform as:

$$(25) \quad P_i(\omega_\sigma) \to -P_i(\omega_\sigma), E_i^{\omega_1} \to -E_i^{\omega_1}, \text{ and } E_j^{\omega_2} \to -E_j^{\omega_2}$$

Therefore, Eq. (23) becomes:

(26) $\quad -P_i^{(2)}(\omega_\sigma) = \chi_{ijk}^{(2)}(-E_j^{\omega_1})(-E_k^{\omega_2})$

or

(27) $\quad P_i^{(2)}(\omega_\sigma) = -\chi_{ijk}^{(2)} E_j^{\omega_1} E_k^{\omega_2}$

As can be seen by comparing Eq. (24) and Eq. (27), the polarization has changed sign and therefore the statement can only be true if:

(28) $\quad \chi_{ijk}^{(2)} = 0$

This means that no second-harmonic generation can come from a centrosymmetric medium. Only noncentrosymmetric media will give a second-order response.

We should keep in mind that this rule is applied on the molecular as well as on the macroscopic scale. At the molecular scale this problem can be solved by using electron donor and acceptor substituted conjugated D-π-A dipolar or octopolar molecules. When all these molecules are also randomly oriented, no second-order signal will occur. A polar order should be induced in the macroscopic scale. This can be achieved by electric poling, noncentrosymmetric crystal growth, the Langmuir-Blodgett technique, etc. An exception to this rule is Hyper-Rayleigh Scattering, where the natural orientational fluctuations of molecules in solution leads to local breaking of the centrosymmetry, allowing the measurement of second-order nonlinear response.

3.2 Second-Harmonic Generation

In Second-Harmonic Generation (SHG) experiments, an input beam of frequency ω incident in the material generates an output beam of frequency 2ω. The response is described by the second-order nonlinear susceptibility:

(29) $\quad P_i(2\omega) = \frac{1}{2}\chi_{ijk}^{(2)}(-2\omega; \omega, \omega)E_j(\omega)E_k(\omega)$

By solving the nonlinear wave equation, under the assumption of undepleted input beams, it is found that the intensity of the output wave at frequency 2ω varies as a function of the interaction length inside of the sample, l, and the wave vector mismatch Δk [26]:

(30) $\quad I_{2\omega} = I_{2\omega}^{\max} \frac{\sin^2(\Delta k \, l/2)}{(\Delta k \, l/2)^2}$

where the wave vector mismatch is defined as:

(31) $\quad \Delta k = 2k_\omega - k_{2\omega}$

From Eq. (30) we can see that by varying the interaction length inside of the sample, the output intensity at frequency 2ω will be changed. This is the basic idea behind most Second-Harmonic Generation measurement techniques.

3.2.1 Second-harmonic generation on crystals

The Maker fringe method is one of the most used methods for determining the second-order nonlinear susceptibility $\chi^{(2)}_{ijk}$ of a crystal [27]. It is a relative method and is only useful when the second-harmonic signal is compared with the signal from a crystal with known $\chi^{(2)}_{ijk}$ values. The experiment is designed as follows.

A light beam at frequency ω and associated with an electric field E_ω is interacting with a nonlinear active crystal. The light is linearly polarized. The electric field E_ω, associated with the wavevector k_ω, will interact with the crystal with thickness l. The incident angle of the light beam on the crystal surface is θ. Then a nonlinear polarization is induced in the crystal. Inside the crystal two waves are present: the first wave is called a forced wave and is at incident frequency ω and progresses at a speed n_ω (n_ω = refractive index at frequency ω). The propagation wavevector is k'_ω. At the surface a second wave is generated, the harmonic wave and is proceeding with a speed $n_{2\omega}$ ($n_{2\omega}$ = refractive index at frequency 2ω). This wave is associated with the wavevector $k'_{2\omega}$. The material is then rotated around an axis perpendicular to the incoming laser beam. The fringes are then caused by the angular dependence of the phase mismatch Δk between the forced and harmonic waves:

$$(32) \quad \Delta k = 2k_\omega - k_{2\omega} = \frac{4\pi}{\lambda \left(n_\omega \cos \theta'_\omega - n_{2\omega} \cos \theta'_{2\omega} \right)}$$

where θ'_ω and $\theta'_{2\omega}$ are the angles of refraction at the two frequencies. The relationship between these quantities can schematically be seen in Figure 1.

The study of the output fringe pattern, as a function of the incident angle, θ, allows to determine $\chi^{(2)}_{ijk}$.

The harmonic power $P_{2\omega}$ varies with the angle θ:

$$(33) \quad P_{2\omega} = I_m(\theta) \sin^2 \psi$$

where ψ is defined as:

$$(34) \quad \psi = \frac{\pi l}{2 l_c(\theta)}$$

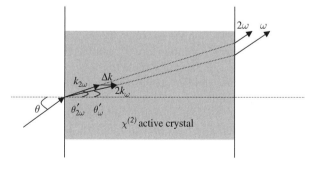

Figure 1. Schematic view of the wave propagation in a second-order active crystal

while the angular dependence of the coherence length, l_c, is given by:

$$(35) \quad l_c(\theta) = \frac{\lambda}{4(n_\omega \cos\theta_\omega - n_{2\omega} \cos\theta_{2\omega})}$$

The quantity $I_m(\theta)$ is an envelope function defined by:

$$(36) \quad I_m(\theta) = G\left(\chi^{(2)}_{ijk}\right)^2 P^2_\omega$$

where $\chi^{(2)}_{ijk}$ is the effective second-order nonlinear coefficient and P_ω the incident power. The proportionality constant G contains reflection and transmission coefficients and other optical factors for determining the magnitude of the optical fields inside the crystal. A more detailed description has been provided by Kurtz [28].

The oscillating pattern can also be obtained by measuring at a constant angle θ and varying the interaction length l inside of the crystal. In this case the crystal is wedge-shaped and the crystal is moved perpendicular to the incoming laser beam. In this case the coherence length is a constant and the fringes result from the variation of l [29]:

$$(37) \quad P_{2\omega} = G\left(\chi^{(2)}_{ijk}\right)^2 P^2_\omega \sin^2\psi$$

where

$$(38) \quad \psi = \frac{2\pi \Delta n \, l}{\lambda}$$

with

$$(39) \quad \Delta n = (n_\omega \cos\theta_\omega - n_{2\omega} \cos\theta_{2\omega})$$

at a specific incoming angle θ.

3.2.2 Second-harmonic generation on films

To determine the $\chi^{(2)}_{ijk}$ in a film, the centrosymmetry of the film should be broken. This can be achieved by poling the polymer film or depositing the film on a glass substrate by the Langmuir-Blodgett technique. In the case of poled polymer films, the film is prepared by adding a polymer layer on the substrate by spincoating. The sample is then heated near the glass transition temperature T_g and the dipoles are aligned by applying an electric field. After cooling the sample to room temperature the electric field is turned off. The nonlinear dipoles are frozen in a noncentrosymmetric condition (see Figure 2).

An often used material is a "guest-host" system, where the dipoles are mixed in a polymer, usually PMMA. This is a relatively easy system to obtain but has the disadvantage of being relatively unstable in time. To avoid relaxation after poling the matrix is often cross-linked after poling. To have better alignment of the chromophores a polymer with semi-crystalline side-chains can be introduced, which

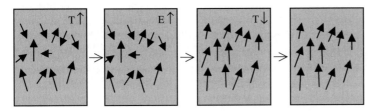

Figure 2. Schematic view of the alignment of dipoles in a polymer matrix by poling. First the sample is heated above the glass transition temperature. Then an external electric field is applied. When the dipoles are aligned the temperature is lowered below the glass transition temperature. The result is a noncentrosymmetric distribution of nonlinear dipoles even in the absence of the external applied field

increases the poling condition by a factor of two. The mayor disadvantage of this "guest-host" system is the relatively low solubility of the dipoles in the polymer matrix. It is common to use a fully functionalized polymer, where the dipole is covalently linked to the side chain of the polymer, with or without crystalline side chains.

Poling induces a polar axis in the polymer film. The z-axis is essentially an infinite-fold rotational axis with an infinite number of mirror planes. This type of symmetry is denominated ∞mm or $C_{\infty v}$. In this case the molecules are distributed cylindrically about the z-axis and the angle α, defined as the angle between the z-axis and the dipole moment of the molecule, varies from molecule to molecule. In the weak poling limit the distribution of α is broad, but with a tendency to orient in the direction of z compared to the unpoled state. The non-vanishing nonlinear coefficients for $C_{\infty v}$ symmetry are $\chi^{(2)}_{xxz} = \chi^{(2)}_{xzx} = \chi^{(2)}_{yyz} = \chi^{(2)}_{yzy}$, $\chi^{(2)}_{zxx} = \chi^{(2)}_{zyy}$ and $\chi^{(2)}_{zzz}$.

A second technique to obtain a noncentrosymmetric film is by the Langmuir-Blodgett deposition technique. This technique is useful for ordering (small) amphiphilic molecules consisting of a polar hydrophobic head group and a long hydrophilic aliphatic chain. On a water layer an ordered condensed monolayer of nonlinear active molecules is formed. The force of the water surface and the lateral surface pressure are used to condense a randomized set of molecules to a highly organized and stabilized monolayer of molecules by the van der Waals forces between the molecules. The film can be transferred to a glass substrate as a film. A schematic view of the classical deposition is presented in Figure 3. A typical amphiphilic molecule to deposit LB films is a steric acid (eg $C_{17}H_{35}COOH$) where the acid group is the polar head group, symbolized by a circle with an attached aliphatic section (the alkyl chain). The first withdrawal of a polar surface induces the addition of the polar head group on the substrate (see step 2 on Figure 3). A re-immersion organizes an additional layer in a tail-to-tail configuration. The next withdrawal is responsible for a head-to-head deposition. This can be repeated several times until films of a thickness of about 1 μm are formed. This type of deposition is referred to as Y-type deposition. Different types of deposition are used to obtain LB films. For instance, one can chose for horizontal or vertical deposition of the substrate on the monolayer.

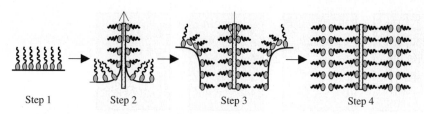

Figure 3. Schematic representation of the deposition of Langmuir-Blodgett layers on a polar substrate. **Step1**: the molecules are aligned on the water surface forming a monolayer (polar head group (circle), aliphatic alkyl chain). **Step 2**: the substrate is removed out of the water surface, depositing an organized monolayer onto the surface. **Step 3**: the surface is immersed into the water layer again depositing a second layer in a tail-to-tail configuration. **Step 4**: the surface is withdrawed from the water depositing a third layer, this time in a head-to-head position, generating a classical Y-type deposition

Since LB films posses the same symmetry as poled polymer films, which is also $C_{\infty v}$, the same tensor components will be non-vanishing: $\chi^{(2)}_{xxz} = \chi^{(2)}_{xzx} = \chi^{(2)}_{yyz} = \chi^{(2)}_{yzy}, \chi^{(2)}_{zxx} = \chi^{(2)}_{zyy}$ and $\chi^{(2)}_{zzz}$.

At this point we should also mention the possibility of introducing chirality into the sample [30]. A chiral molecule is inherently noncentrosymmetric and even a random distribution of enantiomerically pure molecules will never lead to systems with inversion symmetry. In fact, the requirement of noncentrosymmetry is only strictly required in the electric dipole approximation. In the presence of magnetic dipole and electric quadrupole contributions to the nonlinearity, noncentrosymmetry is not a strong requirement anymore for observing second-order processes. Only chiral systems have substantial magnetic dipole and electric quadrupole contribution to nonlinear optics. It can be shown that for a chiral isotropic surface, the non-vanishing tensor components are:

$$(40) \quad \chi^{(2)}_{zzz}, \chi^{(2)}_{zxx} = \chi^{(2)}_{zyy}, \chi^{(2)}_{xxz} = \chi^{(2)}_{yyz} = \chi^{(2)}_{yzy} = \chi^{(2)}_{xzx}, \chi^{(2)}_{xyz} = -\chi^{(2)}_{yxz} = -\chi^{(2)}_{yzx} = \chi^{(2)}_{xzy}$$

Apart from their symmetry properties, the interest for chiral structures was enhanced by the discovery of their optical activity. Chiral molecules have different efficiencies for generating second-harmonic light for left- and right-handed circularly polarized fundamental light. This effect is called Second-Harmonic Generation Circular Dichroism (SHG-CD). The SHG-CD effect is several orders of magnitude higher than circular dichroism in linear optics. Since centrosymmetry is broken at surfaces or interfaces and so SHG is sensitive to surfaces, and SHG-CD is an effect specific to chiral materials, SHG-CD is used to investigate chiral surfaces and interfaces. SHG-CD can be used to probe a chiral surface but can also be used to enhance the second-order nonlinear response.

The second-order susceptibility tensor can be measured by performing polarization experiments (see Figure 4). In this type of experiments a fundamental p-polarized light is passed through a quarter wave plate which is continuously rotated. The second-harmonic signal is detected in reflection and transmission. Only the *s* or *p* polarization is detected because of the presence of an analyzer in front of

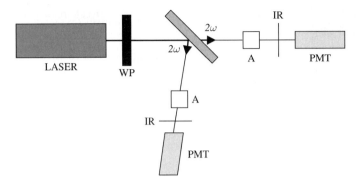

Figure 4. Schematic view of the experiment that characterizes the tensor elements on a thin film. WP = waveplate (quarter wave or half wave), A = analyzers (*p* or *s* polarized depending on the experiment), IR = infra-red blocking filter; PMT = photomultiplier tube

the detector. The same experiments are done by using a half-wave plate. The exact equation and further detailed explanation can be found in [31].

3.2.3 Electric-field-induced second-harmonic generation

The Electric-Field-Induced Second-Harmonic Generation (EFISHG) technique makes it possible to measure the molecular hyperpolarizability, β, on liquids or molecular solutions. The centrosymmetry of the solution is broken by applying a DC electric field to induce an average orientation of the molecules due to interactions of the permanent dipoles of the molecules and the electric field. The energy of a dipole with a permanent dipole $\vec{\mu}$ in an electric field $\vec{E}$ is given by:

(41) $$U = -\vec{\mu} \cdot \vec{E} \cos\theta$$

where θ is the angle between the dipole moment and the applied electric field.

An isolated dipole reduces its energy by orientating in the direction of the electric field. In solution, however, the thermal energy induces spatial distribution on the dipole orientation in a Boltzman distribution. The degree of orientation of the dipoles by the electric field results in an averaged angle $\langle\cos\theta\rangle$. In a non-oriented medium $\langle\cos\theta\rangle = 0$ and in a perfect aligned medium $\langle\cos\theta\rangle = 1$.

Because a strong electric field is required to align the molecules, further restrictions are imposed on the molecules: they should have a permanent dipole moment. For instance, EFISHG can not be applied to measure the second-order nonlinear susceptibilities of octopolar molecules, even though at the molecular level, their molecular hyperpolarizability, β_{ijk}, is non-zero. Also, EFISHG can not be used with ionic molecules or with a polar solvent.

In order to measure the molecular first hyperpolarizability β_{ijk} of a molecule, a strong static electric field and two optical fields are applied on the designed molecules. By the interaction of the two optical fields, coming from a laser beam, and the static electric field we actually measure a third-order effect, related to $\chi^{(3)}(-2\omega;\omega,\omega,0)$ through:

(42) $\quad P_i(2\omega) = \dfrac{3}{2}\chi^{(3)}_{ijkl}(-2\omega;\omega,\omega,0)E_j(\omega)E_k(\omega)E_l(0)$

where $\chi^{(3)}_{ijkl}$ is the macroscopic third-order susceptibility. $E_j(\omega)$ and $E_k(\omega)$ are the optical electric field components at frequency ω, while $E_j(0)$ is the applied static electric field. The macroscopic susceptibility $\chi^{(3)}_{ijkl}$ is linked to the third-order molecular polarizability γ_{ijkl} by the following relation [32]:

(43) $\quad \chi^{(3)}_{ijkl} = N\, f_0\, f_{2\omega}\, f_\omega^2\, \gamma_{ijkl}$

where N equals the number density of the molecules in solution and f_0, $f_{2\omega}$ and f_ω are the local field factors. The first correction for the applied static electric field is given by:

(44) $\quad f_0 = \dfrac{\varepsilon_0(n^2+2)}{(n^2+2\varepsilon_0)}$

where ε_0 is the static dielectric constant. This correction takes into account the orientation of permanent dipoles due to the interaction with neighboring dipoles, and is known as the Onsager local field factor.

The other two correction factors account for the effects of induced dipoles in the medium through electronic polarization, and are known as the Lorentz-Lorenz correction factors. The correction for the optical field at frequency 2ω is given by:

(45) $\quad f_{2\omega} = \dfrac{n^2_{2\omega}+2}{3}$

and the correction factor for the optical field at frequency ω is given by:

(46) $\quad f_\omega = \dfrac{n^2_\omega+2}{3}$

The molecular third-order polarizability γ_{ijkl} has three contributions [33]:

(47) $\quad \gamma_{ijkl} = \gamma^e_{ijkl} + \gamma^v_{ijkl} + \gamma^r_{ijkl}$

γ^e_{ijkl} is the averaged electronic contribution, γ^v_{ijkl} is the averaged vibronic contribution and γ^r_{ijkl} is the averaged dipole rotation contribution. It is this last contribution that is proportional to the second-order polarizability:

(48) $\quad \gamma^r_{ijkl} = \dfrac{\mu_z \beta_z}{5kT}$

where β_z is the vector component of the tensor β_{ijk} projected on the axis of the dipole moment [34].

(49) $\quad \mu_z\,\beta_z = \mu_z\,\beta_{iiz} + \mu_z\,\beta_{izi} + \mu_z\,\beta_{zii}$

This can be represented as $\bar{\mu} \cdot \bar{\beta}$, where $\bar{\beta}$ is the vectorial part of the third-rank tensor β_{ijk} [35];

(50) $\quad \beta_z = \beta_{zzz} + \dfrac{1}{3}\left(\beta_{zxx} + \beta_{zyy} + \beta_{xzx} + \beta_{yzy} + \beta_{xxz} + \beta_{yyz}\right)$

and since $\beta_{ijk} = \beta_{ikj}$ due to the intrinsic permutation symmetry we find:

(51) $\quad \beta_z = \beta_{zzz} + \dfrac{1}{3}\left(\beta_{zxx} + \beta_{zyy} + 2\beta_{xzx} + 2\beta_{yzy}\right)$

Far-off from resonances, where there is no energy dissipation in the material, we can use Kleinman symmetry ($\beta_{ijk} = \beta_{jki} = \beta_{kij}$), which reduces the number of independent tensorial components even more:

(52) $\quad \beta_z = \beta_{zzz} + \beta_{zxx} + \beta_{zyy}$

For conjugated molecular systems [36, 37]: $\gamma^r \gg \gamma^e$ and $\gamma^r \gg \gamma^v$ so we can conclude:

(53) $\quad \chi^{(3)} = N\, f_0\, f_{2\omega}\, f_\omega^2\, \gamma^r$

or equivalently,

(54) $\quad \chi^{(3)} = N\, f_0\, f_{2\omega}\, f_\omega^2\, \dfrac{\mu_z \beta_z}{5kT}$

In EFISHG an oscillating pattern is observed as is done with SHG on a crystal. The theory has been developed by Levine and Bethea [38–41]. The measuring cell is wedge-shaped. An external electric field is applied over the cell. The analysis of the harmonic signal however is a little different than for the crystal. The glass slides will also contribute to the overall signal. The total intensity at the harmonic wavelength can be written as:

(55) $\quad I_{2\omega} = I_{2\omega}(G1) + I_{2\omega}(L) + I_{2\omega}(G2)$

with G1 and G2 referring to the contributions of the first and second glass slide forming the wedge-shaped cell and L to the solution in between. The relation between the macroscopic third-order polarizability $\chi^{(3)}$ and the harmonic intensity is found by solving the Maxwell equations in the separate parts:

(56) $\quad \dfrac{\partial E^{2\omega}}{\partial z^2} + \alpha_{2\omega}\dfrac{\partial E^{2\omega}}{\partial z} + k_f^2 E^{2\omega} = -\dfrac{4\pi(2\omega)^2}{c^2}\chi^{(3)}(E_\omega(z))^2 E_0 e^{ik_b z}$

where z is the propagation direction of the fundamental laser beam, $\alpha_{2\omega}$ is the absorption coefficient at frequency 2ω, $E_\omega(z)$ is the amplitude of the fundamental field and E_0 is the amplitude of the applied external static dc field. The wave vectors of the fundamental and harmonic wave are defined as:

(57) $\quad k_{2\omega} = \dfrac{2\omega n_{2\omega}}{c} \quad$ and $\quad k_\omega = \dfrac{\omega n_\omega}{c}$

In the absence of absorption at fundamental and harmonic frequency, the harmonic intensity is then given by:

(58) $\quad I_{2\omega} = \dfrac{c}{2\pi} \left(t_{2\omega}^G \left(T_G E_\omega^G - T_L E_{2\omega}^L\right)\right)^2 \sin^2\left(\dfrac{\pi l}{2 l_c^L}\right)$

where l is the interaction length inside of the solution and with:

(59) $\quad t_{2\omega}^G = \dfrac{2n_{2\omega}^G}{1 + n_{2\omega}^G}, \; T_G = \dfrac{n_\omega^G + n_{2\omega}^L}{n_{2\omega}^L + n_{2\omega}^G}, \; T_L = \dfrac{n_\omega^L + n_{2\omega}^L}{n_{2\omega}^L + n_{2\omega}^G}, \; l_c^L = \dfrac{\lambda}{4(n_{2\omega}^L - n_\omega^L)}$

(60) $\quad E_\omega^G = \dfrac{4\pi}{(n_{2\omega}^G)^2 - (n_\omega^L)^2} \chi_L E_0 \left(E_\omega \dfrac{2}{1 + n_\omega^G}\right)^2$

(61) $\quad E_\omega^L = \dfrac{4\pi}{(n_{2\omega}^G)^2 - (n_\omega^L)^2} \chi_L E_0 \left(E_\omega \dfrac{2}{1 + n_\omega^G} \dfrac{2 n_\omega^G}{n_\omega^L + n_\omega^G}\right)^2$

The translation of the measuring cell, perpendicular to the laser beam, induces a fringe pattern in a $\sin^2$-function of the harmonic intensity by variation of the path length l (see Fig. 5).

Several reference methods can be used. The first method is using a quartz reference wedge-shaped crystal. The ratio of the harmonic intensities of the cell $I_{2\omega}^L$ and the quarts crystal $I_{2\omega}^Q$ gives $\chi_{ijkl}^{(3)}$. Since the $\chi_{ijkl}^{(3)}$ values from some solvents are described in literature [42], they can be used as a reference. The harmonic intensities of the pure solvent S and the solution L are then measured in the same experimental conditions and the ratio of $I_{2\omega}^L$ and $I_{2\omega}^S$ determine $\chi_{ijkl}^{(3)}$. From the value $\chi_{ijkl}^{(3)}$, the product $\mu_z \beta_z$ can be obtained (Eq. 54). If the molecular dipole moment is known, β_z is calculated.

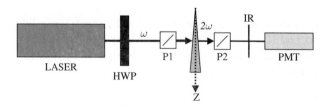

Figure 5. Schematic representation of an EFISHG experiment: HWP = half-wave plate to set the input intensity, P1 = polarizer to set the input s-polarization, wedge-shaped cell translated in the z-direction to vary the pathlength l, P2 = second polarizer to polarize the second-harmonic light, IR = infra-red blocking filter, PMT = photomultiplier tube for detection of second-harmonic light

3.2.4 Hyper-Rayleigh scattering

Hyper-Rayleigh scattering (HRS) is another technique to measure the second-order polarizability or first hyperpolarizability of compounds and is now one of the most used technique to do so. HRS has several advantages over the EFISHG technique. With this technique no knowledge of the dipole moment is required. Because no external electric field is applied, ionic molecules can also be measured. Since alignment is not necessary, octopolar molecules can be investigated as well. The centrosymmetry of the solution is broken by orientational fluctuations. Additionally, the depolarization ratio can give inside into the tensorial character of the molecules.

However, HRS has also some disadvantages. Unlike EFISHG, it is a non-coherent technique and second-order efficiencies are low. The incoherent scattering is also responsible for not being able to distinguish between harmonic hyper-Rayleigh signal and multi-photon fluorescence [43–45], which can result in an overestimation of the hyperpolarizability β. Three-photon fluorescence (3PF) should theoretically be detectable due to the cubic dependence of the fundamental intensity I_ω. Experimentally, it is not always easy to distinguish between a quadratic dependence and a quadratic dependence with a small contribution of cubic dependence [46]. A 3PF band can extend as far as to the second-harmonic wavelength. Two-photon fluorescence (2PF) is also quadratically dependent on the fundamental intensity, and based on intensity dependence no discrimination between HRS and 2PF can be made. 2PF exhibits in most cases a Stokes shift to longer wavelengths which allows discrimination between HRS and 2PF by using a narrow interference filter. However, anti-stokes 2PF has also been reported [45, 47]. In this case, an optical filter has no use. Since there is a difference in spectral width, HRS signal is a small sharp peak while 2PF is a broad background peak, this can be used to discriminate between both signals by detecting the signal at different wavelengths. This technique has been reported but seems to be very time-consuming [45].

A more elegant solution is to use the time difference between the time-delayed MPF and the immediate HRS [46]. Since a typical fluorescence lifetime is in the ns scale, the classical set-up with a Nd^{3+} : YAG laser is unsuited for performing this experiments. A ps or fs pulsed laser would be more appropriate. The use of a narrow and early temporal gate together with fast counting electronics makes the experiment possible. The Fourier transform of this technique in the frequency domain has been implemented in our group [48], which is experimentally easier to achieve than the time-domain approach. The principle of this set-up will be explained later after the introduction to Hyper-Rayleigh scattering and the experimental details of classical "ns" HRS.

A Hyper-Rayleigh scattering experiment is performed by measuring the intensity of the incoherently scattered frequency-doubled light generated by an intense laser beam from an isotropic solution [49, 50]. The scattered intensity of a single molecule at the harmonic wavelength can be calculated by performing an orientational average over β:

(62) $$I_{2\omega} = \frac{32\pi^2}{c\varepsilon_0^3 \lambda^4 r^2} \langle \beta_{HRS}^2 \rangle I_\omega^2$$

The brackets indicate the orientational averaging, λ the fundamental wavelength and r the distance to the scattering molecule, c the speed of light in vacuum ($c = 2.998 \times 10^8$ m/s) and ε_0 the permittivity of free space ($\varepsilon_0 = 8.85 \times 10^{-12}$ F/m).

Since an isotropic solution consists of a large number of molecules, summing the electric fields scattered by the individual molecules in the scattering volume and then squaring the result could provide the total scattering intensity if they are correlated scatterers. However, assuming the molecules in the scattering volume are independent, the total intensity is proportional to the sum of the intensity scattered by the individual molecules:

$$(63) \quad I_{2\omega} = \frac{32\pi^2}{c\varepsilon_0^3 \lambda^4 r^2} N f_\omega^4 f_{2\omega}^2 \langle \beta_{HRS}^2 \rangle I_\omega^2$$

where N is the concentration of chromophores, and f_ω and $f_{2\omega}$ are the local field factors, as defined in Eqs. (45) and (46). Until now, no evidence has been found that individual molecules in solution should be treated as correlated.

The relationship between $\langle \beta_{HRS}^2 \rangle$ and the molecular tensor components β_{ijk} depends on the polarization state of both fundamental and harmonic light and the scattering geometry. In classical HRS experiments the 90° angle geometry is mainly used. This means we build the set-up in such a way that the fundamental light beam is propagating in the X-direction and polarized in the Z-direction, and the scattered light is collected in the Y-direction (see Figure 6). Note that we distinguish between the laboratory coordinate system of reference (X, Y, Z), and the molecular coordinate system of reference (x, y, z).

In such a measuring geometry the relation between the orientationally averaged tensor components and the molecular tensor components are expressed as follows:

$$(64) \quad \langle \beta_{ZZZ}^2 \rangle = \frac{1}{7} \sum_i \beta_{iii}^2 + \frac{6}{35} \sum_{i \neq j} \beta_{iii} \beta_{ijj} + \frac{9}{35} \sum_{i \neq j} \beta_{iij}^2$$
$$+ \frac{6}{35} \sum_{ijk,cyclic} \beta_{iij} \beta_{jkk} + \frac{12}{35} \beta_{ijk}^2$$

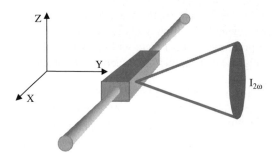

Figure 6. Schematic view of the classical 90° angle HRS geometry. An intense infrared laser beam is brought to focus in a cell containing the isotropic solution and the frequency-doubled light $I_{2\omega}$ is collected and detected under 90°

(65) $$\langle\beta_{XZZ}^2\rangle = \frac{1}{35}\sum_i \beta_{iii}^2 - \frac{2}{105}\sum_{i\neq j}\beta_{iii}\beta_{ijj} + \frac{11}{105}\sum_{i\neq j}\beta_{iij}^2$$
$$- \frac{2}{105}\sum_{ijk,cyclic}\beta_{iij}\beta_{jkk} + \frac{8}{35}\beta_{ijk}^2$$

The first subscript (X or Z) refers to the polarization state of the frequency-doubled light (in the laboratory coordinate system). Since both polarizations are detected with equal sensitivity, and the fundamental light polarized vertically, the orientational average over β is the sum of both equations.

(66) $$\langle\beta_{HRS}^2\rangle = \langle\beta_{ZZZ}^2\rangle + \langle\beta_{XZZ}^2\rangle$$

The orientational averaged hyperpolarizability squared $\langle\beta_{HRS}^2\rangle$ is related to the molecular hyperpolarizability tensor components according to Eq. (64) and Eq. (65). For a molecule of $C_{\infty v}$ symmetry, these equations reduce to:

(67) $$\langle\beta_{ZZZ}^2\rangle = \frac{1}{7}\beta_{zzz}^2 + \frac{6}{35}\beta_{zzz}\beta_{zyy} + \frac{9}{35}\beta_{zyy}^2$$

(68) $$\langle\beta_{XZZ}^2\rangle = \frac{1}{35}\beta_{zzz}^2 - \frac{2}{105}\beta_{zzz}\beta_{zyy} + \frac{11}{105}\beta_{zyy}^2$$

The square root of the orientational averaged hyperpolarizability:

(69) $$\sqrt{\langle\beta_{HRS}^2\rangle} = \sqrt{\langle\beta_{XZZ}^2\rangle + \langle\beta_{ZZZ}^2\rangle}$$

is related to β_{zzz} and reduces to:

(70) $$\langle\beta_{HRS}^2\rangle \approx \left(\frac{1}{7} + \frac{1}{35}\right)\beta_{zzz}^2 = \frac{6}{35}\beta_{zzz}^2$$

under the assumption that β_{zzz} is much larger than β_{zyy} and β_{zxx}.

For an octopolar molecule with D_{3h} symmetry only 4 equal tensor components remain $\beta_{zzz} = -\beta_{zzx} = -\beta_{xzx} = -\beta_{xxz}$ and Eqs. (64) and (65) reduce to:

(71) $$\langle\beta_{ZZZ}^2\rangle = \frac{1}{7}\beta_{zzz}^2 + \frac{6}{35}\beta_{zzz}(-\beta_{zxx}) + \frac{9}{35}\beta_{zxx}^2 = \frac{8}{35}\beta_{zzz}^2$$

(72) $$\langle\beta_{XZZ}^2\rangle = \frac{1}{35}\beta_{zzz}^2 - \frac{2}{105}\beta_{zzz}(-\beta_{zxx}) + \frac{11}{105}\beta_{zxx}^2 = \frac{16}{105}\beta_{zzz}^2$$

and therefore:

(73) $$\langle\beta_{HRS}^2\rangle = \frac{8}{21}\beta_{zzz}^2$$

Note that the orientational averaged hyperpolarizability is dependent on the symmetry of the molecule investigated and that there are different relations to the molecular tensor elements. It is also important to mention that the vectorial part of the hyperpolarizability β_{vec} is constituted out of different tensor elements than the orientational averaged hyperpolarizability measured by hyper-Rayleigh scattering.

The exact experimental details are described elsewhere [51] but basically, the technique works as follows: Since HRS is a forbidden process in isotropic solution, the efficiency is very low. As a consequence the optical fields with high optical power-density are needed together with an efficient collection system to detect the HRS signal. The fundamental light beam is passed between two crossed polarizers. A half-wave plate is place in between the two polarizers to control the intensity of the fundamental beam. Then the fundamental beam is focused in the cell. Part of the intensity is split of and detected by a photodiode (PD) which will read the fundamental signal I_ω. The collection system is constituted out of a concave mirror, an aspherical lens, a planoconvex lens and a photomultiplier. Separation of the fundamental and harmonic light is achieved by an interference filter. A schematic view of the set-up is shown in Figure 7.

For a solution of two components (solvent and solute), the harmonic intensity $I_{2\omega}$ equals:

$$(74) \quad I_{2\omega} = G \left(N_s \langle \beta_{HRS}^2 \rangle_s + N_x \langle \beta_{HRS}^2 \rangle_x \right) I_\omega^2$$

where G includes all experimental factors and the subscripts s and x refer to solvent and chromophores, respectively. From a concentration series $\langle \beta_{HRS}^2 \rangle_x$ can be determined when $\langle \beta_{HRS}^2 \rangle_s$ is known. This method is referred to as the internal reference method. There is also an external reference method where also a concentration series of a reference compound, with known hyperpolarizability, is measured. The ratio of both slopes gives $\langle \beta_{HRS}^2 \rangle_x$.

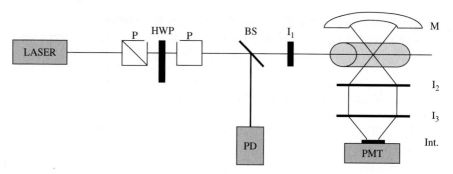

Figure 7. Schematic view of the experimental HRS set-up. (LASER, P = polarizer, HWP = half-wave plate, BS = beam splitter, M = concave mirror, l_1 = focusing lens, l_2 = aspheric lens, l_3 = planoconvex lens, Int. = interference filter, PD = photodiode, PMT = photomultiplier)

In a classical HRS experiment a ns pulsed laser (often Nd^{3+} : YAG laser with a fundamental lightbeam of 1064 nm) is used. Because of the low repetition rate of the laser pulses, gated integrators are used to measure the intensity of the HRS signal. The measurement is computer controlled and I_ω and $I_{2\omega}$ are recorded.

The principle of discriminating between immediate HRS and time-delayed fluorescence is based on the phase shift φ and the demodulation $M = M_F/M_R$ (ratio between the intensity of fluorescence, M_F, and the intensity of excitation M_R) that the fluorescence acquires versus the excitation light at a particular amplitude modulation frequency $\omega = 2\pi f$. The normalized magnitude M and the phase φ of the fluorescence at a particular frequency $\omega = 2\pi f$ are experimentally observable in the time domain and in the real and imaginary part of the Fourier transform. The frequency dependence of φ and M is determined by the fluorescence decay parameters, which is the fluorescence lifetime and its respective amplitude. In the frequency domain the phase shift φ of the fluorescence tends to 90° and the normalized magnitude M tends to zero for long lifetimes τ and/or high modulation frequencies. It is said that than the fluorescence is "out-of-phase" and completely "demodulated" with the excitation. Thus an attractive way is offered to eliminate fluorescence from the scattered light through high frequency modulation of the incident light. A repetitive short pulse in the time domain contains the fundamental repetition frequency and its higher harmonics under an envelope that is determined by the inverse of the pulse width. A Ti^{3+}:sapphire laser with a pulse width of 100 fs, has modulation frequencies available well into the GHz frequency range. The intrinsic high harmonic content of the repetitive femtosecond pulse is used as a source for high-frequency amplitude modulation. The set-up is constructed in the same way as for classical ns HRS. The electronic reading of the signals is somewhat different. Experimental details are described by Olbrechts et al [48].

So far we can say that efforts have been made to improve the measurements of the second-order nonlinear optical response. In time, techniques have been developed for determining the hyperpolarizability of compounds. First they were measured in their crystal structure. The disadvantage of this technique is that the molecules should crystalize in a non-centrosymmetric space group, and that it is not always easy to grow large crystals. The Kurz-powder technique has the advantage that only small crystals are needed. It is not always easy to determine the size of the powder and still the crystal needs to be in a noncentrosymmetric space group. It can be however that a dipolar molecule (which should exhibit second-order effects), is crystalized in a noncentrosymmetric space group and therefore has no harmonic signal. The EFISHG technique can measure the hyperpolarizability of molecules in solution. The molecules are limited to be neutral molecules with a permanent dipole moment. The HRS technique however can measure also octopolar and ionic molecules. This is the mean advantage of the technique and has lead to the recent improvements of the technique. HRS has been combined with electrochemistry to probe the changes in hyperpolarizability on structural change by oxidation or reduction [52]. An overview of different switching mechanism can be found in [53].

4. CHARACTERIZATION TECHNIQUES OF THIRD-ORDER SUSCEPTIBILITIES

4.1 The Intensity-dependent Refractive Index

An extremely useful feature of the third-order nonlinear optical response is the intensity-dependent refractive index, where the refractive index of the medium changes due to the interaction with a light beam. This optically-induced change in the refractive index is essential for all-optical switching applications.

The intensity-dependent refractive index can be defined as [54]:

$$(75) \quad n(\vec{r}, t) = n_0(\vec{r}, t) + \Delta n[I(\vec{r}, t)]$$

where $n_0(\vec{r}, t)$ is the linear refractive index which dominates the response for low intensity fields while $\Delta n[I(\vec{r}, t)]$ represents the component of the refractive index that depends on the intensity $I(\vec{r}, t)$. In optical Kerr-like media, this dependence is linear so it is possible to write $\Delta n[I(\vec{r}, t)]$ as:

$$(76) \quad \Delta n = 2n_2 \left|\vec{E}\right|^2$$

where $\vec{E}$ is the total optical field.

Since the intensity is proportional to the optical field, $I \propto \left|\vec{E}\right|^2$, in optical Kerr-like media:

$$(77) \quad \Delta n = n_2' I$$

Again, it has to be realized that the definitions of $\Delta n[I(\vec{r}, t)]$ are different in different conventions.

In order to successfully substitute the role of the electron by the photon in photonic applications, it is necessary to achieve high processing speeds. For applications of the intensity-dependent refractive index, one could define the following figure of merit that evaluates the optical switching performance of a third-order nonlinear optical material [24]:

$$(78) \quad F(\lambda) = \frac{\chi^{(3)}(\lambda)}{\alpha'(\lambda)\tau}$$

where $\chi^{(3)}$ is the magnitude of the third-order electronic susceptibility corresponding to the third-order nonlinear response of the material, α' is the absorptivity of the material, and τ is the lifetime of the third-order nonlinear optical response. Both $\chi^{(3)}$ and α' are usually dependent on the wavelength of the optical beam, λ.

The reasoning for such figure of merit is the following: as the magnitude of $\chi^{(3)}$ increases, less light intensity is needed to induce the response; as the absorptivity lowers, the longer the propagation length possible for an induced nonlinear response; and for the shorter the lifetime, the faster the processing speed of the response.

However, the figure of merit presented in Eq. (78) is not useful in characterizing third-order nonlinear materials because it is not dimensionless and it does not separate between linear and nonlinear absorption. More appropriate dimensionless figures of merit have been proposed by Stegeman [55, 56].

4.2 Physical Mechanisms of the Third-order Nonlinear Response

In order to select a particular experimental technique to measure $\chi^{(3)}$, it is very important to keep in mind which parameter of the third-order nonlinear response has to be characterized. For example, if one wants to determine the time-response due to molecular reorientation, one cannot choose Third-Harmonic Generation or Electric-Field-Induced Second-Harmonic Generation, since none of these techniques provide time-response information. Depending on the parameter of interest, a specific technique must be chosen. The following physical mechanisms can contribute to the third-order nonlinear response [54]:

4.2.1 Electronic polarization

Through this mechanism, the nonlinear response is produced by the changes on the electronic cloud around the atom or molecule through the optical field. It is related to the microscopic third-order molecular polarizability γ. Typically, nonresonant electronic processes in non-absorbing media yield values of $\chi^{(3)} \sim 10^{-14}$ esu. The time response of nonresonant electronic processes is $\sim 10^{-15}$ s. This is the fastest time response for third-order nonlinear processes.

4.2.2 Raman induced Kerr effect

This effect is related with the Stimulated Raman scattering process. A strong beam (pump) incident on a Raman active medium induces a change of the refractive index, which in turn influences the propagation of a weaker beam (probe). A typical value for Raman susceptibility is $|\text{Re}(\chi^{(3)})| \sim 10^{-12}$ esu, with a time response $\sim 10^{-12}$ s.

4.2.3 Molecular orientational effects

Anisotropic molecules show optically isotropic behavior in the bulk when they are disordered and randomly oriented, for instance in solutions or liquid crystal above the transition temperature. Under the influence of a strong beam, the induced dipole moment of the molecules feels a torque that tends to orient the molecule. The reorientation of the molecular dipoles induces a change in the refractive index. The typical values for molecular susceptibilities and the time-responses vary depending on the type of systems. For small anisotropic molecular systems, $\chi^{(3)} \sim 10^{-12}$ esu, with a time response $\sim 10^{-12}$ s. However, in the nematic phase, liquid crystal molecules are strongly correlated, resulting in much higher values, $\chi^{(3)} \sim 10^{-2}$ esu, with slow time responses $\sim 10^{-3} - 10^{-2}$ s.

4.2.4 Electrostriction

Electrostriction is an effect that requires the existence of inhomogeneities in the intensity of the electric field. The inhomogeneous field creates a force on the molecules proportional to the gradient of the intensity of the electric field. A typical value for electrostriction susceptibility is $\chi^{(3)} \sim 10^{-12}$ esu, with a time response $\sim 10^{-9}$ s.

4.2.5 Population redistribution

This effect occurs when the frequency of the incident beam is near a resonant energy transition of the atom or molecule that is responsible for the nonlinear behavior. Near the resonance the electrons occupy a real excited state for a finite period of time. For low intensity light, the population redistribution results in a change of the index of reflection, since it is mostly determined by the molecules in the ground state. A typical value for population redistribution susceptibility is $\chi^{(3)} \sim 10^{-8}$ esu, with a time response $\sim 10^{-8}$ s.

4.2.6 Thermal contributions

In this case, the change in the refractive index is related to the changes of temperature. In general, as the temperature increases, the density of the material decreases. A change in the density reflects in a change of the refractive index. A typical value for population redistribution susceptibility is $\chi^{(3)} \sim 10^{-4}$ esu, with a time response $\sim 10^{-3}$ s.

4.2.7 Cascade second-order effects

Through cascade second-order effects, the second-order optical nonlinearities result on a third-order optical effect in a multistep or cascade process. This process is a due to the existence of microscopic electric fields that are generated by second-order nonlinear aligning of molecular dipoles.

4.2.8 Photorefractive Effect

The photorefractive effect is a physical mechanism where the change in the intensity-dependent refractive index is dependent on the spatial variations of intensity. It is a non-local process, because unlike most processes, the change in the refractive index is not dependent on the magnitude of the intensity that produces such change.

Comparing results from different techniques is difficult. First, the tensorial character of $\chi^{(3)}$ has to be taken into account. $\chi^{(3)}$ is a fourth-rank tensor, which means that it has 81 components. In isotropic media there are only three independent components, $\chi^{(3)}_{xxxx}$, $\chi^{(3)}_{xyxy}$ and $\chi^{(3)}_{xxyy}$. In the case of purely electronic contributions in the off-resonance regime the components are further related: $\chi^{(3)}_{xxxx} = 3\chi^{(3)}_{xxyy} = 3\chi^{(3)}_{xyxy}$.

Secondly, since $\chi^{(3)}$ is dependent on the input frequencies and near the electronic or vibrational resonances of the material the dependence is very strong, one has to be very careful when results from different techniques are compared since they might operate in the vicinity of different resonances. In fact, only the electronic contribution should be compared. Sometimes the experimental values are extrapolated to the off-resonance regime, which requires the use of a quantum mechanical model for the material response.

Finally, near a resonance, $\chi^{(3)}$ is a complex quantity. This has to be taken into account when the experiments only measure the magnitude of $\chi^{(3)}$. The imaginary part of $\chi^{(3)}$ generally leads to nonlinear absorption and therefore will deplete the beam intensity. Although some experimental setups allow the combined measurement of both the real and imaginary part of $\chi^{(3)}$, in general, different techniques will be needed to completely characterize the complex $\chi^{(3)}$ near the resonance.

The wavelength of operation is very important due to the strong frequency dependence of $\chi^{(3)}$ close to the resonance. Also in order to measure only the electronic contribution to $\chi^{(3)}$ and to avoid the dynamic nonlinearities that occur at time scales longer than ps, a short pulse duration and a low repetition rate is required.

Most measurements are made with respect to a reference material that has to be very well characterized. Each technique might need a different material as reference. For instance, while in Third-Harmonic Generation, glass is used as a reference, in Degenerate Four-Wave Mixing it is usually carbon sulfide.

Each experimental technique is best suited for a particular type of sample and will be more relevant for a particular type of application, but in general, the different experimental techniques complement each other allowing the study of the various parameters that determine $\chi^{(3)}$.

4.3 Third-Harmonic Generation

In the Third-Harmonic Generation (THG) experiment an input beam at frequency ω is incident into the nonlinear sample and an optical signal oscillating at frequency 3ω is generated through the nonlinear interaction inside of the material. This is described by $\chi^{(3)}(-3\omega, \omega, \omega, \omega)$:

(79) $$P_i(3\omega) = \frac{1}{4}\chi^{(3)}_{ijkl}(-3\omega, \omega, \omega, \omega) E_j(\omega) E_k(\omega) E_l(\omega)$$

Through THG only the electronic contribution to $\chi^{(3)}$ is measured, because no other mechanism is fast enough to produce a nonlinear polarization oscillating at the third-harmonic frequency of the incident beam. Harmonic generation is a coherent process that occurs through purely electronic interactions that are almost instantaneous.

The intensity of the third-harmonic signal is related to the path length of the beam inside of the material (in MKS units) through [57]:

(80) $$I_{3\omega} = \frac{(3\omega)^2}{n_{3\omega} n_\omega^3 c^4 \varepsilon_0^2} \frac{\sin^2[\Delta k(l/2)]}{[\Delta k(l/2)]^2} |\chi^{(3)}|^2 l^2 I_\omega^3$$

where $I_{3\omega}$ is the intensity of the third-harmonic generated beam, I_ω is the intensity of the input beam at frequency ω, n_ω is the linear refractive index at frequency ω, ε_0 is the permittivity of the free space, l is the interaction length and Δk is the wave vector mismatch. The above equation assumes non-depletion for the fundamental beam and non-absorbing media. These two conditions can be relaxed.

The third-harmonic signal is maximized when there is exact phase-matching:

(81) $\quad \Delta k = k_{3\omega} - 3k_\omega = 3\omega \dfrac{(n_{3\omega} - n_\omega)}{c} = 0$

Since for most materials $n_\omega < n_{3\omega}$, exact phase-matching is very difficult to obtain. Instead, in third-harmonic generation measurements the signal is maximized by proper change of the interaction length.

For non-phased THG, the Maker fringe or wedge-shaped fringe method is used to determine the coherence length of the material, l_c. The coherence length is defined as:

(82) $\quad l_c = \pi/\Delta k$

By studying the dependence of the third-harmonic signal as a function of the interaction length, the coherence length of the material is obtained.

With this method, absolute values of $\chi^{(3)}$ can be measured. In the simple case described by Eq. (80), the dependence of $I_{3\omega}$ as a function of I_ω could be obtained. Further knowledge of the material parameters (n_ω, $n_{3\omega}$, ε_0) is needed, as well as an accurate measurement of the interaction length. Furthermore, the pump intensity I_ω has to be properly characterized, the beam $1/e^2$ radius has to be determined and in the case of pulsing lasers, the pulse width has to be measured carefully.

In practice, it is customary to perform relative measurements, using a well-characterized reference material. Under the same experimental conditions (same input intensity in sample and reference), the coherence length is measured for both the reference and the sample. The third-order susceptibility is then approximated as:

(83) $\quad \dfrac{I_{3\omega,sample}}{I_{3\omega,reference}} = \left(\dfrac{\chi^{(3)}_{sample}}{\chi^{(3)}_{reference}}\right)^2 \left(\dfrac{l_c^{reference}}{l_c^{sample}}\right)^2.$

In the case of relative measurements, the selection of the wavelength is very crucial, since the reference values of $\chi^{(3)}$ are frequency dependent. Traditionally BK-7 glass has been used as reference material for THG. Heflin, Cai and Garito [58] report a value THG of $\chi^{(3)}_{xxxx} = 5.8 \times 10^{-15}$ esu at frequency $\lambda = 1910$ nm. If one intends to measure $\chi^{(3)}$ of a liquid solution, chloroform is a more convenient reference. However, for liquid samples, the contribution of the container windows to $\chi^{(3)}$ has to be taken into account. Kajzar and Messier report the following values of $\chi^{(3)}$ and l_c for fused silica (window material) and chloroform (reference material) at frequency $\lambda = 1064$ nm [59]:

(84) $\quad$ Fused Silica : $\chi^{(3)}_{xxxx} = 3.1 \times 10^{-14}$ esu and $l_c = 6.6\,\mu$m

(85) $\quad$ Chloroform : $\chi^{(3)}_{xxxx} = 7.0 \times 10^{-14}$ esu and $l_c = 5.2\,\mu$m

In practice, THG measurements are complicated by the fact that all materials show third-order nonlinear effects (unlike second-order nonlinear effects which require non-centrosymmetry). This means that the surroundings of the sample (including air) will contribute to $\chi^{(3)}$. Different ways of minimizing the contributions of the air and film support materials have been proposed [59–62].

Because THG only measures the electronic contribution to $\chi^{(3)}$, the pulse width it is not as relevant as in other type of experiments. Usually a Q-switched pulsed Nd^{3+}:Yag laser with nanosecond pulses is used, at low repetition rate (10–30 Hz). The selection of the wavelength is important in THG, especially when organic materials have to be characterized since organic materials generally absorb in the UV spectral range. For this reason, the fundamental output of the Nd^{3+}:Yag laser is usually shifted to a longer wavelength.

The laser beam is split in two parts. The first one goes through the reference and generates the third-harmonic signal in the reference, while the second one goes through the sample and generates the third-harmonic signal in the sample. By changing the path length of the sample and reference the coherence lengths are obtained, and the value of $\chi^{(3)}$ for the sample is obtained through Eq. (83).

4.4 Electric-Field-Induced Second-Harmonic Generation

This method is used to indirectly compute $\chi^{(2)}$ and the molecular hyperpolarizability, β. By applying a static electric field, one obtains χ_{eff} which has two components, one related to the third-order nonlinearity $\chi^{(3)}$, while the other one is related to the second-order nonlinearity, through dipole orientation (see section 3.2.3 for details). For clarity, we reproduce Eq. (42), which relates the induced polarization $P_i(2\omega)$ with the third-order nonlinear susceptibility $\chi^{(3)}(-2\omega; \omega, \omega, 0)$:

$$(86) \quad P_i(2\omega) = \frac{3}{2}\chi^{(3)}_{ijkl}(-2\omega, \omega, \omega, 0)E_j(\omega)E_k(\omega)E_l(0)$$

The solution of the non-linear wave equation can be approximated as the solution for SHG intensity with the strength of the response given by $\chi^{(3)}$ instead of $\chi^{(2)}$:

$$(87) \quad I_{2\omega} = I_{2\omega}^{max}\frac{\sin^2(\Delta k\, l/2)}{(\Delta k\, l/2)^2}$$

In this case the wave vector mismatch is defined as:

$$(88) \quad \Delta k = 2k_\omega - k_{2\omega} = 2\omega\frac{(n_\omega - n_{2\omega})}{c}$$

The EFISHG technique obtains $\chi^{(3)}$ from both second- and third-order nonlinearities. In the case of solutions, $\chi^{(3)}$ will be related to the microscopic third-order polarizability γ_0 of the dissolved molecules which is given by:

$$(89) \quad \gamma^0 = \langle\gamma\rangle + \frac{\mu\cdot\beta_{vec}}{5kT}$$

where $\mu \cdot \beta_{vec}$ is the scalar product between the molecular dipole moment vector with the vector part of the second-order polarizability tensor and $\langle \gamma \rangle$ is the scalar orientationally averaged part of the third-order polarizability tensor.

Although EFISHG is used to indirectly compute $\chi^{(2)}$ and the molecular first hyperpolarizability, for centrosymmetric structures, when $\mu = 0$ and $\beta = 0$, EFISHG will measure only contributions from the third-order molecular polarizability.

As in the THG measurements, no time response is obtained by EFISHG. Maker fringe or wedge-shaped fringe techniques can be applied to measure $\chi^{(3)}(-2\omega; \omega, \omega, 0)$ of samples. As the interaction length is changed, the second-harmonic signal undergoes maxima and minima. The spacing between the two consecutive maxima is two times the coherence length, l_c. For experimental setup details, refer to Section 3.2.3.

Levine and Bethea have estimated the third-order susceptibilities for solutions in terms of the coherence lengths and second-harmonic signal intensities using quartz as a reference [33, 63]:

$$(90) \quad \chi^{(3)} = \frac{(n_{2\omega}^L + n_{2\omega}^G)l_c^G}{(n_{2\omega}^G + n_\omega^G)l_c^L}\chi_G^{(3)} + \frac{(n_{2\omega}^L + n_{2\omega}^G)d_{11}l_c^Q}{(n_{2\omega}^G + n_\omega^G)E_0 l_c^L}\left(\frac{e^A I_{2\omega}^L}{I_{2\omega}^Q}\right)^{1/2}$$

where L stand for the liquid solution, G for the glass window cell, and Q for the reference quartz; n_ω is the linear refractive index at frequency ω; l_c is the coherence length; E_0 is the amplitude of the applied electric field, A is the attenuation coefficient induced by absorption of harmonic light; $I_{2\omega}$ is the peak intensity of the harmonic light; and d_{11} for quartz is defined as half the value of the second-order polarizability of quartz.

4.5 Degenerate Four-Wave Mixing

Four-Wave Mixing is the optical process where the nonlinear interaction between three beams generates a fourth beam. When the frequencies of the input waves are different, new frequencies are generated. The case of equal frequencies is called degenerate four-wave mixing (DFWM).

DFWM experiments are of particular interest in order to characterize centrosymmetric isotropic materials, where there is no competition with second-order effects and the time-response of the nonlinearity has to be evaluated.

The expression for the induced nonlinear polarization as a function of the input electric fields is given by:

$$(91) \quad P_i(\omega) = \frac{1}{2}\chi_{ijkl}^{(3)}(-\omega; \omega, \omega, -\omega)E_j(\omega)E_k(\omega)E_l^*(\omega)$$

DFWM processes get contributions from both the imaginary and real parts of $\chi^{(3)}$. It also allows measurement of electronic and dynamic contributions (molecular orientational, electrostriction, thermal effects...) to the third-order nonlinear susceptibility.

Two different geometries can be used in DFWM experimental setups: forward-wave geometry and backward-wave geometry.

In the forward-wave geometry, all the waves travel in the forward direction, making this type of geometry very well suited for the study of nonlinearities in thin samples. The typical configuration for the forward-wave DFWM geometry is shown in Fig. 8.

In this configuration, a forward pump E_f is incident in the sample, while a weaker probe beam E_p coming also in the forward direction is incident on the medium, making an angle θ with the forward pump E_f. Probe and pump beam overlap inside the sample and the output waves include a conjugated beam E_c apart from the transmitted pump and probe beams. Actually, in the case of a thin medium, there might be more than one new generated output wave. The nonlinear interaction of the pump and probe beams inside of the sample creates an index grating in the medium. The optically thin created grating self-diffracts these waves through Raman-Nath diffraction. Usually it is enough to consider only the conjugated beam E_c.

For the induced nonlinear polarization given by Eq. (91) the nonlinear wave equation can be solved under the assumption of no pump depletion and using the slowly varying approximation [26]. As in the case of THG or EFISGH, the wave equation yields a phase-matching condition on the wave-vectors of the four beams:

$$(92) \qquad \vec{k}_1 + \vec{k}_2 + \vec{k}_3 + \vec{k}_4 = 0$$

For the forward-geometry configuration, Eq. (92) determines the angle between the transmitted pump beam E_f and the generated conjugated beam E_c.

In the backward-wave geometry two waves, a pump and the probe, travel in the forward direction while the two other travel in the backward direction, one of them being the other pump and the other being the generated conjugated beam. As illustrated in Fig. 9, the sample is hit with two strong pump beams with opposite directions, E_f in the forward direction and E_b in the backward direction. At the same time, the probe beam E_p is incident at an angle θ with respect to the forward pump beam E_f, which results in the creation of a conjugated beam E_c that propagates counter to the probe beam.

The generated conjugated beam is proportional to the third-order polarization oscillating at frequency ω (Eq. 91). Since $\vec{k}_f = -\vec{k}_b$ and $\vec{k}_p = -\vec{k}_c$, the backward-geometry is always phase-matched.

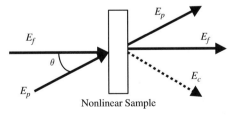

Figure 8. Schematic configuration for the forward-wave DFWM geometry

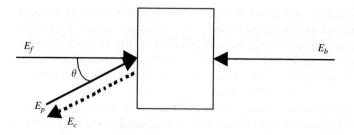

Figure 9. Schematic diagram for the backward-wave DFWM geometry

Backward-wave geometry is preferred when transparent and weakly absorbing samples are going to be studied. Because in this geometry the phase-matching condition (Eq. 92) is satisfied automatically, the method is very sensitive on the alignment of the beams.

4.5.1 Optical phase conjugation

When the backward geometry is used in a DFWM process, the process is also called Optical Phase Conjugation (OPC) [64]. One of the most interesting applications of OPC is the correction of optical (phase) aberrations. In the backward geometry, the conjugated wave travels in the opposite direction of the probe with exactly the time-reversed phase of the probe wave, so it becomes the "reflected" conjugated wave and the material responsible for the effect becomes a "phase-conjugated mirror". In OPC, when the probe beam goes through an aberrating medium, the reflected phase-conjugated beam has to follow exactly the reversed path through the aberration medium. Its reversed phase will then suffer the same aberration but with an opposite sign, so it cancels out with the positive phase aberration. In this way, through OPC an aberration-free output is generated.

Under the assumption of no depletion of the pumps, low reflectivity and no absorption, the intensity of the conjugated beam is related to the intensity of the other three beams [24].

$$(93) \quad I_c = \frac{\omega^3}{4c^2 n_\omega^2} \left|\chi^{(3)}\right|^2 l^2 I_f I_b I_p$$

Usually, the three input beams are obtained from the same fundamental beam, and the conjugated signal shows a cubic dependence with respect to the fundamental intensity. The efficiency of the backward-geometry DFWM is measured through the reflectivity $R = I_c/I_p$. Because the phase-conjugated beam can get energy from the pump beams, reflectivities higher than 100% have been reported [65–67].

Since both electronic and dynamic components can be measured, the choice of the laser pulse width is very important to determine which components will be measured. In organic materials, the peak power is usually in the $10\,\text{MW/cm}^2 - 1\,\text{GW/cm}^2$ range. The laser pulses have to be well resolved temporally and spatially and pulse fluctuations should be minimized. With long laser pulses, on the order of ns, dynamic nonlinearity and thermal contributions due to absorption might dominate the response. In order to measure only the electronic contribution, ps or smaller laser pulses should be used.

In the backward-wave geometry, the fundamental laser beam is usually split into forward, backward and pump beam. It is necessary to use delay lines in the case where the pulses are very short in order to assure a good temporal overlap between the three input beams. The output phase-conjugated intensity is maximized by changing the path length of one of the pump beams. The maximum value for the DFWM signal is measured for both sample (I_{sample}) and a reference ($I_{reference}$). In the case of non-absorbing medium, they are simply related [24]:

$$(94) \quad \frac{\chi^{(3)}_{sample}}{\chi^{(3)}_{reference}} = \left(\frac{n^0_{sample}}{n^0_{reference}}\right)^2 \frac{l_{reference}}{l_{sample}} \sqrt{\frac{I_{sample}}{I_{reference}}}$$

where n^0 is the linear refractive index and l is the path length of the media. Eq. (94) assumes that the interaction lengths of both sample and reference are the same as the path lengths, which means that experimentally the angle between the pump beam and the probe beam should be minimized and strong focusing should be avoided. Eq. (94) also assumes a cubic dependence of the DFWM signal on the input intensity, a fact that has to be monitored in the experimental setup.

Eq. (94) can be corrected for absorbing samples if the linear absorption coefficient is known. In any case, the value of the linear refractive index must be known. For liquid samples an Abe refractometer can be used. For solid samples index matching liquids can be used, although for thin films the most convenient method is the m-line technique [68].

DFWM has several advantages: the conjugated beam is easily distinguished through spatial separation; the intensity dependence of the conjugated beam is easy to check; different sample shapes can be used; all the components of $\chi^{(3)}$ can be determined; it does not depend strongly in the beam shape and it allows to study the time dependence of the nonlinearity.

However, it also has some disadvantages: it does not separate between real and imaginary parts of $\chi^{(3)}$ and it is very sensitive on the alignment of the incident beams, which usually requires short pulses on the order of ps and a good control of the experimental conditions.

4.6 Optical Kerr Gate

The Optical Kerr Gate (OKG) method allows measurement of $\chi^{(3)}$ by studying the polarization change of a probe beam, propagating through the system where the optical birefringence is induced through an intensity-dependent refractive index. The method was described by Ho [69].

An intense linearly polarized beam is used to induce optical birefringence in the media, $\delta n = n_\parallel - n_\perp$, which results on a change of polarization of the weaker probe beam. Monitoring the time evolution of the birefringence provides information for the response time of $\chi^{(3)}$.

The method is better suited for isotropic materials, since the optically induced anisotropy is usually small. The values of δn are obtained by monitoring the phase retardation of the probe beam, $\delta\phi$, as a function of the delay time between the probe and the orienting pulsing:

$$(95) \qquad \delta\phi(t) = \frac{2\pi l}{\lambda} \delta n(t)$$

where l is the sample path length and λ is the wavelength of the probe beam.

When the response is purely electronic the phase retardation can be related to the intensity of the pump beam through:

$$(96) \qquad \delta\phi(t) = \frac{2\pi l}{\lambda} n'_2 I_{pump}$$

where n'_2 is related with $\chi^{(3)}$ by:

$$(97) \qquad n'_2 = \frac{12\pi}{n^0} (\chi^{(3)}_{xxxx} - \chi^{(3)}_{xxyy})$$

4.7 Self-Focusing Methods

When there is a spatial variation of the laser intensity, the beam shape might change as it travels through a nonlinear material. This effect, which relates to the intensity-dependent refractive index, allows measurement of $\chi^{(3)}$ by two simple methods: Power Limiting and Z-scan.

The first method, Power Limiting, was proposed by Soileau et al. [70] and it is based on the idea of studying the intensity of the transmitted beam through a sample. It assumes a positive intensity-dependent refractive index.

When the intensity of the incident beam is low, the nonlinear effects are negligible, and the transmitted intensity (measured by a detector) is linear with the input intensity (see Fig. 10). After the intensity reaches a critical power, P_c, self-focusing occurs. A more focused beam induces other nonlinear effects which levels-off the transmitted power (see Figure 11). Therefore, for powers greater than P_c, the transmitted intensity does not depend linearly on the input intensity.

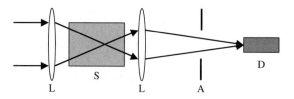

Figure 10. Schematic diagram representing the Power Limiting method at low input intensities. L: focusing lens; S: nonlinear sample; A: optical aperture; D: detector

Characterization Techniques of Nonlinear Optical Materials

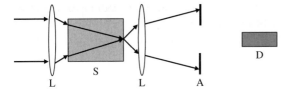

Figure 11. Schematic diagram representing the Power Limiting at high input intensities. L: focusing lens; S: nonlinear sample; A: optical aperture; D: detector

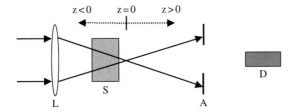

Figure 12. Schematic diagram representing the Z-scan method. L: focusing lens; S: nonlinear sample; A: optical aperture; D: detector. The sampled is moved along the z-axis

By solving the nonlinear equation for a focused Gaussian beam, P_c and n_2 can be related [71]:

$$(98) \quad P_c = \frac{3.72 c \lambda^2}{32 \pi^2 n_2}$$

The second method, Z-scan, was developed by Sheik-bahae et al. [72] Z-scan allows measurement of both the sign and the magnitude of $\chi^{(3)}$ with a simple setup (see Figure 12).

A Gaussian beam is focused and made to pass through the nonlinear medium. The output power is measured as a function of the sample position in the z-direction (with respect to the focal plane). The analysis of the transmitted intensity power profile provides the sign and magnitude of $\chi^{(3)}$.

While the main advantage of Z-scan is the fact that it requires a relatively simple set-up, it suffers from some disadvantages. It requires a high quality Gaussian beam shape (TEM_{00}). It also needs a high power density together with a long interaction length, so it is not recommended for polymers. Thermal effects are expected due to the required long interaction length. Finally, it does not provide time response information.

4.8 Nonlinear Fabry-Perot Methods

With the Fabry-Perot etalon method, $\chi^{(3)}$ is obtained by measuring the intensity-dependent phase shift that results from the intensity-dependent refractive index.

A Fabry-Perot etalon consists of two mirrors separated by a medium of refractive index n and thickness l. When a beam enters the etalon it undergoes successive multiple reflections. The ratio between incident and transmitted intensity is given by:

$$(99) \qquad \frac{I_t}{I_c} = \frac{1-R^2}{(1-R)^2 + 4R\sin^2(\delta/2)}$$

where no losses are assumed, R is the reflectivity of the mirrors and δ is the phase delay between two transmitted waves, differing in pathlength by one round-trip inside of the cavity. For normal incidence:

$$(100) \qquad \delta = \frac{4\pi n l}{\lambda}$$

From Eq. (100), maximum transmittance occurs when $\delta = 2m\pi$, where m is any integer. Thus the length separation between two consecutive maxima (Δl) is given by [54]:

$$(101) \qquad \frac{n\Delta l}{\lambda} = \frac{1}{2}$$

Therefore, measuring the length of separation between two consecutive maxima provides the refractive index. By monitoring the values of the refractive index as a function of the input intensity, n_2 can be obtained and then $\chi^{(3)}$. The nonlinear Fabry-Perot method can be used with liquids and solids.

4.9 Connection Between Microscopic and Macroscopic Quantities in Third-order Nonlinear Optics

Macroscopically the third-order nonlinear susceptibility $\chi^{(3)}$ is defined to correlate the third-order nonlinear polarization with the electric fields:

$$(102) \qquad \vec{P}^{(3)}(\omega) \propto \chi^{(3)}(\omega = \omega_i + \omega_j + \omega_k)\vec{E}(\omega_i)\vec{E}(\omega_j)\vec{E}(\omega_k)$$

At the molecular level the third-order nonlinear polarization $\vec{\mu}$, is correlated to the local electric fields $(\vec{f})$ through the third-order nonlinear molecular susceptibility γ:

$$(103) \qquad \vec{\mu}^{(3)}(\omega) \propto \gamma(\omega = \omega_i + \omega_j + \omega_k)\vec{f}(\omega_i)\vec{f}(\omega_j)\vec{f}(\omega_k)$$

With organic compounds it is very useful to characterize the nonlinear optical response at the molecular level, and it becomes necessary to establish the relationships between the macroscopic and microscopic quantities.

When the macroscopic third-order electric susceptibility $\chi^{(3)}_{IJKL}$ is measured in the laboratory (the subscripts I, J, K, L refering to the axes in the laboratory-fixed coordinate system), it is first related to the microscopic third-order coefficients c_{IJKL} [19]:

$$(104) \qquad \chi^{(3)}_{IJKL} = Nf_I(\omega_1)f_J(\omega_2)f_K(\omega_3)f_L(\omega_4)c_{IJKL}$$

where N is the species number density and $f_I(\omega_i)$ is the local field factor corresponding to the radiation frequency ω_i. The simplest way to calculate the local field factors is to use the Lorentz local field model that applies to dipolar liquid or solid solutions [73]:

$$(105) \quad f_I(\omega_i) = \frac{n(\omega_i)+2}{3}$$

where $n(\omega_i)$ is the refractive index for the liquid at frequency ω_i. The Lorentz local field model is an approximation that assumes that the species occupy a spherical cavity in the material and that the local surroundings of the species are treated as a continuum.

The microscopic third-order nonlinear coefficients c_{IJKL} are expressed in laboratory-fixed coordinates and must be related to the molecule-fixed coordinate system. Instantaneously:

$$(106) \quad c_{IJKL} = \sum_{ijkl} \phi_{Ii}\phi_{Jj}\phi_{Kk}\phi_{Ll}\gamma_{ijkl}$$

where ϕ_{Ii} is the direction cosine between the laboratory-fixed I axis and the molecule-fixed i axis, and γ_{ijkl} is the $ijkl$ component of the third-order polarizability tensor.

In the case of isotropic liquids, the instantaneous coefficients have to be averaged over all direction in order to obtain the orientationally averaged third-order polarizability, showing the statistical contributions of all molecules:

$$(107) \quad \langle\gamma\rangle = \langle c_{IJKL}\rangle = \sum \langle\phi_{Ii}\phi_{Jj}\phi_{Kk}\phi_{Ll}\rangle \gamma_{ijkl}$$

As the third-order electric susceptibility, γ_{ijkl} has 81 components. Fortunately, the symmetry of the molecule allows to reduce the number of nonzero components. As an example, if molecules belonging to the orthorhombic point group are considered there are only 21 nonzero components [74]: γ_{iiii}, where $i = x, y, z$ and γ_{iijj}, γ_{ijij}, γ_{jiji}, γ_{ijji}, γ_{jjii}, γ_{jiij}, where $i, j = x, y$; $i, j = x, z$; and $i, j = y, z$.

Far off from the resonances, under the assumption that there is no energy dissipation through the nonlinear process, Kleinman symmetry can be used to reduce the number of independent components of γ_{ijkl} [75]. In the case of molecules that belong to the orthorhombic point group, the 21 nonzero components get reduced to only 6 independent components: γ_{xxxx}, γ_{yyyy}, γ_{zzzz}, $6\gamma_{xxyy}$, $6\gamma_{xxzz}$ and $6\gamma_{yyzz}$.

Assuming an isotropic liquid of the orthorhombic point group, Eq. (107) reduces to:

$$(108) \quad \langle\gamma\rangle = \frac{1}{5}(\gamma_{xxxx} = \gamma_{yyyy} + \gamma_{zzzz} + 2\gamma_{xxyy} + 2\gamma_{xxzz} + 2\gamma_{yyzz})$$

It is important to realize that experimentally, for amorphous materials, only the averaged third-order polarizability can be measured, so the information about the different components of γ_{ijkl} is lost.

5. CONCLUSIONS

We have reviewed the principles of operation of the most relevant techniques employed to characterize the second- and third-order response of nonlinear media.

While, in principle, the second-order response should have a higher strength than the third-order response, a strong geometrical condition (noncentrosymmetry at the atomic/molecular and at the bulk levels) limits the availability of second-order nonlinear materials. Experimentally, one has to ensure that a noncentrosymmetric configuration is used if one desires to measure the strength of the second-order nonlinear response, characterized by $\chi^{(2)}$.

The measurement of third-order nonlinear response, characterized by $\chi^{(3)}$, is simplified because no geometrical condition in the material is required. The intensity-dependent refractive index, a unique feature of the third-order nonlinear response, allows to characterize $\chi^{(3)}$ by studying the change in the refractive index of the nonlinear material. This effect is exploited in numerous technical applications, and results in different experimental techniques that determine $\chi^{(3)}$. However, the absence of a geometrical condition in the material results in an extra complication when measurements are performed, since all materials (cell walls, glass, air, ...) contribute to $\chi^{(3)}$.

Both $\chi^{(2)}$ and $\chi^{(3)}$ are complex tensorial quantities and in general, the different experimental techniques complement each other in order to fully characterize $\chi^{(2)}$ and $\chi^{(3)}$. In the case of molecular materials, the response of the bulk can be related to the individual molecular response. One has to always keep in mind that the experimental setup has to meet the appropriate conditions imposed by the particular model applied to derive $\chi^{(2)}$ or $\chi^{(3)}$ from the experimental measurements.

Although absolute methods are available, in most cases reference samples are used to simplify the experimental procedure. The general techniques must be adapted to perform measurements on crystals, films, solutions, etc, which adds some complexity to the methods.

Finally, it is important to realize that the development and improvement of characterization techniques rely heavily on instrumental developments, a fact that can not be overemphasized. After all, nonlinear optical effects were not known until the invention of a highly coherent and powerful enough light source, the laser.

REFERENCES

[1] Franken, P.A., Hill, A.E., Peters, C.W., Weinreich, G.: Generation of optical harmonics, Phys. Rev. Lett. **7**, 118–119 (1961)
[2] Giordmaine, J.A., Miller, R.C.: Tunable coherent parametric oscillation in LiNbO$_3$, at optical frequencies, Phys. Rev. Lett. **14**, 973–976 (1965)
[3] Harris, S.E.: Tunable optical parametric oscillators, Proc. IEEE **57**, 2096–2113 (1969)
[4] Edelstein, D.C., Wachman, E.S., Tang, C.L.: Broadly tunable high repetition rate femtosecond optical parametric oscillator, Appl. Phys. Lett. **54**, 1728–1730 (1989)
[5] Bosemberg, W.R., Guyer, D.R.: Single-frequency optical parametric oscillator, Appl. Phys. Lett. **61**, 387–389 (1993)

[6] Mak, G., Fu, Q., van Driel, H.M.: Externally pumped high repetition rate femtosecond infrared optical parametric oscillator, Appl. Phys. Lett. **60**, 542–544 (1992)
[7] Betts, G.E.: Linearized modulators for suboctave-bandpass optical analog links, IEEE Trans. Microwave Theory Tech. **42**, 2642–2649 (1994)
[8] Lee S.S., Udupa A.H., Erlig, H., Zhang, H., Chang, Y., Zhang, C., Chang, D.H., Bhattacharya, D., Tsap, B., Steiner, W.H., Dalton, L.R., Fetterman, H.R.: Demonstration of a photonically controlled RF phase shifter, IEEE Microwave Guided Wave Lett. **9**, 357–359 (1999)
[9] Lee, S.S., Garner, S.M., Chuyanov, V., Zhang, H., Steiner, W.H., Wang, F., Dalton, L.R., Udupa, A.H., and Fetterman, H.R.: Optical intensity modulator based on a novel electrooptic polymer incorporating a high $\mu\beta$ chromophore, IEEE J. Quant. Electron. **36**, 527–532 (2000)
[10] Bloom, D.M., Bjorklund, G.C.: Conjugate wave-front generation and image reconstruction by four-wave mixing, Appl. Phys. Lett. **31**, 592–594 (1997)
[11] Fisher, R.A.: Ed., Optical Phase Conjugation, Academic New York, 1983
[12] Pepper, D.M.: Optical Engineering **21**, 156–183 (1982)
[13] Agrawal, G.P., Nonlinear Fiber Optics, Academic, San Diego, 1995
[14] Mollenauer, L.F., Stolen, R.H., Gordon, J.P.: Experimental observation of picosecond pulse narrowing and solitons in optical fibers, Phys. Rev. Lett. **45**, 1095–1098 (1980)
[15] Zakharov, V.E., Shabat, A.B.: Exact theory of two-dimensional self-focusing and one-dimensional self-modulation of waves in nonlinear media, Sov. Phys. JETP **34**, 62–69 (1972)
[16] Gibbs, H.M.: Optical Bistability, Academic, Orlando, 1985
[17] Gibbs, H.M., McCall, S.L., and Venkatesan, S.L.: Differential gain and bistability using a sodium-filled Fabry-Perot interferometer, Phys. Rev. Lett. **36**, 1135–1138 (1976)
[18] Lugiato, L.A., Theory of Optical Bistability in Progress in Optics XXI, Edited by Wolf. E., North-Holland Physics Publishing, Amsterdam, pp. 69–216 (1984)
[19] Prasad, P.N., and Williams, D.J.: Introduction to Nonlinear Optical Effects in Molecules and Polymers, Wiley, New York, 1991
[20] Zernike, F., Midwinter, J.E.: Applied nonlinear optics, Wiley, New York, 1973
[21] Weis, R.S., Gaylord, T.K.: Lithium niobate: Summary of physical properties and crystal structure, Appl. Phys. A **36**, 191–203 (1985)
[22] Lines, M.E., Glass, A.M.: Principles and Applications of Ferroelectrics and Related Materials, Clarendon, Oxford, 1997
[23] Klein, M.B., Dunning, G.J., Valley, G.C., Lind, R.C., O'Meara, T.R.: Imaging threshold detector using a phase-conjugate resonator in BaTiO3, Opt. Lett. **11**, 575–577 (1986)
[24] Bredas, J.L., Adant, C., Tackx, P, Persoons A.: Third-Order nonlinear optical response in organic materials: theoretical and experimental aspects, Chem. Rev. **94**, 243–278 (1994)
[25] Shi, R.F., Garito, A.F.: Introduction: conventions and standards for nonlinear optical processes, in Characterization techniques and tabulations for organic nonlinear optical materials, edited by Kuzyk, M.G., and Dirk, C.W., Marcel Dekker, Inc. pp. 1–13 (1998)
[26] Boyd, R.W.: Nonlinear optics, Academic Press, San Diego, 1992
[27] Maker, P.D., Terhune, R.W., Nisenoff, M., Savage, C.M.: Effects of dispersion and focusing on the production of optical harmonics, Phys. Rev. Lett. **8**, 21–22 (1962)
[28] Jerphagnon, J., Kurtz, S.K.: Maker fringes: a detailed comparison of theory and experiment for isotropic and uniaxial crystals, J. Appl. Phys. **41**, 1667–1681 (1970)
[29] Jerphagnon, J., Kurtz, S.K.: Optical nonlinear susceptibilities: accurate relative values for quartz, ammonium dihydrogen phosphate, and potassium dihydrogen phosphate, Phys. Rev. B **1**, 1739–1744 (1970)
[30] Verbiest, T., Kauranen, M., Persoons, A.J.: Second-order nonlinear optical properties of chiral thin films, Mater. Chem. **9**, 2005–2012 (1999)
[31] Sioncke, S., Verbiest, T., Persoons, A.: Second-order nonlinear properties of chiral materials, Materials Science and Engineering **R42**, 115–155 (2003)
[32] Oudar, J.L.: Optical nonlinearities of conjugated molecules. Stilbene derivatives and highly polar aromatic compounds, J. Chem. Phys. **67**, 446–457 (1997)

[33] Levine, B.F., Bethea, C.G.: Second and third order hyperpolarizabilities of organic molecules, J. Chem. Phys. **63**, 2666–2682 (1975)
[34] Lamala, S.J., Garito, A.F.: Origin of the nonlinear second-order optical susceptibilities of organic systems, Phys. Rev. A **20**, 1179–1194 (1979)
[35] Kielich, S.: Optical second-harmonic generation by electrically polarized isotropic media, IEEE Journal of Quantum Electronics **5**, 562–568 (1969)
[36] Levine, B.F.: Donor-acceptor charge transfer contributions to the second order hyperpolarizability, Chem. Phys. Lett. **37**, 516–520 (1976)
[37] Puccetti, G.: Electric field induced second harmonic generation/third harmonic generation measurements on molecules with extended charge transfer: Absorption domain and strong resonance effects, J. Chem. Phys. **102**, 6463–6475 (1995)
[38] Levine, B.F., Bethea, C.G.: Molecular hyperpolarizabilities determined from conjugated and nonconjugated organic liquids, Appl. Phys. Lett. **24**, 445–447 (1974)
[39] Levine, B.F., Bethea, C.G.: Absolute signs of hyperpolarizabilities in the liquid state, J. Chem. Phys. **60**, 3856–3858 (1974)
[40] Levine, B.F.: Conjugated electron contributions to the second order hyperpolarizability of substituted benzene molecules, J. Chem. Phys. **63**, 115–117 (1975)
[41] Bethea, C.G.: Experimental technique of dc induced SHG in liquids: measurement of the nonlinearity of CH_2I_2, Applied Optics **14**, 1447–1451 (1975)
[42] Kajzar, F., Ledoux, I., Zyss, J.: Electric-field-induced optical second-harmonic generation in polydiacetylene solutions, Phys. Rev. A **36**, 2210–2219 (1987)
[43] Flipse, M.C., de Jonge, R., Woudenberg, R.H., Marsman, A.W., van Walree, C.A., Jenneskens, L.W.: The determination of first hyperpolarizabilities β using hyper-Rayleigh scattering: a caveat, Chem. Phys. Lett. **245**, 297–303 (1995)
[44] Morrison, I.D., Denning, R.G., Laidlaw, W.M., Stammers, M.A.: Measurement of first hyperpolarizabilities by hyper-Rayleigh scattering, Rev. Sci. Instrum. **67**, 1445–1453 (1996)
[45] Song, N.W., Kang, T.-I., Jeoung, S.C., Joen, S.-J., Cho, B.R., Kim, D.: Improved method for measuring the first-order hyperpolarizability of organic NLO materials in solution by using the hyper-Rayleigh scattering technique, Chem. Phys. Lett. **261**, 307–312 (1996)
[46] Noordman, O.F.J., van Hulst, N.F.: Time-resolved hyper-Rayleigh scattering: measuring first hyperpolarizabilities β of fluorescent molecules, Chem. Phys. Lett. **253**, 145–150 (1996)
[47] Hubbard, S.F., Petschek, R.G., Singer, K.D.: Spectral content and dispersion of hyper-Rayleigh scattering, Opt. Lett. **21**, 1774–1776 (1996)
[48] Olbrechts, G., Trobbe, R., Clays, K., Persoons, A.: High-frequency demodulation of multi-photon fluorescence in hyper-Rayleigh scattering, Rev. Sci. Instrum. **69**, 2233–2241 (1998)
[49] Hendrickx, E., Clays, K., Persoons, A.: Hyper-Rayleigh Scattering in Isotropic Solution, Acc. Chem. Res. **31**, 675–683 (1998)
[50] Clays, K., Persoons, A.: Hyper-Rayleigh scattering in solution, Phys. Rev. Lett. **66**, 2980–2983 (1991)
[51] Clays, K., Persoons, A.: Hyper-Rayleigh scattering in solution, Rev. Sci. Instrum. **63**, 3285–3289 (1992)
[52] Asselberghs, I., Clays, K., Persooons, A., McDonagh, A., Ward, M., McCleverty, J., Chem. Phys. Lett. **368**, 408–411 (2003)
[53] Asselberghs, I., Clays, K., Persooons, A., McDonagh, A., Ward, M., McCleverty, J., Switching of molecular second-order polarisability in solution, J. Mater. Chem. **14**, 2831–2839 (2004)
[54] Sutherland, R.L.: Handbook of Nonlinear optics, Marcel Dekker, Inc., 1996
[55] Stegeman, G.I., Miller, A.: Physics of All-Optical Switching devices: In: Photonic Switching, vol. 1,: Midwinter, J.E. (ed) Academic, Orlando, Fl., pp. 81–146 (1993)
[56] Stegeman, G.I.: Material figures of merit and implications to all-optical waveguide switching, Proc. SPIE vol. **1853**, 75–89 (1993)
[57] Thalhammer, M., Penzkofer, A.: Measurement of third order nonlinear susceptibilities by non-phase matched third harmonic generation, Appl. Phys. B **32**, 137–143 (1983)
[58] Heflin, J.R., Cai, Y.M., Garito, A.F.: J. Opt. Soc. Am. B **8**, 2132–2139 (1991)

[59] Kajzar, F., and Messier, J.: Third-harmonic generation in liquids, Phys. Rev. A **32**, 2352–2363 (1985)
[60] Meredith, G.R., Buchalter, B., Hanzlik, C.: Third-order susceptibility determination by third harmonic generation. II, J. Chem. Phys. **78**, 1543–1551 (1983)
[61] Kajzar, F., Messier, J.: Nonlinear Optical Properties of Organic Molecules and Crystals, Vol. II, edited by Chemla, D.S., and Zyss, J., (Academic, Orlando, Fl), pp. 51–83 (1987)
[62] Kajzar, F., Messier, J.: Original technique for third-harmonic-generation measurements in liquids, Rev. Sci. Instrum. **58**, 2081–2085 (1987)
[63] Huijts, R.A., Hesselink, G.L.J.: Length dependence of the second-order polarizability in conjugated organic molecules, Chem. Phys. Lett. **156**, 209–212 (1989)
[64] Hellwarth, R.W.: Generation of time-reversed wavefront by nonlinear refraction, J. Opt. Soc. Am. **67**, 1–3 (1977)
[65] Bloom, D.M., Liao, P.F., Economou, N.P.: Observation of amplified reflection by degenerate four-wave mixing in atomic sodium vapor, Opt. Lett. **2**, 58–61 (1978)
[66] Pepper, D.M., Fekete, D., Yariv, A.: Observation of amplified phase-conjugate reflection and optical parametric oscillation by degenerate four-wave mixing in a transparent medium, Appl. Phys. Lett. **33**, 41–44 (1978)
[67] Jain, R.K., Klein, M.B., Lind, R.C.: High-efficiency degenerate four-wave mixing of 1.06-Mum radiation in silicon, Opt. Lett. **4**, 328–331 (1979)
[68] Ding, T.N., Garmire, E.: Measuring refractive index and thickness on films: a new technique, Appl. Opt. **22**, 3177–3181 (1983)
[69] Ho, P.P.: Semiconductors Probed by Ultrafast Laser Spectroscopy, Vol. 2, edited by Alfano, R.R., Academic, New York, pp. 410–439 (1984)
[70] Soileau, M., Williams, W., Van Stryland, E.: Optical power limiter with picosecond response time, IEEE J. Quant. Electron. **19**, 731–735
[71] Marburger, J.H.: Progress of Quantum Electronics **4**, edited by Sandom J.H., Stenholm, S., Pergamon, New York, pp. 35–110 (1997)
[72] Sheik-Bahae, M., Said, A.A., Van Stryland, E.W.: High-sensitivity, single-beam n2 measurements, Opt. Lett. **14**, 995–958 (1989)
[73] Lorentz, H.A.: The Theory of Electrons and Its Applications to the Phenomena of Light and Radiant Heat, Leipzig, Germany, Teubner, (1909)
[74] Hellwarth, R.W.: Third-order optical susceptibilities of liquids and solids, Prog. Quant. Elect. **5**, 1–68 (1997)
[75] Kleinman, D.A.: Nonlinear dielectric polarization in optical media, Phys. Rev. **126**, 1977–1979 (1962)

CHAPTER 15

THIRD-ORDER NONLINEAR OPTICAL RESPONSE OF METAL NANOPARTICLES

BRUNO PALPANT
Institut des Nano-Sciences de Paris, Université Pierre et Marie Curie – Paris 6, Université Denis Diderot – Paris 7, CNRS, Campus Boucicaut, 140 rue de Lourmel, 75015 Paris, France

Abstract: We present a review of the main results reported in the literature regarding the third-order nonlinear optical response of nanocomposite media consisting of noble metal nanoparticles surrounded by a dielectric host. This phenomenon, known as optical Kerr effect, can be characterized by the intensity-dependent complex optical index of the material or, equivalently, its complex third-order susceptibility. The theoretical basis of the linear and nonlinear optical properties of metal nanoparticles and nanocomposite media are described first. The different third-order optical phenomena which have been observed in such materials are then examined. The dependence of the nonlinear properties on morphological parameters – nature of the dielectric host, metal concentration, particle size and shape – as well as on laser excitation characteristics – wavelength, intensity, pulsewidth – will be explained and illustrated by selected experimental results. The final part points out the important role played by thermal effects in the nonlinear optical response

Keywords: Noble metals; nanoparticles; nanocomposite materials; surface plasmon resonance; local field enhancement; nonlinear optical response; optical Kerr effect; third-order susceptibility; saturation of absorption; optical limiting; self-focusing; metal concentration; size effects; spectral dispersion; interband transitions; hot electrons; thermal lensing

1. INTRODUCTION

The fascinating optical properties of metal nanoparticles have caught the attention of many researchers from the pioneering and almost parallel works of G. Mie and J.C. Maxwell-Garnett at the beginning of the twentieth century. These original properties, like many other phenomena specifically appearing in matter divided to the nanoscale, are linked with confinement effects, since quasi-free conduction electrons cannot spread beyond the limits of the metal nanoparticle. When particle size ranges from nanometer to a few tens of nanometers, confinement results in the possibility of resonantly exciting the electron gas collectively by coupling with an appropriate oscillating electromagnetic field. This phenomenon is known as the surface plasmon resonance (SPR). Whereas some of its detailed features may be explained by invoking quantum effects, its essential characteristics are understandable through very classical considerations. In the optical response of a material containing metal nanoparticles it manifests itself as an absorption band, which is located in the visible or near ultraviolet spectral domain for noble metal spheres.

As the local electric field in the particles is enhanced at the SPR, the metal nonlinear optical response can be amplified as compared to the bulk solid one. Moreover, the intrinsic nonlinear properties of metals may themselves be modified by effects linked with electronic confinement. These interesting features have led an increasing number of people to devote their research to the study of nonlinear optical properties of nanocomposite media for about two decades. The third-order nonlinear response known as optical Kerr effect have been particularly investigated, both theoretically and experimentally. It results in the linear variation of both the refraction index and the absorption coefficient as a function of light intensity. These effects are usually measured by techniques employing pulsed lasers.

In this chapter, we will present a large but non-exhaustive review of the main results which have been published about the third-order nonlinear optical properties of metal/dielectric nanocomposite materials. These properties depend significantly on many factors regarding both the materials themselves (metal and host medium kinds, metal concentration, particle size, shape and spatial arrangement) and the excitation laser (wavelength, intensity, pulsewidth). The comparison of different experimental results then appears to be a quite difficult task to perform if one aims at highlighting the role played by each of these factors independently. Nevertheless, several general features can be extracted from the abundant literature, as will be established in the different following sections.

In the first part, emphasis will be put on the linear optical properties of dielectric media doped with noble metal nanoparticles. Indeed, the study of the linear response is definitely needed to further explore the nonlinear one. We will then introduce the fundamentals of the theoretical tools required to understand why and how people inquire into the third-order nonlinear properties of nanocomposite materials. In the second part, experimental results will be presented by first examining the different nonlinear optical phenomena which have been observed in these media. We will then focus on the nanoparticle intrinsic nonlinear susceptibility before analysing the influence of the main morphological factors on the nonlinear optical response. The dependence of the latter on laser characteristics will finally be investigated, as well as the crucial role played by different thermal effects.

2. LINEAR OPTICAL PROPERTIES OF NOBLE METAL NANOPARTICLES AND NANOCOMPOSITE MEDIA

The propagation, in the linear regime, of an electromagnetic wave in a homogeneous and isotropic medium is governed by the usual complex optical index $\tilde{n} = n + i\kappa$ of the latter. n is the refractive index and κ the extinction coefficient, proportional to the absorption coefficient $\alpha = 4\pi\kappa/\lambda$ where λ is the wavelength of the incident radiation. The complex index is linked to the medium dielectric function $\varepsilon = \varepsilon_1 + i\varepsilon_2$ through $\varepsilon = \tilde{n}^2$. We will now particularly examine the optical properties of noble metals, first in their bulk phase, then as nanoparticles, before getting insight into those of nanocomposite materials.

2.1 Dielectric Function of the Noble Metals

Noble metals – copper, silver and gold – are monovalent elements with a *fcc*-like crystallographic structure in the bulk phase under normal conditions. Their dielectric function has been the subject of various experimental investigations in the past [1–6]. A compilation and an analyse of the main results can be found in [7]. The response of noble metals to an electromagnetic excitation in the UV–visible range cannot be described, contrarily to the case of alkalis, by the only behaviour of the quasi-free conduction electrons (*sp* band), but must include the influence of the bound electrons of the so-called *d* bands [8]. Hence, the total dielectric function ε_m of noble metals can be written as the sum of two contributions, one due to electronic transitions within the conduction band (intraband transitions) and the other stemming from transitions from the *d* bands to the conduction one (interband transitions):

(1) $\quad \varepsilon = \varepsilon_f + \varepsilon_{ib}.$

ε_f stands for the free-electron contribution, which can be described in a classical way by the Drude model [9]:

(2) $\quad \varepsilon_f(\omega) = 1 - \dfrac{\omega_p^2}{\omega(\omega + i\Gamma)}.$

ω is the applied wave circular frequency, Γ is a phenomenological damping constant characterizing all the collision processes experienced by the electrons in the metal, and ω_p is the volume plasmon circular frequency, given by

(3) $\quad \omega_p = \sqrt{\dfrac{Ne^2}{m^*\varepsilon_0}}$

where N, e and m^* are the density, charge and effective mass of the conduction electrons, respectively, and ε_0 the permittivity of vacuum. The interband contribution to the dielectric function, ε_{ib}, can be calculated from the detailed band structure of the metal [1, 10–13]. Usually, only the imaginary part of ε_{ib} is determined this way, the real part being subsequently deduced using the Kramers-Kronig relations. Due to the Pauli principle, there is a minimum photon energy for which an interband (IB) transition can occur, corresponding to the excitation of an electron from the top of the d band to the Fermi level. This defines an energy threshold under which the imaginary part of ε_{ib} is zero. Whereas this threshold lies in the UV for silver ($\sim 3.9\,\mathrm{eV}$) [1] it is in the visible for gold and copper (at about 2.4 and 2.1 eV, respectively) [1, 4], which explains the specific colour of these metals in the bulk phase.

2.2 The Surface Plasmon Resonance in Nanoparticles

2.2.1 Intuitive description

When dividing bulk metal into very small entities, its optical response changes drastically due to the confinement of the electrons. Indeed, if the size of such an entity is much smaller than the applied radiation wavelength (which is the case of nanoparticles in the near-UV–visible spectral range), all the conduction electrons experience the same homogeneous electromagnetic field and oscillate collectively like, at first order, a giant dipole. This excitation is resonant when the applied wave frequency matches the eigenfrequency of the electron gas motion relative to the ionic core. This phenomenon is known as the *surface plasmon resonance* (SPR). From the more realistic quantum point of view, it corresponds to the excitation of coherent electronic transitions within the conduction band.

2.2.2 Local field factor

The SPR can be simply formalized, in a first approach, by solving Laplace's equation in the case of a single conducting sphere surrounded by a homogeneous transparent medium, with the appropriate continuity relations at the metal–dielectric interface and assuming that the sphere radius is much lower than the wavelength (quasi-static approximation). The homogeneous local electric field inside the particle, $\mathbf{E}_1$, then writes

$$(4) \quad \mathbf{E}_1 = \frac{3\varepsilon_d}{\varepsilon_m + 2\varepsilon_d} \mathbf{E}_0,$$

where $\mathbf{E}_0$ is the applied field, ε_m and ε_d the dielectric functions of metal and host medium, respectively. One then defines the local field factor, f, as the ratio of the local field to the applied one: $f = E_1/E_0$. It is a complex quantity and will be a highly relevant parameter when discussing below the third-order nonlinear response of nanocomposite materials. It can be seen in Eq. (4) that $|f|$ presents a resonance behaviour at the minimum value of $|\varepsilon_m + 2\varepsilon_d|$. The dielectric host being transparent in the spectral range of interest, ε_d is real. If ε_{m2} is negligible, or if $\partial\varepsilon_{m2}/\partial\omega \approx 0$, the resonance condition simplifies into

$$(5) \quad \varepsilon_{m1} = -2\varepsilon_d.$$

For a noble metal nanoparticle the dielectric function of which is given by Eqs. (1) and (2) with $\omega \gg \Gamma$, this condition leads to the following resonance circular frequency:

$$(6) \quad \omega_{sp} = \frac{\omega_p}{\sqrt{\varepsilon_{ib1}(\omega_{sp}) + 2\varepsilon_d(\omega_{sp})}}$$

which simplifies into $\omega_{sp} = \omega_p/\sqrt{3}$ for a sphere of simple metal in vacuum. Unfortunately, Eq. (6) fails in accurately predicting the SPR frequency for gold and especially copper nanoparticles, because the hypotheses leading to Eq. (5) are not valid, due to the spectral proximity of the SPR to the IB transition threshold. Nevertheless, such a simple analytical formula allows to discuss the influence of the bound d electrons or the host matrix refractive index on the SPR spectral location.

Equation (4) is illustrated on Fig. 1 which exhibits the spectral dependence of the local field modulus in different cases, and from which several major features regarding the SPR can be highlighted. First, the SPR of silver particles, located in the UV, has a larger oscillator strength than the gold particle one, lying in the visible. This is mainly due to the coupling between the core d electrons and the conduction electrons, which is significant in gold (as well as in copper) while being negligible in silver. Secondly, the SPR amplitude is as large as the host matrix refractive index is high. Thirdly, the SPR maximum shifts towards red with increasing ε_d.

As the existence of a resonance behaviour can be explained by pure electromagnetic considerations, using as only ingredients macroscopic quantities that are the dielectric functions of the different media, the local field amplification phenomenon is often said to originate from *dielectric confinement*.

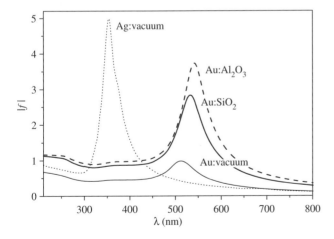

Figure 1. Modulus of the local field factor, $|f|$, calculated by using Eq. (4) for silver nanoparticles in vacuum ($\varepsilon_d = 1$) and gold nanoparticles in different surrounding media: Vacuum, silica ($\varepsilon_d \approx 2.1$) and alumina ($\varepsilon_d \approx 2.6$). The precise dielectric functions are those of [7]

2.2.3 Mie theory

At the beginning of the twentieth century, G. Mie published his results regarding the optical response of a sphere to an electromagnetic wave excitation [14]. A detailed development of his calculations can be found in [15] and [16]. He obtained the analytical expression of the extinction cross section as a multipolar expansion by solving Maxwell's equations in spherical coordinates. Hence, his theory can be applied for particles much bigger than the ones we are dealing with in the present article, where the first dipolar electric term alone largely dominates the optical behaviour; extinction reduces to pure absorption, elastic scattering of light being negligible. In this case, the absorption cross section writes

$$(7) \quad \sigma_{abs}(\omega) = 9\frac{\omega}{c}\varepsilon_d(\omega)^{3/2} V \frac{\varepsilon_{m2}(\omega)}{[\varepsilon_{m1}(\omega) + 2\varepsilon_d(\omega)]^2 + \varepsilon_{m2}(\omega)^2}$$

where c is the speed of light in vacuum and V the particle volume. σ_{abs} exhibits a resonance profile, with the same resonance condition as the one obtained above for the local field factor (Eq. 5).

The SPR is then also called *Mie resonance*. For simple metals, the SPR absorption band has a Lorentzian shape peaked at ω_{sp}, the width of which is directly proportional to the collision constant Γ introduced in the Drude description of the metal dielectric constant (Eq. 2). Of course, for noble metals the absorption due to interband transitions has to be taken into account in order to obtain the complete spectrum.

2.2.4 Finite size effects

In the dipolar approximation of the Mie theory, particle size is involved only insofar as absorption is proportional to particle volume (Eq. 7). It will play an important part as soon as higher order terms in the multipolar expansion of the extinction cross section become significant [17, 18]. In the corresponding size range (a few tens of nanometers to micrometers) the Mie theory is fully suited to describe, for example, the scattering of light by atmospheric dusts, fog, factory smokes, etc... However, it fails in explaining size effects observed in the optical response of nanoparticles whose radius is lower than about a few nanometers. In particular, there are other phenomena, linked to the finite size, which lead to the modification of the SPR spectral location, amplitude and width. The term *electronic confinement* is sometimes used in the literature to design all the finite (quantum) size effects. It has been the subject of a very large number of both theoretical and experimental studies, the description of which would obviously be out of the scope of this article. A very complete review of these works – and, more generally, of the studies devoted to the linear optical properties of metal nanoparticles – has been written by U. Kreibig and M. Vollmer about ten years ago in [19].

The theoretical approaches range from the simplest phenomenological models to complex quantum calculations. They can be split into two main strategies. The first one consists in keeping the classical Mie expression for the absorption cross section and modifying in a proper manner the metal dielectric function by introducing a size dependence in its parameters. In this aim, one can either restrict oneself to a pure classical description of the phenomena involved, or use the results of quantum calculations. Of course, such an approach cannot hold if particles are very small (less than a few hundreds of atoms, i.e. having a sub-nanometric size); in this case, the discrete electronic shell structure governs the optical response [20]. In the size range we are dealing with in the scope of this chapter, the high electron density of noble metals makes the energy level splitting be sufficiently small to assimilate the conduction band to a quasi-continuum. The second strategy consists in calculating directly the nanoparticle optical response through a fully microscopic quantum approach, without using the Mie formulation. Whatever the method chosen, the number and complexity of the physical mechanisms to be taken into account in order to describe an experimental situation in a realistic manner render the calculation a hard task. Very often, the quantum approach is carried out by selecting a phenomenon among all and study independently its influence on the optical response, the others being included in a phenomenological way.

As several works devoted to the nonlinear optical properties of metal nanoparticles include a size dependence of the linear dielectric function, it seems to us relevant to introduce and briefly comment now the most widespread approach used to describe such a dependence. It consists in modifying the phenomenological collision factor Γ in the Drude contribution (Eq. 2) as:

$$(8) \qquad \Gamma(R) = \Gamma_\infty + A \frac{v_F}{R}.$$

Γ_∞ is the bulk collision constant, A is a positive dimensionless factor, v_F is the Fermi velocity and R the particle radius. From a classical point of view, this modification is supported by the fact that, when the radius is smaller than the bulk mean free path of the electrons, there is an additional scattering factor at the particle surface. This phenomenon, known as the *mean free path effect*, is abundantly discussed in [19]. In a quantum approach, the boundary conditions imposed to the electron wave functions lead to the appearance of individual electron-hole excitations (Landau damping) [21] resulting in the broadening of the SPR band proportional to the inverse of the particle radius as in Eq. (8) [22]. A *chemical interface damping* mechanism has also been considered, leading to the $1/R$ dependence of Γ [23].

Whereas Eq. (8) succeeds in explaining qualitatively the broadening and damping of the SPR absorption band with decreasing nanoparticle size, it presents some major drawbacks. First, the parameter A takes different values, from tenths to few units, depending on the theory. The value $A = 1$ is arbitrarily the most often used. Secondly, the introduction of such a $1/R$ dependence in the Drude model results in the red-shift of the SPR with decreasing size, whereas a blue-shift is observed for noble metal nanoparticles [19]. This is due to the influence of bound d electrons which is ignored in the size-dependence considerations that we have described until now [24–27]. However – and even if it cannot of course explain on its own all the size effects – the $1/R$ dependence of different factors is an attractive intuitive idea as the magnitude of the physical mechanisms involved in the nanoparticle properties often amounts to a balance between their volume and surface.

Hence, finite size effects on the optical response of metal nanoparticles are very difficult to take into account in an accurate manner. Moreover, in most experiments carried out on thin nanocomposite films or colloidal solutions the particle size distribution is not mono-dispersed but more or less broad, that can be usually determined by analysis of transmission electronic microscopy images. It should be underlined that the relevant quantity for studying size effects in the optical response of such media can definitely not be the mean cluster radius $<R>$, although it is often used in the literature [28–33], since the contribution of one nanoparticle to the optical response of the whole medium is proportional to its volume, i.e. to R^3 (cf. Eq. 7). The relevant quantity, that we call the "optical mean radius" $<R_{opt}>$, would then rather be the third-order momentum of the size distribution, $<R_{opt}> = <R^3>^{1/3}$.

2.3 Nanocomposite Materials

In the preceding section, we have examined the optical response of a single nanoparticle surrounded by a transparent dielectric medium. In practice, such an approach can be relevant for studying individual particles (using, for instance, near field optical microscopy techniques [34, 35]), or in a pinch for composite materials with a very weak metal concentration (dilute medium limit). As metal volume fraction becomes larger, while keeping particle size constant, electromagnetic interactions between neighbouring particles cannot be neglected any more. In order to include them in the calculation of the linear optical response of nanocomposite materials, different effective medium theories (EMT) have been developed, each of them being suited for a specific morphology and a given concentration range. Whereas such theories generally provide an analytical expression of the whole material effective dielectric susceptibility, other approaches are being developed, which are based on the numerical resolution of the electromagnetic equations governing the interaction of an applied wave with a sampled volume of an inhomogeneous medium. The principles of both strategies are now briefly presented.

2.3.1 Effective medium theories

Heterogeneous media consisting of a mixing of non-miscible materials exhibit specific macroscopic physical properties, different from the ones of their constituents. One is then tempted to define a fictitious homogeneous medium which would have, for one or a few given properties, the same macroscopic response as the heterogeneous material one; this defines the concept of *effective medium*. Such a process is valid only in return for some restriction imposed on the material morphology. In particular, the simulation of its optical properties by those of an effective medium requires constraints on the size of its components as compared to the incident radiation wavelength (light scattering can be neglected), as well as on the volume fraction occupied by these components in the medium. These conditions being fulfilled, one is brought to define the concept of *effective dielectric function*.

The materials we are interested in are made of metallic inclusions (nanoparticles) embedded in a dielectric host (solid matrix or solution). We suppose that the conditions required for using the dipolar and the quasi-static approximations are satisfied. The inclusions can then be likened as dipoles. The calculation of the effective dielectric function can be performed directly through the Mie theory [36, 37], but the usual procedure is rather based on the evaluation of the local field which polarizes each dipole; this field is the sum of the macroscopic applied field and the fields created by all the other dipoles. Its determination can be simplified by using the concept of the Lorentz sphere and is carried out exactly as for the microscopic calculation of the dielectric function of a homogeneous material leading to the Clausius-Mossotti relation [8]. There exists a lot of effective medium theories, adapted for different types of inclusions (spheres, ellipsoids, coated particles,...) and concentration ranges, from the dilute medium to the percolated one [38]. The most famous theories – and, from a historical point of view, the pioneering ones – are those of Maxwell-Garnett [39] and Bruggeman [40]. The former (MG) considers one type of identical spherical inclusions embedded in a continuous matrix. The calculation amounts to determining the volume mean value of the field in a unit cell containing a metal particle and the surrounding dielectric so as to preserve the metal volume fraction, p [41, 42]. One then obtains the MG expression for the effective dielectric function, ε_{eff}:

$$(9) \quad \varepsilon_{eff} = \varepsilon_d \frac{(1+2p)\varepsilon_m + 2(1-p)\varepsilon_d}{(1-p)\varepsilon_m + (2+p)\varepsilon_d}.$$

The MG theory neglects the dipolar interactions between particles; its strict quantitative applicability is thus limited to weak metal concentrations ($p < 10\%$), but it can reasonably be used for a somewhat higher p values. Bruggeman's approach is different: All constituents of the heterogeneous material – that is, here, metal and dielectric – are considered as inclusions embedded in the effective medium and treated the same way in the calculation of the mean field. Hence, unlike MG theory, Bruggeman's one is fully symmetric regarding the different constituents of the medium. It overestimates the dipolar interactions between particles, and is *a priori* better suited for high metal concentrations. From there, numerous refinements have been proposed to adapt these basic EMT to more complex situations [43] like non-spherical inclusions (ellipsoids [44, 45], cylinders [46]), presence of an oxide layer [47], nanoshells [48–51], and effect of a size distribution [52–54].

2.3.2 Beyond EMT

The hypotheses on which effective medium theories lie remain quite restrictive regarding the material morphology. Indeed, whereas they hold for pure random particle spatial distribution or perfectly ordered superlattices, they are unsuited for distributions where partial coalescence, aggregation, or dense packing of individual particles is present [19].

In these cases, both the quasi-static and the dipolar approximations may be invalid. For example, MG theory is unable to reproduce the anomalous IR linear absorption observed in nanocomposite media with large metal amount [55], where field retardation effects as well as multipolar electromagnetic interactions have to be taken into account. Among the numerous approaches which have been developed in this sense, two general alternatives can be distinguished. The first one consists in extending and generalizing effective medium theories [56], sometimes by treating the material morphology through statistical methods [57]; several developments of this kind rest on the formalism of the *spectral density* [58], which is used for nonlinear optics as well (see Section 3.2.5 below). The second alternative consists in calculating the optical response of a given specific nanoparticle arrangement, like a particle pair, triplet, chain, or any more complex distribution, including percolated and fractal structures [59, 60]. This can be performed analytically or numerically with more or less approximations; for example, the linear (and third-order nonlinear) optical properties of a chain of silver spheres coated with a dielectric shell has been recently calculated by numerically solving Maxwell's equations thanks to the 3D finite-difference time-domain method [61]. The resulting polarizability of the complex object may then be possibly inserted in a classical EMT [62]. The *generalized Mie theory* is based on such an approach: An individual particle is described by its polarizability as given by the classical Mie theory, and the interaction between all particles of a chosen sample of inhomogeneous medium is treated by calculating the multiscattered near-fields, including both retardation and multipolar effects [63]. The extinction spectrum of the whole sample can then be deduced by properly summing the contributions of all particles [64]. The medium sample is defined by the number, radius, and spatial location of each particle in the dielectric host. The sample size is limited by computing capacities and degree of approximation authorized. For calculating the field scattered by the interacting spheres of the sample an approach based on a recursive transfer matrix method has been recently proposed, allowing to determine the complex electric field topography [65–67]. Such calculations make available the simulation of the optical response of nanocomposite materials containing spherical nanoparticles with any given spatial arrangement. Moreover, they provide relevant information for understanding the influence of local field enhancement on the material nonlinear optical properties [67].

3. OPTICAL KERR EFFECT IN NANOCOMPOSITE MEDIA

3.1 Third-order Nonlinear Response of Materials

3.1.1 Third-order nonlinear susceptibility

When the electric field associated with the incoming light wave is sufficiently intense (10^3–10^4 V/cm), the relationship between the polarization induced in a material and the field amplitude is no longer linear. The real polarization **P** can be expanded into a power series of the real electric field **E**

$$(10) \quad \mathbf{P} = \varepsilon_0 \left[\chi^{(1)} \mathbf{E} + \chi^{(2)} \mathbf{E}^2 + \chi^{(3)} \mathbf{E}^3 + \ldots + \chi^{(p)} \mathbf{E}^p + \ldots \right].$$

$\chi^{(1)} = \varepsilon - 1$ is the linear susceptibility and $\chi^{(p)}$ is the pth-order nonlinear susceptibility which is a tensor of rank $p+1$. The actual situation is, in fact, more complicated, as the preceding expression holds for isotropic and homogeneous media with nonlinear susceptibilities being real scalar quantities. In the general case, the susceptibilities are complex tensors, defined through the relationship between the spatiotemporal Fourier components of both the polarization and the electric field [68, 69].

In centrosymmetric media (i.e. media exhibiting inversion symmetry) the even-order susceptibilities from electric dipole origin vanish. The first non-zero nonlinear susceptibility is then the third-order one. In the following, all the materials we will deal with present such inversion symmetry. Let us mention, however, that there have yet been experimental results concerning the enhancement of second harmonic generation (SHG) in noble metal particles due to the SPR. This second-order nonlinear optical response is rendered possible, whether when particles are not spherical [70], or when spherical particles are dispersed at the interface between two different media, thus breaking the inversion symmetry [71], or clustered together in low-symmetry aggregates, or even in the centrosymmetric situation by exciting the electric quadrupole contribution [72]. We refer the reader to the article of P.-F. Brevet and co-workers in the present book for more details regarding SHG in gold nanoparticles.

The component $P_i^{(3)}(\omega_m)$ ($i = x, y, z$) of the third-order nonlinear polarization, oscillating at circular frequency ω_m, is expressed as the sum of terms proportional to the product of three Fourier components of the electric field

(11) $$P_i^{(3)}(\omega_m) = \varepsilon_0 \sum_{npq}\sum_{jkl} \chi_{ijkl}^{(3)}(-\omega_m; \omega_n, \omega_p, \omega_q) E_j(\omega_n) E_k(\omega_p) E_l(\omega_q).$$

Of course, the frequencies and wave vectors fulfil the phase-matching conditions. The third-order susceptibility $\chi_{ijkl}^{(3)}$ is a fourth-rank tensor having *a priori* 81 elements. In an isotropic material, there remain 21 non-vanishing elements, among which only three are independent [69]. The simplest case consists in a unique incident plane wave, linearly polarized. Indeed, the third-order polarization vector is then parallel to the electric field and reduces to the sum of two propagating terms, one oscillating at the wave circular frequency ω, and another at the circular frequency 3ω. The amplitudes of these two contributions write, respectively,

(12) $$\begin{cases} P^{(3)}(\omega) = 3\varepsilon_0 \chi^{(3)}(\omega) |E(\omega)|^2 E(\omega) \\ P^{(3)}(3\omega) = \varepsilon_0 \chi^{(3)}(3\omega) E^3(\omega). \end{cases}$$

In these expressions, the simplified susceptibilities denote (if the field oscillates along the x axis)

(13) $$\begin{cases} \chi^{(3)}(\omega) = \chi_{xxxx}^{(3)}(-\omega; \omega, \omega, -\omega) \\ \chi^{(3)}(3\omega) = \chi_{xxxx}^{(3)}(-3\omega; \omega, \omega, \omega). \end{cases}$$

The first contribution to the polarization induces a modification of the wave propagation in the material, for both its amplitude and phase, but without any frequency change. This phenomenon is known as the *optical Kerr effect*, by analogy with the magneto-optic and electro-optic Kerr effects where the medium refractive index varies proportionally with the square of the applied magnetic or electric static field. The second contribution corresponds to the third harmonics generation (THG).

Most of the studies devoted to the nonlinear optical properties of metal nanoparticles use the notation $\chi^{(3)}(\omega)$ to refer to the susceptibility for the optical Kerr effect. Unless otherwise specified, we will also adopt this simplified designation in the following. Let us just recall that it corresponds, in fact, to an experimental situation where a unique plane wave, linearly polarized (or three plane waves with same polarization and frequency), generates the third-order nonlinear optical phenomenon in an isotropic medium at the same frequency, and that the susceptibility is *a priori* a complex quantity.

3.1.2 Intensity-dependent optical coefficients

The optical Kerr effect, introduced in the preceding section through the third-order nonlinear susceptibility, results in the dependence of the complex optical index of the material on wave intensity, I, as

(14) $$\begin{cases} n = n_0 + \gamma I \\ \alpha = \alpha_0 + \beta I. \end{cases}$$

n_0 and α_0 are the linear refractive index and absorption coefficient, respectively (§ 2); γ is the nonlinear refraction coefficient, while β is the nonlinear absorption coefficient. By developing the relation between the electric displacement and the electric field and neglecting the terms proportional to I^2, one easily obtains the link between these coefficients and the complex third-order nonlinear susceptibility, $\chi^{(3)} = \chi^{(3)}_r + i\chi^{(3)}_i$ [73–75]:

(15) $$\begin{cases} \chi^{(3)}_r = \dfrac{2\varepsilon_0 c n_0}{3}\left(2n_0\gamma - \dfrac{\alpha_0 \beta}{2k^2}\right) \\ \chi^{(3)}_i = \dfrac{2\varepsilon_0 c n_0}{3}\left(\dfrac{n_0\beta}{k} + \dfrac{\alpha_0\gamma}{k}\right) \end{cases}$$

where k denotes the modulus of the wave vector. Whereas Eq. (15) holds in the general case, it can be simplified when linear absorption in the material is negligible, i.e. when $\kappa_0 = \alpha_0/2k \ll n_0$. The real and imaginary parts of the nonlinear susceptibility are then proportional, respectively, to the nonlinear refraction and absorption coefficients through

(16) $$\begin{cases} \chi^{(3)}_r = \dfrac{4}{3}\varepsilon_0 c n_0^2 \gamma \\ \chi^{(3)}_i = \dfrac{2}{3k}\varepsilon_0 c n_0^2 \beta. \end{cases}$$

Let us stress that these approximated relations have often been wrongly used in the literature concerning metal nanoparticles. Indeed, linear absorption in materials containing such particles is, most of the time, absolutely not negligible, and Eq. (15) has to be used instead of Eq. (16).

3.1.3 Units

In Eqs. (15) and (16), the different quantities are expressed in SI units. However, the third-order nonlinear susceptibility is often expressed in electrostatic units (esu):

(17) $\quad \chi^{(3)}(\text{esu}) = 10^{-8} \dfrac{c^2}{4\pi} \chi^{(3)}(\text{SI}).$

In this case, Eq. (15) becomes

(18) $\quad \begin{cases} \chi_r^{(3)}(\text{esu}) = \dfrac{cn_0}{240\pi^2}\left(2n_0\gamma - \dfrac{\alpha_0\beta}{2k^2}\right) \\ \chi_i^{(3)}(\text{esu}) = \dfrac{cn_0}{240\pi^2}\left(\dfrac{n_0\beta}{k} + \dfrac{\alpha_0\gamma}{k}\right). \end{cases}$

Moreover, the nonlinear optical coefficients γ and β are often expressed in the submultiple units $\text{cm}^2\,\text{W}^{-1}$ and $\text{cm}\,\text{W}^{-1}$, respectively.

3.1.4 Figure-of-merit for all-optical telecom applications

Due to the possibility of modifying optical absorption or refraction in a medium by using light as command, materials exhibiting sufficiently large nonlinear optical coefficients – among which are nanocomposite media – are thought to be possibly used in all-optical signal processing devices, such as ultrafast switches [76]. Such a functionality requires not only a high nonlinear response in the spectral range of interest (near infrared for fiber-optics communications, for instance), but also a low linear absorption α_0 as well as a fast response time τ. Thus, the following figures-of-merit are often determined in the literature, as, either [76]

(19) $\quad F = \dfrac{\chi^{(3)}}{\alpha_0}$

or

(20) $\quad F_\tau = \dfrac{\chi^{(3)}}{\alpha_0 \tau}.$

However, there is most of the time an ambiguity regarding the nonlinear susceptibility in these formula, since, either the real part, the imaginary part, or the modulus of $\chi^{(3)}$ should be prioritized depending on the application envisaged. Let us finally mention that other figures-of-merit have sometimes been proposed [77, 78].

3.2 The Case of Nanocomposite Materials

As for the linear optical response, different approaches have been proposed to describe the nonlinear optical properties of nanocomposite media. Nevertheless, a few general principles can be identified. First, each component of such a medium possesses its own susceptibility; however, as the typical structure size is much smaller than the wavelength, the observable result of light interaction with the medium is different from a simple combination of the individual responses of the separated constituents (again, we do not treat the case of spatially-resolved studies of the optical response). One is then again led to introduce the concept of *effective medium*, extended to the case of nonlinear optical properties.

Secondly, as stated in the introduction, the local electromagnetic field enhancement in the metal nanoparticles at the SPR is responsible for the large enhancement of the metal nonlinear optical response [79, 80]. This is a direct consequence of the dielectric confinement described above (§ 2.2.2): Metal divided into nanometric entities can present a nonlinear response several orders of magnitude larger than the one of its bulk phase. Several studies have shown that the optical nonlinearities in nanocomposite materials originate from particles and not from their host matrix, but let us qualify this statement: If the metal nonlinear response is indeed much larger than the matrix one, it is further amplified by the SPR phenomenon which depends significantly on the dielectric contrast between the particles and the host medium. Thus, strictly speaking, metal alone cannot explain by itself the high nonlinear response of nanocomposite media.

Thirdly, just like the linear ones, the metal nonlinear properties are affected by electronic confinement. In other words, the intrinsic nonlinear optical susceptibility of a small particle might be different from the bulk metal one, due to finite size effects as already evoked in Section 2.2.4 for the linear susceptibility.

We will now discuss each point through different approaches of the nanocomposite material nonlinear optical response.

3.2.1 A general formulation of the effective third-order nonlinear susceptibility

Following the method of Stroud and Hui [81], Stroud and Wood [82], and later Ma et al. [83], have derived from Maxwell's equations the general expression of the effective $\chi^{(3)}$ in the quasi-static approximation. For this, they have considered that the magnitude of the nonlinear coefficients remains sufficiently small to neglect the nonlinearity in the electric field evaluation. $\chi^{(3)}_{eff}$ then writes

$$(21) \quad \chi^{(3)}_{eff} = \frac{\frac{1}{V}\int \chi^{(3)}(\mathbf{r})\,|\mathbf{E}(\mathbf{r})|^2\,\mathbf{E}(\mathbf{r})^2\,d\mathbf{r}}{|\mathbf{E}_0(\mathbf{r})|^2\,\mathbf{E}_0(\mathbf{r})^2}.$$

V denotes an averaging volume of the inhomogeneous medium and $\mathbf{E}_0$ the spatially averaged applied electric field. The frequency dependence of the different quantities involved has been omitted for sake of clarity. It is easy to deduce from this formula that – at least for its modulus – the effective nonlinear susceptibility will be enhanced as soon as the amplitude of the local electric field $\mathbf{E}(\mathbf{r})$ is enhanced itself somewhere in the medium. Moreover, as Eq. (21) exhibits the dependence of $\chi^{(3)}_{eff}$ on the fourth power of the local field, any enhancement of the latter will lead to an amplification of the nonlinear response relatively much higher than the one of the linear optical response. Of course, the precise result of Eq. (21) depends on both material morphology and intrinsic susceptibilities of the constituents, which are, just like the electric field, complex quantities.

Most of the time, the nanocomposite materials studied in experiments contain metal inclusions the nonlinear susceptibility of which has a much larger modulus than the one of the surrounding host matrix. For example, $|\chi^{(3)}|$ values of the order of 10^{-8} esu and 10^{-14} esu have been reported for bulk gold [73] (or gold particles [80]) and different transparent materials [84–86] respectively, in the visible or near infrared range. Assuming that the third-order nonlinear susceptibility of metal, $\chi^{(3)}_m$, is the same for all inclusions of the medium and constant in each of them, Eq. (21) can then be simplified into

$$(22) \quad \chi^{(3)}_{eff} = \chi^{(3)}_m \frac{\frac{1}{V}\int_{metal} |\mathbf{E}(\mathbf{r})|^2 \mathbf{E}(\mathbf{r})^2 d\mathbf{r}}{|\mathbf{E}_0(\mathbf{r})|^2 \mathbf{E}_0(\mathbf{r})^2}$$

where the integral runs over the volume of all metal inclusions encompassed in the total medium volume V.

This equation underlines the need to know the local field topography in the material if one aims at determining the actual complex value of the effective $\chi^{(3)}$. It can be further simplified if the field inside particles is homogeneous (mean field, or "decoupling", approximation) [82, 83]. Indeed, in this case the normalized volume averaged product (i.e. the quotient in Eq. (22)) reduces to give

$$(23) \quad \chi^{(3)}_{eff} = p \langle f^2(\mathbf{r}) \rangle_m \langle |f(\mathbf{r})|^2 \rangle_m \chi^{(3)}_m.$$

$f(\mathbf{r})$ denotes the local field factor and the brackets indicate volume averaging in metal particles.

3.2.2 Simple approach for dilute media

Equation 23 can be significantly simplified in the limit of dilute media, that is, for low metal concentrations, since in this case the local field factor is the same for all inclusions and is given by Eq. (4). D. Ricard et al. have proposed a straightforward perturbation method to get the expression of $\chi^{(3)}_{eff}$ in the dilute medium approximation [79]. The optical Kerr effect in the material amounts to modifying the effective dielectric function under the action of the applied field E_0 as

$$(24) \quad \delta\varepsilon_{eff} = \chi^{(3)}_{eff} |E_0|^2.$$

This change is due to the modification of the particle dielectric function under the action of the local field E_l

$$(25) \quad \delta\varepsilon_m = \chi^{(3)}_m |E_l|^2.$$

Moreover, the derivation of the Maxwell-Garnett expression of ε_{eff} with respect to ε_m (Eq. (9)) provides

$$(26) \qquad \frac{\partial \varepsilon_{eff}}{\partial \varepsilon_m} = pf^2,$$

$f = E_l/E_0$ being given by Eq. (4). Combining Eqs. (24)–(26), one obtains

$$(27) \qquad \chi_{eff}^{(3)} = pf^2 |f|^2 \chi_m^{(3)}$$

$$= p \left(\frac{3\varepsilon_d}{\varepsilon_m + 2\varepsilon_d} \right)^2 \left| \frac{3\varepsilon_d}{\varepsilon_m + 2\varepsilon_d} \right|^2 \chi_m^{(3)}.$$

This formulation can be also obtained by other approaches [81, 83, 87, 88]. It is extensively used in the literature to analyse the nonlinear optical properties of nanocomposite materials determined experimentally.

3.2.3 Intrinsic nonlinear optical properties of metal nanoparticles

Whatever the degree of approximation used in evaluating the effective nonlinear susceptibility of a composite medium, it can be seen in Eqs. (22), (23) or (27) that the result depends on the product of two complex quantities: One linked with the medium morphology and composition (the local field factor), the other linked with the nonlinear optical properties of the metal inclusions themselves (the intrinsic third-order susceptibility, $\chi_m^{(3)}$) – inasmuch as the own contribution of the host matrix to the whole nonlinear response still remains negligible. We will focus here on the second factor. It is noteworthy that very few theoretical work has been accomplished regarding the value of $\chi_m^{(3)}$ for noble metal nanoparticles after the pioneering studies of Flytzanis and coworkers [79, 80, 89, 90]. Moreover, as will be underlined below, their results may not be used in every experimental situation as they are.

In Flytzanis' group investigations the different electronic contributions to the third-order nonlinear susceptibility of a gold particle, including its tensor aspect, were calculated. They first evaluated, using quantum mechanics density matrix theory, the *intraband* contribution, $\chi_{intra}^{(3)}$, that is the electric dipole nonlinear susceptibility associated with transitions involving the only confined conduction electrons [80]. This contribution is expected to depend on particle size as $1/R^3$ if R is sufficiently small. Rautian proposed an alternative calculation of the intraband contribution and refuted some important points of the approach of Hache et al., leading to discrepancies between their respective results, in particular regarding the frequency dependence of $\chi_{intra}^{(3)}$ [91]. Whatever the theoretical method used, let us emphasize that this contribution is due to electronic confinement, that is, stems from a quantum size effect, and is therefore absent in the metal bulk phase.

The second contribution, on the contrary, is size-independent down to very small R values and already exists in bulk metal. It originates from transitions from the fulfilled core-electron d band to the conduction sp band and is called, therefore, the *interband* contribution, $\chi^{(3)}_{inter}$. As it corresponds to resonant two-levels transitions which saturate, its value is expected to be mainly imaginary and negative. However, let us stress that this statement may only be valid for photon energies at least as large as the IB transition threshold. Moreover, the calculation of Hache et al. is carried out in a particular spectral range, close to both the SPR of gold particles and the IB transition threshold (in the vicinity of the X point of the Brillouin zone) [89]. This means that the imaginary character of $\chi^{(3)}_{inter}$ as well as its sign and magnitude are likely to experience spectral variations, what is rarely considered in the literature.

When using a light source with high instantaneous power, like ultrashort pulse lasers, and at a frequency close to the SPR one, the energy absorbed by the conduction electron gas is sufficiently high to induce a significant rise of their temperature, thus modifying the Fermi-Dirac distribution around the Fermi level. This phenomenon, known as *Fermi smearing*, leads to the modification of the transition probabilities and, consequently, of the linear optical properties [92, 93]. Whereas this is not, strictly speaking, a direct electronic nonlinearity, Hache and coworkers could nevertheless establish, within certain approximations, a proportionality between the incident wave intensity and the induced variation of the metal dielectric function [89]. They thus associated a third contribution to the metal particle nonlinear susceptibility, called the *hot electron* contribution, $\chi^{(3)}_{hot\ electrons}$. However, the expression they obtained exhibits a temperature dependence due to the temperature-dependent heat capacity of electrons (see Eq. (23) in [89]). Now electron temperature varies of course with the energy absorbed; $\chi^{(3)}_{hot\ electrons}$ is then not a real third-order susceptibility, independent of incident intensity. Moreover, the calculation is restricted, not only to a particular photon energy domain close to the SPR, but also to a particular excitation pulsewidth (picosecond pulses), a parameter on which the electron temperature is highly dependent as we will detail deeper in § 8.3.

From the above discussion it can be inferred that the approach of Flytzanis' group has to be considered with care, since its strict applicability imposes constraints regarding light excitation wavelength, intensity and temporal regime. However, within these limited range, it provides a good idea of the nature and order of magnitude of the different contributions. Let us summarize their main results: At the SPR frequency of gold nanoparticles the intraband contribution is found to be negligible against the interband one. The latter is mainly imaginary and has a negative sign: $\text{Im}\chi^{(3)}_{inter} \sim -1.7 \times 10^{-8}$ esu. For picosecond pulse excitation at the SPR, the hot electron contribution is also mainly imaginary but has a positive sign. Its magnitude is higher than the one of $\chi^{(3)}_{inter}$: $\text{Im}\chi^{(3)}_{hot\ electron} \sim 1.1 \times 10^{-7}$ esu. Additionally, the Kerr susceptibility $\chi^{(3)}(\omega;\omega,-\omega,\omega)$ is calculated to be larger by several orders of magnitude than the third-harmonics one, $\chi^{(3)}(3\omega;\omega,\omega,\omega)$ [89].

3.2.4   Consequences and illustration

One of the main consequences of dielectric confinement for the third-order nonlinear optical properties is the fact that the response of a composite medium can be very different in both sign and magnitude from the one of its constituents [73, 89, 94].

This can be easily understood by examining Eq. (27): Both f and $\chi_m^{(3)}$ are complex quantities, and the resulting effective susceptibility may exhibit unexpected values. This is exemplified in Fig. 2, where we have reported the spectral variations of the real and imaginary parts of $\chi_{eff}^{(3)}$, as well as its modulus, calculated using Eq. (27) for a Au:SiO$_2$ medium with $p = 1\%$. The gold particle intrinsic susceptibility has been taken equal to $\chi_m^{(3)} = (-1 + 5i) \times 10^{-8}$ esu, as evaluated by Smith et al. from experiments realised on a gold-glass composite film with 30 ps laser pulses at $\lambda = 532$ nm (see § 6 below) [73]. Let us recall that, as previously discussed in § 3.2.3, the actual complex value of $\chi_m^{(3)}$ is surely not constant over the spectral range under consideration (and, moreover, depends on the pulse duration). However, we use a constant value here since we only aim at emphasizing the effect of both the complex nature of f and its enhancement at the SPR. Hence, it can be observed on Fig. 2 that, whereas $\chi_m^{(3)}$ is mainly imaginary and positive, $\chi_{eff}^{(3)}$ is also mainly imaginary but negative at the SPR. This stems from the local field correction, since f is itself mainly imaginary at the resonance due to phase shift between the external and local electromagnetic fields.

Another consequence of the local field correction is that both the real and imaginary parts of the effective susceptibility undergo strong sign and magnitude spectral variations in the vicinity of the SPR. One can nevertheless deduce from Eq. (27) that the enhancement of the local field – i.e. $|f| > 1$ – at the SPR induces a high enhancement of the global amplitude of the material nonlinear response, since $\left|\chi_{eff}^{(3)}\right|$

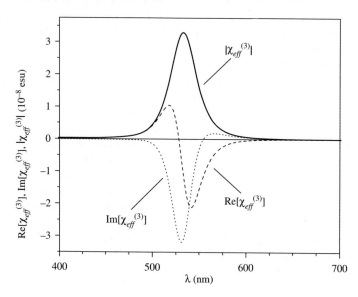

Figure 2. Real part (dashed line), imaginary part (dotted line) and modulus (solid line) of the effective third-order nonlinear susceptibility, $\chi_{eff}^{(3)}$, of a Au:SiO$_2$ nanocomposite medium, calculated by using Eq. (27) with $p = 1\%$ and $\chi_m^{(3)} = (-1 + 5i) \times 10^{-8}$ esu

varies as $|f|^4$. This is again confirmed by Fig. 2, where $\left|\chi_{eff}^{(3)}\right|$ exhibits a resonance behaviour at the SPR (compare with $|f|$ in Fig. 1). This, of course, renders such nanocomposite materials interesting for their remarkable nonlinear optical properties.

Let us finally notice that, investigating the dispersion theory for the effective third-order nonlinear susceptibility of nanocomposite media, Peiponen et al. established that Kramers-Kronig relations are not valid for $\chi_{eff}^{(3)}$, whereas they are valid for other nonlinear processes such as frequency conversion [95].

3.2.5 Nonlinear EMT and other theoretical approaches

As for the linear properties, numerous approaches have been proposed to predict and explain the nonlinear optical response of nanocomposite materials beyond the hypothesis leading to the simple model presented above (§ 3.2.2). Especially, Eq. (27) does not hold as soon as metal concentration is large and, *a fortiori*, reaches the percolation threshold. Several EMT or topological methods have then been developed to account for such regimes and for different types of material morphology, using different calculation methods [38, 81, 83, 88, 96–116]. Let us mention works devoted to ellipsoidal [99, 100, 109] or cylindrical [97] inclusions, effect of a shape distribution [110, 115], core-shell particles [114, 116], layered composites [103], nonlinear inclusions in a nonlinear host medium [88], linear inclusions in a nonlinear host medium [108], percolated media and fractals [101, 104–106, 108]. Attempts to simulate in a nonlinear EMT the influence of temperature have also been reported [107, 113].

As an example among the significant results of such theoretical developments, we would like to underline that a spectral decoupling between the linear absorption maximum and the nonlinear response maximum has been predicted for ellipsoidal metal particles [110]; the longitudinal SPR of such particles being shifted to longer wavelengths as their asymmetry increases, one can thus expect to get a high figure-of-merit (Eq. 19) in the telecom IR spectral range by optimizing particle shape.

In the case where metal concentration is high, that is, close to or above the percolation threshold, there have been predictions for giant spatial fluctuations of the electromagnetic field [105, 106], which have been verified experimentally [117]. Consequently, the possibility of locally obtaining very large enhancements of the third-order nonlinear optical response in such highly-concentrated media is expected. The Sheng EMT also predicts a maximum enhancement of $\chi^{(3)}$ at the percolation threshold for certain wavelengths [83].

Finally, as for the linear optical properties, alternative approaches are being developed to calculate the nonlinear optical response of nanostructured materials. They are most of the time based on the numerical resolution of the equations governing the electromagnetic behaviour of a finite set of nanoparticles in a given spatial arrangement [61].

It is unfortunately worth noticing that, up to now, there has been very few published literature regarding the experimental verification of nonlinear EMT predictions [83, 112]. Only those established within the framework of the simple dilute medium approximation are generally considered.

4. EXPERIMENTAL STUDIES: SOME GENERALITIES

4.1 Experimental Methods

Most of the time, metal/dielectric nanocomposites are studied in the form of solutions or thin solid films on a substrate: Colloids, doped and annealed glasses, sol-gels, surfactant-stabilized nanoparticles, micelles, two- or three-dimension self-assembled nanocomposites, self-organized mesoporous oxides filled with metals, electrochemically-loaded template membranes, metal-ion implanted crystals, nanocomposite films elaborated by laser ablation, cluster-beam deposition, radio-frequency sputtering, or nanolithography.

The usual experimental techniques developed to study the optical Kerr effect in materials have already been described in a preceding chapter of this book. We only mention here the methods which have especially been used for nanocomposite materials as colloidal solutions or thin films: Degenerate four-wave mixing (DFWM) and optical phase conjugation, which provide the modulus of $\chi^{(3)}$ only and may be completed by interferometry techniques to get its phase as well, optical limiting, optical Kerr shutter, and z-scan, which is probably the most common technique used in recent years due to its ability to provide simultaneously the nonlinear refraction and absorption coefficients of the same sample point [118].

4.2 Relevant Parameters and Orders of Magnitude of the Nonlinear Response

It appears to us worthwhile to point out the different parameters relevant to the analysis of the third-order nonlinear optical response of nanocomposite materials, because some of them are sometimes omitted in the literature, rendering the comparison difficult. They can be classified into two main sets. First, some parameters are linked with the optical excitation source, which usually consists of a pulsed laser beam: Wavelength λ, pulse energy E, pulse duration τ, repetition rate ν. Secondly, other relevant parameters concern the material itself: Particle size and shape (and distributions), metal volume fraction p, particle spatial arrangement in the host medium.

There are sometimes ambiguities in the literature regarding "$\chi^{(3)}$", which can be used to denote either $\chi_r^{(3)}$, $\chi_i^{(3)}$ or $|\chi^{(3)}|$. In the following sections the notation "$\chi^{(3)}$" will refer to the *complex* third-order nonlinear susceptibility of the composite material. As a large part of experimental works focus on its modulus, we will also use the notation $\chi^{(3)} = |\chi^{(3)}| e^{i\psi}$ where ψ denotes the phase of $\chi^{(3)}$.

Finally, before closing this general introduction to experimental studies, let us provide orders of magnitude of the third-order nonlinear response of nanocomposite materials. Depending on the composition and morphological characteristics of the latter, as well as on the experimental conditions regarding in particular laser excitation, the nonlinear absorption and refraction coefficient absolute values can reach $|\gamma| \sim 10^{-7}\,\mathrm{cm^2/W}$ and $|\beta| \sim 10^{-2}\,\mathrm{cm/W}$. The magnitude of the third-order susceptibility takes values within $\sim 10^{-13}$ and 10^{-5} esu, whereas the figure-of-merit F defined in Eq. (19) ranges between $\sim 10^{-13}$ and 10^{-10} esu cm.

5. DIFFERENT THIRD-ORDER NONLINEAR OPTICAL PHENOMENA

5.1 Third-harmonic Generation

Several aspects of the nonlinear optical behaviour of metal/dielectric nanocomposites have been reported in the literature. First, very few works have been devoted to the third-harmonics generation (THG) from such materials. The THG from $\lambda = 1064\,\mathrm{nm}$, $\tau = 35\,\mathrm{ps}$ pulses has been measured in colloidal Au, Ag, Pt and Cu in [119]. While no signal could be detected for Au and Ag, the authors found a linear variation of the TH intensity with the pump one for Pt and Cu, providing a maximum conversion efficiency of 7×10^{-7} and 3×10^{-7}, respectively. The corresponding values of $\chi^{(3)}(3\omega)$ (Eq. 13) are worth $(1.5 \pm 0.75) \times 10^{-14}$ esu and $(1.0 \pm 0.5) \times 10^{-14}$ esu, and are of the same order of magnitude as the Kerr susceptibilities $\chi^{(3)}(\omega)$. In [121] the THG in metal nanoparticles is not really studied for itself, but rather used as a probe for measuring the SPR decay time with femtosecond resolution. The advantage of THG over SHG lies in the fact that it allows the study of particles having centrosymmetric shape. Very recently, THG efficiency has been measured on individual gold colloids excited at $\lambda = 1500\,\mathrm{nm}$ with $\tau = 1\,\mathrm{ps}$ pulses [119]. Lastly, let us mention recent investigations regarding the observation of nonlinear magneto-optical Kerr effect (NOMOKE) and magnetization-induced optical THG (and SHG) in Co_xAg_{1-x} nanogranular films (5–20 nm in size) [122]. The SPR in silver nanoparticles at the Co–Ag interface is thought to be involved in these effects.

5.2 Saturation of Absorption

We now focus on the main subject of this contribution, namely the optical Kerr effect. Depending on the material characteristics and experimental conditions – that is, on laser wavelength and power as well as on metal and matrix kinds and relative amounts – the nonlinear absorption coefficient β is found to be either negative or positive. The influence of each of these parameters on the nonlinear response will be examined in details in forthcoming sections.

The case $\beta < 0$ corresponds, in the spectral domain of a resonant transition, to a saturation of absorption (SA) phenomenon. The origin of such a behaviour in metal/insulator nanocomposite media is not described the same way from authors to others. On the one hand, the simple and intuitive explanation generally given is that strong light absorption at the SPR causes the saturation (bleaching) of the corresponding electronic transition [123, 124]. The two levels involved are then the SPR ground and excited states. On the other hand, the hot electron phenomenon is evoked. The calculation of the third-order susceptibility of Au:SiO$_2$ nanocomposites at the SPR performed by Hache et al. leads to such a conclusion. Indeed, as already evoked in § 3.2.3, among the three contributions to $\chi_m^{(3)}$ the hot electron one (imaginary and positive) slightly dominates over the interband one (imaginary and negative), both dominating the intraband contribution (also imaginary and negative). To corroborate their theoretical findings, these authors studied, by optical phase conjugation and SA experiments with $\lambda = 527$ nm, $\tau = 5$ ps and $\lambda = 532$ nm, $\tau = 25$ ps pulses the nonlinear response of glasses doped with gold nanoparticles ($R = 1$–15 nm, $p \sim 10^{-5}$) [89]. The wavelengths used were close to the SPR maximum of the nanoparticles (~ 530 nm). The authors measured a negative value for the imaginary part of $\chi^{(3)}$, that is a negative value of β since the low linear absorption due to the very weak metal volume fraction justifies the use of Eq. (16). The local field factor being mainly imaginary at the SPR, they deduced through Eq. (27) that the imaginary part of the gold particle intrinsic susceptibility is positive (see discussion in § 6 below). This is consistent with their theoretical investigations carried out within the same conditions and using the same parameters as those of their experiments. However, it seems hard to explain that way the persistence of a SA-like behaviour in experiments using long-lasting laser pulses in the vicinity of the SPR [74, 125–132], since in such a temporal excitation regime the hot electron phenomenon should be negligible for usual experimental conditions (see § 8.3.2.2). In this case, an alternative origin of the absorptive nonlinear response could possibly be found in a slow thermo-optical effect, as will be discussed later.

The SA phenomenon can be described within the framework of a two-level atomic model [69]. When the medium can be assimilated to a coherent ensemble of identical two-level systems all having the same response – that is, when the broadening of the transition is purely homogeneous –, the absorption coefficient is related to the incident light intensity I through

$$(28) \quad \alpha(I) = \frac{\alpha_0}{1 + I/I_s}$$

where I_s is defined as the saturation intensity [69]. The second expression in Eq. (14) can then be viewed as the first-order truncation of the power series expansion of Eq. (28) relative to I. Puech et al. examined the dependence of the absorption coefficient on laser intensity in colloidal solutions of gold nanoparticles with $p \ll 1$ [125]. For this, they used $\tau = 500$ ps, $\lambda = 516$ and 522 nm, $\nu = 5$ Hz pulses at a peak intensity I_{00} of about 1 GW/cm^2. It was found that, whereas Eq. (28) cannot correctly account for the experimental results, the latter are well fitted by the following law, suited for inhomogeneously broadened two-level systems

$$(29) \quad \alpha(I) = \frac{\alpha_0}{\sqrt{1 + I/I_s}}.$$

The authors confirmed this finding afterwards by studying the ultrafast dephasing time of the coherent excitation [133]. The inhomogeneous broadening is thought to stem from size and shape distributions. Whereas such distributions are likely to be significant in many experiments, no other group, to our knowledge, has used Eq. (29) rather than Eq. (28) for describing the intensity dependence of the absorption coefficient from the SA point of view.

5.3 Positive Nonlinear Absorption and Optical Limiting

The case $\beta > 0$, opposite to the previous one, can stem in nanocomposite media from reverse saturation of absorption (RSA), multi-photon absorption, or nonlinear scattering. It is of course of high interest for optical limiting applications.

5.3.1 Multi-photon processes

Two-photon absorption may be the dominant effect in a spectral range far from any resonant transition, i.e. where $\alpha_0(\omega)$ is negligible, but with significant absorption at frequency 2ω. This occurs in noble metal nanoparticles for 2ω matching the SPR spectral domain, or for 2ω larger than the IB transition threshold. Such a nonlinear mechanism has been observed, for example, at $\lambda = 1064$ nm in silica glasses doped with Cu nanoparticles ($\lambda_{SPR} \sim 560$ nm) [134]. Both two-photon and three-photon absorption processes have been simultaneously observed in Ag nanorods in borosilicate glass at $\lambda = 800$ nm [135]. They are respectively associated with the transverse SPR ($\lambda_{SPR} \sim 400$ nm) and the interband transitions in silver. Two-photon absorption connected with IB transitions has also been reported in Cu and Ag nanoparticles produced by ion implantation in silica [136, 137].

5.3.2 Reverse saturation of absorption

RSA corresponds to the situation where the absorption cross section of the excited state is larger than the ground state one. RSA was found by Ganeev et al. to be the dominant nonlinear absorption mechanism in Ag and Pt colloidal solutions at $\lambda = 1064$ nm ($\tau = 35$ ps) [120]. In this case, the photon energy is not sufficient to induce two-photon absorption associated either with the SPR or with interband transitions. More surprisingly, RSA was also found to occur at 532 nm in Au colloids, that is close to their SPR ($\lambda_{SPR} = 525$ nm), whereas other experimental studies showed a SA behaviour within similar conditions, as reported in the preceding section. Finally, both RSA and two-photon absorption occurred in the same gold colloids at 1064 nm. Other authors have observed either SA or RSA in the same nanocomposite medium, depending on metal concentration (cf. § 7.3) [138, 139] or on laser intensity (cf. § 8.1) [123, 124, 140, 141].

5.3.3 Nonlinear scattering

Let us now examine the case of a nonlinear scattering process. Ispasoiu et al. deduced from considerations regarding the excited-state lifetime that the optical limiting observed using nanosecond laser pulses at $\lambda = 532$ nm in their silver-dendrimer nanocomposite aqueous solution was due to absorption-induced nonlinear scattering [142]. They suggested that the scattering centres were micro-bubbles produced by local heating. Examining the calculation results of the thermal response of nanocomposite media under pulsed light excitation, as will be discussed below (§ 8.3.2), this assumption is quite probable. François et al. also invoked absorption-induced nonlinear scattering to interpret the positive sign of β in gold nanoparticles, but they distinguished between medium laser fluences, with the generation of solvent bubbles, and high fluences, with fast expansion and vaporization of the particles [143, 144].

5.3.4 Optical limiting

Whatever the nonlinear process involved, the metal/dielectric nanocomposites have been often inquired into for their optical limiting (OL) properties [120, 123, 124, 140–150], owing to the important stake that such a functionality represents for civil and military applications in human eye or detector protection. Indeed, as for telecom applications, metal nanoparticles present the advantages of both intense and fast nonlinear response.

Figure 3 compiles selected results obtained for optical limiting properties of materials containing gold or silver nanoparticles. Note that the ordinate axis shows the normalized transmission – that is, the transmission normalized to unity at low input fluence – and not the absolute one. All kinds of materials roughly present the

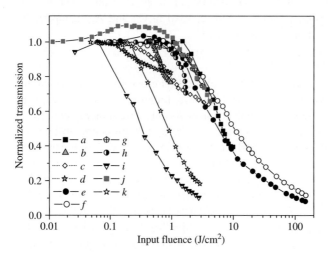

Figure 3. Normalized transmission as a function of laser fluence for material containing gold and silver nanoparticles. *a*: Ag-dendrimer nanocomposites in aqueous solution, $\lambda = 532$ nm and $\tau = 6.5$ ns [142]; *b*, *c* and *d*: octadecanethiol-caped Ag, Au, and AuAg$_{0.75}$ colloids ($R = 3$–4 nm) in toluene, respectively, at $\lambda = 532$ nm with $\tau = 35$ ps [123]; *e* and *f*: Au colloidal solutions ($R = 15$ nm) at $\lambda = 530$ nm and $\lambda = 630$ nm, respectively, with $\tau = 8$ ns [143]; *g* and *h*: colloidal silver and gold, respectively, $\lambda = 1064$ nm and $\tau = 35$ ps [120]; *i*: ligand (C$_{60}$tpy)-protected Au colloids ($R = 5$–15 nm) in chloroform, $\lambda = 532$ nm and $\tau = 8$ ns [147]; *j*: Au nanoparticle ($R = 5$–15 nm) in glass, $\lambda = 532$ nm and $\tau = 8$ ns [124]; and *k*: C$_{60}$ suspension in toluene with identical linear transmission as sample *i* (70%), $\lambda = 532$ nm and $\tau = 8$ ns [147]

same OL efficiency, with varying thresholds, apart from C_{60}tpy-Au in chloroform (sample *i*) which appears better than C_{60} in toluene, a compound which is yet itself known for its high OL ability [147, 148]. A similar result was previously established by the same group for [60]fullerene-silver nanocomposites (DTC_{60}-Ag) in hexane (not shown in the figure) [146]. This strong OL is expected by the authors to originate from the excited state interaction between the [60]fullerene and metal nanoparticles.

Let us also notice the special case of material *j* (Au nanoparticles in glass excited at the SPR by 8 ns pulses): Transmission begins to increase (SA phenomenon) and then decreases as the input fluence rises above $\sim 1 \, \text{J/cm}^2$. Such a behaviour illustrates the fact that, within similar conditions but with different laser powers, either a negative or a positive sign has been found for the nonlinear absorption coefficient, as already evoked above (§ 5.2 and 5.3.2). François et al. studied the influence of particle size on the OL performances of gold colloidal solutions excited by nanosecond pulses at the SPR, in the range $R = 2.5$–$15 \, \text{nm}$ [143]. They found that, as R increases, both the OL threshold and amplitude increase. Moreover, they reported later the influence of wavelength on the OL efficiency, and established that the latter is as high as λ is close to λ_{SPR} [144]. All these observation are fully coherent with the interpretation given by the authors in terms of nonlinear scattering, as described in Section 5.3.3.

5.4 Self-focusing and Self-defocusing

As for the absorption nonlinear properties, the refraction ones have also been shown to present either negative or positive sign. Intensity-induced positive (negative) refraction is usually denoted by self-focusing (self-defocusing). Of course, γ is likely to vary with wavelength in the SPR spectral domain like $\chi_r^{(3)}$ and $\chi_i^{(3)}$ (cf. Fig. 2), but may also vary with laser fluence, pulsewidth, metal concentration, matrix index... The mechanisms involved to explain the self-focusing or self-defocusing are the same as those described for nonlinear absorption, comprising both electronic and thermal effects [132].

6. Determination of $\chi_m^{(3)}$

In order to analyse the microscopic origin of their experimental results regarding the third-order nonlinear optical response of nanocomposite media, several authors have extracted the value of $\chi_m^{(3)}$ by inverting Eq. (27). We want to stress here that this procedure is sometimes applied incorrectly in the following cases: (*i*) the material metal concentration is too large, whereas Eq. (27) has been established assuming low *p* value [128]; (*ii*) moreover, the material complex $\chi^{(3)}$ is often deduced from the measured values of γ and β through the simplified relations 16, valid for low linear absorption, whereas in most cases Eq. (15) should be used instead; (*iii*) the local field factor *f* involved in Eq. (27) is generally estimated from tabulated bulk values of the metal and matrix linear dielectric functions, which do not take into account the SPR maximum spectral shift and quenching due to finite size effects (§ 2.2.4); a simple test of the validity of the f values might be realised by comparing the experimental linear absorption spectrum with the calculated one.

Due to these points, and especially the last one, the applicability of Eq. (27) to measurement results appears quite restricted. Nevertheless, it allows to outline qualitative trends and to provide an estimated order of magnitude for the particle intrinsic nonlinear susceptibility. We now focus on results selected from the literature which appear to be relevant following the preceding clarification.

First, let us recall that, at the SPR, the local field effect results in the opposite signs of the metal inclusion and composite material imaginary third-order susceptibilities, respectively. This was already stated when discussing the results of Hache et al. in § 5.2. This feature was also pointed out later by Smith and co-workers in the case of gold colloids in solution [94]. They also showed that the sign of $\chi_i^{(3)}$ of gold colloids was opposite to the one of a thin gold film [73].

In Table 1 we have reported the value of $\chi_m^{(3)} = \mathrm{Re}\chi_m^{(3)} + i\mathrm{Im}\chi_m^{(3)} = \left|\chi_m^{(3)}\right| e^{i\psi_m}$ for noble metal nanoparticles in different host media and excited at different laser wavelengths and pulsewidths [151–163].

Several features can be deduced from these data. First, Au and Cu nanoparticles excited at or close to the SPR present higher nonlinear response (i.e. higher $\left|\chi_m^{(3)}\right|$) than Ag particles in similar conditions. This may be due to the fact that, in Ag, the photon energy corresponding to the plasmon resonance is lower than the IB transition threshold; hence, at λ_{SPR}, the contribution of interband transitions to $\left|\chi_m^{(3)}\right|$ is weaker in silver than in gold and copper. Puech et al. have experimentally shown, moreover, that for colloidal gold in acetone ($R = 50$ nm, $\lambda_{SPR} = 530$ nm), $\left|\chi_m^{(3)}\right|$ is almost wavelength-independent in the spectral range $\lambda = 562$–606 nm ($\tau = 5$ ps) [155]. Indeed, this range corresponds to photon energies close but lower than the IB transition threshold (2.4 eV, see § 2.1); the dispersion of $\left|\chi_m^{(3)}\right|$ is then expected to be weak. This property was also previously suggested by Hamanaka et al. [161].

Secondly, looking at Table 1, one can notice that there is no significant effect of the surrounding medium on the value of $\chi_m^{(3)}$. Let us just remind that its refractive index governs the SPR wavelength. Thirdly, pulsewidth has a large influence on the value of $\chi_m^{(3)}$, the magnitude of which increases by several orders when increasing pulse duration. This is directly linked with the generation of different thermal effects, that is, hot electron contribution for ultrafast pulses, and thermal lensing for long-lasting pulses [74]. This point will be specifically discussed in Section 8.3. Finally, no clear feature can be extracted from Table 1 regarding either the phase of the metal particle intrinsic susceptibility or the sign of its real component. For its imaginary one, it seems that it is positive for wavelengths close to the SPR, but the precise comparison of the corresponding results is rendered difficult since it should also integrate considerations about the IB transition threshold and the pulsewidth altogether. The influence of particle size on $\chi_m^{(3)}$ will be tackled in the next paragraph.

Table 1. Intrinsic third-order nonlinear susceptibility of noble metal nanoparticles, $\chi_m^{(3)}$, determined through Eq. (27) from $\chi^{(3)}$ values measured on low-p nanocomposite media

Metal	Host medium	Pulse width	λ (nm)	$\|\chi_m^{(3)}\|$ (esu)	$\psi_m(°)$	$\mathrm{Im}\chi_m^{(3)}$ (esu)	$\mathrm{Re}\chi_m^{(3)}$ (esu)	Reference
Au	SiO$_2$	110 fs	560	9×10^{-10}	-169	-1.7×10^{-10}	-8.8×10^{-10}	[74]
Au	SiO$_2$	200 fs	532	34×10^{-10}				[159]
Au	water	5–25 ps	527–532	5.0×10^{-8}	80	4.9×10^{-8}	0.9×10^{-8}	[89]
Au	acetone	5 ps	562–606	2.5×10^{-8}				[155]
Au	water	30 ps	532			11×10^{-8}		[94]
Au	water	30 ps	532	$4-10 \times 10^{-8}$				[151]
Au	glass	30 ps	532	5.1×10^{-8}	-79	5.0×10^{-8}	-1.0×10^{-8}	[73]
Au	SiO$_2$	70 ps	532	4.2×10^{-8}				[159]
Au	water	0.5 ns	516, 522			22×10^{-8}		[125]
Au	water	0.5 ns	522	1.4×10^{-8}				[133]
Au	glass	7 ns	532	2.5×10^{-8}				[152]
Au	SiO$_2$	7 ns	560	1.6×10^{-6}	-62	-1.4×10^{-6}	0.75×10^{-6}	[74]
Au	Al$_2$O$_3$	8 ns	530	7.1×10^{-6}				[158]
			550	8.9×10^{-6}				
			570	20×10^{-6}				
Au	SiO$_2$	ns	530–570	1.1×10^{-6}				[157]
Au	SiO$_2$	20 ns	530	30×10^{-8}				[156]
Ag	glass	150 fs	384–574	1.5×10^{-10}	63	1.3×10^{-10}	0.7×10^{-10}	[162]
Ag	glass	500 fs	388				0.11×10^{-10}	[161]
Ag	water	28 ps	400	2.4×10^{-9}				[79]
Ag	glass	7 ns	415–430	$2-4 \times 10^{-9}$				[154]
Cu	Al$_2$O$_3$	12 ps	585	1.5×10^{-8}				[163]
Cu	Al$_2$O$_3$	30 ps	590				$5-25 \times 10^{-8}$	[160]
Cu	SiO$_2$	35 ps	532	$5-47 \times 10^{-9}$				[153]
Cu	glass	7 ns	565–580	$1-2 \times 10^{-6}$				[154]

7. INFLUENCE OF MATERIAL MORPHOLOGY

As for linear optical properties, the morphology of the material is of course expected to influence its nonlinear response as well [164]. We now examine the most important effects of the morphological parameters as found in the literature.

7.1 Finite Size Effects

A rather large number of experimental studies have been devoted to the influence of nanoparticle size on the optical Kerr effect in nanocomposite media. β and/or γ have been found to present a R-dependence, sometimes even accompanied with a sign change in the size range investigated. The interpretation of such results has to be carried out with care: First, as stated at the end of § 2.2.4, the evaluation of the relevant mean particle size in a material is not straightforward for broad size distributions. Secondly, in the long-lasting pulse domain of laser excitation, or when using high pulse repetition rates, the thermo-optical response of the material may also depend on particle radius, as will be evoked in Section 8.3.2.2. Thirdly, a change in mean particle size from one sample to another often comes with a change in linear absorption spectrum, or even with a change in metal concentration. Hence, some results which have been reported in the literature regarding the size variation of β and γ appear to be quite difficult to analyse [165, 166].

The possible variation of the material third-order susceptibility or nonlinear optical coefficients with particle size can originate from extrinsic effects, as the local field factor and metal concentration, or from intrinsic ones, that is from the size dependence of $\chi_m^{(3)}$. Let us recall that, for Hache et al., the only size dependence of $\chi_m^{(3)}$ lies in the intraband contribution, due to quantum confinement (cf. § 3.2.3). However, they predicted and evidenced experimentally that $\chi_m^{(3)}$ is roughly size independent at the SPR for R varying from 1.4 to 15 nm [89]. This fact was confirmed by other experimental investigations [151, 152, 154]. The intraband contribution was then deduced to be negligible against the others [89]. However, in a similar small size range, a few authors found a size effect on the value of $\chi_m^{(3)}$ for copper nanoparticles at or close to the SPR frequency [153, 160]. Moreover, $|\chi_m^{(3)}|$ was found, in these cases, to roughly follow the $1/R^3$ dependence expected for the intraband contribution only [80], whereas among the three noble metals Cu is likely to exhibit the highest relative contribution of the interband part at the SPR, due to its low IB transition threshold. There has been, up to now, no clear explanation for such discrepancies.

Several authors have underlined the role of the size dependence of the local field factor on the nonlinear response. An increase of the $|\chi^{(3)}|$ value with R has been reported for radii up to ∼ 20 nm, attributed to the increase of $|f|$ [152, 154, 167–170]. Whereas in this small size domain the dipolar approximation holds, field retardation effects and multipolar terms in the polarization cannot be neglected for larger particles [18]. The SPR resonance is then increasingly red-shifted, damped and broadened as R increases above few tens of nanometres. The amplification of the nonlinear response due to the local field enhancement is of course affected by these effects, and $|\chi^{(3)}|$ also begins to decrease [167, 169, 170].

7.2 Nature of the Host Matrix

We now look at the different roles that the surrounding host medium can play in the composite material nonlinear response. The main one is easily understood when examining Eq. (27) and Fig. 1, even if there are valid for weakly concentrated media only: The larger the refractive index of the surrounding dielectric, the larger the local field enhancement at the SPR [167, 171, 172]. If the chemical interaction between particle and matrix is weak, that is, if the modification of the metal electronic properties induced by the presence of the surrounding medium is negligible, the value of $\chi_m^{(3)}$ is independent of the kind of matrix. This already appeared to be demonstrated by different experimental studies in Section 6. However, Gao and Li claimed that the metal/dielectric interactions at the interface largely affect the third-order susceptibility [173]. This fact was, to our knowledge, unfortunately never clearly confirmed elsewhere [174]. The host medium may also present itself a substantial nonlinear response, which can be enhanced by the large local electromagnetic field scattered in the vicinity of the metal particles excited at the SPR. This is the case for certain crystals and/or semiconductor materials, or molecules in solutions [94, 170, 171]. In the same way, the nonlinear optical properties of the composite can be modified by doping the host medium with other species [175, 176]. Let us finally mention that in some cases the host medium may have an influence through slow thermal effects (see Section 8.3.2 below). The relevant parameter is then its thermal conductivity.

Whereas the majority of experimental works has been focused on silica-, glass- or alumina-embedded noble metal nanoparticles, or aqueous colloidal solutions, a few ones have dealt with other kinds of matrices, either amorphous (BaO [177], BaTiO$_3$ [164, 167], Bi$_2$O$_3$ [178], Nb$_2$O$_5$ [179], TiO$_2$ [180, 181], ZrO$_2$...[167]) or crystalline (BaTiO$_3$ [164, 182, 183], Bi$_2$O$_3$ [184], LiNbO$_3$ [185], SrTiO$_3$ [172], ZnO...[186]). A direct comparison of the nonlinear properties from one matrix to another is difficult to carry out, since all other parameters should be kept constant while tuning the wavelength as to match the SPR maximum.

7.3 Metal Concentration

7.3.1 Local field enhancement

When looking at the simple model suited for low metal concentration (Eq. 27), one expects a linear variation of the material effective third-order susceptibility with p. Such a behaviour has been experimentally evidenced in nanocomposite media with very low filling factors [94, 125, 187].

For larger metal volume fractions, the material complex nonlinear optical response may depend on p in a way different from a simple proportionality. For instance, both β and ψ are found in [130] to increase with p in Au:BaTiO$_3$. We also already mentioned in § 5.3.2 that some authors reported the evolution of the nonlinear absorption from RSA-like ($\beta > 0$) into SA-like ($\beta < 0$) with increasing p [138, 139]. This behaviour is possibly accompanied by a change from self-focusing ($\gamma > 0$) to self-defocusing ($\gamma < 0$) [138]. The modification of the complex local field factor linked with the rise of p, due to the increase of electromagnetic interactions between particles within the medium as already evoked in Section 2.3, is of course thought to be responsible for such phenomena [67]. Let us recall that the complex third-order susceptibility of the whole material is very sensitive to the complex value of the local field factor in the SPR spectral domain. Hence, a variation of f may possibly induce a sign and magnitude change of the real and imaginary parts of $\chi^{(3)}$ at a given wavelength [188, 189]. This can occur via the nanoparticle intrinsic nonlinear response (Eq. 22) and, when slow thermal processes are involved like in experiments using long-lasting pulses [130, 138], via the complex thermo-optical response of each of the constituents, as will be evoked at the end of this chapter.

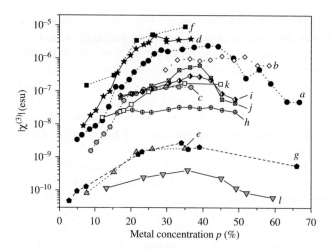

Figure 4. Modulus of the third-order nonlinear susceptibility of different metal/dielectric nanocomposite materials as a function of metal volume fraction, p. a: Au:SiO$_2$, $\lambda = 532$ nm, $\tau = 70$ ps [190]; b: Au:Al$_2$O$_3$, $\lambda = 532$ nm, $\tau = 70$ ps [191]; c and d: Au:SiO$_2$, $\lambda = 532$ nm, $\tau = 200$ fs and 70 ps, respectively [159]; e and f: Au:SiO$_2$, $\lambda = 560$ nm, $\tau = 110$ fs and 6.5 ns, respectively [74]; g: $\left|\chi_i^{(3)}\right|$ of Au:SiO$_2$, $\lambda = 520$–560 nm, $\tau = 150$ fs [193]; h, i and j: Au:TiO$_2$, $\lambda = 532$ nm, 630 nm and 670 nm, respectively, $\tau = 70$ ps [180]; k: Cu:Al$_2$O$_3$, $\lambda = 585$ nm, $\tau = 12$ ps [163]; l: Ag:Bi$_2$O$_3$, $\lambda = 800$ nm, $\tau = 100$ fs [192]

Figure 4 presents the variation of $\left|\chi^{(3)}\right|$ with p for different metal/dielectric composites, measured at different excitation wavelengths and with different pulsewidths, as selected from the literature [74, 139, 163, 180, 190–193]. The influence of both λ and τ on the nonlinear response will be discussed in Section 8. Nevertheless, one can already notice that the concentration dependence of $\left|\chi^{(3)}\right|$ exhibits a similar behaviour whatever the experimental conditions, that is a proportionality to p for very low concentrations ($p < 10\%$), then a stronger increase with p in the intermediate concentration range until about $p = 20$–30%, mainly due to increasing electromagnetic interactions between particles [188, 189, 193], followed by a saturation which will be discussed in the next paragraph. Takeda et al. found a similar p-dependence profile for β of Cu:SiO$_2$ nanocomposites excited with femtosecond pulses [194]. Some authors have established a p^3 dependence of $\left|\chi^{(3)}\right|$ in the second regime [190], but no obvious theoretical support could help assessing this finding. It is unfortunately noteworthy that only very few works have dealt with the complex nature of $\chi^{(3)}$ regarding the influence of metal concentration, that is, have reported the p-dependence of its phase together with the one of its modulus [74, 139]. Our group measured a rather constant value for ψ (around $-135°$) in Au:SiO$_2$ composites excited at 560 nm in the femtosecond pulse regime, while in the nanosecond regime it varies between about $-10°$ and $-105°$ in the concentration range under investigation [74]. These variations were ascribed to a thermo-optical effect.

7.3.2 Reaching the optical percolation threshold

The saturation of $|\chi^{(3)}|$ at high metal concentration – and even more its decrease at higher p values – has been reported in several papers [159, 163, 190, 191, 193], where different interpretations have been given. For Liao et al. [159, 190, 191], it stems from an "optical percolation" effect: At high concentration the particle size becomes large so that only the particle outer layer contributes to the local field enhancement. The explanation of del Coso et al. [163] is rather similar: The decrease of $|\chi^{(3)}|$ at high p stems from the percolated spatial arrangement with particles, the susceptibility of which tends to the bulk metal one. For Hamanaka et al., [193] the saturation and further decrease originate from the suppression of the local field enhancement due to an electron tunnelling effect between very close particles through the thin dielectric spacer. While such interpretations may be plausible, though needing further investigations for being confirmed, we think from our part that the saturation and decrease phenomenon stems from the appearance of significant multipolar electromagnetic interactions between particles. This statement is supported by the results of numerical simulations of the local field enhancement in random nanocomposite media with various metal concentrations. These calculations, recently performed in our group, do not include any tunnelling or optical percolation effect. They consist in evaluating, by using a recursive transfer matrix method, the topography of the complex electromagnetic field due to multi-order Mie scattering from all particles of a sample distribution [67]. Our results regarding large numerical samples of random nanocomposites have not yet been published.

7.4 Particle Shape

As already mentioned in § 3.2.5, non-spherical metal nanoparticles or core-shell particles may exhibit original nonlinear optical properties as predicted by certain theoretical investigations. However, there have been up to now too rare experimental works devoted to such materials. Deviation from spherical shape leads to the modification of the SPR characteristics, and, subsequently, of the local field factor spectral profile. Haus et al. found that the presence of gold spheroids with a random orientation in a colloidal solution has a large influence on the DFWM signal measured [100]. Liao et al. also studied the nonlinear response of spheroidal Au particles, but a clear shape influence is difficult to extract from their results since both particle size and volume fraction increase with anisotropy [195]. Kyoung and Lee obtained interesting results regarding the nonlinear absorption and refraction coefficients in parallel silver nanorods in borosilicate glass [135]. Such anisotropic material exhibits a longitudinal SPR around 800 nm, matching the wavelength of the $\tau = 240$ fs laser pulses used in this study. Positive and negative values were respectively measured for the two- and three-photon absorption coefficients. The two-photon absorption energy matches the transverse SPR while remaining too low to generate interband transitions in silver. Moreover, γ is found to be positive. When exciting the transverse SPR, γ is about ten times lower than with the longitudinal one. Qu et al. compared the nonlinear optical coefficients of gold nanorods with two different aspect ratios excited with 8 ns pulses at $\lambda = 532$ nm, that is at their transverse SPR maximum [149]. Both present self-focusing behaviour and RSA-like nonlinear absorption, which is ascribed to scattering from thermally-induced formation of micro-bubbles (see § 5.3.3). West et al. showed that the optical limiting efficiency is larger for gold nanorods than for gold nanospheres, both excited at $\lambda = 532$ nm with 6.5 ns pulses [196]. They ascribed this to strong light scattering processes in nanorods.

Okada and coworkers investigated the nonlinear optical response of silver triangular nanoprisms by pump-probe femtosecond spectroscopy [197]. They reported a different $\chi_i^{(3)}$ value at the in-plane dipole and quadrupole plasmon resonances, which they showed to correspond to the difference in local field enhancement factors. In both cases, the spectral dispersion of the nonlinear susceptibility exhibits similar behaviour, that is negative and positive values at the low-energy and high-energy sides of the plasmon band, respectively.

Finally, let us mention the original work performed by Yang and co-workers, who studied the nonlinear optical properties of spherical gold nanoparticles coated with a CdS shell at $\lambda = 800$ nm with $\tau = 200$ fs pulses [171]. With increasing shell thickness, the optical Kerr effect signal decreases, which is again explained by local field effects.

8. INFLUENCE OF LASER CHARACTERISTICS

8.1 Intensity Dependence of $\chi^{(3)}$

In Sections 5.2–5.4 we discussed the variation of the complex optical index of nanocomposite materials as a function of laser intensity I. But may the nonlinear coefficients be intensity-dependent themselves? If so, it amounts to go further in the development of Eq. (14). The expression given in Eq. (28) (or 29) is then better suited to account for such a development, at least for the absorptive part of the index. In this sense, β depends itself on I. In optical limiting experiments (§ 5.3.4) the variation of the material transmittance with laser fluence may reflect such features. The identification of a three-photon absorption process (i.e. a fifth-order nonlinearity) has already been reported in § 5.3.1 [135–137]. The sign of the nonlinear absorption coefficient has sometimes been found to change with increasing laser intensity I (with fluence up to several J/cm^2), mostly when using nanosecond pulses: SA → RSA processes have been reported (see, for example, curve j in Fig. 3) [123, 124, 141] as well as RSA → SA [139, 198], and even SA → RSA → SA [140]. Similar phenomena are also sometimes observed for refraction: Change from self-focusing to self-defocusing [198], or the inverse evolution [124, 199]. These results are explained through excited-state theory for the conduction-band electrons [124], i.e. SPR bleach followed by free-carrier absorption [123, 140], thermo-optical effects [198], and photochemical changes [140]. It is likely that slow thermal effects play an important role in this excitation temporal regime, as will be discussed in § 8.3.2.2, associated with pure electronic nonlinear mechanisms as the above-mentioned ones.

Following our work reported in [74], we analysed the dependence of $|\chi^{(3)}|$ and ψ on laser intensity for Au:SiO$_2$ thin films ($\lambda_{SPR} = 495$ to 560 nm, increasing with p) at 560 nm. $|\chi^{(3)}|$ is found to decrease with rising intensity ($I < 150\,\text{GW/cm}^2$) in the femtosecond pulse regime, whereas ψ remains constant. The interpretation lies on both the hot electron effect and the local field enhancement. In the nanosecond regime, $|\chi^{(3)}|$ and ψ of the same samples exhibit a I-dependence which is as obvious as p is large. We assign this trend to the influence of thermo-optical effects. These results have not yet been published. Moreover, we recently found no significant variation of β in Cu:ZnO nanocomposites ($\lambda_{SPR} \sim 600\,\text{nm}$) with varying intensity ($I = 8$–$16\,\text{MW/cm}^2$) when exciting at 532 nm with nanosecond pulses [186]. This is certainly due to the low metal filling fraction and the fact that the measurement is carried out rather off resonance.

8.2 Spectral Dependence of $\chi^{(3)}$

As already stated for other experimental parameters, two factors may account for the nonlinear optical response dependence on excitation wavelength: Local field factor, f, and intrinsic nonlinear properties of the particles, $\chi_m^{(3)}$. The interband contribution to $\chi_m^{(3)}$ is expected to vary only for photon energies at least equal to the IB transition threshold, provided the intraband contribution remains negligible. On the other hand, the hot electron contribution, which accounts for the Fermi smearing mechanism, presents spectral variations for photon energies close to the IB transition threshold, since the electron distribution is modified around the Fermi level by the temperature increase subsequent to light absorption (see § 3.2.3). The wavelength dependence of $\chi_m^{(3)}$ has been already discussed in Section 6.

Regarding the local field factor, a glance at Figs. 1 and 2 allows to easily understand its influence on the spectral dispersion of $\chi^{(3)}$. Two important facts have been highlighted in the literature. First, the nonlinear response is as large as λ is close to λ_{SPR}, due to the maximum enhancement of $|f|$. This has been experimentally established in numerous metal/dielectric composites [100, 136, 154, 156–159, 162, 163, 173, 181, 193, 200]. Let us emphasize that in the long-lasting pulse temporal regime this behaviour is thought to originate from the enhancement of the thermo-optical response at the SPR rather than from the one of electronic nonlinear properties. Secondly, as predicted with a simple model in § 3.2.4 and exemplified in Fig. 2, both $\chi_r^{(3)}$ and $\chi_i^{(3)}$ undergo strong sign changes in the SPR spectral domain, due to f dispersion. To our knowledge, two recent papers only have up to now addressed this point, by using the z-scan technique. Hamanaka et al. measured the complex optical properties of Ag:glass nanocomposites with very low volume fraction ($\lambda_{SPR} = 419\,\text{nm}$) as a function of photon energy, from 2.16 to 3.23 eV (574–384 nm), with $\tau = 150\,\text{fs}$ pulses at $\nu = 1\,\text{kHz}$ repetition rate [162]. They assumed a constant $\chi_m^{(3)}$ value over this spectral range, since the IB transition threshold is located at higher energy in silver. They then extracted the complex value of $\chi_m^{(3)}$ by fitting the model (Eq. 27) on the measured $\chi^{(3)}$. Good agreement was found up to $\sim 2.9\,\text{eV}$ photon energy, but discrepancies were significant above. They envisaged them to stem from higher-order Mie resonance or frequency-dependent $\chi_m^{(3)}$ when approaching the IB threshold. This last statement looks rather plausible. Takeda et al. studied $\chi^{(3)}$ dispersion in Cu:SrTiO$_3$ ($p = 30\%$, $\lambda_{SPR} = 608$ nm) using $\tau = 200$ fs pulses at $\nu = 1$ kHz in the spectral range 2.0–2.3 eV (610–540 nm) [172]. Their results qualitatively agree with calculations of $\chi_r^{(3)}$ and $\chi_i^{(3)}$ through Eq. (27), but the use of the latter is doubtful since p is too high to ensure the validity of the simple expression of f. Moreover, the host medium is found to exhibit a significant nonlinear response itself, not taken into account in the calculation.

8.3 Influence of the Excitation Pulsewidth

8.3.1 Experimental facts

Among the numerous works related to the third-order nonlinear optical response of nanocomposite media, some have been especially devoted to the influence of the excitation laser pulsewidth on the value of $\chi^{(3)}$ [74, 89, 120, 159]. Liao et al. have shown by DFWM experiments that, as this pulsewidth decreases from 70 ps to 200 fs, $|\chi^{(3)}|$ of Au:SiO$_2$ composites undergoes strong reduction of about one order of magnitude, whatever metal concentration (curves c and d in Fig. 4) [159]. The authors invoke thermal effects to explain the differences observed between their excitation temporal regimes. Ganeev et al. have also demonstrated, thanks to z-scan measurements, that the drastic reduction of the nonlinear refraction coefficient of noble metal colloidal solutions observed when decreasing the pulsewidth from nanosecond to picosecond can be mainly ascribed to a thermal lensing effect present in the former temporal regime [120]. Our group has recently reported that $|\chi^{(3)}|$ of Au:SiO$_2$ materials with different gold concentration exhibits a significant dependence on pulse duration τ (see Fig. 4, curves e and f) [74]: When using 6.5 ns pulses, the value of $|\chi^{(3)}|$ is at least three orders of magnitude larger than with 110 fs pulses, the ratio globally increasing with p as for the results of Liao et al. (curves c and d of Fig. 4). As we used the z-scan technique, we could also measure the phase of $\chi^{(3)}$. We found that its variations with p has a different behaviour depending on τ, as reported at the end of § 7.3.1.

8.3.2 Role of thermal effects in the third-order nonlinear optical response

When the pulse duration is much shorter than the nonlinear response time of the material (transient case), the third-order nonlinear susceptibility can be expressed in a phenomenological time-independent form similar to the one corresponding to the long-lasting pulse regime – the adiabatic case [68]. It is then *a priori* possible to compare the measurement results obtained in such opposite temporal regimes, provided the physical processes involved in the optical response remain exactly the same and, especially, are not intensity-dependent. This condition being fulfilled, the modulus of the susceptibility measured in a short pulse regime, $\left|\chi^{(3)}_{pulse}\right|$, is linked with the adiabatic one $\left|\chi^{(3)}_{cw}\right|$ through [201]

$$\left|\chi^{(3)}_{cw}\right| = \left|\chi^{(3)}_{pulse}\right| \xi \frac{T_1}{\tau}. \quad (30)$$

ξ is a dimensionless factor accounting for the pulse temporal shape and T_1 denotes the material nonlinear response time. Eq. (30) holds for $\tau \ll T_1$. When the decay is multi-exponential, T_1 may be replaced by an effective lifetime as in the work of Li et al., who have successfully applied such an analysis to the case of $\left|\chi^{(3)}\right|$ in semiconducting CuBr nanocrystals [201].

In nanocomposite media, T_1 is worth about a few picoseconds (see § 8.3.2.3 below). Eq. (30) then helps explaining the fact that, as noticed in the preceding section, $\left|\chi^{(3)}\right|$ values measured with femtosecond pulses are smaller than those obtained with longer pulsewidths. However, dynamical thermal effects are likely to play a crucial role in the material nonlinear optical response, as will be shown in the following. As their influence depends on the excitation temporal regime, the measurement analysis is not as simple as one could expect from the only characteristic time comparison of Eq. (30). We now go deeper into these thermal effects.

8.3.2.1 Thermal effects in the dynamics of the optical response. As the physical mechanisms involved in the optical response of nanocomposite media exhibit specific temporal behaviour, many time-resolved optical studies using pump-probe techniques have been reported in the literature; all of them converge on the link between the dynamics of these physical mechanisms and the dynamics of thermal effects [202–211]. The latter have different origins. The first one has already been evoked in § 3.2.3: Ultrafast pulse absorption by metal nanoparticles leads to a high rise of the electron gas temperature, thus modifying the optical properties through the Fermi smearing mechanism [89, 202–206, 208–211]. This dynamical process is of course linked with the hot electron contribution to the third-order nonlinear susceptibility. Additionally, the electron-electron and electron-phonon scattering processes may also exhibit a possible electron temperature dependence [202, 204, 206, 208, 209, 211]; moreover, the electron-surface interaction has been shown to become a significant relaxation channel in small nanoparticles [204]. Second, once the metal lattice has been heated by coupling with the hot electron gas, the subsequent metal density modification affects the optical properties [210, 211]. Finally, as thermal energy diffuses towards the surrounding host, the whole material is heated.

In conventional nonlinear optical experiments a single laser pulse both excites the nonlinear response and probes it. The high sensitivity of $\chi^{(3)}$ on the excitation pulsewidth evoked in Section 8.3.1 in then very likely to partially stem from thermal effects, as it has already been suggested in the literature [132, 159, 207, 212].

8.3.2.2 Long-lasting pulse excitation. Following the work of Hamanaka et al [210], a numerical method has been recently proposed by our group to solve, whatever the excitation temporal regime, the three coupled differential equations governing the time-dependence of the electron, metal lattice and surrounding host temperatures for a single nanosphere [74, 213]. Moreover, this model has been improved so as to include the influence of thermal exchanges between neighbouring particles in the medium, and has further been especially adapted for the case of thin nanocomposite films [213]. In a regime of long-lasting pulses (τ longer than a few hundreds of picoseconds), or continuous-wave excitation, the electron and lattice baths in metal nanoparticles are in quasi-equilibrium at the timescale of a pulse within usual power ranges. Electron temperature then reaches much weaker values than in ultrashort pulse experiments. Indeed, the energy being provided slowly as compared with the characteristic relaxation times, heat is released to the surrounding matrix during the optical absorption process itself. This implies that, while the hot electron phenomenon can be neglected in this temporal regime, the whole material is significantly heated during the pulse passage. This is of course not the case in ultrashort pulse regime with low repetition rate. We also established that the material temperature increases with both R and p due to, respectively, the decrease of the particle surface/volume ratio and the rising overheating stemming from thermal diffusion from neighbouring particles [213]. The size dependence was already demonstrated, for instance, in [214] and [215].

8.3.2.3 Ultrashort pulse excitation and relaxation time of the nonlinear response. In the ultrashort pulse temporal regime (τ smaller than a few picoseconds) the electron temperature can reach several thousand Kelvin under usual power conditions. The contribution of the hot electron phenomenon to $\chi_m^{(3)}$ is then expected to be significant. It is also possible to rest with an athermal (non Fermi-Dirac) electron distribution for the whole pulse duration under low power excitation conditions [209]. Many different phenomena can affect the ultrafast dynamics of the optical response – and then the hot electron contribution to $\chi_m^{(3)}$ –, the description of which is largely out of the scope of this chapter [208, 209].

Among the numerous works concerned with this dynamics, a few ones were directly focused on the nonlinear response, that is, on $\chi^{(3)}$ [79, 158, 161, 163, 177, 197, 207, 216, 217]. It is usually found that the nonlinear response relaxation time presents two components, a fast one ($\sim$ 1–5 ps) corresponding to electron-phonon coupling in metal nanoparticles, and a slow one ($\sim$ 10–500 ps) linked with the thermal energy diffusion towards the surrounding medium. Some authors additionally showed that the slow process characteristic time is as large as metal concentration is high [159, 163, 180, 214], which confirms our own theoretical results [213]. In the particular configuration of optical Kerr-shutter experiments, the Kerr signal has been found to present a response time of the order of 200–400 fs [177, 206, 217].

8.3.2.4 Thermal lensing contribution to the measured nonlinear optical properties. If the pulse duration is longer than the characteristic time of the heat diffusion in the medium, or if this time is itself longer than the delay between successive pulses, material heating may lead to an observable transient thermal lens phenomenon [120, 165, 212, 218, 219]. This can show itself, in experiments, with characteristics similar to those of a pure (electronic) Kerr effect. There have been some attempts to extract the respective values of the thermal and electronic contributions to γ from z-scan measurements [136, 160, 165, 166, 175, 220]. However, de Nalda et al. proved later that this method was not reliable enough to get quantitative results [219].

The sign and magnitude of the resulting "thermal nonlinear refraction" coefficient (which is, actually, a pure linear effect [219]) depend on the thermo-optical coefficient $\partial n/\partial T$ of the material. This coefficient has sometimes been assimilated to the one of the surrounding host only [132, 218], but we have recently shown that, due to local field enhancement at the SPR, they can be very different – even for weakly concentrated media –, exactly as for the pure electronic nonlinear properties as demonstrated in Section 3.2.4. Moreover, an absorptive thermo-optical effect, which is always disregarded in the literature, can occur parallel to the refractive one. These conclusions will be published soon.

In the ultrashort excitation regime the pulse energy is too weak and its duration too short for a thermo-optical effect to be induced by a single pulse, but cumulative heating can occur when using high pulse repetition rates (above a few tens of kHz) [155, 165, 212, 219]. In the long-lasting pulse regime, on the contrary, the thermal lens effect may be easily excited due to the possibility of generating a significant material overheating during each pulse passage. Moreover, as discussed in § 8.3.2.2, the material temperature increases with metal concentration and needs to be calculated accurately by accounting for the composite morphology of the medium, the time-dependence of the excitation, and the possible finite dimensions of the sample [213].

9. CONCLUSION

In this chapter we have shown that the third-order nonlinear optical response of metal/dielectric nanocomposite media varies in a complex manner with many parameters. These parameters are difficult to control independently in experimental investigations. We have distinguished the roles respectively played by the local field enhancement and the metal particle intrinsic nonlinearity. We have highlighted the influence of different parameters, and emphasized the significant links between optical properties and thermal effects.

Due to their high and ultrashort nonlinear response, nanocomposite materials are expected, at least when reading most of the numerous papers published on the subject, to open promising outlooks for technological applications in optical limiting or all-optical telecommunications (for signal processing, switching, reshaping), as already mentioned in Section 3.1.4. However, no such concrete achievement has, to our knowledge, been up to now reported.

The specific optical properties of metal nanoparticles have yet found many applications [221], as in labelling of biological molecules, biosensors, surface-enhanced Raman scattering, or even commercial paintings. One of the most fascinating realisations is the recent use of core-shell nanoparticles with tailored SPR wavelength as local heat sources for targeted cancer tumoral cell destruction [222]. The principle is simply based on the thermal energy release from the particles to the surrounding medium after light absorption in metal nanoshells. We believe such nano-objects – as well as metal nanorods or ellipsoids (see § 3.2.5) – to be also good candidates for the conception of ultrafast all-optical signal processing devices. Indeed, the SPR wavelength can be easily tuned to the telecom infrared by modifying the particle geometrical characteristics. Another interesting outlook may consist in inserting nanocomposite materials in photonic band gap structures, as recently proposed [223]. Whatever the application envisaged, fundamental investigations need to be further carried out for better understanding the physics inherent to the metal nanoparticle nonlinear optical properties.

ACKNOWLEDGEMENTS

The author thanks S. Debrus, B. Perrin, N. Pinçon, D. Prot, M. Rashidi-Huyeh, and A. Ryasnyanskiy for abundant and fruitful discussions.

REFERENCES

[1] Ehrenreich, H., Philipp, H.R.: Optical properties of Ag and Cu. Phys. Rev. **128**, 1622–1629 (1962)
[2] Thèye, M.L.: Investigation of the optical properties of Au by means of thin semitransparent films. Phys. Rev. B **2**, 3060–3078 (1970)
[3] Johnson, P.B., Christy, R.W.: Optical constants of the noble metals. Phys. Rev. B **6**, 4370–4379 (1972)
[4] Aspnes, D.E., Kinsbron, E., Bacon, D.D.: Optical properties of Au: Sample effects. Phys. Rev. B **21**, 3290–3299 (1980)
[5] Innes, R.A., Sambles, J.R.: Optical characterisation of gold using surface plasmon-polaritons. J. Phys. F: Metal Phys. **17**, 277–287 (1987)
[6] Nash, D.J., Sambles, J.R.: Surface plasmon-polariton study of the optical dielectric function of silver. J. Mod. Opt. **43**, 81–91 (1996)
[7] Palik, E.D. (ed.): Handbook of Optical Constants of Solids, vols. I and II, Academic, New York (1985/1991)
[8] Ashkroft, N.W., Mermin, N.D.: Solid State Physics. Saunders College Publishing, Philadelphia (1976)
[9] Kittel, C.: Introduction to Solid State Physics. Wiley, New York (1983)
[10] Cooper, B.R., Ehrenreich, H., Philipp, H.R.: Phys. Rev. **138**, A494–A507 (1965)
[11] Rosei, R., Lynch, D.W.: Thermomodulation Spectra of Al, Au, and Cu. Phys. Rev. B **5**, 3883–3894 (1972)
[12] Rosei, R.: Temperature modulation of the optical transitions involving the Fermi surface in Ag: Theory. Phys. Rev. B **10**, 474–483 (1974)
[13] Rosei, R., Culp, C.H., Weaver, J.H.: Temperature modulation of the optical transitions involving the Fermi surface in Ag: Experimental. Phys. Rev. B **10**, 484–489 (1974)

[14] Mie, G.: Beitrage zur Optik truber Medien, speziell kolloidaler Metallosungen. Ann. Phys. (Leibzig) **25**, 377 (1908)
[15] Born, M., Wolf, E.: Principles of Optics. University Press, Cambridge (1999)
[16] Bohren, C.F., Huffman, D.P.: Absorption and Scattering of Light by Small Particles. Wiley, New York (1983)
[17] Russell, B.K., Mantovani, J.G., Anderson, V.E., Warmack, R.J., Ferrell, T.L.: Experimental test of the Mie theory for microlithographically produced silver spheres. Phys. Rev. B **35**, 2151–2154 (1987)
[18] Kolwas, K., Demianiuk, S., Kolwas, M.: Optical excitation of radius-dependent plasmon resonances in large metal clusters. J. Phys. B: At. Mol. Opt. Phys. **29**, 4761–4770 (1996)
[19] Kreibig, U., Vollmer, M.: Optical Properties of Metal Clusters. Springer Verlag, Berlin, Heidelberg (1995)
[20] Fedrigo, S., Harbich, W., Buttet, J.: Collective dipole oscillations in small silver clusters embedded in rare-gas matrices. Phys. Rev. B **47**, 10706–10715 (1993)
[21] Kawabata, A., Kubo, R.: Electronic properties of fine metallic particles. II Plasma resonance absorption. J. Phys. Soc. Jpn. **21**, 1765–1772 (1966)
[22] Yannouleas, C., Broglia, R.A.: Landau damping and wall dissipation in large metal clusters. Ann. Phys. N. Y. **217**, 105–141 (1992)
[23] Persson, B.N.J.: Polarizability of small spherical metal particles: influence of the matrix environment. Surf. Sci. **281**, 153–162 (1993)
[24] Kasperovich, V., Kresin, V.V.: Ultraviolet photoabsorption spectra of silver and gold nanoclusters. Philos. Mag. B **78**, 385–396 (1998)
[25] Palpant, B., Prével, B., Lermé, J., Cottancin, E., Pellarin, M., Treilleux, M., Perez, A., Vialle, J.L., Broyer, M.: Optical properties of gold clusters in the size range 2–4 nm. Phys. Rev. B **57**, 1963–1970 (1998)
[26] Lermé, J., Palpant, B., Prével, B., Pellarin, M., Treilleux, M., Vialle, J.L., Perez, A., Broyer, M.: Quenching of the size effects in free and matrix-embedded silver clusters. Phys. Rev. Lett. **80**, 5105–5108 (1998)
[27] Celep, G., Cottancin, E., Lermé, J., Pellarin, M., Arnaud, L., Huntzinger, J.R., Broyer, M., Palpant, B., Boisron, O., Mélinon, P.: Size evolution of the optical properties of copper clusters embedded in alumina: an experimental and theoretical study of size dependence. Phys. Rev. B **70**, 165409 (2004)
[28] Quinten, M.: Optical constants of gold and silver clusters in the spectral range between 1.5 eV and 4.5 eV. Z. Phys. B **101**, 211–217 (1996)
[29] Link, S., El-Sayed, M.A.: Size and temperature dependence of the plasmon absorption of colloidal gold nanoparticles. J. Phys. Chem. B **103**, 4212–4217 (1999)
[30] Lee, M., Chae, L., Lee, K.C.: Microstructure and surface plasmon absorption of sol-gel-prepared Au nanoclusters in TiO_2 thin films. Nanostruct. Mater. **11**, 195–201 (1999)
[31] Dalacu, D., Martinu, L.: Spectroellipsometric characterization of plasma-deposited Au/SiO_2 nanocomposite films. J. Appl. Phys. **87**, 228–235 (2000)
[32] Cai, W., Hofmeister, H., Dubiel, M.: Importance of lattice contraction in surface plasmon resonance shift for free and embedded silver particles. Eur. Phys. J. D **13**, 245–253 (2001)
[33] Sasai, J., Hirao, K.: Relaxation behavior of nonlinear optical response in borate glasses containing gold nanoparticles. J. Appl. Phys. **89**, 4548–4553 (2001)
[34] Klar, T., Perner, M., Grosse, S., von Plessen, G., Spirkl, W., Feldmann, J.: Surface-plasmon resonances in single metallic nanoparticles. Phys. Rev. Lett. **80**, 4249–4252 (1998)
[35] Grésillon, S., Aigouy, L., Boccara, A.C., Rivoal, J.C., Quelin, X., Desmarest, C., Gadenne, P., Shubin, V.A., Sarychev, A.K., Shalaev, V.M.: Experimental observation of localized optical excitations in random metal-dielectric films. Phys. Rev. Lett. **82**, 4520–4523 (1999)
[36] Stroud, D., Pan, F.P.: Self-consistent approach to electromagnetic wave propagation in composite media: Application to model granular metals. Phys. Rev. B **17**, 1602–1610 (1978)
[37] Niklasson, G.A., Granqvist, C.G.: Optical properties and solar selectivity of coevaporated $Co-Al_2O_3$ composite films. J. Appl. Phys. **55**, 3382–3410 (1984)

[38] See, for a large review of the advances in this field, the proceedings of the conferences on Electrical Transport and Optical Properties of Inhomogeneous Media: ETOPIM 1, edited by Garland, J.C., Tanner, D.B. (AIP, New York, 1978); ETOPIM 2, edited by Lafait, J., Tanner, D.B.: Physica A **157** (1) (1989); ETOPIM 3, edited by Mochan W.L., Barrera, R.G.: Physica A **207** (1–3) (1994); ETOPIM 4, edited by Dykhne, A.M., Lagarkov, A.N., Sarychev, A.K.: Physica A **24** (1–2) (1997); ETOPIM 5, edited by Hui, P.M., Sheng, P., Tang, L.-H.: Physica B **279** (1–3) (2000); ETOPIM 6, edited by Milton, G.W., Golden, K.M., Dobson, D., Vardeny, A.Z.: Physica B **338** (1–3) (2003)
[39] Maxwell-Garnett, J.C.: Colours in metal glasses and in metallic films. Philos. Trans. R. Soc. London **203**, 385 (1904)
[40] Bruggeman, D.A.G.: The calculation of various physical constants of heterogeneous substances. I. The dielectric constants and conductivities of mixtures composed of isotropic substances. Ann. Phys. (Leipzig) **24**, 636–664 (1935); 665-679 (1935)
[41] Marton, J.P., Lemon, J.R.: Optical properties of aggregated metal systems. I. Theory. Phys. Rev. B **4**, 271–280 (1971)
[42] Marton, J.P., Lemon, J.R.: Optical properties of aggregated metal systems: Real metals. J. Appl. Phys. **44**, 3953–3959 (1973)
[43] Gehr, R.J., Boyd, R.W.: Optical properties of nanostructured optical materials. Chem. Mater. **8**, 1807–1819 (1996)
[44] Polder, D., van Santen, J.H.: The effective permeability of mixtures of solids. Physica **12**, 257 (1946)
[45] Cohen, R.W., Cody, G.D., Coutts, M.D., Abeles, B.: Optical properties of granular silver and gold films. Phys. Rev. B **8**, 3689–3701 (1973)
[46] Foss, Jr., C.A., Hornyak, G.L., Stockert, J.A., Martin, C.R.: Template-synthesized nanoscopic gold particles: Optical spectra and the effects of particle size and shape. J. Phys. Chem. **98**, 2963–2971 (1994)
[47] Granqvist, C.G., Hunderi, O.: Optical properties of ultrafine gold particles. Phys. Rev. B **16**, 3513–3534 (1977)
[48] Neeves, A.E., Birnboim, M.H.: Composite structures for the enhancement of nonlinear optical materials. Opt. Lett. **13**, 1087–1089 (1988); Composite structures for the enhancement of nonlinear-optical susceptibility. J. Opt. Soc. Am. B **6**, 787–796 (1989)
[49] Zhang, X., Stroud, D.: Numerical studies of the nonlinear properties of composites. Phys. Rev. B **49**, 944–955 (1994)
[50] Fedotov, V.A., Emel'yanov, V.I., MacDonald, K.F., Zheludev, N.I.: Optical properties of closely packed nanoparticle films: spheroids and nanoshells. J. Opt. A: Pure Appl. Opt. **6**, 155–160 (2004)
[51] Hui, P.M., Xu, C., Stroud, D.: Second-harmonic generation for a dilute suspension of coated particles. Phys. Rev. B **69**, 014203 (2004)
[52] Chýlek, P., Srivastava, V.: Dielectric constant of a composite inhomogeneous medium. Phys. Rev. B **27**, 5098–5106 (1983)
[53] Barrera, R.G., Villaseñor–González, P., Mochán, W.L., Monsivais, G.: Effective dielectric response of polydispersed composites. Phys. Rev. B **41**, 7370–7376 (1990)
[54] Spanoudaki, A., Pelster, R.: Effective dielectric properties of composite materials: The dependence on the particle size distribution. Phys. Rev. B **64**, 064205 (2001)
[55] Ducourtieux, S., Gresillon, S., Boccara, A.C., Rivoal, J.C., Quelin, X., Gadenne, P., Drachev, V.P., Bragg, W.D., Safonov, V.P., Podolskiy, V.A., Ying, Z.C., Armstrong, R.L., Shalaev, V.M.: Percolation and fractal composites: optical studies. J. Nonlinear Opt. Phys. Mater. **9**, 105–116 (2000)
[56] Ping Sheng: Theory for the dielectric function of granular composite media. Phys. Rev. Lett. **45**, 60–63 (1980)
[57] Andraud, C., Lafait, J., Beghdadi, A.: Entropic model for the optical properties of heterogeneous media: Validation on granular gold films near percolation. Phys. Rev. B **57**, 13227–13234 (1998)

[58] Bergman, D.: The dielectric constant of a composite material – A problem in classical physics. Phys. Rep. C **43**, 377–407 (1978); Exactly solvable microscopic geometries and rigorous bounds for the complex dielectric constant of a two-component composite material. Phys. Rev. Lett. **44**, 1285–1287 (1980)
[59] Markel, V.A., Shalaev, V.M., Stechel, E.B., Kim, W., Armstrong, R.L.: Small-particle composites. I. Linear optical properties. Phys. Rev. B **53**, 2425–2436 (1996)
[60] Pustovit, V.N., Niklasson, G.A.: Observability of resonance optical structure in fractal metallic clusters. J. Appl. Phys. **90**, 1275–1279 (2001)
[61] Panoiu, N.-C., Osgood, Jr., R.M.: Subwavelength nonlinear plasmonic nanowire. Nano Lett. **4**, 2427–2430 (2004)
[62] Granqvist, C.G., Hunderi, O.: Conductivity of inhomogeneous materials: Effective-medium theory with dipole-dipole interaction. Phys. Rev. B **18**, 1554–1561 (1978); Optical properties of Ag-SiO_2 cermet films: A comparison of effective-medium theories. Phys. Rev. B **18**, 2897–2906 (1978)
[63] Quinten, M., Kreibig, U.: Optical properties of aggregates of small metal particles. Surf. Sci. **172**, 557–577 (1986)
[64] Gérardy, J.M., Ausloos, M.: Absorption spectrum of clusters of spheres from the general solution of Maxwell's equations. The long-wavelength limit. Phys. Rev. B **22**, 4950–4959 (1980); II. Optical properties of aggregated metal spheres. ibid. **25**, 4204–4229 (1982)
[65] Auger, J.C., Stout, B., Lafait, J.: Dependent light scattering in dense heterogeneous media. Physica B **279**, 21–24 (2000)
[66] Stout, B., Auger, J.-C., Lafait, J.: Individual and aggregate scattering matrices and crosssections: conservation laws and reciprocity. J. Mod. Opt. **48**, 2105–2128 (2001)
[67] Prot, D., Stout, D.B., Lafait, J., Pinçon, N., Palpant, B., Debrus, S.: Local electric field enhancements and large third-order optical nonlinearity in nanocomposite materials. J. Opt. A: Pure and Appl. Opt. **4**, S99–S102 (2002)
[68] Butcher, P.N., Cotter, D.: The Elements of Nonlinear Optics. University Press, Cambridge (1991)
[69] Boyd, R.W.: Nonlinear Optics. Academic, San Diego (1992)
[70] Lambrecht, B., Leitner, A., Aussenegg, F.R.: Femtosecond decay-time measurement of electron-plasma oscillation in nanolithographically designed silver particles. Appl. Phys. B **64**, 269–272 (1997); SHG studies of plasmon dephasing in nanoparticles. ibid. **68**, 419–423 (1999)
[71] Antoine, R., Brevet, P.F., Girault, H.H., Bethell, D., Schiffrin, D.J.: Surface plasmon enhanced non-linear optical response of gold nanoparticles at the air/toluene interface. Chem. Commun. **1997** (1901–1902)
[72] Antoine, R., Pellarin, M., Palpant, B., Broyer, M., Prével, B., Perez, A., Galetto, P., Brevet, P.F., Girault, H.H.: Surface plasmon enhanced second harmonic response from gold clusters embedded in an alumina matrix. J. Appl. Phys. **84**, 4532–4536 (1998)
[73] Smith, D.D., Yoon, Y., Boyd, R.W., Campbell, J.K., Baker, L.A., Crooks, R.M., George, M.: z-scan measurement of the nonlinear absorption of a thin gold film. J. Appl. Phys. **86**, 6200–6205 (1999)
[74] Palpant, B., Prot, D., Mouketou-Missono, A.-S., Rashidi-Huyeh, M., Sella, C., Debrus, S.: Evidence for electron thermal effect in the third-order nonlinear optical response of matrix-embedded gold nanoparticles. Proc. of SPIE **5221**, 14–23 (2003)
[75] del Coso, R., Solis, J.: Relation between nonlinear refractive index and third-order susceptibility in absorbing media. J. Opt. Soc. Am. B **21**, 640–644 (2004)
[76] Ando, M., Kadono, K., Haruta, M., Sakaguchi, T., Miya, M.: Large third-order optical nonlinearities in transition-metal oxides. Nature **374**, 625–627 (1995)
[77] Cattaruzza, E., Battaglin, G., Gonella, F., Polloni, R., Mattei, G., Maurizio, C., Mazzoldi, P., Sada, C., Montagna, M., Tosello, C., Ferrari, M.: On the optical absorption and nonlinearity of silica films containing metal nanoparticles. Philos. Mag. B **82**, 735–744 (2002)
[78] Battaglin, G., Cattaruzza, E., Gonella, F., Polloni, R., Scremin, B.F., Mattei, G., Mazzoldi, P., Sada, C.: Structural and optical properties of Cu:silica nanocomposite films prepared by co-sputtering deposition. Appl. Surf. Sci. **226**, 52–56 (2004)

[79] Ricard, D., Roussignol, P., Flytzanis, C.: Surface-mediated enhancement of optical phase conjugation in metal colloids. Opt. Lett. **10**, 511–513 (1985)
[80] Hache, F., Ricard, D., Flytzanis, C.: Optical nonlinearities of small metal particles: Surface-mediated resonance and quantum size effects. J. Opt. Soc. Am. B **3**, 1647–1655 (1986)
[81] Stroud, D., Hui, P.M.: Nonlinear susceptibilities of granular matter. Phys. Rev. B **37**, 8719–8724 (1988)
[82] Stroud, D., Van E. Wood: Decoupling approximation for the nonlinear-optical response of composite media. J. Opt. Soc. Am. B **6**, 778–786 (1989)
[83] Ma, H., Xiao, R., Sheng, P.: Third-order optical nonlinearity enhancement through composite microstructures. J. Opt. Soc. Am. B **15**, 1022–1029 (1998)
[84] Hellwarth, R.W.: Third-order optical susceptibilities of liquids and solids. Prog. Quantum Electron. **5**, 1–68 (1979)
[85] Buchalter, B., Meredith, G.R.: Third-order optical susceptibility of glasses determined by third harmonic generation. Appl. Opt. **21**, 3221–3224 (1982)
[86] Santran, S., Canioni, L., Sarger, L., Cardinal, T., Fargin, E.: Precise and absolute measurements of the complex third-order optical susceptibility. J. Opt. Soc. Am. B **21**, 2180-2190 (2004)
[87] Hache, F., Ricard, D., Girard, C.: Optical nonlinear response of small metal particles: A self-consistent calculation. Phys. Rev. B **38**, 7990–7996 (1988)
[88] Sipe, J.E., Boyd, R.W.: Nonlinear susceptibility of composite optical materials in the Maxwell Garnett model. Phys. Rev. A **43**, 1614–1629 (1992)
[89] Hache, F., Ricard, D., Flytzanis, C., Kreibig, U.: The optical Kerr effect in small metal particles and metal colloids: the case of gold. Appl. Phys. A **47**, 347–357 (1988)
[90] Flytzanis, C., Hache, F., Klein, M.C., Ricard, D., Roussignol, Ph.: Nonlinear optics in composite materials. In: Wolf, E. (ed.) Progress in Optics XXIX, Elsevier Science Publishers B.V., pp. 321–411 (1991)
[91] Rautian, S.G.: Nonlinear spectroscopy of the degenerate electron gas in spherical metallic particles. JETP **85**, 451–461 (1997)
[92] Eesley, G.L.: Generation of nonequilibrium electron and lattice temperatures in copper by picosecond laser pulses. Phys. Rev. B **33**, 2144–2151 (1986)
[93] Schoenlein, R.W., Lin, W.Z., Fujimoto, J.G., Eesley, G.L.: Femtosecond studies of nonequilibrium electronic processes in metals. Phys. Rev. Lett. **58**, 1680–1683 (1987)
[94] Smith, D.D., Fischer, G., Boyd, R., Gregory, D.A.: Cancellation of photoinduced absorption in metal nanoparticle composites through a counterintuitive consequence of local field effects. J. Opt. Soc. Am. B **14**, 1625–1631 (1997)
[95] Peiponen, K.-E., Vartiainen, E.M., Asakura, T.: Dispersion theory of effective meromorphic nonlinear susceptibilities of nanocomposites. J. Phys.: Condens. Matter **10**, 2483–2488 (1998)
[96] Agarwal, G.S., Dutta Gupta, S.: T-matrix approach to the nonlinear susceptibilities of heterogeneous media. Phys. Rev. A **38**, 5678–5687 (1988)
[97] Zeng, X.C., Bergman, D.J., Hui, P.M., Stroud, D.: Effective-medium theory for weakly nonlinear composites. Phys. Rev. B **38**, 10970–10973 (1988)
[98] Blumenfeld, R., Bergman, D.J.: Exact calculation to second order of the effective dielectric constant of a strongly nonlinear inhomogeneous composite. Phys. Rev. B **40**, 1987–1989 (1989)
[99] Haus, J.W., Inguva, R., Bowden, C.M.: Effective-medium theory of nonlinear ellipsoidal composites. Phys. Rev. A **40**, 5729–5734 (1989)
[100] Haus, J.W., Kalyaniwalla, N., Inguva, R., Bloemer, M., Bowden, C.M.: Nonlinear-optical properties of conductive spheroidal particle composites. J. Opt. Soc. Am. B **6**, 797–807 (1989)
[101] Shalaev, V.M., Stockman, M.I.: Resonant excitation and nonlinear optics of fractals. Physica A **185**, 181–186 (1992)
[102] Levy, O., Bergman, D.J.: Clausius-Mossotti approximation for a family of nonlinear composites. Phys. Rev. B **46**, 7189–7192 (1992)
[103] Boyd, R.W., Gehr, R.J., Fischer, G.L., Sipe, J.E.: Nonlinear optical properties of nanocomposite materials. Pure Appl. Opt. **5**, 505–512 (1996)

[104] Shalaev, V.M., Poliakov, E.Y., Markel, V.A.: Small-particle composites. II. Nonlinear optical properties. Phys. Rev. B **53**, 2437–2449 (1996)
[105] Shalaev, V.M., Sarychev, A.K.: Nonlinear optics of random metal-dielectric films. Phys. Rev. B **57**, 13265–13288 (1998)
[106] Sarychev, A.K., Shalaev, V.M.: Electromagnetic field fluctuations and optical nonlinearities in metal-dielectric composites. Phys. Rep. **375**, 275–371 (2000)
[107] Gao, L., Li, Z.-Y.: Temperature dependence of nonlinear optical response in metal/dielectrics composite media. Sol. State Commun. **107**, 751–755 (1998); Temperature dependence of nonlinear optical properties in metal/dielectric composites. Phys. Stat. Sol. (b) **218**, 571–582 (2000)
[108] Stockman, M.I., Kurlayev, K.B., George, T.F.: Linear and nonlinear optical susceptibilities of Maxwell Garnett composites: Dipolar spectral theory. Phys. Rev. B **60**, 17071–17083 (1999)
[109] Wu, Y.M., Gao, L., Li, Z.-Y.: The influence of particle shape on nonlinear optical properties of metal-dielectric composites. Phys. Stat. Sol. (b) **220**, 997–1008 (2000)
[110] Gao, L., Yu, K.W., Li, Z.Y., Bambi Hu, Effective nonlinear optical properties of metal-dielectric composite media with shape distribution. Phys. Rev. E **64**, 036615-1–8 (2001)
[111] Mackay, T.G., Lakhtakia A., Weiglhofer, W.S.: Homogeneisation of isotropic, cubically nonlinear, composite mediums by the strong-permittivity-fluctuation theory: third-order considerations. Opt. Commun. **204**, 219–228 (2002)
[112] Ma, H., Sheng, P., Wong, G.K.L.: Third-order nonlinear properties of Au clusters containing dielectric thin films. Topics Appl. Phys. **82**, 41–62 (2002)
[113] Gao, L., Li, Z.-Y.: Effect of temperature on nonlinear optical properties of composite media with shape distribution. J. Appl. Phys. **91**, 2045–2050 (2002)
[114] Neeves, A.E., Birnboim, M.H.: Composite structures for the enhancement of nonlinear optical materials. Opt. Lett. **13**, 1087–1089 (1988); Composite structures for the enhancement of nonlinear-optical susceptibility. J. Opt. Soc. Am. B **6**, 787–796 (1989)
[115] Goncharenko, A.V., Popelnukh, V.V., Venger, E.F.: Effect of weak nonsphericity on linear and nonlinear optical properties of small particle composites. J. Phys. D: Appl. Phys. **35**, 1833–1838 (2002)
[116] Pinchuk, A.: Optical bistability in nonlinear composites with coated ellipsoidal nanoparticles. J. Phys. D: Appl. Phys. **36**, 460–464 (2003)
[117] Ducourtieux, S., Podolskiy, V.A., Grésillon, S., Buil, S., Berini, B., Gadenne, P., Boccara, A.C., Rivoal, J.C., Bragg, W.D., Banerjee, K., Safonov, V.P., Drachev, V.P., Ying, Z.C., Sarychev, A.K., Vladimir M. Shalaev, Near-field optical studies of semicontinuous metal films. Phys. Rev. B **64**, 165403 (2001)
[118] Sheik-Bahae, M., Said, A.A., Wei, T.-H., Hagan, D.J., Van Stryland, E.W.: Sensitive measurement of optical nonlinearities using a single beam. IEEE J. Quant. Electron. **26**, 760–769 (1990)
[119] Lippitz, M., van Dijk, M.A., Orrit, M.: Third-harmonic generation from single gold nanoparticles. Nano Lett. **5**, 799–802 (2005)
[120] Ganeev, R.A., Ryasnyansky, A.I., Kamalov, Sh.R., Kodirov, M.K., Usmanov, T.: Nonlinear susceptibilities, absorption coefficients and refractive indices of colloidal metals. J. Phys. D: Appl. Phys. **34**, 1602–1611 (2001)
[121] Lamprecht, B., Krenn, J.R., Leitner, A., Aussenegg, F.R.: Resonant and off-resonant light-driven plasmons in metal nanoparticles studied by femtosecond-resolution third-harmonic generation. Phys. Rev. Lett. **83**, 4421–4424 (1999)
[122] Aktsipetrov, O.A., Murzina, T.V., Kim, E.M., Kapra, R.V., Fedyanin, A.A., Inoue, M., Kravets, A.F., Kuznetsova, S.V., Ivanchenko, M.V., Lifshits, V.G.: Magnetization-induced second- and third-harmonic generation in magnetic thin films and nanoparticles. J. Opt. Soc. Am. B **22**, 138–147 (2005)
[123] Philip, R., Kumar, G.R., Sandhyarani N., Pradeep, T.: Picosecond optical nonlinearity in monolayer-protected gold, silver, and gold-silver alloy nanoclusters. Phys. Rev. B **62**, 13160–13166 (2000)
[124] Qu, S., Zhao, C., Jiang, X., Fang, G., Gao, Y., Zeng, H., Song, Y., Qiu, J., Zhu, C., Hirao, K.: Optical nonlinearities of space selectively precipitated Au nanoparticles inside glasses. Chem. Phys. Lett. **368**, 352–358 (2003)

[125] Puech, K., Henari, F., Blau, W., Duff, D., Schmid, G.: Intensity-dependent optical absorption of colloidal solutions of gold nanoparticles. Europhys. Lett. **32**, 119–124 (1995)
[126] Huang, H.H., Yan, F.Q., Kek, Y.M., Chew, C.H., Xu, G.Q., Ji, W., Oh, P.S., Tang, S.H.: Synthesis, characterization, and nonlinear optical properties of copper nanoparticles. Langmuir **13**, 172–175 (1997)
[127] Danilova, Y.E., Lepeshkin, N.N., Rautian, S.G., Safonov, V.P.: Excitation localization and nonlinear optical processes in colloidal silver aggregates. Physica A **241**, 231–235 (1997)
[128] Debrus, S., Lafait, J., May, M., Pinçon, N., Prot, D., Sella, C., Venturini, J.: Z-scan determination of the third-order optical nonlinearity of gold:silica nanocomposites. J. Appl. Phys. **88**, 4469–4475 (2000)
[129] Wang, W.-T., Yang, G., Chen, Z.-H., Zhou, Y.-L., Lü, H.-B., Yang, G.-Z.: Large third-order optical nonlinearity in Au nanometer particle doped $BaTiO_3$ composite films near the resonant frequency. Chinese Phys. **11** 1324–1327 (2002)
[130] Wang, W., Yang, G., Chen, Z., Lu, H., Zhou, Y., Yang, G., Kong, X.: Nonlinear refraction and saturable absorption in $Au:BaTiO_3$ composite films. Appl. Opt. **42**, 5591–5595 (2003)
[131] Wang, P., Lu, Y., Tang, L., Zhang, J., Ming, H., Xie, J., Ho, F.-H., Chang, H.-H., Lin, H.-Y., Tsai, D.-P.: Surface-enhanced optical nonlinearity of a gold film. Opt. Commun. **229**, 425–429 (2004)
[132] Ganeev, R.A., Ryasnyansky, A.I., Stepanov, A.L., Usmanov, T.: Saturated absorption and nonlinear refraction of silicate glasses doped with silver nanoparticles at 532 nm. Opt. Quant. Electron. **36**, 949–960 (2004)
[133] Puech, K., Henari, F., Blau, W., Duff, D., Schmid, G.: Investigation of the ultrafast dephasing time of gold nanoparticles using incoherent light. Chem. Phys. Lett. **247**, 13–17 (1995)
[134] Ganeev, R.A., Ryasnyansky, A.I., Stepanov, A.L., Usmanov, T.: Characterization of nonlinear optical parameters of copper- and silver-doped silica glasses at $\lambda = 1064$ nm. Phys. Stat. Sol. (b) **241**, 935–944 (2004)
[135] Kyoung, M., Lee, M.: Nonlinear absorption and refractive index measurements of silver nanorods by the Z-scan technique. Opt. Commun. **171**, 145–148 (1999)
[136] Haglund, Jr., R.F., Yang, L., Magruder III, R.H., Wittig, J.E., Becker, K., Zuhr, R.A., Picosecond nonlinear optical response of a Cu:silica nanocluster composite. Opt. Lett. **18**, 373–375 (1993)
[137] Magruder, III, R.H., Osborne, Jr., D.H., Zuhr, R.A.: Non-linear optical properties of nanometer dimension Ag—Cu particles in silica formed by sequential ion implantation. J. Non-Cryst. Solids **176**, 299–303 (1994)
[138] Yang, G., Wang, W., Zhou, Y., Lu, H., Yang, G., Chen, Z.: Linear and nonlinear optical properties of Ag nanocluster/$BaTiO_3$ composite films. Appl. Phys. Lett. **81**, 3969–3971 (2002)
[139] Scalisi, A.A., Compagnini, G., D'Urso, L., Puglisi, O.: Nonlinear optical activity in Ag–SiO_2 nanocomposite thin films with different silver concentration. Appl. Surf. Sci. **226**, 237–241 (2004)
[140] Unnikrishnan, K.P., Nampoori, V.P.N., Ramakrishnan, V., Umadevi, M., Vallabhan, C.P.G.: Nonlinear optical absorption in silver nanosol. J. Phys. D: Appl. Phys. **36**, 1242–1245 (2003)
[141] Qu, S., Gao, Y., Jiang, X., Zeng, H., Song, Y., Qiu, J., Zhu, C., Hirao, K.: Nonlinear absorption and optical limiting in gold-precipitated glasses induced by a femtosecond laser. Opt. Commun. **224**, 321–327 (2003)
[142] Ispasoiu, R.G., Balogh, L., Varnavski, O.P., Tomalia, D.A., Goodson, III, T.: Large optical limiting from novel metal-dendrimer nanocomposite materials. J. Am. Chem. Soc. **122**, 11005–11006 (2000)
[143] François, L., Mostafavi, M., Belloni, J., Delouis, J.F., Delaire, J., Feneyrou, P.: Optical limitation induced by gold clusters. 1. Size effect. J. Phys. Chem. B **104**, 6133–6137 (2000)
[144] François, L., Mostafavi, M., Belloni, J., Delaire, J.A.: Optical limitation induced by gold clusters: Mechanism and efficiency. Phys. Chem. Chem. Phys. **3**, 4965–4971 (2001)
[145] Zhang, H., Zelmon, D.E., Deng, L., Liu, H.-K., Teo, B.K.: Optical limiting behavior of nanosized polyicosahedral gold-silver clusters based on third-order nonlinear optical effects. J. Am. Chem. Soc. **123**, 11300–11301 (2001)
[146] Sun, N., Wang, Y., Song, Y., Guo, Z., Dai, L., Zhu, D.: Novel [60]fullerene-silver nanocomposite with large optical limiting effect. Chem. Phys. Lett. **344**, 277–282 (2001)

[147] Qu, S., Du, C., Song, Y., Wang, Y., Gao, Y., Liu, S., Li, Y., Zhu, D.: Optical nonlinearities and optical limiting properties in gold nanoparticles protected by ligands. Chem. Phys. Lett. **356**, 403–408 (2002)
[148] Fang, H., Du, C., Qu, S., Li, Y., Song, Y., Li, H., Liu, H., Zhu, D.: Self-assembly of the [60]fullerene-substituted oligopyridines on Au nanoparticles and the optical nonlinearities of the nanoparticles. Chem. Phys. Lett. **364**, 290–296 (2002)
[149] Qu, S., Li, H., Peng, T., Gao, Y., Qiu, J., Zhu, C.: Optical nonlinearities from transverse plasmon resonance in gold nano-rods. Mat. Lett. **58**, 1427–1430 (2004)
[150] Zhan, C., Li, D., Zhang, D., Xu, W., Nie, Y., Zhu, D.: The excited-state absorption and third-order optical nonlinearity from 1-dodecanethiol protected gold nanoparticles: Application for optical limiting. Opt. Mat. **26**, 11–15 (2004)
[151] Bloemer, M.J., Haus, J.W., Ashley, P.R.: Degenerate four-wave mixing in colloidal gold as a function of particle size. J. Opt. Soc. Am. B **7**, 790–795 (1990)
[152] Fukumi, K., Chayahara, A., Kadono, K., Sakaguchi, T., Horino, Y., Miya, M., Fujii, K., Hayakawa, J., Satou, M.: Gold nanoparticles ion implanted in glass with enhanced nonlinear optical properties. J. Appl. Phys. **75**, 3075–3080 (1994)
[153] Yang, L., Becker, K., Smith, F.M., Magruder III, R.H., Haglund, Jr., R.F., Yang, L., Dorsinville, R., Alfano, R.R., Zuhr, R.A.: Size dependence of the third-order susceptibility of copper nanoclusters investigated by four-wave mixing. J. Opt. Soc. Am. B **11**, 457–461 (1994)
[154] Uchida, K., Kaneko, S., Omi, S., Hata, C., Tanji, H., Asahara, Y., Ikushima, A.J., Tokizaki, T., Nakamura, A.: Optical nonlinearities of a high concentration of small metal particles dispersed in glass: copper and silver particles. J. Opt. Soc. Am. B **11**, 1236–1243 (1994)
[155] Puech, K., Blau, W., Grund, A., Bubeck, C., Gardenas, G.: Picosecond degenerate four-wave mixing in colloidal solutions of gold nanoparticles at high repetition rates. Opt. Lett. **20**, 1613–1615 (1995)
[156] Tanahashi, I., Manabe, Y., Tohda, T., Sasaki, S., Nakamura, A.: Optical nonlinearities of Au/SiO$_2$ composite thin films prepared by a sputtering method. J. Appl. Phys. **79**, 1244–1249 (1996)
[157] Lee, M., Kim, T.S., Choi, Y.S.: Third-order optical nonlinearities of sol–gel-processed Au–SiO$_2$ thin films in the surface plasmon absorption region. J. Non-Cryst. Solids **211**, 143–149 (1997)
[158] Hosoya, Y., Suga, T., Yanagawa, T., Kurokawa, Y.: Linear and nonlinear optical properties of sol-gel-derived Au nanometer-particle-doped alumina. J. Appl. Phys. **81**, 1475–1480 (1997)
[159] Liao, H.B., Xiao, R.F., Fu, J.S., Wang, H., Wong, K.S., Wong, G.K.L.: Origin of third-order optical nonlinearity in Au:SiO$_2$ composite films on femtosecond and picosecond time scales. Opt. Lett. **23**, 388–390 (1998)
[160] Ballesteros, J.M., Solis, J., Serna, R., Afonso, C.N.: Nanocrystal size dependence of the third-order nonlinear optical response of Cu:Al$_2$O$_3$ thin films. Appl. Phys. Lett. **74**, 2791–2793 (1999)
[161] Hamanaka, Y., Nakamura, A., Omi, S., Del Fatti, N., Vallée, F., Flytzanis, C.: Ultrafast response of nonlinear refractive index of silver nanocrystals embedded in glass. Appl. Phys. Lett. **75**, 1712–1714 (1999)
[162] Hamanaka, Y., Nakamura, A., Hayashi, N., Omi, S.: Dispersion curves of complex third-order optical susceptibilities around the surface plasmon resonance in Ag nanocrystal–glass composites. J. Opt. Soc. Am. B **20**, 1227–1232 (2003)
[163] del Coso, R., Requejo-Isidro, J., Solis, J., Gonzalo, J., Afonso, C.N.: Third order nonlinear optical susceptibility of Cu:Al$_2$O$_3$ nanocomposites: From spherical nanoparticles to the percolation threshold. J. Appl. Phys. **95**, 2755–2762 (2004)
[164] Wang, W., Yang, G., Wu, W., Chen, Z.: Effects of the morphology and nanostructure on the optical nonlinearities of Au:BaTiO$_3$ nanocomposite films. J. Appl. Phys. **94**, 6837–6840 (2003)
[165] Magruder, III, R.H., Haglund, Jr., R.F., Yang, L., Wittig, J.E., Zuhr, R.A.: Physical and optical properties of Cu nanoclusters fabricated by ion implantation in fused silica. J. Appl. Phys. **76**, 708–715 (1994)
[166] Serna, R., Ballesteros, J.M., Solis, J., Afonso, C.N., Osborne, D.H., Haglund, Jr., R.F., Petford-Long, A.K.: Laser-induced modification of the nonlinear optical response of laser-deposited Cu:Al$_2$O$_3$ nanocomposite films. Thin Sol. Films **318**, 96–99 (1998)

[167] Ma, G., Sun, W., Tang, S.-H., Zhang, H., Shen, Z., Qian, S.: Size and dielectric dependence of the third-order nonlinear optical response of Au nanocrystals embedded in matrices. Opt. Lett. **27**, 1043–1045 (2002)

[168] Selvan, S.T., Hayakawa, T., Nogami, M., Kobayashi, Y., Liz-Marzán, L.M., Hamanaka, Y., Nakamura, A.: Sol-gel derived gold nanoclusters in silica glass possessing large optical nonlinearities. J. Phys. Chem. B **106**, 10157–10162 (2002)

[169] Liao, H.B., Wen, W., Wong, G.K.L.: Preparation and optical characterization of Au/SiO_2 composite films with multilayer structure. J. Appl. Phys. **93**, 4485–4488 (2003)

[170] Liao, H., Wen, W., Wong, G.K.L., Yang, G.: Optical nonlinearity of nanocrystalline Au/ZnO composite films. Opt. Lett. **28**, 1790–1792 (2003)

[171] Yang, Y., Nogami, M., Shi, J., Chen, H., Ma, G., Tang, S.: Enhancement of third-order optical nonlinearities in 3-dimensional films of dielectric shell capped Au composite nanoparticles. J. Phys. Chem. B **109**, 4865–4871 (2005)

[172] Takeda, Y., Lu, J., Plaksin, O.A., Kono, K., Amekura, H., Kishimoto, N.: Control of optical nonlinearity of metal nanoparticle composites fabricated by negative ion implantation. Thin Sol. Films **464–465**, 483–486 (2004)

[173] Gao L., Li, Z.-Y.: Third-order nonlinear optical response of metal dielectric composites. J. Appl. Phys. **87**, 1620–1625 (2000)

[174] Ishizaka, T., Muto, S., Kurokawa, Y.: Nonlinear optical and XPS properties of Au and Ag nanometer-size particle-doped alumina films prepared by the sol–gel method. Opt. Commun. **190**, 385–389 (2001)

[175] Yang, L., Osborne, D.H., Haglund, Jr., R.F., Magruder, R.H., White, C.W., Zuhr, R.A., Hosono, H.: Probing interface properties of nanocomposites by third-order nonlinear optics. Appl. Phys. A **62**, 403–415 (1996)

[176] Wang, W.T., Chen, Z.H., Yang, G., Guan, D.Y., Yang, G.Z., Zhou, Y.L., Lu, H.B.: Resonant absorption quenching and enhancement of optical nonlinearity in $Au:BaTiO_3$ composite films by adding Fe nanoclusters. Appl. Phys. Lett. **83**, 1983–1985 (2003)

[177] Zhang, Q.F., Liu, W.M., Xue, Z.Q., Wu, J.L., Wang, S.F., Wang, D.L., Gong, Q.H.: Ultrafast optical Kerr effect of Ag–BaO composite thin films. Appl. Phys. Lett. **82**, 958–960 (2003)

[178] Zhou, P., You, G., Li, J., Wang, S., Qian, S., Chen, L.: Annealing effect of linear and nonlinear optical properties of $Ag:Bi_2O_3$ nanocomposite films. Opt. Expr. **13**, 1508–1514 (2005)

[179] Cotell, C.M., Schiestel, S., Carosella, C.A., Flom, S., Hubler, G.K., Knies, D.L.: Ion-beam assisted deposition of Au nanocluster/Nb_2O_5 thin films with nonlinear optical properties. Nucl. Instr. and Meth. in Phys. Res. B **127/128**, 557–561 (1997)

[180] Liao, H.B., Xiao, R.F., Wang, H., Wong, K.S., Wong, G.K.L.: Large third-order optical nonlinearity in $Au:TiO_2$ composite films measured on a femtosecond time scale. Appl. Phys. Lett. **72**, 1817–1819 (1998)

[181] Zhang, C.-F., You, G.-J., Dong, Z.-W., Liu, Y., Ma, G.-H., Qian, S.-X.: Off-resonant third-order optical nonlinearity of an $Ag:TiO_2$ composite film. Chin. Phys. Lett. **22**, 475–477 (2005)

[182] Yang, G., Wang, W.-T., Yang, G.-Z., Chen, Z.-H.: Enhanced nonlinear optical properties of laser deposited $Ag/BaTiO_3$ nanocomposite films. Chin. Phys. Lett. **20**, 924–927 (2003)

[183] Wang, W., Qu, L., Yang, G., Chen, Z.: Large third-order optical nonlinearity in $BaTiO_3$ matrix-embedded metal nanoparticles. Appl. Surf. Sci. **218**, 24–28 (2003); Wang, W., Yang, G., Chen, Z., Lu, H., Zhou, Y., Yang, G., Kong, X.: Nonlinear refraction and saturable absorption in $Au:BaTiO_3$ composite films. Appl. Opt. **42**, 5591–5595 (2003)

[184] Zhou, P., You, G., Li, J., Wang, S., Qian, S., Chen, L.: Annealing effect of linear and nonlinear optical properties of $Ag:Bi_2O_3$ nanocomposite films. Opt. Expr. **13**, 1508–1514 (2005)

[185] Williams, E.K., Ila, D., Sarkisov, S., Curley, M., Cochrane, J.C., Poker, D.B., Hensley, D.K., Borel, C.: Study of the effects of MeV Ag and Au implantation on the optical properties of $LiNbO_3$. Nucl. Instr. and Meth. in Phys. Res. B **141**, 268–273 (1998); Sarkisov, S.S., Williams, E., Curley, M., Ila, D., Venkateswarlu, P., Poker, D.B., Hensley, D.K.: Third order optical nonlinearity of colloidal metal nanoclusters formed by MeV ion implantation. Nucl. Instr. and Meth. in Phys. Res. B **141**, 294–298 (1998)

[186] Ryasnyansky, A., Palpant, B., Debrus, S., Ganeev, R., Stepanov, A., Can, N., Buchal, C., Uysal, S.: Nonlinear optical absorption of ZnO doped with copper nanoparticles in the picosecond and nanosecond pulse laser field. Appl. Opt. **44**, 2839–2845 (2005)
[187] Huang, H.H., Yan, F.Q., Kek, Y.M., Chew, C.H., Xu, G.Q., Ji, W., Oh, P.S., Tang, S.H.: Synthesis, characterization, and nonlinear optical properties of copper nanoparticles. Langmuir **13**, 172–175 (1997)
[188] Pinçon-Roetzinger, N., Prot, D., Palpant, B., Charron, E., Debrus, S.: Large optical Kerr effect in matrix-embedded metal nanoparticles. Mat. Sci. Eng. C **19**, 51–54 (2002)
[189] Pinçon, N., Palpant, B., Prot, D., Charron, E., Debrus, S.: Third-order nonlinear optical response of Au:SiO$_2$ thin films: Influence of gold nanoparticle concentration and morphologic parameters. Eur. Phys. J. D **19**, 395–402 (2002)
[190] Liao, H.B., Xiao, R.F., Fu, J.S., Yu, P., Wong, G.K.L., Ping Sheng, Large third-order optical nonlinearity in Au:SiO$_2$ composite films near the percolation threshold. Appl. Phys. Lett. **70**, 1–3 (1997)
[191] Liao, H.B., Xiao, R.F., Fu, J.S., Wong, G.K.L.: Large third-order nonlinear optical susceptibility of Au–Al$_2$O$_3$ composite films near the resonant frequency. Appl. Phys. B **65**, 673–676 (1997)
[192] Zhou, P., You, G.J., Li, Y.G., Han, T., Li, J., Wang, S.Y., Chen, L.Y., Liu, Y., Qian, S.X.: Linear and ultrafast nonlinear optical response of Ag:Bi$_2$O$_3$ composite films. Appl. Phys. Lett. **83**, 3876–3878 (2003)
[193] Hamanaka, Y., Fukuta, K., Nakamura, A., Liz-Marzán, L.M., Mulvaney, P.: Enhancement of third-order nonlinear optical susceptibilities in silica-capped Au nanoparticle films with very high concentrations. Appl. Phys. Lett. **84**, 4938–4940 (2004)
[194] Takeda, Y., Lu, J., Plaksin, O.A., Amekura, H., Kono, K., Kishimoto, N.: Optical properties of dense Cu nanoparticle composites fabricated by negative ion implantation. Nucl. Instr. and Meth. in Phys. Res. B **219–220**, 737–741 (2004)
[195] Liao, H.B., Wen, W., Wong, G.K.L.: Preparation and characterization of Au/SiO$_2$ multilayer composite films with nonspherical Au particles. App. Phys. A: Mat. Sci. Proc. **80**, 861–864 (2005)
[196] West, R., Wang, Y., Goodson III, T.: Nonlinear absorption properties in novel gold nanostructured topologies. J. Phys. Chem. B **107**, 3419–3426 (2003)
[197] Okada, N., Hamanaka, Y., Nakamura, A., Pastoriza-Santos, I., Liz-Marzán, L.M.: Linear and nonlinear optical response of silver nanoprisms: Local electric fields of dipole and quadrupole plasmon resonances. J. Phys. Chem. B **108**, 8751–8755 (2004)
[198] Zhao, C.-J., Qu, S.-L., Gao, Y.-C., Song, Y.-L., Qiu, J.-R., Zhu, C.-S.: Preparation and nonlinear optical properties of Au colloid. Chin. Phys. Lett. **20**, 1752–1754 (2003)
[199] Ganeev, R.A., Ryasnyansky, A.I., Stepanov, A.L., Usmanov, T.: Saturated absorption and nonlinear refraction of silicate glasses doped with silver nanoparticles at 532 nm. Opt. Quant. Electron. **36**, 949–960 (2004)
[200] Faccio, D., Di Trapani, P., Borsella, E., Gonella, F., Mazzoldi, P., Malvezzi, A.M.: Measurement of the third-order nonlinear susceptibility of Ag nanoparticles in glass in a wide spectral range. Europhys. Lett, **43**, 213–218 (1998)
[201] Li, Y., Takata, M., Nakamura, A.: Size-dependent enhancement of nonlinear optical susceptibilities due to confined excitons in CuBr nanocrystals. Phys. Rev. B **57**, 9193–9200 (1998)
[202] Bigot, J.-Y., Merle, J.-C., Cregut, O., Daunois, A.: Electron dynamics in copper metallic nanoparticles probed with femtosecond optical pulses. Phys. Rev. Lett. **75**, 4702–4705 (1995)
[203] Hamanaka, Y., Hayashi, N., Omi, S., Nakamura, A.: Ultrafast relaxation dynamics of electrons in silver nanocrystals embedded in glass. J. Lumin. **76&77**, 221–225 (1997)
[204] Nisoli, M., Stagira, S., De Silvestri, S., Stella, A., Tognini, P., Cheyssac, P., Kofman, R.: Ultrafast electronic dynamics in solid and liquid gallium nanoparticles. Phys. Rev. Lett. **78**, 3575–3578 (1997)
[205] Perner, M., Bost, P., Becker, U., Mennig, M., Schmitt, M., Schmidt, H.: Optically induced damping of the surface plasmon resonance in gold colloids. Phys. Rev. Lett. **78**, 2192–2195 (1997); Hartland, G.V., Hodak, J.H., Martini, I.: Comment. ibid. **82**, 3188 (1999); Perner, M., von Plessen, G., Feldmann, J.: Reply. ibid. **82**, 3189 (1999)

[206] Inouye, H., Tanaka, K., Tanahashi, I., Hirao, K.: Ultrafast dynamics of nonequilibrium electrons in a gold nanoparticles system. Phys. Rev. B **57**, 11334–11340 (1998)
[207] Inouye, H., Tanaka, K., Tanahashi, I., Hirao, K.: Femtosecond optical Kerr effect in the gold nanoparticles system. Jpn. J. Appl. Phys. **37**, L1520–L1522 (1998)
[208] Bigot, J.-Y., Halté, V., Merle, J.-C., Daunois, A.: Electron dynamics in metallic nanoparticles. Chem. Phys. **251**, 181–203 (2000)
[209] Del Fatti, N., Vallée, F.: Ultrafast optical nonlinear properties of metal nanoparticles. Appl. Phys. B **73**, 383–390 (2001)
[210] Hamanaka, Y., Kuwabata, J., Tanahashi, I., Omi, S., Nakamuka, A.: Ultrafast electron relaxation via breathing vibration of gold nanocrystals embedded in a dielectric medium. Phys. Rev. B **63**, 104302 (2001)
[211] Voisin, C., Del Fatti, N., Christofilos, D., Vallée, F.: Ultrafast electron dynamics and optical nonlinearities in metal nanoparticles. J. Phys. Chem. B **105**, 2264–2280 (2001)
[212] Falconieri, M.: Thermo-optical effects in Z-scan measurements using high-repetition-rate lasers. J. Opt. A: Pure Appl. Opt. **1**, 662–667 (1999)
[213] Rashidi-Huyeh, M., Palpant, B.: Thermal response of nanocomposite materials under pulsed laser excitation. J. Appl. Phys. **96**, 4475–4482 (2004)
[214] Halté, V., Bigot, J.-Y., Palpant, B., Broyer, M., Prével, B., Pérez, A.: Size dependence of the energy relaxation in silver nanoparticles embedded in dielectric matrices. Appl. Phys. Lett. **75**, 3799–3801 (1999)
[215] Hu, M., Hartland, G.V.: Heat dissipation for Au particles in aqueous solution: Relaxation time versus size. J. Phys. Chem. B **106**, 7029–7033 (2002)
[216] Heilweil, E.J., Hochstrasser, R.M.: Nonlinear spectroscopy and picosecond transient grating study of colloidal gold. J. Chem. Phys. **82**, 4762–4770 (1985)
[217] Inouye, H., Tanaka, K., Tanahashi, I., Hattori, T., Nakatsuka, H.: Ultrafast optical switching in a silver nanoparticle system. Jpn. J. Appl. Phys. **39**, 5132–5133 (2000)
[218] Mehendale, S.C., Mishra, S.R., Bindra, K.S., Laghate, M., Dhami, T.S., Rustagi, K.C.: Nonlinear refraction in aqueous colloidal gold. Opt. Commun. **133**, 273–276 (1997)
[219] de Nalda, R., del Coso, R., Requejo-Isidro, J., Olivares, J., Suarez-Garcia, A., Solis, J., Afonso, C.N.: Limits to the determination of the nonlinear refractive index by the z-scan method. J. Opt. Soc. Am. B **19**, 289–296 (2002)
[220] Ballesteros, J.M., Serna, R., Solís, J., Afonso, C.N., Petford-Long, A.K., Osborne, D.H., Haglund, Jr., R.F.: Pulsed laser deposition of $Cu:Al_2O_3$ nanocrystal thin films with high third-order optical susceptibility. Appl. Phys. Lett. **71**, 2445–2447 (1997)
[221] Corti, C.W., Holliday, R.J., Thompson, D.T.: Developing new industrials applications for gold: gold nanotechnology. Gold Bulletin. **35**, 111–117 (2002)
[222] Hirsch, L.R., Stafford, R.J., Bankson, J.A., Sershen, S.R., Rivera, B., Price, R.E., Hazle, J.D., Halas, N.J., West, J.L.: Nanoshell-mediated near-infrared thermal therapy of tumors under magnetic resonance guidance. PNAS **100**, 13549–13554 (2003); Loo, C., Lin, A., Hirsch, L., Lee, M.H., Barton, J., Halas, N., West, J., Drezek, R.: Nanoshell-enabled photonics-based imaging and therapy of cancer. Technol. Cancer Res. Treat. **3**, 33–40 (2004)
[223] Inouye, H., Kanemitsu, Y.: Direct observation of nonlinear effects in a one-dimensional photonic crystal. Appl. Phys. Lett. **82**, 1155–1157 (2003); Inouye, H., Kanemitsu, Y., Hirao, K.: Nonlinear optical response in a total-reflection-type one-dimensional photonic crystal with gold nanoparticles. Physica E **17**, 414–417 (2003)

CHAPTER 16

FROM DIPOLAR TO OCTUPOLAR PHTHALOCYANINE DERIVATIVES: THE EXAMPLE OF SUBPHTHALOCYANINES

CHRISTIAN G. CLAESSENS, GEMA DE LA TORRE AND TOMÁS TORRES
Universidad Autónoma de Madrid, Departamento de Química Orgánica C-I, Campus de Cantoblanco, 28049 Madrid, Spain

Abstract: Boron-subphthalocyanines (SubPcs)—cone-shaped 14-π electron aromatic macrocycles—are lower phthalocyanine analogues which consist of three isoindole units N-fused around a central boron atom which fourth valency is occupied by an axial ligand. SubPcs have been shown to possess very interesting features for NLO. Although they present a permanent dipole moment along the boron-axial substituent axis, their optical response is essentially associated to charge transfer inside the macrocycle π-surface. Moreover, due to the C_{3v} symmetry of the SubPc core, its NLO behavior is mostly octupolar. It will be shown that the application to SubPcs of the design criteria that have been successful in phthalocyanines and porphyrins led to high-performance second-harmonic generators

Keywords: Phthalocyanines – Subphthalocyanines–Boron–Nonlinear Optics-Dipolar Compounds – Octupolar Compounds

1. INTRODUCTION

The last few decades have witnessed a spectacular growth of multidisciplinary research activity involving materials that exhibit nonlinear optical (NLO) behavior [1–3]. NLO activity was first found in semiconductors and inorganic crystals [2, 3] followed by the coming out of organic materials [4–10] by the mid-1980's. The latter have fascinated most scientists in this area because they present advantages over conventional inorganics for practical applications. First, they are, in general, easier to process and to integrate into micro-optoelectronic devices. Moreover, the inherent tailorability of organic compounds renders them suitable for achieving a fine-tuning of the NLO properties by rational modification of the

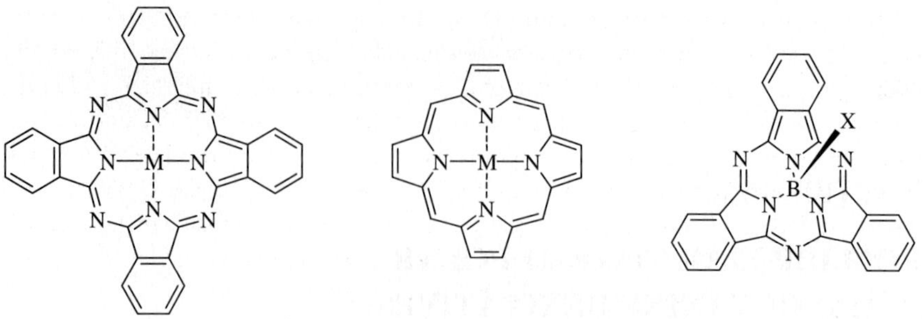

Figure 1. Molecular structures of from left to right: phthalocyanine; porphyrins and subphthalocyanine

chemical structure. Typically, organic molecules for nonlinear optics possess highly polarizable delocalized π-electrons, i.e. polyenes. Among organic compounds, phthalocyanines (**1**, Figure 1, left) [11–14] and other related aromatic macrocycles such as porphyrins (**2**, Figure 1, center) [14] and subphthalocyanines (**3**, Figure 1, right) [15], stand out because of the fast and large nonlinearities they exhibit and their stability and processability features. In addition, their chemical versatility allows the manipulation of the electronic distribution of the macrocyclic core and, therefore, the fine-tuning of the NLO response [16–19].

In recent years there has been a growing interest in the search for materials with large macroscopic second-order nonlinearities [20–22] because of their practical utility as frequency doublers, frequency converters and electro-optic modulators [23] by means of second-harmonic generation, parametric frequency conversion (or mixing) and the electro-optic (EO) effect. They are described by $\chi^{(2)}(2\omega;\omega,\omega)$, $\chi^{(2)}(\omega;\omega_1 \pm \omega_2)$, $\chi^{(2)}(\omega;0,\omega)$, respectively. In order to optimize these effects, highly efficient materials have to be engineered. Second-order NLO effects are usually observed from noncentrosymmetric materials which are built up, for example, by incorporating donor-acceptor substituted organic molecules that have nonvanishing molecular hyperpolarizability (β). Thus, molecular engineering of one-dimensional (1-D) chromophores has been particularly active, leading to push-pull derivatives displaying huge first-order hyperpolarizabilities [24–26]. However, such chromophores display an intense absorption band in the visible region, due to a strong intramolecular charge transfer (ICT) transition, which is an important drawback especially in the context of effective materials for SHG in the visible. Moreover, the polarity of such molecules is a disadvantage for obtaining non-centrosymmetric crystals because dipole-dipole interactions favor an antiparallel alignment of the chromophores. For these reasons, another type of molecules having planar or quasi-planar structures and an octupolar charge distribution have been developed; in this way they offer additional parameters for the optimization of the NLO response. So, have been brought into consideration molecular symmetries such as D_{3h} which, lacking a permanent dipole moment, can exhibit non-zero β due to the non-diagonal contribution to the β tensor. The octupolar route, pioneered by Lehn, Zyss and coworkers [27, 28], allows the optimization of the efficiency-transparency trade-off and hold potential for non-centrosymmetric crystallization.

In this chapter we report on some novel strategies that have been pursued to obtain efficient second-order nonlinear molecules starting from the well-known phthalocyanines. In principle, these planar centrosymmetric molecules do not present second-order activity and have been extensively studied for third-order applications. In order to induce asymmetry, two main approaches have been followed: a) peripheral substitution of the macrocycle with donor and acceptor groups and b) structural modifications of the Pc core to reduce the symmetry, the resulting-noncentrosymmetric compounds (i.e. subphthalocyanines) presenting variable degrees of dipolarity/octupolarity in the nonlinear response.

2. FROM DIPOLAR TO OCTUPOLAR APPROACHES: PHTHALOCYANINES AND RELATED MACROCYCLES

As mentioned above, in many organic compounds the second-order optical nonlinearity arises from one-dimensional highly polarizable π-conjugated systems capped with groups of different electron affinities. Such dipolar polarizable molecules exhibit one dominant hyperpolarizability component lying in the direction of the charge transfer axis. There are also examples of 2D and 3D systems displaying second-order NLO behavior, but the available data are still scarce. Although high values have been determined or calculated for β, in only a few cases they approach the best 1D values. Phthalocyanines and other related analogues such as porphyrins and subphthalocyanines are among the most relevant 2D/3D targets for second-order NLO. The main advantage of these multidimensional compounds is that they offer the possibility of investigating the role of dimensionality on the NLO response; the correlation between structure and NLO response is much richer for these than for 1D molecules and offers more variables of optimization.

In the case of molecules having inversion symmetry, as non-substituted phthalocyanines, all the components of the first hyperpolarizability β are zero. For this reason, appropriately substituted phthalocyanines have to be designed if one wishes to obtain efficient second-order NLO responses. Thus, theoretical calculations developed at the end of the 80's by T. J. Marks and coworkers [29] suggested that push-pull unsymmetrically substituted phthalocyanines with suitable electron-donor and acceptor groups and efficient intramolecular charge transfer should yield interesting compounds for second-order applications.

Therefore, prediction of the β values for a given chemical structure is essential in the second-order NLO field in order to prepare appropriate systems for a specific application and constitutes the basis for the optimization of the microscopic NLO performance. The most widely used theoretical model to discuss the NLO performance of organic molecules consists of a summation over the electronic states of the system (SOS: sum over states). The general SOS expressions for the components of the β tensors are well-known [30]. In many cases, only a few electronic levels contribute significantly to the NLO response and the general expressions simplify. For 1D charge-transfer molecules (such as p-nitroaniline) a two-level model has yielded satisfactory results for the quadratic hyperpolarizability. For planar (2D) systems, as most phthalocyanines and analogous compounds, three levels (or even more) are generally required to account for β. However, for strongly push-pull phthalocyanines with a dominant optical absorption band, a two-level model may still be a reasonable approximation. In principle, it is convenient to consider the second-order response of phthalocyanines in a three level model (the ground 0 and the excited 1 and 2 levels, responsible for the Q band).

In accordance with their electronic structure, phthalocyanines present intense π-π bands at the visible (Q band) and UV (B or Soret band) spectral regions that mostly determine the NLO response (Figure 2). The Q band corresponds to transitions to the lowest excited state orbitals (e_g) from the highest occupied orbital (a_u). For metal-free phthalocyanines, the Q band is split into two main components whilst metal-containing phthalocyanines with D_{4h} symmetry exhibit only a single Q band in their visible spectra (Figure 2). The exact position of these bands depends on the particular structure, metal complexation and peripheral substituents [11]. For peripherally substituted metallic Pcs (e.g. push-pull compounds) the degenerate Q bands show some splitting due to the reduction in symmetry. The broad absorption valley between the B and Q bands is the zone used for frequency doubling into the green spectral region. The two- and three-level models mostly used to account for the second-order NLO behavior of phthalocyanines include either the doubly degenerate or split excited levels responsible for the sharp Q band.

Since the former theoretical predictions of "push-pull" substituted phthalocyanines as candidates for second-order NLO properties, some work has been devoted to prepare and study different substituted derivatives with the aim to establish the key structural parameters affecting the NLO response. Some revisions have already been done on the second-order NLO behavior of phthalocyanines [16, 17, 31]. For most of the unsymmetrically substituted push-pull compounds (planar conjugated π-electron systems in the XZ plane), C_{2v} symmetry may be assumed due to the

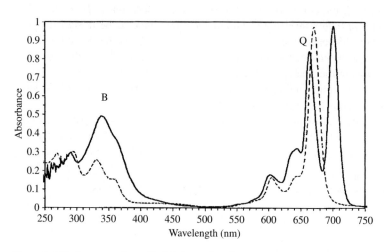

Figure 2. UV-vis spectra of metallophthalocyanines (dotted line) and metal-free phthalocyanines (continuous line).

presence of the charge-transfer axis z. In this case, three non-zero and independent components of the β tensor can be considered, namely β_{zzz}, β_{zxx} and β_{xxz}.

(1) $$\beta_{zzz} = \beta_{33} = \frac{(\mu_{01}^z)^2 \Delta\mu_{01}^z}{\hbar^2} d_{33}$$

$$\beta_{zxx} = \beta_{31} = \frac{(\mu_{02}^x)^2 \Delta\mu_{02}^z}{\hbar^2} d_{31} + \frac{2\mu_{01}^z \mu_{02}^x \mu_{12}^x}{\hbar^2} d_{31}'$$

$$\beta_{xxz} = \beta_{15} = \frac{(\mu_{02}^x)^2 \Delta\mu_{02}^z}{\hbar^2} d_{15} + \frac{\mu_{02}^x \mu_{12}^x \mu_{01}^z}{\hbar^2} d_{15}'$$

where the dispersion factors are:

(2) $$d_{33} = \frac{3\omega_{01}^2}{2(\omega_{01}^2 - 4\omega^2)(\omega_{01}^2 - \omega^2)}, \quad d_{31} = \frac{1}{2(\omega_{02}^2 - \omega^2)}$$

$$d_{31}' = \frac{\omega_{01}\omega_{02} + 2\omega^2}{2(\omega_{01}^2 - 4\omega^2)(\omega_{02}^2 - \omega^2)}, \quad d_{15} = \frac{\omega_{02}^2 + 2\omega^2}{2(\omega_{02}^2 - 4\omega^2)(\omega_{02}^2 - \omega^2)}$$

$$d_{15}' = \frac{1}{2(\omega_{01}^2 - \omega^2)} \left\{ \frac{\omega_{01}\omega_{02} + 2\omega^2}{(\omega_{02}^2 - 4\omega^2)} + \frac{\omega_{01}\omega_{02} - \omega^2}{(\omega_{02}^2 - \omega^2)} \right\}$$

In these expressions ω_{01} and ω_{02} stand for the peak frequencies of the two relevant optical transitions, μ_{01}, μ_{02} and μ_{12} refer to the optical transition moments and $\Delta\mu_{01}$ and $\Delta\mu_{02}$ to the change in electrical dipole moment when the molecule is excited from the ground to the 1 or 2 excited states, respectively. Damping factors are neglected. When states 1 and 2 coincide one goes from the three to the two-level model and the corresponding expressions write

(3) $$\beta_{zzz} = \beta_{33} = \frac{(\mu_{01}^z)^2 \Delta\mu_{01}^z}{\hbar^2} d_{33}$$

$$\beta_{zxx} = \beta_{31} = \frac{(\mu_{01}^x)^2 \Delta\mu_{01}^z}{\hbar^2} d_{31}$$

$$\beta_{xxz} = \beta_{15} = \frac{(\mu_{01}^z)^2 \Delta\mu_{01}^z}{\hbar^2} d_{15}$$

Most of the studies on push-pull phthalocyanines have been carried out by means of Electric Field Induced Second Harmonic (EFISH) Generation experiments in solution. Since only one experimental condition is favorable for EFISH, namely, parallel polarizations for all optical and static fields, these experiments lead to only one observable: the vector component along the charge transfer axis ($\beta_{zzz} = \beta_z$, assuming that the dipole moment vector and the vector part of the third-rank tensor along the molecular z-axis are collinear).

EFISH studies on push-pull phthalocyanines have been focused on determining the role of the donor-acceptor substitution pattern, the electron donor or acceptor

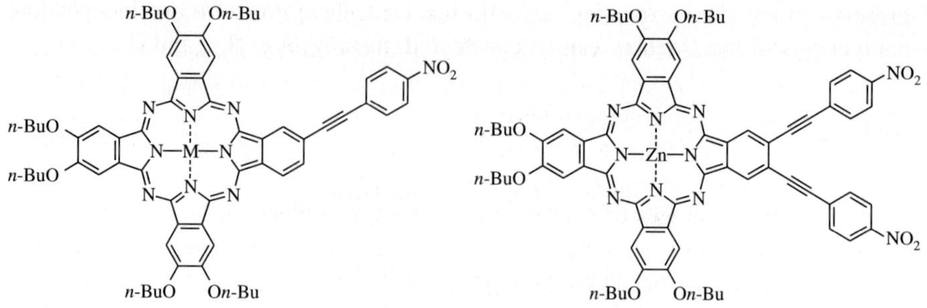

Figure 3. Highly conjugated "push-pull" phthalocyanines

strength of the substituents or the central metal while keeping the peripheral groups. The approach followed by some authors in order to enhance the quadratic hyperpolarizabilities is the extension of the conjugation pathway by the introduction of π-delocalized electron-acceptor substituents [32–37], Particularly, EFISH experiments have been performed on solutions of phthalocyanines bearing one and two 4-nitrophenylethynyl moieties as the acceptor component of the push-pull system (Figure 3, top) [37]. Moreover, Hyper-Raleigh Scattering (HRS) experiments have been also carried out which lead to the experimental determination of an additional off-diagonal tensor element, β_{zxx}. The β_{HRS} values obtained from the experiments are exceptionally high, particularly for the dinitrophenylethynyl-Pc (right in Figure 3), whose β_{HRS} value can be estimated in *ca.* 550×10^{-30} esu. These values are the largest found for push-pull phthalocyanines and offer a route for the optimization of the SHG response.

Beyond the classical approach, one has to consider the concept of octupolar nonlinearity in order to optimize the NLO response. Thus, it is sometimes convenient to decompose the β tensor into irreducible spherical multipolar components [27, 28]. When Kleinman symmetry applies i.e. under off-resonant conditions, the decomposition for β is as follows,

(4) $\quad \beta = \beta^{J=1} \oplus \beta^{J=3}$

where $\beta^{J=1}(\beta^{1,-1}, \beta^{1,0}, \beta^{1,1})$ and $\beta^{J=3}(\beta^{3,-2}, \beta^{3,-1}, \beta^{3,0}, \beta^{3,1}, \beta^{3,2})$ stand for the *vector (dipolar)* and *octupolar* components of the β tensor. For 2D molecules in the XZ plane

(5) $\quad \beta^{1,0} = [2/(15)^{1/2}] \, (\beta_{zzz} + \beta_{zxx})$

$\beta^{1,1} = -\beta^{1,-1} = -[3/(30)^{1/2}](\beta_{xxx} + \beta_{xzz})$

$\beta^{3,\pm 3} = [\pm 1/2(2)^{1/2}](-\beta_{xxx})$

$\beta^{3,\pm 2} = [2/(3)^{1/2}]\beta_{zxx}$

$\beta^{3,\pm 1} = [\pm 3/2(30)^{1/2}](\beta_{xxx} - 4\beta_{xzz})$

$\beta^{3,0} = [1/(10)^{1/2}] \, (2\beta_{zzz} - 3\beta_{zxx})$

In the case of $\beta^{J=1} = 0$ (octupolar symmetry) the molecule does not possess any permanent dipole moment even if it still presents second-order activity associated to $\beta^{J=3}$. Since this type of molecules present multidirectional charge transfer, the simple two-level model is clearly non valid [38]. Therefore, for 2D molecules one can go from purely dipolar to purely octupolar behavior and so examine the role of multipolarity on the response.

One of the advantages of octupolar molecules in comparison to dipolar systems is the improved nonlinearity-transparency trade-off. The archetype of the octupolar structure is a cube with alternating donor and acceptor groups at the edges. Pure octupolar symmetries are derived from this cubic T_d structure, either by projection along a C_3 axis, giving rise to D_{3h} or D_3 symmetry (the so-called "TATB route"), or by fusion of one type of charge in the barycenter, leading to D_{3h}, D_3, T_d or D_{2d} symmetry (the so-called "guanidinium route"). Some of these octupolar charge distributions are illustrated in Figure 4.

Most octupolar systems developed to date are organic molecules. Molecular engineering of these octupoles is based on a spatially controlled organization of charge transfers within a molecule in order to reach the desired symmetry. They are usually designed by chemical functionalization of a central core and can be grouped into: a 2D molecules of global D_{3h} symmetry, obtained by 1,3,5 functionalization of aromatic cores such as phenyl [27, 39–43], or triazine [44, 45]; b D_{3h} or slightly twisted D_3 molecules such as appropriately functionalized trivalent carbons [46, 47]; and c three-dimensional tetrahedral molecules, such as tetrasubstituted carbon [48], phosphonium [49, 50] or tin derivatives [51]. Coordination chemistry can also offer a way to build up octupolar arrangements [52–57].

In the phthalocyanine field, the octupolar route provides additional degrees of freedom to help in the design of efficient nonlinear molecules. One of the possible methodologies to reach Pc-based octupolar architectures is the arrangement of the Pc cores into D_{3h} or T_d structures by means of attaching the macrocycles to benzene [58] or to a tetravalent atom such as phosphorus. Thus, for example, aryl trisphthalocyanine phosphonium salt (Figure 5) has been prepared and the second-order NLO response at the molecular level has been measured by HRS [59]. The β_{HRS} values at $\lambda = 1.06\,\mu\text{m}\,(189 \times 10^{-30}\,\text{esu})$ is superior to those available for other related unsymmetrically substituted phthalocyanines with dipolar characteristics.

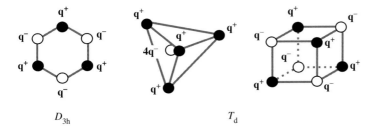

Figure 4. Archetypes of octupolar structures

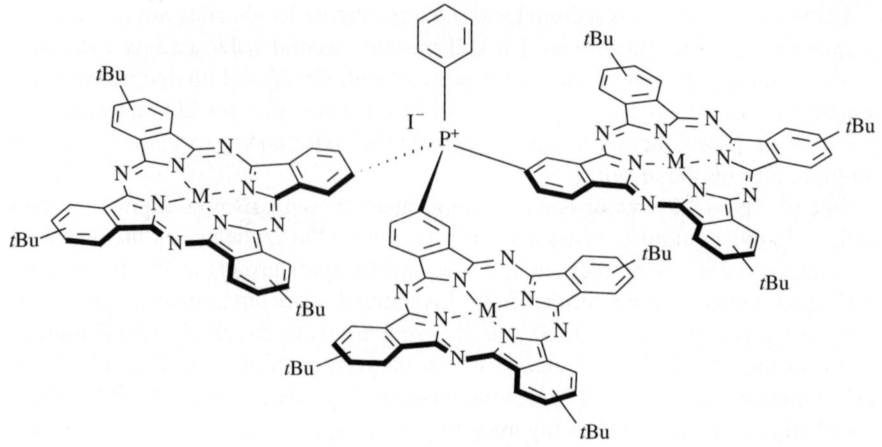

Figure 5. Octupolar aryl trisphthalocyanine phosphonium salt

This fact points out the role of "central acceptor" played by the positively charged phosphorus atom in tailoring the multipolar character of the hyperpolarizability [49, 50].

Another possible approach to the octupolar route is the axial substitution of appropriate metallic phthalocyanines. The introduction of a substituent in the axial position breaks the centrosymmetry of the Pc macrocycle and adds to the molecule a non-negligible octupolar character and a dipole moment perpendicular to the surface of the phthalocyanine. It is well-known that the combination of dipolar and octupolar features yields interesting NLO properties. Some examples of axially substituted phthalocyanines have been reported, namely, titanium(IV) [60], gallium(III) [60] and indium(III) derivatives [61] axially substituted by different ligands. Particularly, titanium and gallium phthalocyanines have been measured by EFISH and HRS.

Among the formal structural modifications of the phthalocyanine core to reduce the symmetry, the most exciting one is the subphthalocyanine route (Figure 1, right) [15]. These compounds have shown to be excellent high-performance second-order molecules and therefore constitute an optimal route to practical NLO applications. Hence, we will focus now on the outstanding optical properties that these compounds exhibit, giving a detailed description of the NLO measurements performed in solution and condensed phases of these fascinating molecules.

3.   SUBPHTHALOCYANINES

Subphthalocyanines are cone-shaped 14-π electron aromatic macrocycles, which consist of three isoindole units *N*-fused around a central boron atom. Preparation of subphthalocyanines is carried out by condensation reaction of the appropriate phthalonitriles in the presence of BCl_3 or BBr_3. The halogen atom linked to the boron core defines the z molecular axis along which a dipole moment may

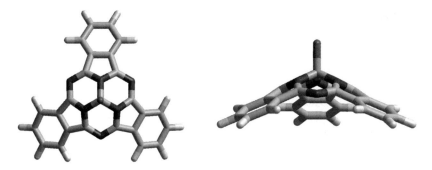

Figure 6. Top (left) and side (right) views of chlorosubphthalocyanine

exist. The molecular topology of these molecules (Figure 6) (exact C_{2v} symmetry with a T_d-like charge distribution) provides them with a multipolar character; the combination of dipolar and octupolar features makes them very attractive for practical NLO applications, since it allows the preparation of ordered macromolecular structures, i.e. *via* corona poling and the achievement of high NLO responses.

Despite being non planar, subphthalocyanines exhibit aromatic behavior associated with the delocalized 14 π-electron system. The UV-visible spectra of SubPcs are comparable to those of Pcs in that they both show a Q-band and Soret B band as in other aza aromatic macrocyclic compounds. In the case of SubPcs there is a tendency for both the Soret band (300 nm) and the Q-band (560 nm) to shift to shorter wavelength with respect to Pcs as a consequence of the decrease of the π-conjugation system (Figure 7). Absorption coefficients (ε) in both the Soret band and the Q-band also decrease on going from Pcs to SubPcs. For example, the ε values of the Q-band of SubPcs are ca. 5-6 × 10^4 dm^3 mol^{-1} cm^{-1}, and for most Pcs are in the range 8-24 × 10^4 dm^3 mol^{-1} cm^{-1}. The smaller Q-band intensity of SubPcs compared to that of Pcs may be attributed to their nonplanar structure. Peripheral

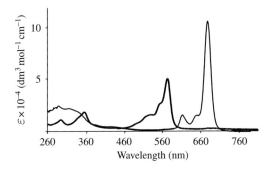

Figure 7. UV-vis spectrum of chlorosubphthalocyanine (thick line) compared to that of nickel phthalocyanine (thin line) in chloroform

donor and acceptor substituents tend to shift the Q-band of SubPcs towards longer wavelengths, while axial substituents have no or only a very small effect on the position of the bands.

4. NONLINEAR OPTICAL PROPERTIES OF SUBPHTHALOCYANINES IN SOLUTION

4.1 Influence of the Peripheral Substituents

The influence of the substituent at the periphery of the macrocycle on the NLO properties was studied in solution. Thus, various subphthalocyanines bearing representative electron-donor or acceptor groups were synthesized and their NLO properties were measured employing HRS, EFISH and Third Harmonic Generation (THG) methods [62, 63].

Not that many functional groups could be introduced at the periphery of the SubPc macrocycle as a consequence of their incompatibility with boron trihalides (Figure 8). Thus, subphthalocyanines **3**, **6**, **8** and **9** that bear three or six donor substituents in the form of thioether groups were synthesized. Subpcs **1**, **5**, **7**, **10**, and **11** possess three or six electron withdrawing groups in their peripheral positions. Tri-iodoSubPc **4** is a borderline case that can not be assigned to any category since iodine is π-donor and σ-acceptor. Thus SubPc **4** will be considered as neutral along with **2**.

The experimental NLO data for all investigated molecules are summarized in Table 1 along with the experimental values for the permanent dipole moment (μ_0).

It appears that, as a general trend, subphthalocyanines are moderately polar, whereas **5**, **8**, and **10** are more strongly polar ($\mu_0 \geq 10\,\mathrm{D}$).

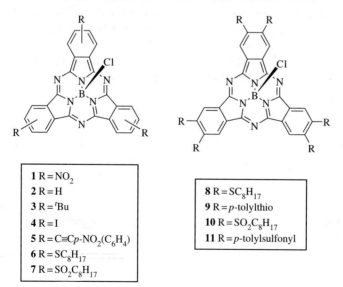

1 R = NO$_2$
2 R = H
3 R = ^{t}Bu
4 R = I
5 R = C≡Cp-NO$_2$(C$_6$H$_4$)
6 R = SC$_8$H$_{17}$
7 R = SO$_2$C$_8$H$_{17}$

8 R = SC$_8$H$_{17}$
9 R = p-tolylthio
10 R = SO$_2$C$_8$H$_{17}$
11 R = p-tolylsulfonyl

Figure 8. Molecular structures of subphthalocyanines **1–11**

Table 1. Experimental data obtained from EFISH, THG, and HRS experiments for compounds **1–11**

Compd.	λ_{max} (nm)	μ_0 (D)	γ_{EFISH} (10^{-34} esu)		γ_{THG} (10^{-34} esu)	β_{HRS} (10^{-30} esu)	
			1.9 μm	1.34 μm	1.34 μm	1.34 μm	1.46 μm
1[a]	586	5.5	−8.5	16	−13	2000	144.3
2	565	0	−6	–	–	296	–
3	570	1.3	−3	–	–	380	–
4	573	5.3	−7.3	4.3	−5.8	–	164.5
5	590	10.1	−6.0	7.0	−14	–	38
6	584	6.7	−3.7	9.7	−23.4	–	76.5
7	570	7.6	−6.6	22.3	−18	–	168.5
8	603	15	−3.0	13.0	−106	–	40
9	607	4.8	−8.0	13.6	−70	–	64.3
10	579	14.8	−15.0	80	−27	–	260
11	587	8.6	−13.4	130	−1.6	–	211.5

[a] The measurements were performed on a 1:3 isomeric mixture **1a:1b** (see part 4.3)

γ_{EFISH} values at 1.9 μm are essentially negative and do not indicate large differences among the various molecules. On the other hand, at 1.34 μm, the EFISH hyperpolarizability becomes positive, and the measured values become strongly dependent on the specific compound. These values are rather high for the molecules possessing acceptor substituents and reach a remarkable value for **10** (80×10^{-34} esu) and particularly for SubPc **11** (130×10^{-34} esu). On the other hand, molecules bearing donor substituents (**6**, **8**, and **9**) show somewhat small γ_{EFISH}(1.34 μm) values ($\sim 10 \times 10^{-34}$ esu). These are in the same range as those reported for some unsymmetrically substituted Pcs [64].

The γ_{THG} data presented in Table 1 indicate that the electronic contribution γ_e to γ_{EFISH} can not be neglected. Thus, the determination of the β_{EFISH} values is not feasible with enough precision in this case. It would be necessary, as will be commented below, to evaluate γ_e from theoretical models and then infer the $\beta_{EFISH} = <\beta(2\omega,\omega,\omega)>_{ZZZ}$ values following equation (6).

$$(6) \quad \gamma_{EFISH}(-2\omega;\omega,\omega,0) = \langle\gamma(2\omega;\omega,\omega,0))\rangle_{ZZZZ} + \frac{\langle\beta(2\omega;\omega,\omega)\rangle_{ZZZ}}{E_Z(0)}$$

The β_{HRS} values of subphthalocyanines **1–11** depend very much on the donor/acceptor character of the substituents and show the same trend as that described for γ_{EFISH} at $\lambda = 1.34$ μm. The highest values (ca. 2000×10^{-30} esu) are reached for the molecules containing the electron acceptor groups (**1**, **4**, **7**, **10**, and **11**), whereas much lower values are measured in the case of electron-donor groups (**3**, **6**, **8**, and **9**). It is noteworthy that these lower values correspond to more resonant conditions than those of acceptor groups. One should observe that the β_{HRS} value, measured at $\lambda = 1.46$ μm, for trinitro-chlorosubphthalocyanine **1** is remarkably lower than that at $\lambda = 1.34$ μm. This is most probably due to residual fluorescence contamination at the latter wavelength that could not be avoided in the 1.34 μm setup.

With regard to the cubic hyperpolarizability data, γ_{THG}, the values obtained are negative and markedly lower than those reported for symmetric phthalocyanines but quite comparable to those of some unsymmetric Pcs. This could indicate a decrease in π-conjugation related to the smaller aromaticity of SubPcs in comparison to Pcs. On the other hand, they show a completely opposite tendency with the substituents to that observed for the β_{HRS}. In fact, the highest γ_{THG} values are obtained for those subphthalocyanines, **6**, **8**, and **9**, containing donor substituents. This behavior may be due, in part, to the large red shift of the Q-band that most probably increases the resonant behavior at 2ω. Furthermore, these compounds show an additional absorption band at ca. 400 nm, due to the n-π donation of the sulfur atoms, which is strongly resonant at the third-harmonic frequency and so should also contribute to the enhancement of the NLO response.

In order to suppress any multiphoton fluorescence that often contributes to the detected second harmonic signal in the HRS experiment, Clays et al. devised an experimental setup that effectively excludes any fluorescence phenomenon [65–67]. Thus, by high-frequency demodulation of the fluorescence contribution it was demonstrated unambiguously that multiphoton fluorescence does indeed participate in the overall second harmonic signal (see compound **1** for example). Previously described results dealing with subphthalocyanines were shown to overestimate greatly the purely second order response. This technique was applied to a series of hexasubstituted SubPcs **8**, **12**–**14** (Figure 9) that were shown to possess interesting liquid crystalline properties. At $\lambda = 1.3\,\mu$m, for the lowest modulation frequency (1.6 kHz chopper frequency) the total response includes the modulation frequency independent HRS signal and the entire fluorescence signal. Such a setup at this low chopper frequency is the equivalent of a standard HRS experiment without any action taken to exclude fluorescence contribution, resulting in an overestimation of the apparent β_{HRS} value. For higher modulation frequencies (higher harmonics of the laser repetition frequency of 80 MHz), the total signal

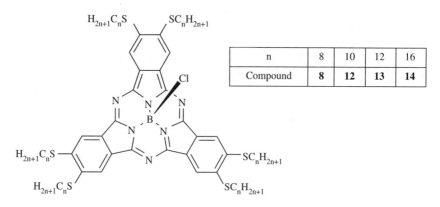

Figure 9. Hexathioalkyl chlorosubphthalocyanines

n	8	10	12	16
Compound	8	12	13	14

decreases. This is caused by the decrease in amplitude (demodulation) of the multi-photon fluorescence contribution.

The resulting fluorescence-free β_{HRS} value for the hexa-octyl-subphthalocyanine **8** amounts to $190 \pm 30 \times 10^{-30}$ esu in chloroform at 1.3 μm. This is clearly much smaller than the value that would have been obtained without the fluorescence demodulation $1760 \pm 30 \times 10^{-30}$ esu. By studying the other members of the series **12–14**, it was found that the nonlinear optical properties are not affected by the length of the alkyl chain. The same values (within experimental error) for the hyperpolarizability were retrieved for the four homologues **8, 12–14**. This demonstrates that the absence of multiphoton fluorescence to the Hyper-Rayleigh scattering signal appears indispensable when reporting values for the first hyperpolarizability of fluorescent molecules in general and subphthalocyanines in particular.

4.2 Influence of the Expansion of the π-surface

Bearing in mind the results obtained with the various peripheral substituents described in part 4.1, the next logical step is to check the effect of the expansion of the subphthalocyanine π-conjugated system without touching its intrinsic cone-shaped geometry.

This study has been achieved by comparing second-order NLO properties of chlorosubphthalocyanine **2** and chlorosubnaphthalocyanine (SubNc, **15**) that is composed of three *N*-fused 2,3-dicyanonaphthalene units in the same fashion as in the case of subphthalocyanines (Figure 10) [68]. The calculated and experimental β_{HRS} data at 1.064 μm and the γ_{EFISH} susceptibilities at 1.064 and 1.9 μm for unsubstituted SubPc **2** and SubNc **15** are presented in Tables 2 and 3, respectively.

For a purely octupolar molecule, the orientational contribution to γ_{EFISH} should be 0, so the measured value should correspond as a whole to the electronic contribution. Therefore, the information on the β tensor is almost entirely expressed in the β_{HRS}. For the unsubstituted SubPc **2**, the β_{HRS} at $\lambda = 1.06\,\mu$m are in the range measured for donor subphthalocyanines **6, 8** and **9** at $\lambda = 1.46\,\mu$m. The β_{HRS} value measured for the unsubstituted SubNc **15** is about a factor 2 smaller.

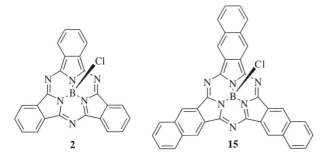

Figure 10. Chlorosubphthalocyanine (left) and chlorosubnaphthalocyanine (right)

Table 2. Calculated and experimental quadratic hyperpolarizability, β_{HRS} values in units of 10^{-30} esu

Compound	Calculated		Experimental	
	$\beta_{xxx}(1.06)$	$\beta_{xxx}(1.06)$	$\beta_{HRS}(0)$	$\beta_{HRS}(1.06)$
SubPc 2	164	101	38	92
SubNc 15	109	67	35	41

Table 3. Calculated and experimental γ_{EFISH} values in units of 10^{-33} esu

Compound	$\gamma_{calculd}(1.06)$	$\gamma_{exptl}(1.06)$	$\gamma_{calculd}(1.9)$	$\gamma_{exptl}(1.9)$
SubPc 2	1.1	−9.7	1.2	−2.9
SubNc 15	0.5	9.5	10.5	−6.0

However, the two values become approximately identical when the SHG response is extrapolated at $\omega = 0$. The essentially octupolar nature of **2** and **15** requires at least a three-level model. In general, due to the predominance of the Q band in the absorption spectrum, it appears reasonable to consider the ground and the two lowest degenerate excited states. Assuming a D_{3h} molecular symmetry, the nonzero components of the β tensor are $\beta_{xxx} = -\beta_{xyy} = -\beta_{yyx} = -\beta_{yxy}$ with

$$\text{(7)} \quad \beta_{xxx} = \frac{3\omega_{01}^2 \left[\Delta\mu_{01}^x \left((\mu_{01}^x)^2 - (\mu_{02}^y)^2 \right) - 2\mu_{01}^x \mu_{12}^x \right]}{2\hbar^2 \left(\omega_{01}^2 - \omega^2 \right) \left(\omega_{01}^2 - 4\omega^2 \right)}$$

where damping corrections, the effects of uncertainty, Doppler, collisional, and other mechanisms that lead to the broadening of the electronic transitions, have been ignored. In expression 7, $\Delta\mu_{01}^x$ represents the x component of the change in dipole moment between ground and first excited states. For a planar molecule with D_{3h} symmetry, β_{HRS} is given by:

$$\text{(8)} \quad \langle \beta_{HRS} \rangle = 2\sqrt{\frac{2}{21}} \beta_{xxx}$$

in which the x axis is defined as perpendicular to one of σ planes of the C_{3v} molecules. β_{xxx} and β_{HRS} (Table 2) were obtained by introducing into expressions 7 and 8 the parameters obtained from the INDO semiempirical calculations. Thus, INDO parameters seem to describe correctly the experimental trend and account for superior β_{HRS} value for the SubPc **2** over that of the SubNc **15** at $\lambda = 1.06\,\mu\text{m}$. The theoretical values are larger by 10% (for SubPc **2**) and 60% (for SubNc **15**) than the experimental ones. These differences may be considered reasonable taking into account the neglect of damping factors and higher electronic levels in Equation (7), as well as the approximations involved in the quantum-chemistry semiempirical methods. Finally, this theoretical framework allows the determination of β_{HRS} values at $\omega = 0$ that are also included in Table 2. It comes out that at low frequencies (i.e., off-resonance) SubPc **2** and SubNc **15** show an essentially similar SHG performance. The differences found at $\lambda = 1.06\,\mu m$ are mostly caused by frequency dispersion.

The analysis of γ_{EFISH} experimental and theoretical values is more complicated. The electronic contribution to γ_{EFISH} was calculated using the corresponding formula for D_{3h} symmetry. The expression for the electronic contribution to γ_{EFISH} is

$$(9) \quad \gamma_{elec} = \frac{8}{30\hbar^3}\left[|\bar{\mu}_{01}|^2\,|\bar{\mu}_{12}|^2\right]\sum_{i=1}^{6}D^{i}_{111} - \frac{7}{30\hbar^3}\left[|\bar{\mu}_{01}|^4\right]\sum_{i=1}^{3}d^{i}_{11}$$

Where the D^{i}_{111} and d^{i}_{11} are the appropriate resonance factors given next:

$$(10) \quad \sum_{i=1}^{6} D^{i}_{111} = \frac{6\omega_{01}^3\,(2\omega_{01}^2 - 5\omega^2)}{(\omega_{01}^2 - 4\omega^2)^2\,(\omega_{01}^2 - 4\omega^2)^2}$$

$$\sum_{i=1}^{3} d^{i}_{11} = \frac{12\omega_{01}}{(\omega_{01}^2 - 4\omega^2)\,(\omega_{01}^2 - 4\omega^2)}$$

Results are given in Table 3. The agreement with the experiment is very poor. In particular, the theoretical nearly off-resonant values (at $\lambda = 1.90\,\mu m$) are very small and positive at variance with the negative sign found in experiment. So, γ_{EFISH} values obtained from INDO calculations under a restrictive D_{3h} symmetry are much less valid. Of course, this is not surprising since subphthalocyanines are nonplanar molecules with C_{3v} symmetry with a non-zero dipole moment along the three-fold z axis. This fact introduces additional nonzero components of the β tensor involving the z axis. Consequently, a dipolar (vector) component for β should contribute to the orientational term in γ_{EFISH} that has not been included in the calculations.

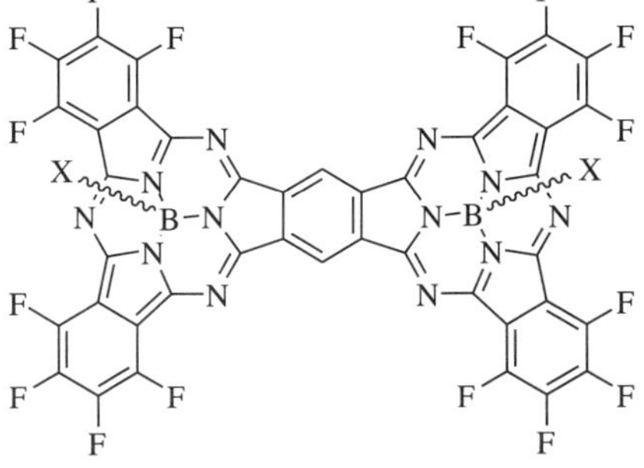

Figure 11. Fused subphthalocyanine dimer **16**

As a conclusion for this section, it appears that the expansion of the π-surface of subphthalocyanines into subnaphthalocyaines does not modify significantly its second and third order NLO response.

Another interesting π-extended subphthalocyanine derivative that would be worthwhile to compare with standard SubPc is the fused subphthalocyanine dimer **16** whose NLO properties have not been yet studied (Figure 11) [69]. The presence of two isomers possessing very different dipole moment but identical π-surfaces should give rise to interesting differences that may contribute to the understanding of the NLO behavior of this type of π-extended molecules.

4.3   Influence of the Substitution Pattern

In order to go even deeper into the understanding of the nonlinear optical properties of subphthalocyanines in solution, and in particular of its mixed octupolar/dipolar character, HRS and EFISH experimental studies were performed on four structural isomers of trinitro chlorosubphthalocyanine **1a-d** (Figure 12). These studies were reinforced by theoretical calculation at high to very high level [70].

The experimental β_{HRS} values (Table 4) for all four isomers are rather similar and comparable to those of the trinitro chlorosubphthalocyanine isomeric mixture **1a-b** (in a 1 to 3 ratio, respectively, see part 4.2), showing the very little influence of the substitution pattern on the "octupolar" NLO response. This trend is further confirmed by the similarity between the (i) HOMO-LUMO energy differences (Table 4) and (ii) the SOS-derived γ_{elec} values (Table 5) between all four compounds.

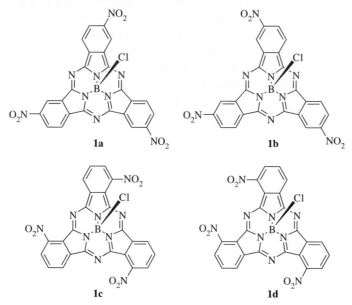

Figure 12. Four trinitro chlorosubphthalocyanine isomers

Table 4. Experimental and theoretical values for $\beta_{HRS}(\times 10^{-30}$ esu$)^a$

	$\Delta E_{HOMO \to LUMO}$ (eV)	Exper. β_{HRS}	Theoretical $\beta_{SOS}{}^b$	$\beta_{SOS}(0)$	$\beta_{FF}{}^b$
$m-C_3$ (1a)	2.5	92	196	36	48
$m-C_1$ (1b)	2.5	94	236	42	40
$o-C_3$ (1c)	2.7	104	145	31	40
$o-C_1$ (1d)	2.6	78	129	33	36

[a] $\lambda = 1.34\,\mu$m.
[b] SOS and FF stand for sum over states and finite field calculation methods, respectively.

On the contrary, the γ_{EFISH} experimental values (Table 5) are extremely sensitive to the substitution pattern at the periphery of the subphthalocyanine core and also depend strongly on the orientational contribution γ_{orient} to γ_{EFISH}. As observed in Table 5, γ_{EFISH} depends on the permanent dipole moment of the molecule, and in particular on its component along the axis defined by the B-Cl bond (z axis). The correlation is such that sign reversal of the dipole moment value along the B-Cl axis induces a change in the sign of γ_{EFISH} at both theoretical and experimental levels.

The observed independence of the β_{HRS} values with respect to the substitution pattern at the periphery of the SubPc macrocycle may be understood as a signature of the dominant octupolar character of the subphthalocyanine core, as further confirmed by FF calculations. As seen in Table 6, both C_3-symmetrical SubPcs

Table 5. $\gamma_{EFISH}(\times 10^{-34}$ esu) experimental and theoretical values for SubPcs 1a-d

		Exper.	Calculated SOS				Calc. FF
	μ_0 (D)	$\gamma_{EFISH}{}^a$	$\gamma_{EFISH}{}^a$	$\gamma_{elec}{}^a$	$\gamma_{orient}{}^a$	$\gamma_{total}{}^b$	$\gamma_{total}{}^b$
$m-C_3$ (1a)	0.4	−7.7	−2.5	−1.5	−1.0	−1.5	−1.7
$m-C_1$ (1b)	4.1	−9.0	−10.3	−2.2	−8.1	−6.1	−8.4
$o-C_3$ (1c)	4.6	17.9	6.6	−1.4	8.0	5.9	3.6
$o-C_1$ (1d)	9.2	11.6	9.8	−1.7	11.5	6.2	9.4

[a] Values at $\lambda = 1.9\,\mu$m.
[b] Values at $\omega = 0$.

Table 6. Dipolar and octupolar spherical components and molecular anisotropy ratio, ρ, for compounds 1a-d

	m-C_3 (1a)	m-C_1 (1b)	o-C_3 (1c)	o-C_1 (1d)
$\|\beta^{J=1}\|^2$	0.4	11140	0.02	14040
$\|\beta^{J=3}\|^2$	306700	131510	61750	23620
ρ	840	3	1690	1

1a and **1c** present a strong octupolar character ($\beta_{J=3}$), with a weak $\beta_{J=1}$ dipolar component. For C_1-symmetrical SubPcs **1b** and **1d**, dipolar to octupolar ratio of the NLO response is close to unity, showing that especially for $o\text{-}C_1$ (**1d**) these molecules present a relatively smaller octupolar behavior. This is however expected, since **1d** possesses the highest permanent dipole moment. Still, in all the four isomers the octupolar component is stronger than the dipolar one, confirming the robust octupolar character of all four SubPc isomers.

4.4 Degenerate Four Wave Mixing Experiments

The third-order optical nonlinearity of bromosubphthalocyanine **17** (Figure 13) in absolute ethanol was measured by the 3D DFWM technique which provides information on the magnitude, speed and origin of the third order susceptibility $\chi^{(3)}$ [71, 72]. It was observed that the phase conjugate signal is symmetrical with respect to the zero time delay, indicating a response time limited by the pulse duration of the laser (35 ps). At the high excitation intensities employed in the experiment, the broadening of the phase conjugate response were absent and the signals become very similar to those observed for nonresonant $\chi^{(3)}$ materials. The shortening of the lifetime of the excited state, observed at high excitation intensities, is attributed to exciton-exciton annihilation mechanism. Since the working wavelength (532 nm) lies in the region of strong absorption, the observed effects should involve resonance processes.

The third-order nonlinear optical susceptibility $\chi^{(3)}$ was obtained by comparing the measured signals for the sample with that of carbon disulfide as reference under the same experimental condition. The measured $\chi^{(3)}$ value is 6.2×10^{-14} esu for the subphthalocyanine at a concentration 1.25×10^{-4} M. Considering an isotropic media under the Lorenz-Lorentz approximation, the second hyperpolarizability $<\gamma>$ value was found to be 3.0×10^{-31} esu. Furthermore, $\chi^{(3)}_{pure}$ corresponding to the pure subphthalocyanine **17** was calculated to be 6.9×10^{-10} esu, about four times higher

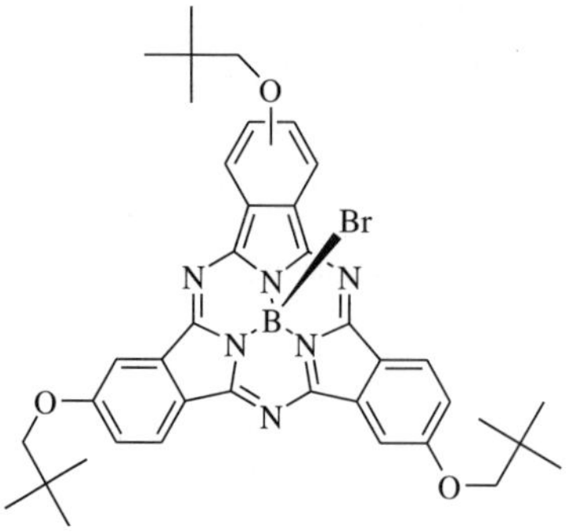

Figure 13. Trineopentyloxi-bromosubphthalocyanine

than $\chi^{(3)}$ values obtained with SubPc thin films (see section 5). The experimental errors are estimated to be 20%.

The 3D DFWM signal intensity was also measured as a function of the incident fluence. At low incident fluences a linear correlation was found in the log-log plot with a slope of ca. 3, which demonstrates that a third order nonlinear optical process occurs, and saturates at higher intensities which is indicative of an absorption saturation effect.

Thus, subphthalocyanine **17** exhibits a large resonant third-order optical nonlinearity and fast response. The measured resonant third-order NLO response is two orders of magnitude larger than the off-resonant values of chloroboron subphthalocyanine, and of the same order of magnitude as the off-resonant values of polydiacetylene due to resonant enhancement.

4.5 Remarks on Theoretical Calculations with Subphthalocyanines

4.5.1 THE SOS and FF methods

Most of the NLO experiments performed with subphthalocyanines could not be interpreted without theoretical calculations. Thus, it is crucial to possess the knowledge of the minimum requirements for obtaining reliable theoretical values that may be compared to the experimental ones in the case of SubPcs. The molecular hyperpolarizabilities are generally calculated employing two parallel theoretical approaches: (i) the finite field (FF) method and (ii) the sum over states (SOS) formalism as stated earlier.

In the FF procedure, the dipole moment (μ) of a molecule in its ground state, in the presence of a static external electric field (E), is expanded as a Taylor series:

$$(11) \quad \mu_{induced} = \mu_i + \alpha_{ij} \cdot E_i + 1/2! \beta_{ijk} \cdot E_i \cdot E_j + 1/3! \gamma_{ijkl} \cdot E_i \cdot E_j \cdot E_k$$

The tensor component of the molecular polarizability (α_{ij}) and hyperpolarizabilities (β_{ijk} and γ_{ijkl}) may be calculated by taking the appropriate derivatives of the total electronic energy or dipole moment with respect to the external electric field:

$$(12) \quad \mu_i = \frac{\partial U}{\partial E_i}; \quad \alpha_{ij} = \frac{\partial \mu_i}{\partial E_j}; \quad \beta_{ijk} = \frac{\partial^2 \mu_i}{\partial E_j \partial E_k}; \quad \gamma_{ijkl} = \frac{\partial^3 \mu_i}{\partial E_j \partial E_k \partial E_l}$$

In the SOS procedure, the theoretical description of hyperpolarizability tensors β_{ijk} and γ_{ijkl} requires a three-level model. Within this formalism, the more general expression for β_{ijk} is given by:

$$(13) \quad \beta_{ijk} = \beta_{ijk}\left(2lev, |1\rangle\right) + \beta_{ijk}\left(2lev, |2\rangle\right) + \frac{1}{\hbar^2}\left[\mu^i_{01}\left(\mu^j_{12}\mu^k_{02} + \mu^j_{02}\mu^k_{12}\right)D^{(a)}_{12}\right.$$
$$\left. + \mu^i_{02}\left(\mu^j_{12}\mu^k_{01} + \mu^j_{01}\mu^k_{12}\right)D^{(a)}_{21} + \mu^i_{12}\left(\mu^j_{02}\mu^k_{01} + \mu^j_{01}\mu^k_{02}\right)D^{(b)}_{12}\right]$$

where:

(14) $$\beta_{ijk}(2lev,|n\rangle = \frac{1}{\hbar^2}\left[\mu^i_{0n}\left(\Delta\mu^j_{0n}\mu^k_{0n}+\Delta\mu^k_{0n}\mu^j_{0n}\right)D^{(a)}_{nn}+\Delta\mu^i_{0n}\mu^j_{0n}\mu^k_{0n}D^{(b)}_{nn}\right]$$

is the theoretical expression in a two-level system. The dispersion factors D are given by:

(15) $$D^{(a)}_{nm} = \frac{\omega_{0n}\omega_{0m}+2\omega^2}{2\left(\omega^2_{0n}-4\omega^2\right)\left(\omega^2_{0m}-\omega^2\right)}$$

$$D^{(b)}_{nm} = \frac{\omega_{0n}\omega_{0m}-\omega^2}{2\left(\omega^2_{0n}-\omega^2\right)\left(\omega^2_{0m}-\omega^2\right)}$$

Those expressions are of great relevance since they allow the evaluation of the dispersive contribution to the signal when the harmonic wavelength is close to the linear absorption wavelength.

The accuracy of the polarizability and hyperpolarizabilities SOS calculations for subphthalocyanines are strongly dependent on both: (i) the active space selected for the calculations and (ii) the method employed in deriving the ground state properties. As a rule the active space in the best calculations consists of the highest occupied molecular orbital (HOMO) and the two lowest unoccupied ones (LUMO and LUMO+1) which are energetically degenerated in the C_3-symmetrical SubPcs. This active space very well reproduces the electronic low-energy transitions of the SubPcs due to the breakdown of the Gouterman four-level model observed in these compounds. Nevertheless, the expression for γ_{ijkl} in a three-level model is appreciably complex, and only in the case of reduced symmetry the final expression is manageable. Molecular symmetry in the studied compounds is C_1 or C_3, which means that a mixed x,y,z character of the transitions is present and SOS calculations become rather tiresome. Thus, theoretical treatment with a two-level model is usually employed so as to calculate γ_{ijkl} under Kleinman symmetry conditions for a C_3 system. The permanent dipolar moments, if not experimentally available, employed in the SOS calculations are calculated at semiempirical level by the INDO method using the DFT optimized geometries.

4.5.2 β_{HRS} and γ_{EFISH} calculations

Using the tensor components of the molecular hyperpolarizabilities (β_{ijk} and γ_{ijkl}) obtained by SOS and FF procedures, it is possible to infer the contributions to the HRS and EFISH signals. Depending on the C_1 or C_3 molecular symmetry, the expressions for β_{HRS} and γ_{EFISH} are given by:

(16) $$\langle\beta^2_{HRS}\rangle_{C_3} = \frac{6}{35}\beta^2_{zzz}+\frac{32}{105}\beta_{zzz}\beta_{zxx}+\frac{40}{105}\beta^2_{xxx}+\frac{40}{105}\beta^2_{yyy}$$
$$+\frac{108}{105}\beta^2_{zxx}+\frac{60}{35}\beta^2_{xyz}$$

(17) $$\langle\beta^2_{HRS}\rangle_{C_1} = \frac{6}{35}(\beta^2_{xxx}+\beta^2_{yyy}+\beta^2_{zzz})$$
$$+\frac{16}{105}(\beta_{xxx}\beta_{xyy}+\beta_{xxx}\beta_{xzz}+\beta_{yyy}\beta_{yxx}+\beta_{yyy}\beta_{yzz}+\beta_{zzz}\beta_{zxx}+\beta_{zzz}\beta_{zyy})$$
$$+\frac{38}{105}(\beta^2_{xyy}+\beta^2_{xzz}+\beta^2_{yxx}+\beta^2_{yzz}+\beta^2_{zxx}+\beta^2_{zyy})$$
$$+\frac{16}{105}(\beta_{xxy}\beta_{yzz}+\beta_{xxz}\beta_{zyy}+\beta_{yyx}\beta_{xzz}+\beta_{yyz}\beta_{zxx}+\beta_{zzx}\beta_{xyy}+\beta_{zzy}\beta_{yxx})$$
$$+\frac{20}{35}(3\beta^2_{xyz})$$

And the general expression for γ_{EFISH} is given by Equation (1). Where, assuming Kleinman symmetry conditions,

(18) $$\langle\beta\rangle^{Kleinman}_{\Omega,ZZZ} = \frac{\mu_0 E_z(0)}{5kT}(\beta_{zzz}+\beta_{zxx}+\beta_{zyy})$$

and,

$$\gamma^{Kleinman}_{electronic} = \frac{1}{5}\sum_{i\neq j}^{x,y,z}(\gamma_{iiii}+\gamma_{iijj})$$

In the case of the electronic contribution to the γ_{EFISH}, the expression for a C_3 symmetry gives:

(19) $$\gamma^{C_3}_{elec} = \frac{1}{5}[\gamma_{zzzz}+2\gamma_{xxxx}+4\gamma_{xxzz}+2\gamma_{xxyy}]$$

These expressions allow the comparison between theoretical and experimental results, as well as the determination of the relative strength of the electronic contribution to γ_{EFISH}.

4.5.3 Calculations of the dipolar and octupolar contributions to β

In order to quantify the dipolar to octupolar contribution to the NLO response, two spherical components of β related to the dipolar and octupolar molecular anisotropy are employed. They represent the dipolar and octupolar contributions to the NLO response of the molecules. The expressions which define them are, respectively:

(20) $$\|\beta^{J=1}\|^2 = \frac{3}{5}(\beta_{xxx}+\beta_{xyy})^2$$

(21) $$\|\beta^{J=3}\|^2 = \frac{1}{20}[3(\beta_{xxx}+\beta_{xyy})^2+5(\beta_{xxx}-3\beta_{xyy})^2]$$

The nonlinear molecular anisotropy ratio ρ is then defined from the previous equations in the following manner:

(22) $$\rho = \frac{\|\beta^{J=3}\|}{\|\beta^{J=1}\|}$$

Whose values run from 0 (pure dipole) to ∞ (pure octupole).

5. NONLINEAR OPTICAL PROPERTIES OF SUBPHTHALOCYANINES IN CONDENSED PHASES

Establishing the relationship between microscopic and macroscopic second-order behavior of subphthalocyanines still remains a challenging target. Consequently, a number of studies have been performed on SubPc systems in condensed phases.

A detailed study of the SHG from both spin-coated and evaporated films of trinitro-(**1**) [73, 74], triiodo-(**4**) [74] and trioctylsulfonyl-(**7**) [74] subphthalocyanines has been carried out. In order to induce molecular ordering in the spin coated film, a corona poling technique was used. Films were prepared by spin-coating from a solution of the corresponding SubPc in poly(methylmethacrylate) (PMMA). For comparison purposes, evaporated films were also prepared by vacuum sublimation onto amorphous silica plates.

Three polarizations of a 1.064 μm fundamental light source are needed to determine the three non-zero components of the $\chi^{(2)}_{ijk}$ tensor. The polarizations are: (i) along the x axis, (ii) along the y axis, (iii) at 45° from either the x or y axis (see Figure 14).

Poling the spin-coated films causes a slight decrease in the Q-band intensity of the chromophores. This behavior can be rationalized assuming that the transition dipole moment responsible for the Q-band absorption is aligned, as the permanent ground-state dipole moment, along the B-Cl axis. Molecular modeling at DFT level confirmed this fact [70]. The molecular ordering prompted by the poling field partially aligns these moments perpendicularly to the film faces and hence, for light propagation along this z-axis, the absorption decreases in comparison with that observed for a purely isotropic distribution. It is worthwhile to mention that evaporated films show similar ordering, as can be deduced from the dependence of the SHG yield on rotating angle θ for incident x and y polarization. It can be assumed that, during deposition, SubPc molecules preferentially arrange with their macrocycle plane parallel to the glass substrate.

Taking into account the $C_{\infty v}$ symmetry expected for the spin-coated and evaporated films and fitting the experimental theoretical expressions [75] to the experimental data (SHG yield vs rotation angle θ) for the three different polarizations, the values for the three non-zero independent components of the $\chi^{(2)}$ tensor ($\chi_{31}^{(2)}$, $\chi_{15}^{(2)}$, $\chi_{33}^{(2)}$, using Voigt notation: 1 ↔ 11, 2 ↔ 22, 3 ↔ 33, 4 ↔ 23, 5 ↔ 31, 6 ↔ 12) have been determined (see Table 7). The values are about one order

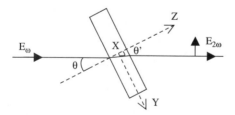

Figure 14. Orientation of the spin-coated films of compounds **1, 4, 7** with respect to the incident beam

Table 7. Components of the $\chi_{ij}^{(2)}$ tensor, molecular density (N) and thickness of the films

SubPc	Type of film	N (10^{20} cm^{-3})	Thickness (μm)	$\chi_{31}^{(2)}$ (10^{-10} esu)	$\chi_{15}^{(2)}$ (10^{-10} esu)	$\chi_{33}^{(2)}$ (10^{-10} esu)
1	Spin coated	0.4	2.7	4.3	−2.0	0.3
4	Spin coated	0.8	1.5	5.7	−2.6	0.3
7	Spin coated	0.2	1.9	7.7	−3.1	1.1
1	Evaporated	6.8	0.1	14.5	−5.0	0.5
4	Evaporated	13	0.08	25	−4.5	4.4

of magnitude smaller than those reported [75, 76] for guest-host PMMA systems containing the well known dye DR1, but the molecular concentrations are also significantly lower. It is worth mentioning that, for the two types of films, $\chi_{31}^{(2)}$ and $\chi_{15}^{(2)}$ come out with opposite (but undetermined) sign, whereas $\chi_{33}^{(2)}$ has the same sign as $\chi_{31}^{(2)}$. The values given in Table 7 come from assuming $\chi_{31}^{(2)} > 0$.

One should note that $\chi_{31}^{(2)}$, $\chi_{15}^{(2)} \geq \chi_{33}^{(2)}$, this situation being clearly at variance with that found for linear molecules, where the relation $\chi_{31}^{(2)} = \chi_{33}^{(2)}/3$ should be obeyed. Another conclusion is that Kleinman symmetry implying $\chi_{31}^{(2)} = \chi_{15}^{(2)}$ is not obeyed. As expected, subphthalocyanine bearing the stronger acceptor SO$_2$C$_8$H$_{17}$ (7) shows the highest $\chi_{ij}^{(2)}$.

The microscopic SHG response has been also evaluated from the determined $\chi_{ij}^{(2)}$ tensor for the spin-coated films. The correlation between $\chi_{ij}^{(2)}$ and β_{ij} tensor component, assuming non-interacting molecules in thermal equilibrium under an applied field is well known [77]. Only the components β_{31}, β_{15} and β_{33} common to C_3, C_{3v} and $C_{\infty v}$ symmetry groups can be determined from the measured $\chi_{31}^{(2)}$, $\chi_{15}^{(2)}$ and $\chi_{33}^{(2)}$. The calculated β_{ij} values are listed in Table 8. In all cases, the highest component is β_{31} and the most efficient response corresponds to the trioctylsulfonylsubphthalocyanine 7. This result confirms that peripheral substitution with electron acceptor groups enhances the second-order NLO response.

One should note that the predominance of the β_{31} and β_{15} components in relation to β_{33} is consistent with a charge flow pattern during light excitation from (or to) the peripheral substituents into (or from) the capping Cl [78].

Another study on the macroscopic SHG response of subphthalocyanines is that carried out on Langmuir-Blodgett (LB) films of compounds substituted with long

Table 8. Calculated quadratic molecular hyperpolarizabilities β_{ij} from the spin-coated films data

SubPc	β_{31} (10^{-30} esu)	β_{15} (10^{-30} esu)	β_{33} (10^{-30} esu)
1	9	−4	0.4
4	5	−2	−0.3
7	29	−9	≈ 0

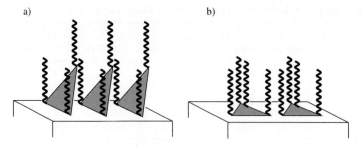

Figure 15. Schematic representations of a) an edge-on and b) a side-on disposition for subphthalocyanines at the air-water interface

alkyl chains. The Langmuir-Blodgett technique is one of the best ways for obtaining well-defined multilayers of an organic compound on a solid substrate [79].

Subphthalocyanines **3**, **6**, **7** and **10** have been organized into Langmuir films at the air-water interface. From the compression isotherms, an edge-on disposition may be assumed for compounds **3**, **6**, and **7** (see Figure 15a). However, SubPc **10** is forced to lie flat on the water surface (see Figure 15b). These results open a new way for the chemical modulation of the SubPcs organization, and hence, for the modulation of the second-order macroscopic response: by increasing the number of the polar groups in the periphery or adding a bulky group in the axial position, we can achieve respectively a somewhat edge-on or side-on disposition.

Only SubPcs **3**, **6**, and **7** have been successfully transferred onto hydrophobic glass substrates. Further structural characterizations and preliminary SHG measurements were done exclusively in the case of compound **7**. Infrared linear dichroism suggests that various transition dipoles of SubPcs are tilted versus the normal to the substrate with an angle of *ca.* 60°, thus indicating that the macrocycle is tilted in the LB film. Preliminary SHG experiments have been also carried out at $\lambda = 1.064\,\mu m$ on LB films made of 100 layers of SubPc **7**. For an s-polarized fundamental beam, the data show small but appreciable SHG yield when the observation and illumination directions lie at an angle of $\approx 50°$ with the normal to the film. However, no significant yield could be measured for a p-polarized fundamental beam. These results are consistent with a non-centrosymmetric ordering restricted to the first layers on the glass substrate. On the other hand, a strong increase of the SHG response was obtained from alternated films of SubPc **7** and behenic acid. From this data, one can conclude that layer alternation induces non-centrosymmetrical molecular ordering through the whole film thickness.

Even if it is out of the scope of the present chapter, it is worth mentioning that the third-order NLO behavior of SubPcs-based condensed phases has been also studied. Resonant and off-resonant third-order NLO properties of spin-coated films of SubPc **17** were determined *via* Z-scan technique [80]. THG experiments have been also performed on sublimated boron SubPc **2** thin films [81].

6. CONCLUDING REMARKS

Chemical variations on Pcs can alter the electronic structure of their macrocyclic core, thus allowing the fine tuning of their nonlinear response. Pc-related molecules present very attractive features for fundamental NLO studies. They offer the possibility of investigating the role of dimensionality on the NLO response. By core modification one can alter the point-group symmetry and the ratio between off-diagonal and diagonal tensor components. In fact, for 2D molecules one can go from purely dipolar to purely octupolar behavior and so examine the role of multipolarity on the NLO response. Moreover, it is possible to obtain three-dimensional (3D) structures with pyramidal shape and examine the effect of the third dimension on the NLO response. In any case, NLO measurements provide sensitive and meaningful tests that serve to assess theoretical models both on geometrical and electronic points of views. In this sense, subpthalocyanines represent a very good example of versatile three-dimensional NLO chromophores. One of the main advantages of these nearly octupolar molecules is the existence of a dipole moment along the B-Cl axis that is fundamental for their organization in condensed phases.

ACKNOWLEDGEMENTS

We are grateful for the financial support of the CICYT and Comunidad de Madrid (Spain) through grants BQU2002-04697 and GR/MAT/0513/2004, respectively. CGC thanks CICYT for a Ramon y Cajal fellowship.

REFERENCES

[1] Boyd, R.W.: Nonlinear Optics; Academic, San Diego, CA (1992)
[2] Zyss, J. (ed.): Molecular Nonlinear Optics: Materials, Physics and Devices, Academic, New York (1994)
[3] Saleh, B.E.A., Teich, M.C.: Fundamentals of Photonics, Wiley, New York (1991)
[4] Chemla, D.S., Zyss, J. (eds.): Nonlinear Optical Properties of Organic Molecules and Crystals, vols. **1, 2**. Academic, Orlando (1987)
[5] Prasad, P.N., Williams, D. (eds.): Introduction to Nonlinear Effects in Molecules and Polymers, Wiley, New York (1991)
[6] Nalwa, H.S., Miyata, S. (eds.): Nonlinear Optics of Organic Molecules and Polymers, CRC Press: Boca Ratón, FL (1997)
[7] Nalwa, H.S.: Adv. Mater. **5**, 341 (1993)
[8] Long, N.J.: Angew. Chem. Int. Ed. Engl. **34**, 21 (1995)
[9] Coe, B.J.: Comprehensive Coordination Chemistry II **9**, 621 (2004)
[10] Mark, T.J., Ratner, M.A.: Angew. Chem. Int. Ed. Engl. **34**, 155 (1995)
[11] Leznoff, C.C., Lever, A.B.P. (eds.): Phthalocyanines. Properties and Applications; vols. **1–4** VCH Publishers (LSK) Ltd.: Cambridge (1989, 1993, 1996)
[12] McKeown, N.B.: Phthalocyanine Materials: Synthesis, Structure and Function; Cambridge University Press: Cambridge (1998)
[13] de la Torre, G., Nicolau, M., Torres, T.: In: Nalwa, H.S. (ed.) Supramolecular Photosensitive and Electroactive Materials; Academic: San Diego (2001)
[14] Kadish, K.M., Smith, K.M., Guilard, R. (eds.): Porphyrin Handbook, vol. **1–20**, Academic, Boston, (2003)

[15] Claessens, C.G., González-Rodriguez, D., Torres, T.: Chem. Rev. **102**, 835 (2002)
[16] de la Torre, G., Vázquez, P., Agulló-López, F., Torres, T.: J. Mater. Chem. **8**, 1671 (1998)
[17] de la Torre, G., Torres, T., Agulló-López, F.: Adv. Mater. **9**, 265 (1997)
[18] Nalwa, H.S., Shirk, J.S.: In: Leznoff, C.C., Lever, A.B.P., (eds.) Phthalocyanines. Properties and Applications; vol. **4**, p. 79. VCH Publishers (LSK) Ltd (1996)
[19] Flom, S.R.: In: Kadish, K.M., Smith, K.M., Guilard, R. (eds.) Porphyrin Handbook; vol. **19**, p. 179. Academic, Boston (2003)
[20] Verbiest, T., Houbrechts, S., Kauranen, M., Clays, K., Persoons, A.: J. Mater. Chem. **7**, 2175 (1997)
[21] Coe, B.J.: Chem. Eur. J. **5**, 2464 (1999)
[22] Dagani, R.: Chem. Eng. March 4, 22 (1996)
[23] Agulló-López, F.: Cabrera, J.M., Agulló-Rueda, F.: Electrooptics: Phenomena, Materials and Applications, Academic, San Diego, CA (1994)
[24] Craig, G.S., Cohen, R.E., Silbey, R.R., Pucceti, G., Ledoux, I., Zyss, J.: J. Am. Chem. Soc. **115**, 860 (1993)
[25] Blanchard-Desce, M., Alain, V., Bedworth, P.V., Marder, S.R., Fort, A., Runser, C., Barzoukas, M., Lebus, S., Wortmann, R.: Chem. Eur. J. **3**, 1091 (1997)
[26] Alain, V., Blanchard-Desce, M., Ledoux-Rak, I., Zyss, J.: Chem. Commun. 353 (2000)
[27] Ledoux, I., Zyss, J., Siegel, J.S., Brienne, J., Lehn, J.-M.: Chem. Phys. Lett. **172**, 440 (1990)
[28] Zyss, J., Ledoux, I.: Chem. Rev. **94**, 77 (1994)
[29] Li, D.A., Ratner, M.A., Marks, T.J.: J. Am. Chem. Soc. **110**, 1707 (1998)
[30] Orr, B.J., Ward, F.J.: Mol. Phys. **20**, 513 (1971)
[31] de la Torre, G., Vázquez, P., Agulló-López, F., Torres, T.: Chem. Rev. **104**, 3723 (2004)
[32] de la Torre, G., Torres, T.: J. Porphyrins Phthalocyanines **1**, 221 (1997)
[33] Torres, T., de la Torre, G., García-Ruiz, J.: Eur. J. Org. Chem. 2323 (1999)
[34] Tian, M., Wada, T., Sasabe, H.: J. Heterocyclic Chem. **34**, 171 (1997)
[35] Tian, M., Wada, T., Kimura-Suda, H., Sasabe, H.: Mol. Cryst. Liq. Cryst. **294**, 271 (1997)
[36] Tian, M., Wada, T., Kimura-Suda, H., Sasabe, H.: J. Mat. Chem. **7**, 861 (1997)
[37] Maya, E.M., García-Frutos, E.M., Vázquez, P., Torres, T., Martín, G., Rojo, G., Agulló-López, F., González-Jonte, R.H., Ferro, V.R., García de la Vega, J.M., Ledoux, I., Zyss, J.: J. Phys. Chem. A **107**, 2110 (2003)
[38] Zyss, J., Brasselet, S.: J. Nonlinear Phys. Mater. **7**, 397 (1998)
[39] Verbiest, T., Clays, K., Semyr, C., Wolff, J., Reihoudt, D., Persoons, A.: J. Am. Chem. Soc. **116**, 9320 (1994)
[40] Bredas, J.L., Meyers, F., Pierce, B.M., Zyss, J.: J. Am. Chem. Soc. **114**, 4928 (1992)
[41] McDonagh, A.M., Humphrey, M.G., Samoc, M., Luther-Davies, B., Houbrechts, S., Wada, T., Sasabe, H., Persoons, A.: Adv. Mater. **11**, 1292 (1999)
[42] Cho, B.R., Lee, S.J., Lee, S.H., Son, K.H., Kim, Y.H., Doo, I.-Y., Lee, G.J., Kang, T.I., Lee, Y.K., Cho, M., Jeon, S.-J.: Chem. Mater. **13**, 1438 (2001)
[43] Cho, B.R., Park, S.B., Lee, S.J., Son, K.H., Lee, S.H., Lee, M.-J., Yoo, J., Lee, Y.K., Lee, G.J., Kang, T.I., Cho, M., Jeon, S.-J.: J. Am. Chem. Soc. **123**, 6421 (2001)
[44] Brasselet, S., Cherioux, F., Audebert, P.: J. Zyss. Chem. Mater. **11**, 1915 (1999)
[45] Wolff, J.J., Siegler, F., Matschiner, R., Wortmann, R.: Angew. Chem. Int. Ed. **39**, 1436 (2000)
[46] Verbiest, A., Clays, K., Samyn, C., Wolff, J., Reinhoudt, D., Persoons, A.: J. Am. Chem. Soc. **116**, 9320 (1994)
[47] Lee, Y.-K., Jeon, S.-J., Cho, M.: J. Am. Chem. Soc. **120**, 10921 (1998)
[48] Maker, P.D.: Phys. Rev. A **1**, 923 (1970)
[49] Bourgogne, C., Le Fur, Y., Juen, P., Masson, P., Nicoud, J.-F., Masse, R.: Chem. Mater. **12**, 1025 (2000)
[50] Lambert, C., Schmälzlin, E., Meerholz, K., Bräuchle, C.: Chem. Eur. J. **4**, 512 (1998)
[51] Lequan, M., Branger, C., Simon, J., Thami, T., Chauchard, E., Persoons, A.: Adv. Mater. **6**, 851 (1994)

[52] Dhenaut, C., Ledoux, I., Samuel, I.D.W., Zyss, J., Bourgault, M., Le Bozec, H.: Nature **374**, 339 (1995)
[53] Lin, W., Wang, Z., Ma, L.: J. Am. Chem. Soc. **121**, 11249 (1999)
[54] Le Bozec, H., Le Bouder, T., Maury, O., Bondon, A., Zyss, J., Ledoux, I.: Adv. Mater. **13**, 1677 (2001)
[55] Sénéchal, K., Maury, O., Le Bozec, H., Ledoux, I., Zyss, J.: J. Am. Chem. Soc. **124**, 4561 (2002)
[56] Le Bouder, T., Maury, O., Le Bozec, H., Bondon, A., Costuas, K., Amouyal, E., Zyss, J., Ledoux, I.: J. Am. Chem. Soc. **125**, 12884 (2003)
[57] Maury, O., Viau, L., Sénéchal, K., Corre, B., Guégan, J.-P., Renouard, T., Ledoux, I., Zyss, J., Le Bozec, H.: Chem. Eur. J. **10**, 4454 (2004)
[58] Bottari, G., Torres, T.: Chem. Commun. 2668 (2005)
[59] de la Torre, G., Gouloumis, A., Vázquez, P., Torres, T.: Angew. Chem. Int. Ed. **40**(15), 2895 (2001)
[60] Claessens, C.G., Gouloumis, A., Barthel, M., Chen, Y., Martin, G., Agulló-López, F., Ledoux-Rak, I., Zyss, J., Hanack, M., Torres, T.: J. Porphyrins Phthalocyanines **7**, 291 (2003)
[61] Rojo, G., Martin, G., Agulló-López, F., Torres, T., Heckman, H., Hanack, M.: J. Phys. Chem. B **104**, 7066 (2000)
[62] del Rey, B., Keller, U., Torres, T., Rojo, G., Agulló-López, F., Nonell, S., Marti, C., Brasselet, S., Ledoux, I., Zyss, J.: J. Am. Chem. Soc. **120**(49), 12808 (1998)
[63] Sastre, A., Torres, T., Díaz-García, M.A., Agulló-López, F., Dhenaut, C., Brasselet, S., Ledoux, I., Zyss, J.: J. Am. Chem. Soc. **118**(11), 2746 (1996)
[64] Maya, E.M., Garcia-Frutos, E.M., Vazqucz, P., Torres, T., Martin, G., Rojo, G., Agullo-Lopez, F., Gonzalez-Jonte, R.H., Ferro, V.R., Garcia de la Vega, J.M., Ledoux, I., Zyss, J.: J. Phys. Chem. A **107**(12), 2110 (2003)
[65] Olbrechts, G., Wostyn, K., Clays, K., Persons, A., Kang, S.H., Kim, K.: Chem. Phys. Lett. **308**, 173 (1999)
[66] Kang, S.H., Kang, Y.-S., Zin, W.-C., Olbrechts, G., Wostyn, K., Clays, K., Persoons, A., Kim, K.: Chem. Commun. 1661 (1999)
[67] Olbrechts, G., Clays, K., Wostyn, K., Persoons, A.: Synth. Met. **115**, 207 (2000)
[68] Martín, G., Rojo, G., Agulló-López, F., Ferro, V.R., García de la Vega, J.M., Martínez-Díaz, M.V., Torres, T.: J. Phys. Chem. B **106**, 13139 (2002)
[69] Claessens, C.G., Torres, T.: Angew. Chem. Int. Ed. **41**(14), 2561 (2002)
[70] Claessens, C.G., González-Rodríguez, D., Torres, T., Martín, G., Agulló-López, F., Ledoux, I., Zyss, J., Ferro, V.R., García de la Vega, J.M.: J. Phys. Chem. B **109**, 3800 (2005)
[71] Gu, Y., Wang, Y., Gan, F.: Mat. Lett. **52**, 404 (2002)
[72] According to the expertise of the authors of this chapter in subphthalocyanine synthesis, SubPc 17 is probably fairly unstable in ethanol. Most probably, the measured $\chi^{(3)}$ corresponds to the trineopentyloxi-ethyloxisubphthalocyanine or any other non-subphthalocyanine decomposition product. The UV-vis presented in the article does not show any of the characteristic features of SubPcs
[73] Rojo, G., Hierro, A., Díaz-García, M.A., Agulló-López, F., del Rey, B., Sastre, A., Torres, T.: Appl. Phys. Lett. **70**(14), 1802 (1997)
[74] Rojo, G., Agulló-López, F., del Rey, B., Torres, T.: J. Appl. Phys. **84**(12), 6507 (1998)
[75] Hayden, L.M., Sauter, G.F., Ore, F.R., Pasillas, P.L., Hoover, J.M., Lindsay, G.A., Henry, R.A.: J. Appl. Phys. **68**, 456 (1990)
[76] Meredith, G.R., Van Dusen, J.G., Williams, D.J.: Macromolecules **15**, 1358 (1982)
[77] Kielich, S.: IEEE J. Quantum Electron. **QE-5**, 562 (1969)
[78] Ferro, V.R., Poveda, L.A., Claessens, C.G., Gonzalez-Jonte, R.H., Garcia de la Vega, J.M.: Int. J. Quant. Chem. **91** (3), 369 (2003)
[79] Roberts, G. (ed.) Langmuir-Blodgett Films, Plenum, New York (1990)
[80] Gu, Y., Wang, Y., Gan, F.: Mater. Lett. **52**, 404 (2002)
[81] Díaz-García, M.A., Agulló-López, F., Sastre, A., Torres, T., Torruellas, W.E., Stegeman, G.I.: J. Phys. Chem. **99**, 14988 (1995)

CHAPTER 17

NLO PROPERTIES OF METAL ALKYNYL AND RELATED COMPLEXES

JOSEPH P.L. MORRALL[1,2], MARK G. HUMPHREY[1],
GULLIVER T. DALTON[1,2], MARIE P. CIFUENTES[1] AND MAREK SAMOC[2]

[1] *Department of Chemistry, Australian National University, Canberra ACT 0200, Australia*
[2] *Laser Physics Centre, Research School of Physical Sciences and Engineering, Australian National University, Canberra ACT 0200, Australia*

Abstract: The NLO properties of iron, ruthenium, osmium, nickel, and gold alkynyl complexes and some related compounds prepared in the authors' laboratories are reviewed. Structure-property relationships for both quadratic and cubic NLO merit for these complexes have been developed from hyper-Rayleigh scattering studies at $1.064\,\mu m$ and Z-scan studies at $0.800\,\mu m$, respectively

Keywords: Organometallics, Alkynyl, Vinylidene, Quadratic Hyperpolarizability, Cubic Hyper-polarizability

1. INTRODUCTION

Many different types of molecular and bulk material have been examined as possible nonlinear optical (NLO) materials, one very important category being organic molecules containing conjugated π-systems with unsymmetrical charge distributions. Organometallic complexes are similar to organic molecules in that they can possess large NLO responses, fast response times, and ease of fabrication and integration into composites, but they possess greater flexibility at the design stage; variation in the metal oxidation state, ligand environment, and geometry can permit tuning of NLO response in ways not possible for organic molecules. As a result, organometallic complexes have come under considerable scrutiny [1–13]. However, reports from various laboratories employing different techniques at varying measurement wavelengths, with different pulse lengths, (internal) standards, and indeed definitions of NLO properties render comparisons and thereby development of structure-property relationships difficult, if not impossible. To achieve reliable structure-property trends, standardization of most or all of these potential variables is required.

This Chapter summarizes our studies of specific organometallic complexes (largely alkynyl compounds) over the past decade. In almost all instances, second-order NLO properties have been measured by hyper-Rayleigh scattering (HRS) at 1.064 μm as part of an ongoing collaboration with the Leuven group of Prof. André Persoons, and third-order NLO properties have been measured by Z-scan at 0.800 μm in our labs at ANU, although electric field-induced second-harmonic generation (EFISH, at Leuven) and degenerate four-wave mixing (DFWM, at ANU) studies have been undertaken to provide complementary information in limited cases (for a thorough discussion of these experimental techniques, see the Chapter by Asselberghs et al). This standardization facilitates development of structure-NLO property relationships.

The molecular composition that we explored in our earlier studies is depicted in Figure 1. Metal alkynyl complexes are organometallic compounds that have attracted considerable interest[14]. In metal alkynyl complexes, the metal is in the plane of the organic π-system, an important molecular design concept that has been suggested to optimize NLO properties[15], and in contrast to probably the most extensively-studied organometallic system, namely metallocenyl complexes, for which the metal-to-ligand charge-transfer (MLCT) axis is perpendicular to the plane of the organic π-system (Figure 2). The molecular compositions in Figures 1 and 2 permit a systematic examination of the effect of varying metal, co-ligands, d-electron count, and nature of the π-bridge and acceptor group on quadratic and cubic NLO properties.

One shortcoming of the donor-bridge-acceptor design illustrated in Figures 1 and 2 is that increasing the nonlinearity is correlated with a red shift of the charge-transfer absorption band (the so-called nonlinearity-transparency trade-off).

A possible means to increase the nonlinearity without sacrificing transparency is to replace the dipolar with an octupolar composition[16], so more recently we have explored trigonally-branched metal alkynyl complexes as potential NLO materials,

$L_nM\!\equiv\!\boxed{\pi\text{-bridge}}\text{-}\boxed{\text{acceptor}}$

Figure 1. Molecular composition of metal alkynyl complexes explored in our dipolar studies. $L_nM \equiv$ ligated metal

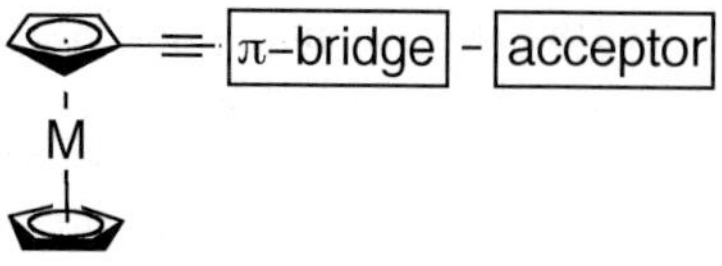

Figure 2. Metallocenyl complex composition for second-order NLO application. $M \equiv$ metal (commonly Fe)

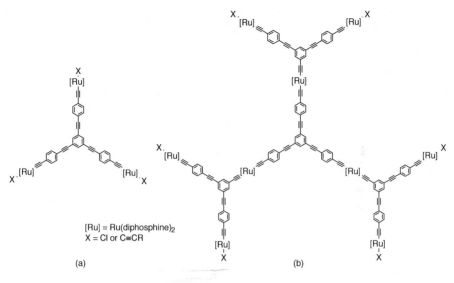

Figure 3. Trigonally-branched molecular composition explored in our studies

which has lead to a study of metal alkynyl octupolar complexes (Figure 3 (a)) and dendrimers (Figure 3 (b)). The results from our quadratic and cubic NLO studies of dipolar and octupolar complexes are described in this Chapter.

2. SECOND-ORDER NLO STUDIES

As was mentioned above, one important variable in our molecular composition is the nature of the metal. We therefore carried out a study of iron, ruthenium, and osmium alkynyl complexes by HRS at 1.064 μm, the results from which are tabulated in Table 1; examples of efficient molecules are depicted in Figure 4. The β_{HRS} values for the metal alkynyl complexes [M(C≡C-4-$C_6H_4NO_2$)(dppe) (η^5-C_5H_5)](M = Fe, Ru, Os) and *trans*-[M(C≡C-4-$C_6H_4NO_2$)(Cl{(R, R) − diph}$_2$] [diph ≡ 1,2-bis(methylphenylphosphino)benzene] increase as Fe ≤ Ru ≤ Os. The iron alkynyl complexes have absorption bands closer to the second harmonic wavelength of 532 nm than either the ruthenium or osmium homologues (with the exception of the iron and ruthenium carbonyl complexes), suggesting that the β_{HRS} values for the iron complexes contain a larger resonance contribution than those of the ruthenium and osmium homologues. The β_{HRS} values for the osmium alkynyl complexes are, in each case, greater than the values for the ruthenium-containing complexes. Absorption bands for the osmium complexes are closer to the second-harmonic than are those of the ruthenium homologues. For the cyclopentadienyl complexes, if static β values are calculated from the experimental β_{HRS} values using a two-level model, then the trend remains the same as for the experimental values. The values for the complexes containing the diph ligand show that ruthenium has the higher calculated static value. While this suggests some ambiguity, the two-level

Table 1. Molecular second-order NLO results for selected iron, ruthenium and osmium alkynyl complexes[a]

Complex	λ_{max} (nm)	β[b] (10^{-30} esu)	β_0[c] (10^{-30} esu)	Ref.
[Fe(C≡C-4-C$_6$H$_4$NO$_2$)(CO)$_2$(η^5-C$_5$H$_5$)]	370	49	22	[17]
[Ru(C≡C-4-C$_6$H$_4$NO$_2$)(CO)$_2$(η^5-C$_5$H$_5$)]	364	58	27	[17]
[Fe(C≡C-4-C$_6$H$_4$NO$_2$)(dppe)(η^5-C$_5$H$_5$)]	498	665	64	[17]
[Ru(C≡C-4-C$_6$H$_4$NO$_2$)(dppe)(η^5-C$_5$H$_5$)]	447	664	161	[17]
[Os(C≡C-4-C$_6$H$_4$NO$_2$)(dppe)(η^5-C$_5$H$_5$)]	461	929	188	[17]
(−)$_{436}$-trans-[Fe(C≡C-4-C$_6$H$_4$NO$_2$)Cl{(R, R) − diph}$_2$]	543	440	−14	[18]
(−)$_{589}$-trans-[Ru(C≡C-4-C$_6$H$_4$NO$_2$)Cl{(R, R) − diph}$_2$]	467	530	97	[18]
(−)$_{365}$-trans-[Os(C≡C-4-C$_6$H$_4$NO$_2$)Cl{(R, R) − diph}$_2$]	490	620	74	[18]
[Ru(C≡C-4-C$_6$H$_4$NO$_2$)(PPh$_3$)$_2$(η^5-C$_5$H$_5$)]	460	468	96	[17]
[Os(C≡C-4-C$_6$H$_4$NO$_2$)(PPh$_3$)$_2$(η^5-C$_5$H$_5$)]	474	1051	174	[17]

[a] All measurements in THF solvent. All complexes are optically transparent at 1.064 μm.
[b] Hyper-Rayleigh scattering studies at 1.064 μm; values ±10 %.
[c] HRS at 1.064 μm corrected for resonance enhancement at 532 nm using the two-level model with $\beta_0 = \beta[1 - (2\lambda_{max}/1064)^2][1 - (\lambda_{max}/1064)^2]$; damping factors not included.

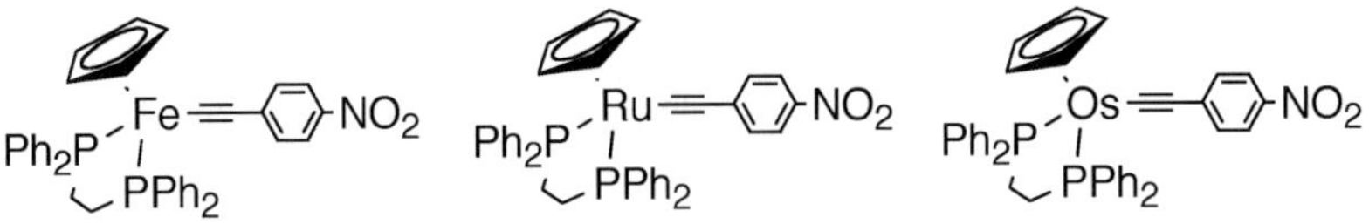

Figure 4

Fe complex: Ph$_2$P–Fe(dppe)(η-C$_5$H$_5$)–C≡C–C$_6$H$_4$–NO$_2$
$\beta_{1.064} = 665 \times 10^{-30}$ esu
$\beta_0 = 64 \times 10^{-30}$ esu

Ru complex: Ph$_2$P–Ru(dppe)(η-C$_5$H$_5$)–C≡C–C$_6$H$_4$–NO$_2$
$\beta_{1.064} = 664 \times 10^{-30}$ esu
$\beta_0 = 161 \times 10^{-30}$ esu

Os complex: Ph$_2$P–Os(dppe)(η-C$_5$H$_5$)–C≡C–C$_6$H$_4$–NO$_2$
$\beta_{1.064} = 929 \times 10^{-30}$ esu
$\beta_0 = 188 \times 10^{-30}$ esu

Figure 4. The effect of varying group 8 metal on frequency-dependent and two-level-corrected β values for selected alkynyl complexes

model may have limited applicability with organometallic complexes of this type. The present data is suggestive of β_{HRS} values for this type of complex following the ordering Fe ≤ Ru ≤ Os.

Comparison of the β_{HRS} values in Table 1 permits assessment of the effect of varying co-ligands. Replacement of the electron-donating diphosphine ligand with the relatively strongly electron-withdrawing carbonyls results in a significant reduction of the second-order NLO response.

This is readily rationalized because the amount of electron density available to the donating metal centre, and hence its donor strength, is reduced on replacing diphosphine by two carbonyl groups. This is an example of one strength of organometallic systems, *viz.* the possibility of tuning donor strength by ligand modification.

The metal centres in the iron, ruthenium, and osmium alkynyl complexes listed in Table 1 possess 18 valence electrons. Table 2 contains HRS data at 1.064 μm and two-level-corrected values for similar 18 valence electron alkynyl and chloro nickel complexes, and a particularly efficient example is illustrated in Figure 5. These data are substantially resonance enhanced, although the relative orderings are maintained with two-level-corrected values.

The experimental data for the nickel complexes with "extended-chain" alkynyl ligands reveal an efficiency sequence *E*-ene-linkage > yne-linkage ≈ imino-linkage > biphenylene unit, but intense ($\varepsilon > 10000$) linear optical absorption maxima for all "extended-chain" alkynyl complexes within 100 nm of 2ω are consistent with substantial dispersion enhancement of the observed nonlinearities. It has been suggested that the two-state model is appropriate in the limited cases where structural change is restricted to the molecular component responsible for the charge-transfer band contributing to the hyperpolarizability [19]. It is likely for these nickel complexes that the higher energy bands are associated with transitions involving other ligands, with little change in dipole moment between ground and excited states, and hence only a small contribution to the optical nonlinearity.

Although the absolute value of β_{HRS} for the biphenyl-linked [Ni(C≡C-4-C$_6$H$_4$-4-C$_6$H$_4$NO$_2$)(PPh$_3$)(η^5-C$_5$H$_5$)] is less than that for the phenyl-containing chromophore [Ni(C≡C-4-C$_6$H$_4$NO$_2$)(PPh$_3$)(η^5-C$_5$H$_5$)], this relative ordering is reversed with two-level-corrected values. Two pyridyl-containing complexes were examined, their quadratic nonlinearities being similar in magnitude to their phenyl-containing analogues.

Table 2. Molecular second-order NLO results for nickel alkynyl and chloro complexes[a]

Complex	λ_{max} (nm)	β[b] (10^{-30} esu)	β_0[c] (10^{-30} esu)	Ref.
[NiCl(PPh$_3$)(η^5-C$_5$H$_5$)]	330	89	45	[20]
[Ni(C≡CPh)(PPh$_3$)(η^5-C$_5$H$_5$)]	307	24	15	[20, 21]
[Ni(C≡C-2-C$_5$H$_4$N)(PPh$_3$)(η^5-C$_5$H$_5$)]	415	25	8	[22]
[Ni(C≡C-2-C$_5$H$_4$N-5-NO$_2$)(PPh$_3$)(η^5-C$_5$H$_5$)]	456	186	41	[22]
[Ni(C≡C-4-C$_6$H$_4$NO$_2$)(PPh$_3$)(η^5-C$_5$H$_5$)]	439	221	59	[20]
[Ni(C≡C-4-C$_6$H$_4$-4-C$_6$H$_4$NO$_2$)(PPh$_3$)(η^5-C$_5$H$_5$)]	413	193	65	[20]
[Ni{C≡C-4-C$_6$H$_4$-(E)-CH=CH-4-C$_6$H$_4$NO$_2$}(PPh$_3$)(η^5-C$_5$H$_5$)]	437	445	120	[20]
[Ni{C≡C-4-C$_6$H$_4$-(Z)-CH=CH-4-C$_6$H$_4$NO$_2$}(PPh$_3$)(η^5-C$_5$H$_5$)]	417	145	47	[20]
[Ni(C≡C-4-C$_6$H$_4$C≡C-4-C$_6$H$_4$NO$_2$)(PPh$_3$)(η^5-C$_5$H$_5$)]	417	326	106	[20]
[Ni(C≡C-4-C$_6$H$_4$N=CH-4-C$_6$H$_4$NO$_2$)(PPh$_3$)(η^5-C$_5$H$_5$)]	448	387	93	[20]
[1-(HC≡C)-3,5-C$_6$H$_3${(C≡C)Ni(PPh$_3$)(η^5-C$_5$H$_5$)}$_2$]	316	94	55	[21]

[a] All measurements in THF solvent. All complexes are optically transparent at 1.064 µm.
[b] Hyper-Rayleigh scattering studies at 1.064 µm; values ±10%.
[c] HRS at 1.064 µm corrected for resonance enhancement at 532 nm using the two-level model with $\beta_0 = \beta[1-(2\lambda_{max}/1064)^2][1-(\lambda_{max}/1064)^2]$; damping factors not included.

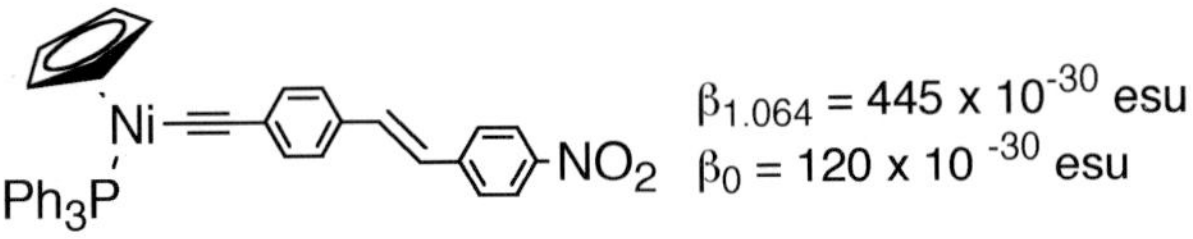

Ph₃P–Ni≡C–C₆H₄–CH=CH–C₆H₄–NO₂ $\beta_{1.064} = 445 \times 10^{-30}$ esu
                                  $\beta_0 = 120 \times 10^{-30}$ esu

Figure 5. The most efficient nickel-containing complex from these studies

Ph₃P·Au–C≡C–C₆H₄–CH=CH–C₆H₄–NO₂ $\beta_{1.064} = 180 \times 10^{-30}$ esu
                                    $\beta_0 = 68 \times 10^{-30}$ esu

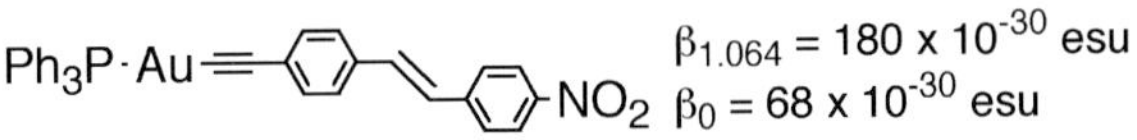

Figure 6. The most efficient gold-containing complex from these studies

The studies summarized above involve complexes with 18 valence electron metal centres. We also examined related 14 electron gold alkynyl complexes. Molecular quadratic hyperpolarizabilities of gold alkynyl complexes by HRS measurements at 1.064 μm are given in Table 3, and the most efficient gold-containing complex is depicted in Figure 6.

Complex [Au(C≡CPh)(PPh₃)] has a large $\beta_{1.064}$ for a compound that can be considered as a phenyl group containing a donor substituent (Ph₃PAuC≡C) only. Replacement of the 4-arylalkynyl H in [Au(C≡CPh)(PPh₃)] by a nitro substituent to generate the donor-acceptor alkynyl complex [Au(C≡C-4-C₆H₄NO₂)(PPh₃)] leads to a substantial increase in the nonlinearity, with an efficiency similar to that of 4-nitroaniline (21.4 × 10⁻³⁰ esu, THF solvent) [19].

Examination of the effect of π-system lengthening on optical nonlinearity of donor-acceptor gold alkynyl complexes reveals an order [Au(C≡C-4-C₆H₄NO₂)(PPh₃)] < [Au(C≡C-4-C₆H₄-4-C₆H₄NO₂)(PPh₃)] < [Au{C≡C-4-C₆H₄-(Z)-CH=CH-4-C₆H₄NO₂}(PPh₃)] ≈ [Au(C≡C-4-C₆H₄C≡C-4-C₆H₄NO₂)(PPh₃)] < [Au(C≡C-4-C₆H₄N=CH-4-C₆H₄NO₂)(PPh₃)] < [Au{C≡C-4-C₆H₄-(E)-CH=CH-4-C₆H₄NO₂}(PPh₃)] < [Au{C≡C-4-C₆H₄-(E)-N=N-4-C₆H₄NO₂}(PPh₃)]. These data are consistent with an increase in nonlinearity for "extended chain" two-ring organometallic alkynyl chromophores versus one-ring complexes, confirming the observation in the nickel system. Examination of the effect of varying carbon-containing bridges in the "extended" two-ring gold alkynyl complexes reveals an efficiency sequence C₆H₄C₆H₄ ≈ C₆H₄C≡CC₆H₄ < C₆H₄CH=CHC₆H₄ for C-containing bridges; the linear optical absorption bands for the gold alkynyl complexes are significantly removed from the harmonic frequency, suggesting that this relative ordering accurately reflects off-resonance nonlinearities.

Torsion effects at the phenyl-phenyl linkage (for diphenyl compounds) and orbital energy mismatch of *p* orbitals of *sp*-hybridized alkynyl carbons with *p* orbitals of *sp²*-hybridized phenyl carbons (for diphenylacetylene compounds) have been suggested as reasons for lower β values for C₆H₄C₆H₄- and C₆H₄C≡CC₆H₄- linked organic compounds, compared with their *trans*-stilbene analogues [28]; it is likely that the same factors influence relative nonlinearities for "extended-chain" alkynyl complexes, as the trend in the nonlinear optical merit for the gold alkynyl

Table 3. Molecular second-order NLO results for gold alkynyl complexes[a]

Complex	λ_{max} (nm)	β^b (10^{-30} esu)	β_0^c (10^{-30} esu)	Ref.
[Au(C≡CPh)(PPh$_3$)]	296	6	4	[23]
[Au(C≡C-4-C$_6$F$_4$OMe)(PMe$_3$)]	324	44	25	[24]
[Au(C≡C-4-C$_6$F$_4$OMe)(PPh$_3$)]	292	20	13	[24]
[Au(C≡C-2-C$_5$H$_4$N)(PPh$_3$)]	300	7	4	[22]
[Au(C≡C-2-C$_5$H$_4$N-5-NO$_2$)(PPh$_3$)]	339	38	20	[22]
[Au(C≡C-2-C$_5$H$_4$N-5-NO$_2$)(PMe$_3$)]	340	12	6	[22]
[Au(C≡C-4-C$_6$H$_4$CHO)(PPh$_3$)]	322	14	8	[25]
[Au{C≡C-4-C$_6$H$_4$CHO(CH$_2$)$_3$O}(PPh$_3$)]	296	15	4	[25]
[Au{C≡C-4-C$_6$H$_4$CHO(CH$_2$)$_3$O}(PMe$_3$)]	292	48	13	[25]
[Au(C≡C-4-C$_6$H$_4$NO$_2$)(PMe$_3$)]	339	50	27	[26]
[Au(C≡C-4-C$_6$H$_4$NO$_2$)(PCy$_3$)]	342	31	16	[26]
[Au(C≡C-4-C$_6$H$_4$NO$_2$)(PPh$_3$)]	338	22	12	[23]
[Au(C≡C-4-C$_6$H$_4$-4-C$_6$H$_4$NO$_2$)(PPh$_3$)]	350	39	20	[23]
[Au{C≡C-4-C$_6$H$_4$-(E)-CH=CH-4-C$_6$H$_4$NO$_2$}(PPh$_3$)]	386	120	49	[23]
[Au{C≡C-4-C$_6$H$_4$-(Z)-CH=CH-4-C$_6$H$_4$NO$_2$}(PPh$_3$)]	362	58	28	[23]
[Au(C≡C-4-C$_6$H$_4$C≡C-4-C$_6$H$_4$NO$_2$)(PPh$_3$)]	362	59	28	[23]
[Au{C≡C-4-C$_6$H$_4$-(E)-N=N-4-C$_6$H$_4$NO$_2$}(PPh$_3$)]	398	180	68	[27]
[Au(C≡C-4-C$_6$H$_4$N=CH-4-C$_6$H$_4$NO$_2$)(PPh$_3$)]	392	85	34	[23]
[1,3,5-C$_6$H$_3${(C≡C)Au(PPh$_3$)}$_3$]	298	6	4	[21]

[a] All measurements in THF solvent. All complexes are optically transparent at 1.064 μm.
[b] Hyper-Rayleigh scattering studies at 1.064 μm; values ± 10%.
[c] HRS at 1.064 μm corrected for resonance enhancement at 532 nm using the two-level model with $\beta_0 = \beta[1-(2\lambda_{max}/1064)^2][1-(\lambda_{max}/1064)^2]$; damping factors not included.

complexes mirrors that of the organic compounds. Observed nonlinearities suggest that $\beta[(Z)] < \beta[(E)]$ for bridge stereochemistry variation in the $C_6H_4CH=CHC_6H_4$ linked complexes; although it is tempting to ascribe this variation to dipole moment differences (the molecular geometry of the E isomer leads to an increased charge separation compared to the Z isomer), the relevant optical transition in the E isomer is almost twice as intense as that for the Z isomer, and a combination of these effects is likely. The difference in intrinsic nonlinearity between the Z and E isomers is likely to be substantially greater than that observed experimentally; the Z form fluoresces significantly at the frequency-doubled wavelength, inflating its observed nonlinearity compared to that of the E isomer. Effects of bridge atom variation on observed nonlinearity reveal the experimentally determined nonlinearity for the (E)-imino-complex [Au(C≡C-4-C_6H_4N=CH-4-$C_6H_4NO_2$)(PPh$_3$)] is about two-thirds that of the (E)-ene-linked complex [Au(C≡C-4-C_6H_4-(E)-CH=CH-4-$C_6H_4NO_2$)(PPh$_3$)]. In the gold system, it is evident that ene-linkage is more effective than imino-linkage at maximizing nonlinearity in these organometallic alkynyl chromophores.

Table 3 also includes the two-level corrected values, with β_0 values about half those of $\beta_{1.064}$ values for the gold alkynyl complexes. The two-state model may not be adequate for donor-acceptor organometallic systems where two dominant optical transitions are close to 2ω, as is the case for the "extended-chain" complexes (by analogy with previous work on ethynylgold(I) complexes [29, 30], the higher energy bands ($\lambda < 310$ nm) are probably due to $\sigma(Au \leftarrow P) \rightarrow \pi^*(PPh)$ transitions; the low nonlinearity for [Au(C≡CPh)(PPh$_3$)] suggests that these transitions do not significantly influence the observed nonlinearities for [Au(C≡C-4-$C_6H_4NO_2$)(PPh$_3$)] and the "extended-chain" complexes, and any contribution that they make is likely to be consistent across the series of complexes). For these gold complexes, the relative ordering for observed and two-level corrected β are the same, and all complexes are optically transparent at 2ω; it is therefore almost certain that the effects of structural modification on observed nonlinearity reflect their effect upon intrinsic nonlinearity.

Table 3 also affords the possibility of assessing acceptor group variation, phenyl substitution resulting in β values increasing as 4-H < 4-CH{C(O)Me}$_2$, 4-$\overline{CHO(CH_2)_3O}$ < 4-CHO < 4-NO$_2$, the expected trend for increasing acceptor strength in these dipolar molecules. Replacing co-ligand PPh$_3$ by PMe$_3$ in proceeding from [Au{4-C≡CC$_6$H$_4\overline{CHO(CH_2)_3O}$}(PPh$_3$)] to [Au{4-C≡CC$_6H_4\overline{CHO(CH_2)_3O}$}(PMe$_3$)] results in a three-fold increase in $\beta_{1.064}$ and β_0. PMe$_3$ is a more basic phosphine, resulting in a more electron-rich gold donor, but PPh$_3$ provides for more extensive π-delocalization; these data suggest that donor strength is the more important factor influencing the magnitude of β in these complexes. The most efficient complex from the series is [Au{C≡C-4-C_6H_4-(E)-N=N-4-$C_6H_4NO_2$}(PPh$_3$)], suggesting that azo-linked compounds may be viable alternatives to ene-linked complexes if possible photoisomerization at the azo linkage is not a problem.

Data for 18 valence electron ruthenium vinylidene and alkynyl complexes are collected in Table 4. Cyclopentadienylbis(phosphine)ruthenium complexes containing the same alkynyl ligands as those listed above for nickel and gold complexes have been examined; one of the most efficient of these complexes is depicted in Figure 7. The linear optical absorption spectra for these donor-acceptor alkynyl complexes have a low-energy MLCT transition that undergoes a red shift on alkynyl ligand chain-lengthening, with the lowest energy absorption for the imine-linked alkynyl complex. The nonlinearities have been determined experimentally by HRS and confirmed for two complexes, [Ru(C≡C-4-C_6H_4N=CH-4-$C_6H_4$$NO_2$) ($PPh_3$)$_2$ (η^5-C_5H_5)] and [Ru{C≡C-4-C_6H_4-(E)-CH=CH-4-$C_6H_4$$NO_2$}($PPh_3$)$_2$($\eta^5$-$C_5H_5$)], by EFISH. The experimentally obtained $\mu \cdot \beta_{1.064}$ values, 9700×10^{-48} esu and 5800×10^{-48} esu, respectively, are resonance enhanced, but extremely large in magnitude compared to previously reported organometallic data (the $\mu \cdot \beta$ product is the relevant parameter to assess poled polymer potential) [31]. Experimentally determined dipole moments of 6.7 D and 7.6 D, respectively, afford β_{vec} values consistent with those obtained by HRS. As the values of β measured by EFISH and HRS are equal, and considering that these techniques determine different combinations of tensor components, one can conclude that there is only one dominating tensor component, and that therefore the β as determined by HRS for these complexes can also be considered as β_{vec} ($\beta_{EFISH}=\beta_{HRS}=\beta_{vec}=\beta_{zzz}$). Not surprisingly, chain-lengthening in proceeding from [Ru(C≡C-4-$C_6H_4$$NO_2$)($PPh_3$)$_2$($\eta^5$-$C_5H_5$)] to [Ru{C≡C-4-$C_6H_4$-(E)-X=CH-4-$C_6H_4$$NO_2$}($PPh_3$)$_2$($\eta^5$-$C_5H_5$)](X=CH or N) leads to an increased nonlinearity.

Phosphine replacement (in proceeding from [Ru(C≡C-4-$C_6H_4$$NO_2$)($PMe_3$)$_2$ (η^5-C_5H_5)] to [Ru(C≡C-4-$C_6H_4$$NO_2$)($PPh_3$)$_2$($\eta^5$-$C_5H_5$)]) leads to an increase in corrected quadratic nonlinearity, but further data is needed to corroborate this result. Although it might be expected that the more strongly electron-donating PMe_3 ligand should give rise to increased nonlinearity (cf. the observation with gold alkynyl complexes summarized above), Mulliken analyses of charge density suggest that the ruthenium in the PPh_3 complex is more electron rich than in the PMe_3 complex (+1.40 and +1.47, respectively); this may be responsible for its higher-than-expected nonlinearity.

The lack of contribution from transitions involving other ligands is also suggested by the low nonlinearity of the precursor chloride [RuCl(PPh_3)$_2$(η^5-C_5H_5)]. Given the preceding, it is possible that the two-level corrected values may have some significance. The corrected nonlinearities suggest an efficiency series E-ene-linkage ≥ yne-linkage > azo-linkage > biphenyl > imino-linkage [28].

Both HRS and EFISH data suggest a substantial increase in two-level-corrected β_{vec} on atom replacement of N by CH in the bridging group, there being a three-fold increase in quadratic optical nonlinearity at 1.064 μm in proceeding from [Ru(C≡C-4-C_6H_4N=CH-4-$C_6H_4$$NO_2$)($PPh_3$)$_2$($\eta^5$-$C_5H_5$)] to [Ru{C≡C-4-$C_6$$H_4$-(E)-CH=CH-4-$C_6H_4$$NO_2$}($PPh_3$)$_2$($\eta^5$-$C_5H_5$)]. The significant difference between [Ru{C≡C-4-C_6H_4-(E)-CH=CH-4-$C_6H_4$$NO_2$}($PPh_3$)$_2$($\eta^5$-$C_5H_5$)] and [Ru(C≡C-4-$C_6H_4$N=CH-4-$C_6H_4$$NO_2$)($PPh_3$)$_2$($\eta^5$-$C_5H_5$)] is the strength of the

Table 4. Molecular second-order NLO results for ruthenium vinylidene and alkynyl complexes[a]

Complex	λ_{max} (nm)	β[b] (10^{-30} esu)	β_0[c] (10^{-30} esu)	Technique	Ref.
[Ru(C≡CPh)(PPh$_3$)$_2$(η^5-C$_5$H$_5$)]	310	16	10	HRS	[32]
[Ru(C≡C-4-C$_6$H$_4$NO$_2$)(PPh$_3$)$_2$(η^5-C$_5$H$_5$)]	460	468	96	HRS	[33]
[Ru(C≡C-4-C$_6$H$_4$NO$_2$)(CO)$_2$(η^5-C$_5$H$_5$)]	364	58	27	HRS	[17]
[Ru(C≡C-4-C$_6$H$_4$NO$_2$)(dppe)(η^5-C$_5$H$_5$)]	447	664	161	HRS	[17]
[Ru(C≡C-4-C$_6$H$_4$NO$_2$)(PMe$_3$)$_2$(η^5-C$_5$H$_5$)]	477	248	38	HRS	[32]
[Ru(C≡C-4-C$_6$H$_4$-4-C$_6$H$_4$NO$_2$)(PPh$_3$)$_2$(η^5-C$_5$H$_5$)]	448	560	134	HRS	[32]
[Ru(C≡C-2-C$_5$H$_3$N)(PPh$_3$)$_2$(η^5-C$_5$H$_5$)]	331	18	10	HRS	[22]
[Ru(C≡C-2-C$_5$H$_4$N-5-NO$_2$)(PPh$_3$)$_2$(η^5-C$_5$H$_5$)]	468	622	113	HRS	[22]
[Ru(C≡C-4-C$_6$H$_4$CHO)(PPh$_3$)$_2$(η^5-C$_5$H$_5$)]	400	120	45	HRS	[25]
[Ru(C≡C-4-C$_6$H$_4$CH{OC(O)Me}$_2$)(PPh$_3$)$_2$(η^5-C$_5$H$_5$)]	326	68	38	HRS	[25]
[Ru{C≡C-4-C$_6$H$_4$N=CCH=CBu^{t}C(O)CBu^t=CH}(PPh$_3$)$_2$(η^5-C$_5$H$_5$)]	622	658	159	HRS	[34]
[Ru(C≡C-4-C$_6$H$_4$C≡C-4-C$_6$H$_4$NO$_2$)(PPh$_3$)$_2$(η^5-C$_5$H$_5$)]	446	865	212	HRS	[32]
[Ru{C≡C-4-C$_6$H$_4$-(E)-CH=CH-4-C$_6$H$_4$NO$_2$}(PPh$_3$)$_2$(η^5-C$_5$H$_5$)]	476	1455	232	HRS	[33]
		1464	234	EFISH	
[Ru{C≡C-4-C$_6$H$_4$-(E)-N=CH-4-C$_6$H$_4$NO$_2$}(PPh$_3$)$_2$(η^5-C$_5$H$_5$)]	496	840	86	HRS	[33]
		740	78	EFISH	
[Ru{C≡C-4-C$_6$H$_4$-(E)-N=N-4-C$_6$H$_4$NO$_2$}(PPh$_3$)$_2$(η^5-C$_5$H$_5$)]	565	1627	149	HRS	[27]
trans-[Ru(C≡CHPh)Cl(dppm)$_2$](PF$_6$)	320	24	16	HRS	[35]
trans-[Ru(C≡CPh)Cl(dppm)$_2$]	308	20	12	HRS	[36]
trans-[Ru{C≡CH-4-C$_6$H$_4$C≡CPh}Cl(dppm)$_2$](PF$_6$)	380	64	31	HRS	[35]
trans-[Ru{C≡C-4-C$_6$H$_4$C≡CPh}Cl(dppm)$_2$]	381	101	43	HRS	[35]
trans-[Ru{C≡CH-4-C$_6$F$_4$OMe}Cl(dppm)$_2$](PF$_6$)	334	32	17	HRS	[24]
trans-[Ru{C≡C-4-C$_6$F$_4$OMe}Cl(dppm)$_2$]	337	26	14	HRS	[24]
trans-[Ru{C≡C-2-C$_5$H$_3$N}Cl(dppm)$_2$]	351	35	19	HRS	[36]
trans-[Ru{C≡C-2-C$_5$H$_3$N-5-NO$_2$}Cl(dppm)$_2$]	490	468	56	HRS	[36]
trans-[Ru{C≡CH-4-C$_6$H$_4$CHO(CH$_2$)$_3$O}Cl(dppm)$_2$](PF$_6$)	317	64	38	HRS	[25]
trans-[Ru{C≡C-4-C$_6$H$_4$CHO(CH$_2$)$_3$O}Cl(dppm)$_2$]	320	61	35	HRS	[25]
trans-[Ru{C≡CH-2-C$_6$H$_4$CHO}Cl(dppm)$_2$](PF$_6$)	555	27	2	HRS	[25]
trans-[Ru{C≡CH-3-C$_6$H$_4$CHO}Cl(dppm)$_2$](PF$_6$)	320	45	26	HRS	[25]
trans-[Ru{C≡C-3-C$_6$H$_4$CHO}Cl(dppm)$_2$]	321	58	34	HRS	[25]

(Continued)

Table 4. (Continued)

Complex	λ_{max} (nm)	β[b] (10^{-30} esu)	β_0[c] (10^{-30} esu)	Technique	Ref.
trans-[Ru(C≡CH-4-C$_6$H$_4$CHO)Cl(dppm)$_2$](PF$_6$)	403	108	39	HRS	[35]
trans-[Ru(C≡C-4-C$_6$H$_4$CHO)Cl(dppm)$_2$]	405	106	38	HRS	[35]
trans-[Ru(C≡CH-4-C$_6$H$_4$NO$_2$)Cl(dppm)$_2$](PF$_6$)	470	721	127	HRS	[35]
trans-[Ru(C≡C-4-C$_6$H$_4$NO$_2$)Cl(dppm)$_2$]	473	767	129	HRS	[36]
trans-[Ru(C≡CH-4-C$_6$H$_4$C≡C-4-C$_6$H$_4$NO$_2$)Cl(dppm)$_2$](PF$_6$)	326	424	122	HRS	[35]
trans-[Ru(C≡C-4-C$_6$H$_4$C≡C-4-C$_6$H$_4$NO$_2$)Cl(dppm)$_2$]	464	833	161	HRS	[35]
trans-[Ru(C≡C-4-C$_6$H$_4$C≡C-4-C$_6$H$_4$NO$_2$)Cl(dppm)$_2$]	439	1379	365	HRS	[35]
trans-[Ru(C≡C-4-C$_6$H$_4$N=CCH=CBu^{t}C(O)CBu^t=CH)Cl(dppm)$_2$]	645	417	124	HRS	[34]
trans-[Ru(C≡C-4-C$_6$H$_4$-(E)-N=N-4-C$_6$H$_4$NO$_2$)Cl(dppm)$_2$]	583	1649	232	HRS	[27]
trans-[Ru(C≡C-4-C$_6$H$_4$-(E)-CH=CH-4-C$_6$H$_4$NO$_2$)Cl(dppm)$_2$](PF$_6$)	369	1899	314	HRS	[35]
trans-[Ru(C≡C-4-C$_6$H$_4$-(E)-CH=CH-4-C$_6$H$_4$NO$_2$)Cl(dppm)$_2$]	490	1964	235	HRS	[35]
trans-[Ru(C≡CPh)Cl(dppe)$_2$]	319	6	3	HRS	[35]
trans-[Ru(C≡CH-4-C$_6$H$_4$CHO)Cl(dppe)$_2$](PF$_6$)	412	181	61	HRS	[35]
trans-[Ru(C≡C-4-C$_6$H$_4$CHO)Cl(dppe)$_2$]	413	120	40	HRS	[35]
trans-[Ru(C≡CH-4-C$_6$H$_4$NO$_2$)Cl(dppe)$_2$](PF$_6$)	476	1130	180	HRS	[35]
trans-[Ru(C≡C-4-C$_6$H$_4$NO$_2$)Cl(dppe)$_2$]	477	351	55	HRS	[35]
trans-[Ru[C≡CH-4-C$_6$H$_4$-(E)-CH=CH-4-C$_6$H$_4$NO$_2$]Cl(dppe)$_2$](PF$_6$)	473	441	74	HRS	[35]
trans-[Ru[C≡C-4-C$_6$H$_4$-(E)-CH=CH-4-C$_6$H$_4$NO$_2$]Cl(dppe)$_2$]	489	2676	342	HRS	[35]
trans-[Ru(C≡CPh)(C≡C-4-C$_6$H$_4$C≡CPh)(dppe)$_2$]	383	34	–	HRS	[37]
(−)$_{578}$-trans-[Ru(C≡CPh)Cl{(R, R)-diph}$_2$]	292	~0	~0	HRS	[18]
(−)$_{578}$-trans-[Ru(C≡C-4-C$_6$H$_4$NO$_2$)Cl{(R, R)-diph}$_2$]	467	530	97	HRS	[18]
(−)$_{589}$-trans-[Ru[C≡C-4-C$_6$H$_4$-(E)-CH=CH-4-C$_6$H$_4$NO$_2$]Cl{(R, R)-diph}$_2$]	481	2795	406	HRS	[18]
1-(HC≡C)-3,5-C$_6$H$_3$[trans-(C≡C)RuCl(dppm)$_2$]$_2$	323	42[d]	24	HRS	[21]
1,3,5-{trans-[RuCl(dppe)$_2$(C≡C-4-C$_6$H$_4$C≡C)]}$_3$C$_6$H$_3$	414	94	–	HRS	[37]
1,3,5-{trans-[Ru(C≡CPh)(dppe)$_2$(C≡C-4-C$_6$H$_4$C≡C)]}$_3$C$_6$H$_3$	411	93	–	HRS	[37]

[a] All measurements in THF solvent. All complexes are optically transparent at 1.064 μm.
[b] Hyper-Rayleigh scattering studies at 1.064 μm; values ±10%. EFISH studies at 1.064 μm; values ±15%.
[c] HRS or EFISH data at 1.064 μm corrected for resonance enhancement at 532 nm using the two-level model with $\beta_0 = \beta[1-(2\lambda_{max}/1064)^2][1-(\lambda_{max}/1064)^2]$; damping factors not included.
[d] Upper bound; demodulation of SHG and fluorescence contributions could not be effected.

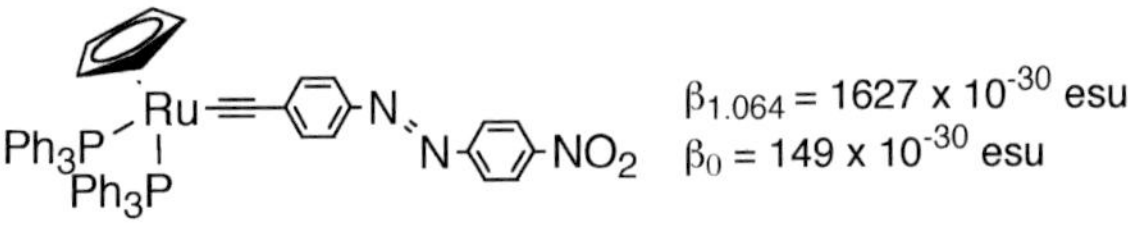

Ph₃P–Ru≡⟨⟩–N=N–⟨⟩–NO₂ $\beta_{1.064} = 1627 \times 10^{-30}$ esu
Ph₃P $\beta_0 = 149 \times 10^{-30}$ esu

Figure 7. One of the most efficient cyclopentadienylbis(triphenylphosphine)ruthenium-containing complexes from these studies

MLCT optical transition; the ene-linked alkynyl complex has an oscillator strength for this transition twice that of the imino-linked analogue. The two-level corrected β_{vec} is dependent on the difference in dipole moments between ground and excited states, and extinction coefficients of the relevant transition. Assuming that the excited state dipoles for these complexes are similar, and knowing dipole moments for these complexes are comparable, the three-fold enhancement in β_0 in proceeding from [Ru(C≡C-4-C₆H₄N=CH-4-C₆H₄NO₂)(PPh₃)₂(η^5-C₅H₅)] to [Ru{C≡C-4-C₆H₄-(E)-CH=CH-4-C₆H₄NO₂}(PPh₃)₂(η^5-C₅H₅)] arises largely from differences in oscillator strength for the dominant transition. Replacement of ene-linkage by yne-linkage to give [Ru(C≡C-4-C₆H₄C≡C-4-C₆H₄NO₂)(PPh₃)₂(η^5-C₅H₅)] affords a complex with a large nonlinearity, suggesting that yne-linkage may be worthy of consideration for the preparation of longer chromophores (while the nonlinearities of [Ru{C≡C-4-C₆H₄-(E)-CH=CH-4-C₆H₄NO₂}(PPh₃)₂(η^5-C₅H₅)] and [Ru(C≡C-4-C₆H₄C≡C-4-C₆H₄NO₂)(PPh₃)₂(η^5-C₅H₅)] are comparable, the latter involves less trade-off in optical transparency).

We have also studied a further ruthenium-containing series of complexes with a different ligand set, namely chlorobis(bidentate phosphine)ruthenium alkynyl complexes, the results from which are summarized in Table 4, and efficient examples illustrated in Figure 8.

Chain-lengthening in proceeding from *trans*-[Ru(C≡C-4-C₆H₄NO₂)Cl(dppm)₂] and *trans*-[Ru(C≡C-4-C₆H₄C≡C-4-C₆H₄NO₂)Cl(dppm)₂] to *trans*-[Ru(C≡C-4-C₆H₄C≡C-4-C₆H₄C≡C-4-C₆H₄NO₂)Cl(dppm)₂] results in increased β and β_0, as expected (Figure 9). Surprisingly, though, proceeding from *trans*-[Ru(C≡C-4-C₆H₄NO₂)Cl(dppm)₂] to *trans*-[Ru(C≡C-4-C₆H₄C≡C-4-C₆H₄NO₂)Cl(dppm)₂] does not result in a significant increase in β or β_0 value. It has been shown with organic compounds that "chain-lengthening" arylalkynes leads to a saturation of the β response for two repeat units, whereas the β response for oligo-phenylenevinylene compounds does not saturate until the complex contains approximately twenty repeat

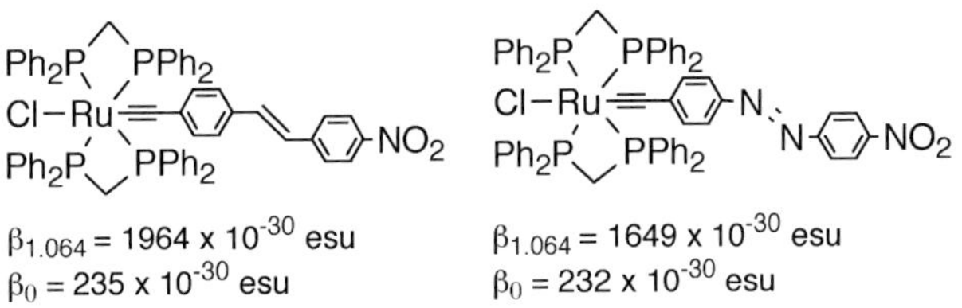

$\beta_{1.064} = 1964 \times 10^{-30}$ esu
$\beta_0 = 235 \times 10^{-30}$ esu

$\beta_{1.064} = 1649 \times 10^{-30}$ esu
$\beta_0 = 232 \times 10^{-30}$ esu

Figure 8. Efficient chlorobis(bidentate phosphine)ruthenium alkynyl complexes from these studies

units [38]. These ruthenium-containing compounds afford an unusual series for which minimal increase in β or β_0 (within the error margin of $\pm 10\%$) is seen on progressing from n = 0 to 1 for trans-[Ru{C≡C-4-(C$_6$H$_4$-4-C≡C)$_n$C$_6$H$_4$NO$_2$}Cl(dppm)$_2$] but a significant increase is seen on progressing to n = 2. For this series of complexes, increasing β is not correlated with a red-shift in λ_{max}; chain-lengthening is accompanied by a blue-shift in optical absorption maxima. Replacing an yne-linkage by an ene-linkage results in increased β and β_0 as demonstrated in proceeding from trans-[Ru(C≡C-4-C$_6$H$_4$C≡C-4-C$_6$H$_4$NO$_2$)Cl(dppm)$_2$] to trans-[Ru{C≡C-4-C$_6$H$_4$-(E)-CH=CH-4-C$_6$H$_4$NO$_2$}Cl(dppm)$_2$] [23, 39].

These alkynyl complexes can be protonated to afford vinylidene complexes, which can in turn be deprotonated to give the starting alkynyl complex, reactions that are spectroscopically quantitative. The tabulated data also provide the opportunity to assess the effect of this protonation, in proceeding from alkynyl complex to vinylidene derivative. One would perhaps expect that replacing the electron-rich ruthenium donor in the alkynyl complexes with a (formally) cationic ruthenium centre in the vinylidene complexes would result in a significant decrease in nonlinearity.

However, some vinylidene/alkynyl complex pairs have similar nonlinearities (e.g. trans-[Ru(C=CHPh)Cl(dppm)$_2$](PF$_6$), trans-[Ru(C≡CPh)Cl(dppm)$_2$]; trans-[Ru(C=CH-4-C$_6$H$_4$NO$_2$)Cl(dppm)$_2$](PF$_6$), trans-[Ru(C≡C-4-C$_6$H$_4$NO$_2$)Cl(dppm)$_2$]), while in some instances the "expected" trend β(vinylidene) < β(alkynyl complex) is observed (e.g. trans-[Ru(C=CH-4-C$_6$H$_4$C≡CPh)Cl(dppm)$_2$](PF$_6$), trans-[Ru(C≡C-4-C$_6$H$_4$C≡CPh)Cl(dppm)$_2$]; trans-[Ru(C=CH-4-C$_6$H$_4$C≡C-4-C$_6$H$_4$NO$_2$)Cl(dppm)$_2$](PF$_6$), trans-[Ru(C≡C-4-C$_6$H$_4$C≡C-4-C$_6$H$_4$NO$_2$)Cl(dppm)$_2$]). If β_{exp} and β_0 values for vinylidene/alkynyl complex pairs differ sufficiently to readily distinguish β signals into bi-stable "off" and "on" states, the alkynyl complexes can be reprotonated to afford the precursor vinylidene complex, and this sequence can be repeated. These complex pairs can therefore provide a protically-switchable NLO-active system where the "on" signal is the alkynyl complex.

The alkynyl ligands that were coupled to the nickel, gold, and cyclopentadienylbis(triphenylphosphine)ruthenium centres described above were also attached to the chlorobis(bidentate phosphine)ruthenium moiety, with a similar outcome in relative quadratic NLO merit upon alkynyl ligand variation. Further variation

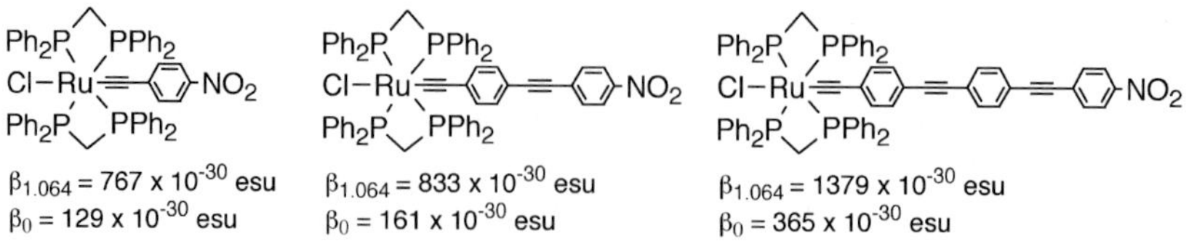

Figure 9. Effect of phenyleneethynylene chain-lengthening on quadratic nonlinearities for selected ruthenium complexes

in alkynyl ligand was pursued; phenyl substituent location affects β, in replacing 3-CHO by 4-CHO (proceeding from trans-[Ru(C≡C-3-C$_6$H$_4$CHO)Cl(dppm)$_2$] to trans-[Ru(C≡C-3-C$_6$H$_4$CHO)Cl(dppm)$_2$]), with the magnitude increasing upon formal conjugation of the metal centre with the acceptor formyl unit; however, this result does not translate to increased corrected nonlinearities, experimentally indistinguishable β_0 values being observed.

A shortcoming with the arylalkynyl ligands described so far is that the polarization of the electron density is rendered energetically unfavourable by the loss of aromatic stabilization energy. To circumvent this problem, indoaniline compounds have been suggested by Marder et al. [40]. The relatively low β values recorded for the indoaniline compounds trans-[Ru(C≡C-4-C$_6$H$_4$N=CCHCBu^{t}C(O)CBu^t=CH)Cl(dppm)$_2$] and trans-[Ru(C≡C-4-C$_6$H$_4$N=CCHCBu^{t}C(O)CBu^t=CH)(PPh$_3$)$_2$(η^5-C$_5$H$_4$)] may be due to the presence of the *tert*-butyl groups on the quinone ring (Figure 10). These relatively electron rich groups may be expected to reduce the electron acceptor properties of the quinone ring, consistent with the results of Marder et al. [40].

As was mentioned above, replacing the traditional dipolar composition with alternative multipolar geometries is of significant interest. The molecular hyperpolarizability of [1-(HC≡C)C$_6$H$_3$-3,5-{C≡CNi(PPh$_3$)(η^5-C$_5$H$_5$)}$_2$] was determined to be 96×10^{-30} esu, with the two-level corrected value $\beta_0 = 57 \times 10^{-30}$ esu amongst the largest values for an organometallic complex lacking a donor-acceptor composition. Significantly, attaching two Ni(PPh$_3$)(η^5-C$_5$H$_5$) units to the central 1,3,5-triethynylbenzene core to afford [1-(HC≡C)C$_6$H$_3$-3,5-{C≡CNi(PPh$_3$)(η^5-C$_5$H$_5$)}$_2$] leads to a nonlinearity four times that of the monometallic analogue [Ni(C≡CPh)(PPh$_3$)(η^5-C$_5$H$_5$)], but with negligible shift to low energy of λ_{max}. With this exception, though, our focus has been on octupolar ruthenium alkynyl complexes (Tables 4, 5 and 6). The ligated ruthenium centre is sufficiently large that insertion of phenylethynyl or phenylethenyl "spacer" groups is required to accommodate three such units about a 1,3,5-trisubstituted benzene core.

The complex [1,3,5-C$_6$H$_3$-trans-{C≡C-4-C$_6$H$_4$C≡C[RuCl(dppe)$_2$]}$_3$] possesses a significant β_{HRS}. Extending the delocalized π-system through the metal in progressing to [1,3,5-C$_6$H$_3$-trans-{C≡C-4-C$_6$H$_4$C≡C[Ru(C≡CPh)(dppe)$_2$]}$_3$] (Figure 11), though, is ineffective in increasing β, indicating that the trans-phenylalkynyl ligand is acting largely as a π-donor ligand (it has been reported that phenylalkynyl ligands are pseudo-halides in complexes of this type) [41]; a similar lack of β enhancement on extending the π-system through a metal has been

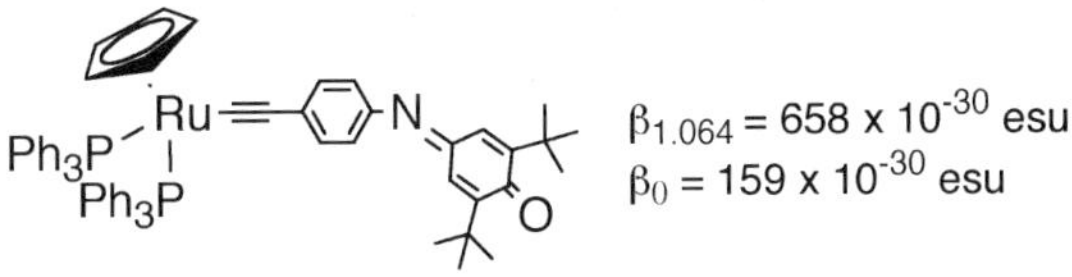

$\beta_{1.064} = 658 \times 10^{-30}$ esu
$\beta_0 = 159 \times 10^{-30}$ esu

Figure 10. Indoanilinoalkynylruthenium complex from these studies

Table 5. Molecular second-order NLO results for selected linear and octupolar alkynyl complexes[a]

Complex	λ_{max} (nm)	$\sqrt{\langle\beta^2\rangle}$[b] ($10^{-30}$ esu)	Ref.
trans-[Ru(C≡CPh)(dppe)$_2$Cl]	319	6	[37]
trans-[Ru(C≡C-4-C$_6$H$_4$C≡CPh)(C≡CPh)(dppe)$_2$]	383	34	[37]
[1,3,5-(trans-[(dppe)$_2$ClRu(C≡C-4-C$_6$H$_4$C≡C)]$_3$C$_6$H$_3$]	414	94	[37]
[1,3,5-(trans-[(dppe)$_2$(PhC≡C)Ru(C≡C-4-C$_6$H$_4$C≡C)])$_3$C$_6$H$_3$]	411	93	[37]

[a] All measurements in THF solvent. All complexes are optically transparent at 1.064 μm.
[b] Hyper-Rayleigh scattering studies at 1.064 μm; values ±10%.

Table 6. Molecular second-order NLO results for ruthenium alkynyl and vinylidene complexes measured at two wavelengths[a]

Complex	λ_{max} (nm)	$\beta^{b}_{1.064}$ (10^{-30} esu)	$\beta^{c}_{0.800}$ (10^{-30} esu)	Ref.
trans-[Ru{C≡CH-4-C$_6$H$_4$-(E)-CH=CHPh}Cl(dppm)$_2$](PF$_6$)	360	70	33	[43, 44]
trans-[Ru{C≡C-4-C$_6$H$_4$-(E)-CH=CHPh}Cl(dppm)$_2$]	397	200	≤920	[43, 44]
[1,3,5-(trans-[(dppm)$_2$ClRu{C≡CH-4-C$_6$H$_4$-(E)-CH=CH}]$_3$C$_6$H$_3$](PF$_6$)$_3$	396	165	483	[43, 44]
[1,3,5-(trans-[(dppm)$_2$ClRu{C≡C-4-C$_6$H$_4$-(E)-CH=CH}]$_3$C$_6$H$_3$]	415	244	.935	[43, 44]

[a] All measurements in THF solvent. All complexes are optically transparent at 1.064 and 0.800 μm.
[b] HRS studies at 1.064 μm; values ±10%.
[c] HRS studies at 0.800 μm; values ±10%.

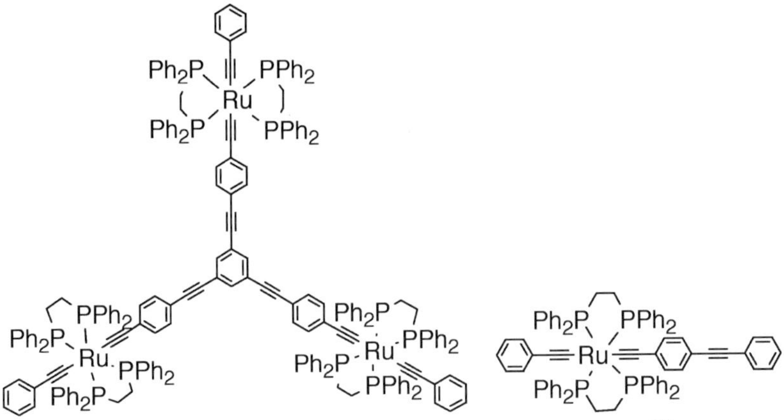

Figure 11. Example of an octupolar alkynyl complex and its linear alkynyl complex analogue from these studies

reported recently in a dipolar system [42]. trans-[Ru(C≡C-4-C_6H_4C≡CPh)(C≡CPh)(dppe)$_2$] (Figure 11) is a linear fragment of the octupolar complex [1,3,5-C_6H_3-trans-{C≡C-4-C_6H_4C≡C[Ru(C≡CPh)(dppe)$_2$]}$_3$]; progressing from the linear fragment to the octupolar complex results in a three-fold increase in oscillator strength of the UV-vis band assigned to the MLCT transition, and a three-fold increase in quadratic NLO merit.

The absolute value of β_{HRS} for [1,3,5-C_6H_3-trans-{C≡C-4-C_6H_4C≡C[Ru(C≡CPh)(dppe)$_2$]}$_3$] is amongst the largest thus far for a multipolar compound optically transparent at the second-harmonic, for which resonance enhancement is much less important. It is also amongst the largest thus far for a multipolar compound lacking a formal acceptor moiety (results with organic compounds suggest that a further increase in β is likely upon replacing the arene ring with an electron acceptor such as 2,4,6-trinitroaryl or 2,4,6-triazine groups).

HRS measurements of trans-[Ru{C≡C-4-C_6H_4-(E)-CH=CHPh}Cl(dppm)$_2$] and [1,3,5-(trans-[(dppm)$_2$ClRu{C≡C-4-C_6H_4-(E)-CH=CH}])$_3C_6H_3$] and their protio-vinylidene derivatives are listed in Table 6. These complexes are transparent at the second-harmonic wavelength of 532 nm, permitting assessment of the impact of structural variation on quadratic NLO merit. Nonlinearities for [1,3,5-(trans-[(dppm)$_2$ClRu{C=CH-4-C_6H_4-(E)-CH=CH}])$_3C_6H_3$](PF$_6$)$_3$ and [1,3,5-(trans-[(dppm)$_2$ClRu{C≡C-4-C_6H_4-(E)-CH=CH}])$_3C_6H_3$] are large for octupolar complexes without polarizing acceptor substituents, but data for trans-[Ru{C≡C-4-C_6H_4-(E)-CH=CHPh}Cl(dppm)$_2$], [1,3,5-(trans-[(dppm)$_2$ClRu{C=CH-4-C_6H_4-(E)-CH=CH}])$_3C_6H_3$](PF$_6$)$_3$ and [1,3,5-(trans-[(dppm)$_2$ClRu{C≡C-4-C_6H_4-(E)-CH=CH}])$_3C_6H_3$] are experimentally indistinguishable within the error margins; the ene-linked alkynyl complexes have $\beta_{1.064}$ values larger that those of the related yne-linked complexes [1,3,5-(trans-[(dppm)$_2$

ClRu{C≡C-4-C_6H_4C≡C}])$_3C_6H_3$] described above. HRS measurements of these complexes at 800 nm with femtosecond pulses are also listed in Table 6; fluorescence contributions could not be completely eliminated from the data for *trans*-[Ru{C≡C-4-C_6H_4-(*E*)-CH=CHPh}Cl(dppm)$_2$], but are absent from the other three complexes (there is no demodulation of the signals as modulation frequency is varied). Proceeding from vinylidene complex [1,3,5-(*trans*-[(dppm)$_2$ClRu{C=CH-4-C_6H_4-(*E*)-CH=CH}])$_3C_6H_3$](PF$_6$)$_3$ to alkynyl complex [1,3,5-(*trans*-[(dppm)$_2$ClRu{C≡C-4-C_6H_4-(*E*)-CH=CH}])$_3C_6H_3$] results in a two-fold increase in $\beta_{0.800}$, the latter suggestive of a similar increase in $\beta_{1.064}$ obscured by error margins. These complexes are rare examples of organometallics for which quadratic optical nonlinearities have been determined at more than one wavelength. The $\beta_{0.800}$ values are mostly significantly larger than their $\beta_{1.064}$ values, consistent with significant resonance enhancement for the former resulting from close proximity of the optical absorption maxima to the second-harmonic wavelength (400 nm).

3. THIRD-ORDER NLO STUDIES

Our studies of the third-order NLO properties of alkynyl metal complexes have focused on iron, ruthenium, nickel and gold complexes. Most of the iron-containing examples are ferrocene derivatives, listed in Table 7. The real components γ_{real} of the nonlinearities for most of the ferrocenyl complexes are negative, and the imaginary components γ_{imag} for most are significant, consistent with two-photon absorption contributions to the observed molecular nonlinearities $|\gamma|$; comment on the effect of structural variation on the magnitude of $|\gamma|$ is therefore cautious, particularly in the light of the significant error margins. Replacing the 14 electron gold centre with an 18 electron ligated ruthenium centre (in proceeding to [Fe{η^5-C_5H_4-(*E*)-4-CH=CHC$_6H_4$CH=CRuCl(dppm)$_2$}$_2$](PF$_6$)$_2$ and [Fe{η^5-C_5H_4-(*E*)-4-CH=CHC$_6H_4$C≡CRuCl(dppm)$_2$}$_2$]) (Figure 12) results in intense transitions in the UV-vis spectra close to the second-harmonic wavelength of our Ti-sapphire laser (400 nm) and, as a consequence, these complexes possess large negative γ_{real} and large γ_{imag} values.

Third-order NLO data for nickel complexes are collected in Table 8 and the most efficient example depicted in Figure 13. Similar to the ferrocenyl complexes above, the real component of the γ values for the nitro-containing nickel complexes are large, negative and incorporate substantial error margins; the negative real components and presence of significant imaginary components of the nonlinearities indicate two-photon dispersion is contributing to the observed responses.

As noted above, the two-photon states become important for an 800 nm irradiating wavelength when complexes contain $\lambda_{max} > 400$ nm, and real components for the nickel complexes considered here become negative when the optical absorption maximum fulfils this criterion. Despite the large error margins, an increase in the real component of the nonlinearity upon chain-lengthening is evident, and an increase in efficiency upon replacing *Z* by *E* stereochemistry, in

Table 7. Molecular third-order NLO data for iron alkynyl and ferrocenyl complexes[a]

Complex	λ_{max} (nm)	γ_{real} (10^{-36} esu)	γ_{imag} (10^{-36} esu)	$|\gamma|$ (10^{-36} esu)	Ref.
[Fe(C≡C-4-C$_6$H$_4$NO$_2$)(dppe)(η^5-C$_5$H$_5$)]	497	-410 ± 200	580 ± 200	710 ± 280	[45]
[Fe{C≡C-4-C$_6$H$_4$-(E)-CH=CH-4-C$_6$H$_4$NO$_2$}(dppe)(η^5-C$_5$H$_5$)]	499	-2200 ± 600	1200 ± 300	2500 ± 670	[45]
[Fe{η^5-C$_5$H$_4$-(E)-CH=CH-4-C$_6$H$_4$C≡CAu(PCy$_3$)}$_2$]	468	-400 ± 500	500 ± 100	640 ± 390	[46]
[Fe{η^5-C$_5$H$_4$-(E)-CH=CH-4-C$_6$H$_4$C≡CAu(PPh$_3$)}$_2$]	465	-1100 ± 300	300 ± 60	1140 ± 310	[46]
[Fe{η^5-C$_5$H$_4$-(E)-CH=CH-4-C$_6$H$_4$C≡CAu(PMe$_3$)}$_2$]	463	200 ± 150	0 ± 30	200 ± 150	[46]
[Fe{η^5-C$_5$H$_4$-(E)-CH=CH-4-C$_6$H$_4$CH=CRuCl(dppm)$_2$}$_2$](PF$_6$)$_2$	383	-3000 ± 1200	2300 ± 800	3800 ± 1400	[46]
[Fe{η^5-C$_5$H$_4$-(E)-CH=CH-4-C$_6$H$_4$C≡CRuCl(dppm)$_2$}$_2$]	396	-7100 ± 3000	10600 ± 2000	13000 ± 3000	[46]

[a] All complexes are optically transparent at the fundamental frequency corresponding to the measurement wavelength of 800 nm. Measured in THF by the Z-scan technique. Results are referenced to the nonlinear refractive index of silica $n_2 = 3 \times 10^{-16}$ cm^2 W^{-1}.

Table 8. Molecular third-order NLO data for nickel alkynyl complexes[a]

Complex	λ_{max} (nm)	γ_{real} (10^{-36} esu)	γ_{imag} (10^{-36} esu)	Ref.
[Ni(C≡CPh)(PPh$_3$)(η^5-C$_5$H$_5$)]	307	15 ± 10	<10	[20]
[Ni(C≡C-4-C$_6$H$_4$NO$_2$)(PPh$_3$)(η-C$_5$H$_5$)]	439	−270 ± 100	70 ± 50	[20]
[Ni(C≡C-4-C$_6$H$_4$-4-C$_6$H$_4$NO$_2$)(PPh$_3$)(η^5-C$_5$H$_5$)]	413	−580 ± 200	300 ± 60	[20]
[Ni{C≡C-4-C$_6$H$_4$-(E)-CH=CH-4-C$_6$H$_4$NO$_2$}(PPh$_3$)(η^5-C$_5$H$_5$)]	437	−420 ± 100	480 ± 150	[20]
[Ni{C≡C-4-C$_6$H$_4$-(Z)-CH=CH-4-C$_6$H$_4$NO$_2$}(PPh$_3$)(η^5-C$_5$H$_5$)]	417	−230 ± 50	160 ± 80	[20]
[Ni(C≡C-4-C$_6$H$_4$C≡C-4-C$_6$H$_4$NO$_2$)(PPh$_3$)(η^5-C$_5$H$_5$)]	417	−640 ± 300	720 ± 300	[20]
[Ni(C≡C-4-C$_6$H$_4$N=CH-4-C$_6$H$_4$NO$_2$)(PPh$_3$)(η^5-C$_5$H$_5$)]	448	<120	360 ± 100	[20]

[a] All complexes are optically transparent at the fundamental frequency corresponding to the measurement wavelength of 800 nm. Measured in THF by the Z-scan technique. Results are referenced to the nonlinear refractive index of silica $n_2 = 3 \times 10^{-16}$ cm^2 W^{-1}.

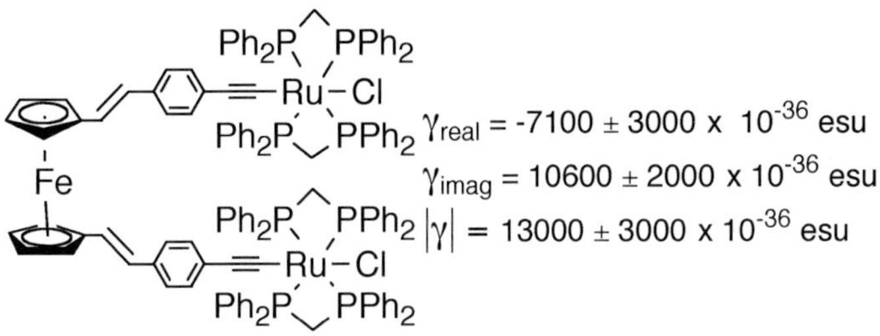

Figure 12. The most efficient iron-containing alkynyl complex in these third-order NLO studies

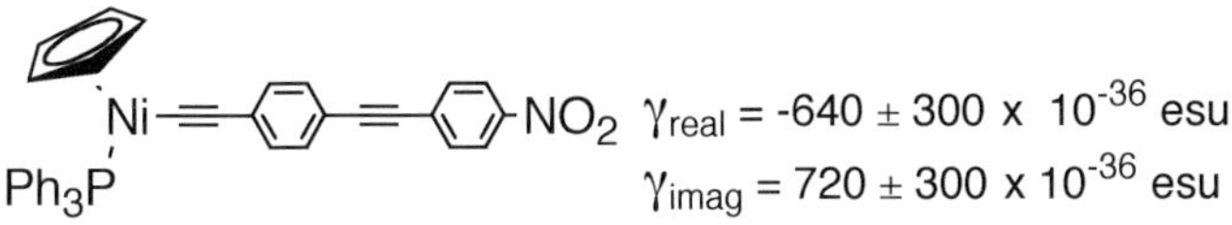

Figure 13. The most efficient nickel-containing complex from these studies

progressing from [Ni{C≡C-4-C_6H_4-(E)-CH=CH-4-$C_6H_4NO_2$}(PPh_3)(η^5-C_5H_5)] to [Ni{C≡C-4-C_6H_4-(Z)-CH=CH-4-$C_6H_4NO_2$}(PPh_3)(η^5-C_5H_5)], is suggested.

The linear optical absorption maxima for all gold complexes excepting the ferrocenyl-containing examples lie to higher energy of 2ω, reducing problems arising from two-photon states, and permitting comment on the effect of alkynyl ligand variation on cubic NLO merit. Third-order NLO data for gold complexes are listed in Table 9, and efficient examples depicted in Figure 14.

A number of conclusions from comparisons across these data can be made. Replacement of 4-H by 4-NO_2 in proceeding from [Au(C≡CPh)(PPh_3)] to [Au(C≡C-4-$C_6H_4NO_2$)(PPh_3)] results in a substantial increase in nonlinearity; unlike the related nitro-containing nickel alkynyl complexes above, for which negative nonlinearities were obtained, γ_{real} values for almost all nitro-containing gold alkynyl compounds considered here are positive, consistent with greatly diminished two-photon dispersion [it should be emphasized, though, that the presence of imaginary components in the γ values for most examples (associated with nonlinear absorption) suggests that electronic resonance enhancement still exists, though diminished with respect to the nickel complexes]. Chain-lengthening of the alkynyl chromophore leads to a dramatic increase in nonlinearity, with nonlinearities

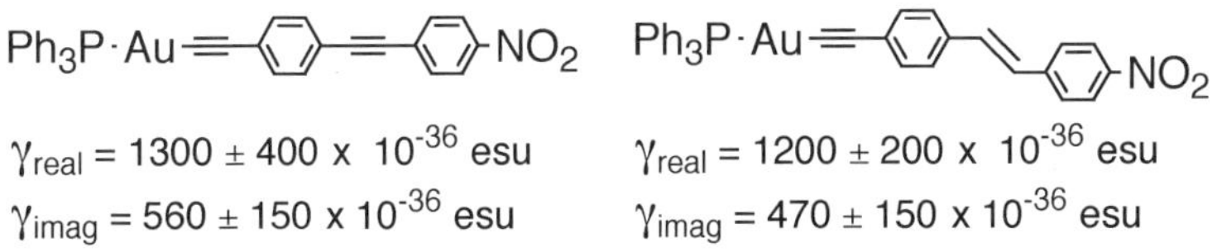

Figure 14. Efficient gold-containing complexes from these studies

Table 9. Molecular third-order NLO data for gold alkynyl complexes[a]

Complex	λ_{max} (nm)	Solvent	γ_{real} (10^{-36} esu)	γ_{imag} (10^{-36} esu)	$|\gamma|$(10^{-36} esu)	Ref.
[Au(C≡CPh)(PPh$_3$)]	296	THF	39 ± 20	–		[47, 39]
[Au(C≡C-4-C$_6$H$_4$C≡CPh)(PMe$_3$)]	335	THF	−200 ± 150	0 ± 50	200 ± 150	[44]
[Au(C≡C-4-C$_6$H$_4$C≡CPh)(PPh$_3$)]	336	THF	−900 ± 400	0 ± 100	900 ± 400	[44]
[Au{C≡C-4-C$_6$H$_4$-(E)-CH=CHPh}(PPh$_3$)]	338	THF	0 ± 300	0 ± 50	0	[44]
[Au(C≡C-4-C$_6$H$_4$CHO)(PMe$_3$)]	322	THF	35 ± 20	45 ± 30	60 ± 35	[25]
[Au(C≡C-4-C$_6$H$_4$CHO)(PPh$_3$)]	322	THF	300 ± 150	0	300 ± 150	[25]
[Au(C≡C-4-C$_6$H$_4${CHO[CH$_2$]$_3$O})(PPh$_3$)]	296	THF	210 ± 100	0	210 ± 100	[25]
PPN[Au(C≡C-4-C$_6$H$_4$NO$_2$)$_2$]	376	CH$_2$Cl$_2$	−800 ± 400	115 ± 50		[47]
NPr$_4$[Au(C≡C-4-C$_6$H$_4$NO$_2$)$_2$]	374	CH$_2$Cl$_2$	90 ± 150	190 ± 50		[47]
[Au(C≡C-4-C$_6$H$_4$NO$_2$)(CNBu^t)]	332	CH$_2$Cl$_2$	≤130	≤50		[47]
[Au(C≡C-4-C$_6$H$_4$-4-C$_6$H$_4$NO$_2$)(CNBu^t)]	343	CH$_2$Cl$_2$	20 ± 100	70 ± 50		[47]
[Au(C≡C-4-C$_6$H$_4$-4-C$_6$H$_4$NO$_2$)(PPh$_3$)]	338	THF	120 ± 40	20 ± 50		[47, 39]
[Au(C≡C-4-C$_6$H$_4$C≡C-4-C$_6$H$_4$NO$_2$)(PPh$_3$)]	350	THF	540 ± 150	120 ± 50		[47, 39]
[Au{C≡C-4-C$_6$H$_4$-(E)-CH=CH-4-C$_6$H$_4$NO$_2$}(PPh$_3$)]	362	THF	1300 ± 400	560 ± 150		[47, 39]
[Au{C≡C-4-C$_6$H$_4$-(Z)-CH=CH-4-C$_6$H$_4$NO$_2$}(PPh$_3$)]	386	THF	1200 ± 200	470 ± 150		[47, 39]
[Au{C≡C-4-C$_6$H$_4$-(E)-N=CH-4-C$_6$H$_4$NO$_2$}(PPh$_3$)]	362	THF	420 ± 150	92 ± 30		[47, 39]
[Au{C≡C-4-C$_6$H$_4$-(E)-CH=CH-4-C$_6$H$_4$NO$_2$}(CNBu^t)]	392	THF	130 ± 30	330 ± 60		[47, 39]
[Au{C≡C-4-C$_6$H$_4$-(E)-CH=CH-4-C$_6$H$_4$NO$_2$}{C(NHBu^t)(NEt$_2$)}]	381	CH$_2$Cl$_2$	390 ± 200	1050 ± 300		[47]
[Au{C≡C-4-C$_6$H$_4$-4-C$_6$H$_4$NO$_2$}{C(NHBu^t)(NEt$_2$)}]	354	CH$_2$Cl$_2$	10 ± 100	160 ± 40		[47]
[Au{C≡C-4-C$_6$H$_4$-(E)-CH=CH-4-C$_6$H$_4$NO$_2$}{C(NHBu^t)(NEt$_2$)}]	389	CH$_2$Cl$_2$	−200 ± 80	610 ± 200		[47, 39]
[Fe{η^5-C$_5$H$_4$-(E)-4-CH=CHC$_6$H$_4$C≡CAu(PCy$_3$)}$_2$]	468	THF	−400 ± 500	500 ± 100	640 ± 390	[46]
[Fe{η^5-C$_5$H$_4$-(E)-4-CH=CHC$_6$H$_4$C≡CAu(PPh$_3$)}$_2$]	465	THF	−1100 ± 300	300 ± 60	1140 ± 310	[46]
[Fe{η^5-C$_5$H$_4$-(E)-4-CH=CHC$_6$H$_4$C≡CAu(PMe$_3$)}$_2$]	463	THF	200 ± 150	0 ± 30	200 ± 150	[46]

[a] All complexes are optically transparent at the fundamental frequency corresponding to the measurement wavelength of 800 nm. Measured in THF by the Z-scan technique. Results are referenced to the nonlinear refractive index of silica $n_2 = 3 \times 10^{-16}$ cm^2 W^{-1}.

for [Au(C≡C-4-C$_6$H$_4$-4-C$_6$H$_4$NO$_2$)(PPh$_3$)], [Au{C≡C-4-C$_6$H$_4$-(E)-CH=CH-4-C$_6$H$_4$NO$_2$}(PPh$_3$)], [Au{C≡C-4-C$_6$H$_4$-(Z)-CH=CH-4-C$_6$H$_4$NO$_2$}(PPh$_3$)], and [Au(C≡C-4-C$_6$H$_4$C≡C-4-C$_6$H$_4$NO$_2$)(PPh$_3$)] significantly larger than that for [Au(C≡C-4-C$_6$H$_4$NO$_2$)(PPh$_3$)]. The present data permit discrimination of the merit of the linking unit: ene-linkage [Au{C≡C-4-C$_6$H$_4$-(E)-CH=CH-4-C$_6$H$_4$NO$_2$}(PPh$_3$)] and yne-linkage [Au(C≡C-4-C$_6$H$_4$C≡C-4-C$_6$H$_4$NO$_2$)(PPh$_3$)] are much more effective than biphenyl linkage [Au(C≡C-4-C$_6$H$_4$-4-C$_6$H$_4$NO$_2$)(PPh$_3$)], torsional effects at polyphenyl groups probably leading to difficulty in enforcing coplanarity of rings and efficient π-delocalization. The effect of bridging unit stereochemistry for the ene-linked complexes has also been probed: γ_{real} for the E complex [Au{C≡C-4-C$_6$H$_4$-(E)-CH=CH-4-C$_6$H$_4$NO$_2$}(PPh$_3$)] is about three times larger than that for the Z complex [Au{C≡C-4- C$_6$H$_4$-(Z)-CH=CH-4-C$_6$H$_4$NO$_2$}(PPh$_3$)].

Although the π-system is the same length for [Au{C≡C-4-C$_6$H$_4$-(E)-CH=CH-4-C$_6$H$_4$NO$_2$}(PPh$_3$)] and [Au{C≡C-4-C$_6$H$_4$-(Z)-CH=CH-4-C$_6$H$_4$NO$_2$}(PPh$_3$)], the E geometry leads to a greater charge separation for the important MLCT transition (difference in ground and excited state dipole moment is an important contributor to one of three terms influencing γ in the perturbation theory-derived three-level model [48–50], but validity of this model for organometallic complexes of the type considered here is unclear. Contributions from the oscillator strength (that of [Au{C≡C-4-C$_6$H$_4$-(E)-CH=CH-4-C$_6$H$_4$NO$_2$}(PPh$_3$)] is twice that of [Au{C≡C-4-C$_6$H$_4$-(Z)-CH=CH-4-C$_6$H$_4$NO$_2$}(PPh$_3$)]) may also be important]. The effect of bridging atom variation was also probed, replacing CH in [Au{C≡C-4-C$_6$H$_4$-(E)-CH=CH-4-C$_6$H$_4$NO$_2$}(PPh$_3$)] by N to afford [Au{C≡C-4-C$_6$H$_4$N=CH-4-C$_6$H$_4$NO$_2$}(PPh$_3$)]. Surprisingly, there is an order of magnitude difference in γ_{real} between [Au(C≡C-4-C$_6$H$_4$-(E)-CH=CH-4-C$_6$H$_4$NO$_2$)(PPh$_3$)] and [Au(C≡C-4-C$_6$H$_4$N=CH-4-C$_6$H$_4$NO$_2$)(PPh$_3$)], consistent with electronegative atoms in the π-system diminishing electron delocalization. There are broad similarities in relative efficiency for γ with those for β.

The effect on refractive nonlinearity γ_{real} of phosphine ligand replacement in the dipolar series [Au(C≡C-4-C$_6$H$_4$NO$_2$)(PCy$_3$)], [Au(C≡C-4-C$_6$H$_4$NO$_2$)(PPh$_3$)], and [Au(C≡C-4-C$_6$H$_4$NO$_2$)(PMe$_3$)] is negligible, all γ_{real} data being equivalent within the error margins; unlike [Au(C≡C-4-C$_6$H$_4$NO$_2$)(PPh$_3$)], no detectable γ_{imag} component is present for [Au(C≡C-4-C$_6$H$_4$NO$_2$)(PCy$_3$)] and [Au(C≡C-4-C$_6$H$_4$NO$_2$)(PMe$_3$)]. In contrast, replacing PMe$_3$ by PPh$_3$ in proceeding from [Au(C≡C-4-C$_6$H$_4$CHO)(PMe$_3$)] to [Au(C≡C-4-C$_6$H$_4$CHO)(PPh$_3$)] results in increased γ_{real} and $|\gamma|$.

By far the largest class of complexes we have examined as third-order NLO materials are ruthenium alkynyl and vinylidene compounds, the results being summarized in Table 10. The observed cubic responses are not simply the sums of nonlinearities for the molecular fragments; γ for [Ru(C≡C-4-C$_6$H$_4$NO$_2$)(PMe$_3$)$_2$(η^5-C$_5$H$_5$)] is much larger than that for [RuCl(PMe$_3$)$_2$(η^5-C$_5$H$_5$)] and 4-nitrophenylethyne ($\leq 80 \times 10^{-36}$ esu and 20×10^{-36} esu, respectively), indicating that electronic communication between the ligated metal and alkynyl fragments is important. Only one pair of

Table 10. Molecular third-order NLO data for ruthenium alkynyl and vinylidene complexes[a]

Complex	λ_{max} (nm)	Tech.	Solvent	γ_{real} (10^{-36} esu)	γ_{Imag} (10^{-36} esu)	$|\gamma|$ (10^{-36} esu)	Ref.
[Ru(C≡CPh)(PPh$_3$)$_2$(η^5-C$_5$H$_5$)]	310	Z-scan	THF	≤150			[32, 51]
[Ru(C≡C-4-C$_6$H$_4$Br)(PPh$_3$)$_2$(η^5-C$_5$H$_5$)]	325	Z-scan	THF	≤150			[32, 51]
[Ru(C≡C-4-C$_6$H$_4$CH{OC(O)Me}$_2$)(PPh$_3$)$_2$(η^5-C$_5$H$_5$)]	326	Z-scan	THF	100±100	0	100±100	[25]
[Ru(C≡C-4-C$_6$H$_4$CHO)(PPh$_3$)$_2$(η^5-C$_5$H$_5$)]	400	Z-scan	THF	−75±50	210±50	220±60	[25]
[Ru(C≡C-4-C$_6$H$_4$NO$_2$)(PPh$_3$)$_2$(η^5-C$_5$H$_5$)]	460	DFWM	THF	−260±60			[32, 51]
[Ru(C≡C-4-C$_6$H$_4$NO$_2$)(PMe$_3$)$_2$(η^5-C$_5$H$_5$)]	477	Z-scan	THF	−210±50	≤10		[32, 51]
				−230±70	74±30		
[Ru(C≡C-4-C$_6$H$_4$-4-C$_6$H$_4$NO$_2$)(PPh$_3$)$_2$(η^5-C$_5$H$_5$)]	448	Z-scan	THF	−380±200	320±160		[32]
[Ru{C≡C-4-C$_6$H$_4$-(E)-CH=CH-4-C$_6$H$_4$NO$_2$}(PPh$_3$)$_2$(η^5-C$_5$H$_5$)]	476	Z-scan	THF	−450±100	210±100		[32, 51]
[Ru(C≡C-4-C$_6$H$_4$C≡C-4-C$_6$H$_4$NO$_2$)(PPh$_3$)$_2$(η^5-C$_5$H$_5$)]	446	Z-scan	THF	−450±100	≤20		[32, 51]
[Ru(C≡C-4-C$_6$H$_4$N=CH-4-C$_6$H$_4$NO$_2$)(PPh$_3$)$_2$(η^5-C$_5$H$_5$)]	496	Z-scan	THF	−850±300	360±200		[32]
trans-[Ru(C≡CHPh)Cl(dppm)$_2$](PF$_6$)	833	Z-scan	CH$_2$Cl$_2$	1300±500	−2200±1000	2600±1000	[52, 53]
	320	Z-scan	THF	<440	<50	<440	[35]
trans-[Ru(C≡CPh)Cl(dppm)$_2$]	318	Z-scan	CH$_2$Cl$_2$	<300	<200	≈0	[52, 53]
	308	Z-scan	THF	<120	0	<120	[35]
trans-[Ru(C≡CH-4-C$_6$H$_4$C≡CPh)Cl(dppm)$_2$](PF$_6$)	380	Z-scan	THF	<500	0	<500	[35]
trans-[Ru(C≡C-4-C$_6$H$_4$C≡CPh)Cl(dppm)$_2$]	381	Z-scan	THF	65±40	520±200	520±200	[35]
trans-[Ru(4-C≡CHC$_6$H$_4$NO$_2$)Cl(dppm)$_2$](PF$_6$)	470	Z-scan	THF	<50	<30	<50	[35]
trans-[Ru(C≡C-4-C$_6$H$_4$NO$_2$)Cl(dppm)$_2$]	466[b]	Z-scan	CH$_2$Cl$_2$	170±34	230±46	290±60	[54]
	448[b]	Z-scan	CH$_2$Cl$_2$	140±28	64±13		[54]
trans-[Ru(C≡CH-4-C$_6$H$_4$C≡C-4-C$_6$H$_4$NO$_2$)Cl(dppm)$_2$](PF$_6$)	326	Z-scan	THF	<500	420±60	420±60	[35]
trans-[Ru(C≡C-4-C$_6$H$_4$C≡C-4-C$_6$H$_4$NO$_2$)Cl(dppm)$_2$]	464	Z-scan	THF	−160±80	160±60	230±100	[35]
trans-[Ru(C≡C-4-C$_6$H$_4$C≡C-4-C$_6$H$_4$NO$_2$)Cl(dppm)$_2$]	439	Z-scan	THF	−920±200	970±200	1300±300	[35]
trans-[Ru{C≡C-4-C$_6$H$_4$-(E)-CH=CH-4-C$_6$H$_4$NO$_2$}Cl(dppm)$_2$]	471[b]	Z-scan	CH$_2$Cl$_2$	200±40	1100±220	1100±220	[54]
trans-[Ru(C≡C-4-C$_6$H$_4$NO$_2$)$_2$(dppm)$_2$]	474[b]	Z-scan	CH$_2$Cl$_2$	300±60	490±98		[54]

(Continued)

Table 10. (Continued)

Complex	λ_{max} (nm)	Tech.	Solvent	γ_{real} (10^{-36} esu)	γ_{imag} (10^{-36} esu)	$|\gamma|$ (10^{-36} esu)	Ref.
trans-[Ru{C≡C-4-C$_6$H$_4$-4-C$_6$H$_4$NO$_2$}$_2$(dppm)$_2$]	453[b]	Z-scan	CH$_2$Cl$_2$	≤800	2500±500		[54]
trans-[Ru{C≡C-4-C$_6$H$_4$-(E)-CH=CH-4-C$_6$H$_4$NO$_2$}$_2$(dppm)$_2$]	367[b]	Z-scan	CH$_2$Cl$_2$	≤1100	3400±680		[54]
trans-[Ru{C≡CH-4-C$_6$H$_4$CHO(CH$_2$)$_3$O}Cl(dppm)$_2$](PF$_6$)	317	Z-scan	THF	75±75	0	75±75	[25]
trans-[Ru{C≡C-4-C$_6$H$_4$CHO(CH$_2$)$_3$O}Cl(dppm)$_2$]	320	Z-scan	THF	50±50	0	50±50	[25]
trans-[Ru{C≡CH-2-C$_6$H$_4$CHO}Cl(dppm)$_2$](PF$_6$)	555	Z-scan	THF	450±150	150±60	470±160	[25]
trans-[Ru{C≡CH-3-C$_6$H$_4$CHO}Cl(dppm)$_2$](PF$_6$)	320	Z-scan	THF	200±200	0	200±200	[25]
trans-[Ru{C≡C-3-C$_6$H$_4$CHO}Cl(dppm)$_2$]	321	Z-scan	THF	150±150	0	150±150	[25]
trans-[Ru{C≡CH-4-C$_6$H$_4$CHO}Cl(dppm)$_2$](PF$_6$)	403	Z-scan	THF	0	<20	<20	[35]
trans-[Ru{C≡C-4-C$_6$H$_4$CHO)Cl(dppm)$_2$]	405	Z-scan	THF	<120	210±60	210±60	[35]
[Fe{η^5-C$_5$H$_4$-(E)-4-CH=CHC$_6$H$_4$CH=CHC$_6$H$_4$CH=CRuCl(dppm)$_2$}$_2$][PF$_6$]$_2$	383	Z-scan	THF	-3000±1200	2300±800	3800±1400	[46]
[Fe{η^5-C$_5$H$_4$-(E)-4-CH=CHC$_6$H$_4$C≡CRuCl(dppm)$_2$}$_2$]	396	Z-scan	THF	-7100±3000	10600±2000	13000±3000	[46]
trans-[Ru(C≡CHPh)Cl(dppe)$_2$](PF$_6$)	317	Z-scan	THF	380±400	<50	380±400	[35]
trans-[Ru(C≡CPh)Cl(dppe)$_2$]	319	Z-scan	THF	-170±40	71±20	18±45	[11]
trans-[Ru{C≡CH-4-C$_6$H$_4$C≡CPh}Cl(dppe)$_2$](PF$_6$)	893	Z-scan	CH$_2$Cl$_2$	2900±1000	-1200±600	3100±1000	[52, 53]
trans-[Ru{C≡C-4-C$_6$H$_4$C≡CPh}Cl(dppe)$_2$]	387	Z-scan	CH$_2$Cl$_2$	-100±100	450±200	460±200	[52, 53]
trans-[Ru{C≡CPh)(C≡C-4-C$_6$H$_4$C≡CPh)(dppe)$_2$]	383	Z-scan	CH$_2$Cl$_2$	-670±300	1300±300	1500±500	[52, 53]
trans-[Ru{C≡CH-4-C$_6$H$_4$CHO)Cl(dppe)$_2$](PF$_6$)	412	Z-scan	THF	<260	0	<260	[35]
trans-[Ru{C≡C-4-C$_6$H$_4$CHO)Cl(dppe)$_2$]	413	Z-scan	THF	-300±500	<200	300±500	[35]
trans-[Ru{C≡CH-4-C$_6$H$_4$NO$_2$)Cl(dppe)$_2$](PF$_6$)	476	Z-scan	THF	250±300	<50	250±300	[35]
trans-[Ru{C≡C-4-C$_6$H$_4$NO$_2$)Cl(dppe)$_2$]	477	Z-scan	THF	320±55	<50	320±55	[35]
trans-[Ru{C≡CH-4-C$_6$H$_4$-(E)-CH=CH-4-C$_6$H$_4$NO$_2$}Cl(dppe)$_2$]PF$_6$	473	Z-scan	THF	650±500	<50	650±500	[35]
trans-[Ru{C≡C-4-C$_6$H$_4$-(E)-CH=CH-4-C$_6$H$_4$NO$_2$}Cl(dppe)$_2$]	489	Z-scan	THF	40±200	<100	40±200	[35]
trans-[Ru{C≡CPh)(C≡C-4-C$_6$H$_4$C≡CPh)(dppe)$_2$]	383	Z-scan	CH$_2$Cl$_2$	-670±300	1300±300	1500±500	[37]
trans-[Ru{C≡C-4-C$_6$H$_4$-(E)-CH=CHPh)Cl(dppe)$_2$]	404	Z-scan	THF	300±400	300±100	420±350	[44]

(Continued)

Table 10. (Continued)

Complex	λ_{max} (nm)	Tech.	Solvent	γ_{real} (10^{-36} esu)	γ_{imag} (10^{-36} esu)	$|\gamma|$ (10^{-36} esu)	Ref.
[1,3,5-(trans-[(dppm)$_2$ClRu{C≡CH-4-C$_6$H$_4$-(E)-CH=CH}])$_3$C$_6$H$_3$][PF$_6$)$_3$	396	Z-scan	THF	−900±500	700±400	1100±700	[43, 44]
[1,3,5-(trans-[(dppm)$_2$ClRu{C≡C-4-C$_6$H$_4$-(E)-CH=CH}])$_3$C$_6$H$_3$]	415	Z-scan	THF	−640±500	2000±500	2100±600	[43, 44]
[1,3,5-(trans-[(dppe)$_2$ClRu{C≡C-4-C$_6$H$_4$C≡C}]$_3$C$_6$H$_3$](PF$_6$)$_3$	893	Z-scan / DFWM	CH$_2$Cl$_2$ / CH$_2$Cl$_2$	13500±3000	−4700±500	14000±3000 / 2000	[52] [55]
[1,3,5-(trans-[(dppe)$_2$ClRu{C≡C-4-C$_6$H$_4$C≡C}]$_3$)C$_6$H$_3$]	414	Z-scan / DFWM	CH$_2$Cl$_2$ / CH$_2$Cl$_2$	−330±100	2200±500	22000±600 / 2000	[37, 52] [55]
[1,3,5-{trans-[Ru(C≡CPh)(C≡C-4-C$_6$H$_4$C≡C)(dppe)$_2$]}$_3$C$_6$H$_3$]	411	Z-scan	THF	−600±200	2900±500	3000±600	[37]
[1-(Me$_3$SiC≡C)C$_6$H$_3$-3,5-{C≡C-4-C$_6$H$_4$C≡C-trans-[RuCl(dppe)$_2$]}$_2$]	411	Z-scan	THF	−510±500	4700±1500	4700±2000	[56]
[1-(Me$_3$SiC≡C)C$_6$H$_3$-3,5-{C≡C-4-C$_6$H$_4$C≡C-trans-[Ru(C≡CPh)(dppe)$_2$]}$_2$]	407	Z-scan	THF	−700±100	2270±300	2400±300	[56]
[1-(HC≡C)C$_6$H$_3$-3,5-{C≡C-4-C$_6$H$_4$C≡C-trans-[Ru(C≡CPh)(dppe)$_2$]}$_2$]	408	Z-scan	THF	−830±100	2200±300	2400±300	[56]
1,3,5-C$_6$H$_3$(C≡C-4-C$_6$H$_4$C≡C-trans-[Ru(dppe)$_2$]C≡C-3,5-C$_6$H$_3$-1-(C≡C-4-C$_6$H$_4$C≡C-trans-[Ru(C≡CPh)(dppe)$_2$]}$_2$)$_3$	402	Z-scan	THF	−5050±500	20100±2000	20700±2000	[56]
[1,3,5-(trans-[(dppe)$_2$ClRu{C≡C-4-C$_6$H$_4$-(E)-CH=CH}])$_3$C$_6$H$_3$]	426	Z-scan	THF	−4600±2000	4200±800	6200±2000	[43, 44]
[1,3,5-(trans-[(dppe)$_2$(PhC≡C)Ru{C≡C-4-C$_6$H$_4$-(E)-CH=CH}])$_3$C$_6$H$_3$]	421	Z-scan	THF	−11200±3000	8600±2000	14000±4000	[43, 44]

[a] All complexes are optically transparent at the fundamental frequency corresponding to the measurement wavelength of 800 nm. Measured in THF by the Z-scan technique. Results are referenced to the nonlinear refractive index of silica $n_2 = 3 \times 10^{-16}$ cm^2 W^{-1}.

complexes varying in trialkyl-vs triaryl-phosphine ([Ru(C≡C-4-$C_6H_4NO_2$)(PPh$_3$)$_2$(η^5-C_5H_5)], [Ru(C≡C-4-$C_6H_4NO_2$)(PMe$_3$)$_2$(η^5-C_5H_5)]) give sufficiently large nonlinearities to permit comparison. Proceeding from [Ru(C≡C-4-$C_6H_4NO_2$)((PPh$_3$)$_2\eta^5$-C_5H_5)] to [Ru(C≡C-4-$C_6H_4NO_2$)(PMe$_3$)$_2$(η^5-C_5H_5)] makes little difference to γ_{real}, although a 50% decrease in response was noted with the corresponding β_{HRS} values. Minor variation in the alkynyl ligand (replacement of 4-H by 4-Br in proceeding from [Ru(C≡CPh)(PPh$_3$)$_2$(η^5-C_5H_5)] to [Ru(C≡C-4-C_6H_4Br)(PPh$_3$)$_2$(η^5-C_5H_5)]) has no discernible effect on γ_{real}.

However, replacement of H by the strongly-withdrawing NO$_2$ ([Ru(C≡CPh)(PPh$_3$)$_2$(η^5-C_5H_5)] to [Ru(C≡C-4-$C_6H_4NO_2$)(PPh$_3$)$_2$(η^5-C_5H_5)]) makes a significant difference to the cubic nonlinearity; a similar increase was observed in proceeding from styrene to its 4-nitro derivative.[57] Not surprisingly, extension from a one-ring chromophore [Ru(C≡C-4-$C_6H_4NO_2$)(PPh$_3$)$_2$(η^5-C_5H_5)] to "extended-chain" two-ring chromophores [Ru(C≡C-4-C_6H_4-4-$C_6H_4NO_2$)(PPh$_3$)$_2$(η^5-C_5H_5)], [Ru{C≡C-4-C_6H_4-(E)-CH=CH-4-$C_6H_4NO_2$}(PPh$_3$)$_2$(η^5-C_5H_5)], and [Ru(C≡C-4-C_6H_4C≡C-4-$C_6H_4NO_2$)(PPh$_3$)$_2$(η^5-C_5H_5)] leads to a large increase in γ_{real}. Complexes [Ru(C≡CPh)(PPh$_3$)$_2$(η^5-C_5H_5)] and [Ru(C≡C-4-C_6H_4Br)(PPh$_3$)$_2$(η^5-C_5H_5)] have positive γ_{real}, whereas the nitro-containing cyclopentadienylruthenium complexes incorporating a strong donor-acceptor interaction have negative γ_{real}. It was of interest to ascertain the origin of the negative responses. Investigations on [Ru(C≡C-4-$C_6H_4NO_2$)(PPh$_3$)$_2$(η^5-C_5H_5)] and [Ru(C≡C-4-$C_6H_4NO_2$)(PMe$_3$)$_2$(η^5-C_5H_5)] by femtosecond time-resolved DFWM confirm that the negative γ_{real} are not due to thermal lensing; the observed response of solutions shows a concentration dependence characteristic for a negative real part of γ of the solute while the DFWM signal retains its femtosecond response. The observation of a negative real component of γ together with a complex component (as evident for [Ru(C≡C-4-$C_6H_4NO_2$)(PMe$_3$)$_2$(η^5-C_5H_5)] and [Ru{C≡C-4-C_6H_4-(E)-CH=CH-4-$C_6H_4NO_2$}(PPh$_3$)$_2$(η^5-C_5H_5)]) is suggestive of electronic resonance enhancement, with the imaginary part relating to nonlinear absorption. It is perhaps significant that negative γ is observed for complexes with λ_{max} longer than 400 nm, and positive γ for complexes with λ_{max} shorter than 400 nm, consistent with the dispersion effect of two-photon states contributing to the observed responses.

Nonlinearities of the bis(bidentate phosphine)ruthenium vinylidene and alkynyl complexes are characterized by large error margins in many instances, rendering extraction of structure-property relationships difficult, and negative real components and significant imaginary components for many complexes, indicative of two-photon resonance effects. Nevertheless, as observed with β_{exp} and β_0 trends, the effect of chain lengthening on $|\gamma|$ is insignificant within error margins on proceeding from trans-[Ru(C≡C-4-$C_6H_4NO_2$)Cl(dppm)$_2$] to trans-[Ru(C≡C-4-C_6H_4C≡C-4-$C_6H_4NO_2$)Cl(dppm)$_2$], but there is a dramatic increase in $|\gamma|$ in proceeding to trans-[Ru(C≡C-4-C_6H_4C≡C-4-C_6H_4C≡C-4-$C_6H_4NO_2$)Cl(dppm)$_2$]. The vinylidene/alkynyl pair trans-[Ru(C=CH-4-$C_6H_4NO_2$)Cl(dppm)$_2$](PF$_6$) and trans-[Ru(C≡C-4-$C_6H_4NO_2$)Cl(dppm)$_2$] have significantly different γ_{imag} values.

Since γ_{imag} is related to the two-photon absorption (TPA) cross-section σ_2, the significant variation in γ_{imag} values for this pair provides protically-switchable materials in which the TPA response can be alternatively switched "on" and "off". Replacing NO_2 in the dipolar examples with *trans*-[(C≡C)RuCl(dppm)$_2$] to afford *trans*, *trans*-[RuCl(dppm)$_2$(C≡C-4-C$_6$H$_4$C≡C)RuCl(dppm)$_2$] or *trans*, *trans*-[RuCl(dppm)$_2$(C≡C-4-C$_6$H$_4$-4-C$_6$H$_4$C≡C)RuCl(dppm)$_2$] results in significant increases in $|\gamma|$, the presence of the second electron-rich metal centre being more important than dipolar composition in enhancing cubic NLO merit. These data suggest that extending π-delocalization is the critical factor. Significant extension of the π-system, in proceeding from *trans*, *trans*-[RuCl(dppm)$_2$(C≡C-4-C$_6$H$_4$C≡C)RuCl(dppm)$_2$] and *trans*, *trans*-[RuCl(dppm)$_2$(μ-C≡C-4-C$_6$H$_4$-4-C$_6$H$_4$C≡C)RuCl(dppm)$_2$] to [Fe{η^5-C$_5$H$_4$-(*E*)-CH=CH-4-C$_6$H$_4$C≡CRuCl(dppm)$_2$}$_2$], results in a further considerable increase in $|\gamma|$.

The dendritic ruthenium alkynyl complexes have very large cubic NLO coefficients. Inspection of γ values for *trans*-[Ru(C≡CC$_6$H$_4$-4-C≡CPh(C≡CPh)(dppe)$_2$] and [1, 3, 5-C$_6$H$_3$-*trans*-{C≡C-4-C≡CC$_6$H$_4$C≡C[Ru(C≡CPh)(dppe)$_2$]}$_3$]

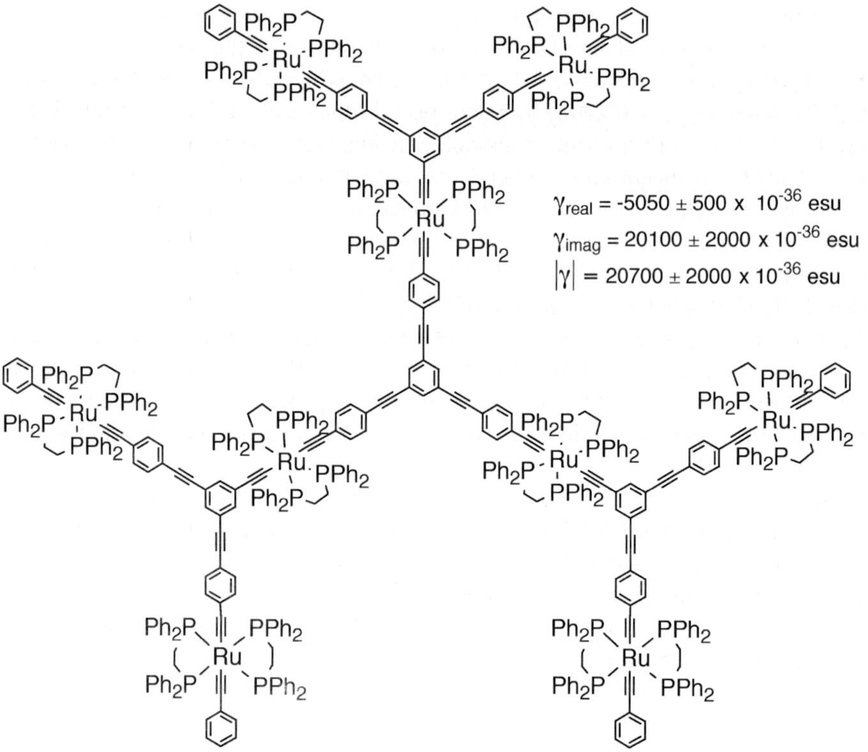

γ_{real} = -5050 ± 500 x 10^{-36} esu
γ_{imag} = 20100 ± 2000 x 10^{-36} esu
$|\gamma|$ = 20700 ± 2000 x 10^{-36} esu

Figure 15. Alkynylruthenium dendrimer from these studies

NLO Properties of Metal Alkynyl and Related Complexes

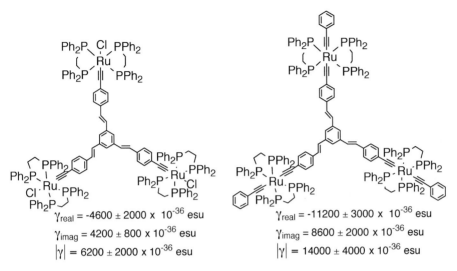

Figure 16. Octupolar alkynylruthenium complexes incorporating phenylenevinylene groups from these studies

reveals a significant increase in the imaginary component on progressing from the linear to the multipolar complex, but no increase in γ_{real}. Both real and imaginary components of the third-order hyperpolarizability for the dendrimer [1, 3, 5-C_6H_3- trans- {C≡C-4-C≡CC$_6$H$_4$C≡C[Ru(dppe)$_2$]C≡C-3,5-C_6H_3-trans-{C≡C-4-C≡CC$_6$H$_4$C≡C[Ru(C≡CPh)(dppe)$_2$]}$_2$}$_3$)] (Figure 15) are much larger than those of its components [HC≡C-3, 5-C_6H_3-trans-{C≡C-4-C≡CC$_6$H$_4$C≡C[Ru(C≡CPh)(dppe)$_2$]}$_2$] and [1, 3, 5-C_6H_3-trans-{C≡C-4-C≡CC$_6$H$_4$C≡C[RuCl(dppe)$_2$]}$_3$] or the related complex 1, 3, 5-C_6H_3-trans-{C≡C-4-C≡CC$_6$H$_4$C≡C[Ru(C≡CPh)(dppe)$_2$]}$_3$]. In particular, progressing from [1, 3, 5-C_6H_3-trans-{C≡C-4-C≡CC$_6$H$_4$C≡C[Ru(C≡CPh)(dppe)$_2$]}$_3$] to [1, 3, 5-C_6H_3-trans-{C≡C-4-C≡CC$_6$H$_4$C≡C[Ru(dppe)$_2$]C≡C-3, 5-C_6H_3-trans-{C≡C-4-C≡CC$_6$H$_4$C≡C[Ru(C≡CPh)(dppe)$_2$]}$_2$}$_3$)] results in increases in both γ_{real} and γ_{imag} proportionately greater than either the increase in the number of phenylethynyl groups or the extinction coefficient.

Comparison of the complexes [1, 3, 5-(trans-[(dppe)$_2$ClRu{C≡C-4-C_6H_4-(E)-CH=CH}])$_3$C$_6$H$_3$] and [1, 3, 5-(trans-[(dppe)$_2$(PhC≡C)Ru{C≡C-4-C_6H_4-(E)-CH=CH}])$_3$C$_6$H$_3$] (Figure 16) with the analogous alkynyl complexes above demonstrates enhancement of third-order NLO properties by replacement of acetylene linkages with (E)-ene linkages.

Two-photon absorption cross-sections have been calculated, selected examples of which are listed in Table 11. The σ_2 values increase upon increase in dendrimer size and π-system, with the value for the dendrimer depicted in Figure 15 amongst the largest for organometallic compounds.

Table 11. Two-photon absorption cross-sections of selected alkynylruthenium complexes from these studies

Complex	λ_{max} (nm)	σ_2 (10^{-50} cm^4 s)[b]	Ref.
[1,3,5-{trans-[Ru(C≡CPh)(C≡C-4-C$_6$H$_4$C≡C)(dppe)$_2$]}$_3$C$_6$H$_3$]	411	700 ± 120	[37]
[1-(Me$_3$SiC≡C)C$_6$H$_3$-3,5-{C≡C-4-C$_6$H$_4$C≡C-trans-[RuCl(dppe)$_2$]}$_2$]	411	1100 ± 360	[56]
[1-(Me$_3$SiC≡C)C$_6$H$_3$-3,5-{C≡C-4-C$_6$H$_4$C≡C-trans-[Ru(C≡CPh)(dppe)$_2$]}$_2$]	407	550 ± 70	[56]
[1-(HC≡C)C$_6$H$_3$-3,5-{C≡C-4-C$_6$H$_4$C≡C-trans-[Ru(C≡CPh)(dppe)$_2$]}$_2$]	408	530 ± 70	[56]
1,3,5-C$_6$H$_3$(C≡C-4-C$_6$H$_4$C≡C-trans-[Ru(dppe)$_2$]C≡C-3,5-C$_6$H$_3$-{C≡C-4-C$_6$H$_4$C≡C-trans-[Ru(C≡CPh)(dppe)$_2$]}$_2$)$_3$	402	4800 ± 500	[56]

[a] All complexes are optically transparent at the fundamental frequency corresponding to the measurement wavelength of 800 nm. Measured in THF by the Z-scan technique. Results are referenced to the nonlinear refractive index of silica $n_2 = 3 \times 10^{-16}$ cm^2 W^{-1}.
[b] Calculated using the equation $\sigma_2 = \hbar\omega\beta/2\pi N$, where β is the two-photon absorption coefficient [58].

4. CONCLUSION

The studies described above have resulted in identification of the key structural components for quadratic NLO response in alkynylmetal complexes and the identification of compounds with some of the largest cubic NLO coefficients and particularly TPA cross-sections thus far. Quadratic hyperpolarizabilities increase upon similar structural modifications to the alkynyl ligands as those that increase NLO response in organic chromophores but, significantly, also upon increasing metal valence electron count, ease of oxidation, and ligand substitution at the metal centre. Cubic hyperpolarizabilities increase substantially upon increasing π-system length and "dimensionality" (progressing to octupolar complexes and dendrimers). Recently, attention has turned to reversibly "switching" such responses using a variety of stimuli, and this Chapter contains examples of protic switching with reversible interconversion of alkynyl and vinylidene ligands. The ruthenium complexes also possess readily accessible and fully reversible $Ru^{II/III}$ redox couples; our experiments demonstrating electrochemically induced NLO "switching" of these complexes are summarized in the Chapter by Coe.

REFERENCES

[1] Ashitaka, H.: Organometallic News 81–86 (1991)
[2] Di Bella, S.: Chem. Soc. Rev. **30**, 355–366 (2001)
[3] Humphrey, M.G.: Gold Bulletin **33**, 97–102 (2000)
[4] Long, N.J.: In Fackler, J.P., Roundhill, D.M. (eds) Optoelectronic Properties of Inorganic Compounds, pp 107–167, Plenum, New York (1999)
[5] Marder, S.R.: In Bruce, D.W., O'Hare, D. (eds) Inorg. Mater., pp 115–164, Wiley Sussex, England (1992)
[6] Matsubayashi, G.: Organometallic News 80–83 (1992)
[7] Meng, X.R., Zhao, J.A., Hou, H.W., Mi, L.W.: Chin. J. Inorg. Chem. **19**, 15–19 (2003)
[8] Nalwa, H.S.: Appl. Organomet. Chem. **5**, 349–377 (1991)
[9] Uno, M.: Recent trends, Organometallic News 42–47 (1995)
[10] Whittall, I.R., McDonagh, A.M., Humphrey, M.G., Samoc, M.: In Stone, F.G.A. (ed.) Advances in Organometallic Chemistry, vol. 42, pp 291–362, Harcourt Publishers Ltd, (1998)
[11] Whittall, I.R., McDonagh, A.M., Humphrey, M.G., Samoc, M.: In Stone, F.G.A., Hill, A.F., West, R. (eds.) Advances in Organometallic Chemistry, vol. 43, pp 349–405, Academic press (1999)
[12] Powell, C.E., Humphrey, M.G.: Coord. Chem. Rev. **248**, 725–756 (2004)
[13] Cifuentes, M.P., Humphrey, M.G.: J. Organomet. Chem. **689**, 3968–3981 (2004)
[14] Humphrey, M.G., Mingos, D.M.: J. Organomet. Chem., vol. 670, Elsevier, London, UK (2004)
[15] Marder, S.R., Beratan, D.N., Tiemann, B.G., Cheng, L.T., Tam, W.: Structure/property relationships for organic and organometallic materials with second-order optical nonlinearities, Spec. Publ. – R.S.C. **91**, 165–175 (1991)
[16] Zyss, J., Dhenaut, C., Chauvan, T., Ledoux, I.: Chem. Phys. Lett. **206**, 409–414 (1993)
[17] Powell, C.E., Cifuentes, M.P., McDonagh, A.M., Hurst, S.K., Lucas, N.T., Delfs, C.D., Stranger, R., Humphrey, M.G., Houbrechts, S., Asselberghs, I., Persoons, A., Hockless, D.C.R.: Inorg. Chim. Acta **352**, 9–18 (2003)
[18] McDonagh, A.M., Cifuentes, M.P., Humphrey, M.G., Houbrechts, S., Maes, J., Persoons, A., Samoc, M., Luther-Davies, B.: J. Organomet. Chem. **610**, 71–79 (2000)

[19] Cheng, L.-T., Tam, W., Stevenson, S., Meredith, G.R., Rikken, G., Marder, S.R.: J. Phys. Chem. **95**, 10631–10652 (1991)
[20] Whittall, I.R., Cifuentes, M.P., Humphrey, M.G., Luther-Davies, B., Samoc, M., Houbrechts, S., Persoons, A., Heath, G.A., Bogsanyi, D.: Organometallics **16**, 2631–2637 (1997)
[21] Whittall, I.R., Humphrey, M.G., Houbrechts, S., Maes, J., Persoons, A., Schmid, S., Hockless, D.C.R.: J. Organomet. Chem. **544**, 277–283 (1997)
[22] Naulty, R.H., Cifuentes, M.P., Humphrey, M.G., Houbrechts, S., Boutton, C., Persoons, A., Heath, G.A., Hockless, D.C.R., Luther-Davies, B., Samoc, M.: J. Chem. Soc., Dalton Trans. 4167–4174 (1997)
[23] Whittall, I.R., Humphrey, M.G., Houbrechts, S., Persoons, A., Hockless, D.C.R.: Organometallics **15**, 5738–5745 (1996)
[24] Hurst, S.K., Lucas, N.T., Humphrey, M.G., Asselberghs, I., Van Boxel, R., Persoons, A.: Aust. J. Chem. **54**, 447–451 (2001)
[25] Hurst, S.K., Lucas, N.T., Cifuentes, M.P., Humphrey, M.G., Samoc, M., Luther-Davies, B., Asselberghs, I., Van Boxel, R., Persoons, A.: J. Organomet. Chem. **633**, 114–124 (2001)
[26] Hurst, S.K., Cifuentes, M.P., McDonagh, A.M., Humphrey, M.G., Samoc, M., Luther-Davies, B., Asselberghs, I., Persoons, A.: J. Organomet. Chem. **642**, 259–267 (2002)
[27] McDonagh, A.M., Lucas, N.T., Cifuentes, M.P., Humphrey, M.G., Houbrechts, S., Persoons, A.: J. Organomet. Chem. **605**, 193–201 (2000)
[28] Cheng, L.-T., Tam, W., Marder, S.R., Stiegman, A.E., Rikken, G., Spangler, C.W.: J. Phys. Chem. **95**, 10643–10652 (1991)
[29] Muller, T.E., Wing-Kin Choi, S., Mingos, D.M., Murphy, D., Williams, D.J., Wing-Wah Yam, V.: J. Organomet. Chem. **484**, 209–224 (1994)
[30] Yam, V.W.W., Choi, S.W., Cheung, K.: Organometallics **15**, 1734–1739 (1996)
[31] Di Bella, S., Fragala, I., Ledoux, I., Marks, T.J.: J. Am. Chem. Soc. **116**, 10089–10085 (1995)
[32] Whittall, I.R., Cifuentes, M.P., Humphrey, M.G., Luther-Davies, B., Samoc, M., Houbrechts, S., Persoons, A., Heath, G.A., Hockless, D.C.R.: J. Organomet. Chem. **549**, 127–137 (1997)
[33] Whittall, I.R., Humphrey, M.G., Persoons, A., Houbrechts, S.: Organometallics **15**, 1935–1941 (1996)
[34] McDonagh, A.M., Cifuentes, M.P., Lucas, N.T., Humphrey, M.G., Houbrechts, S., Persoons, A.: J. Organomet. Chem. **605**, 184–192 (2000)
[35] Hurst, S.K., Cifuentes, M.P., Morrall, J.P.L., Lucas, N.T., Whittall, I.R., Humphrey, M.G., Asselberghs, I., Persoons, A., Samoc, M., Luther-Davies, B., Willis, A.C.: Organometallics **20**, 4664–4675 (2001)
[36] Naulty, R.H., McDonagh, A.M., Whittall, I.R., Cifuentes, M.P., Humphrey, M.G., Houbrechts, S., Maes, J., Persoons, A., Heath, G.A., Hockless, D.C.R.: J. Organomet. Chem. **563**, 137–146 (1998)
[37] McDonagh, A.M., Humphrey, M.G., Samoc, M., Luther-Davies, B., Houbrechts, S., Wada, T., Sasabe, H., Persoons, A.: J. Am. Chem. Soc. **121**, 1405–1406 (1999)
[38] Kanis, D.R., Ratner, M., Marks, T.J.: Chem. Rev. **94**, 195–242 (1994)
[39] Whittall, I.R., Humphrey, M.G., Samoc, M., Luther-Davies, B.: Ange. Chem.-Int. Ed. Engl. **36**, 370–371 (1997)
[40] Marder, S.R., Cheng, L.-T., Tiemann, B.G.: J. Chem. Soc., Chem. Commun, 672–674 (1992)
[41] McGrady, J.E., Lovell, T., Stranger, R., Humphrey, M.G.: Organometallics **16**, 4004–4011 (1997)
[42] Coe, B.J., Harris, J.A., Harrington, L.J., Jeffery, J.C., Rees, L.H., Houbrechts, S., Persoons, A.: Inorg. Chem. **37**, 3391–3399 (1998)
[43] Hurst, S.K., Humphrey, M.G., Isoshima, T., Wostyn, K., Asselberghs, I., Clays, K., Persoons, A., Samoc, M., Luther-Davies, B.: Organometallics **21**, 2024–2026 (2002)
[44] Hurst, S.K., Lucas, N.T., Humphrey, M.G., Isoshima, T., Wostyn, K., Asselberghs, I., Clays, K., Persoons, A., Samoc, M., Luther-Davies, B.: Inorg. Chim. Acta **350**, 62–76 (2003)
[45] Garcia, M.H., Robalo, M.P., Dias, A.R., Duarte, M.T., Wenseleers, W., Aerts, G., Goovaerts, E., Cifuentes, M.P., Hurst, S., Humphrey, M.G., Samoc, M., Luther-Davies, B.: Organometallics **21**, 2107–2118 (2002)

[46] Hurst, S.K., Humphrey, M.G., Morrall, J.P., Cifuentes, M.P., Samoc, M., Luther-Davies, B., Heath, G.A., Willis, A.C.: J. Organomet. Chem. **670**, 56–65 (2003)
[47] Vicente, J., Chicote, M.T., Abrisqueta, M.D., de Arellano, M.C.R., Jones, P.G., Humphrey, M.G., Cifuentes, M.P., Samoc, M., Luther-Davies, B.: Organometallics **19**, 2968–2974 (2000)
[48] Dirk, C.W., Cheng, L.-T., Kuzyk, M.G.: Int. J. Quantum Chem. **43**, 27–36 (1992)
[49] Dirk, C.W., Caballero, N., Kuzyk, M.G.: Chem. Mater. **5**, 733–737 (1993)
[50] Kuzyk, M.G., Dirk, C.W.: Phys. Rev. A **41**, 5098–5109 (1990)
[51] Whittall, I.R., Humphrey, M.G.: Organometallics **14**, 5493–5495 (1995)
[52] Cifuentes, M.P., Powell, C.E., Humphrey, M.G., Heath, G.A., Samoc, M., Luther-Davies, B.: J. Phys. Chem. A **105**, 9625–9627 (2001)
[53] Powell, C.E., Cifuentes, M.P., Morrall, J.P., Stranger, R., Humphrey, M.G., Samoc, M., Luther-Davies, B., Heath, G.A.: J. Am. Chem. Soc. **125**, 602–610 (2003)
[54] McDonagh, A.M., Cifuentes, M.P., Whittall, I.R., Humphrey, M.G., Samoc, M., Luther-Davies, B., Hockless, D.C.R.: J. Organomet. Chem. **526**, 99–103 (1996)
[55] Powell, C.E., Humphrey, M.G., Cifuentes, M., P., Morrall, J.P., Samoc, M., Luther-Davies, B.: J. Phys. Chem. A **107**, 11264–11266 (2003)
[56] McDonagh, A.M., Humphrey, M.G., Samoc, M., Luther-Davies, B.: Organometallics **18**, 5195–5197 (1999)
[57] Myers, L.K., Langhoff, C., Thompson, M.E.: J. Am. Chem. Soc. **114**, 7560–7561 (1992)
[58] Sutherland, R.L. Handbook of Nonlinear Optics, vol. 52, Marcel Dekker (1996)

CHAPTER 18

RUTHENIUM COMPLEXES AS VERSATILE CHROMOPHORES WITH LARGE, SWITCHABLE HYPERPOLARIZABILITIES

BENJAMIN J. COE

School of Chemistry, The University of Manchester, Oxford Road, Manchester M13 9PL, UK

Abstract: This work provides a relatively comprehensive review of studies involving ruthenium coordination and organometallic complexes as nonlinear optical (NLO) compounds/materials, including both quadratic (second-order) and cubic (third-order) effects, as well as dipolar and octupolar chromophores. Such complexes can display very large molecular NLO responses, as characterised by hyperpolarizabilities, and bulk effects such as second harmonic generation have also been observed in some instances. The great diversity of ruthenium chemistry provides an unparalleled variety of chromophoric structures, and facile $Ru^{II} \to Ru^{III}$ redox processes can allow reversible and very effective switching of both quadratic and cubic NLO effects

Keywords: ruthenium complexes; hyperpolarizabilities; redox-switching; protic-switching; second harmonic generation

1. INTRODUCTION: THE CASE FOR RUTHENIUM

Amongst transition metal complexes with nonlinear optical (NLO) properties [1, 2], ruthenium compounds have been studied particularly intensively for both their quadratic (second-order) and cubic (third-order) behaviour. This situation has arisen partly because of the very extensive coordination and organometallic chemistry of ruthenium, which involves an attractive balance between stability and reactivity towards ligand substitutions. However, more importantly, electron-rich d^6 ruthenium(II) centres are especially well-suited for incorporation into NLO chromophores because their highly polarizable d orbitals can engender effective π-electron-donating properties when coordinated to ligands with low-lying π^* orbitals. A wide variety of Ru(II) complexes has been studied over the past 15 or so years, such as ruthenocene derivatives, σ-acetylide, ammine and polypyridyl complexes. These chromophores include neutral and charged complexes, possessing both dipolar and octupolar electronic structures. As such, ruthenium complexes display an unsurpassed degree of chromophoric diversity, combined with extensive opportunities for tuning, switching and optimisation of molecular hyperpolarizabilities and associated optical properties. This review includes almost all of the main primary literature citations concerning ruthenium-based NLO compounds and materials, with an emphasis on the specific and often seminal contributions that such studies have brought to the broader NLO research field.

2. RUTHENOCENYL DERIVATIVES AND OTHER COMPLEXES OF η^5-CYCLOPENTADIENYL AND RELATED LIGANDS

The first reported studies concerning the NLO properties of Ru^{II} complexes of η^5-cyclopentadienyl (Cp) ligands involve the amphiphilic derivative **1** in multilayer Langmuir-Blodgett (LB) films [3]. Such thin films of a cyanoterphenyl complex give 532 nm second harmonic generation (SHG), allowing Richardson et al. to estimate first molecular hyperpolarizabilities β, and mixed films also containing the free cyanoterphenyl ligand show 50% increased SHG intensities [3]. Many related complexes with σ-bonded acetylide ligands are known to display NLO properties, but these will be discussed separately in section 5 below.

Ferrocenyl derivatives have historically attracted considerable interest for their NLO properties, and ruthenium analogues of certain of the most promising chromophores have also been investigated [4, 5]. Such d^6 metallocenyl units behave as relatively strong electron donors, and 1907 nm electric-field-induced SHG (EFISHG) experiments by Calabrese et al. show that replacement of iron by ruthenium causes β to decrease [5]. This observation is attributed to the higher ionisation potential of a ruthenocenyl group when compared with ferrocenyl, i.e. the Ru^{II} unit is a weaker electron donor than its Fe^{II} counterpart, also evidenced by a blue-shifting of the lowest energy metal-to-ligand charge-transfer (MLCT) absorption in ruthenocenyl chromophores [5]. Detailed analyses of the electronic structures of donor-acceptor metallocenyl dyes involving density-functional theory (DFT) calculations and electronic Stark effect (electroabsorption) spectroscopy have been reported by Barlow et al. [6] The results of these studies indicate that both the MLCT and higher energy (predominantly intraligand charge-transfer (ILCT), with some metal character) transitions contribute to the β responses [6, 7].

The introduction of molecular chirality is one means to favour the formation of noncentrosymmetric materials, and studies on some chiral ruthenocene derivatives by Yamazaki et al. reveal a relatively high 1064 nm powder SHG efficiency of 27 × urea for the (+) enantiomer of one complex (**2**) [8]. 1340 nm EFISHG measurements by Bourgault et al. on the vinylruthenocene **3** with a pendent 2,2′-bipyridyl (bpy) unit show that coordination to octahedral *fac*-Re^{I}Cl(CO)$_3$ or tetrahedral Zn^{II}(OAc)$_2$ centres red-shifts the MLCT band and approximately doubles β [9]. Ruthenocenyl polyenes with powerful heterocyclic electron acceptor groups show very large $\mu\beta$ values (μ = dipole moment), as determined via 1907 nm EFISHG studies by Alain et al. [10]. In keeping with expectations, extension of the polyene chain causes red-shifting of the MLCT bands, accompanied by increasing NLO responses, and the superior donor strength of a ferrocenyl unit is again apparent [10]. The largest $\mu\beta$ (1900 × 10^{-48} esu in chloroform) is shown by the complex **4** which contains a 3-(dicyanomethylidene)-2,3-dihydrobenzo-thiophene-1,1-dioxide acceptor group [10]. In an early report by Kimura et al., a series of asymmetric sandwich complexes with Ru^{II}Cp units coordinated to η^5- or η^6-aryl ligands were found to display 1064 nm powder SHG efficiencies of 0.7-1.0 × urea [11].

Heck and co-workers have investigated a range of mono- and dinuclear sesquifulvalene-based and related complexes, a number of which contain ruthenocenyl units [12–14]. 1064 nm hyper-Rayleigh scattering (HRS) experiments conducted in both nitromethane and dichloromethane afford relatively large β responses (*ca.* 350-650 × 10^{-30} esu) for the mononuclear complex in salt **5** with a vinyltropylium electron acceptor group [13], and these NLO properties are maintained when the latter is complexed to a {Ru^{II}(η^5-C$_5$R$_5$)}$^+$ (R = H or Me) moiety [14]. However, the complexity of their electronic structures, combined with extensive resonance-enhancement, serve to limit the extent to which meaningful structure-activity relationships can be derived from these studies. Further related investigations by Heck and colleagues describe dinuclear complexes in which a ferrocenyl group is attached to a guaiazulenylium-$\{Ru^{II}Cp\}^+$ or a borabenzene-$\{Ru^{II}(\eta^6\text{-}C_6H_6)\}^+$ unit [15, 16]. The former compound (6) displays an intense charge-transfer (CT) band at the visible-NIR boundary ($\lambda_{max} = 698$ nm) and possesses a β response of 326×10^{-30} esu, as determined by 1064 nm HRS (data in dichloromethane) [15]. In contrast, the borabenzene complex shows only weak visible absorption which is associated with the absence of a detectable HRS signal at 532 nm, and its BPh_4^- salt unfortunately adopts a centrosymmetric crystal structure [16].

Besides the original work of Richardson et al. [3], a number of other complexes containing $Ru^{II}Cp$ electron donor centres coordinated to aryl nitrile derivatives are known to display NLO properties [17–22]. Dias et al. have reported a series of salts of the complex $[Ru^{II}Cp\{(+)\text{-}(DIOP)\}(nbn)]^+$ [DIOP = 2,3-O-isopropylidene-2,3-dihydroxy-1,4-bis(diphenylphosphino)butane, nbn = 4-nitrobenzonitrile] which show powder SHG activity at 1064 nm, with the largest signal of $10 \times$ urea being for the $CF_3SO_3^-$ salt (7) [17]. Subsequent studies with further related compounds do not reveal any materials with greater SHG activities, but 7 is found to crystallise noncentrosymmetrically in the monoclinic space group $P2_1$ (Figure 1), whilst its PF_6^- counterpart (SHG = $2.7 \times$ urea) crystallises in the triclinic $P1$ [18]. Estimates of the angle between the molecular CT axis and the polar crystal axis are 70.3° for 7 and 83.5° for the PF_6^- salt, which are both far from the optimum values for phase-matched SHG in the relevant space groups (54.74° for $P2_1$ and 35.26° for $P1$) [18].

Salt 7, together with its PF_6^- and $MeC_6H_4SO_3^-$ analogues and several related species containing dppe [1,2-bis(diphenylphosphino)ethane] instead of (+)-DIOP also show third harmonic generation (THG) when doped in polymethylmethacrylate (PMMA) thin films, as probed by Dias et al. using the Maker fringe technique [19]. As expected, extension of the conjugated π-system of the electron-accepting nitrile ligand causes the THG intensity to increase, and the introduction of a second phenylene ring to give a 4-(4-nitrophenyl)benzonitrile (npbn) ligand produces a relatively large second molecular hyperpolarizability γ of 2.3×10^{-33} esu for the PF_6^- salt of the dppe complex (8) [19]. Wenseleers et al. have also subjected 8, its nbn analogue and related compounds of Co^{III}, Ni^{II} and Fe^{II} to 1064 nm HRS experiments [20]. Although the results indicate that β increases as the metal changes in the order

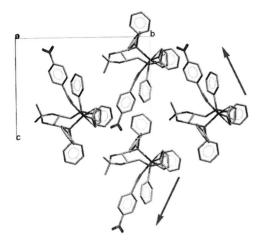

Figure 1. Crystal packing diagram of the complex salt **7** (mono-diethyl ether solvate), showing the noncentrosymmetric arrangement of the chromophoric complex cations (solvent and anions omitted) [18]. The arrows indicate the approximate directions of the molecular dipolar axes

Co < Ni < Ru < Fe, resonance effects play an important part in this apparent trend, with the complex $[Fe^{II}Cp(dppe)(nbn)]^+$ absorbing strongly at 532 nm. Furthermore, the applicability of the widely used two-state model to such chromophores is restricted since $[Ru^{II}Cp(dppe)(nbn)]^+$ shows two strongly overlapping absorptions in the region 300–500 nm. The observations that the complexes of nbn ligands show red-shifted MLCT absorptions and larger β values when compared with their npbn analogues may be attributable to torsion between the phenylene rings in npbn which decreases donor–acceptor electronic coupling (Figure 2) [20].

1064 nm HRS measurements have also been carried out by Mata et al. on the heterobimetallic complex **9** which contains a N-coordinated 1-ferrocenyl-2-E-(4-cyanophenyl)ethylene ligand [21]. As expected, variation of the counter-anion does not affect the linear absorption spectrum, but the BF_4^- salt apparently shows a considerably larger β response than its PF_6^- counterpart, although a convincing explanation for such unusual behaviour remains elusive [21]. Similar studies by Garcia et al. on complexes featuring a $Cr(phenyl)(CO)_3$ unit connected to a

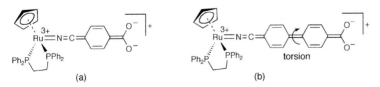

Figure 2. Canonical representations of the MLCT excited state in (a) nbn and (b) npbn ruthenium complexes, showing how torsion between the phenylene rings can act to diminish electronic coupling

nitrile-coordinated {Ru^{II}Cp(dppe)}$^+$ centre confirm the superior electron donating ability of an Fe^{II} fragment with respect to an analogous Ru^{II}-containing moiety [22]. Lee et al. have applied 1064 nm HRS and time-dependent DFT (TD-DFT) calculations to mononuclear sandwich complexes with aryl Ru^{II} units coordinated to a range of E-styryl-2-thiophene derivatives [23]. Such chromophores combine relatively high β responses (up to 389×10^{-30} esu for **10**) with good visible transparency, and display predictable enhancements in β with increasing electron accepting strength of substituents at the 4-position of the styryl ring [23]. Furthermore, the TD-DFT-calculated UV-visible maxima generally agree very closely with those measured in nitromethane solutions [23].

3. AMMINE COMPLEXES

In one of the first ever studies of ruthenium complexes with NLO properties, waveguides comprising LB films of the binuclear Ru^{III} complexes **11** and **12** were considered for SHG of a 1064 nm laser [24]. Although it was anticipated that phase-matching might be achievable due to anomalous dispersion arising from the intense CT absorptions at *ca.* 800 nm, these materials unfortunately proved to be insufficiently photostable for such an application [24]. A few years later, some particularly notable results were disclosed by Laidlaw et al. who used 1064 nm HRS to determine unusually large β values for two mixed-valence bimetallics (e.g. **13**) having intense, low energy intervalence CT transitions [25]. However, these data were subsequently found to be markedly overestimated due to a 6-fold error in the β used for the solvent reference [26], leading to results which are more in keeping with the rather short conjugation pathlengths of the chromophores. In these complexes, the electron-deficient d^5 Ru^{III} and electron-rich d^6 Ru^{II} ions are respectively stabilised by strongly basic ammine and π-accepting cyanide ligands. Static first hyperpolarizabilities β_0 for **13** have also been derived using the two-state model equation $\beta_0 = 3\Delta\mu_{12}(\mu_{12})^2/(E_{max})^2$ (where $\Delta\mu_{12}$ = dipole moment change,

μ_{12} = transition dipole moment and E_{max} = CT energy): Laidlaw et al. used a $\Delta\mu_{12}$ value calculated for transfer of 1 electron over the geometric Ru^{II}–Ru^{III} distance to afford $\beta_0 = 41 \times 10^{-30}$ esu [26], whereas Vance et al. obtained $\beta_0 = 28 \times 10^{-30}$ esu by using a $\Delta\mu_{12}$ derived from Stark spectroscopy [in 1:1 (v:v) ethylene glycol-water at 77 K] [27]. The discrepancy between the latter and that obtained from HRS (81×10^{-30} esu in water) may be due to errors arising from resonance-enhancement in the HRS data [27].

Coe and colleagues have carried out a number of investigations into the NLO properties of ruthenium ammine complexes of (mostly) pyridyl ligands, early studies focusing largely on the development of structure-activity relationships for β values derived via HRS at 1064 nm [28–34]. Results for **14** and **15** show that the β_0 response of the laser dye coumarin-510 (33×10^{-30} esu from EFISHG in chloroform) is not significantly affected by complexation or by $Ru^{II} \to Ru^{III}$ oxidation, because the 3-substitution at the pyridyl ring allows only weak coumarin-metal π-electronic coupling [28]. In contrast, the significant increase in β_0 for **16** (49×10^{-30} esu in acetonitrile, MeCN) when compared with that of the free dye coumarin-523 (32×10^{-30} esu from EFISHG in chloroform) is ascribed to the inductive electron-withdrawing effect of the Ru^{III} centre [28].

n = 2 (**14**), 3 (**15**)

16

Systematic studies by Coe et al. with salts of the form trans-$[Ru^{II}(NH_3)_4(L^D)(L^A)][PF_6]_n$ (L^D = an electron-rich ligand; L^A = an electron acceptor-substituted ligand; n = 2 or 3) initially focused on species where L^A = 4-acetylpyridine, ethylisonicotinate or N-methyl-4,4′-bipyridinium (MeQ$^+$) [29]. The intense, visible $Ru^{II} \to L^A$ MLCT absorptions of these complexes are strongly solvatochromic and dominate the β responses [29]. HRS-derived β_0 values in the range 10–130×10^{-30} esu were obtained, with the largest for trans-$[Ru^{II}(NH_3)_4(MeQ^+)(dmap)][PF_6]_3$ [dmap = 4-(dimethylamino)pyridine] [29]. As expected, complexes of the MeQ$^+$ ligand have larger β_0 values when compared with their analogues with monopyridyl L^A ligands due to extended π-conjugation [29]. The salt trans-$[Ru^{II}(NH_3)_4(MeQ^+)(PTZ)][PF_6]_3 \cdot Me_2CO$ (acetone solvate of **17**, PTZ = phenothiazine) crystallizes in the noncentrosymmetric hexagonal space group $P6_3$, with an almost planar MeQ$^+$ ligand (pyridyl-pyridyl torsion angle = 9.6°) [29]. Although the dipolar cations exhibit a strong projected component along the z axis, crystal twinning precludes bulk NLO effects [29].

17

18

Subsequent studies by Coe and co-workers have permitted further tuning and enhancement of the molecular NLO responses of complex salts *trans*-[Ru^{II}(NH$_3$)$_4$(L^D)(L^A)][PF$_6$]$_3$ [30–35], affording the important conclusion that *N*-arylation of 4,4′-bipyridinium cations is an effective approach to red-shifting MLCT bands and increasing β_0. The most active chromophore to be discovered during these investigations is that in salt **18** ($\beta_0 = 410 \times 10^{-30}$ esu in MeCN) [31]. Arguably the most significant and far-reaching result to emerge from these early studies is the demonstration that the MLCT absorptions and β_0 responses of the pentaammine complexes can be very effectively decreased (10–20-fold for β) by Ru$^{II} \rightarrow$ Ru^{III} oxidation using H$_2$O$_2$ (Figure 3) [33–37]. This unprecedented and facile redox-induced switching of NLO responses is fully reversible and provides an important incentive for incorporating Ru^{II} centres into NLO chromophores. Indeed, similar effects have subsequently been reported by other research groups using

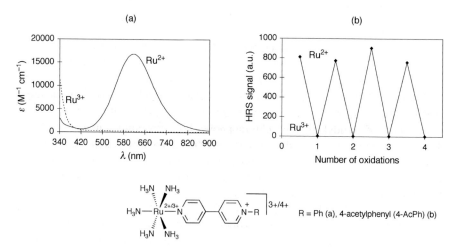

Figure 3. Redox-switching of (a) MLCT absorption and (b) molecular quadratic NLO response in ruthenium pentaammine complexes [33–37]

several different types of organotransition metal complexes, including those studied by Humphrey and co-workers (see Section 5 later).

A later report by Coe et al. describes detailed Stark spectroscopic studies carried out in butyronitrile (PrCN) glasses at 77 K which provide verification and rationalisation of the previously obtained HRS results [38]. Remarkably good agreement is found between the β_0 values calculated according to the two-state model equation $\beta_0 = 3\Delta\mu_{12}(\mu_{12})^2/(E_{max})^2$ and the HRS-derived data [38]. Red-shifting of the MLCT band and increases in both μ_{12} and $\Delta\mu_{12}$ are generally associated with enhancements in β_0, and placing a E-ethylene unit between the pyridyl and pyridinium rings of L^A increases β_0 by up to 50% [38]. ZINDO computations on the pentaammine complexes predict the dipole properties fairly accurately, but do not reliably reproduce E_{max} or β_0 [38]. In contrast, Lin et al. have found that the TD-DFT method gives reasonably good agreement with the β_0[HRS] values for some of Coe's complexes (e.g. $\beta_0 = 31 \times 10^{-30}$ esu for **19**, as opposed to 27×10^{-30} esu in MeCN [29], and $\beta_0 = 326 \times 10^{-30}$ esu for **18**), and is more reliable than the *ab initio* Hartree-Fock approach in treating such species [39]. Additional Stark and HRS studies by Coe et al. show that replacement of a neutral L^D ligand such as dmap with a (presumably *N*-coordinated) thiocyanate anion increases the electron-donating strength of the Ru^{II} centre [40]. This simple structural change logically translates into enhanced β_0 responses (e.g. β_0[HRS] = 513×10^{-30} esu for **20** in MeCN; β_0[Stark] = 553×10^{-30} esu), the magnitude of the observed increase with respect to the analogous *N*-methylimidazole (mim) complexes varying over a range of ca. 25–120%, depending on L^A [40].

Further studies by Coe and colleagues involving extended dipolar Ru^{II} ammine systems have uncovered some very unusual optical behaviour [41, 42]. Within three series of pyridyl polyene chromophores (e.g. the mim-containing **21–24**), the MLCT energy unexpectedly increases as the conjugated system extends from $n = 1$ to 3 [41, 42]. The β_0 values obtained via HRS and Stark spectroscopy are very large (ca.$100–600 \times 10^{-30}$ esu) and maximize at $n = 2$, in marked contrast to other known electron donor-acceptor polyenes in which β_0 increases steadily with n [41, 42]. TD-DFT and finite field (FF) calculations generally predict the empirical trends, both in terms of the blue-shifting of the MLCT bands and the unusual behaviour of β_0 [42]. The TD-DFT results show that the HOMO gains in π character as n increases and consequently the lowest energy transition usually considered as purely MLCT in character has some ILCT contribution which increases with the conjugation pathlength [42]. Hence, the E_{max} and $\Delta\mu_{12}$ values are respectively larger and smaller than expected, causing β_0 to decrease [42]. Notably, these studies show that metal-containing NLO chromophores can show very different optical behaviour when compared with their more thoroughly-studied purely organic counterparts [41, 42].

$n = 0$ (**21**), 1 (**22**), 2 (**23**), 3 (**24**)

The differences in optical properties between Ru^{II} ammine complexes and closely related purely organic chromophores have been thoroughly investigated in three further publications by Coe et al. [43–45] Initial studies with compounds featuring -C_6H_4-4-NMe_2 or pyridyl-coordinated $\{Ru^{II}(NH_3)_5\}^{2+}$ units connected directly to pyridinium electron acceptors show that the Ru^{II} centre is more effective (in terms of enhancing β_0) than the organic group as a π-electron donor [43]. This difference arises because the higher HOMO energy of the metal centre more than offsets the more effective π-orbital overlap in the analogous purely organic chromophores [43]. More extensive studies including TD-DFT and FF calculations confirm that extension of polyene chains in purely organic pyridinium chromophores leads to normal optical behaviour, i.e. red-shifting of the intramolecular CT (ICT) bands and increasing $\Delta\mu_{12}$ and β_0 [44, 45]. The contrasting dependencies of the optical properties on polyene chain length for the Ru^{II} and -C_6H_4-4-NMe_2 compounds (Figure 4) are attributable to the degree of donor–acceptor electronic coupling [44, 45]. Electrochemical, ^{1}H NMR and Stark spectroscopic data all show that π-orbital overlap is more effective in the purely organic compounds than in their Ru^{II} counterparts [44, 45]. The less effective donor–acceptor communication in the complexes becomes increasingly evident over long distances, so that β_0, μ_{12} and H_{ab} (the electronic coupling matrix element) all decrease above $n = 2$ [44, 45]. Although the Ru^{II} pyridyl ammine centres are generally more effective than a -C_6H_4-4-NMe_2 group as π-electron donors, in terms of enhancing β_0, such benefits are lost when $n = 3$ [44, 45].

Coe and co-workers have also investigated a number of complexes featuring redox-switchable cis-$\{Ru^{II}(NH_3)_4\}^{2+}$ centres coordinated to monodentate bipyridinium ligands and related chelating derivatives of 2, 2' : 4, 4" : 4', 4'''- quaterpyridyl [46–48]. Such unusual charged 2-dimensional NLO chromophores can display multiple MLCT bands, the energies (E_{max}) of which decrease in the order R = Me > Ph > 4-AcPh > 2-Pym (2-Pym = 2-pyrimidyl), as the electron-accepting ability of the

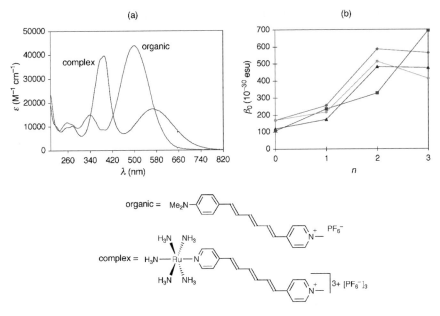

Figure 4. Illustrative examples of the contrasting electronic absorption (a) and molecular quadratic NLO responses (b) of ruthenium ammine complexes and related purely organic pyridinium polyene chromophores [45]. In (b), the squares refer to the organic series, while the diamonds refer to the complex salts **21–24**, the circles refer to the analogous series with trans pyridine (py) ligands, and the triangles refer to the pentaammine series

pyridinium groups increases [46–48]. This trend mirrors that observed previously in related 1-dimensional dipolar species [32]. β values have been measured by using HRS at 800 nm and Stark spectroscopy, incorporating Gaussian fitting to deconvolute the MLCT transitions [46–48]. These dipolar pseudo-C_{2v} chromophores exhibit two substantial components of the β tensor, β_{zzz} and β_{zyy}, with the difference between them being most marked for the non-chelated systems [48]. For example, **25** has $\beta_{zzz} = 110 \times 10^{-30}$ esu and $\beta_{zyy} = 298 \times 10^{-30}$ esu from Stark data (in PrCN at 77 K) [48]. The orbital structures have been elucidated by using TD-DFT which shows that the lowest energy MLCT transition is associated with the β_{zyy} response and the higher energy transition with β_{zzz}, with the former "off-diagonal" component being the larger [48].

Perhaps surprisingly, the cubic NLO properties of ruthenium ammine complexes have not yet been studied experimentally, but a single theoretical report of such behaviour has appeared [49]. Application of a linear algebraic method afforded remarkably high cubic bulk NLO susceptibilities $\chi^{(3)}$ for linear, mixed-valence $Ru^{III/II}$ "Creutz-Taube" oligomeric chain complexes having pyrazine-bridged *trans*-$\{Ru(NH_3)_4\}^{n+}$ (n = 2 or 3) centres with terminal $\{Ru(NH_3)_5\}^{n+}$ (n = 2 or 3) units [49].

4. COMPLEXES OF 2,2'-BIPYRIDYL AND OTHER CHELATING POLYPYRIDYL LIGANDS

Derivatives of the famous complex $[Ru^{II}(bpy)_3]^{2+}$ (RTB) have been investigated for their quadratic NLO properties by Sakaguchi and co-workers [50–55], early reports noting SHG from complex salts (e.g. **26**) in alternate Y-type LB films [50] or crystalline powders [51]. For **26**, a β value of 70×10^{-30} esu was estimated (at 1064 nm) [50], the presence of the electron-withdrawing amide substituents being expected to give an increase in β when compared with RTB [51]. Related studies have involved the use of photoexcitation to allow rapid modulation of the SHG from LB films (Figure 5) of **26** [52–55]. Upon 378 nm irradiation, the SHG from a 590 nm dye laser decreases by 30% in under 2 ps and is restored within several hundred ps [54]. Similar effects are observed following excitation at 355 or 460 nm with a 1064 nm probe [52, 53, 55]. The SHG time-profile and the decay of the MLCT-derived luminescence correlate quantitatively, indicating that changes in β on MLCT excitation are responsible for the SHG switching [52–55]. Matsuo and co-workers have shown that impregnation of the n-hexadecyl analogue of **26** into ultrathin PVC films affords SHG active materials [56, 57], and this group have also described SHG from amphiphilic RTB derivatives in LB films [58–60]. More recently, incorporation of the 1,10-phenanthroline (phen)-containing complex salt **27** into hybrid LB multilayers of a clay mineral has allowed probing of bulk

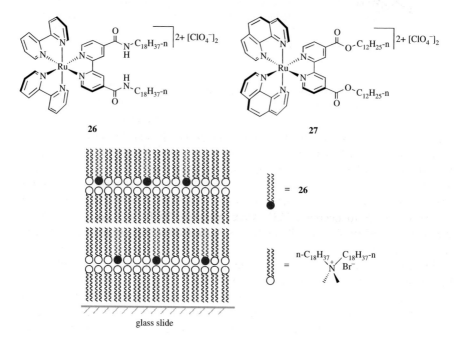

Figure 5. Schematic representation of alternating LB films containing the complex salt **26**. The actual samples feature 30 bilayers on each side of the support [52–55]

structures [61]. The SHG from a 1064 nm fundamental is enhanced when using only the Λ enantiomer of **27** rather than its racemic form, demonstrating the usefulness of the fabrication method for creating noncentrosymmetric ultrathin films [61].

R = N(*n*-Bu)$_2$ (**28**), NEt$_2$ (**29**), OOct (**30**)

Zyss et al. first pointed out the octupolar electronic structures of D_3 tris-chelate complexes, using HRS and a three-state model to obtain respective β_0 values of 53×10^{-30} and 47×10^{-30} esu for the salts [RTB]Br$_2$ and [Ru^{II}(phen)$_3$]Cl$_2$ [62]. Around the same time, Persoons et al. reported somewhat smaller β_0 values for these chromophores, and showed that chemical oxidation to the corresponding Ru^{III} species causes β to decrease [63]. More recently, Hache and co-workers have published several studies describing the observation of nonlinear circular dichroism in solutions of the resolved Δ and Λ isomers of RTB [64–67]. This cubic NLO effect is of interest for use in ultrafast time-resolved dynamic studies of chiral molecules. The two-photon absorption (TPA) properties of [Ru^{II}(4, 7-Me$_2$phen)$_3$][ClO$_4$]$_2$ have also been described [68]. Other NLO studies have involved more elaborately substituted derivatives of RTB and related complexes. A particularly significant report by Dhenaut et al. quotes an extremely large β_0 (2200×10^{-30} esu in chloroform) from 1340 nm HRS for a derivative of RTB with electron-donating styryl substituents (**28**) [69]. However, this β_0 value is apparently overestimated due to luminescence

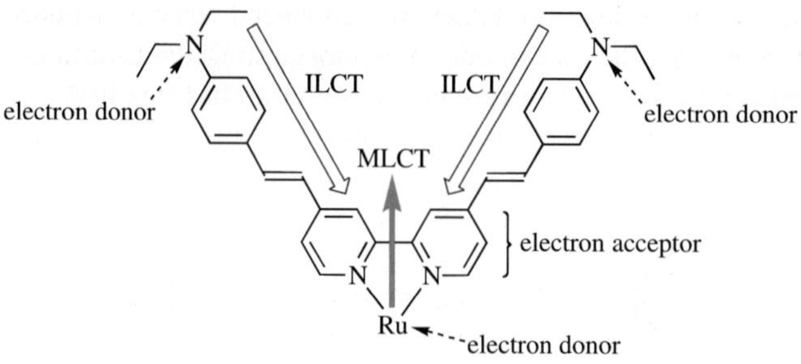

Figure 6. Schematic representation of the opposing charge-transfer excitation processes in complex salts such as **29** [72]

contributions [70], and subsequent studies have afforded a smaller (but still large) β_0 of 380×10^{-30} esu for **28** [71]. The RTB unit itself is octupolar, but polarized HRS and Stark spectroscopic experiments by Vance and Hupp show that the β response of **29** is more accurately described as being associated with several degenerate dipolar CT processes, rather than a truly octupolar transition [72]. Notably, the dominant transitions in such chromophores are of an ILCT nature, red-shifted on Ru^{II} coordination, and opposing the MLCT processes (Figure 6) [71, 72].

The effects of variation of the donor substituents and the metal centre have also been investigated in 1320 nm HRS studies with compounds related to **28** and **29** by Le Bozec and colleagues [71, 73]. Even though the relatively complicated electronic structures of such tris-(α-diimine) complexes hinders the development of clear structure-activity relationships, the smaller β_0 of **30** (298×10^{-30} esu in chloroform) clearly corresponds with a higher ILCT energy when compared with **28** [71, 73]. These differences are attributable to the weaker electron-donating ability of an alkoxy as opposed to a dialkylamino unit [71]. In similar fashion, the presence of a more electron-rich Fe^{II} centre gives rise to a blue-shifted ILCT absorption and red-shifted MLCT band when compared with **28** [71, 73], and both of these changes are consistent with the observed β decrease. Notably, the Zn^{II} analogue of **28** displays a similarly large β_0 value, but possesses a wider visible transparency range, due to the absence of MLCT transitions for the d^{10} ion [71]. Further data from 1907 nm HRS experiments with **28**, its *n*-octyloxy analogue and a range of related chromophores have recently been reported [74], and these results confirm that the Ru^{II} species have larger β responses than their Fe^{II} analogues.

Interestingly, Le Bozec and co-workers have recently incorporated the chromophoric units of **28** and **29** into macromolecular systems which may lead to useful NLO materials [75–78]. A polyimide derivative of **28** (**31**) with high thermal stability has a β of 1300×10^{-30} esu (from HRS at 1907 nm in dichloromethane), several times larger than that of the corresponding monomer [75]. Even more remarkably, a dendrimeric assembly comprising 7 units derived from **29** shows a very large β of 1900×10^{-30} esu (also from HRS at 1907 nm in dichloromethane) [76]. It is apparent that quasi-optimized octupolar ordering of the individual complex moieties in the latter structure is responsible for its enhanced β response when compared with the linear polymer which contains on average twice as many chromophoric units [76–78]. Each monomeric unit hence contributes coherently to the HRS response in a highly ordered dendritic architecture, whereas the linear polymer has a fully disordered structure [77, 78]. Another linear polymer derived from poly(disilanylene-2, 2′-bipyridine-5, 5′-diyl) and incorporating cis-$\{Ru^{II}(bpy)_2\}^{2+}$ units into the backbone has been subjected to SHG, nonlinear absorption and optical limiting studies by Zeng and co-workers [79–81], and 532 nm Z-scan measurements have been applied to a related polydiacetylene-based material by Camacho et al. [82].

Further cubic NLO measurements on Ru^{II} tris-(α-diimine) complexes have been reported by Ji and colleagues, using Z-scan at 540 or 532 nm; all of the compounds investigated display both NLO absorption and self-defocusing properties [83–87]. The $\chi^{(3)}$ and γ responses of complex salts such as **32** are relatively large and increase with the position of the -NO_2 substituent in the order 3- < 2- < 4- [83]. Furthermore, these NLO responses decrease if the imidazolyl nitrogens are deprotonated (but the effect is only statistically significant for **32**), an effect attributable to the concomitant reduction in π-accepting ability of the phen-based ligand [83]. Experiments with complex salts such as **33** predictably show that γ decreases with contraction of the π-system by the sequential removal of aryl rings from the benzo[i]dipyrido[3,2-a: 2′,3′-c]phenazine ligand ($\gamma = 47.8 \times 10^{-30}$ esu for the phen analogue of **33** in dimethylformamide, DMF) [84]. The $\chi^{(3)}$ and γ values of the tetranuclear compound **34** are larger than those of its mono- or dinuclear relatives, and scale with the number of metal centres (e.g. $\gamma = 268.5 \times 10^{-30}$ esu for **34** as opposed to 73.4×10^{-30} esu for the monomer in MeCN), indicating that the individual chromophoric units contribute additively to the NLO responses [85]. γ values in the range $4.15\text{-}4.86 \times 10^{-29}$ esu (in DMF), increasing with n, are observed for the complex salts $[\text{Ru}^{\text{II}}(\text{bpy})_{3-n}(\text{PIP})_n][\text{ClO}_4]_2$ (PIP = 2-phenylimidazo[4,5-f]-1,10-phenanthroline) [86]. Lahiri and co-workers have recently carried out Z-scan measurements of γ on some trinuclear Ru^{II} bpy/phen/arylazopyridine complexes with 1,3,5-triazine-2,4,6-trithiolato cores (and a related mononuclear compound) [88, 89]. 532 nm degenerate four-wave mixing (DFWM) and nonlinear absorption studies on several binuclear complexes with two cis-$\{\text{Ru}^{\text{II}}(\text{bpy})_2\}^{2+}$ units linked via chelating bridging ligands have been described by Sun et al. [90] Of these complexes, **35** displays the largest

γ value of 1.1×10^{-29} esu (in MeCN), which is *ca.* 2.5 times larger than that of the corresponding Ru^{II}-Ru^{III} mixed-valence species [90].

In the light of the relatively extensive NLO studies involving Ru^{II} tris-(α-diimine) species, it is perhaps surprising that only two investigations with related complexes of tridentate ligands have been reported [91, 92]. 532 nm Z-scan studies by Konstantaki et al. on several bis-(tpy)-based (tpy = 2,2':6',2''-terpyridyl) Ru^{II} or Os^{II} complexes, reveal very large cubic NLO responses for compounds including **36** ($\gamma = 228 \times 10^{-30}$ esu in MeCN) [91]. In addition, on 10 ns laser pulse excitation, the Ru^{II} complexes show reverse saturable absorption, but their Os^{II} counterparts show saturable absorption at low incident intensities, turning into reverse saturable absorption above a concentration-dependent intensity threshold. However, all of the complexes show only saturable absorption under faster (500 fs) laser excitation [91]. HRS studies by Uyeda et al. on Zn^{II} porphyrins with appended $[M^{II}(tpy)_2]^{2+}$ (M = Ru or Os) groups reveal unusual frequency dispersion behaviour of β, with a particularly large (albeit resonance enhanced) response of 5100×10^{-30} esu being observed for one Ru^{II} compound (**37**, in dichloromethane) at the technologically important wavelength of 1300 nm [92]. Roberto et al. have disclosed the first NLO studies with mono-tpy species, subjecting complexes of 4'-(4-dibutylaminophenyl)-tpy with several different metal centres to 1340 nm EFISHG [93]. Enhancements of β are observed in all cases (when compared with the free ligand), being positive for Zn^{II} species with trigonal bipyramidal structures, but negative for octahedral Ru^{III} or Ir^{III} complexes [93].

36

37

5. COMPLEXES OF σ-BONDED ACETYLIDE AND RELATED LIGANDS

The NLO properties of organometallic Ru^{II} complexes of σ-acetylide ligands have been extensively studied, most notably by Humphrey and co-workers [94–127], who have also published several recent reviews [128–131]. Many of these complexes also feature Cp co-ligands. Compounds of this type are particularly attractive due to their facile syntheses, high stability and potential for incorporation into polynuclear assemblies.

Whittall et al. carried out 1064 nm powder SHG studies on a series of aryldiazovinylidene Ru^{II} complex salts and found relatively modest activities of up to about that of urea [94]. Surprisingly, the most active compound (**38**) crystallises centrosymmetrically in $P\bar{1}$, and the SHG activity was attributed to surface effects or small deviations from centrosymmetry [94]. Subsequent early reports from this group describe the NLO properties of a range of dipolar σ-acetylide complexes featuring either $\{Ru^{II}Cp(PR_3)_2\}^+$ (R = Ph or Me) [95–97, 102] or trans-$\{Ru^{II}(dppm)_2\}^{2+}$ [dppm = bis(diphenylphosphino)methane] [98–101, 103] electron donor groups. ZINDO calculations on **39** afford a β_{vec} of 31×10^{-30} esu (at 1907 nm) and indicate that replacement of the -NO_2 substituent with H and/or substituting PPh_3 for PMe_3 cause the NLO response to decrease, as the strengths of the electron donating and accepting groups are diminished [95]. Complex **39** and its analogue lacking a -NO_2 substituent crystallise in the noncentrosymmetric space groups $Pca2_1$ and Cc, respectively [95]. In keeping with normal design principles for NLO molecules, the extended dipolar complex **40** has much larger β_0 values of 232 and 234×10^{-30} esu, determined at 1064 nm via HRS and EFISHG, respectively (both in tetrahydrofuran, THF) [97]. The corresponding ZINDO-derived value is 45×10^{-30} esu (essentially non-resonant at 1907 nm), but **40** unfortunately crystallises centrosymmetrically in $P2_1/n$ [97].

The cubic NLO properties of **40**, which has a low energy MLCT absorption at 476 nm (in THF), together with those of some related complexes, have been assessed via DFWM and Z-scan measurements at 800 nm, revealing moderate γ values [96]. Similar results have been obtained from Z-scan studies with related trans-$\{Ru^{II}(dppm)_2\}^{2+}$ complexes [100], and further ZINDO calculations [98, 99] reveal the expected increases in β_{vec} on extension of conjugated chains, with a value of 60×10^{-30} esu (at 1907 nm) for **41** [98]. HRS at 1064 nm was subsequently used to derive a β_0 value of 235×10^{-30} esu for the latter compound (in THF), which is about 30% larger than that of the analogous species lacking an E-ethylene bridge in the acetylide ligand [103]. A binuclear complex of C_{2v} symmetry with two trans-$\{Ru^{II}Cl(dppm)_2\}^+$ centres (**42**) was found by 1064 nm HRS to display a β response approximately twice that of its mononuclear counterpart trans-$Ru^{II}Cl(C\equiv CPh)(dppm)_2$ [101].

Particularly notable reports from McDonagh et al. describe the molecular quadratic and cubic NLO properties of the 3-fold symmetric octupolar Ru^{II} complexes **43** and **44** [104, 105]. According to the results of 1064 nm HRS experiments, moving from the dipolar complex **45** to its octupolar analogue **44** produces a ca. 3-fold enhancement in the uncorrected β response, with only a small accompanying loss of visible transparency (the MLCT λ_{max} changes from 383 to 411 nm in THF) [105]. Z-scan studies at 800 nm on these same compounds afford the first such cubic NLO data for organometallic complexes, and despite large experimental errors, indicate a ca. 2-fold increase in γ on moving from **45** to **44** [105]. By extending the molecular structure of such octupolar species in 3 directions, McDonagh et al. have also prepared and studied a nonanuclear dendrimeric complex (**46**) [106]. 800 nm Z-scan measurements show that progression from **44** to **46** results in no loss of visible transparency, but a very large increase in γ of almost an order of magnitude (to 207×10^{-34} esu in THF), together with a dramatic enhancement of the TPA cross-section (σ_2) [106]. Remarkably, this observed increase in γ is proportionately several times greater than either the increase in the number of phenylethynyl groups or the extinction coefficient for the MLCT band which is expected to dominate the NLO response [106].

McDonagh et al. have also used HRS at 1064 nm to study dipolar {Ru^{II}Cp(PR$_3$)$_2$}$^+$ and trans-{Ru^{II}Cl(dppm)$_2$}$^+$ complexes of ligands featuring imino or diazo units within the electron-acceptor-substituted σ-acetylide ligands [107, 108]. The complexes containing indoanilino units (e.g. **47**) have especially low energy MLCT bands in the 620–650 nm region (in THF) and large, albeit resonantly enhanced, β responses [107]. Related HRS studies with $\{Ru^{II}Cp(PPh_3)_2\}^+$ aryldiazovinylidene complexes reveal that such chromophores have smaller responses than similar σ-acetylide species [109], in keeping with their higher energy MLCT transitions. A number of vinylidene complexes with $\{Ru^{II}Cp(PPh_3)_2\}^+$ or trans-$\{Ru^{II}Cl(P–P)_2\}^+$ [P–P = dppm or dppe] electron donor units have also been prepared and compared with their σ-acetylide counterparts by using HRS and 800 nm Z-scan measurements [111–113]. The extended complex **48** has an especially large β_0 of 365×10^{-30} esu, a γ of 13×10^{-34} esu and MLCT maximum at 439 nm (all in THF) [112]. These vinylidine species are converted into their analogous σ-acetylides simply via deprotonation; although in some cases this change causes substantial modulation of β_0 (for example, for the dppe analogue of **41**, β_0 decreases from 342 to 74×10^{-30} esu on deprotonation), such variations in β_0 are overall inconsistent, meaning that this is not a general approach to switching first hyperpolarizabilities [112]. However, the σ_2 values are about an order of magnitude larger for the σ-acetylides in two such pairs of complexes, providing a potential mechanism for protic-switching of TPA effects (Figure 7) [112].

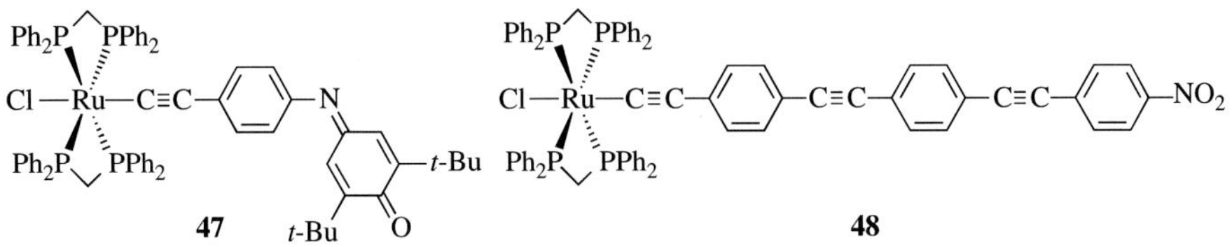

47, **48**

Ru^{II} complexes containing optically active 1,2-bis(methylphenylphosphino)-benzene [(R,R)-diph] ligands in trans-[Ru^{II}Cl{(R, R)-diph}$_2$]$^+$ centres have also been studied, together with their Fe^{II} and Os^{II} analogues in the case of **49** [110]. Although the HRS β_0 values indicate that the Ru^{II} and Os^{II} complexes have similar responses, whilst that of the corresponding Fe^{II} species is apparently smaller [110], resonance effects complicate this comparison because the MLCT band for the Fe^{II} complex is very close to 532 nm. These Ru^{II} and Os^{II} complexes show no or little powder SHG activity at 1064 nm, but the Fe^{II} analogue has an efficiency of ca. twice that of urea, indicating the adoption of a noncentrosymmetric crystal structure [110]. Notably, the complex **50** has a β_0 of 406×10^{-30} esu, apparently the largest for such a mononuclear Ru^{II} σ-acetylide, and a MLCT maximum at 481 nm (in THF) [110]. Comparison of this NLO response with that for **40** indicates that a trans-[Ru^{II}Cl{(R, R)-diph}$_2$]$^+$ centre is a rather more effective electron donor group than a {Ru^{II}Cp(PPh$_3$)$_2$}$^+$ unit.

Recent reports from Humphrey and colleagues have included several demonstrations of redox-switching of quadratic and cubic NLO responses in Ru^{II} σ-acetylides [114, 115, 118, 123]. Such compounds generally display reversible Ru$^{III/II}$ processes at potentials similar to those of the ammine species studied by Coe et al. [36]. The first switching of nonlinear absorption (a cubic effect) was achieved with two

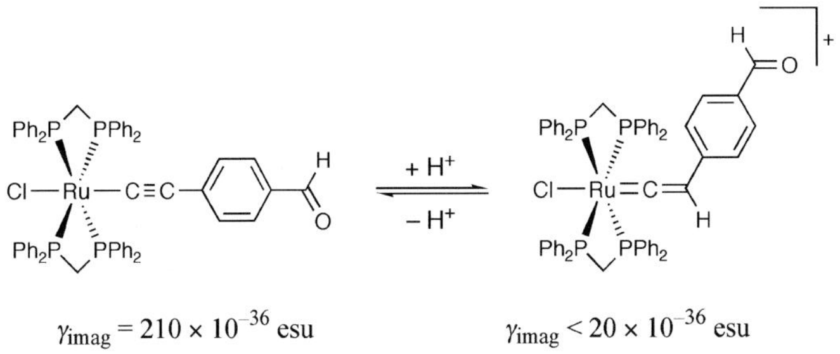

$\gamma_{imag} = 210 \times 10^{-36}$ esu $\gamma_{imag} < 20 \times 10^{-36}$ esu

Figure 7. Protic-switching of cubic NLO responses in ruthenium σ-acetylide/vinylidene complexes. The imaginary component of the second hyperpolarizability γ_{imag} is related to σ_2 [112]

[Structures 49 and 50 shown at top]

dipolar trans-{Ru^{II}Cl(P–P)$_2$}$^+$ (P–P = dppm or dppe) complexes and the octupolar **43** by using 800 nm Z-scans in an OTTLE cell [114, 115]. Three-electron oxidation of the trinuclear **43** causes a large increase in γ and a transition from TPA to saturable absorption behaviour [114, 115], and these changes have also been monitored by using a combination of femtosecond DFWM and transient-absorption experiments at 800 nm [123]. The UV-vis-NIR absorption spectral changes upon oxidation of **43** are depicted in Figure 8 [115]. Related detailed redox-switching studies with a series of dipolar trans-{Ru^{II}Cl(dppe)$_2$}$^+$ complexes have included further UV-vis-NIR spectroelectrochemical experiments and TD-DFT calculations as a means to assist rationalisation of the optical absorption spectra [118]. The Ru^{II} species are transparent below about 460 nm, whilst their Ru^{III} analogues show NIR

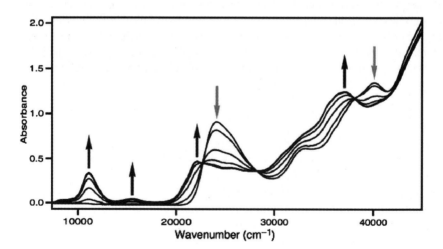

Figure 8. UV-vis-NIR spectra of a dichloromethane solution of complex **43** in a 0.5 mm pathlength OTTLE cell during oxidation with an applied potential of ca. 0.80 V vs. Ag–AgCl at 248 K (all three Ru^{II} centres oxidise at identical potentials of ca. 0.5 V) [115]

absorption bands, assigned to transitions from either a chloride or an ethynyl p orbital to the partially occupied HOMO [118].

Hurst et al. and Powell et al. have described 800 nm Z-scan studies on some symmetrical binuclear σ-acetylide complexes of trans-$\{Ru^{II}Cl(P-P)_2\}^+$ (P–P = dppm or dppe) centres [116, 119]. At the time of writing, complex **51** had the largest σ_2 to molecular weight ratio reported for an organometallic compound [116]. The related species **52** containing a ferrocenyl bridging unit has been characterised crystallographically, and a four-fold decrease in σ_2 and γ on conversion of **52** into its vinylidene analogue shows that these compounds have potential for protic-switching of NLO effects [119]. Further NLO studies with C_3 octupolar $Ru^{II}\sigma$-acetylides have been reported recently [117, 120, 121]. 1064 nm HRS data shows that the β responses of such chromophores with stilbenyl linkages are large, and especially so given the absence of strongly electron-accepting substituents [117, 120, 121]. The cubic molecular NLO parameters σ_2 and γ increase on moving from linear dipolar to octupolar species, showing what has been termed as a "dimensional evolution" [117, 120]. These octupoles possess notably large cubic NLO properties as assessed via Z-scan experiments, and bulk $\chi^{(3)}$ values determined by using Stark spectroscopy with doped PMMA thin films also scale with the number of metal ions [117, 120, 121].

Powell et al. have reported 1064 nm HRS and TD-DFT results for the carbonyl σ-acetylide complex **53** and its analogues in which PPh_3 or dppe replace the CO ligands [122]. These phosphine complexes have considerably larger β_0 responses than their CO counterpart, consistent with their lower $Ru^{III/II}$ potentials and MLCT energies [122]. Comparisons with the analogous Fe^{II} and Os^{II} dppe complexes show that β_0 increases in the order Fe ≤ Ru ≤ Os, although the observed MLCT energies follow the order Fe < Os < Ru and the corrresponding calculated trend is Fe < Ru < Os [122]. Further studies describing the use of Z-scan to probe the dispersion of γ in a hexanuclear dendrimeric complex (**54**) have recently been disclosed [124, 125]. The observed NLO behaviour is rationalised in terms of an interplay between TPA and absorption saturation, and can be reproduced by using a simple dispersion equation [125]. A collaboration with Adams and colleagues has included Z-scan measurements on several mixed σ-acetylide-bpy complexes of the form $Ru^{II}(C\equiv CR)_2(4,4'-Me_2bpy)(PPh_3)_2$ (R = alkyl or aryl), including **55**, for which only modest γ values were determined [126].

Several other research groups have described NLO studies with $Ru^{II}\sigma$-acetylide complexes [132–139]. Gimeno and co-workers have prepared a number of indenyl complexes, including heterobinuclear species, with $\{Ru^{II}(\eta^5-C_9H_7)(PPh_3)_2\}^+$

electron donor groups and studied them by using HRS at 1064 nm [132–134]. The complex **56** which features a nitrile-coordinated W(CO)$_5$ electron-accepting unit was studied as a mixture of its E and Z stereoisomers, giving a MLCT λ_{max} of 456 nm and a β_0 of 150×10^{-30} esu (in dichloromethane), which is notably large for a binuclear complex [132–134]. The mononuclear Ru^{II} precursor has a smaller β_0 of 71×10^{-30} esu and a blue-shifted λ_{max} of 427 nm, demonstrating the inductive electron-withdrawing effect of the W(CO)$_5$ moiety [132–134]. The ferrocenyl complex **57** also has a large β_0 response, despite the presence of two electron-donating metal centres [134].

Lin and colleagues have carried out HRS studies with {Ru^{II}Cp(PPh$_3$)$_2$}$^+$ complexes having a N-methylpyridinium [135] or nitroaryl [136] electron acceptor group. Measurements with a 1560 nm laser show that incorporation of thienyl rings into such chromophores acts to enhance quadratic NLO responses, as also noted in purely organic species [136]. The extended complex **58** has $\beta_0 = 195 \times 10^{-30}$ esu and λ_{max}[MLCT] = 536 nm (in dichloromethane), which compares favourably with the β_0 of 105×10^{-30} esu determined for **40** under the same conditions [136]. By using off-resonance 1907 nm EFISHG data for two {Ru^{II}(η^5-C$_5$Me$_5$)(dppe)}$^+$ complexes, Paul et al. have concluded that the analogous Fe^{II} centre is a more effective electron donor [138], at odds with some other studies [110, 122]. A recent report by Fillaut et al. describes DFWM results and also acoustically-induced SHG from PMMA-included trans-{Ru^{II}Cl(dppe)$_2$}$^+$ complexes with oligothienyl spacers and aldehyde acceptor groups, including **59** [139]. Interestingly, π-stacking interactions between thiophene carboxaldehyde units cause the latter complex to adopt the centrosymmetric space group $P\bar{1}$ (Figure 9) [139].

The first NLO studies on Ru^{II} σ-vinyl complexes have been reported recently by Xia et al. [140] The C_3 symmetric species **60** and **61** have structures reminiscent of some of the octupolar σ-acetylide complexes studied by Humphey and co-workers [104, 105, 117, 120, 121], but 1064 nm HRS experiments in chloroform reveal only relatively small β_0 values of 9 and 14×10^{-30} esu, for **60** and **61**, respectively [140].

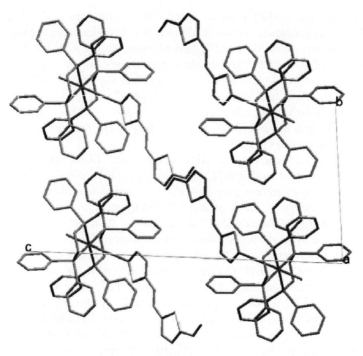

Figure 9. Crystal packing diagram of the complex **59** (bis-dichloromethane solvate), showing the centrosymmetric arrangement of the complexes (solvent omitted) and the π-stacking interactions [139]

6. MISCELLANEOUS COMPLEXES

The cubic NLO properties of a Ru^{II} phthalocyanine (Pc) and 3,4-naphthalocyanine (NPc) complexes have been investigated [141, 142]. Hanack and co-workers have carried out 1064 nm DFWM and THG studies on spin-cast thin films of soluble $Ru^{II}(t\text{-Bu})_4Pc$ oligomers with 4-diisocyanobenzene bridges (**62**) [141]. The DFWM wavelength was varied over the complete range of the Pc Q-band absorption, showing the dispersion of $\chi^{(3)}$ to a maximum value of 11.5×10^{-8} esu [141]. The optical properties of this oligomer are primarily determined by the planar macrocylic units (which show saturable absorption behaviour), due to only weak axial electronic coupling via the bridging ligands [141]. Nalwa et al. have reported THG measurements on a tetra-*t*-Bu substituted NPc complex (**63**), deriving a $\chi^{(3)}$ value of 1.0×10^{-12} esu when using a 2100 nm laser [142]. Similar cubic NLO susceptibilities were found for several related NPcs, showing that the optical properties are dominated by the NPc organic macrocycle [142]. Wang et al. have investigated the ultrafast cubic NLO responses of Ru^{III} and Fe^{III} complexes of the derivatised fullerene $C_{60}(NH_2CN)_5$ by using the femtosecond optical Kerr gate technique in dimethylsulfoxide at 830 nm [143]. The presence of two *N*-coordinated Ru^{III} centres enhances the γ of the fullerene unit by more than an order of magnitude, whilst

Fe^{III} has an opposite effect (although only one metal ion appears to be coordinated in this case) [143]. Despite speculation regarding bridging of several C$_{60}$ units by Ru^{III} [143], the origin of the observed optical effects is unclear.

Coe et al. have studied a series of Ru^{II} complexes of the chelating ligand 1,2-phenylenebis(dimethylarsine) (pdma), in part for purposes of comparison with the related ammine compounds of the same pyridinium-substituted ligands [144–146]. A major advantage of these arsine species is their improved crystallisation behaviour. Although many phosphine-containing complexes have been investigated (see section 5 above), no other arsine species appear to have been studied for NLO effects. Initial studies showed that the lower electron-richness of a trans-{Ru^{II}Cl(pdma)$_2$}$^+$ centre when compared with a {Ru^{II}(NH$_3$)$_5$}$^{2+}$ group leads to increased Ru$^{III/II}$ potentials and MLCT energies, although the same trends are observed in both arsine and ammine species as the structure of the pyridinium-substituted ligand is varied [144]. An analysis of bond lengths and dihedral angles obtained from crystallographic studies provides no evidence for ground state charge-transfer, despite the strongly dipolar, polarizable nature of these complexes [144]. Stark spectroscopic studies reveal that the β responses of the pdma complexes are only a little smaller than those of their {Ru^{II}(NH$_3$)$_5$}$^{2+}$ analogues, but this result is partly attributable to unexpected changes in the relative μ_{12} values on freezing [145, 146]. The complex of a N-(4-acetylphenyl)-4,4'-bipyridinium ligand as its PF$_6^-$ salt (**64**) adopts the noncentrosymmetric space

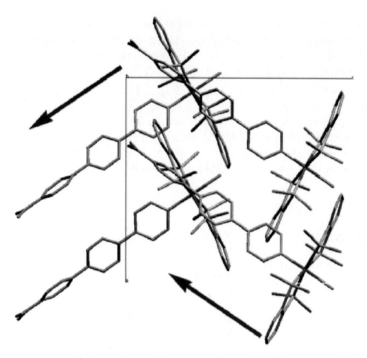

Figure 10. Crystal packing diagram of the complex salt **64** (mono-acetone solvate), showing the noncentrosymmetric arrangement of the chromophoric complex cations (solvent and anions omitted) [144]. The arrows indicate the approximate directions of the molecular dipolar axes

group $Pna2_1$ (Figure 10) [144], and with a β of 170×10^{-30} esu (in PrCN at 77 K) can be expected to show substantial bulk NLO effects [145, 146].

7. CONCLUSIONS AND OUTLOOK

There is no doubt that the studies described herein form an important contribution to the field of molecular NLO materials. A very wide range of Ru complexes has been prepared and investigated, many of which are relatively easily synthesised and highly stable. Chromophores possessing very large molecular NLO responses, which can compete with those of all but the very best known purely organic species, have been identified in several instances. Although the design criteria for Ru-containing NLO chromophores have generally been found to mirror those already established for metal-free organic compounds, some unusual effects have been noted, such as the decreasing of β_0 values with π-conjugation extension in pyridyl polyene complexes. Furthermore, the interesting and potentially useful phenomena of redox-switching of molecular hyperpolarizabilities and associated properties were first demonstrated in Ru complexes, and have subsequently been observed in other metal complexes. Whilst most of the work to date has focused on the molecular level, i.e. tuning and optimisation of chromophoric properties, a number of studies have uncovered crystalline materials showing bulk NLO effects, and several different types of polymeric Ru^{II}-containing materials have also been created. Although future studies can be expected to feature an increased emphasis upon such materials aspects, fundamental molecular engineering investigations are still valuable and desirable and can be expected to afford further insights. Furthermore, the potential for applications of transition metal complex dyes in a wide range of electronic/photonic areas [147], means that studies with NLO properties as their primary focus may have spin-off relevance, e.g. for photovoltaic devices. In a relatively short time period, Ru complexes have become established as being amongst the most versatile and fascinating NLO compounds, and their future is surely promising.

ACKNOWLEDGEMENT

The author would like to thank Dr Simon J. Coles (University of Southampton, UK) for assistance with the preparation of Figures 1, 9 and 10.

REFERENCES

[1] Di Bella, S.: Second-order nonlinear optical properties of transition metal complexes. Chem. Soc. Rev. **30**, 355–366 (2001)

[2] Coe, B.J.: Non-linear optical properties of metal complexes: In: Comprehensive Coordination Chemistry II; McCleverty, J.A., Meyer, T.J., (eds) Elsevier Pergamon: Oxford, United Kingdom, vol. 9, pp. 621–687 (2004)

[3] Richardson, T., Roberts, G.G., Polywka, M.E.C., Davies, S.G.: Preparation and characterization of organotransition metal Langmuir-Blodgett films. Thin Solid Films **160**, 231–239 (1988)

[4] Cheng, L.-T., Tam, W., Meredith, G.R.: Quadratic hyperpolarizabilities of some organometallic compounds. Mol. Cryst. Liq. Cryst. **189**, 137–153 (1990)

[5] Calabrese, J.C., Cheng, L.-T., Green, J.C., Marder, S.R., Tam, W.: Molecular second-order optical nonlinearities of metallocenes. J. Am. Chem. Soc. **113**, 7227–7232 (1991)

[6] Barlow, S., Bunting, H.E., Ringham, C., Green, J.C., Bublitz, G.U., Boxer, S.G., Perry, J.W., Marder, S.R.: Studies of the electronic structure of metallocene-based second-order nonlinear optical dyes. J. Am. Chem. Soc. **121**, 3715–3723 (1999)

[7] Barlow, S., Marder, S.R.: Electronic and optical properties of conjugated group 8 metallocene derivatives. Chem. Commun. 1555–1562 (2000)

[8] Yamazaki, Y., Hosono, K., Matsuda, H., Minami, N., Asai, M., Nakanishi, H.: Enzymic preparation of organometallic compounds with second-order optical nonlinearities. Biotechnol. Bioeng. **38**, 1218–1222 (1991)

[9] Bourgault, M., Baum, K., Le Bozec, H., Pucetti, G., Ledoux, I., Zyss, J.: Synthesis and molecular hyperpolarizabilities of donor-acceptor bipyridyl metal complexes (M = Re, Zn, Hg). New J. Chem. 517–522 (1998)

[10] Alain, V., Blanchard-Desce, M., Chen, C.-T., Marder, S.R., Fort, A., Barzoukas, M.: Large optical nonlinearities with conjugated ferrocene and ruthenocene derivatives. Synth. Met. **81**, 133–136 (1996)

[11] Kimura, M., Abdel-Halim, H., Robinson, D.W., Cowan, D.O.: Synthesis and nonlinear optical properties of some substituted ruthenium(II) η^5-cyclopentadienyl η^5- or η^6-arene salts. J. Organomet. Chem. **403**, 365–372 (1991)

[12] Heck, J., Dabek, S., Meyer-Friedrichsen, T., Wong, H.: Mono- and dinuclear sesquifulvalene complexes, organometallic materials with large nonlinear optical properties. Coord. Chem. Rev. **190–192**, 1217–1254 (1999)
[13] Wong, H., Meyer-Friedrichsen, T., Farrell, T., Mecker, C., Heck, J.: Second harmonic generation and two-photon fluorescence as nonlinear optical properties of dipolar mononuclear sesquifulvalene complexes. Eur. J. Inorg. Chem. 631–646 (2000)
[14] Meyer-Friedrichsen, T., Wong, H., Prosenc, M.H., Heck, J.: Vinylogue mono- and bimetallic cationic sesquifulvalene and monohydro sesquifulvalene complexes for second harmonic generation. Eur. J. Inorg. Chem. 936–946 (2003)
[15] Farrell, T., Meyer-Friedrichsen, T., Malessa, M., Haase, D., Saak, W., Asselberghs, I., Wostyn, K., Clays, K., Persoons, A., Heck, J., Manning, A.R.: Azulenylium and guaiazulenylium cations as novel accepting moieties in extended sesquifulvalene type D–π–A NLO chromophores. J. Chem. Soc., Dalton Trans. 29–36 (2001)
[16] Behrens, U., Meyer-Friedrichsen, T., Heck, J.: Cationic borabenzene complexes as electron accepting groups for molecular material with nonlinear optical properties. Z. Anorg. Allg. Chem. **629**, 1421–1430 (2003)
[17] Dias, A.R., Garcia, M.H., Rodrigues, J.C., Green, M.L.H., Kuebler, S.M.: Synthesis and characterization of η^5-monocyclopentadienyl (p-nitrobenzonitrile)ruthenium(II) salts: second harmonic generation powder efficiencies. J. Organomet. Chem. **475**, 241–245 (1994)
[18] Garcia, M.H., Rodrigues, J.C., Dias, A.R., Piedade, M.F.M., Duarte, M.T., Robalo, M.P., Lopes, N.: Second harmonic generation of η^5-monocyclopentadienyl ruthenium p-benzonitrile derivatives by Kurtz powder technique. Crystal and molecular structure determinations of $[Ru(\eta^5\text{-}C_5H_5)((+)\text{-DIOP})(p\text{-NCC}_6H_4NO_2)][X]$, $X = PF_6^-, CF_3SO_3^-$ and $[Ru(\eta^5\text{-}C_5H_5)((+)\text{-DIOP})(NCCH_3)][PF_6]$. J. Organomet. Chem. **632**, 133–144 (2001)
[19] Dias, A.R., Garcia, M.H., Rodrigues, J.C., Petersen, J.C., Bjørnholm, T., Geisler, T.: Third-harmonic generation in organometallic ruthenium(II) derivatives containing coordinated p-substituted benzonitriles. J. Mater. Chem. **5**, 1861–1865 (1995)
[20] Wenseleers, W., Gerbrandij, A.W., Goovaerts, E., Garcia, M.H., Robalo, M.P., Mendes, P.J., Rodrigues, J.C., Dias, A.R.: Hyper-Rayleigh scattering study of η^5-monocyclopentadienyl–metal complexes for second order non-linear optical materials. J. Mater. Chem. **8**, 925–930 (1998)
[21] Mata, J.A., Peris, E., Uriel, S., Llusar, R., Asselberghs, I., Persoons, A.: Preparation and properties of new ferrocenyl heterobimetallic complexes with counterion dependent NLO responses. Polyhedron **20**, 2083–2088 (2001)
[22] Garcia, M.H., Royer, S., Robalo, M.P., Dias, A.R., Tranchier, J.-P., Chavignon, R., Prim, D., Auffrant, A., Rose-Munch, F., Rose, E., Vaissermann, J., Persoons, A., Asselberghs, I.: Synthesis, characterisation of (arene)tricarbonylchromium complexes linked to cationic Fe and Ru derivatives and studies of first hyperpolarisabilities by hyper-Rayleigh scattering. Eur. J. Inorg. Chem. 3895–3904 (2003)
[23] Lee, I.S., Choi, D.S., Shin, D.M., Chung, Y.K., Choi, C.H.: Preparation of thiophene-coordinated ruthenium complexes for nonlinear optics. Organometallics **23**, 1875–1879 (2004)
[24] Cahill, P.A.: Nonlinear optical waveguides containing a transition metal-based dye molecule. Mater. Res. Soc. Symp. Proc. **109**, 319–322 (1988)
[25] Laidlaw, W.M., Denning, R.G., Verbiest, T., Chauchard, E., Persoons, A.: Large second-order optical polarizabilities in mixed-valency metal complexes. Nature **363**, 58–60 (1993)
[26] Laidlaw, W.M., Denning, R.G., Verbiest, T., Chauchard, E., Persoons, A.: Second-order nonlinearity in mixed-valence metal chromophores. Proc. SPIE, Int. Soc. Opt. Eng. **2143**, 14–19 (1994)
[27] Vance, F.W., Karki, L., Reigle, J.K., Hupp, J.T., Ratner, M.A.: Aspects of intervalence charge transfer in cyanide-bridged systems: modulated electric field assessment of distances, polarizability changes, and anticipated first hyperpolarizability characteristics. J. Phys. Chem. A **102**, 8320–8324 (1998)
[28] Coe, B.J., Chadwick, G., Houbrechts, S., Persoons, A.: Molecular linear and quadratic non-linear optical properties of pentaammineruthenium complexes of coumarin dyes. J. Chem. Soc., Dalton Trans. 1705-1711 (1997)

[29] Coe, B.J., Chamberlain, M.C., Essex-Lopresti, J.P., Gaines, S., Jeffery, J.C., Houbrechts, S., Persoons, A.: Large molecular quadratic hyperpolarizabilites in donor/acceptor-substituted trans-tetraammineruthenium(II) complexes. Inorg. Chem. **36**, 3284–3292 (1997)
[30] Coe, B.J., Essex-Lopresti, J.P., Harris, J.A., Houbrechts, S., Persoons, A.: Ruthenium(II) ammine centres as efficient electron donor groups for quadratic non-linear optics. Chem. Commun. 1645–1646 (1997)
[31] Coe, B.J., Harris, J.A., Harrington, L.J., Jeffery, J.C., Rees, L.H., Houbrechts, S., Persoons, A.: Enhancement of molecular quadratic hyperpolarizabilites in ruthenium(II) 4,4′-bipyridinium complexes by N-phenylation. Inorg. Chem. **37**, 3391–3399 (1998)
[32] Coe, B.J., Harris, J.A., Asselberghs, I., Persoons, A., Jeffery, J.C., Rees, L.H., Gelbrich, T., Hursthouse, M.B.: Tuning of charge-transfer absorption and molecular quadratic nonlinear optical properties in ruthenium(II) ammine complexes. J. Chem. Soc., Dalton Trans. 3617–3625 (1999)
[33] Houbrechts, S., Asselberghs, I., Persoons, A., Coe, B.J., Harris, J.A., Harrington, L.J., Essex-Lopresti, J.P.: Large molecular hyperpolarizabilities in donor/acceptor-substituted trans-tetraammine ruthenium(II) complexes. Mol. Cryst. Liq. Cryst. Science & Tech., Sect. B: Nonlinear Opt. **22**, 161-164 (1999)
[34] Houbrechts, S., Asselberghs, I., Persoons, A., Coe, B.J., Harris, J.A., Harrington, L.J., Chamberlain, M.C., Essex-Lopresti, J.P., Gaines, S.: Tuning the hyperpolarizabilities of asymmetrically substituted trans-tetraammineruthenium(II) complexes. Proc. SPIE-Int. Soc. Opt. Eng. **3796**, 209-218 (1999)
[35] Asselberghs, I., Houbrechts, S., Persoons, A., Coe, B.J., Harris, J.A.: Hyper-Rayleigh scattering of trans-tetraammineruthenium(II) complexes. Synth. Met. **124**, 205–207 (2001)
[36] Coe, B.J., Houbrechts, S., Asselberghs, I., Persoons, A.: Efficient, reversible redox-switching of molecular first hyperpolarizabilities in ruthenium(II) complexes possessing large quadratic optical nonlinearities. Angew. Chem. Int. Ed. **38**, 366–369 (1999)
[37] Coe, B.J.: Molecular materials possessing switchable quadratic nonlinear optical properties. Chem. Eur. J. **5**, 2464–2471 (1999)
[38] Coe, B.J., Harris, J.A., Brunschwig, B.S.: Electroabsorption spectroscopic studies of dipolar ruthenium(II) complexes possessing large quadratic nonlinear optical responses. J. Phys. Chem. A **106**, 897–905 (2002)
[39] Lin, C.-S., Wu, K.-C., Snijders, J.G., Sa, R.-J., Chen, X.-H.: TDDFT and ab initio study on the quadratic hyperpolarizabilities of trans-tetraammineruthenium(II) complexes. Acta Chim. Sinica **60**, 664–668 (2002)
[40] Coe, B.J., Jones, L.A., Harris, J.A., Sanderson, E.E., Brunschwig, B.S., Asselberghs, I., Clays, K., Persoons, A.: Molecular quadratic non-linear optical properties of dipolar trans-tetraammineruthenium(II) complexes with pyridinium and thiocyanate ligands. Dalton Trans. 2335–2341 (2003)
[41] Coe, B.J., Jones, L.A., Harris, J.A., Brunschwig, B.S., Asselberghs, I., Clays, K., Persoons, A.: Highly unusual effects of π-conjugation extension on the molecular linear and quadratic nonlinear optical properties of ruthenium(II) ammine complexes. J. Am. Chem. Soc. **125**, 862–863 (2003)
[42] Coe, B.J., Jones, L.A., Harris, J.A., Brunschwig, B.S., Asselberghs, I., Clays, K., Persoons, A., Garín, J., Orduna, J.: Syntheses and spectroscopic and quadratic nonlinear optical properties of extended dipolar complexes with ruthenium(II) ammine electron donor and N-methylpyridinium acceptor groups. J. Am. Chem. Soc. **126**, 3880–3891 (2004)
[43] Coe, B.J., Harris, J.A., Clays, K., Persoons, A., Wostyn, K., Brunschwig, B.S.: A comparison of the pentaammine(pyridyl)ruthenium(II) and 4-(dimethylamino)phenyl groups as electron donors for quadratic non-linear optics. Chem. Commun. 1548–1549 (2001)
[44] Coe, B.J., Jones, L.A., Harris, J.A., Asselberghs, I., Wostyn, K., Clays, K., Persoons, A., Brunschwig, B.S., Garín, J., Orduna, J.: Quadratic nonlinear optical properties of novel pyridinium salts. Proc. SPIE-Int. Soc. Opt. Eng. **5212**, 122–136 (2003)
[45] Coe, B.J., Harris, J.A., Brunschwig, B.S., Garín, J., Orduna, J., Coles, S.J., Hursthouse, M.B.: Contrasting linear and quadratic nonlinear optical behavior of dipolar pyridinium chromophores

with 4-(dimethylamino)phenyl or ruthenium(II) ammine electron donor groups. J. Am. Chem. Soc. **126**, 10418–10427 (2004)

[46] Coe, B.J., Harris, J.A., Brunschwig, B.S.: Determination of the molecular quadratic non-linear optical responses of V-shaped metallochromophores by using Stark spectroscopy. Dalton Trans. 2384–2386 (2003)

[47] Harris, J.A., Coe, B.J., Jones, L.A., Brunschwig, B.S., Asselberghs, I., Clays, K., Persoons, A.: Quadratic nonlinear optical properties of transition metal quaterpyridyl complexes. Proc. SPIE-Int. Soc. Opt. Eng. **5212**, 341–350 (2003)

[48] Coe, B.J., Harris, J.A., Jones, L.A., Brunschwig, B.S., Song, K., Clays, K., Garín, J., Orduna, J., Coles, S.J., Hursthouse, M.B.: Syntheses and properties of two-dimensional charged nonlinear optical chromophores incorporating redox-switchable cis-tetraammineruthenium(II) centers. J. Am. Chem. Soc. **127**, 4845–4859 (2005)

[49] Ferretti, A., Lami, A., Villani, G.: Third-harmonic generation in mixed-valent Ru-pyrazine chains: a theoretical study. J. Phys. Chem. A **101**, 9439–9444 (1997)

[50] Sakaguchi, H., Nakamura, H., Nagamura, T., Ogawa, T., Matsuo, T.: Second harmonic generation by the use of metal to ligand charge-transfer transition of a ruthenium(II)-bipyridine metal complex in Langmuir-Blodgett film. Chem. Lett. 1715–1718 (1989)

[51] Sakaguchi, H., Nagamura, T., Matsuo, T.: Quadratic nonlinear optical properties of ruthenium(II)-bipyridine complexes in crystalline powders. Appl. Organomet. Chem. **5**, 257–260 (1991)

[52] Sakaguchi, H., Nagamura, T., Matsuo, T.: Laser-induced modulation of second-harmonic light emission from ruthenium(II)-bipyridine metal complex in Langmuir-Blodgett film. Jpn. J. Appl. Phys., Part 2 **30**, L377–L379 (1991)

[53] Nagamura, T., Sakaguchi, H., Matsuo, T.: Photochemical modulation of second order nonlinear optical properties of alternate Langmuir-Blodgett films containing ruthenium(II)-bipyridine complexes. Thin Solid Films **210–211**, 160–162 (1992)

[54] Sakaguchi, H., Gomez-Jahn, L.A., Prichard, M., Penner, T.L., Whitten, D.G., Nagamura, T.: Subpicosecond photoinduced switching of second-harmonic generation from a ruthenium complex in supported Langmuir-Blodgett films. J. Phys. Chem. **97**, 1474–1476 (1993)

[55] Sakaguchi, H., Nagamura, T., Penner, T.L., Whitten, D.G.: Ultrafast optical modulation of quadratic nonlinearity from an Ru(II)-bipyridine complex in Langmuir-Blodgett assemblies. Thin Solid Films **244**, 947–950 (1994)

[56] Matsuo, T., Nakamura, H., Nakao, T., Kawazu, M.: Optical second harmonic generation from ultrathin polymer films impregnated with ruthenium polypyridine complexes. Chem. Lett. 2363–2366 (1992)

[57] Yamada, S., Kawazu, M., Matsuo, T.: Second-order nonlinear optical effects in stacked assemblies of ultrathin polymer films with amphiphilic ruthenium complex. J. Phys. Chem. **98**, 3573–3574 (1994)

[58] Yamada, S., Nakano, T., Matsuo, T.: Second harmonic generation from Langmuir-Blodgett monolayers of amphiphilic ruthenium(II) complexes: structural characterization with polarization measurements. Thin Solid Films **245**, 196–201 (1994)

[59] Yamada, S., Yamada, Y., Nakano, T., Matsuo, T.: In-situ observation of second harmonic light from amphiphilic ruthenium(II) tris(2, 2′-bipyridine) complex at glass/liquid interface. Chem. Lett. 937–940 (1994)

[60] Nakano, T., Yamada, Y., Matsuo, T., Yamada, S.: In-situ second-harmonic generation and luminescence measurements for structural characterization of ruthenium-polypyridine complex monolayers with two and four aliphatic tails at the air/water interface. J. Phys. Chem. B **102**, 8569–8573 (1998)

[61] Umemura, Y., Yamagishi, A., Schoonheydt, R., Persoons, A., De Schryver, F.: Langmuir-Blodgett films of a clay mineral and ruthenium(II) complexes with a noncentrosymmetric structure. J. Am. Chem. Soc. **124**, 992–997 (2002)

[62] Zyss, J., Dhenaut, C., Chauvan, T., Ledoux, I.: Quadratic nonlinear susceptibility of octupolar chiral ions. Chem. Phys. Lett. **206**, 409–414 (1993)

[63] Persoons, A., Clays, K., Kauranen, M., Hendrickx, E., Put, E., Bijnens, W.: Characterization of nonlinear optical properties by hyper-scattering techniques. Synth. Met. **67**, 31–38 (1994)
[64] Mesnil, H., Hache, F.: Experimental evidence of third-order nonlinear dichroism in a liquid of chiral molecules. Phys. Rev. Lett. **85**, 4257–4260 (2000)
[65] Mesnil, H., Schanne-Klein, M.C., Hache, F., Alexandre, M., Lemercier, G., Andraud, C.: Experimental observation of nonlinear circular dichroism in a pump-probe experiment. Chem. Phys. Lett. **338**, 269–276 (2001)
[66] Mesnil, H., Schanne-Klein, M.C., Hache, F., Alexandre, M., Lemercier, G., Andraud, C.: Wavelength dependence of nonlinear circular dichroism in a chiral ruthenium-tris(bipyridyl) solution. Phys. Rev. A **66**, 013802/1–013802/9 (2002)
[67] Mesnil, H., Schanne-Klein, M.-C., Hache, F., Alexandre, M., Lemercier, G., Andraud, C.: Nonlinear circular dichroism in a chiral ruthenium-tris(bipyridyl) solution. Trends Opt. Photon. **79**, 354–356 (2002)
[68] Kawamata, J., Ogata, Y., Yamagishi, A.: Two-photon fluorescence property of tris(4,7-diphenyl-1, 10-phenanthroline)ruthenium(II)perchlorate. Mol. Cryst. Liq. Cryst. Science & Tech., Sect. A: Mol. Cryst. Liq. Cryst. **379**, 389–394 (2002)
[69] Dhenaut, C., Ledoux, I., Samuel, I.D.W., Zyss, J., Bourgault, M., Le Bozec, H.: Chiral metal complexes with large octupolar optical nonlinearities. Nature **374**, 339–342 (1995)
[70] Morrison, I.D., Denning, R.G., Laidlaw, W.M., Stammers, M.A.: Measurement of first hyperpolarizabilities by hyper-Rayleigh scattering. Rev. Sci. Instrum. **67**, 1445–1453 (1996)
[71] Le Bozec, H., Renouard, T., Bourgault, M., Dhenaut, C., Brasselet, S., Ledoux, I., Zyss, J.: Molecular engineering of octupolar tris(bipyridyl) metal complexes. Synth. Met. **124**, 185–189 (2001)
[72] Vance, F.W., Hupp, J.T.: Probing the symmetry of the nonlinear optic chromophore Ru(trans-4,4′-diethylaminostyryl-2, 2′-bipyridine)$_3^{2+}$: insight from polarized hyper-Rayleigh scattering and electroabsorption (Stark) spectroscopy. J. Am. Chem. Soc. **121**, 4047–4053 (1999)
[73] Le Bozec, H., Renouard, T.: Dipolar and non-dipolar pyridine and bipyridine metal complexes for nonlinear optics. Eur. J. Inorg. Chem. 229–239 (2000)
[74] Maury, O., Viau, L., Sénéchal, K., Corre, B., Guégan, J.-P., Renouard, T., Ledoux, I., Zyss, J., Le Bozec, H.: Synthesis, linear, and quadratic-nonlinear optical properties of octupolar D_3 and D_{2d} bipyridyl metal complexes. Chem. Eur. J. **10**, 4454–4466 (2004)
[75] Le Bouder, T., Maury, O., Le Bozec, H., Ledoux, I., Zyss, J.: Synthesis of a highly thermally stable octupolar polyimide for nonlinear optics. Chem. Commun. 2430–2431 (2001)
[76] Le Bozec, H., Le Bouder, T., Maury, O., Bondon, A., Ledoux, I., Deveau, S., Zyss, J.: Supramolecular octupolar self-ordering towards nonlinear optics. Adv. Mater. **13**, 1677–1681 (2001)
[77] Le Bozec, H., Le Bouder, T., Maury, O., Ledoux, I., Zyss, J.: Coordination chemistry for nonlinear optics: a powerful tool for the design of octupolar molecules and supramolecules. J. Opt. A: Pure Appl. Opt. **4**, S189–S196 (2002)
[78] Le Bouder, T., Maury, O., Bondon, O., Costuas, K., Amouyal, E., Ledoux, I., Zyss, J., Le Bozec, H.: Synthesis, photophysical and nonlinear optical properties of macromolecular architectures featuring octupolar tris(bipyridine) ruthenium(II) moieties: evidence for a supramolecular self-ordering in a dentritic structure. J. Am. Chem. Soc. **125**, 12284–12299 (2003)
[79] Liu, C.-Y., Zeng, H.-P., Segawa, Y., Kira, M.: Optical limiting performance of a novel σ-π alternating polymer. Opt. Commun. **162**, 53–56 (1999)
[80] Zeng, H.-P., Liu, C.-Y., Tokura, S., Kira, M., Segawa, Y.: Scanning second-harmonic microscopy of a thin film of σ-π copolymer with 2, 2′-bipyridine in the backbone. J. Phys.: Condensed Matter **11**, L333–L340 (1999)
[81] Zeng, H.-P., Liu, C.-Y., Tokura, S., Kira, M., Segawa, Y.: Optical limiting and bistability of a σ-π photoconductive copolymer. Chem. Phys. Lett. **331**, 71–77 (2000)
[82] Camacho, M.A., Kar, A.K., Lindsell, W.E., Murray, C., Preston, P.N., Wherrett, B.S.: Synthesis and nonlinear optical properties of functionalized polydiacetylenes and their complexes with transition metals. J. Mater. Chem. **9**, 1251–1256 (1999)
[83] Chao, H., Li, R.-H., Ye, B.-H., Li, H., Feng, X.-L., Cai, J.-W., Zhou, J.-Y., Ji, L.-N.: Syntheses, characterization and third order non-linear optical properties of the ruthenium(II) complexes

containing 2-phenyl-imidazo[4,5-f][1,10]phenanthroline derivatives. J. Chem. Soc., Dalton Trans. 3711–3717 (1999)

[84] Jiang, C.-W., Chao, H., Li, R.-H., Li, H., Ji, L.-N.: Syntheses, characterization and third-order nonlinear optical properties of ruthenium(II) complexes containing 2-phenylimidazo-[4,5-f][1,10]phenanthroline and extended diimine ligands. Polyhedron **20**, 2187–2193 (2001)

[85] Chao, H., Li, R.-H., Jiang, C.-W., Li, H., Ji, L.-N., Li, X.-Y.: Mono-, di- and tetra-nuclear ruthenium(II) complexes containing 2, 2′-p-phenylenebis(imidazo[4,5-f]phenanthroline): synthesis, characterization and third-order non-linear optical properties. J. Chem. Soc., Dalton Trans. 1920–1926 (2001)

[86] Jiang, C.-W., Chao, H., Li, R.-H., Li, H., Ji, L.-N.: Ruthenium(II) complexes of 2-phenylimidazo[4,5-f][1,10]phenanthroline. Synthesis, characterization and third order nonlinear optical properties. Trans. Met. Chem. **27**, 520–525 (2002)

[87] Chao, H., Yuan, Y.-X., Ji, L.-N.: Synthesis, characterization and third-order nonlinear optical properties of ruthenium(II) complexes containing 2-(4-nitrophenyl)imidazo[4,5-f][1,10]phenanthroline. Trans. Met. Chem. **29**, 774–779 (2004)

[88] Kar, S., Miller, T.A., Chakraborty, S., Sarkar, B., Pradhan, B., Sinha, R.K., Kundu, T., Ward, M.D., Lahiri, G.K.: Synthesis, mixed valence aspects and non-linear optical properties of the triruthenium complexes [{(bpy)$_2$Ru^{II}}$_3$(L)]$^{3+}$ and [{(phen)$_2$Ru^{II}}$_3$(L)]$^{3+}$ (bpy = 2, 2′-bipyridine, phen = 1, 10-phenanthroline and L^{3-} = 1,3,5-triazine-2,4,6-trithiol). Dalton Trans. 2591–2596 (2003)

[89] Kar, S., Pradhan, B., Sinha, R.K., Kundu, T., Kodgire, P., Rao, K.K., Puranik, V.G., Lahiri, G.K.: Synthesis, structure, redox, NLO and DNA interaction aspects of [{(L′-‴)$_2$Ru^{II}}$_3$(μ_3 – L)]$^{3+}$ and [(L′)$_2$Ru^{II}(NC$_5$H$_4$S$^-$)]$^+$[L^{3-} =1,3,5-triazine-2,4,6-trithiolato, L′-‴ = arylazopyridine]. Dalton Trans. 1752–1760 (2004)

[90] Sun, W.-F., Patton, T.H., Stultz, L.K., Claude, J.P.: Resonant third-order nonlinearities of tetrakis(2, 2′-dipyridyl)diruthenium complexes. Opt. Commun. **218**, 189–194 (2003)

[91] Konstantaki, M., Koudoumas, E., Couris, S., Laine, P., Amouyal, E., Leach, S.: Substantial non-linear optical response of new polyads based on Ru and Os complexes of modified terpyridines. J. Phys. Chem. B **105**, 10797–10804 (2001)

[92] Uyeda, H.T., Zhao, Y.-X., Wostyn, K., Asselberghs, I., Clays, K., Persoons, A., Therien, M.J.: Unusual frequency dispersion effects of the nonlinear optical response in highly conjugated (polypyridyl) metal-(porphinato)zinc(II) chromophores. J. Am. Chem. Soc. **124**, 13806–13813 (2002)

[93] Roberto, D., Tessore, F., Ugo, R., Bruni, S., Manfredi, A., Quici, S.: Terpyridine Zn(II), Ru(III) and Ir(III) complexes as new asymmetric chromophores for nonlinear optics: first evidence for a shift from positive to negative value of the quadratic hyperpolarizability of a ligand carrying an electron donor substituent upon coordination to different metal centres. Chem. Commun. 846–847 (2002)

[94] Whittall, I.R., Cifuentes, M.P., Costigan, M.J., Humphrey, M.G., Goh, S.C., Skelton, B.W., White, A.H.: Organometallic materials for nonlinear optics. Second harmonic generation by (aryldiazovinylidene)ruthenium complexes. X-ray structure of [Ru(C=CPhN=NC$_6$H$_4$OMe-4)(PPh$_3$)$_2$(η-C$_5$H$_5$)][BF$_4$]•CH$_2$Cl$_2$. J. Organomet. Chem. **471**, 193–199 (1994)

[95] Whittall, I.R., Humphrey, M.G., Hockless, D.C.R., Skelton, B.W., White, A.H.: Organometallic complexes for nonlinear optics. 2. Syntheses, electrochemical studies, structural characterization, and computationally-derived molecular quadratic hyperpolarizabilities of ruthenium σ-arylacetylides : X-ray crystal structures of Ru(C≡CPh)(PMe$_3$)$_2$(η-C$_5$H$_5$) and Ru(C≡CC$_6$H$_4$NO$_2$-4)(L)$_2$(η-C$_5$H$_5$)(L = PPh$_3$, PMe$_3$). Organometallics **14**, 3970–3979 (1995)

[96] Whittall, I.R., Humphrey, M.G., Samoc, M., Swiatkiewicz, J., Luther-Davies, B.: Organometallic complexes for nonlinear optics. 4. Cubic hyperpolarizabilities of (cyclopentadienyl)bis(phosphine)ruthenium σ-arylacetylides. Organometallics **14**, 5493–5495 (1995)

[97] Whittall, I.R., Humphrey, M.G., Persoons, A., Houbrechts, S.: Organometallic complexes for nonlinear optics. 3. Molecular quadratic hyperpolarizabilities of ene-, imine-, and azo-linked ruthenium σ-acetylides: X-ray Crystal Structure of Ru((E)-4, 4′-C≡CC$_6$H$_4$CH=CHC$_6$H$_4$NO$_2$)(PPh$_3$)$_2$(η-C$_5$H$_5$). Organometallics **15**, 1935–1941 (1996)

[98] McDonagh, A.M., Whittall, I.R., Humphrey, M.G., Skelton, B.W., White, A.H.: Organometallic complexes for nonlinear optics. V. Syntheses and computationally derived quadratic nonlinearities of trans-[Ru(C≡CC$_6$H$_4$R-4)Cl(dppm)$_2$][R = H, NO$_2$, C$_6$H$_4$NO$_2$-4, CH=CHC$_6$H$_4$NO$_2$-4, (E)]; X-ray crystal structure of trans-[Ru(C≡CC$_6$H$_4$C$_6$H$_4$NO$_2$-4,4′)Cl(dppm)$_2$]. J. Organomet. Chem. **519**, 229–235 (1996)

[99] McDonagh, A.M., Whittall, I.R., Humphrey, M.G., Hockless, D.C.R., Skelton, B.W., White, A.H.: Organometallic complexes for nonlinear optics. VI: Syntheses of rigid-rod ruthenium σ-acetylide complexes bearing strong acceptor ligands; x-ray crystal structures of trans-[Ru(C≡CC$_6$H$_4$NO$_2$-4)$_2$(dppm)$_2$] and trans-[Ru(C≡CC$_6$H$_4$C$_6$H$_4$NO$_2$-4,4′)$_2$(dppm)$_2$]. J. Organomet. Chem. **523**, 33–40 (1996)

[100] McDonagh, A.M., Cifuentes, M.P., Whittall, I.R., Humphrey, M.G., Samoc, M., Luther-Davies, B., Hockless, D.C.R.: Organometallic complexes for nonlinear optics. VII. Cubic optical nonlinearities of octahedral trans-bis{bis(diphenylphosphino)methane}ruthenium acetylide complexes; x-ray crystal structure of trans-[Ru(C≡CPh)(4-C≡CC$_6$H$_4$NO$_2$)(dppm)$_2$]. J. Organomet. Chem. **526**, 99–103 (1996)

[101] Whittall, I.R., Humphrey, M.G., Houbrechts, S., Joachim, M., Persoons, A., Schmid, S., Hockless, D.C.R.: Organometallic complexes for nonlinear optics. 14. Syntheses and second-order nonlinear optical properties of ruthenium, nickel and gold σ-acetylides of 1,3,5-triethynylbenzene: X-ray crystal structures of 1-(HC≡)-3, 5-C$_6$H$_3$(trans-C≡CRuCl(dppm)$_2$)$_2$ and 1, 3, 5-C$_6$H$_3$(C≡CAu(PPh$_3$))$_3$. J. Organomet. Chem. **544**, 277–283 (1997)

[102] Whittall, I.R., Cifuentes, M.P., Humphrey, M.G., Luther-Davies, B., Samoc, M., Houbrechts, S., Persoons, A., Heath, G.A., Hockless, D.C.R.: Organometallic complexes for nonlinear optics. X. Molecular quadratic and cubic hyperpolarizabilities of systematically varied (cyclopentadienyl)bis(phosphine)ruthenium σ-arylacetylides: X-ray crystal structure of Ru((E)-4, 4′-C≡CC$_6$H$_4$CH=CHC$_6$H$_4$NO$_2$)(PPh$_3$)$_2$(η-C$_5$H$_5$). J. Organomet. Chem. **549**, 127–137 (1997)

[103] Naulty, R.H., McDonagh, A.M., Whittall, I.R., Cifuentes, M.P., Humphrey, M.G., Houbrechts, S., Maes, J., Persoons, A., Heath, G.A., Hockless, D.C.R.: Organometallic complexes for nonlinear optics. 15. Molecular quadratic hyperpolarizabilities of trans-bis{bis(diphenylphosphino)methane}ruthenium σ-aryl- and σ-pyridyl-acetylides: x-ray crystal structure of trans [Ru(2-C≡CC$_5$H$_3$N-5-NO$_2$)Cl(dppm)$_2$]. J. Organomet. Chem. **563**, 137–146 (1998)

[104] Houbrechts, S., Wada, T., Sasabe, H., Morrall, J.P.L., Whittall, I.R., McDonagh, A.M., Humphrey, M.G., Persoons, A.: Novel organometals for nonlinear optics: octupolar alkynylmetal complexes. Mol. Cryst. Liq. Cryst. Science & Tech., Sect. B: Nonlinear Opt. **22**, 165–168 (1999)

[105] McDonagh, A.M., Humphrey, M.G., Samoc, M., Luther-Davies, B., Houbrechts, S., Wada, T., Sasabe, H., Persoons, A.: Organometallic complexes for nonlinear optics. 16. Second and third order optical nonlinearities of octopolar alkynylruthenium complexes. J. Am. Chem. Soc. **121**, 1405–1406 (1999)

[106] McDonagh, A.M., Humphrey, M.G., Samoc, M., Luther-Davies, B.: Organometallic complexes for nonlinear optics. 17. Synthesis, third-order optical nonlinearities, and two-photon absorption cross section of an alkynylruthenium dendrimer. Organometallics **18**, 5195–5197 (1999)

[107] McDonagh, A.M., Cifuentes, M.P., Lucas, N.T., Humphrey, M.G., Houbrechts, S., Persoons, A.: Organometallic complexes for nonlinear optics. Part 19. Syntheses and molecular quadratic hyperpolarizabilities of indoanilino-alkynyl-ruthenium complexes. J. Organomet. Chem. **605**, 184–192 (2000)

[108] McDonagh, A.M., Lucas, N.T., Cifuentes, M.P., Humphrey, M.G., Houbrechts, S., Persoons, A.: Organometallic complexes for nonlinear optics. Part 20. Syntheses and molecular quadratic hyperpolarizabilities of alkynyl complexes derived from (E)-4, 4′-HC≡CC$_6$H$_4$N=NC$_6$H$_4$NO$_2$. J. Organomet. Chem. **605**, 193–201 (2000)

[109] Cifuentes, M.P., Driver, J., Humphrey, M.G., Asselberghs, I., Persoons, A., Samoc, M., Luther-Davies, B.: Organometallic complexes for nonlinear optics. Part 18. Molecular quadratic

and cubic hyperpolarizabilities of aryldiazovinylidene complexes. J. Organomet. Chem. **607**, 72–77 (2000)

[110] McDonagh, A.M., Cifuentes, M.P., Humphrey, M.G., Houbrechts, S., Maes, J., Persoons, A., Samoc, M., Luther-Davies, B.: Organometallic complexes for nonlinear optics Part 21. Syntheses and quadratic hyperpolarizabilities of alkynyl complexes containing optically active 1,2-bis(methylphenylphosphino)benzene ligands. J. Organomet. Chem. **610**, 71–79 (2000)

[111] Hurst, S.K., Lucas, N.T., Cifuentes, M.P., Humphrey, M.G., Samoc, M., Luther-Davies, B., Asselberghs, I., Van Boxel, R., Persoons, A.: Organometallic complexes for nonlinear optics Part 23. Quadratic and cubic hyperpolarizabilities of acetylide and vinylidene complexes derived from protected and free formylphenylacetylenes. J. Organomet. Chem. **633**, 114–124 (2001)

[112] Hurst, S.K., Cifuentes, M.P., Morrall, J.P.L., Lucas, N.T., Whittall, I.R., Humphrey, M.G., Asselberghs, I., Persoons, A., Samoc, M., Luther-Davies, B., Willis, A.C.: Organometallic complexes for nonlinear optics. 22. Quadratic and cubic hyperpolarizabilities of trans-bis(bidentate phosphine)ruthenium σ-arylvinylidene and σ-arylalkynyl complexes. Organometallics **20**, 4664–4675 (2001)

[113] Hurst, S.K., Lucas, N.T., Humphrey, M.G., Asselberghs, I., Van Boxel, R., Persoons, A.: Organometallic complexes for non-linear optics. XXVI. Quadratic hyperpolarizabilities of some 4-methoxytetrafluorophenylalkynyl gold and ruthenium complexes. Aust. J. Chem. **54**, 447–451 (2001)

[114] Samoc, M., Humphrey, M.G., Cifuentes, M.P., McDonagh, A.M., Powell, C.E., Heath, G.A., Luther-Davies, B.: Third-order optical nonlinearities of organometallics: influence of dendritic geometry on the nonlinear properties and electrochromic switching of nonlinear absorption. Proc. SPIE–Int. Soc. Opt. Eng. **4461**, 65–77 (2001)

[115] Cifuentes, M.P., Powell, C.E., Humphrey, M.G., Heath, G.A., Samoc, M., Luther-Davies, B.: Organometallic complexes for nonlinear optics. 24. Reversible electrochemical switching of nonlinear absorption. J. Phys. Chem. A **105**, 9625–9627 (2001)

[116] Hurst, S.K., Cifuentes, M.P., McDonagh, A.M., Humphrey, M.G., Samoc, M., Luther-Davies, B., Asselberghs, I., Persoons, A.: Organometallic complexes for nonlinear optics Part 25. Quadratic and cubic hyperpolarizabilities of some dipolar and quadrupolar gold and ruthenium complexes. J. Organomet. Chem. **642**, 259–267 (2002)

[117] Hurst, S.K., Humphrey, M.G., Isoshima, T., Wostyn, K., Asselberghs, I., Clays, K., Persoons, A., Samoc, M; Luther-Davies, B.: Organometallic complexes for nonlinear optics. 28. Dimensional evolution of quadratic and cubic optical nonlinearities in stilbenylethynylruthenium complexes. Organometallics **21**, 2024–2026 (2002)

[118] Powell, C.E., Cifuentes, M.P., Morrall, J.P., Stranger, R., Humphrey, M.G., Samoc, M., Luther-Davies, B., Heath, G.A.: Organometallic complexes for nonlinear optics. 30. Electrochromic linear and nonlinear optical properties of alkynylbis(diphosphine)ruthenium complexes. J. Am. Chem. Soc. **125**, 602–610 (2003)

[119] Hurst, S.K., Humphrey, M.G., Morrall, J.P., Cifuentes, M.P., Samoc, M., Luther-Davies, B., Heath, G.A., Willis, A.C.: Organometallic complexes for nonlinear optics. Part 31. Cubic hyperpolarizabilities of ferrocenyl-linked gold and ruthenium complexes. J. Organomet. Chem. **670**, 56–65 (2003)

[120] Humphrey, M.G., Cifuentes, M.P., Samoc, M., Isoshima, T., Persoons, A.: Hyper-structured alkynylruthenium complexes: effect of dimensional evolution of NLO properties. Spec. Pub.–Roy. Soc. Chem. **287** (Perspectives in organometallic chemistry), 100–110 (2003)

[121] Hurst, S.K., Lucas, N.T., Humphrey, M.G., Isoshima, T., Wostyn, K., Asselberghs, I., Clays, K., Persoons, A., Samoc, M., Luther-Davies, B.: Organometallic complexes for nonlinear optics. Part 29. Quadratic and cubic hyperpolarizabilities of stilbenylethynyl-gold and -ruthenium complexes. Inorg. Chim. Acta **350**, 62–76 (2003)

[122] Powell, C.E., Cifuentes, M.P., McDonagh, A.M., Hurst, S.K., Lucas, N.T., Delfs, C.D., Stranger, R., Humphrey, M.G., Houbrechts, S., Asselberghs, I., Persoons, A., Hockless, D.C.R.: Organometallic complexes for nonlinear optics. Part 27. Syntheses and optical properties of some iron, ruthenium and osmium alkynyl complexes. Inorg. Chim. Acta **352**, 9–18 (2003)

[123] Powell, C.E., Humphrey, M.G., Morrall, J.P., Samoc, M., Luther-Davies, B.: Organometallic complexes for nonlinear optics. 33. Electrochemical switching of the third-order nonlinearity observed by simultaneous femtosecond degenerate four-wave mixing and pump-probe measurements. J. Phys. Chem. A **107**, 11264–11266 (2003)
[124] Humphrey, M.G., Powell, C.E., Cifuentes, M.P., Morrall, J.P., Samoc, M.: Syntheses and nonlinear optical properties of alkynylruthenium dendrimers. Polym. Preprints (Am. Chem. Soc., Div. Polym. Chem.) **45**, 367–368 (2004)
[125] Powell, C.E., Morrall, J.P., Ward, S.A., Cifuentes, M.P., Notaras, E.G.A., Samoc, M., Humphrey, M.G.: Dispersion of the third-order nonlinear optical properties of an organometallic dendrimer. J. Am. Chem. Soc. **126**, 12234–12235 (2004)
[126] Adams, C.J., Bowen, L.E., Humphrey, M.G., Morrall, J.P.L., Samoc, M., Yellowlees, L.J.: Ruthenium bipyridyl compounds with two terminal alkynyl ligands. Dalton Trans. 4130–4138 (2004)
[127] Houbrechts, S., Boutton, C., Clays, K., Persoons, A., Whittall, I.R., Naulty, R.H., Cifuentes, M.P., Humphrey, M.G.: Novel organometallic compounds for nonlinear optics: metal σ-phenyl and pyridyl acetylide complexes. J. Nonlinear Opt. Phys. Mater. **7**, 113–120 (1998)
[128] Whittall, I.R., McDonagh, A.M., Humphrey, M.G., Samoc, M.: Organometallic complexes in nonlinear optics I: second-order nonlinearities. Adv. Organomet. Chem. **42**, 291–362 (1998)
[129] Whittall, I.R., McDonagh, A.M., Humphrey, M.G., Samoc, M.: Organometallic complexes in nonlinear optics II: third-order nonlinearities and optical limiting studies. Adv. Organomet. Chem. **43**, 349–405 (1998)
[130] Powell, C.E., Humphrey, M.G.: Nonlinear optical properties of transition metal acetylides and their derivatives. Coord. Chem. Rev. **248**, 725–756 (2004)
[131] Cifuentes, M.P., Humphrey, M.G.: Alkynyl compounds and nonlinear optics. J. Organomet. Chem. **689**, 3968–3981 (2004)
[132] Houbrechts, S., Clays, K., Persoons, A., Cadierno, V., Gamasa, M.P., Gimeno, J.: Large second-order nonlinear optical properties of novel organometallic (σ-aryl-enynyl)ruthenium complexes. Organometallics **15**, 5266–5268 (1996)
[133] Houbrechts, S., Clays, K., Persoons, A., Cadierno, V., Gamasa, M.P., Gimeno, J., Whittall, I.R., Humphrey, M.G.: New organometallic materials for nonlinear optics: Metal σ-arylacetylides. Proc. SPIE-Int. Soc. Opt. Eng. **2852**, 98–108 (1996)
[134] Cadierno, V., Conejero, S., Gamasa, M.P., Gimeno, J., Asselberghs, I., Houbrechts, S., Clays, K., Persoons, A., Borge, J., Garcia-Granda, S.: Synthesis and second-order nonlinear optical properties of donor-acceptor σ-alkynyl and σ-enynyl indenylruthenium(II) complexes. X-ray crystal structures of [Ru{C≡CCH=C(C$_6$H$_4$NO$_2$-3)$_2$}(η^5- C$_9$H$_7$)(PPh$_3$)$_2$] and (*EE*)-[Ru{C≡C(CH=CH)$_2$-C$_6$H$_4$NO$_2$-4}(η^5-C$_9$H$_7$)(PPh$_3$)$_2$] Organometallics **18**, 582–597 (1999)
[135] Wu, I.-Y., Lin, J.T., Luo, J., Sun, S.-S., Li, C.-S., Lin, K.J., Tsai, C., Hsu, C.-C., Lin, J.-L.: Syntheses and reactivity of ruthenium σ-pyridylacetylides. Organometallics **16**, 2038–2048 (1997)
[136] Wu, I.-Y., Lin, J.T., Luo, J., Li, C.-S., Tsai, C., Wen, Y.S., Hsu, C.-C., Yeh, F.-F., Liou, S.: Syntheses and second-order optical nonlinearity of ruthenium σ-acetylides with an end-capping organic electron acceptor and thienyl entity in the conjugation chain. Organometallics **17**, 2188–2198 (1998)
[137] Tamm, M., Jentzsch, T., Werncke, W.: Complexes with (2,4,6-cycloheptatrien-1-ylidene) ethenylidene ligands: strongly polarized ruthenium(II) allenylidene complexes. Organometallics **16**, 1418–1424 (1997)
[138] Paul, F., Costuas, K., Ledoux, I., Deveau, S., Zyss, J., Halet, J.-F., Lapinte, C.: Redox-switchable second-order molecular polarizabilities with electron-rich iron σ-aryl acetylides. Organometallics **21**, 5229–5235 (2002)
[139] Fillaut, J.-L., Perruchon, J., Blanchard, P., Roncali, J., Golhen, S., Allain, M., Migalsaka-Zalas, A., Kityk, I.V., Sahraoui, B.: Design and synthesis of ruthenium oligothienylacetylide complexes. New materials for acoustically induced nonlinear optics. Organometallics **24**, 687–695 (2005)

[140] Xia, H.-P., Wen, T.B., Hu, Q.Y., Wang, X., Chen, X.-G., Shek, L.Y., Williams, I.D., Wong, K.S., Wong, G.K.L., Jia, G.-C.: Synthesis and characterization of trimetallic ruthenium and bimetallic osmium complexes with metal-vinyl linkages. Organometallics **24**, 562–569 (2005)

[141] Grund, A., Kaltbeitzel, A., Mathy, A., Schwarz, R., Bubeck, C., Vermehren, P., Hanack, M.: Resonant nonlinear optical properties of spin-cast films of soluble oligomeric bridged (phthalocyaninato)ruthenium(II) complexes. J. Phys. Chem. **96**, 7450–7454 (1992)

[142] Nalwa, H.S., Kobayashi, S., Kakuta, A.: Third-order nonlinear optical properties of processable metallo-naphthalocyanine dyes. Mol. Cryst. Liq. Cryst. Sci. Technol., Sect. B **6**, 169–179 (1993)

[143] Wang, S.-F., Huang, W.-T., Liang, R.-S., Gong, Q.-H., Li, H.-B., Chen, H.-Y., Qiang, D.: Transient nonlinear optics of organometallic fullerene: research on iron(III) and ruthenium(III) derivatives of C_{60}. J. Phys. Chem. B **105**, 10784–10787 (2001)

[144] Coe, B.J., Beyer, T., Jeffery, J.C., Coles, S.J., Gelbrich, T., Hursthouse, M.B., Light, M.E.: A spectroscopic, electrochemical and structural study of polarizable, dipolar ruthenium(II) arsine complexes as models for chromophores with large quadratic non-linear optical responses. J. Chem. Soc., Dalton Trans. 797–803 (2000)

[145] Coe, B.J., Harries, J.L., Harris, J.A., Brunschwig, B.S.: Molecular quadratic nonlinear optical properties of dipolar ruthenium(II) arsine complexes. Proc. SPIE-Int. Soc. Opt. Eng. **5212**, 332–340 (2003)

[146] Coe, B.J., Harries, J.L., Harris, J.A., Brunschwig, B.S., Coles, S.J., Light, M.E., Hursthouse, M.B.: Syntheses, spectroscopic and molecular quadratic nonlinear optical properties of dipolar ruthenium(II) complexes of the ligand 1,2-phenylenebis(dimethylarsine). Dalton Trans. 2935–2942 (2004)

[147] Coe, B.J., Curati, N.R.M.: Metal complexes for molecular electronics and photonics. Comments Inorg. Chem. **25**, 147–184 (2004)

CHAPTER 19

LINEAR AND NONLINEAR OPTICAL PROPERTIES OF SELECTED ROTAXANES AND CATENANES

JACEK NIZIOL, KAMILA NOWICKA AND FRANCOIS KAJZAR
Commissariat à l'Energie Atomique, DRT – LITEN, DSEN/GENEC/L2C, CEA Saclay, 91191 Gif sur Yvette, France

Abstract: Linear and nonlinear optical properties of catenanes and rotaxanes in thin films and in solution are reviewed and discussed. The compounds represent a new class of molecules, with mobile subparts. It offers a new kind of applications, particularly for optical switching. The rotational mobility of the subparts of these molecules was studied by the electro-optic Kerr effect. Both catenanes and rotaxanes can be processed into partly ordered thin films by vacuum sublimation. The degree of order may be controlled by an adequate chemical modification of the molecules, as it was observed in a series of substituted rotaxanes. Methods for controlling the motion of the components using light and electric fields are presented. The linear optical properties were studied by UV-Vis spectrometry and m-lines technique. The nonlinear optical properties were studied in solution and/or in thin films by the optical second and third harmonic generation techniques and by the quadratic electro-optic Kerr effect. The knowledge on the rotaxanes and the catenanes linear and nonlinear optical properties obtained by theses studies is important for the future construction of synthetic molecular machines and optical switching elements

Keywords: catenanes, rotaxanes, molecular switching, electro-optic effect, Kerr effect, photoisomerization, second harmonic generation, third harmonic generation, refractive index dispersion, absorption, molecular motors

1. INTRODUCTION

Catenanes and rotaxanes (for a review see [1, 2, 3, 4, 5, 6, 7, 8, 9, 10, 11, 12, 13]), belong to a new class of supramolecules composed from, mechanically bond, constituent smaller molecules, able to move independently. Due to this unique property these new class of organic molecules represent a great interest for potential applications in photonics, particularly in all optical and electro-optic switching [14, 15, 16, 17, 18].

The name of catenanes originates from latin *catena* which means a chain. Indeed these supramolecules are fundamentally made from interlocked macrocycles (Figure 1(a)) with, as already mentioned, ability of a relative movement of one macrocycle with respect to the another one(s) (pirouetting). The number of macrocycle is included in the used notation: [n] catenanes denote n interlocked chains. Up to now supramolecules of up to 4 macrocycles were synthesized. Large catenanes ($M_w = 10^5$) are present in nature in DNA as intermediates during the replication, transcription, and recombination process. Since the first two-ring catenane was obtained in early sixties, smaller synthetic catenanes ($M_w = 10^3$) have attracted the interest of chemists and physicists.

The name of rotaxanes takes its origin in latin too. *Rota* and *axis* mean, respectively wheel and axle. In the case of rotaxanes the macrocycle (or more of them) is locked onto a linear thread terminated on both sides with bulky stoppers (cf. Figure 1(b)). Rotaxanes exhibit more degrees of freedom than catenanes. The macrocycle can not only rotate along the thread (piroutetting as in the case of catenanes, cf. Fig. 2a) but can also move along it (shuttling, Fig. 2b). The shuttling movement of macrocycle is limited by stoppers located at the both ends of thread (two phenyl rings in the case of nitrone [2] rotaxane, cf. Figure 1(b)). Another degree of freedom consists on a "bending" movement of rotaxane thread which may lead to clipping. This could be realized by an appropriate design of thread (e.g. a photo isomerising one). It may be leading to a reversible (or irreversible) transformation of a [2] rotaxane into [2] catenane under an external stimulus (e.g. light), as shown in Figure 2c. This could be possible with threads containing alternate e.g. photoisomerizable –C = C– or –N = N– segments [19, 20, 21, 22]. Because of these different degrees of freedom in these molecules they represent a particular interest for the fabrication of the nanoscale (nano motors) devices [6, 9, 14, 16, 23, 24, 25, 26, 27, 28, 29].

The interest in the development of these supramolecules is triggered not only by the possible applications in photonics, but also in medicine, biology and everyday life. Indeed many phenomena of biological interest originate directly from the light induced and/or controlled mechanical motions at the molecular level. One of the very well known exemple is our vision which exploits the trans-cis isomerisation of

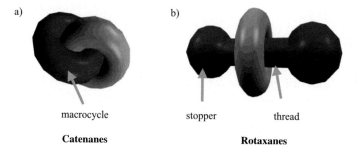

Figure 1. Schematic representation of a [2] catenane (a) and a [2] rotaxane (b)

Linear and Nonlinear Optical Properties

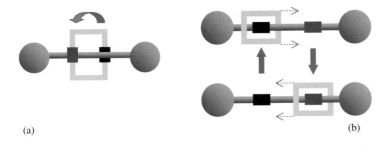

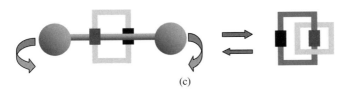

Figure 2. Schematic representation of different possible movements in rotaxanes: a) pirouetting, b) shuttling, and c) clipping, the last leading to the reversible (or irreversible) transformation of a [2] rotaxane into a [2] catenane

retinal molecule. Another famous example is the energy conversion in cells based on the rotary motion of the enzyme F_1-ATPase. But there are also other aspects of the use of these molecules to trap atoms, virus, transport atoms, molecules and as memory elements. Therefore it is important to know how different external stimulis like low (AC fields) and high frequency optical fields may control and/or induce some well defined mechanical motions.

Both catenanes and rotaxanes can be functionalized and their physico-chemical properties can be tailored by an adequate substitution. Some of them can be processed into good optical quality thin films by vacuum sublimation.

In this paper we review the linear and nonlinear optical properties studies performed on solutions or thin film of rotaxanes and catenanes. The linear optical properties were studied by UV-Vis spectrometry and m-line techniques. The nonlinear optical properties were studied by second (SHG) and third (THG) harmonic generation in thin films and by electro-optic Kerr effect measurements in solutions.

2. CHEMICAL SYNTHESIS AND MATERIAL PROCESSING

2.1 Synthesis of Catenanes and Rotaxanes

The early syntheses of catenanes and rotaxanes were mainly based on statistical threading approaches or on directed methodologies involving chemical conversion [30]. To obtain a catenated molecule, one ring must be closed in the presence of a

second ring ("clipping"). In the case of rotaxanes, there are three routes leading to their formation:
 (i) synthesis of a macrocycle and then of thread, followed of its capping on the ends
 (ii) slipping of a preformed ring over the stoppers of a preformed dumbbell-stopped component into a thermodynamically favorable site on the rod part of the dumbbell, and
 (iii) clipping of a preformed dumbbell with a suitable u-type component that is subsequently cyclized.

Catenanes and rotaxanes were first synthesized by Stoddart and coworkers [31, 32], Ashton and coworkers [33, 34, 35, 36] and by Sauvage and coworkers [37, 38, 39]. The strategies chosen by Stoddart and his co-workers are based on electron donor-acceptor interactions and hydrogen bonding between crown ethers and ammonium ions [32]. In the case of the donor-acceptor interactions, their work is mainly based on the combination of π-electron deficient bipyridinium and π-electron rich hydroquinone moieties. By employing a supramolecularly assisted synthetic methodology based on π-π stacking and [C-H···O] hydrogen-bonding interactions, they have self-assembled [2] catenanes [35]. Also rotaxanes based on π-donor-acceptor interactions have been prepared via both the threading and slippage procedure [34]. The rotaxanes in which the most important interaction is hydrogen bonding, are based on the inclusion complexes between ammonium ions and crownethers [33]. Among the strongest type of interactions, which are used in the synthesis of interlocked molecules, is the metal coordination of organic ligands. The synthesis and the studies of rotaxanes and catenanes, based on these interactions, have been performed by the group of Sauvage [37, 39]. They exploited the coordination of suitable ligands around a tetrahedral copper(I) to template the formation of interlocked molecules [38]. Once the Cu(I) complex is formed,

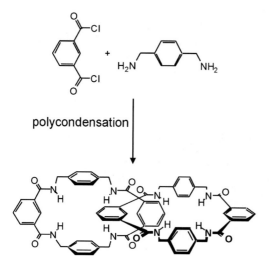

Figure 3. Synthetic route for catenanes as proposed by D. Leigh group (after [40, 41])

conventional organic reactions are employed to close the rings of catenanes and to attach the stoppers to rotaxanes.

Later on the group of D. Leigh, presently at the University of Edinburgh, proposed a simpler and more efficient route for the chemical synthesis of catenanes giving a high yield [40, 41]. The proposed method uses the commercially available para-xylene diamine and isophthaloyl chloride compounds and catenanes are obtained by their condensation in an appropriate solvent, as it is shown schematically in Fig. 3. As amphasized by the authors this reaction is very versatile and may be used for the synthesis of a large number of catenanes with different functional groups [41]. The chemical synthesis of the molecules whose studies are describe in this paper was done by the D. Leigh group. In the case of benzylic amide catenane it was derived from isophthaloyl dichloride and xylylene diamine with purity greater than 95% [40, 41].

The rotaxanes studied by us were also synthesied by the D. Leigh group using the third method. First the thread was prepared, as it's shown in Fig. 4. Than, by a condensation of triethylamine, isophthaloyl dichloride and xylylene diamine in chloroform, the rotaxanes were obtained [42, 43, 44]. We have studied more particularly the rotaxanes with the fumaric (fumrot) and nitrone (norot) threads (cf. Fig. 5).

2.2 Thin Film Processing

The simplest catenanes and rotaxanes such as benzylic amide [2] catenanes, fumrots and norots were successfully processed into thin films by using 3 techniques:
(i) vacuum sublimation
(ii) spinning
(iii) drawing

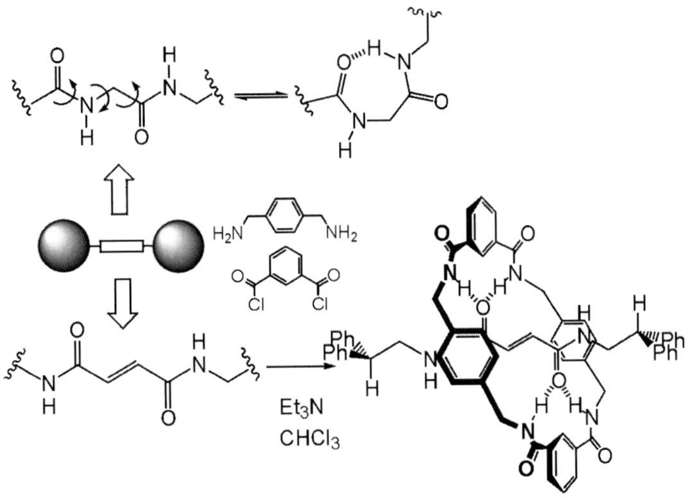

Figure 4. Synthetic route for rotaxanes as proposed by D. Leigh group [42, 43]

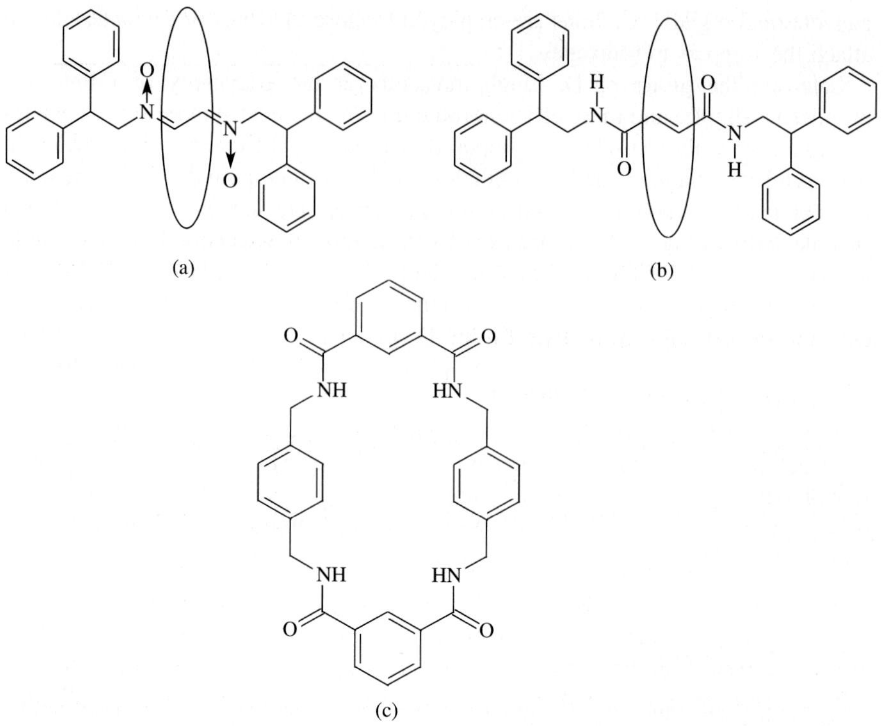

Figure 5. Chemical structures of norot (a), fumrot (b) and of macrocycle (c)

The first techniques yields good optical quality thin films with thickness controlled by the deposition time. Thin films with thickness varying from ca. 100 to 1000 nm were easily obtained. The conditions for thin film deposition were as follows:

Initial and final level of vacuum: -10^{-6} and 10^{-5} Torr, respectively
Crucible temperatures: $-220°$ C
Target temperature: $-25°$ C.
Deposition rate: – from 10 to 30 A/s.

The films were deposited on both glass and fused silica substrates. As example, Table 1 gives the thicknesses of vacuum deposited thin films determined by 3 different techniques:

Profilometry
Fabry Perrot intereference
m-lines technique

Within the experimental accuracy these techniques give the same result, showing the good homogeneity of the deposited films.

Both solution cast techniques led to very thin films, because of limited solubility. Due to their polycrystalline structure these films scatter highly light.

Linear and Nonlinear Optical Properties 615

Table 1. Comparison of the thin films thickness values as measured by use of different techniques

Sample	d (±30 nm) profilometry	d (±20 nm) m-line spectroscopy	d (nm) Fabry Perot interferometry
BAC-A	756	800	820 ± 20
BAC-B	350	388	336 ± 25
BAC-W	826	829	844 ± 40

3. LINEAR OPTICAL PROPERTIES

3.1 Optical Absorption Spectra

The absorption spectra of solution and thin films of studied catenanes and rotaxanes were recorded in transmission mode using a Perkin Elmer Lambda 19 spectrometer. The solution optical absorption of benzylic amide [2] catenane and of fumrot as well as of norot are shown in Figs. 6 and 7, respectively. The first macromolecule exhibit a large transparency range. Due to the absorption of silica cell we were able to measure only the edge of its absorption, which is defined by the absorption of phenyl rings of macrocycle. When the concentration is increasing the tail of absorption band is shifting towards larger wavelengths, as it is seen in Fig. 6.

In the case of fumrot and norot rotaxanes the transparency range is slightly smaller, due to the absorption by the conjugated thread. Its absorption band is

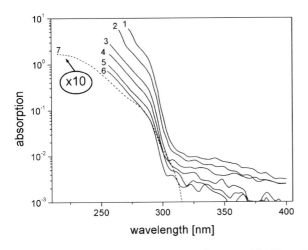

Figure 6. Edge of the absorption spectrum (in log scale) of benzylic amide [2] catenane as function of solution concentration. The increasing number correspond to decreasing concentration. The spectra 1-5 were recorded in solutions with a concentration of 7.5, 4.0, 1.5, 0.88, and 0.36 g/l respectively. The dashed line (7) is the spectrum of a solution in methanol with a concentration of about 0.04 g/l. Its optical density of 7 is multiplied by a factor of 10

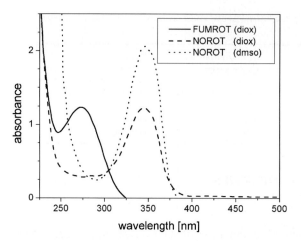

Figure 7. Absorption spectra of fumrot and norot. In the case of Norot two different solvents were used: dioxane and DMSO

located around 274 nm in the case of fumrot and around 346 nm in the case of norot, respectively. No solvatochromic effect was aserved when changing the solvent polarity (cf. Fig. 7).

The absorption spectra of vacuum sublimated thin films of catenanes and rotaxanes on fused silica substrates are shown in Figs. 8 and 9 respectively. Similarly as in the solution case the catenane thin films exhibit a large transparency range, as it is seen in Fig. 8. In this figure the optical absorption spectrum of a solution cast (spinning) thin film is also shown. This film exhibits a larger light scattering, due to its polycrystalline structure, as compared to the vacuum evaporated thin films and as it is seen from the tail of optical absorption spectrum. Due to the poor solubility of these molecules only very thin films were obtained.

We have deposited also thin films of a modified fumrot by functionalizing the macrocycle with NO_2 group, as it is seen in Fig. 10. The optical absorption spectra of modified in this way fumrot with mono and di-substitution are shown in Fig. 11. Later we will show that such modification, which doesn't alter significantly the absorption band, influence significantly the structural order and the value of refractive index through the influence of the substitution on the molecular packing in solid state.

3.2 m-lines Spectroscopy

The refractive indices of vacuum deposited thin films of the studied catenanes and rotaxanes were determined by the m-lines technique. It consists on finding solutions of an eigen equation for modes propagating in thin film. For a planar waveguide,

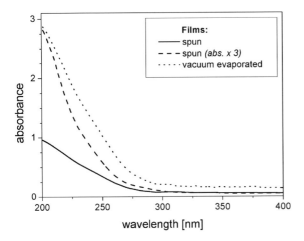

Figure 8. Absorption spectra of thin films the benzylic amide [2] catenane obtained by vacuum evaporation and by solution cast

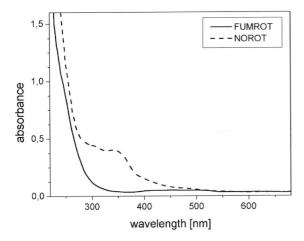

Figure 9. Absorption spectra of vaccum evaporated thin films fumrot (solid line) and norot (dashed line)

shown schematically in Fig. 12, this equation, derived from boundary condition for optical wave propagation in a thin layer, has the following form [45, 46]:

(1) $\quad 2kn_2 d \sin \Theta_m - 2\Phi_{23} - 2\Phi_{21} = 2j\pi$

where $j = 0, 1, 2 \ldots$ is the mode number, d is the waveguide thickness, k is the wavector, n_i's are refractive indices of the guiding layer (n_2) and of the buffer layers (n_1, n_2), respectively (cf. Fig. 12) and Φ_{ij} are the phase factors, which for

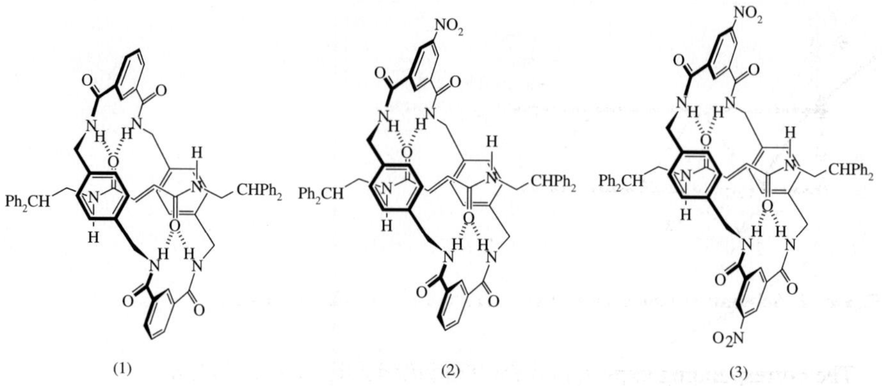

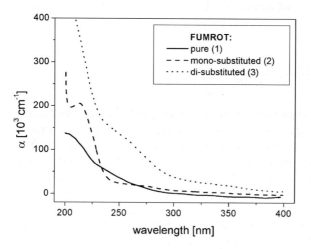

Figure 10. Chemical structure of pure (1), mono – (2), di – (2) substitued fumrot

Figure 11. Absorption spectra of vacuum deposited thin films of pure (1), mono – (2), di – (2) substitued fumrot. The observed modulations are most likely due to the Fabry–Perrot interferences

the TE polarization of propagating wave are given by

$$\Phi_{23} = tg^{-1}\left[\frac{(n_2^2 \sin^2 \theta_2 - n_3^2)^{1/2}}{n_2 \cos \theta_2}\right] \quad (2)$$

and

$$\Phi_{21} = tg^{-1}\left[\frac{(n_2^2 \sin^2 \theta_2 - n_1^2)^{1/2}}{n_2 \cos \theta_2}\right] \quad (3)$$

where θ_i's are propagation angles in corresponding media (cf. Fig. 12).

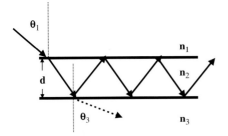

Figure 12. Schematic representation of wave propagation in a planar waveguide

The corresponding expressions for TM polarization are as follows

(4) $$\Phi_{23} = tg^{-1}\left[n_2^2\frac{(n_2^2\sin^2\theta_2 - n_3^2)^{1/2}}{n_3^2 n_2 \cos\theta_2}\right]$$

and

(5) $$\Phi_{21} = tg^{-1}\left[n_2^2\frac{(n_2^2\sin^2\theta_2 - n_1^2)^{1/2}}{n_1^2 n_2 \cos\theta_2}\right]$$

By measuring the coupling angles for a series of modes j with the experimental set up shown schematically in Fig. 13 and by solving the eigen equation (1) for a given mode propagation (or rather a set of modes) one can determine both the thin film thickness d and its refractive index for a given polarization of propagating wave (TE or TM). The precision depends on the number of modes propagating in thin film, thus on the thin film thickness and on the differences of refractive indices between the substrate and the measured film. Usually the upper buffer layer is air or vacuum ($n_1 = 1$). In this determination a precise knowledge of the substrate refractive index is required.

The m – lines technique allows also to measure the anisotropy of refractive index. Figure 14 shows, as example, the measured intensity dependence of reflected laser beam (no coupling into waveguide) on the interface coupling prism – air gap controlled by the point pressure on the substrate (e.g. a bolt, (cf. Fig. 13)) as function of the incidence angle. The coupling is manifesting by a dip occurring at the output light intensity. The deepness of this dip depends on the quality of thin film. A sharp, narrow and dip coupling curve is a finger print of good optical quality of thin films.

3.3 Refractive Index Dispersion

The measured refractive index dispersions for benzylic amide [2] catenane, fumrot and norot are displayed in Figs. 15–17, respectively. In all cases we observe a birefringence due to the order created during the thin film deposition and favored by

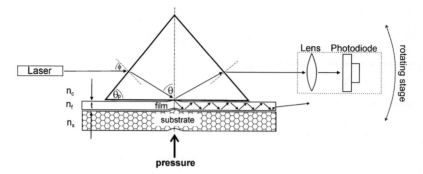

Figure 13. Schematic representation of experimental set-up for refractive index and thin film thickness determination by using the m-lines technique

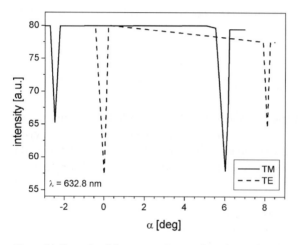

Figure 14. Example of the measured output intensity, as function of incidence angle using the experimental set up shown schematically in Fig. 13. The wave coupling corresponds to the dips in the measured output intensity

the structure of molecules. Apparently there is more phenyl rings oriented parallel than perpendicular to the substrate plane, as the ordinary index of refraction is larger than the extraordinary.

In the case of benzylic amide [2] catenane (cf. Fig. 15) we observe a large refractive index, larger than for common, organic and nonconjugated polymeric thin films. This is most likely due to a good packing, as it follows from Clausius-Masotti formula, linking index of refraction to the material density. The refractive indices of fumrot and norot are large too. It reflects influence not only of the conjugated thread, but also of packing. Indeed, when functionalizing the macrocycle by NO_2 substitution we observe the decrease of refractive index, thus worse packing and

Linear and Nonlinear Optical Properties

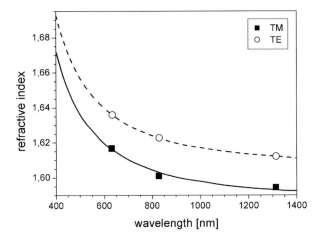

Figure 15. Refractive index dispersion in a vacuum evaporated thin film of benzylic amide [2] catenane

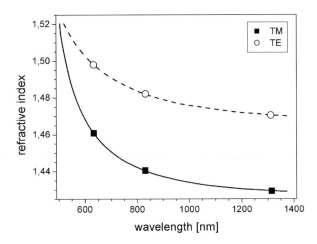

Figure 16. Refractive index dispersion in evaporated thin film of norot

decrease of order. For the di-substituted fumrot the vacuum evaporated thin films are isotropic, as it is seen in Fig. 18.

The dispersions of refractive indices of these materials can be well fitted by the Sellmeier formula

$$(6) \quad n^2 = n_0^2 + \frac{A}{\lambda_0^2 - \lambda^2}$$

where n_0 is refractive index at zero frequency and λ is the wavelength. In Table 2 we reported the values of least square fit parameters n_0, λ and A for selected BAC, fumrot and norot thin films.

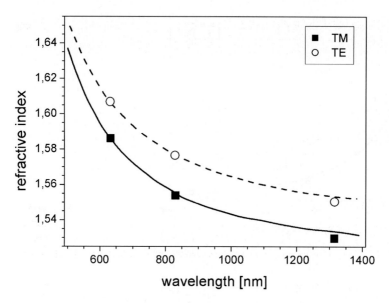

Figure 17. Refractive index dispersion in evaporated thin film of fumrot

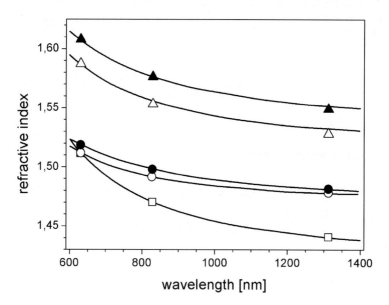

Figure 18. Refractive index dispersion in evaporated thin films of pure (triangles), monosubstitued (circles) and disubstitued fumrot (squares) (cf. Fig. 10). Full figures show ordinary whereas the open the extraordinary index of refraction, respectively. Solid lines are Sellmeier fits

Linear and Nonlinear Optical Properties

Table 2. Values for the Sellmeier's fit parameters (cf. Eq. 6) for selected thin films

Molecule	Polarization	n_0	$A(nm^2)$	λ_0 (nm)
BAC	TE	1.610	34500.00	198.00
	TM	1.588	33300.00	198.00
NOROT	TE	1.4624	35780.71	245.34
	TM	1.4229	39021.21	201.19
FUMROT	TE	1.5383	76728.1	201.01
	TM	1.5199	73538.5	208.2

3.4 Propagation Losses

One of the important factor determining the use of thin films in optics, particularly in waveguiding configuration are propagation losses defined as

$$(7) \quad PL = \frac{10}{d} \log_{10} \frac{I_d}{I_0}$$

where d is the propagation length, I_0 and I_d are the light intensities at the entrance to the medium and at the distance d, respectively. Usually d is expressed in cm giving propagation losses in commonly used units dB/cm.

The optical propagation losses in the vacuum evaporated benzylic amide [2] catenane thin films, measured in planar waveguide configuration [47, 48] were found to be PL = 2.8 ± 0.1 dB/cm at $\lambda = 1.32\,\mu m$ and PL = 4.0 ± 0.1 dB/cm at $\lambda = 1.55\,\mu m$, respectively. These values were determined by a two prism method [49]. As for polycrystalline thin films these value are significantly smaller han usually observed. It shows the ability of these molecules to form good optical quality thin films by using these technologically friendly technique. It shows also that the crystallites are very small, tens to a few hundreds of nanometers size.

4. NONLINEAR OPTICAL PROPERTIES

4.1 Definitions

Under the action of a strong electric field (DC or AC) the polarization of a medium is changing and its variation can be expanded, in dipolar approximation, into the power series of the forcing external field strength E giving

$$(8) \quad \Delta P = P(\omega_\sigma) - P_0 = K_1 \chi^{(1)}_{IJ}(-\omega_\sigma; \omega_\sigma) E_J^{\omega_\sigma}$$
$$+ K_2 \chi^{(2)}_{IJK}(-\omega_\sigma; \omega_1, \omega_2) E_J^{\omega_1} E_K^{\omega_2}$$
$$+ K_3 \chi^{(3)}_{IJKL}(-\omega_\sigma; \omega_1, \omega_2, \omega_3) E_J^{\omega_1} E_K^{\omega_2} E_L^{\omega_3} + \ldots$$

where P_0 is the static polarization (in absence of external field), $\chi^{(n)}$ is a three dimensional (n+1) rank tensor describing linear (n = 0) and nonlinear (n > 1) optical properties of a given material, K's are coefficients depending on the conventions used. For the Fourrier transform of electric field we use

$$(9) \quad E(r,t) = \frac{1}{2}E(r)\left[e^{i(\omega t - kr)} + c.c\right]$$

and similarly for the polarization field. We include also all degeneracy factors into K coefficients for a given process.

Similar expansion is valid on the molecule level for its dipole moment variation under the applied external electric field:

$$(10) \quad \Delta\mu_i(\omega_\sigma) = \mu_i(\omega_\sigma) - \mu_{0i} = K_1 \alpha_{ij}(-\omega_\sigma; \omega_\sigma)E_j$$
$$+ K_2 \beta_{ijk}(-\omega_\sigma; \omega_1, \omega_2)E_j E_k$$
$$+ K_3 \gamma_{ijkl}(-\omega_\sigma; \omega_1, \omega_2)E_j E_k E_l + \ldots$$

Due to the screening of external field by molecular field the field intervening in Eq. (10) is given by

$$(11) \quad E(\omega) = f_\omega E(\omega)$$

where f_ω is the local field factor giving the corresponding correction. For a cylindrical shape molecule this factor is equal to 1, while for a molecule with spherical symmetry it is given by

$$(12) \quad f_\omega = \frac{\varepsilon(\omega) + 2}{3}$$

where $\varepsilon(\omega)$ is the dielectric constant of material at optical frequency ω.

4.2 Second-order NLO Properties

The second order NLO effects are described by $\chi(2)$ susceptibility on macroscopic level and by first hyperpolarizability β on microscopic one. From symmetry consideration and within the dipolar approximation for a centrosymmetric molecule or a bulk material with center of inversion the corresponding quantities describing the NLO response are equal to zero. For a single crystal and for noninteracting dipole moments the macroscopic NLO susceptibilities can be obtained by transformation of β hyperpolarizability from the molecule reference frame to the laboratory system:

$$(13) \quad \chi^{(2)}_{IJK}(-\omega_\sigma; \omega_1, \omega_2) = f_I^{\omega_\sigma} f_J^{\omega_1} f_K^{\omega_2} \sum_n N^{(n)} \sum_{ijk} a_{iI}^{(n)} a_{jJ}^{(n)} a_{kK}^{(n)} \beta_{ijk}^{(n)}(-\omega_\sigma; \omega_1, \omega_2)$$

Where $N^{(n)}$ is the density of (n) molecular specie and a_{iJ} are Wigner's rotation matrices. Often this transformation leads to centrosymmetric bulk materials with

$\chi^{(2)} \equiv 0$ because of dipole–dipole interaction favoring usually their antiparallel alignment. Therefore a lot of efforts is done in order to get a noncentrosymmetric arrangement of molecules. In the case of partly ordered materials the relation (13) is replaced by an orientational average:

(14) $\quad \chi^{(2)}_{IJK}(-\omega_\sigma; \omega_1, \omega_2) = f_I^{\omega_\sigma} f_J^{\omega_1} f_K^{\omega_2} \sum_n N^{(n)} < \beta^{(n)}_{ijk}(-\omega_\sigma; \omega_1, \omega_2) >_{IJK}$

In the case of thin films with point symmetry ∞mm there are 2 nonzero $\chi^{(2)}$ tensor components: $\chi_{ZZZ}^{(2)}$ and $\chi_{XXZ}^{(2)}$, were Z is the symmetry axis. For historical reasons often in the literature "d" tensor is used to describe the second order NLO properties, with corresponding components defined as $d_{sp} = \frac{1}{2}\chi_{XXZ}^{(2)}$ and $d_{pp} = \frac{1}{2}\chi_{ZZZ}^{(2)}$. In practice "s" denotes the polarization of fundamental beam and "p" of the harmonic one, respectively.

Although the studied catenanes and rotaxanes are presumably centrosymmetric the vacuum deposited thin films of benzylic amide [2] catenane exhibit SHG ability with the susceptibilities given in Table 2. The $\chi^{(2)}(-2\omega; \omega, \omega)$ susceptibility was measured, using the experimental set up shown in Fig. 19 for 3 films with different thicknesses. The measurements were done at 1064.2 nm fundamental wavelength with 13 ns pules and 10 Hz operation rate. The films were rotated along an axis perpendicular to the propagation direction. The measurements were performed for two fundamental – harmonic beam polarization configurations: p-p and s-p, allowing to determine the nonzero diagonal and off diagonal $\chi^{(2)}(-2\omega; \omega, \omega)$ tensor components.

Figure 20 shows the incidence angle dependence of SHG intensity at pp fundamental – harmonic beam polarization configuration for a vacuum evaporated BAC thin film, showing very similar dependence as poled polymers with point symmetry ∞mm. This dependence can be well described by the formulas derived for this symmetry [50]. The fact that the measured quadratic NLO susceptibilities

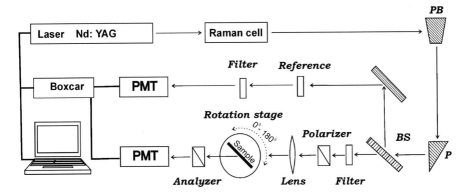

Figure 19. Experimental set-up used for harmonic generation measurements: BS – beam splitter, P – prism, PB – Pellin –Broca prism, PMT – photomultiplier

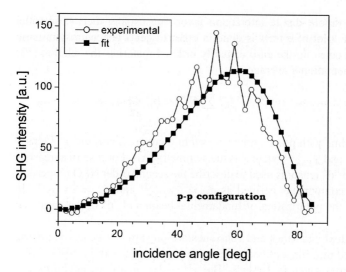

Figure 20. Incidence angle dependence of SHG intensity for a thin film of benzylic amide [2] catenane. Full squares depict a least square fit to experimental data with $\chi^{(2)}(-2\omega; \omega, \omega)$ suceptibility tensor components corresponding to ∞mm point symmetry of thin film

(cf. Table 2) don't depend on the thin films thickness is in favor of a bulk response and not a surface effect. The data were calibrated with SHG measurements on a single crystal plate of α-quartz ($\chi_{111}^{(2)}(-2\omega; \omega, \omega) = 0.6\,\text{pm/V}$ [51]). The average $\chi_{ZZZ}^{(2)}$ susceptibility found for these films is of about 0.025 pm/V.

The diagonal component of the linear electro-optical tensor was measured for a pristine vacuum evaporated BAC thin film by the modulated ellipsometry technique [52, 53, 54] at the wavelength $\lambda = 633\,\text{nm}$. It was found to be $r_{33} = (1.2 \pm 0.2)\,\text{pm/V}$ [47]. This is significantly larger than measured by the second harmonic generation technique (cf. Table 3) at the fundamental wavelength of $\lambda_f = 1064\,\text{nm}$ with $\chi^{(2)}_{ZZZ}(-2\omega; \omega, \omega) = 0.024\,\text{pm/V}$ for a film 106 nm thick.

Table 3. Diagonal $\chi_{ZZZ}^{(2)}(-2\omega; \omega, \omega)$ and off diagonal $\chi_{XXZ}^{(2)}(-2\omega; \omega, \omega)$ components of quadratic susceptibility for vacuum evaporated BAC thin films. The data were calibrated with SHG measurements on a single crystal plate of α-quartz carried out at the same conditions with uadratic susceptibility $\chi_{XXX}^{(2)}(-2\omega; \omega, \omega) = 0.6\,\text{pm/V}$ [51]

Sample	Thicknees (nm)	$\chi_{XXZ}^{(2)}(-2\omega; \omega, \omega)$	$\chi_{ZZZ}^{(2)}(-2\omega; \omega, \omega)$
		(pm/V)	(pm/V)
A	754	0.0086 ± 0.001	0.034 ± 0.004
B	354	0.0050 ± 0.0005	0.016 ± 0.002
C	106	0.0008 ± 0.00001	0.024 ± 0.002

From theoretical considerations a ratio of $r_{33}/\chi_{ZZZ}^{(2)} = 0.3$ is expected by taking into account only electronic contributions to the electro-optic coefficient, in a large disagreement with the observed ratio of 60 [47, 48].

Although the origin of the observed nonzero $\chi_{ZZZ}^{(2)}$ susceptibility is not very clear, it may be either quadrupolar, or of higher order. It may be also due to an intrinsic noncentrosymmetry in vacuum deposited thin films of benzylic acid [2] catenane. The observed large linear electro-optic (Pockels) effect is also in favor of an intrinsic noncentrosymmetry. Moreover, it was also shown that the linear electro-optic effect depends on the applied external field, increasing with its strength. This observation was tentatively interpreted by the field induced mobility of, most likely, macrocycles.

SHG measurements were performed also on vacuum deposited thin films of fumrot. Within the experimental sensitivity, no SHG was observed on pristine sublimed film, as it was the case of catenane films. However these films have shown the ability to be poled by an external DC field (corona poling). Significantly larger values of $\chi_{ZZZ}^{(2)}$ susceptibility, of up to 6.8 pm/V at 1064 nm fundamental wavelength, were observed. The poling kinetics was very similar to that observed in poled polymers, as it is shown in Fig. 21. The poling efficiency depends on poling temperature, thus on the thermal mobility of molecules. There is a temperature range (cf. Fig. 22) at which the poling efficiency is increasing. However, above a certain temperature this efficiency starts to decrease. It's also a very similar behaviour to that observed in polymers, where the limiting temperature corresponding to the maximum poling temperature is the glass transition temperature.

Also, similarly as in functionalized polymers the induced orientation is unstable in time due to the relaxation of induced polar order. The kinetics of relaxation

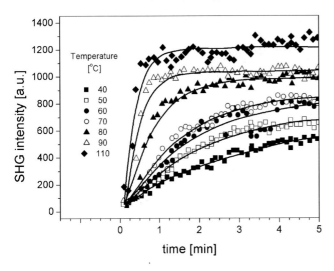

Figure 21. Temporal growth of SHG intensity for fumrot during the corona poling at different temperatures

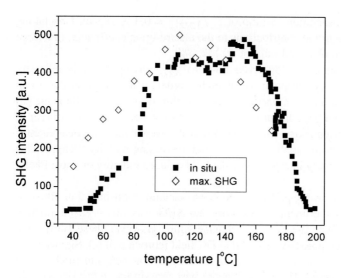

Figure 22. Temperature variation of corona poling efficiency as measured by SHG: in situ (squares) or by taking maximum of SHG signal after poling (diamonds)

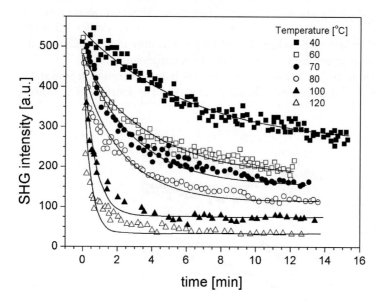

Figure 23. Temporal decay of SHG intensity for corona poled fumrot at different temperatures [20]

depends also on temperature, as it is seen in Fig. 23. The time constants dependence corresponding to the poling and to the relaxation as function of temperature exhibit very similar behavior, as it is seen in Fig. 24. However the relaxation time constants are larger than that of for orientation, what has interesting practical implications.

Linear and Nonlinear Optical Properties

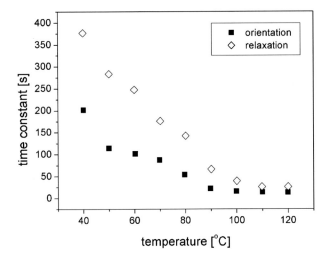

Figure 24. Temperature variation of time constants of polar orientation (squares) and relaxation (diamonds) in corona poled rotaxane [20]

4.2.1 Optical depoling

Recently it was observed that shining with light in the absorption band of poled polymers, functionalized with NLO chromophores leads to a reversible destruction of polar order [55, 56]. The order is restored (under the DC field) when the light is switched off. The amount of polar order remained constant in the case of PMMA functionalized with Disperse Red 1 for a large number of on – off cycles with the light, whereas in the case of zwiterionic chromophores an increase in polar order, with its better temporal stability was observed [56]. We have applied similar treatment to the poled norot films. At the beginning we observed a reversible destruction of polar order as in preceding cases, however the amplitude of variation was decreasing with the number of cycle (cf. Fig. 25), leading to a constant amount of polar order. Most likely the light induced some mobility to the rotaxane molecule (or rather its subparts) and after some number of cycles the system is locked. Again it's an interesting result concerning the practical application of these molecules as such a behavior leads to stable in time noncentrosymmetric structure.

4.3 Third-order NLO Properties

The third-order NLO effects are described by $\chi(3)$ susceptibility on macroscopic level and by second hyperpolarizability γ tensor on microscopic one. In contrary to the second order NLO effects the third order effects are present in all molecules and in bulk materials. There exists also a similar relationship between the corresponding bulk susceptibilities and the molecular hyperpolarizabilities as in the

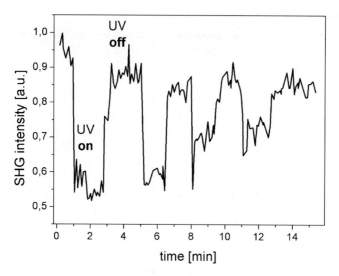

Figure 25. Temporal variation of the SHG intensity from the corona poled rotaxane thin film subjected to the action of light in its absorption band

case of 2nd order NLO effects. For a single crystal it is given by the following expression:

$$(15) \quad \chi^{(3)}_{IJKL}(-\omega_\sigma; \omega_1, \omega_2, \omega_3) = f_I^{\omega_\sigma} f_J^{\omega_1} f_K^{\omega_2} f_L^{\omega_3} \sum_n N^{(n)}$$

$$\times \sum_{ijk} a^{(n)}_{iI} a^{(n)}_{jJ} a^{(n)}_{kK} a^{(n)}_{lL} \gamma^{(n)}_{ijkl}(-\omega_\sigma; \omega_1, \omega_2, \omega_3)$$

(for notations cf. Eq. (13)).

Similarly, for a disordered system the cubic susceptibility is given as a configurational average over all non zero γ tensor components:

$$(16) \quad \chi^{(2)}_{IJKL}(-\omega_\sigma; \omega_1, \omega_2, \omega_3) = f_I^{\omega_\sigma} f_J^{\omega_1} f_K^{\omega_2} f_L^{\omega_3}$$

$$\times \sum_n N^{(n)} < \gamma^{(n)}_{ijkl}(-\omega_\sigma; \omega_1, \omega_2, \omega_3) >_{IJKL}$$

The nonlinear optical properties of rotaxanes and catenanes were studied mainly by three techniques: the optical second and third harmonic generation and the electro-optic Kerr effect. As already mentioned, the harmonic generation techniques give the fast, electronic in origin, molecular and bulk hyperpolarizabilities, whereas the electro-optic methods are sensitive to all effects which induce optical birefringence, such as e.g. the rotation of molecules. Therefore the last technique is very useful to study the rotational mobility of molecules and/or their parts.

4.3.1 Third harmonic generation measurements

The THG measurements were done in solution of catenanes and in thin films of catenanes and rotaxanes using the experimental setup shown in Fig. 18. The measurements were performed at 1 064.2 and 1 907 nm. The last wavelength was obtained by first Stock shift of the fundamental 1 064.2 nm beam in a Raman cell. Similarly as in SHG experiments the TH intensities were collected as function of incidence angle when rotating the film along an axis perpendicular to the beam propagation direction. Figure 26 shows an example of such a dependence for a vacuum evaporated catenane thin film. The nonzero minima are due to the contribution from thin film. All THG experiments were performed in vacuum in order to avoid the air contribution [57]. The harmonic intensity, $I_{3\omega}$, generated by a thin film deposited on a substrate is given by the following expression [58, 59].

$$(17) \quad I_{3\omega}(\theta) = \frac{64\pi^4}{c^2} \left| \frac{\chi^{(3)}(-3\omega;\omega,\omega,\omega)}{\Delta\varepsilon} \right|_s^2$$
$$\left| \exp\left[i\left(\varphi_\omega^s + \varphi_{3\omega}^p\right)\right] \{T_1\left[1 - \exp(-i\Delta\varphi_s)\right] + \rho T_2 \exp(i\Delta\varphi_f)\left[\exp(i\Delta\varphi_f) - 1\right]\} \right|^2 I_\omega^3$$

where I_ω is the intensity of the fundamental beam, $\chi^{(3)}(-3\omega;\omega,\omega,\omega)$ is the cubic susceptibility describing the THG process

$$(18) \quad \rho = \left(\frac{\chi^{(3)}}{\Delta\varepsilon}\right)_f / \left(\frac{\chi^{(3)}}{\Delta\varepsilon}\right)_s$$

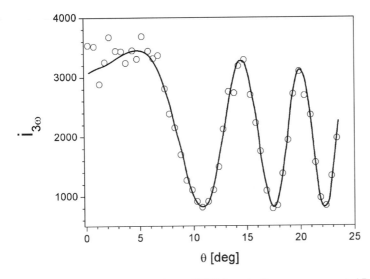

Figure 26. Incidence angle dependence of THG intensity for vacuum evaporated BAC film at 1064 nm fundamental wavelength. Circles are measured values whereas solid line depicts the calculated ones [47]

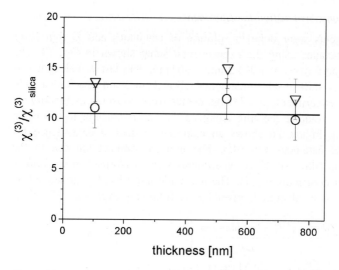

Figure 27. Thickness dependence of the ratio $\chi^{(3)}_{BAC}(-3\omega;\omega,\omega,\omega)/\chi^{(3)}_{silica}(-3\omega;\omega,\omega,\omega)$ at 1064 nm (circles) and 1907 nm (triangles) fundamental wavelength [76]

is the ratio of the cubic susceptibility of the film (f) to that of substrate (s)

(19) $\quad \Delta\varepsilon = \varepsilon_{3\omega} - \varepsilon_{\omega}$

is the dielectric constant dispersion of substrate

(20) $\quad \Delta\varphi = \varphi_{\omega} - \varphi_{3\omega} \quad (i = f, s)$

is the phase mismatch between the fundamental and the harmonic beams, respectively, and is given by

(21) $\quad \varphi_{\omega(3\omega)} = \dfrac{6\pi d n_{\omega(3\omega)}}{\lambda_{\omega}} \cos\theta_{\omega(3\omega)}$

where d is the medium thickness and λ_{ω} is the fundamental beam wavelength.

Figure 26 shows an example of the fit of Eq. (17) to experimental data. A very good agreement is obtained. Such a fit allows to determine precisely the ratio ρ (cf. Eq. (18)) and the phase of thin film third order NLO susceptibility if independent THG measurements are done on substrate at the same conditions. At the same time the substrate serves to calibrate the data if its THG susceptibility is known. In our case we used silica as substrate and for calibration the $\chi^{(3)}(-3\omega;\omega,\omega,\omega)_{silica}$ values of 2.8×10^{-14} esu at 1 907 nm [60] and 3.1×10^{-14} esu at 1064 nm [57].

The measured values for representative molecules are collected in Table 4. As expected they can be well interpreted in terms of the bond additivity model [47],

Table 4. Third order NLO susceptibilities and molecular second hyperpolarizabilities of simple catenane and rotaxane molecules as determined by the optical third harmonic generation technique

Molecule	Wavelength (nm)	n_ω	$n_{3\omega}$	l_c (nm)	$\chi^{(3)}(-3\omega; \omega; \omega; \omega)$ 10^{-13} esu	$\gamma(-3\omega; \omega; \omega; \omega)$ 10^{-34} esu
BAC	1064			.	2.9	0.65
Fumrot	"	1.5601	1.8174	689.39	3.77	1.49
Norot	"	1.4732	1.6376	1082.16	5.79	1.67
BAC	1907			.	2.9	0.65
Fumrot	«	1.5452	1.6062	5210.38	4.17	1.99
Norot	"	1.4658	1.4978	9895.8	6.86	2.13

showing no enhancement in ultra fast $\chi^{(3)}$ susceptibility owing to the pecular structure and mobility of these supramolecules.

4.3.2 Quadratic electro-optic effect

Electro-optic effects refer to the changes in the refractive index of a material induced by the application of an external electric field, which "modulates" their optical properties [61, 62]. Application of an applied external field induces in an optically isotropic material, like liquids, isotropic thin films, an optical birefringence. The size of this effect is represented by a coefficient B, called Kerr constant. The electric field induced refractive index difference is given by

(22) $$\Delta n = n_{II} - n_\perp = B\lambda E^2$$

where n_{II}, $n_\perp$ are refractive indices parallel and perpendicular to the applied external field, respectively, λ is the wavelength of the incident light, E is the strength of the applied electric field, and B is the Kerr constant. This electric field induced birefringence induced a phase change $\Delta\Phi$ of the propagating beam in a Kerr cell of thickness l under an applied electric field E which is given by

(23) $$\Delta\Phi = \Delta k l = \frac{2\pi}{\lambda} l \Delta n = 2\pi l B E^2$$

where k is the wave vector and l is the interaction length of the light with wavelength λ within the material. The applied electrical field E is given by $E(t) = E_\sim \sin \omega t$, where $E_\sim$ is its amplitude. Thus, Equation (23) takes a special form

(24) $$\Delta\Phi = 2\pi l B \left[\frac{1}{2} E_\sim^2 - \frac{1}{2} E_\sim^2 \cos 2\omega_{el} t \right]$$

That means that, beside a time-independent phase shift, there exists a phase shift of the frequency $2\omega_{el}$, where ω_{el} is the circular frequency of the electric field. The Kerr cell is a set of two parallel electrodes placed between crossed polarizers. Thus the optical signal s behind the crossed polarizers is given by

$$(25) \quad s(t) = s_{\max} \sin^2\left(\frac{\Phi(t)}{2}\right)$$

where $\Phi(t)$ is the time dependent phase difference between components of electric field vector of the laser beam parallel and perpendicular to the electric field generated by the metal electrodes; $s_{\max}$ is the maximum signal going through the setup at a phase difference $\Phi = \pi$. When we replace the phase change $\Delta\Phi$ by the signal amplitude $s_\sim$ ($s_\sim = \frac{1}{2} s_{\max} \pi l B E^2$) at the phase shift $\Phi = \pi/2$, the electrode length l by the distance d of the gap between the ITO layers, the Kerr constant in the case of parallel glass plates is [61]:

$$(26) \quad B = \frac{G}{\pi} \frac{s_\sim}{s_{\max}} \left(\frac{d}{U^2}\right)$$

where U is the effective value of the applied voltage and G is a geometrical factor [62]. Figure 28 depicts the experimental set up for electro-optic Kerr effect measurements in solution. The light source is a cw He-Ne laser operating at the wavelength of 633 nm. The laser beam is propagating through a polarizer (P) and than a Soleil-Babinet compensator (SBC). The incident beam polarization is fixed at 45 degrees with respect to the incidence plane. We used SBC to introduce an extra phase shift between s and p polarizations of incident wave in order to control the phase shift between the components of the electric field vector perpendicular and parallel to the electric field generated by the electrodes. Afterwards, the beam is sent into the Kerr cell, whose normal is tilted by an angle of ca. 45° with respect to the beam propagation direction. The Kerr cell consists of the material being examined, placed in an insulating container with two electrodes attached to supply the necessary electric field. Next, the light passes through another polarizer (the analyzer (A)). Finally, the light is filtered from background and scattered light by using appropriate interference filter and sent into a photodetector. Initially, the

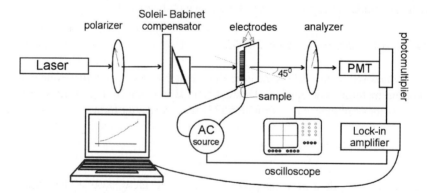

Figure 28. Schematic representation of the experimental set up for electro-optic Kerr effect measurements. The normal to the sample is tilted by 45 degrees with respect to the beam propagation direction and the incident polarization makes an angle of 45° with incidence plane

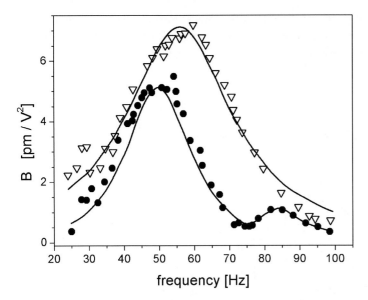

Figure 29. Frequency dependence of Kerr constant B (in pmV^{-2}) for fumrot and norot in dioxan solution 1 and 2. a, Kerr constant of 1 (triangles) and 2 (circles) as a function of the AC frequency of the electric field in dioxane at 296 K. Applied voltage U = 7 V; separation between electrodes = 20 μm (that is, an electric field strength of 0.35 Vμm^{-1}). Solid lines are Lorentz fits (maxima at 57.7 Hz, 53.3 and 82 Hz [63]

polarizer-Kerr cell-analyzer combination is adjusted so that no light is transmitted in the absence of the external electric field. By applying the external electric field is to the Kerr cell one creates optical birefringence (cf. Eq. (22)) and consequently a phase mismatch between the s and p polarized waves resulting in a small signal in detector. By using the lock-in detection, tuned at $2\omega_{el}$ frequency with the use of appropriate filters very weak Δn values can be measured.

As already mentioned, the electro-optic Kerr effect measurements in solution permits to test the rotational mobility of molecules, or their parts, under the applied low frequency AC field. Applying this technique for rotaxane solutions [63] a resonance enhancement in Kerr constant B at the frequency of 57.7 for norot at the external field strength 0.35 V μm^{-1} (cf. Fig. 29) was observed the first time. For similar fumrot solutions a more complex frequency dependence of Kerr constant is seen, with a strong resonance at 53.3 Hz, accompanied by a shoulder at 83 Hz (cf. Fig. 29). The generated signals are large due to large birefringences created by rotation of these big molecules. At the concentration of 10^{-7} mol dm^{-3} the measured Kerr constant was comparable to that of liquid nitrobenzene (44 000 10^{-16} m V^{-2}) [64]. The resonance frequencies depend on the strength of applied electric field and on temperature as its is seen in Figs. 30 and 31, respectively. It shifts toward lower values when the strength of electric field is increasing. It shows that it is possible to control the speed of rotation of macrocycle by electric field. The decrease of the pirouetting rate with the field strength shows braking effect of electric field, as expected. The increase of temperature (cf. Fig. 31) leads to the increase of resonance frequency, which is also expected because of temperature dependence of solution viscosity.

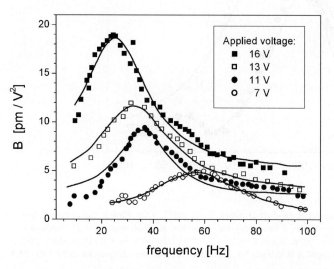

Figure 30. Field-strength-dependence of the Kerr constant a function of the AC frequency of the electric field at 296 K. Applied voltage U = 7 V (open circles, maxima at 56 Hz), 11 V (full circles, maxima at 36 Hz), 13 V (open squares, maxima at 33 Hz) and 16 V (bfull squares, maxima at 25 Hz)

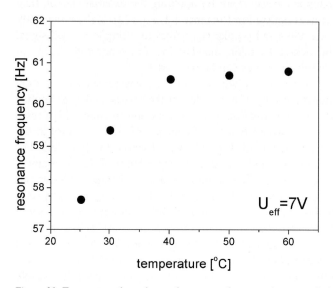

Figure 31. Temperature dependence of resonance frequency in norot solutions in dioxan

The rotation of macrocycle around the thread in fumrot and norot was also observed by the ^{1}H NMR spectroscopy [63], however at frequencies significantly larger than those observed in the Kerr effect measurements: 405 Hz and 310 Hz for fumrot and norot, respectively. As already mentioned (cf. Fig. 30) the resonance frequency is increasing with decreasing field strength. The measurements of resonance frequency at different field strength values permitted to extrapolate the resonance frequency to zeroth field strength, finding it slightly larger than that observed by NMR spectroscopy.

Large electro-optic Kerr effect was also observed in BAC solutions in dioxan (cf. Fig. 32). At the concentration of 10^{-4} mol dm^{-3} and at the field strength of 1.6 V μm^{-1} a resonance in B constant at the frequency value close to 2000 Hz was observed [65]. This frequency is significantly larger than that observed in rotaxanes, but smaller than measured by NMR. Again, we believe, this difference can be explained by the braking effect of external field on the macrocycle rotation. Indeed it was also found [65] that the resonance frequency depends on the external field strength, shifting to lower values with its increasing strength.

4.3.3 Trans-cis izomerization (clipping movement)

The molecular structure of the fumaramide and nitrone thread predispose them to bending under the light illumination and/or through heating through the already mentioned trans-cis izomerization process. This is expected to lead to a clipping

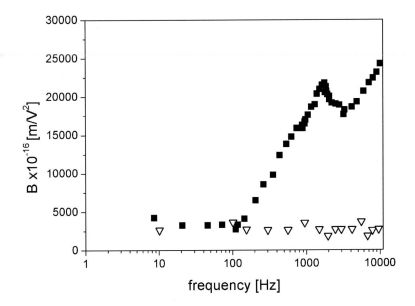

Figure 32. Frequency dependence of electro-optic Kerr constant B for BAC (squares) macrocycle (triangles) solutions in dioxan. The applied voltage was equal to 40 V and the distance between electrodes of 25 μm [65]

movement. The first studies made on silica deposited thin films indicated such a movement through 3 observations due to the irradiation at the tread single – double band absorption band:
  (i) reversible shift of optical absorption spectrum
  (ii) reversible change of the thin film Fabry – Perot fringes pattern
  (iii) reversible variation of the $\chi^{(3)}(-3\omega; \omega, \omega, \omega)$ susceptibility, after its initial decrease

However, the IR studies made on thin films of pure trans and cis forms of fumrot, deposited on CaF_2 substrates show a little variation of their IR specrum after UV irradiation. It means that the energy barrier for this transformation is very high. The IR spectra in the $(1000–4000)\,cm^{-1}$ region of the studied rotaxane films are shown in Figure 33. The bands due to the ν hydrogen stretching vibration of the trans isomers appears at 3326, 3246 and $3049\,cm^{-1}$, whereas bands associated with the ν carbon-carbon double stretching vibrations appear at 1667, 1627, 1539 and $1438\,cm^{-1}$. For the cis form there are two hydrogen bands at 3279 and $3063\,cm^{-1}$, and carbon-carbon double bands at 1647 and $1526\,cm^{-1}$. In order to check the photoisomerization process the infrared absorption spectra were collected in function of the irradiation time. The rotaxane film was irradiated with an UV mercury vapors lamp at 365 nm. The absorbance was measured irradiating the sample and than during the relaxation. Figure 34 shows the variation of the $1539\,cm^{-1}$ band during and after UV irradiation. These measurements were performed over a long period of time, to study in detail the thermal cis-trans back-relaxation process. After few minutes the absorbance increases. This behavior is essentially due to the angular redistribution of trans and cis isomers according to their capability to rotate by random thermal interactions. Six hours after irradiation the absorption

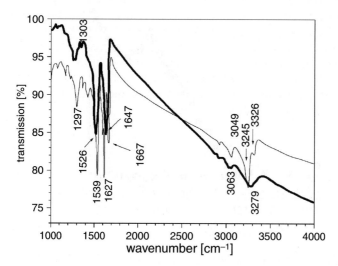

Figure 33. IR asorption spectra of the pure trans (thinner line) and pure cis (thicker line) forms of fumrot

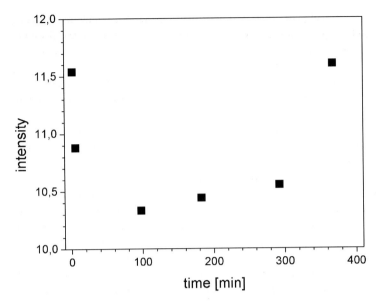

Figure 34. Variation of the transmission intensity of the carbon-carbon double band at 1539 cm^{-1} as a function of time

spectra almost recover their initial shape. The observation of the transient changes in the optical absorption spectrum after photoexcitation allows direct quantitative monitoring of the submolecular translational process.

Recently, Gatti et al. have shown [19] that the rate of rotation of the interlocked components of fumaric-derived [2] rotaxanes can be accelerated by isomerizing them to the corresponding maleamide [2] rotaxanes. Using light, the researchers were able to disrupt the hydrogen-bonding motif between the ring and the thread, drastically reducing energy barriers and allowing the components to move with respect to each other at far greater speeds. As a result, the rate of the pirouetting motion was increased approximately by six orders of magnitude. In the second type of rotaxane, intermolecular photoinduced electron transfer was used to induce a reversible shuttling motion on a time scale of microseconds.

5. CONCLUSIONS

In conclusion, we have reviewed the linear and nonlinear optical properties of simplest benzylic acide [2] catenanes and benzylic amide [2] rotaxanes. These molecules can be processed into good optical quality thin films with a high refractive index, which can be modified by a subsequent functionalization, as it was shown in the case of rotaxanes.

We have shown that the electro—optic Kerr effect in solution is a useful tool to study the relative motion of these molecules. We observed a strong birefringence at frequencies that correspond to the rate at which the molecular cycle "pirouettes" around another component. The rotation rate of macrocycle in catenane solutions is significantly larger (more than one order of magnitude) than in rotaxanes. In both cases the rate of rotation depends on the strength of applied field and decreases with its increase. The rotation rate depends also on temperature and increases with its increase as it was observed in the case of rotaxanes. It shows that these molecules re interesting, functional materials, whose properties can be controlled under external excitation, like electric field and light.

Rotaxanes were proposed as challenging materials in areas covering molecular electronics like for example switches [66], molecular wires [14], and finally, logic gates [14]. These molecules are already known for their binary reversible states arising from shuttling of the macrocycle along the thread.

Many physical phenomena have been exploited to distinguish between two positions of the shuttling macrocycle. In particular the interest was drawn by optical properties. For example the UV-vis absorption spectroscopy and circular dichroism measurements were applied [67]. The authors studied peptido[2]rotaxanes consisting of an intrinsically achiral benzylic amide macrocycle locked onto various chiral dipeptide. Their results, although very interesting required lot of experimental skill since the dichroic induced changes manifested in the wavelength region below 300 nm, where most of solvents start to absorb the light. A competitive, convenient and cost effective methods relies on fluorescence output. The reason is accessibility of high performance laser able to excite in a tiny volume without producing sideproducts. Some materials adapted for this technique have already been reported [68, 69, 70]. The light is therein used both to induce macrocycle movement and "read" its position relative to the thread terminations. For example, the authors [70, 71] fabricated [2]rotaxanes terminated by two different fluorescent stoppers. The thread is terminated by 4-amino-1,8-naphtalimide-3,6-disulphonic disodium salt ("N" stopper) and 1,8-naphtalimide-5-sulphonic sodium salt ("S" stopper). The macrocycle in bell shape was α-cyclodextrin. Under illumination with the 360 nm line, the molecule changes its isomer form from cis to trans. Excitation with the 430 nm line reverses the process. The maximum of fluorescence originating from "N" stopper is around of 520 nm and that of its counterpart in the vicinity of 395 nm. It was shown that in the presence of the macrocycle enhanced characteristic fluorescence of the neighbouring stopper. In that way the total fluorescence spectrum can be altered by shuttling movement of the macrocycle. The experiment was done in solution.

Fluorescence spectroscopy was used also as a detection tool in case of a peptide based [2]rotaxane [72]. It was observed that in polar solvents like dimethyl sulfoxide (DMSO) fluorescence emission spectra were virtually same like those of the thread alone. Since fluorescence origins in this rotaxane from anthracene group, it was concluded that hydrogen bonds linking the macrocycle to the anthracene-based stopper were broken and the macrocycle resided on the alkyl side of the thread. When the solvent was hardly polar like 1,4-dioxane, the macrocycle rest bonded to the anthracene stopper and the spectrum broadened and red-shifted. By mean of time-resolved fluorescence spectroscopy, the authors demonstrated that it was possible to induce the macrocycle motion on nanosecond scale without any other external assistance but photons.

Finally, the anthracene-based [2] rotaxanes were studied in single layer OLED architecture [73, 74]. The measurements revealed new features in electroluminescence spectra comparing to photoluminescence both in solid state and in solution. However, the origin of the phenomena is not still well understood and far to be exploited in applications.

Polymerized [2]rotaxanes were claimed as good chemical vapour sensors [75]. In thin film form they were sensible to phenol vapours as well as to other H-bond donors such as p-nitrophenol or 2,2,2-trifluoroethanol. The observed phenomena was reversible and resulted in fluorescence quenching accompanied by a slight bathochromic shift. It was also found that the polymers were apt to metal bonding due to the presence of the tetrahedral pockets. This fact manifested itself in the appearance of an additional absorption band. The sensitivity of a given polymer thin film was proportional to the film porosity defined by steric properties of the R-substituant. The studies and applications of these molecules are just at the beginning.

ACKNOWLEDGEMENTS

This work was performed within the European Community projects ENBAC, EMMMA (contract n°: HPRN-CT-2002-00168) and MechMol (IST-2001-35504). The authors would like to thank Drs V. Bermudez, D. Grando and T. Gase, who participated in these studies as post-docs funded by European project ENBAC, and to Dr. P.-A. Chollet for their contributions to these studies. Thanks are also due to the other partners in this projects, and more particularly to the group of Prof. D. Leigh at University of Edinburgh for chemical synthesis of the studied compounds

REFERENCES

[1] Gomez-Lopez, M., Preece, J.A., Stoddart, J.F.: Nanotechnology, **7**, 183–192 (1996)
[2] Schill, G.: Catenanes, Rotaxanes and Knots, Academic, New York (1971)
[3] Raymo, F.M., Stoddart, J.F.: Templated Organic Synthesis. (Diederich, F., Stang, P.J., (eds.) Weinheim: Wiley-VCH (2000)
[4] Credi, A.: Molecular – Level Machines and Logic Gates, PhD Thesis, Bologna (1998)
[5] Molecular Machines: Special Issue of Acc. Chem. Res. **34(6)**, 409–522 (2001)
[6] Sauvage, J.-P.: Acc. Chem. Res. **31**, 611–619 (1998)
[7] Sauvage, J.-P. (ed.), Molecular Machines and Motors. Springer, New York (2001)
[8] Balzani, V., Gómez-López, M., Stoddart, J.F.: Acc. Chem. Res. **31**, 405–414 (1998)
[9] Balzani, V., Credi, A., Venturi, M.L.: Molecular Devices and Machines; A Journey into the Nano World, Willey, Weinham (2003)
[10] Flood, A.H., Ramirez, R.J.A., Deng, W.-Q., Muller, R.P., Goddard, W.A., Stoddart, J.F.: Aust. J. Chem. **57**, 301–322 (2004)
[11] Balzani, V., Credi, A., Raymo, F.M., Stoddart, J.F.: Angew. Chem. Int. Ed. **39**, 3348–3391 (2000)
[12] Kidd, T.J., Leigh, D.A., Wilson, A.J.: J. Am. Chem. Soc. **121**, 1599–1600 (1999)
[13] Spencer, J.F., Stoddart, C., Vicent, Williams, D.J.: Angew. Chem. Int. Ed. Engl. **28**, 1396 (1989)

[14] Collier, C.P., Wong, E., Belohradsky, M., Raymo, F.M., Stoddart, J.F., Kuekes, P.J., Williams, R.S., Heath, J.R.: Science **285**, 391–394 (1999)
[15] Pease, A., Jeppesen, R., Stoddart, J.F., Luo, Y., Collier, C.P., Heath, J.R.: Acc. Chem. Res. **34**, 433–444 (2001)
[16] Collier, C.P., Mattersteig, G., Wong, E.W., Luo, Y., Beverly, K., Sampaio, J., Raymo, F.M., Stoddart, J.F., Heath, J.R.: Science **289**, 1172–1175 (2000)
[17] Bissell, R.A., Cordowa, E., Kaifer, A.E., Stoddart, J.F.: Nature **369**, 133–137 (1994)
[18] Bermudez, V., Gase, T., Kajzar, F., Capron, N., Zerbetto, F., Gatti, F.G., Leigh, D.A., Zhang, S.: Opt. Materials, **21**, 39–44 (2002)
[19] Gatti, F.G., Leon, S., Wong, J.K.Y., Bottari, G., Altieri, A., Morales, A.F.M., Teat, S.J., Frochot, C., Leigh, D.A., Brouwer, A.M., Zerbetto, F.: PNAS **100**, 10–14 (2003)
[20] Bermudez, V., Chollet, P.-A., Kajzar, F., Lorin, A., Bottari, G., Gatti, F.G., Leigh, D.A.: Mol. Cryst. Liq. Cryst. **374**, 343–356 (2002)
[21] Brouwer, A.M., Frochot, C., Gatti, F.G., Leigh, D.A., Mottier, L., Paolucco, F., Roffia, S., Wurpel, G.W.H.: Nature **406**, 608–611 (2001)
[22] Buffeteau, T., Lagugné Labarther, F., Pézolet, M., Sourisseau, M.C.: Macromolecules, **34**, 7514–7521 (2001)
[23] Joachim, C., Gimzewski, J.K., Aviram, A.: Nature **408**, 541–548 (2000)
[24] Heath, J.R.: Pure Appl. Chem. **72**, 11–20 (2000)
[25] Ballardini, R., Balzani, V., Credi, A., Gandolfi, M.T., Venturi, M.L.: Acc. Chem. Res. **34**, 445–455 (2001)
[26] Grando, D., Gase, T., Kajzar, F., Fanti, M., Zerbetto, F., Murphy, A., Leigh, D.A.: Mol. Crys. Liq. Cryst. **353**, 545–559 (2000)
[27] Leigh, D.A., Murphy, A., Smart, J.P., Deleuze, M.S., Zerbetto, F.: J. Am. Chem. Soc. **120**, 6458–6467 (1998)
[28] Lane, A.L., Leigh, D.A., Murphy, A.: J. Am. Chem. Soc. **119**, 11092–11093 (1997)
[29] Deleuze, M.S., Leigh, D.A., Zerbetto, F.: J. Am. Chem. Soc. **121**, 2364–2379 (1999)
[30] Amabilino, D.B., Stoddart, J.F.: Chem. Rev. **95**, 2725 (1995)
[31] Fyfe, M.C.T., Stoddart, J.F.: Acc. Chem. Res. **30**, 393 (1997)
[32] Philip, D., Stoddart, J.F.: Angew. Chem. Int. Ed. Engl. **35**, 1154 (1996)
[33] Ashton, P.R., Ballardini, R., Balzani, V., Baxter, I., Credi, A., Fyfe, M.C.T., Gandolfi, M.T., Gómez-López, M., Martínez-Díaz, M.V., Piersanti, A., Spencer, N., Stoddart, J.F., Venturi, M., White, A.P.J., Williams, D.J.: J. Am. Chem. Soc. **120**, 11932 (1998)
[34] Ashton, P.R., Ballardini, R., Balzani, V., Belohradsky, M., Gandolfi, M.T., Philp, D., Prodi, L., Raymo, F.M., Reddington, M.V., Spencer, N., Stoddart, J.F., Venturi, M., Williams, D.J.: J. Am. Chem. Soc. **118**, 4931 (1996)
[35] Ashton, P.R., Ballardidni, R., Balzani, V., Constable, E.C., Credi, A., Kocian, O., Langford, S.J., Preece, J.A., Prodi, L., Schofield, E.R., Spencer, N.: Chem. Eur. J. **4**, 2413–22 (1998)
[36] Ashton, P.R., Ballardidni, R., Balzani, V., Credi, A., Dress, K.R., Ishow, E., Klervaan, C.J., Kocian, O., Preece, J.A., Spencer, N., Stoddart, J.F., Venturi, M., Wenger, S.A.: Chem. Eur. J. **6**, 3558–74 (1998)
[37] Dietrich-Buchecker, C.O., Sauvage, J.-P., Kintzinger, J.-P.: Tetrahedron Lett. **24**, 5095 (1983)
[38] Dietrich-Buchecker, C.O., Sauvage, J-P.: Bioorganic Chemistry Frontiers, **2**, 197, Springer-Verlag, Berlin (1991)
[39] Sauvage, J.-P.: Acc. Chem. Res. **23**, 319 (1990)
[40] Johnston, A.G., Leigh, D.A., Pritchard, R.J., Deegan, M.D.: Angew. Chem. Int. Ed. Engl. **34**, 1209–12 (1995)
[41] Johnston, A.G., Leigh, D.A., Nezhat, L., Smart, J.P., Deegan, M.D.: Angew. Chem. Int. Ed. Engl. **34**, 1212–16 (1995)
[42] Johnston, A.G., Leigh, D.A., Murphy, A., Smart, J.P., Deegan, M.D.: J. Am. Chem. Soc. **118**, 10662–10663 (1996)
[43] Gatti, F.G., Leigh, D.A., Nepogodiev, S.A., Slawin, A.M.Z., Teat, S.J., Wong, J.K.Y.: J. Am. Chem. Soc. **123**, 5983–5989 (2001)

[44] Leigh, D.A., Murphy, A., Smart, J.P., Slawin, A.M.: Angew. Chem. Int. Ed. Engl. **36**, 728–732 (1997)
[45] Hunsperger, R.G.: Integrated Optics: Theory and Technology, Springer–Verlag, Berlin (1991)
[46] Kogelnik, H.: Theory of Dielectric Waveguides, 15–81, in Integrated Optics, Tamir, T. Ed. Springer Verlag, Berlin (1985)
[47] Gase, T., Grando, D., Chollet, P.-A., Kajzar, F., Murphy, A., Leigh, D.A.: Adv. Mat. **11**, 1303–1306 (1999)
[48] Gase, T., Grando, D., Chollet, P.-A., Lorin, A., Tetard, D., Leigh, D.A.: Photonic Science News **3**, 16–21 (1999)
[49] Weber, H.-P., Dunn, F.A., Leibolt, W.N.: Appl. Opt. **12**, 755 (1973)
[50] Swalen, J., Kajzar, F.: Introduction, in Organic Thin Films for Waveguiding Nonlinear Optics, Eds. Kajzar, F., Swalen, J.: Gordon & Breach Sc. Publ., Amsterdam, pp. 1–44 (1996)
[51] Roberts, D.A.: IEEE J. Quantum Electron. **28**, 2057 (1992)
[52] Teng, C.C., Man, H.T.: Appl. Phys. Lett. **56**, 1735 (1990)
[53] Schildkraut, J.S.: Appl. Optics **29**, 2839 (1990)
[54] Kajzar, F., Chollet, P.-A.: in Poled Polymers and their Applications to SHG and EO Devices, Eds. Miyata, S. and Sasabe, H.: Gordon and Breach Science Publishers, New Jersey (1997)
[55] Large, M., Kajzar, F., Raimond, P.: Applied Phys. Lett. **73**, 3635–7 (1998)
[56] Combellas, C., Kajzar, F., Mathey, G., Petit, M.A., Thiebault, A.: Chem. Phys. **252**, 165–177 (2000)
[57] Kajzar, F., Messier, J.: Phys. Rev. A, **32**, 2352–2363 (1985)
[58] Kajzar, F., Messier, J., Rosilio, C.: J. Appl. Phys. **60**, 3040–3044 (1986)
[59] Kajzar, F.: Third Harmonic Generation, 767–839, in Characterization Techniques and Tabulations for Organic NLO Materials, Eds. Kuzyk, M.G. and Dirk, C.W. Mercel Dekker Inc. (1998)
[60] Meredith, G.R., Buchalter, B., Hanzlik, C.: J. Chem. Phys. **78**, 1533–42 (1983)
[61] Gase, T.: Photonics Sc. News **3**, 22–28 (1998)
[62] Kippelen, B., Meerholz, S.K., Peyghambarian, N.: Appl. Phys. Lett. **68**, 1748–1750 (1996)
[63] Bermudez, V., Capron, N., Gase, T., Gatti, F.G., Kajzar, F., Leigh, D.A., Zerbetto, F., Zhang, S.: Nature **406**, 608–611 (2000)
[64] O'Konski, Ed., Molecular Electro-Optic Mercel-Dekker, New York (1967)
[65] Nowicka, K., Chollet, P.-A., Kajzar, F., Bottari, G., Gatti, F.G., Leigh, D.A.: Nonl. Optics & Quant. Optics **32**, 175–86 (2004)
[66] Bottari, G., Leigh, D.A., Perez, E.M.: J. Am. Chem. Soc. **125**, 13360 (2003)
[67] Asakawa, M., Brancato, G., Fanti, M., Leigh, D.A., Shimizu, T., Slawin, A.M.Z., Wong, J.K.Y., Zerbetto, F., Zhang, S.: J. Am. Chem. Soc. **124**, 2939 (2002)
[68] Leon, M.C., Marchioni, F., Silvi, S., Credi, A.: Synth. Met. **139**, 773 (2003)
[69] Perez, E.M., Dryen, D.T.F., Leigh, D.A., Teobaldi, G., Zerbetto, F.: J. Am. Chem. Soc. **126**, 12210 (2004)
[70] Wang, Q.-C., Qu, D.-H., Ren, J., Chen, K., Tian, H.: Angew. Chem. Int. Ed. **43**, 2661 (2004)
[71] Qu, D.-H., Wang, Q.-C., Ren, J., Tian, H.: Org. Lett. **6**, 2085 (2004)
[72] Wurpel, G.W.H., Brouwer, A.M., van Stokkum, I.H.M., Farran, A., Leigh, D.A.: J. Am. Chem. Soc. **123**, 11327 (2001)
[73] Garnet, G., Ruani, G., Cavallini, M., Biscarini, F., Murgia, M., Zamboni, R., Giro, G., Cocchi, M., Fattori, V., Loontjens, T., Thies, J., Leigh, D.A., Morales, A.F., Mahrt, R.F.: Appl. Surf. Sci. **369**, 175–176 (2001)
[74] Giro, G., Cocchi, M., Fattori, V., Gadret, G., Ruani, G., Murgia, M., Cavallini, M., Biscarini, F., Zamboni, R., Loonjes, T.: Synth. Met. **122**, 27–29 (2001)
[75] Kwan, P.H., MacLachan, M.J., Swanger, T.M.: J. Am. Chem. Soc. **126**, 8638 (2004)
[76] Gase T., Grando D., Chollet P.-A., Kajzar F., Lorin A., Leigh D.A. and Tetard D.: Nonlinear Opt., 22, 491(1999)

CHAPTER 20

SECOND HARMONIC GENERATION FROM GOLD AND SILVER NANOPARTICLES IN LIQUID SUSPENSIONS

JÉRÔME NAPPA, GUILLAUME REVILLOD, GAELLE MARTIN,
ISABELLE RUSSIER-ANTOINE, EMMANUEL BENICHOU,
CHRISTIAN JONIN AND PIERRE-FRANÇOIS BREVET
Laboratoire de Spectrométrie Ionique et Moléculaire, UMR CNRS 5579, Université Claude Bernard Lyon 1, 43 Boulevard du 11 Novembre 1918, 69622 Villeurbanne cedex, France

Abstract: A general review of the second harmonic (SH) light scattered from aqueous suspensions of small gold and silver metallic particles is presented. The first part is devoted to the general theory of the SH generation from particles in order to discuss the incidence of the shape of the particles and the retardation effects of the electromagnetic fields on the total SH scattered field. The next part focuses on the problem of the polarization fields, a problem rather specific to metallic particles. Because of their strong polarizability and the possibility of resonance enhancements through surface plasmon (SP) excitations, the exciting field cannot be taken as the incident field only but rather as the superposition of the incident and the polarization fields. An illustration of this problem is presented with the determination of the exact origin of the SH response from metallic particles. This experimental section presents two different sets of data: the size dependence of the SH intensity and the polarization patterns recorded in the geometrical configuration of Hyper Rayleigh Scattering. SP resonance enhancements of the absolute values of the hyperpolarizabilities are then discussed before a presentation of the SH response of aggregating suspensions of particles and their possible applications is given

Keywords: Second Harmonic Generation – Hyper Rayleigh Scattering – Surface Plasmon Resonances – Hyperpolarizability – Retardation effects – Local fields – Core-Shell particles – Aggregation – Bioassay

1. INTRODUCTION

The synthesis and application of small metallic particles date back from the very early days of history when the stability of gold metal particles conferred them a prominent role in medicine against ailments. During the 1990's, dramatic efforts have been undertaken to develop new routes for the synthesis of metal particles with different sizes, shapes, morphologies or compositions [1]. A particular attention has been paid to particles made from noble metals like silver and gold because of their chemical stability, their use in the elaboration of complex structures and most notably their optical properties. Indeed, their photo-absorption spectra in the visible region of the spectrum between 400–800 nm is dominated by the surface plasmon resonance (SPR), namely the collective oscillation of the conduction band electrons [2]. This resonance confers to aqueous suspensions of gold metal particles a ruby red colour and to that of silver metal particles a bright yellow colour. In recent years, the developments of new synthesis routes have been pursued with different motivations: for instance the grafting of organic compounds at the surface of the particles to develop a refined chemistry based on these nano-objects, the synthesis of a whole range of shapes, from spheres and rods to prisms and cubes or the assembling of the particles into organized films by self-assembling or templating techniques. In parallel, the study of fundamental physical phenomena has been devoted to the understanding of the quantum confinement on their optical and electronic properties and of the relaxation mechanisms following a light pulse perturbation or to the electromagnetic response of the assemblies of such particles [3–5]. In particular, experimental methods at the single particle level or on two-dimensional assembly films have been investigated [6–8].

It is not surprising therefore that the optical properties of small metal particles have received a considerable interest worldwide. Their large range of applications goes from surface sensitive spectroscopic analysis to catalysis and even photonics with microwave polarizers [9–15]. These developments have sparked a renewed interest in the optical characterization of metallic particle suspensions, often routinely carried out by transmission electron microscopy (TEM) and UV-visible photo-absorption spectroscopy. The recent observation of large SP enhancements of the non linear optical response from these particles, initially for third order processes and more recently for second order processes has also initiated a particular attention for non linear optical phenomena [16–18]. Furthermore, the paradox that second order processes should vanish at first order for perfectly spherical particles whereas experimentally large intensities were collected for supposedly near-spherical particle suspensions had to be resolved. It is the purpose of the present review to describe the current picture on the problem.

2. GENERAL THEORY

We are interested in a first stage in the description of the SH response from arbitrary particles, with minimum restrictions on their shape or their size. Before discussing the particular case of small gold and silver metallic particles, we wish to recast the problem in its generality. Originally, second harmonic generation (SHG), the process through which two photons at a fundamental frequency are converted into one photon at the harmonic frequency, has been used to investigate planar interfaces [19–21]. The advantage of this second order nonlinear technique as compared to linear optical methods is that the conversion process is forbidden in

Second Harmonic Generation from Gold and Silver Nanoparticles

media possessing a centre of inversion. Hence, interfaces between two such media are readily accessible through the technique. Buried interfaces like solid/liquid or liquid/liquid interfaces are therefore within reach and many studies have been reported on this topic in the literature to date [22–25]. It is thus of interest to see the possibilities of the technique in the case of non planar interfaces.

To do so, we consider an interface between two media possessing a centre of inversion. Within the electric dipole approximation, no signal may arise from the volume of both media and the spatial location of the source to the conversion process is reduced to a thin layer of material at the surface of the particle. One of the two media is the external medium and the other one the inner medium or the particle medium. Disregarding polarization fields, the incoming exciting fundamental field at any point $\vec{r}'$ at the surface of the particle is simply $\vec{E}(\omega, \vec{r}')$, which taken as a plane monochromatic wave, writes [26]:

$$\vec{E}(\omega, \vec{r}') = E\hat{e}^{(\omega)} \exp[-i\omega t] \exp[-i\vec{k}^{(\omega)} \cdot \vec{r}'] \qquad (1)$$

where $\hat{e}^{(\omega)}$ is a unit vector, E is the field amplitude and $\vec{k}^{(\omega)}$ is the wave vector of the plane wave. For simplicity, we have dropped the complex conjugate and we assume a geometrical configuration where the incoming field propagates along the Oz axis and the incoming field is polarized in the Oxy plane with a polarization angle γ, namely:

$$\hat{e}^{(\omega)} = \cos\gamma\hat{x} + \sin\gamma\hat{y} \qquad (2)$$

see Figure 1.

The induced local nonlinear polarization $\vec{p}(\Omega, \vec{r}')$ at the harmonic frequency $\Omega = 2\omega$ at the location $\vec{r}'$ on the surface of the particle is then [27]:

$$\vec{p}(\Omega, \vec{r}') = \overleftrightarrow{T}(\hat{r}')\overleftrightarrow{\beta}(\hat{r}') : \vec{E}(\omega, \hat{r}')\vec{E}(\omega, \hat{r}') \qquad (3)$$

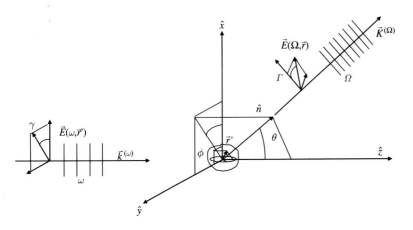

Figure 1. Schematics of the geometrical configuration used for the HRS from small particles

written in the laboratory frame. Hence, in Eq. (3), $\overleftrightarrow{\beta}(\hat{r}')$ is the local hyperpolarizability tensor at location $\vec{r}'$ in the surface local frame and $\overleftrightarrow{T}(\hat{r}')$ is the frame transformation tensor allowing passage from the local surface frame to the laboratory frame. Some considerations on the local hyperpolarizability tensor $\overleftrightarrow{\beta}(\hat{r}')$ can be discussed at that point. For instance, the elements of this local tensor vanish according to the local symmetry of the surface. If an isotropic surface is considered, then the local surface normal must be distinguished from the two in-plane axes, the latter two axes being undistinguishable. Thus, the local hyperpolarizability tensor possesses only three independent elements, namely $\beta_{zzz}(\hat{r}'), \beta_{zxx}(\hat{r}')$ and $\beta_{xzx}(\hat{r}')$, and seven non vanishing elements:

(4a) $\quad \beta_{zzz}(\hat{r}')$

(4b) $\quad \beta_{zxx}(\hat{r}')$

(4c) $\quad \beta_{xxz}(\hat{r}') = \beta_{xzx}(\hat{r}')$

(4d) $\quad \beta_{zyy}(\hat{r}')$

(4e) $\quad \beta_{yyz}(\hat{r}') = \beta_{yzy}(\hat{r}')$

with the added relationships:

(4f) $\quad \beta_{xxz}(\hat{r}') = \beta_{yyz}(\hat{r}')$

(4g) $\quad \beta_{zxx}(\hat{r}') = \beta_{zyy}(\hat{r}')$

For metallic surfaces, the three non vanishing and independent elements have been recast into three parameters, also named the three Rudnick and Stern parameters a, b and d, corresponding to the three nonlinear currents induced at the harmonic frequency, respectively two surface currents perpendicular and parallel to the surface and one volume current perpendicular to the surface [28]. Theoretical expressions for these parameters are known within some approximations and for perfect surfaces [29, 30]. For anisotropic surfaces, the elements of the hyperpolarizability tensor would follow different relationships from Eq. (4) and would be taken from the usual Tables [31]. For metallic particles, the polarization field inside the particle is non negligible and the local exciting field should be taken as the superposition of the incoming and polarization fields. It is also still debatable whether the sheet of polarization is located inside or outside the particle. This problem has been discussed for planar interfaces and a similar discussion could be developed here. The problem of the exciting field will be discussed further below for metallic particles.

At this stage, each single point $\vec{r}'$ radiates a spherical SH wave and the total field amplitude collected at the harmonic frequency at a position $\vec{r}$ is given by the coherent superposition of these spherical waves:

(5) $\quad \vec{E}(\Omega, \vec{r}) = \oint_S \frac{e^{iK^{(\Omega)}|\vec{r}-\vec{r}'|}}{|\vec{r}-\vec{r}'|} \left(\left[\hat{n} \times \overleftrightarrow{T}(\hat{r}') \overleftrightarrow{\beta}(\hat{r}') \vec{E}(\vec{r}', \omega) \vec{E}(\vec{r}', \omega) \right] \times \hat{n} \right) d\hat{r}'$

where we have introduced the direction of collection $\hat{n}$ through $\hat{n} = \vec{r}/r$. Equation (5) is rather general, except that the polarization field at the SH frequency is disregarded, similarly to that at the fundamental frequency. This simple model is though applicable to many systems like molecular systems where the interactions between the scattering centres are weak enough to be neglected. For instance, this model is valid for dye molecules at the surface of liposomes or micelles [32, 33]. Equation (5) is also valid for particles of arbitrary sizes and shapes. Actually, the domain of validity covered by Eq. (5) is larger than that of the SH generation from the surface of particles and could be extended to the case of the SH generation from a collection of point sources located within a known volume. In Eq. (5), a $d\vec{r}'$ volume integration would be inserted in place of the $d\hat{r}'$ surface integration. Retardation effects are taken into account through the full expression of the exponential phase factors. In the approximation of small particles, with a radius much smaller than λ the wavelength of light, the spatial phases can be expanded and truncated to their first order in r'/λ. In general, the field amplitude does not vanish except if the surface of integration is centrosymmetrical. In particular, the total field vanishes altogether for spherical surfaces. On the opposite, if the particles are no longer small, the phase retardation must be taken into account and in that case an SH signal can be collected in the forward direction irrespective of the shape of the particle if the condition:

(6) $\quad \Delta k \cdot a = \left(2k^{(\omega)} - K^{(\Omega)}\right) a >> 1$

is fulfilled where a is the radius of the particle.

One final point must be underlined in the case of rather large particles where the spatial variation of the fields cannot be neglected anymore. For these particles, the field gradients inside to particles no longer vanish and therefore we should not disregard a volume contribution to the nonlinear polarization of the form:

(7) $\quad \vec{p}(\Omega, \vec{r}') = \gamma \vec{\nabla}\left[\vec{E}^2(\omega, \vec{r}')\right] + \beta \vec{E}(\omega, \vec{r}') \vec{\nabla} \cdot \vec{E}(\omega, \vec{r}')$

where γ and β are two complex parameters. Since we are interested in rather small metallic particles in the remainder of the text, we will not discuss this contribution although, originally, it has been introduced and discussed as the source to the SH generation [34, 35].

3. SMALL METALLIC PARTICLES

The problem of metallic particles like gold and silver particles is similar to the previous case except that the material is now highly polarizable. Hence, the polarization sheet is excited by the local field which cannot be taken as the incoming field only. It must be taken as the superposition of the incoming field and the polarization field. This problem is rather difficult in general and several theories have been proposed in the past [35–40]. For arbitrary shapes, one may directly use a numerical approach like the discrete dipole approximation (DDA) for instance. It has however been solved analytically for spherical particles by G. Mie and H. Chew et al. in

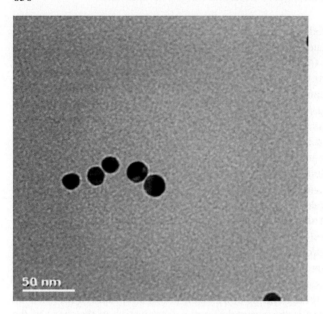

Figure 2. Transmission Electron Microscopy (TEM) picture of small gold metallic particles. The shape of the particles approaches that of perfect spheres

different circumstances for different boundary conditions [41–43]. If we assume that the particles do not depart too far from the spherical shape, this model of the spherical expansion may still be valid. As observed from transmission electron microscopy pictures, this is often the case, see Figure 2.

We therefore start assuming that the particles are compact and define the radius a of the smallest sphere containing the particle [44]. Thus, the condition $r' \approx a$ is always fulfilled. The incoming plane wave can be expanded into spherical waves through a multipole expansion with respect to the parameter $x = a/\lambda$ [27]:

$$(8) \quad \vec{E}(\omega, \vec{r}') = \sum_l \left\{ \vec{E}_l(\omega, \vec{r}') + \vec{M}_l(\omega, \vec{r}') \right\}$$

The first order term of Eq. (8) is reduced to the field $\vec{E}_1(\omega, \vec{r}')$, the electric dipole field, whereas the second order term, linear with the parameter order $x = a/\lambda$, is the sum of the fields $\vec{E}_2(\omega, \vec{r}')$ and $\vec{M}_1(\omega, \vec{r}')$, respectively the electric quadrupole and the magnetic dipole fields. The surface nonlinear polarization of the form of Eq. (3) is now a series expansion with respect to the parameter $x = a/\lambda$ too. Its general expression is:

$$(9) \quad \vec{p}(\Omega, \vec{r}') = \sum_l \vec{p}_l(\Omega, \vec{r}')$$

but because of the tensorial product of the fundamental field $\vec{E}(\omega, \vec{r}')$ its first order term is:

$$(10) \quad \vec{p}_1(\Omega, \vec{r}') = \overleftrightarrow{T}(\vec{r}')\overleftrightarrow{\beta}(\vec{r}') : \vec{E}_1(\omega, \vec{r}')\vec{E}_1(\omega, \vec{r}')$$

and its second order term:

(11) $$\vec{p}_2(\Omega,\vec{r}') = 2\overset{\leftrightarrow}{T}(\vec{r}')\overset{\leftrightarrow}{\beta}(\vec{r}') : \vec{E}_1(\omega,\vec{r}')\left(\vec{E}_2(\omega,\vec{r}') + \vec{M}_1(\omega,\vec{r}')\right)$$

The nonlinear polarization $\vec{p}_1(\Omega,\vec{r}')$ of Eq. (10) is a purely local surface polarization because it only depends on the value of the fields at location $\vec{r}'$. Because it depends on the fundamental field inside the particle, it is subjected to resonances when the quantity $\varepsilon(\omega) + 2\varepsilon_m$ vanishes, where $\varepsilon(\omega)$ is the complex dielectric function of the metal and ε_m that of the surrounding medium. On the opposite, the nonlinear polarization $\vec{p}_2(\Omega,\vec{r}')$ of the form of Eq. (11) is a non local nonlinear polarization since it depends on spatially varying fields within the particle. Similarly to the first order contribution, it is subjected to resonances when the quantity $2\varepsilon(\omega) + 3\varepsilon_m$ vanishes. These resonances are the usual surface plasmon resonances of the particle. Within the condition of non magnetic media, the magnetic dipole field does not introduce any resonances. We neglect higher order terms.

At the harmonic wavelength, the source to the radiated fields is the sheet of polarization. However, the total field radiated at location $\vec{r}$ in the outer medium cannot be taken as the simple superposition of the spherical waves generated by each point $\vec{r}'$ at the surface of the particle because the metal particle is also highly polarizable at the harmonic frequency. The total field in the particle must be the superposition of the radiated field and the polarization field at the harmonic wavelength and the total field at the detector in the outer medium is obtained by solving the boundary conditions at the surface of the particle. In spherical coordinates, the harmonic field at location $\vec{r}$ radiated by a nonlinear polarization source at location $\vec{r}'$ is now also an expansion of the parameter $x = a/\lambda$ of the form:

(12) $$\vec{E}(\Omega,\vec{r},\vec{r}') = \sum_l \left\{\vec{E}_l(\Omega,\vec{r},\vec{r}') + \vec{M}_l(\Omega,\vec{r},\vec{r}')\right\}$$

where the first order term, the electric dipole field $\vec{E}_l(\Omega,\vec{r},\vec{r}')$, is equivalent to the radiation at the harmonic frequency by a point-like dipole located at the centre of the particle and the second order term, the sum of the electric quadrupole field $\vec{E}_2(\Omega,\vec{r},\vec{r}')$ and the magnetic dipole field $\vec{M}_1(\Omega,\vec{r},\vec{r}')$, is the first-order correction term accounting for the non-vanishing size of the particle. In a way similar to the case at the fundamental frequency, surface plasmon resonances can occur at the harmonic frequency. Hence, the first order term has a resonance whenever the quantity $\varepsilon(\Omega) + 2\varepsilon_m$ vanishes and the second order term whenever the quantity $2\varepsilon(\Omega) + 3\varepsilon_m$ vanishes. It therefore appears that the SP resonances may be excited through either the fundamental or the harmonic fields. Because the nonlinear polarization is a surface nonlinear polarization, see Eq. (3), the total field collected at the detector, located at position $\vec{r}$, is a surface integral:

(13) $$\vec{E}(\Omega,\vec{r}) = \oint_s \vec{E}(\Omega,\vec{r},\vec{r}')d\vec{r}'$$

In order to have a better insight into the problem, we first discuss the case of the first term $\vec{E}_1(\Omega, \vec{r})$ of the multipole expansion of the total field $\vec{E}(\Omega, \vec{r})$ which arises from the insertion of Eq. (12) into Eq. (13). The latter SH field $\vec{E}_1(\Omega, \vec{r})$ is generated by a nonlinear polarization source itself the sum of a local and a non local contribution, as seen from Eq. (9)–(11), two successive order of the scaling parameter $x = a/\lambda$. The purely local contribution to the total SH field generated thus stems from the local source to the nonlinear polarization and the local SH harmonic field $\vec{E}_1(\Omega, \vec{r})$. This purely local contribution to the total field scales with the second power of the parameter $x = a/\lambda$ through the surface integral of Eq. (13). It also has SP resonances at the fundamental or the harmonic frequency whenever the quantities $\varepsilon(\omega) + 2\varepsilon_m$ or $\varepsilon(\Omega) + 2\varepsilon_m$ vanishes. Interestingly, the SP resonance at the fundamental frequency should be sharper because it involves a higher power of the fundamental field. The non local contribution to the SH harmonic field on the opposite stems from two origins: the first one is a non local contribution to the nonlinear polarization through the electric quadrupole and magnetic dipole fields $\vec{E}_2(\omega, \vec{r}')$ and $\vec{M}_1(\omega, \vec{r}')$ at the fundamental frequency and the second one through the non local contribution to the SH field $\vec{E}_2(\Omega, \vec{r})$ with the local nonlinear polarization $\vec{p}_1(\Omega, \vec{r}')$. Interestingly, the first of these two non local contributions has its SP resonance frequencies determined by the conditions $2\varepsilon(\omega) + 3\varepsilon_m = 0$ and $\varepsilon(\Omega) + 2\varepsilon_m = 0$ whereas the second one has its SP resonance frequencies determined by the conditions $\varepsilon(\omega) + 2\varepsilon_m = 0$ and $2\varepsilon(\Omega) + 3\varepsilon_m = 0$, according to their different origins. The non local contribution to the total SH field, whether originating from the retardation effects at the fundamental or the harmonic frequency, scales with the third power of the parameter x.

In agreement with the considerations developed above for spherical particles, the first purely local contribution vanishes for perfectly centrosymmetrical particles, and in particular for spheres. In that case, the model developed above coincides with that proposed by J.I. Dadap et al. [39, 40]. In particular, the two non local contributions correspond to the effective dipole $\vec{p}_{eff}$ and effective quadrupole $\vec{Q}_{eff}$ contributions introduced in that work. The effective dipole arises from retardation effects taken into account at the fundamental frequency and the effective quadrupole to retardation effects at the harmonic frequency only. Contributions with retardation effects at both the fundamental and the harmonic frequencies would be of higher orders of the parameter x. As we shall see later, the different contributions may be distinguished through their angular and polarization patterns.

4. HYPER RAYLEIGH SCATTERING

SH generation from aqueous suspensions of particles is not a coherent process because of random arrangement of the particles. No phase relationships occurs between the SH waves produced by each single particle. However the best agreement is obtained with theoretical models because for a liquid suspension of particles, the environment is rather homogeneous. The liquid medium can be described with

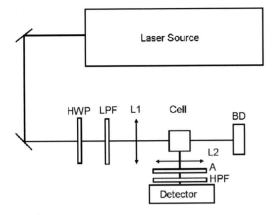

Figure 3. Schematics of an experimental arrangement for the measurement of the Hyper Rayleigh Scattering intensity (HWP: half-wave plate; LPF: low-pass filter, L1 and L2: lens; BD: beam dump; A: analyzer, HPF: high-pass filter)

continuous models. A general experimental set-up of this method, called hyper Rayleigh scattering (HRS), is presented in Figure 3 [45–47].

The total scattered intensity is the mere superposition of the intensities scattered from each particles:

(14) $$I_{HRS}^{\Gamma} = GN \langle E_{\Gamma}(\Omega, \vec{r}) E_{\Gamma}^{*}(\Omega, \vec{r}) \rangle$$

where G is a constant and N is the number of particles in the volume of interaction. The Hyper Rayleigh Scattering (HRS) intensity collected for the light vertically polarized is obtained setting $\Gamma = 0$ or (V) and for the light horizontally polarized setting $\Gamma = \pi/2$ or (H) in Eq. (14).

The intensity therefore scales with the number density of the particles in the solution and it is common practice to verify this behaviour in experiments, see Figure 4.

As expected from the theoretical developments previously available for non centrosymmetrical molecules, the HRS intensity for the two polarization states as a function of the input fundamental wave polarization angle γ is given by [44]:

(15) $$I_{HRS}^{\Gamma} = a^{(\Gamma)} \cos^4 \gamma + b^{(\Gamma)} \cos^2 \gamma \sin^2 \gamma + c^{(\Gamma)} \sin^4 \gamma$$

when the retardation effects at the fundamental and the harmonic frequencies are neglected. In that case, the only contribution to the total intensity is the purely local contribution, similarly to molecules. The real coefficients $a^{(\Gamma)}$, $b^{(\Gamma)}$ and $c^{(\Gamma)}$ are then defined by:

(16a) $$a^{(\Gamma)} = GNa^4 K_a^{(\Gamma)} \frac{1}{|\varepsilon(\omega) + 2\varepsilon_m|^4} \frac{1}{|\varepsilon(\Omega) + 2\varepsilon_m|^2} \langle |\beta_{\Gamma XX}|^2 \rangle$$

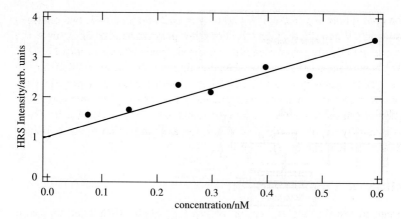

Figure 4. Dependence of the HRS intensity with the number density of the particles for an aqueous suspension of gold particles of 20 nm diameter. The HRS intensity is normalized with the neat water HRS signal

$$(16b) \quad b^{(\Gamma)} = GNa^4 K_b^{(\Gamma)} \frac{1}{|\varepsilon(\omega)+2\varepsilon_m|^4} \frac{1}{|\varepsilon(\Omega)+2\varepsilon_m|^2} \\ \times \langle \beta_{\Gamma XX}\beta^*_{\Gamma YY} + \beta^*_{\Gamma XX}\beta_{\Gamma YY} + 2|\beta_{\Gamma XY}|^2 \rangle$$

$$(16c) \quad c^{(\Gamma)} = GNa^4 K_c^{(\Gamma)} \frac{1}{|\varepsilon(\omega)+2\varepsilon_m|^4} \frac{1}{|\varepsilon(\Omega)+2\varepsilon_m|^2} \langle |\beta_{\Gamma YY}|^2 \rangle$$

where $K_a^{(\Gamma)}$, $K_b^{(\Gamma)}$ and $K_c^{(\Gamma)}$ are constants and where the elements of the hyperpolarizability tensor have been reported into the laboratory frame. These coefficients are not independent and they obey the following relationships [48]:

$$(17a) \quad a^{(0)} + c^{(0)} = b^{(0)}$$

$$(17b) \quad 2a^{(\pi/2)} = 2c^{(\pi/2)} = b^{(\pi/2)}$$

The HRS intensity given through Eqs. (15)–(17) thus possesses a dependence with the power of four of the radius of the particles a and also SP resonances at both the fundamental and the harmonic frequency.

Formally, the expression of Eq. (15) of the HRS intensity is similar to the one usually given in the literature as:

$$(18) \quad I_{HRS} = G' \langle N_0 \beta_0^2 + N\beta^2 \rangle I^2 \exp[-(2A(\omega) + A(\Omega))]$$

where we have furthermore introduced the linear absorption coefficients at the fundamental and the harmonic intensity. In expression Eq. (18), we have made explicit the non vanishing signal arising from the solvent molecules. It is straightforward to see that expression Eq. (15) reduces to expression Eq. (18) for the polarization configuration $\gamma = 0$ and $\Gamma = 0$ or $\gamma = \pi/2$ and $\Gamma = \pi/2$, two configurations known as vV and hH respectively. For other polarization configurations, the connection between the hyperpolarizability tensor elements of expressions Eq. (15) and Eq. (18) is not simple and involves linear combinations. From the slope of the line fit of the HRS intensity as a function of the number density of particles, see Figure 4, its normalization with the extrapolated value of the HRS intensity for a vanishing concentration of particles, taken the usual value of $\beta_0 = 0.56 \times 10^{-30}$ esu for the molecular hyperpolarizability of water, the absolute value of the hyperpolarizability of a particle can be obtained [17].

5. ORIGIN OF THE HRS SIGNAL

It is important to see that the expression of Eq. (15) of the HRS intensity given above vanishes altogether if the particles are perfect spheres. In that case, only the non local contribution to the HRS intensity must be considered, a contribution scaling with the power of six of the radius of the particles. The experimental determination of the vanishing character of the first order term is available: on one hand it is possible to determine the power dependence of the HRS intensity with the radius of the particle. A power of four indicates a pure local response arising from non perfect spheres whereas a power of six indicates a volume contribution from centrosymmetrical particles like perfect spheres. On the other hand, it is possible to investigate the dependence of the HRS intensity with the angle of polarization of the incoming wave. From J.I. Dadap et al. work, the following dependence for prefect spheres should be observed [39]:

(19) $\quad I_{HRS}^{V} = K|\alpha_2|^2 \sin^2 2\gamma I^2$

(20) $\quad I_{HRS}^{H} = K|\alpha_1|^2 I^2$

where K is a constant, I is the fundamental intensity and α_1, α_2 are the components in the laboratory frame of the effective electric dipole $\vec{p}_{eff}$ and vector part of the electric quadrupole $\overleftrightarrow{Q}_{eff}(\hat{y})$ along the $\hat{y}$ axis defined through:

(21) $\quad \vec{p}_{eff} = \alpha_1 \hat{z}$

(22) $\quad \overleftrightarrow{Q}_{eff}(\hat{y}) = \alpha_2 \sin\gamma (\cos\gamma \hat{x} + \sin\gamma \hat{y})$

The experimental work has been recently performed for aqueous suspensions of small gold particles at a fundamental wavelength of 800 nm with a femtosecond laser in a range of diameters between 12 nm and 150 nm [44]. The dependence of the logarithm of the square root of the HRS intensity with the particle diameter is reported in Figure 5. The slope of the line fit is in agreement with a power of four dependence of the HRS intensity with the radius of the particle. This result alone ensures that the origin of the SH process in the particles is indeed of electric dipole origin, stemming from the breaking of the centrosymmetry of the particle shape.

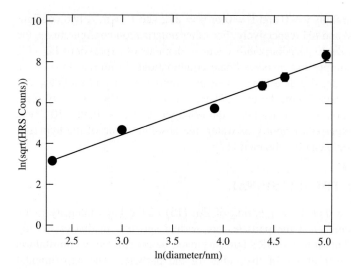

Figure 5. Logarithmic plot of the square root of the HRS intensity, recorded for both the input fundamental and the output harmonic intensities polarized along the Ox direction, versus the diameter of the particles. The slope is 1.9+/−0.2 in agreement with a surface origin of the signal (Reprinted from 44 with permission from the American Physical Society)

A possible origin arising from volume defects in the cubic crystalline structure of the material is also ruled out since such a hypothesis would lead to a dependence of the HRS intensity with the power of six of the particle radius because this is a volume effect. It was however emphasized that the breaking of the particle centrosymmetry can arise from a genuine geometrical shape effect or an inhomogeneous distribution of surface states or adsorbates at the surface of the particle.

It is interesting to note that a similar work has been performed in the past at a fundamental energy of 1064 nm with a nanosecond laser for similar aqueous suspensions of gold particles [18]. The absolute value of the hyperpolarizablity of the gold metallic particles was reported in terms of its value per number of metal atoms in the particle as a function of the particle diameter. This reduced hyperpolarizability $\overline{\beta}$ of the particle, is simply defined by:

$$(23) \quad \overline{\beta} = \frac{\beta}{n}$$

The experimental values are in fact in agreement with such a pure surface electric dipole origin, see Figure 6. This fact was not immediately recognized at that time although the incompatibility with the model of perfect spheres was noticed.

The analysis of the HRS intensity as a function of the angle of polarization of the fundamental wave has also been performed recently [44]. Figure 7 reports the polar plot of the HRS intensity as a function of this angle of polarization for an aqueous suspension of 20 nm diameter gold particles for the output SH light vertically and horizontally polarized. The patterns presented are in agreement with Eq. (15). Unambiguously, the plots do not agree with the dependence expected for perfect

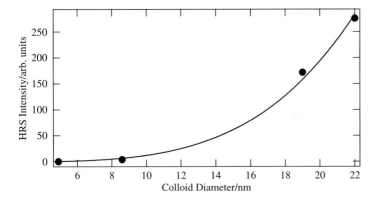

Figure 6. HRS intensity of gold particles in aqueous suspensions as a function of the particle diameter. The continuous line is a fit to a a^4 curve. Data taken from 18

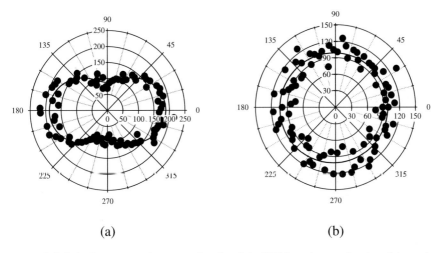

Figure 7. Polarization pattern, given as a polar plot of the HRS intensity as a function of the angle of polarization of the fundamental wave, for an aqueous suspension of 20 nm diameter gold. (a) Vertically polarized SH scattered light and (b) horizontally polarized SH scattered light (Reprinted from 44 with permission from the American Physical Society)

spheres, see Eqs. (19)–(20). The observed patterns are rather in agreement with the expected patterns of point-like sources of SH waves, like dipolar or octupolar molecules.

All these results confirm that the origin of the SH process in small metallic particles arises from the surface of the particle owing to the deviation of the particle shape from that of a perfect sphere. The possibility of a non vanishing hyperpolarizability for small metallic particles has already been discussed with the introduction of shape distortions [49]. All these results underline the sensitivity of the SH process to the breaking of the centrosymmetry.

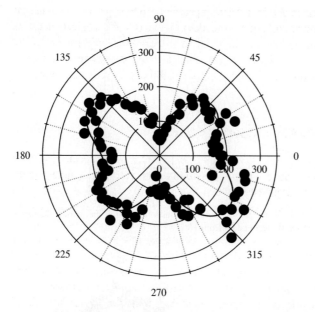

Figure 8. Polarization pattern, given as a polar plot of the HRS intensity as a function of the angle of polarization of the fundamental wave, for an aqueous suspension of 80 nm diameter gold for vertically polarized SH scattered light collected at 400 nm (Reprinted from 44 with permission from the American Physical Society)

Figure 8 gives the pattern obtained for an aqueous suspension of 80 nm gold metallic particles. The pattern no longer follows the form of expression Eq. (15) and now resembles more closely that of a quadrupolar pattern with four lobes. In fact, for 80 nm diameter particles, retardation effects must be taken into account. The total scattered field at the harmonic frequency cannot be described with the pure local contribution only. One has to introduce the non local contribution too. The polarization pattern of the HRS intensity takes the following general form [44]:

$$(24) \quad I_{HRS}^{\Gamma} = a^{(\Gamma)} \cos^4 \gamma \uparrow + b^{(\Gamma)} \cos^2 \gamma \sin^2 \gamma + c^{(\Gamma)} \sin^4 \gamma \\ + d^{(\Gamma)} \cos^3 \gamma \sin \gamma + e^{(\Gamma)} \cos \gamma \sin^3 \gamma$$

with the insertion of two new coefficients $d^{(\Gamma)}$ and $e^{(\Gamma)}$. The relationships of Eqs. (17) followed by the coefficients $a^{(\Gamma)}$, $b^{(\Gamma)}$ and $c^{(\Gamma)}$ are no longer valid, neither are Eqs. (16).

In fact, these three coefficients have an expression of the form:

$$(25) \quad a^{(\Gamma)} = a'^{(\Gamma)} x^4 + a''^{(\Gamma)} x^5$$

whereas the two newly introduced coefficients directly scale as:

$$(26) \quad d^{(\Gamma)} = d''^{(\Gamma)} x^5$$

since they do not appear in the first order expressions. These results demonstrate that as the particles grow in diameter, the deviation from the perfect shape of a sphere is weaker. More mathematically, the local parameter $\eta(\hat{r}') = |1 - r'/a|$ describing the extent of the deviation of the particle from the shape of a perfect sphere decreases for all values of $\hat{r}'$ and the response more closely resembles that of a perfect sphere.

6. ABSOLUTE HYPERPOLARIZABILITIES

Hyperpolarizabilities of gold and silver metallic particles are expected to have rather large magnitudes because of the possibility of resonance enhancements through SP excitations. This possibility has been clearly identified above, see Eq. (16) for instance. Such resonance enhancements have also been observed for the second harmonic generation collected from particles deposited on surfaces [50, 51]. However, in that particular case, the breaking of the particle symmetry is induced by the substrate and even a small static polarization of the particles will lead to a pure electric dipole origin for the signal. Hence, we concentrate in this review on aqueous suspensions of dispersed particles in order to avoid any complications arising from the substrates. In the solution, the surrounding medium is considered homogeneous and modelled through its dielectric function ε_m. In these conditions, large values for the hyperpolarizability tensor magnitudes have been measured for silver and gold particles for diameters in the range of 5 nm to 150 nm.

Gold and silver metal particles have been investigated principally since copper particles are not stable in aqueous solutions [17, 18, 52]. Copper particles require a careful passivation to prevent the re-oxidation of copper. Gold particles with diameters of 22 nm were found to have a hyperpolarizability magnitude of about 16.6×10^{-25} esu and particles with diameters of 5 nm only 0.60×10^{-25} esu when studied with nanosecond light pulses at a harmonic frequency in resonance with the electric dipole SP resonance, namely with a fundamental wavelength of 1064 nm and a harmonic wavelength of 532 nm. Off SP resonances, experiments were also performed and magnitudes of the hyperpolarizability tensor also obtained, for instance 43×10^{-25} esu for 20 nm diameter gold particles at a fundamental wavelength of 820 nm with femtosecond light pulses. In this case, the harmonic frequency with a wavelength of 410 nm is in resonance with the interband transitions and not the SP resonance. Silver particles of 21 nm diameter exhibited similar strong hyperpolarizability magnitudes of about 7×10^{-25} esu off resonance at 532 nm for the harmonic wavelength and nanosecond pulses. Not surprisingly, platinum particles did not give any detectable signal, a feature attributed to the absence of an SP resonance in the vicinity of the excitation frequency [53, 54].

Because the HRS intensity from metallic particles may be enhanced through SP resonances, wavelength dependence measurements were also performed. These experiments allow for a direct confrontation of the theoretical developments with the experimental data. The first wavelength dependence of the HRS intensity for metallic particles dispersed in a homogeneous environment has been reported for

32 nm diameter silver particles by E.C Hao et al. [55]. Figure 9 clearly displays the electric dipole and the electric quadrupole SP resonances. According to the discussion given above, this indicates that in these 32 nm diameter particles retardation effects are already present and non negligible. Furthermore, the spectral location of the two SP resonances are in agreement with the conditions $\varepsilon(\Omega)+2\varepsilon_m = 0$ and $2\varepsilon(\Omega)+3\varepsilon_m = 0$. In Fig. 9, the continuous curves are theoretical models adjusted to the experimental data points for two different ratios of the weights A and B determined by:

$$(27) \quad I_{HRS} = \frac{AC_A(\gamma)}{|\varepsilon(\Omega)+\varepsilon_m|^2} + \frac{BC_B(\gamma)}{|2\varepsilon(\Omega)+3\varepsilon_m|^2}$$

where the functions $C_A(\gamma)$ and $C_B(\gamma)$ of the fundamental polarization angle must be introduced to account for the solid angle integration of the total intensity detected [56]. The adjustment of the model yields an idea of the different contributions to the total SH scattered intensity. However, it has to be noted that according to the previous discussion above, the first term contains the purely local contribution as well as the non local contribution at the fundamental frequency whereas the second term only contains the non local contribution at the harmonic wavelength.

In Eq. (27), the expression is valid for a detection at right angle from the direction of propagation of the incoming exciting wave. The ratio of the two parameters A and B can be modified using a slit arrangement. This possibility arises from the different angular patterns exhibited by the electric dipole and electric quadrupole

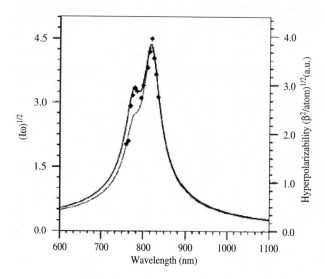

Figure 9. Wavelength dependence of the HRS intensity for an aqueous solution of silver particles the diameter of which is 32 nm for a disk solid angle. Solid lines are (1) fit to the model with $A = 1$ and $B = 6.2$ and (2) G.S. Agarwal and S.S. Jha's model [35] (Reprinted from 54 with permission from the American Institute of Physics)

contributions to the total SH intensity. The slit indeed modifies the solid angle of the collection.

A similar wavelength dependence of the HRS intensity has been reported for an aqueous suspension of the smaller 12 nm diameter gold particles and has yielded only one single electric dipole SP resonance, see Figure 10 [56]. For these particles, only the electric dipole SP resonance is expected. Indeed, even for the larger 20 nm diameter particles, no retardation effects where observed through the polarization pattern. Hence, in Eq. (27), the coefficient B vanishes and the first term only involves the purely local contribution. Similar theoretical curves are also presented for different ratios of the coefficients A and B in order to indicate where the electric quadrupole SP resonance is expected to appear.

One question that has not been discussed so far concerns the metal dielectric functions used in the modelling. It appears in fact that the dielectric functions of the bulk metals, either gold or silver dielectric functions as taken from P.B. Johnson and R.W Christie for example, do not yield a good agreement of the models with the experimental data [57]. Hence, the models used in Figures 9 and 10 are calculated with the use of Drude type dielectric functions of the form $\varepsilon(\omega) = \varepsilon_1(\omega) + i\varepsilon_2(\omega)$ such that [58]:

$$(28) \quad \varepsilon_1(\omega) = \varepsilon_\infty - \frac{\omega_P^2}{\omega^2}$$

$$(29) \quad \varepsilon_2(\omega) = -\frac{\varepsilon_\infty - \varepsilon_1(\omega)}{\omega\tau}$$

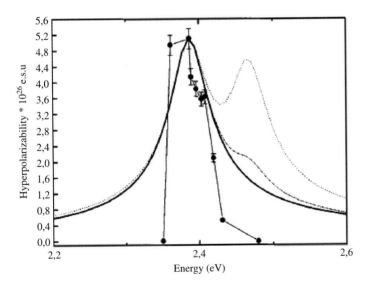

Figure 10. Wavelength dependence of the HRS intensity for an aqueous solution of gold particles the diameter of which is 12 nm. Solid line is a fit with $A = 1$ and $B = 0$, dash-dotted line is a fit with $A = 1$ and $B = 6.2$ and dotted line is a fit with $A = 1$ and $B = 10$ (Reprinted from 56 with permission from the American Institute of Physics)

The interband transitions in particular are only described through the constant ε_∞ which may be adjusted for the best agreement between theory and experiments. It therefore appears that the interband transitions do not lead to strong resonances, as this is experimentally observed on Figure 10 at frequencies above the electric dipole SP resonance with the absence of strong HRS intensities. To further emphasize this problem, gold core – silver shell particles have been investigated, albeit at the water-1,2-dichloroethane interface [59]. The SH intensity I_{SHG} collected with the fundamental light p-polarized and the SH light P-polarized has the following form:

$$(30) \quad I_{SHG} = \frac{\omega^2}{8\varepsilon_0 c^3} \frac{\sqrt{\varepsilon_o(\Omega)}}{\left[\varepsilon_o(\Omega) - \varepsilon_o(\omega)\sin^2\theta_o^\omega\right]} \left|\chi_S^{(2)}\right| I^2$$

where $\varepsilon_o(\omega)$ and $\varepsilon_o(\Omega)$ are the optical dielectric constants of the organic medium at the fundamental and the harmonic frequency, θ_o^ω is the incidence angle from the organic medium and I is the fundamental intensity. The surface tensor $\chi_S^{(2)}$ is the product of the number of particles per unit surface N and the hyperpolarizability of a single nanoparticle β. The SH light collected is thus a coherent signal from the assembly of particles adsorbed at the interface but the wavelength dependence of the signal is determined by the wavelength dependence of the hyperpolarizability of the particles, apart from a factor ω^2. Because a liquid/liquid interface has a minimal dielectric constant mismatch, it is expected that the induced polarization of the particles will be minimum. It is though highly likely that the origin of the SH signal arising from the particles has an electric dipole origin stemming from the surface

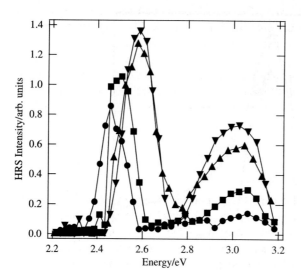

Figure 11. Surface SHG spectra of gold core – silver shell nanoparticles at the water/1,2-dichloroethane interface for different % silver contents: (disks) 20% silver content, (squares) 30% silver content, (triangles) 40% silver content and (inverted triangles) 50% silver content (Reprinted from 59 with permission from the American Institute of Physics)

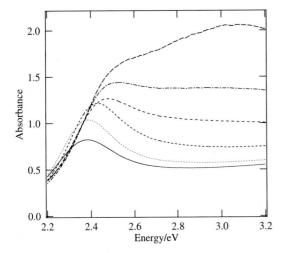

Figure 12. Photo-absorption spectra of aqueous solutions of gold core – silver shell particles for different silver molar contents: (solid) 0%, (dots) 10%, (large dots) 20%, (dashed) 30%, (dashed-dotted) 40% and (large dash-dotted) 50% (Reprinted from 59 with permission from the American Institute of Physics)

induced breaking of the symmetry of the particles. Figure 11 gives the wavelength dependence of the surface SHG signal for gold core – silver shell particles at the liquid-liquid interface.

Two electric dipole SP resonances are observed for such composite particles, in agreement with the theory for the linear optical absorption from core-shell particles, see Figure 12. Furthermore, as the amount of silver in the shell of the particles is increased, the amplitude of the high energy SP resonances increases. However, it is clearly seen that the interband transitions are observed in the linear optical photo-absorption spectroscopy of liquid suspensions of these gold core – silver shell particles but not observed on the corresponding SHG spectra. This question of the role of the interband transitions is still open.

7. AGGREGATION

Aqueous suspensions of metallic particles are not very stable if no stabilizing agents or passivation layers are used. Hence, aggregation may be initiated in these suspensions by simply modifying the ionic strength of the solution thereby decreasing the screening Debye length. This has been recently performed by adding sodium chloride to an aqueous suspension of gold particles by F.W. Vance et al. [17]. The aggregation was followed by Rayleigh and hyper Rayleigh scattering as a function of the amount of NaCl introduced into the solution. A colourful change of the solution from red to blue was observed with an increase of both the Rayleigh and hyper Rayleigh signals, see Figure 13.

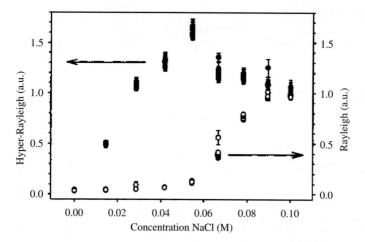

Figure 13. Effect on the Rayleigh and hyper Rayleigh signal intensities recorded at 400 nm of the NaCl induced aggregation of an aqueous suspension of gold particles with a diameter of 13 nm (Reprinted from 17 with permission from the American Chemical Society)

The model of linear chain-like aggregates has been demonstrated in the past to be the initial geometry taken by the aggregates [60]. Thus, the appearance of a longitudinal SP resonance mode in the particle aggregates at longer wavelengths, between 550 nm and 800 nm depending on the aspect ratio of the aggregates, besides the initial SP resonance mode of the spheres located at 525 nm is responsible for the colour changes. In the meantime, this latter transverse SP resonance mode at 525 nm is damped. The HRS intensity during the NaCl induced aggregation was reported for a harmonic wavelength of 410 nm and a fundamental wavelength of 820 nm. The HRS intensity exhibited pronounced enhancements reaching eventually an order of magnitude as compared to the non aggregated solution, for concentrations of NaCl as small as 20 mM, concentrations at which the linear Rayleigh scattering signal did not show any sign of enhancement. This feature was attributed to the greater sensitivity of the HRS signal to small aggregates. One reason for this sensitivity is the occurrence of the longitudinal SP resonance at longer wavelengths, closer to the fundamental wavelength of 820 nm used in this experiments. Large enhancements are indeed expected at the fundamental frequency. Another reason is the possibility that the aggregates have non centrosymmetrical shapes and that this strong breaking of the centrosymmetry enhances the HRS signal intensities.

In order to investigate further these questions, similar experiments were conducted on the aggregation process of gold particles suspensions with pyridine. In this case, pyridine displaces citrate, the charged organic compound stabilizing the suspension. Aggregation is then quickly initiated but minute amounts of pyridine are required to yield a slow kinetics necessary for HRS experiments [61]. Similar measurements have been performed for silver particles [62]. The latter measurements indeed usually require long acquisition times. Large enhancements by a factor of about 10

Second Harmonic Generation from Gold and Silver Nanoparticles 665

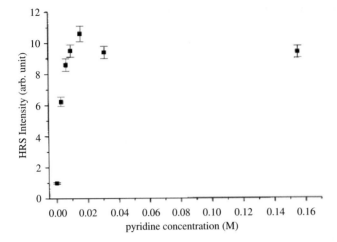

Figure 14. HRS intensity as a function of the pyridine concentration for an aqueous solution of gold particles with a diameter of 22 nm (Reprinted from 61 with permission from the American Chemical Society)

of the HRS intensity were observed for a harmonic wavelength in resonance with the transverse SP mode at 532 nm, see Figure 14. Since the aggregates are expected to take a linear geometry, a simple model based on ellipsoidal particles can be used to describe the data. Obviously, as the aggregation proceeds, larger aggregates are formed which eventually reach radii in excess of the size of the wavelength of light. In that case, the particles should be described with the full multipole expansion of the fields, provided the spherical expansion remains valid.

The enhancement for elliptical particles can be defined as compared to spherical particles:

$$(31) \quad \eta = \frac{|f_{agg}^2(\omega) f_{agg}(\Omega)|^2}{|f_{sph}^2(\omega) f_{sph}(\Omega)|^2}$$

where the field enhancement factors of the aggregates and the spherical particles at the fundamental and the harmonic frequencies are defined using the longitudinal and the transverse polarizabilities $\alpha_i^l(\omega)$ and $\alpha_i^t(\omega)$ of the ellipsoidal particles:

$$(32) \quad f(\omega) = \frac{4\pi}{V} \sum_i w_i \frac{\text{Im}\left[\alpha_i^l(\omega) + 2\alpha_i^t(\omega)\right]}{\text{Im}\varepsilon(\omega)}$$

with the introduction of the weights w_i of the linear aggregates with a well defined ratio present in the suspension. The model is in good agreement with the experimental data and it was concluded that the enhancement of the HRS intensity could be attributed to the excitation of the longitudinal SP resonance mode at the fundamental frequency. These results initiated an interest in the control and the design of dense aqueous suspensions of particles [63]. Synthesis of molecularly-bridged dimers and trimers of gold particles was also achieved and the HRS intensity from aqueous solutions of these aggregates compared for particle diameters ranging between 5 nm and 10 nm. Dimers had large hyperpolarizabilities, larger than that of monomers but trimers had much larger hyperpolarizabilities. This was attributed to the symmetry change operated from the centrosymmetrical monomers and dimers to the non centrosymmetrical trimers. The distance between the different particles of a common aggregate is also an important parameter determining the hyperpolarizability of the ensemble in this case since the size of the aggregate ultimately controls the retardation effects [64].

More recently, HRS from aggregated particles has been used as a possible route to develop new techniques for bio-assay [65]. Protein-modified gold particles were dispersed in solution and their aggregation state was modified by seeding the solution with antigens, see Figure 15. The resulting advancement of the aggregation was monitored by HRS, see Figure 16.

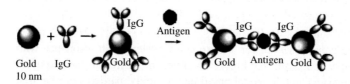

Figure 15. Graphical scheme of Gold-IgG conjugation and antigen induced aggregation (Reprinted from 65 with permission from the American Chemical Society)

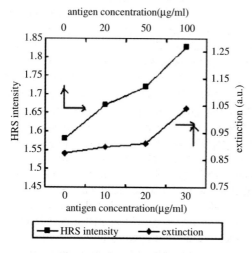

Figure 16. HRS intensity and extinction of gold-IgG conjugates as a function of antigen concentration for an aqueous solution of gold particles with 22 nm diameter (Reprinted from 65 with permission from the American Chemical Society)

The gold particles were first coated by goat-anti-human IgG proteins and as expected, the HRS signal from the protein-aggregated particle solution was larger than that of the non aggregated particle solution. Interestingly, the aging of the solution was described in terms of the time evolution of the HRS intensity after the bio-conjugation of the particles by the proteins, a phenomenon attributed to the rise of a contribution from the surface of the particles. Upon addition of human IgG, aggregation of the gold particles was further induced, in a manner very close to the aggregation reaction observed with addition of NaCl or pyridine.

Comparison with UV-visible photoabsorption spectroscopy showed that HRS was a more sensitive technique revealing the aggregation process at smaller antigen concentration. For instance, at an antigen concentration of $20\,\mu g/ml$, the HRS signal had already increased by 6% from its initial value whereas the extinction coefficient has not increased by more than 1%.

The reason for such a sensitivity of HRS as compared to a linear optical method was attributed to the weak protein-protein interactions. Aggregation was so weak between the particles that linear spectroscopy could not resolve any change in the SP resonances due to the weak electromagnetic coupling between the particles. The possibility of sensing very small aggregates with HRS was thus proven. HRS is then potentially a sensitive method to follow the antigen concentration by conjugation in aqueous samples.

8. CONCLUSION

From the size dependence and the polarization analysis of the HRS intensity from small metallic particle suspensions, it has been demonstrated that the origin of the nonlinear optical response stems from the breaking of the centrosymmetry at the surface of the particles. The origin of the breaking of the centrosymmetry, in turn, should be clarified too but the deviation of the shape of the particles from that of the perfect sphere is observed on TEM images. For large particles, retardation effects become dominant. A spectral analysis also shows that it is possible to distinguish between the electric dipole and electric quadrupole surface plasmon resonances but only in the absence of the latter does the electric dipole resonance characterize a pure local origin of the response. Anyhow, these surface plasmon resonances confer a large magnitude to the hyperpolarizability of the particles. The problem of the role of the interband transitions is though still open.

A great sensitivity of the HRS signal to the aggregation state of the particles has also been noticed. In this case, the exact origin of the nonlinear polarization is still very much in debate. It seems that at the initial stages of the aggregation, the geometry of the aggregates is linear. A model based on the local fields in ellipsoidal particles then correctly accounts for the enhancement of the HRS intensities. At later stages of the aggregation, the exact geometry of the aggregates becomes an important parameter. These concepts have been explored in recent experiments to investigate the potentiality of the HRS technique for immunoassay with interesting perspectives.

REFERENCES

[1] Daniel, M.C., Astruc, D.: Chem. Rev. **103**, 293 (2003)
[2] Kreibig, U., Vollmer, M.: Optical Properties of Metal Clusters, Springer, Berlin, (1995)
[3] Voisin, C., Christofilos, D., Vallée, F.: Phys. Rev. Lett. **85**, 2200 (2000)
[4] Bigot, J.Y., Merle, J.C., Cregeut, O., Daunois, A.: Phys. Rev. Lett. **75**, 4702 (1995)
[5] Shalaev, V.M.: Nonlinear Optics of Random Media, Springer tracts in Modern Physics, vol. **158**, Springer, Berlin (2000)
[6] Boyer, D., Tamarat, P., Maali, A., Lounis, B., Orrit, M.: Science **297**, 1160 (2002)
[7] Arbouet, A., Christofilos, D., Del Fatti, N., Vallée, F., Huntzinger, J.R., Arnaud, L., Billaud, P., Broyer, M.: Phys. Rev. Lett. **93**, 127401 (2004)
[8] Remacle, F.R., Collier, C.P., Heath, J.R., Levine, R.D.: Chem. Phys. Lett. **291**, 453 (1998)
[9] Zhu, T., Yu, H.Z., Wang, J., Wang, Y.Q., Cai, S.M., Liu, Z.F.: Chem. Phys. Lett. **265**, 334 (1997)
[10] Tian, Z.-Q., Ren, B., Wu, D.-Y.: J. Phys. Chem. B **106**, 9463 (2002)
[11] Yoa, J.L., Pan, G.P., Xue, K.H., Wu, D.Y., Ren, B., Sun, D.M., Tang, J., Xu, X., Tian, Z.-Q.: Pure. Appl. Chem. **72**, 221 (2000)
[12] Tarabara, V.V., Nabiev, I.R., Feofanov, A.V.: Langmuir **14**, 1092 (1998)
[13] Vlckova, B., Gu, X.J., Moskovits, M.: J. Phys. Chem. B **101**, 1588 (1997)
[14] Fulishima, A., Honda, K.: Nature **37**, 238 (1972)
[15] Pang, Y.T., Meng, G.W., Zhang, Y., Fang, Q., Zhang, L.D.: Appl. Phys. A **76**, 533 (2003)
[16] Ricard, D., Roussignol, P., Flytzanis, C.: Opt. Lett. **10**, 511 (1985)
[17] Vance, F.W., Lemon, B.I., Hupp, J.T.: J. Phys. Chem. B **102**, 10091 (1998)
[18] Galletto, P., Brevet, P.F., Girault, H.H., Antoine, R., Broyer, M.: Chem. Commun. 581 (1999)
[19] Shen, Y.R.: The Principles of Nonlinear Optics, Wiley, New York, (1984)
[20] Heinz, T.F.: In: Agranovich, V.M., Maradudin, A.A. (eds.), Nonlinear Surface Electromagnetic Phenomena, North Holland, Amsterdam (1991)
[21] Brevet, P.F.: Surface Second Harmonic Generation, Presses Polytechniques Universitaires Romandes, Lausanne (1997)
[22] Shen, Y.R.: Annu. Rev. Phys. Chem. **40**, 329 (1989)
[23] Eisenthal, K.B.: Annu. Rev. Phys. Chem. **43**, 627 (1992)
[24] Higgins, D.A., Corn, R.M.: Chem. Rev. **94**, 107 (1994)
[25] Brevet, P.F.: In: Volkov, A.G., (ed.), Liquid Interfaces in Chemical, Biological and Pharmaceutical Applications, Surface Science Series, vol. **95**, Marcel Dekker, New York (2001)
[26] Nappa, J., Revillod, G., Martin, G., Russier-Antoine, I., Benichou, E., Jonin Ch., Brevet, P.F.: Recent Res. Devel. Physical Chem. **7**, 439 (2004)
[27] Jackson, J.D.: Classical Electrodynamics, Wiley, New York (1975)
[28] Rudnick, J., Stern, E.A.: Phys. Rev. B **4**, 4274 (1971)
[29] Schmickler, W., Urback, M.: Phys. Rev. B **47**, 6644 (1993)
[30] Mendoza, B.S., Mochan, W.L.: Phys. Rev. B **53**, 4999 (1996)
[31] Boyd, R.W.: Nonlinear Optics, Academic, New York (1984)
[32] Wang, H., Yan, E.C.Y., Borguet, E., Eisenthal, K.B.: Chem. Phys. Lett. **259**, 15 (1996)
[33] Revillod, G., Russier-Antoine, I., Benichou, E., Jonin Ch., Brevet, P.F.: J. Phys. Chem. B **109**, 5383 (2005)
[34] Bloembergen, N., Chang, R.K., Jha, S.S., Lee, C.H.: Phys. Rev. **174**, 813 (1968)
[35] Agarwal, G.S., Jha, S.S.: Solid State Commun. **41**, 499 (1982)
[36] Hua, X.M., Gersten, J.I.: Phys. Rev. B **33**, 3756 (1986)
[37] Östling, D., Stampfli P., Bennemann, K.H.: Z. Phys. D **28**, 169 (1993)
[38] Brudny, V.L., Mendoza, B.S., Mochan, W.L.: Phys. Rev. B **62**, 11152 (2000)
[39] Dadap, J.I., Shan, J., Eisenthal, K.B., Heinz, T.F.: Phys. Rev. Lett. **83**, 4045 (1999)
[40] Dadap, J.I., Shan, J., Heinz, T.F.: J. Opt. Soc. Am. B **21**, 1328 (2004)
[41] Mie, G.: Ann. Phys. **25**, 377 (1908)
[42] Chew, H., McNulty P.J., Kerker, M.: Phys. Rev. A **13**, 396 (1976)
[43] Chew, H., McNulty, P.J., Kerker, M.: Phys. Rev. A **16**, 2379 (erratum) (1976)

[44] Nappa, J., Revillod, G., Russier-Antoine, I., Jonin, Ch., Benichou E., Brevet, P.F.: Phys. Rev. B, **71**, 165407 (2005)
[45] Bersohn, R., Pao, Y.H., Frisch, H.L.: J. Chem. Phys. **45**, 3184 (1966)
[46] Cyvin, S.G., Rauch, J.E., Decius, J.C.: J. Chem. Phys. **43**, 4083 (1965)
[47] Clays, K., Persoons, A.: Phys. Rev. Lett. **66**, 2980 (1991)
[48] Brasselet S., Zyss, J.: J. Opt. Soc. Am. B **15**, 257 (1998)
[49] Aktsipetrov, O.A., Elyutin, P.V., Fedyanin, A.A., Nikulin, A.A., Rubtsov, A.N.: Surf. Sci. **325**, 343 (1995)
[50] Antoine, R., Brevet, P.F., Girault, H.H., Bethell, D., Schiffrin, D.J.: Chem. Commun. 1901 (1997)
[51] Antoine, R., Pellarin, M., Prével, B., Palpant, B., Broyer, M., Galletto, P., Brevet, P.F., Girault, H.H.: J. Appl. Phys. **84**, 4532 (1998)
[52] Clays, K., Hendrickx, E., Triest, M., Persoons, A.: J. Mol. Liq. **67**, 133 (1999)
[53] Johnson, R.C., Li, J.T., Hupp, J.T., Schatz, G.C.: Chem. Phys. Lett. **356**, 534 (2002)
[54] Kim, Y., Johnson, R.C., Li, J.T., Hupp J.T., Schatz, G.C.: Chem. Phys. Lett. **352**, 421 (2002)
[55] Hao, E.C., Schatz, G.C., Johnson, R.C., Hupp, J.T.: J. Chem. Phys. **117**, 5963 (2002)
[56] Russier-Antoine, I., Jonin, C., Nappa, J., Benichou, E., Brevet, P.F.: J. Chem. Phys. **120**, 10748 (2004)
[57] Johnson, P.B., Christy, R.W.: Phys. Rev. B **6**, 4370 (1972)
[58] Innes, R.A., Sambles, J.R.: J. Phys. F: Met. Phys. **17**, 277 (1987)
[59] Abid, J.P., Nappa, J., Girault, H.H., Brevet, P.F.: J. Chem. Phys. **121**, 12577 (2004)
[60] Blatchford, C.G., Campbell, J.R., Creighton, J.A.: Surface Sci. **120**, 435 (1982)
[61] Galletto, P., Brevet, P.F., Girault, H.H., Antoine, R., Broyer, M.: J. Phys. Chem. B **103**, 8706 (1999)
[62] Wang, G., Zhang, Y., Cui, Y., Duan, M., Liu, M.: J. Phys. Chem. B **109**, 1067 (2005)
[63] Makeev, E.V., Skipetrov, S.E.: Optics Commun. **224**, 139 (2003)
[64] Novack, J.P., Brousseau, L.C., III, Vance, F.W., Johnson, R.C., Lemon, B.I., Hupp, J.T., Feldheim, D.L.: J. Am. Chem. Soc. **122**, 12029 (2000)
[65] Zhang, C.X., Zhang, Y., Wang, X., Tang Z.M., Lu, Z.H.: Anal. Biochem. **320**, 136 (2003)

INDEX

2n+1 rule, 39, 56, 62, 66, 123, 175–176, 178, 373
[4-(1-pyridinium-1-yl)phenolate, 311
[4-trans-[p-(N,N-Di-n-butylamino)-p-stilbenyl vinyl] pyridine (DBASVP), 311
4-[N-(2-hydroxyethyl)-N-(methyl)amino phenyl]-4′-(6-hydroxyhexyl sulphonyl) (APSS), 222
4-nitro-4′-aminostilbene (ANS), 307

Absolute configuration, 361, 369
Absorption, 25, 193, 481, 483, 615
Achiral CNTs, 321–326, 332
Adiabatic approximation, 28, 40, 102, 153, 159, 262–264
Adiabatic local-density approximation (ALDA), 152
Aggregates, 398
Aggregation, 663
Aminobenzodifuranon (ABF), 304
Amplification-laser field, 8
Amplified spontaneous emission (ASE), 222ff
　left propagating ASE pulse, 236ff
　reabsorption, 242
　right propagating ASE pulse, 235, 236ff, 242
　temporal width, 236
Analytic derivative approach, 36
Armchair CNTs, 321, 322, 332
Aromatic macrocyclic compounds, 509–533
Atomic mean-field (AMFI), 200f
Atomistic molecular modeling, 337
Automatic code generation, 183

Basis set convergence, 62, 67, 74, 79
Bathochromic shift, 302, 641
Benzene, 186
Best harmonic approximation, 262
Betaine-30, 311
Betaine dye, 308
　Pyridinium-N-phenolate, 311
　Reichardt's dye, 308
Binaphthol, 367, 369, 376
BioCARS, 366
Birefringence, 198
Bistability, 274f, 278

Bloch functions, 329
Bond-length alternation (BLA), 143, 305, 310
Bond order alternation (BOA), 305
Born-Oppenheimer approximation, 27–28
Boron compounds, 516, 518, 532
Buckingham effect, 54, 73, 78
Bulk properties, 46

C_3 symmetrical compounds, 525, 528–529, 595
C_{60} fullerene, 189
Carbon nanotubes (CNT), 319, 321, 326, 329
Cascade second-order effect, 444
Catenanes, 609–615, 617, 625, 630, 631, 639
Characterization techniques, 419, 427, 442, 456
Charge crossover, 274, 275, 278
Charge resonance coupling, 252
Charge-transer excited state, 300–314
Chiral CNTs, 320–326, 328, 331, 332–333
CNT hyperpolarizability, 326
CNTs, see Carbon nanotubes (CNT)
Collective phenomena, 251ff
Configuration Interaction (CI), 58, 114, 130ff
　CI singles and doubles (CISD, SDCI), 135
　CI singles only (CIS), 135, 371
　Full (FCI), 34, 55, 134
Conjugation length, 395, 400, 409
Convergence of perturbation expansions, 114
Cooperative phenomena, 251ff, 267ff
Core shell particles, 479, 491, 645, 663
Cotton–Mouton effect, 54, 73, 84–86, 91
Coupled cluster (CC)
　CC2, 53–54, 63, 68, 69, 74–75, 80, 87, 90, 93, 192–3
　CC3, 53–55, 58, 60–66, 68–72, 75, 76, 80, 83–92
　CCSD(T), 53, 55, 60, 65, 66, 69, 71, 76, 83, 86
　CCSD, 53–55, 58, 60, 62–72, 74–78, 80–84, 86, 87, 90–2
　CCSDT(Q), 55, 86
Cubic hyperpolarizability, 520, 537, 567
Cubic NLO properties, 538, 581, 588, 593, 596

671

Davydov shift, 253, 256
Davydov splitting, 253
Degenerate four wave mixing, 322, 366, 422, 445, 448, 480, 526, 538, 586
Delocalization, 252f, 271, 545, 559, 564
Density functional theory, 53, 151ff, 154, 189, 374, 572
Density matrix, 216f, 224
Dephasing rate, electronic, 220, 232
Dexter exchange, 253
Dielectric constant, 189, 203, 215, 260, 301, 384, 385, 386, 394, 421, 434, 624, 632, 662
Dielectric continuum model, 31, 47, 296
Dielectric function, 463ff, 651, 659, 661
Difference-frequency generation, 6–8, 386
Di-phenyl-amino-nitro-stilbene, 211, 243
Dipolar ruthenium complexes, 571ff
Dipole fluctuation operator, 18, 132
Direct inversion in the iterative subspace (DIIS), 157
Dispersion, 303, 375
Donor-acceptor π conjugated molecules (D-"π"-A), 299ff
Doppler effect, 216, 224
Drude model, 463, 467
Duschinsky rotation, 102, 120, 121
Dynamics, 235, 342–346, 398–399, 495–6

Effective medium, 468–470, 474
Ehrenfest equation, 174, 290–291
Ehrenfest theorem, 44, 152, 163f, 174
Elastic scattering, 3–4, 466
Electric field induced second harmonic generation, 308, 378, 422, 426, 433, 443, 447, 513, 538
Electric-dipole approximation, 213, 362, 363, 365
Electro-optic coefficient, 340–341, 346–350, 355, 627
Electro-optical Kerr effect, 10, 105, 472, 497, 611, 630, 634, 635, 637, 639
Electro-optical Pockels effect, 41, 627
Electron paramagnetic resonance (EPR), 199ff
Electron-phonon coupling (e-ph), 262
Electrooptic effect, 214, 419, 420, 627
Electrostatic interactions, 253, 257–258, 261, 264, 266, 284, 286, 291, 296, 302, 308, 340, 341, 347
Electrostriction, 444
Enantiomeric excess, 367
Exchange-correlation (xc), 152–153, 159–162, 165, 167–168, 170, 174, 179ff

Excited-state polarizabilities, 183, 191ff
  of pyrimidine, 183, 191f
  of s-tetrazine, 183, 191f
Excited state properties in DFT, 170f
Exciton, 253
Exciton hopping, 256, 266, 272
Exciton-exciton interaction, 258, 266, 272
Excitonic model (EM), 253, 254, 264, 266, 271, 278
Extended Hückel approximation, 139

Faraday effect, 54, 73, 74
Fermi liquid, 332
Few-states approximation, 140, 300
Few-states model, 130, 140, 141, 144
Finite field, 110, 124–125, 130, 190, 373, 527, 579
Finite field approach, 33, 36, 108, 110
Five-state model, 227
Fluorescence lifetime, 236, 437, 441
Force-field, 338, 339, 341, 343–344, 348, 354
Forster mechanism, 253
Four-wave mixing, 214, 366, 422, 445
Frank-Condon factors, 262
Free energy, 286
Frenkel-like excitation, 253
Frequency-dependent properties, 52–54, 56–57, 59, 60, 61, 63, 66, 68–69, 73, 92, 297, 412
Frequency-upconversion, 211, 222

Gain-narrowing, 236
Generalized gradient approximation (GGA), 151ff
Gold complexes, 545–546, 554, 557
GRINDOL (method, program), 135
G-tensor, 200ff
  di-t-butyl nitric oxide, 202
  diamagnetic term, 200
  diphenyl nitric oxide, 202
  environmental effects, 202, 301, 314
  paramagnetic contribution, 200
  in polarizable continuum model, 191, 202, 304, 407
  relativistic mass velocity correction, 200
  of transition metal compounds, 201ff
Guest-host, 337–340, 347, 430–431, 531

H-aggregates, 256, 272
Hamiltonian
  electronic, 28, 40, 261
  molecular, 11, 36, 40, 43, 260
  nuclear, 28

Index

Hardly polarizable molecules, 256f, 279
Helicene, 371, 374–379
Hellmann-Feynman theorem, 32–33, 37, 45–46, 134, 144
Hexamethylphosphoramide (HMPA), 304
Holstein model, 262
Hot electrons, 477, 482, 486, 493, 495–496
Hyperfine coupling, 157, 183, 199, 203–204
   tensor, 183, 199, 203
Hyperpolarizabilities, 25, 61, 66, 183, 393, 402, 403, 409, 571, 659
Hyperpolarizability
   electronic, 64, 300, 327, 333, 371
   first-order, 6, 15, 17–20, 23, 25–26, 38, 41, 130ff, 140, 306ff, 404, 510
   fourth-order, 66
   higher-order, 25
   imaginary part, 15, 17, 118, 133, 310
   pure vibrational, 28, 103
   residues, 19, 23, 170, 297
   second-order, 19, 22ff, 39, 42, 130ff, 166, 300, 309, 399
   zero-point vibrational average, 28
Hyper-Rayleigh scattering, 388, 422, 428, 437, 440, 521, 538, 573
Hypsochromic, 302

Infinite optical frequency approximation, 108, 110, 114
Infinite periodic polymers, 121
Infinite polymers, 102, 122–124
Intensity-dependent refractive index, 105, 118, 214, 442, 444, 451–453, 456
Interband transitions, 463, 466, 483, 486, 491, 659, 662–663, 667
Intermolecular interactions, 253ff, 342, 352
Intramolecular charge-transfer (CT), 299, 314, 510–511
Ionic NLO materials, 387, 388, 399
Ionic octupolar, 403, 404, 406, 407
Iron complexes, 539
Irreducible representation, 323–325, 328–329

J-aggregates, 256, 272
Jones birefringence, 54, 73

Kerr constant, 633–637
Kerr effect, 73, 105, 383, 443, 462, 470, 472, 475, 480, 481, 487, 492, 497, 611, 630, 634–635, 637, 639

Kohn–Sham theory, 151ff, 159, 174
   spin-restricted, 155ff, 201, 203
   spin-unrestricted, 155, 201, 204
   time dependent, 159, 160
Kurz powder technique, 441

Langevin dipoles, 311
Langmuir-Blodgett films, 387, 531, 572
Lasing, 227–228, 235–238, 242
LiH, 141
Line broadening
   collisional, 219, 220
   emission, 219
   homogeneous, 219, 221, 232
   inhomogeneous, 219, 263
   inverse, 226
   solvent effect, 219
   static, 219
Linear electrooptic effect, 419, 420, 627
Liquids, 359, 361, 362, 366–369, 375, 378
Local density approximation (LDA), 151f, 181, 186ff
Local field
   enhancement, 470, 488, 489, 491–493, 497
   factor, 464ff
Local field factor, 464ff
Local field, 47, 259, 270, 363, 649, 667
Lorentzian decay, 221

Madelung energy, 265
Magnetic circular dichroism, 54, 73
Magneto-electric birefringence, 73, 89, 91, 198
Maxwell equations, 214f, 422, 423, 435
   paraxial, 223, 225
Mean-field approximation (mf), 255, 259, 265
Merocyanine dyes, 304, 308, 411–412
Metal concentration, 462, 468, 469, 475, 479, 483, 485, 489–491, 494, 496, 497
Metallic particles, 646, 648–650, 656–659, 663
Metalloorganic compounds, 537
Mode-mode coupling, 102, 115–116
Molecular crystals, 252, 253, 257, 259
   charge transfer (CT) molecular crystals, 252, 274
Molecular design, 383ff, 409
Molecular dynamics, 339, 342, 343–344, 346, 354
Molecular electronics, 258
Molecular functional materials, 251ff
Molecular hyperpolarizabilities, 409, 421

Molecular materials (mm), 251ff
  calculation of linear and nonlinear
    response of, 253
Molecular mechanics, 283ff
Molecular switching, 609
Monson and McClain formalism, 309
Monte Carlo, 311, 337–340, 341, 342, 344,
  346–347, 348
Multiconfigurational self-cosistent field –
  Molecular mechanics (MCSCF/MM), 283ff
  linear response, 292f
  linear response, 292f
  quadratic response, 294f
  residue, 297
  third-order molecular properties, 283
  time-dependent molecular properties, 290
Multielectron transfer, 277f
Multi-photon absorption, 144, 193, 211, 216
  coherent, 212
  incoherent, 212
  saturation, 212, 228, 234

Nanocomposite materials, 462–464, 468, 470,
  474–476, 479, 480, 490, 492, 497, 498
Nanoparticle, 461–474, 476, 477, 479–489, 491,
  492, 495, 496, 498
Net conversion coefficient, 238–239
Neumann principle, 427
Nickel complexes, 541, 554, 557
Nitrogen, 186
Noble metal, 461
Non-Heitler-London term, 257f, 266
Non-interacting rigid gas, 340, 347ff
Nonlinear optical activity, 360, 361, 365, 366
Nonlinear optical materials, 383ff, 419
Nonlinear optical polymers, 337ff, 354
Nonlinear optical properties (NLO), 3, 51, 101,
  124, 299, 359, 518, 530, 623
Nonlinear optical response, 2, 130, 143, 300,
  301, 303, 314, 320, 323, 373, 384, 421–422,
  426, 441, 442, 454, 461ff, 494
Nonlinear optics, 1ff, 52, 385, 419, 427, 454
Nonlinear polarization, 247, 366, 384, 429, 445,
  448–449, 454, 471, 647, 649–652
Nonlinear spectroscopy, 319
Nonlinear susceptibility, 386–387
Non-resonant NLO processes, 101
Nuclear relaxation hyperpolarizabilities, 105,
  106, 108, 110, 112

Octupolar compounds, 404, 405
Octupolar ruthenium complexes, 551

One-photon absorption (OPA), 232, 233, 235,
  242, 244, 245, 257
  between excited states, 139, 172, 173
  ground-to-excited states, 19
One-photon transition, 19, 40, 234
  off-resonant, 234, 514, 527
  resonant, 118–121
  two-level approximation, 140, 228, 232,
    301, 306
Onsager reaction field theory, 371
Open-shell systems, 153
Optical depoling, 629
Optical index, 463, 472
Optical Kerr effect, 470–480
Optical Kerr Gate, 422, 451–452
Optical limiting, 483–485
Optical phase conjugation, 450–451
Orbital excitation operator, 291
Orbital perturbation theory, 139
Organometallics, 399–402
Oriented gas model (OGM), 252
Osmium complexes, 539

Parity, 325, 361
Pauli matrices, 264
Periodic polymers, 121–124
Perturbation treatment of pure vibrational
  NLO, 104
Perturbative approach, 300
Phosphorescence, 172
Photoabsorption cross section, 221
  effective TPA, 243
  three-photon, 229
Photoisomerization, 545, 638
Photon occupation number
  representation, 217
Photon-echo, 220, 232
Photonic crystals, 215
Photorefractive effect, 444
Phthalocyanines, 511–516
PNA, 40, 183, 303, 310
P-Nitroaniline (PNA), 135, 140, 303, 339,
  347, 370
Polar-polarizable chromophores (pp), 261
Polar solvation, 263
Polarizability
  electronic, 122–123
  excited state, 191–193
  linear, 13–15, 270
  nonlinear, 30, 305, 366, 454
  pure vibrational, 29, 52, 104, 370
  residues, 170–174
  zero-point vibrational average, 125

Index

Polarization interactions, 284, 291, 297
  solvent, 295
Polarization propagator, 14, 43–45
Polarized continuum model (PCM), 311
Polyacenes, 260
Post-VSCF methods, 116
Potential energy surface, 28, 262
Propagation losses, 623
Protic-switching, 591
Pseudoscalars, 361–366
Pulse propagation, 211ff
Pulsewidth, 462, 477, 485, 486, 493–497
Push-pull chromophores (pp), 260–264, 299
  optical spectra in solution, 257, 262, 278

Quadratic electrooptic effect, 214
Quadratic hyperpolarizability, 406, 511, 522
Quadratic NLO properties, 582
Quantum mechanical (QM), 365
Quantum-mechanical-Langevin dipoles/Monte Carlo method (QM/LD/MC), 311
Quantum mechanics, 284, 285
Quasi-energy, 43, 45, 151, 174–179
Quasi-energy ansatz, 151
Quasi-energy Lagrangian, 56, 57

Rabi frequencies, 217, 224, 225
Redox-switching, 578, 592, 598
Refractive index, 442–443, 488, 620-3
Relaxation matrix, 216
Renormalization, 116, 263, 264
Resonant (hyper)polarizabilities, 103, 118–121
Response functions
  cubic, 53, 58, 61, 73, 86, 154, 169, 172–174
  linear, 14, 17, 27, 43, 92, 134, 158, 165, 170–171
  quadratic, 58, 59, 61, 72, 74, 88, 134, 167, 171, 172, 176–177
Response theory, 42–46, 54–61
  cubic, 191
Restricted-unrestricted approach, 157, 204
Retardation effects, 470, 488, 649, 652, 658, 660
Reverse saturation of absorption, 483
Rotaxanes, 609ff
Ruthenium 2,2′-bipyridyl complexes, 582–587
Ruthenium ammine complexes, 581
Ruthenium arsine complexes, 597
Ruthenium complexes, 571–599
Ruthenium phthalocyanine complexes, 509–533

Ruthenium^{II} σ-acetylide complexes, 594
Ruthenium η^5-cyclopentadienyl complexes, 572–576
Ruthenocenyl derivatives, 572–576

Safe transfer operator, 291, 293
Saturation of absorption, 481–483
Scattering duration time, 218–221
Second-harmonic generation (SHG), 4, 308, 422–441, 447–448
Second-order nonlinear optical response, 441, 471
Second-order nonlinear optics, 383ff, 427–428
Self-defocusing, 485
Self-focusing, 214, 215, 247, 452–453, 485
Size effects, 466–468, 487–8
Slowly-varying envelope approximation (SVEA), 215, 224
Solute-solvent interaction, 263, 300, 301, 304, 307
Solvation coordinate, 263, 306
Solvation, 296
  energy of water, 296
  enthalpy of water, 296
Solvatochromic shift, 301, 302, 304, 308
  dispersion contribution, 303
  electrostatic contribution, 301, 302–303
  H-bonds contribution, 303–304
  short-range specific interaction, 301
Solvatochromism, 299ff
  negative, 141, 302, 303, 304, 308
  positive, 302, 303, 308, 309, 310, 311
Solvent effects, 93, 300, 302
Solvent polarity, 308, 313
Spectral dispersion, 492, 493
Spectral line broadening, see Line broadening
Spin contamination, 154, 155, 157, 201, 202
Spin Hamiltonian parameter –molecular structure relationships, 199
Spin–orbit coupling, 172
Spontaneous noise, 228
Spontaneously generated noise photons (SP), 228
Stark spectroscopy, 191, 577, 579, 581
Stilbazolium, 389–399
Stimulated emission, 212, 216, 222, 236, 247
Subphthalocyanines, 509–533
Sudden approximation, 242
Sum-frequency generation, 364–365
Sum-over-orbitals (SOO), 137

Sum-over-states (SOS), 39–42
  first-order hyperpolarizability, 15–19, 42, 135
  modified (MSOS), 310
  polarizability, 13–14, 192
  second-order hyperpolarizability, 19–25, 136, 137, 310
Sum-over-states perturbation theory, 101
Super-radiance, 257
Supermolecule approach, 254–255
Supramolecular interaction, 266, 269, 270
Surface plasmon resonance, 464–468
Susceptibility, 470, 474–475, 490
  density matrix approach, 216–218
  first-order, 15–19, 42, 135
  kinetic equations, 217
  n^{th} order, 110, 213, 340
  simulations, 226
Symmetry, 319ff, 361–366, 386–387, 427–8

Tamm-Dancoff approximation, 135
Thermal effect, 494–495
Thermal lensing, 486, 497
Thin Film, 433, 480, 572, 614–615, 617, 621, 622, 626
Third-harmonic generation, 445–447
Third-order nonlinear optical response, 442, 461ff
Third-order susceptibility, 489, 526
Three-photon absorption (3PA), 193–198, 229–230, 231–3
  ab-initio computations, 230–231
  coefficient, 239, 472
  coherent, 242–246
  cross section, 229–231
  incoherent, 212, 216
  saturation, 234
  two-level approximation, 140, 141
Three-photon matrix element, 23, 193–198
Three-wave mixing, 365
Time-dependent density functional theory (TDDLFT), 151ff, 398
Time-dependent DFT (TDDFT), 152, 182, 199, 205, 576
Time-dependent Hartree-Fock (TDHF), 57
Time-dependent perturbation theory, 10–27
  open-shell, 92, 153, 155
Time-dependent polarization, 5, 7, 12
Time-dependent self-consistent field Hartree-Fock (TDHF) equation, 34, 36, 123, 152
Time-reversal, 362
Transition dipole moment, 275, 277, 306, 307
Transition dipole moments in DFT, 197

Transition moment, 14, 19, 130
  third order, 27, 442–445
Transmission, 233, 650
Transmission electron microscopy, 646, 650
Two-form model, 141–144
Two-photon absorption (TPA), 23, 26, 133, 172, 193, 242, 483, 566
  coherent, 234, 242–246
  cross section, 229, 230, 245, 246, 309–314
  excited-to-excited state, 19
  ground-to-excited-state, 19
  Monson and McClain formalism, 309
  and second-order hyperpolarizability, 19–25, 133, 137
  solvent effects on, 300–301, 313
  and static first-order hyperpolarizability, 135, 310, 313
  tensor, 200, 201, 203–204, 531
  two-level approximation, 140, 141, 232, 301, 306
  two-step, 242–246
  vibrational interaction, 242
Two-photon matrix element, 19, 26, 242–246, 309
Two-state model, 42, 190, 307, 309
  for "alpha", "gamma", 309, 314–315
  valence-bond (VB), 304

Ultraexcitonic correction, 271, 272
Uncoupled Hartree-Fock (UCHF), 130, 137–140
Urbach-like decay, 221

Valence-bond charge transfer model (VB-CT), 143, 305
Valence-bond state model (VBSM), 305
Van der Waals interactions, 253, 284, 343, 344, 392, 420
Van der Waals radius, 220, 252
Verdet constant, 75, 76
Vibrational hyperpolarizabilities, 110, 116, 370
Vinylidene, 560, 567, 591, 593
Virtual excited state, 3
VSCF method, 115

Weisskopf radius, 220
Wrapping procedure, 319, 320

Zero-point vibrational average, 125
Zigzag CNTs, 322, 332
ZPVA contribution, 106, 112, 113
Z-scan, 452, 453, 532, 538, 585
Zwitterion, 141, 143, 305, 308, 398, 408–414

Non-Linear Optical Properties of Matter

CHALLENGES AND ADVANCES IN COMPUTATIONAL CHEMISTRY AND PHYSICS

Volume 1

Series Editor:

JERZY LESZCZYNSKI
Department of Chemistry, Jackson State University, U.S.A.